Sold to H. Bornard
for a rediculus price
M. G. Rekoff Jr

SYNTHESIS
of
PASSIVE NETWORKS

SYNTHESIS
of
PASSIVE NETWORKS

Theory and Methods Appropriate to the Realization and Approximation Problems

Ernst A. Guillemin
PROFESSOR OF ELECTRICAL COMMUNICATION
DEPARTMENT OF ELECTRICAL ENGINEERING
MASSACHUSETTS INSTITUTE OF TECHNOLOGY

NEW YORK · JOHN WILEY & SONS, INC.
London · Chapman & Hall, Limited

Library of Congress Catalog Card Number: 57-8886

Printed in the United States of America

To Otto Brune

who laid the mathematical foundation for modern realization theory

PREFACE

My first thought as I turn to the task of writing this preface is one of great relief that the more or less coherent efforts of many years have reached or at least have come close to their goal. At the same time I am somewhat amazed that it should have taken so long to accomplish this end. Actual writing of the book, although a considerable task in itself, is by no means accountable for much of the total time, for it was done in essentially the space of two summer vacations. The total effort which involves much more than the final documentation, however, extends over a considerably longer period and includes numerous related activities.

I have spent practically a lifetime in electric circuit theory, mostly puttering around in a very restricted area defined by the five adjectives: lumped, linear, finite, passive, and bilateral, yet not once have I felt that the subject was reaching a state where new ideas and things to do were becoming scarce—the feeling that a saturation point was being approached. To be sure, network theory has reached a state of maturity as compared with its infancy during the period prior to about the middle twenties, but this maturity manifests itself primarily in a sense of understanding what network theory really is as a coherent entity with well-defined mathematical foundations. It does not imply by any means that the subject can now be wrapped up in a neat package of known dimensions and arrayed on a shelf along with other "finished" items in the spectrum of past activities.

One reason for this fact is that electrical network theory is essentially the theory of linear systems and hence finds application in numerous fields other than that in which it originated and grew to its present state of respectability. A second reason of no less significance is owing to the importance of linear passive network theory as a tool in dealing with a broader class of network problems defined by the adjectives: nonpassive, nonbilateral, nonlinear, and statistical. Practically all that we know how to accomplish in these areas at present is based upon a manipulative procedure so chosen that linear passive bilateral theory again becomes applicable.

These facts are both good and bad for the state of network theory. They are good because network theory, being an essential tool in a very broad field of engineering activity, attracts the attention of many more individuals than merely those who specialize in it for its own sake. They are bad because many mere "users" of network theory are satisfied with a very superficial understanding of it. They are interested primarily in learning "how things go," and are content with only a smattering knowledge of "why things are that way." They do not realize that even a *tool* must be well understood to be most useful, and that it can betray the unwary who misuses it.

I have always held that there are no short cuts to the acquisition of a really adequate understanding of network theory whether for the user or the specialist. In engineering teaching we cannot overemphasize the importance of having our students grasp the significance and the thrill of acquiring some depth of understanding in at least a few topics, so that they will be aware of the process by which we get to the bottom of things and what we mean by that term. It matters little whether the topic in question promises to be of immediate practical value. The educational value lies in the experience of following a line of reasoning to its logical conclusion; and in the thrill of suddenly realizing that our topic has thereby gained a much wider significance than initially seemed possible. Unless our students are given this kind of experience in at least some spots in their curriculum ("rat race" is the more popular term), I feel that we have failed to give them an education in the true sense of the word.

Most educators will agree, of course, that depth of understanding is an objective we should strive to have our students ultimately achieve, but many feel that thoroughness in the presentation of topics is not effective in the introductory phases of our curriculum—that depth of understanding can come only later on, in graduate work perhaps. They argue that an introductory subject should more appropriately present a wide variety of topics in a shallow manner so as to give a sort of panoramic view of the whole field. Subsequent repetition of these topics (which is essential anyway as an aid to assimilation of new ideas) provides an opportunity for achieving thoroughness.

The fallacy in this argument is that in the second time over the ground the primary interest usually lies in an application to some practical problem and it is taken for granted that the means by which its solution is carried out are "old hat" and do not need further separate consideration which at this point would interfere with the main trend of thought. And so thoroughness yields to expediency, since there never seems to be any time for it.

If anyone doubts the essential validity of these statements, let him ask an average senior engineering student what is meant by an integral. Presumably this topic was "covered" in the student's sophomore calculus (along with innumerable other items) and he has become quite familiar with its use (handbook fashion) in various subsequent engineering applications. In fact, his familiarity with the use of integrals misleads him into believing that he understands what is meant by the process—until he runs into a situation where he really has to know. He then suddenly realizes that what he subconsciously mistook for understanding was not that at all; he had merely gotten accustomed to seeing the thing around. And the mathematics teacher on his part undoubtedly felt that the use to which integration would be put in the subsequent physics and engineering subjects could be relied upon eventually to provide a thorough understanding.

Electric circuit theory is largely in the same category, for, like mathematics, it is a tool in the analysis of many engineering problems. Every electrical engineer thinks he knows circuit theory because he uses it "all over the place." But if you probe for depth, it turns out to be a disappointing business. I do not believe it matters how much or how little circuit theory a student is presented with during his college career so long as the presentation is logical, meaningful, and thorough enough to be intellectually challenging and stimulating.

There is a second very potent reason why thoroughness is commonly postponed or avoided; it requires competence and experience on the part of the teacher. This fact is particularly true at the freshman and sophomore levels where motivation complicates the picture. Here it is much easier to hold the student's attention by parading many topics before him in rapid succession, like keeping a child amused by turning the pages in a picture book. In this way the scene changes before the student becomes bored with each successive item. This kind of performance is the mark of an inexperienced teacher, who has no difficulty presenting a dozen different topics in a single lecture but who is completely stymied if asked to spend the entire lecture talking about a single item in a thorough and interesting fashion.

The relevance of these remarks to the preface I am writing here is to provide at least a contributing reason for the relatively long time it took me to arrive at the documentation represented by this book, which should more appropriately be regarded as a chapter in a much longer story—a story that began with the conception and the writing of *The Mathematics of Circuit Analysis* and continued with *Introductory Circuit Theory*.

From the very start I did not like the idea of providing the mathematical and the introductory circuit theory background along with topics in network synthesis, for the inevitable limitations upon size must then lead to a superficial result, quite apart from other disadvantages arising out of an attempt to cope with such an extensive array of material within the confines of a single volume (or even two). In the present book, the other two (referred to as *MCA* and *ICT*) are regarded as an integral part of the work, and results given there are drawn upon freely, mostly without further comment and with the expectation that the reader has reached a state of familiarity with this essential background, just as any writer would naturally not feel called upon to rediscuss the results of Chapter 1 in Chapter 2, and so forth.

Even so, the job is not finished with the appearance of this volume. When it initially became evident that the material planned for the synthesis book would more likely require two volumes for its accommodation, the whole of it was tentatively divided at a point where the discussion of the *RLC* driving-point impedance was finished and that dealing with the realization of transfer functions began, this division yielding two parts of approximately equal size.

I soon concluded, however, that this "logical" division was not good at all, for it would delay the most interesting and "waited-for" part of the treatment for another two years or more. I therefore separated out all topics having to do with further applications of synthesis methods (such as the broadbanding problem, optimum input and output networks, more detailed material on the design of filters as well as more general types of filter groups, and so on) and decided to include in the present volume all fundamental theory and methods with just sufficient examples to make their presentation clear and meaningful. The rest will appear later on under a title such as "Special Problems in Network Synthesis." Although this separation of the total material does not represent a division into equal parts, since the present volume carries a lion's share, it is one which I feel makes pretty good sense from a practical standpoint.

The job is furthermore not finished because there exists a gap between the *ICT* book and the present volume that needs to be filled. For obvious reasons I decided to skip this intermediate book and bend all efforts to the present task even though some topics (principally those having to do with Fourier theory) had to be presented more fully here for the time being. There are times when one cannot proceed with one's plans in a completely logical manner. For one thing, some of my former students were threatening to "beat me to the draw" in the production of a book on network synthesis, and I just could not let that happen. As it is, the race is a close one.

It is not out of place here to call attention to some of the early work in network synthesis following the initial contributions of O. J. Zobel, G. A. Campbell, and R. M. Foster. It was through the interest and encouragement of Vannevar Bush and Norbert Wiener that the necessary climate for a deeper study of network analysis and synthesis was established and fostered among our group in those early days. The outstanding member of that younger group, and in my opinion the one primarily responsible for establishing a very broad and mathematically rigorous basis for realization theory generally, was Otto Brune, to whose clear and penetrating insight we owe much more than is generally credited to him by those who know of his work only through his very meager publications.

Everyone who studies network synthesis knows about Brune's contributions to the *RLC* driving-point impedance problem, but only those who have listened to my lectures know that it was his creative thinking that also laid the rigorous mathematical foundations for all realization theory. It is for this reason that I enthusiastically dedicate this volume to him.

Regarding the subject with which this book is concerned, it is significant to recognize at the outset that synthesis differs from analysis in two major respects: (*a*) a solution to a stated problem does not necessarily exist, and (*b*) when it does exist it is not unique, there being theoretically an infinite number of circuits all having the same response function or functions at specified points of access. For these reasons the approach to a synthesis problem must begin by establishing necessary conditions for the existence of a solution and effective means for determining whether these conditions are met in a given situation. The nonuniqueness aspect leads us into the byways of equivalent network determination, a subsidiary problem whose solution is far from satisfactory at present.

The approach to a synthesis problem is further complicated by the fact that the specified data regarding excitation and response can be presented in a variety of forms First of all, we recognize that these data may be specified either in terms of time functions or frequency functions; and secondly, that in either of these situations the required network performance may be described graphically or analytically or piecewise analytically or in various combinations of these terms.

Known realization methods for lumped finite networks can accommodate specified performance requirements only when these are expressed in the form of rational functions of the complex frequency variable such as driving-point and/or transfer impedances. Hence it is necessary to translate the stated requirements into this form before a solution to the synthesis problem can be carried through.

This step, which may involve time- as well as frequency-domain considerations, is essentially an approximation procedure with specified tolerances but usually with freedom in other respects so that again an unlimited number of possible solutions exist. In order to choose intelligently between these it is necessary to be thoroughly familiar with the realization methods since one can then control also such important practical matters as economy in element numbers, the structural form of the resulting network, and the probable effect of parasitic elements.

Concerning the arrangement of material in this book, the plan has been to follow as nearly as possible a logical development. The first thirteen chapters deal with realization theory and methods, the last two with the approximation problem in the frequency and time domains respectively. The discussion of realization theory begins with a study of the properties of impedance functions, a formulation of necessary conditions for realization, and methods of testing a given function to determine whether or not these conditions are met. These matters occupy two chapters, of which the first deals with general *RLC* functions and the second with the more particular properties pertinent to two-element-kind networks. Driving-point impedance synthesis for lossless (*LC*) networks is the topic for Chapter 3 and the analogous methods applicable to *RC* and *RL* networks follow in Chapter 4. These results amply demonstrate the nonuniqueness of the realization methods as well as the related reciprocal property arising from the fact that a realizable driving-point function is just as appropriate right side up or upside down; and so these matters are elaborated upon in the next chapter.

Consideration of the general *RLC* driving-point realization problem would seem to be the next logical topic. However, its discussion involves much more than a reasonable extension of the procedures applicable to two-element-kind networks, as one might expect also from the fact that solution of this problem by Brune did not chronologically follow immediately upon the heels of the Foster and Cauer results. The general *RLC* problem involves subtleties of a deeper nature; and the consequent variety of possible procedures for its solution makes it necessary first to become familiar with a number of additional topics including some properties of two terminal-pair networks, realization procedures pertinent to these in the lossless case, and the construction of an impedance from its j-axis real part, which are discussed respectively in Chapters 6, 7, and 8. The latter also includes the related methods of constructing an impedance from its j-axis magnitude or angle as well as a discussion of the uniqueness feature with which these procedures can be endowed if one admits the possibility of impulses in a given j-axis real part.

Chapter 9 presents the classical methods of Brune and Darlington for driving-point impedance synthesis, and Chapter 10 deals with the Bott and Duffin and the Miyata methods of realization so as to avoid mutual inductive coupling. The latter method, which in contrast with the Bott and Duffin method is not generally applicable, is extended and discussed in considerably greater detail than has been done by the originator of this scheme; and the Bott and Duffin method is then further clarified in the light produced by these elaborations.

Chapters 11 and 12 deal with transfer function synthesis, the first giving the various classical procedures and the second presenting a number of alternate methods and more general points of view. *RC* transfer function synthesis is separately discussed in Chapter 13.

At this point an adequate background is laid for consideration of the approximation problem. Its frequency-domain aspects, discussed in Chapter 14, include use of Butterworth, Tschebyscheff, and elliptic functions as well as Fourier and semi-infinite slope methods, and potential analog interpretations. Time-domain synthesis, which differs essentially only in the additional techniques needed to solve the appropriate approximation problem, is dealt with in the final chapter, Chapter 15, which includes a Fourier procedure enabling one to use familiar techniques for direct tolerance control in the time domain.

Throughout the book, but especially in the last chapter and in the ones dealing with transfer function synthesis, it was necessary to omit much existing material which amplifies and enhances the discussions of these topics but which was considered less fundamental or less practical or less essential in providing a general background of understanding. The material assembled in this volume, it is felt, should in any case provide the reader with a good ability to cope with the many articles available in the technical literature on the subject of synthesis. In short, it should make him literate in the field.

As in the case of my introductory book, I do not attempt to provide a bibliography, for I have neither an interest in this sort of thing nor the time for it. Besides, bibliographies are more discouraging than helpful. What good is an endless list of titles except to give one that helpless and frustrated feeling. Certainly, without some discussion and guidance it is practically useless except as an index to find articles on specific problems. My good friend and colleague Professor Ronald M. Foster, of Brooklyn Polytechnic Institute, has compiled such an index of titles. There would certainly be no point in my including an unabridged version of it here.

In closing, I would like to thank all of my past and present students who have taught me many things about network theory and have helped to ferret out errors of all sorts in the manuscript. I wish in the same sense to extend thanks to the staff members of Massachusetts Institute of Technology who, through the years that many of the things recorded here came into being, have indirectly made this book more effective.

E. A. Guillemin

Wellesley Hills, Mass.
June 1957

C O N T E N T S

Chapter

Chapter

C H A P T E R O N E

Properties of Driving-Point and Transfer Impedances

1 Physical Considerations

From our familiarity with the transient or natural behavior of networks (their response to an impulse excitation, for example), we know that the zeros and poles of a driving-point impedance function are identifiable with the complex natural frequencies of the pertinent network for appropriate short-circuit or open-circuit constraints. Thus, if

$$Z(s) = \frac{H(s - s_1)(s - s_3) \cdots}{(s - s_2)(s - s_4) \cdots} \tag{1}$$

is the impedance of a lumped, linear, passive network N (Fig. 1) at a

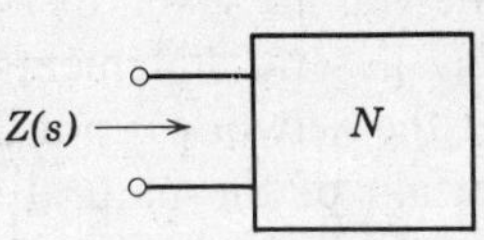

Fig. 1. Linear passive network with single driving point.

selected terminal pair, then the complex frequencies $s_1, s_3, \cdots$ are the natural frequencies of this network when the input terminals are short-circuited, whereas $s_2, s_4, \cdots$ are those for the same network with these terminals open-circuited. More specifically, if we were to excite the network somehow (for example, by having a little boy toss coulombs at the capacitances) with the input terminals short-circuited, then the expression for the current at this point has the form

$$i_1(t) = A_1 e^{s_1 t} + A_3 e^{s_3 t} + \cdots \tag{2}$$

and with the terminals open-circuited the voltage there is given by

$$e_1(t) = B_2 e^{s_2 t} + B_4 e^{s_4 t} + \cdots \tag{3}$$

We call the $s_1, s_3, \cdots$ short-circuit natural frequencies and the $s_2, s_4, \cdots$ open-circuit natural frequencies. The zeros and poles of an impedance $Z(s)$ are thus simply the short-circuit and open-circuit natural frequencies, respectively, of the network at the pertinent driving point. The current expression 2 could be regarded as the response for an applied voltage impulse and the voltage expression 3 as the response for an applied current impulse.

From this physical interpretation of the critical frequencies of a driving-point impedance it follows at once that if we create any other points of access to this network through soldering pairs of leads to any chosen node pairs (the so-called soldering-iron type of entry) then all pertinent impedances have the same poles $s_2, s_4 \cdots$ because they are the natural frequencies of the same network under the same conditions. Similarly, if we short-circuit the given terminal pair and create any other one through cutting a branch (the so-called pliers type of entry) then the impedance at any such other terminal pair has the same zeros $s_1, s_3, \cdots$; and if we make such pliers-type entries but do not short-circuit the original terminal pair then the zeros of the pertinent impedances are $s_2, s_4, \cdots$, the same as the poles of the impedances at the soldering-iron–type entries made before.

When several soldering-iron–type entries are made simultaneously, then we may consider all the open-circuit transfer impedances as well as the driving-point impedances, and *all of these have the same poles.* When several pliers-type entries are made simultaneously, it is more appropriate to consider the short-circuit driving-point and transfer admittances, and *all of their poles are the same* and are given by $s_1, s_3, \cdots$.

These statements (like any general statements about network properties) assume, of course, that the network is not degenerate in one way or another so as to render certain of the natural frequencies nonexcitable from some of its branches. Such degenerate forms can always be constructed, but in a sense the above statement holds true even for these forms, for if a given impedance does not happen to have a certain pole (because the pertinent natural frequency is not excitable from the driving point in question) we can stoutly maintain that indeed it does contain this pole, however, with a zero residue in it. Zero is a very respectable value for a residue to have.

What interests us at the moment about all this is the fact that the passive character of the network at once tells us that expressions 2 and 3 must decay with time and hence that all the critical frequencies of impedance 1 must have a negative real part. The zeros and poles of a driving-point impedance must lie in the left half of the complex frequency plane. This important property of driving-point impedances is thus at once evident from physical considerations regarding network behavior.

With regard to the *transfer* (open-circuit) impedances or (short-circuit) admittances, the above argument says nothing about their zeros. These are in no way related to any natural frequencies of the network. They are values of complex frequency for which the output terminal pair is virtually decoupled from the input terminal pair. It is only the poles of a transfer function (open-circuit impedance or short-circuit admittance, the ratio of output to input) that are related to the network's natural frequencies. These poles, like those of the pertinent driving-point functions, must lie in the left half of the complex frequency plane. This condition, which as mentioned above follows from the fact that the transient response of a passive network must die out, is also referred to as the *stability requirement*. A system (whether it be passive or not) cannot have a stable response (one that will not take off on its own) unless the pertinent system function (driving point or transfer, impedance, admittance, or dimensionless ratio) is analytic in the right half of the complex frequency plane.

With regard to transfer functions, this statement (except for the obvious requirement of boundedness on the j-axis wherever losses are involved) is all that is required. A driving-point impedance or admittance, on the other hand, is not sufficiently circumscribed merely by restricting its zeros and poles to the left half-plane. First of all, the numbers of finite zeros and poles can differ at most by one, which is equivalent to saying that the asymptotic behavior for very small or for very large frequencies can reduce to only one of three things, namely: a constant; a constant times the frequency s; and a constant divided by the frequency s. Physically the reason for this restriction upon the asymptotic behavior follows from the fact that at very large or at very small frequencies the behavior of a driving-point function degenerates into that of one of the three circuit constants R, L, or C. A transfer function by contrast may have many more finite poles than zeros (the boundedness requirement rules out the reverse situation). In fact, its numerator may reduce to a constant and its denominator may have any finite degree. All of its zeros then lie at infinity while its poles are finite. A driving-point function does not have this freedom.

In this connection it is also essential to recall that a driving-point function is a response function or system function (ratio of output to input) whether right side up or upside down, whereas a transfer function is no longer a response function when turned upside down. (The roles of excitation and response may be interchanged in a driving-point situation but not when these two dependent variables refer to different points of entry.) A transfer function in which the degree of the denominator polynomial is several times higher than that in the numerator, when inverted, would seriously violate the boundedness condition. A driving-point function whose numerator and denominator polynomials differ in degree by unity also violates the boundedness condition in one orientation or another, but at most it has a *simple* pole on the j-axis which in a limiting sense (regarding losses) is permissible. In this same sense, simple poles on the j-axis are also permitted in a transfer function but such a situation is even less meaningful practically than it is in a driving-point function.*

Since the latter must be a response function right side up or upside down, the stability condition obviously requires both the zeros as well as the poles to have negative real parts, but the zeros of a transfer function are not so restricted.

A driving-point function is further restricted by the passivity of the network. For all j-axis points, corresponding to sinusoidal behavior with steady amplitude, the average energy absorbed by the circuit must be positive. Hence the real part of a driving-point function must be positive on the entire j-axis of the complex frequency plane. (Note that the negativeness of the real part of a transfer function does not imply negative average energy absorption either by the network or by the load impedance.) If the real part of an impedance is to remain positive, then it is necessary and sufficient that the angle of that impedance be restricted to the range $-\pi/2$ to $+\pi/2$. It is not difficult to appreciate that this angular restriction can be violated even though the finite zeros and poles are alike in number and both are confined to the left half-plane. Visualize in this connection a cluster of zeros on the negative real axis near the j-axis and a cluster of poles far removed. The angles of the frequency factors due to the zeros then rapidly add up with increasing frequency, whereas those due to the poles do not amount to anything before much higher frequencies are reached. Although the net angular increment (for the entire frequency spectrum) is zero for equal numbers of finite zeros and poles, the angle may exceed the value $\pi/2$ by many times at some finite nonzero frequencies.

Restriction of the zeros and poles to the left half-plane and restriction of the relative degrees of numerator and denominator polynomials of a driving-point function are, therefore, not enough. It is additionally necessary to require either that the real part remain positive or that the angle remain within the range $-\pi/2$ to $+\pi/2$ for all j-axis points. As will be seen later on, it turns out that these necessary conditions for the realizability of driving-point impedances or admittances are also sufficient.

2 Analytic Basis for Studying the Properties of Impedance or Admittance Functions

An effective analytic approach to a study of the fundamental properties of driving-point and transfer functions is based upon expressing these in

* This question is considered in greater detail in Art. 6.

terms of the energy functions associated with the network.† On an impedance basis, we have

$$\sum_{i,k=1}^{p} z_{ik} I_i \bar{I}_k = sT_0 + F_0 + \frac{V_0}{s} \tag{4}$$

in which the z_{ik} are the open-circuit driving-point and transfer impedances of a network having l geometrically independent loops in which loop currents with the complex amplitudes $I_1 \cdots I_l$ circulate; the number of accessible terminal pairs formed by appropriately cutting into selected loops are p in number; and the "energy functions" so-called‡ are defined mathematically by

$$T_0 = \sum_{i,k=1}^{l} L_{ik} I_i \bar{I}_k \tag{5}$$

$$F_0 = \sum_{i,k=1}^{l} R_{ik} I_i \bar{I}_k \tag{6}$$

$$V_0 = \sum_{i,k=1}^{l} S_{ik} I_i \bar{I}_k \tag{7}$$

Note that the double summations here extend over all l loops, whereas the summation sign in Eq. 4 extends only over the accessible terminal pairs to which the z_{ik} refer. The L_{ik}, R_{ik}, S_{ik} are the inductance, resistance, and elastance parameters on the chosen loop basis.

When the equilibrium is formulated by the node method, we obtain analogous expressions on an admittance basis that read

$$\sum_{i,k=1}^{p} y_{ik} E_i \bar{E}_k = sV^*_0 + F^*_0 + \frac{T^*_0}{s} \tag{8}$$

in which the y_{ik} are the short-circuit driving-point and transfer admittances with reference to the p-accessible terminal pairs or node pairs which are any or all of the n geometrically independent node pairs of the network. The variables are a set of node-pair voltages with the complex amplitudes $E_1 \cdots E_n$, and the energy functions here are defined by the expressions

$$V^*_0 = \sum_{i,k=1}^{n} C_{ik} E_i \bar{E}_k \tag{9}$$

† See Ch. 10, Arts. 6, 7, 8, *Introductory Circuit Theory*, E. A. Guillemin, John Wiley & Sons, New York, 1953. This approach to the problem was first formulated in the manner presented here by O. Brune in 1930. (Hereafter *Introductory Circuit Theory* will be referred to as *ICT*.)

‡ As pointed out on p. 525 of *ICT*, only T_0 has the dimension of energy.

$$F^*_0 = \sum_{i,k=1}^{n} G_{ik} E_i \bar{E}_k \tag{10}$$

$$T^*_0 = \sum_{i,k=1}^{n} \Gamma_{ik} E_i \bar{E}_k \tag{11}$$

These summations extend over all node pairs from 1 to n, whereas the summation in Eq. 8 extends only over the p node pairs that are selected as points of entry. The C_{ik}, G_{ik}, Γ_{ik} are the capacitance, conductance, and reciprocal inductance parameters on the pertinent node basis.

The complex frequency variable s, although appearing explicitly in only two places in either Eq. 4 or 8, is implicitly more extensively involved because the z_{ik} and y_{ik} are functions of s, and because of the fact that the complex current amplitudes $I_1 \cdots I_l$ or voltage amplitudes $E_1 \cdots E_n$, found from the solution of a pertinent set of equilibrium equations, have values that depend upon the complex frequency assumed for the source or sources of excitation. Hence when we regard the expressions involved in Eqs. 4 or 8 as functions of the frequency variable s, both their implicit as well as their explicit dependence upon s must be kept in mind. Of prime importance in this connection also is the fact (demonstrated in the reference given above) that the energy functions 5, 6, and 7, as well as 9, 10, and 11, have values that are both real and positive for all complex values of the frequency variable s.

It is thus rather easy to see that either function defined by Eq. 4 or Eq. 8, regarded as a function of the complex frequency variable s, has the following general properties:

It is real for real values of s
It has a positive real part for values of s with a positive real part (12)

A function of a complex variable with this basic property (first introduced into network synthesis by O. Brune) is referred to as a *positive real* function. The adjective "real" is intended to indicate the realness of the function for real values of s. The adjective "positive" indicates the positiveness of its real part for values of s with a positive real part. Thus the alternate term "positive" function is used to indicate one having a positive real part for any s with a positive real part but not necessarily one that is real for real values of s. A "positive real" (abbreviated p.r.) function is thus more restricted than is a mere "positive" function. The functions defined in Eqs. 4 and 8 are p.r. We shall see that driving-point impedances and admittances of linear, passive networks are p.r. functions, and that this property is sufficient as well as necessary to insure their realizability. We shall also see that all the detailed properties of these network functions are derivable from their p.r. character alone.

From the fact that the double sums in Eqs. 4 and 8 are p.r. functions, we wish now to extract the properties of the driving-point and transfer impedances z_{ik} and admittances y_{ik}. This process is hampered by the presence in these sums of the complex variables I_i and E_i. Our first task will be to remove this difficulty. The details of the method whereby we accomplish this objective will be discussed with reference to the double sum in Eq. 4. In form they apply likewise to the one in Eq. 8.

Let the complex quantity I_i be written

$$I_i = a_i + jb_i \tag{13}$$

in which a_i and b_i are real quantities but are otherwise unrestricted. The conjugate quantity $\bar{I}_k$ then has the representation

$$\bar{I}_k = a_k - jb_k \tag{14}$$

and we get

$$I_i\bar{I}_k = (a_ia_k + b_ib_k) + j(a_kb_i - a_ib_k) \tag{15}$$

Substitution into the double sum in Eq. 4 gives

$$\sum_{i,\,k=1}^{p} z_{ik}I_i\bar{I}_k = \sum_{i,\,k=1}^{p} z_{ik}(a_ia_k + b_ib_k) + j\sum_{i,\,k=1}^{p} z_{ik}(a_kb_i - a_ib_k) \tag{16}$$

Use of the symmetry condition $z_{ik} = z_{ki}$ shows that the second sum on the right-hand side of this equation is zero; and the fact that the function on the left-hand side must be p.r., even though the currents are caused by appropriately chosen excitation forces to have any arbitrary values, leads us to recognize that it is neither more nor less binding than the original condition to require that the function

$$Z(s) = \sum_{i,\,k=1}^{p} z_{ik}x_ix_k \tag{17}$$

be p.r. for all values of the independent real variables $x_1 \cdots x_p$.

The function $Z(s)$ is a quadratic form in these variables, with the matrix

$$\begin{bmatrix} z_{11} & \cdots & z_{1p} \\ \cdot & \cdots & \cdot \\ z_{p1} & \cdots & z_{pp} \end{bmatrix} \tag{18}$$

which characterizes the open-circuit driving-point and transfer impedances of a p terminal-pair network. Since the coefficients z_{ik} of this matrix are functions of the complex frequency s, the quadratic form $Z(s)$ is a function of s in addition to being a function of the real variables $x_1 \cdots x_p$. It must remain a p.r. function of s for all real finite values of these variables. This requirement imposes conditions upon the z_{ik} which we shall presently describe. In the meantime it is helpful to present a physical interpretation of the function $Z(s)$.

Figure 2 shows schematically a p terminal-pair network N. The volt-

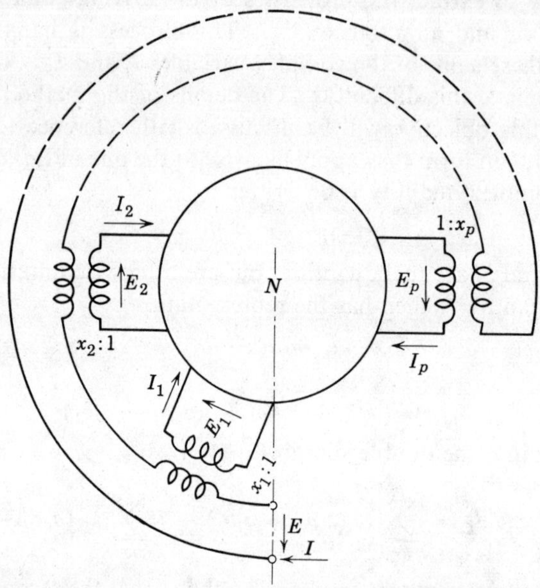

Fig. 2. Relevant to the physical interpretation of Eq. 17.

ages and currents at the terminal pairs are denoted by $E_1, I_1; E_2, I_2 \cdots E_p, I_p$. At each terminal pair an ideal transformer is placed, and all the secondaries of these are connected in series. The secondary-to-primary ratios of the transformers are denoted respectively by the real quantities $x_1 \cdots x_p$ as indicated. Since the voltages on the secondary sides are $x_1E_1, x_2E_2, \cdots x_pE_p$, we see that the net voltage across their series combination is given by

$$E = x_1E_1 + x_2E_2 + \cdots + x_pE_p \tag{19}$$

The common secondary current is I and is related to the various terminal-pair currents by means of the transformer ratios as indicated by

$$I_1 = x_1I, \qquad I_2 = x_2I, \qquad \cdots I_p = x_pI \tag{20}$$

In matrix notation we may rewrite the last two equations in the form

$$E = [x_1 \quad x_2 \cdots x_p] \times \begin{bmatrix} E_1 \\ E_2 \\ \cdot \\ \cdot \\ \cdot \\ E_p \end{bmatrix} \tag{21}$$

and

$$\begin{bmatrix} I_1 \\ I_2 \\ \cdot \\ \cdot \\ \cdot \\ I_p \end{bmatrix} = \begin{bmatrix} x_1 \\ x_2 \\ \cdot \\ \cdot \\ \cdot \\ x_p \end{bmatrix} \times I \tag{22}$$

The equations relating the terminal-pair voltages and currents in terms of the impedance matrix 18 read

$$\begin{bmatrix} z_{11} \cdots z_{1p} \\ \cdots\cdots \\ \cdot \\ \cdot \\ \cdot \\ z_{p1} \cdots z_{pp} \end{bmatrix} \times \begin{bmatrix} I_1 \\ I_2 \\ \cdot \\ \cdot \\ \cdot \\ I_p \end{bmatrix} = \begin{bmatrix} E_1 \\ E_2 \\ \cdot \\ \cdot \\ \cdot \\ E_p \end{bmatrix} \tag{23}$$

If we substitute for the column matrix of currents the equivalent expression in Eq. 22 and then premultiply both sides by the row matrix

$$[x_1 \quad x_2 \cdots x_p] \tag{24}$$

we get, after noting Eq. 21

$$[x_1 \quad x_2 \cdots x_p] \times \begin{bmatrix} z_{11} \cdots z_{1p} \\ \\ \cdots\cdots \\ \\ z_{p1} \cdots z_{pp} \end{bmatrix} \times \begin{bmatrix} x_1 \\ x_2 \\ \cdot \\ \cdot \\ \cdot \\ x_p \end{bmatrix} = \frac{E}{I} = Z(s) \tag{25}$$

The left-hand side of this equation is recognized as being the matrix equivalent of the double sum in Eq. 17, and the ratio E/I is the net impedance looking into the series combination of all the transformer secondaries in Fig. 2. Thus the function $Z(s)$ of Eq. 17 is given the simple physical interpretation (first pointed out by Brune in 1930) of being the net impedance formed through series connection of the p terminal pairs of the network, each provided with an ideal transformer having an independently controllable real ratio. If we interpret Eq. 17 for $p = 1$ it tells us that any driving-point impedance must be p.r. In Fig. 2 a p terminal-pair situation is effectively reduced to a one terminal-pair network, and $Z(s)$ is the driving-point impedance at this single terminal pair. It is physically clear that unless this single driving-point impedance remains p.r. for all choices of the real transformer ratios $x_1 \cdots x_p$, the set of impedances z_{ik} which determine $Z(s)$ cannot represent a physical network. Essentially this circumstance is implied by saying that the function 17 must be p.r. for all real values of the x_k in order that a given set of z_{ik} may be the open-circuit driving-point and transfer impedances of a p terminal-pair network.

So far we have, of course, shown only that this condition is a necessary one for physical realizability. Its sufficiency still remains to be demonstrated by showing that a physical network can actually be constructed when no condition other than this one is met. It is truly remarkable that the necessary and sufficient conditions for physical realizability on the part of linear, passive, bilateral networks can be stated in such simple and compact terms. O. Brune, to whom we owe much in the development of synthesis procedures, was the originator of this elegant approach to the problem.

Entirely analogous reasoning on an admittance basis leads to the result that an equivalent statement may be made in terms of the function

$$Y(s) = \sum_{i,\,k=1}^{p} y_{ik} x_i x_k \tag{26}$$

in which the y_{ik} are the short-circuit driving-point and transfer admittances of the p terminal-pair network. The same kind of physical interpretation as that shown in Fig. 2 is applicable to this expression. If the same network N is involved, then the admittance matrix

$$\begin{bmatrix} y_{11} & \cdots & y_{1p} \\ \cdot & \cdots & \cdot \\ y_{p1} & \cdots & y_{pp} \end{bmatrix} \tag{27}$$

is the inverse of the impedance matrix 18.

This result shows that if the function $Z(s)$ in Eq. 17 is p.r. for all real $x_1 \cdots x_p$, then so is the same function with the z-matrix replaced by its inverse. This inverse matrix, if again regarded as an impedance matrix, characterizes a second p terminal-pair network (which we may denote as a network R in contrast to the original network N) that is referred to as being *reciprocal* to the given network N (as discussed in Ch. 10, Art. 9, of *ICT*). Hence this result is the basis for showing that the reciprocal network with respect to any given one can always be constructed although its dual may not be constructable. (For a detailed comparison of dual and reciprocal networks see *ICT*, p. 539.)

3 Properties of P.R. Functions

Through use of appropriate portions of the theory of functions of a complex variable, we may interpret the detailed properties implied by the p.r. character of a function. Thus it is rather easy to show that a p.r. function can have no poles in the right half of the complex frequency plane. We do this by showing that the real part of a function changes its algebraic sign at least twice as one passes completely around a pole in its immediate vicinity. Since the right half-plane is a region where the real part of a p.r. function remains positive, it is clearly a region where the real part cannot change sign. To assume that a p.r. function has poles in the right half-plane will, therefore, lead to a contradiction, and so we must conclude that this state of affairs is not possible. The detailed analysis supporting this argument follows.

If the function $Z(s)$ has a pole of order n at a point $s = s_1$, then its Laurent expansion about this point reads

$$Z(s) = \frac{a_{-n}}{(s - s_1)^n} + \frac{a_{-(n-1)}}{(s - s_1)^{n-1}} + \cdots + \frac{a_{-1}}{s - s_1} + a_0 + a_1(s - s_1) + \cdots \tag{28}$$

For the immediate vicinity of this point we may replace this series by its dominant term, thus

$$Z(s) \approx \frac{a_{-n}}{(s - s_1)^n} \tag{29}$$

For the complex coefficient a_{-n} let us write

$$a_{-n} = Ae^{j\phi} \tag{30}$$

and in order to facilitate passage around the pole on a concentric circular contour, we let

$$(s - s_1) = \rho e^{j\theta} \tag{31}$$

where ρ is arbitrarily small but nonzero. We then have for the real part of $Z(s)$

$$\operatorname{Re}\left(\frac{A}{\rho^n} e^{(j\phi - n\theta)}\right) = \frac{A}{\rho^n} \cos(n\theta - \phi) \tag{32}$$

The angle ϕ is fixed and θ varies from zero to 2π while ρ remains constant as we pass completely around the pole on the circular contour. It is clear that the real part 32 assumes arbitrarily large positive and negative values on this contour, changing its algebraic sign $2n$ times for a complete circuit around the pole. It is thus seen that a p.r. function must be analytic in the right half-plane where its real part is required to remain positive.

We next show that a function cannot be p.r. unless its reciprocal is p.r., or that the reciprocal is p.r. if the given function is p.r. Suppose we write

$$Z(s) = R(\sigma, \omega) + jX(\sigma, \omega) \tag{33}$$

with

$$s = \sigma + j\omega$$

The reciprocal function is then given by

$$Y(s) = \frac{1}{Z(s)} = \frac{R - jX}{R^2 + X^2} \tag{34}$$

from which it is clear that the function and its reciprocal have identical algebraic signs for their real parts. The statement made at the beginning of this paragraph, therefore, follows at once.

Since the zeros of a function are the poles of its reciprocal, and since poles cannot occur in the right half-plane, zeros cannot occur there either. That is to say, a p.r. function can have neither zeros nor poles in the right half-plane, or the function and its reciprocal must be analytic in the right half of the complex frequency plane. This property of driving-point functions, which was arrived at earlier on the basis of physical reasoning alone, is now seen to be derivable also as a property of p.r. functions.

It is logical at this point to inquire whether poles of a p.r. function are permitted to lie upon the j-axis, which is the boundary between the left and right half-planes. To answer this question we look again at expression 32 for the real part in the vicinity of a pole and observe that if the pole is simple ($n = 1$) and if the coefficient $a_{-n} = a_{-1}$ (which, since $n = 1$, becomes the residue of the function in this pole) is real and positive so that, according to Eq. 30, $\phi = 0$, then the real part in the immediate vicinity of the pole at least remains positive in the right half-plane where $-\pi/2 < \theta < \pi/2$. This argument does not answer the question decisively as to whether poles are or are not permitted to lie upon the j-axis, but it shows that if j-axis poles are simple and if the residues there are real and positive, then the real part is assuredly positive (and incidentally arbitrarily large in absolute value except at the points $\theta = \pm\pi/2$) upon a semicircular contour of sufficiently small radius lying wholly in the right half-plane. We shall see presently that these circumstances are just sufficient to permit such j-axis poles, but to this end we need first to make use of another bit of complex function theory.

A p.r. function, as its title implies, has a real part that is positive over the entire right half of the complex frequency plane. It is also a function that is analytic over this entire half-plane. There is a theorem about the real part of a function, analytic over a given region, and upon the boundary of that region, to the effect that the largest and the smallest values which the real part assumes throughout this region inclusive of its boundary must lie upon the boundary.* To apply this theorem to the p.r. function that may have j-axis poles as well as left half-plane poles,

* See *The Mathematics of Circuit Analysis*, E. A. Guillemin, John Wiley & Sons, New York, 1949, p. 334. (Hereafter *The Mathematics of Circuit Analysis* will be referred to as *MCA*.)

we choose as the boundary of the right half-plane the j-axis exclusive of the immediate vicinities of the poles, and close the gaps at these points by concentric semicircular paths lying wholly within the right half-plane as shown in Fig. 3.

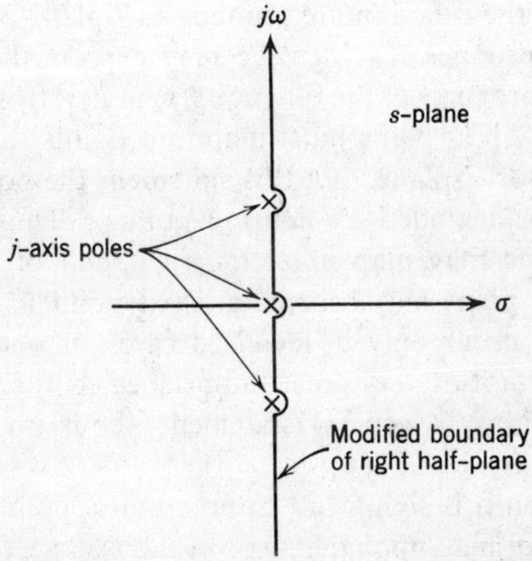

Fig. 3. Modified j-axis on which real part must be nonnegative when the pertinent impedance has j-axis poles.

Since the smallest of all values assumed by the real part throughout the right half-plane occurs upon this boundary, its positiveness on the boundary assures its positiveness over the entire right half-plane, and hence, taking for granted the realness of the function for real values of s, it assures the p.r. character of the function. In other words, a function may have j-axis poles and still be p.r. if these poles are simple and the residues there are real and positive so that the real part of the function is nonnegative on the semicircular arcs, and if on the linear portions of the j-axis exclusive of the immediate vicinities of these poles the real part is likewise nonnegative.

The earlier discussion based upon physical reasoning also showed that the real part must be nonnegative on the j-axis, but it did not yield the more specialized information about j-axis poles. The present analytic approach is thus seen to bring into sharper focus the detailed characteristics of impedance and admittance functions.

We may now summarize the necessary conditions for a function to be p.r. For lumped, finite networks it is a rational function (quotient of finite polynomials), as has amply been demonstrated in our earlier considerations. The requirement that it be real for real values of s merely demands that the polynomials have real coefficients. In turn, the zeros of these polynomials are symmetrically located with respect to the real axis of the s-plane. The pole-zero pattern of a p.r. rational function must first of all display such symmetry.

In addition, its real part must be positive over the entire right half of the s-plane; or if we consider in addition to the s-plane one in which complex values of the function are plotted (an f-plane or a Z-plane if the function is an impedance Z), then we may express this condition as a required mapping property of the function $f(s)$ or $Z(s)$ by stating that points in the right half of the s-plane must map into points that lie in the right half of the f-plane or Z-plane. In this statement the boundaries of these half-planes may be included as limiting points. Thus a point on the j-axis of the s-plane may map upon the right half of the f-plane, or a j-axis point in one plane may become a j-axis point in the other. Such limiting situations occur only in idealized cases of one sort or another and are practically not of very great importance, but, since one can construct these situations, a general statement about p.r. functions must admit their possibility.

In this connection it is significant to note that a point in the right half of the s-plane cannot map upon the j-axis of the f-plane, for it is possible to express the p.r. property in an equivalent polar form (in contrast to the rectangular form in which we have stated it in all of our present discussions) that obviously excludes this possibility. The equivalent polar form reads*

$$|\arg Z| \leq |\arg s| \text{ for } |\arg s| \leq \frac{\pi}{2} \tag{35}$$

In other words, the angle of the function is always smaller in absolute value than the angle of the independent variable, for values of the variable within its right half-plane. Although apparently stronger than the statement about the real part of $f(s)$ being positive for all values of s with a positive real part, the two forms are found mathematically to be entirely equivalent in the sense that one can be derived from the other. The polar form is more convenient for the demonstration of certain detailed properties than the rectangular one, and vice versa. Hence it is well to be familiar with both.

Through use of the above-mentioned theorem about the real part of a complex function and of the analysis relating to its behavior in the vicinity of a pole, we can formulate alternate conditions (necessary and sufficient) that a function (which is real for real s) have a positive real part over the entire right half of the s-plane. Thus, if the function is analytic in the right half-plane and on the j-axis, we merely have to assure ourselves that its real part is nonnegative on the j-axis in order to be sure that it remains

* For a detailed derivation and discussion of this form see *MCA*, p. 419.

positive over the entire right half-plane. An investigation of the behavior of the real part along the j-axis is evidently a less formidable task than an investigation of it over the entire right half-plane. The practical value resulting from use of the pertinent theorem in this situation is easily apparent. Note, however, that we must be assured first of all that the function is analytic over the right half of the s-plane inclusive of the j-axis, otherwise the theorem does not apply.

If j-axis poles are present, we modify the boundary of the right half-plane as is shown in Fig. 3. On this modified boundary the function is analytic, and to it the theorem can be applied, yielding the statement that if the real part of the function is positive everywhere on this new boundary then it is positive everywhere in the right half-plane. In considering the real part along this modified boundary we separate the latter into two portions: the linear sections and the semicircular arcs. Nonnegativeness of the real part on the semicircular arcs is assured by the simplicity of the poles and the positive realness of the residues of the function in these poles. Nonnegativeness of the real part along the linear sections may be investigated (as will presently be shown) by the same methods that are used in the absence of j-axis poles. Hence in the event that the function has j-axis poles, the statement regarding necessary and sufficient conditions for p.r. character is different from those statements applicable to functions analytic on the j-axis only in the additional demand regarding the simplicity of these poles and the positive realness of the residues in them.

We may condense all of the above discussion relating to the necessary and sufficient conditions for p.r. character into a single statement consisting of three parts, which may be referred to hereafter as the ABC's of p.r. functions. This statement reads:

A rational function which is real for real values of s is a p.r. function if:

A. It is analytic in the right half-plane.

B. Its real part is nonnegative on the j-axis. (36)

C. Any j-axis poles are simple and have positive real residues.

The following discussion relates to methods whereby these characteristics may conveniently be investigated.

4 Methods of Testing for P.R. Character

The first major step in testing a given rational function for p.r. character (assuming that its polynomials have real coefficients) involves the question of analyticity in the right half-plane or the investigation of the denominator polynomial to determine whether its zeros have negative real parts. If it does have only such zeros, it is referred to as a Hurwitz polynomial, and the test for this property is also known as a Hurwitz test.* Before going into the details of this procedure we shall recall for the sake of emphasis some of the important properties of Hurwitz polynomials which are indispensable to our present discussions.

Let the polynomial be written

$$Q(s) = a_0 + a_1 s + a_2 s^2 + \cdots + a_n s^n \tag{37}$$

with the even and odd parts

$$\begin{aligned} m(s) &= a_0 + a_2 s^2 + a_4 s^4 + \cdots + a_n s^n \\ n(s) &= a_1 s + a_3 s^3 + a_5 s^5 + \cdots + a_{n-1} s^{n-1} \end{aligned} \tag{38}$$

for n even, and with the last terms interchanged if the degree n is odd. Regarding necessary conditions that can be checked by inspection, it is useful to know that $Q(s)$ cannot be a Hurwitz polynomial unless all coefficients are positive and no intermediate terms are missing. We can easily appreciate these requirements by noting that a Hurwitz polynomial can result only from the product of the following three kinds of factors

$$\begin{aligned} &(s + a) \text{ for } a \text{ real and positive} \\ &(s^2 + b^2) \text{ for } b \text{ real} \\ &(s^2 + 2as + a^2 + b^2) \text{ for } a \text{ and } b \text{ as above} \end{aligned} \tag{39}$$

The first of these represents a zero on the negative real axis. The second represents a pair of j-axis zeros (which we will include as limiting possibilities, although the mathematical definition of a Hurwitz polynomial does not include such zeros), and the third factor represents the pair of conjugate complex zeros: $s = -a \pm jb$. The important point about these factors is that no minus signs occur in them. Hence it is clear that products of any combination of such factors can yield polynomials with plus signs only and no terms can cancel so as to yield a polynomial in which some powers of s are absent. The only exception to the last statement is the degenerate case in which the polynomial consists solely of a product of j-axis factors and becomes an even or an odd function of s (odd if the zero $s = 0$ is present).

Assuming that all the coefficients in the polynomial are positive and none are missing, we must proceed to use more subtle methods for establishing its Hurwitz character. In this connection we make use of some of the more detailed properties of the even and odd part polynomials $m(s)$ and $n(s)$. These properties are stated by saying that the zeros of

* For a detailed discussion of this topic the reader is referred to *MCA*, Ch. VI, Art. 26, p. 395. Only some of the more pertinent aspects of this discussion are presented here.

these polynomials are all simple and are restricted to lie on the j-axis where they occur in conjugate pairs and alternate with each other. Thus $n(s)$, being odd, contains s as a factor and hence is zero at $s = 0$. The next largest pair of zeros in absolute value belong to the polynomial $m(s)$, the next largest are a pair of j-axis zeros contained in $n(s)$, and so forth. We speak of this situation as the *alternation property* of the zeros of m and n along the j-axis. Sometimes one also refers to these sets of zeros as being *interlaced* on the j-axis. If it can be proved that the even and odd parts of an arbitrary polynomial have such simple interlaced zeros, then that polynomial must have Hurwitz character, and vice versa.

Figure 4 illustrates the general character of the behavior of the pair of

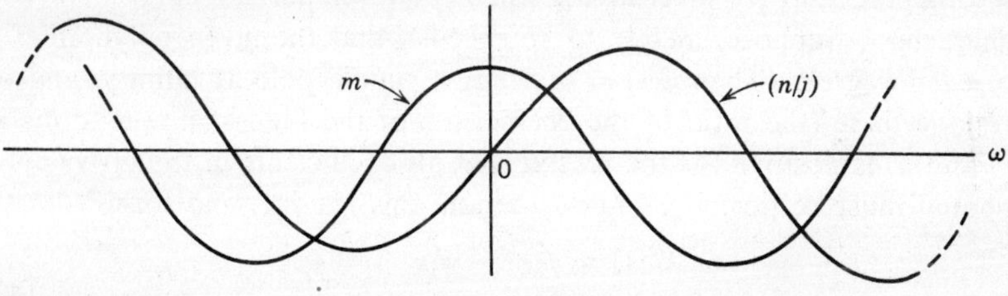

Fig. 4. Behavior of the even and odd parts of a Hurwitz polynomial along the j-axis.

functions $m(j\omega)$ and $n(j\omega)/j$ along the ω axis. The rational function

$$\psi(s) = \frac{m(s)}{n(s)} \tag{40}$$

has simple j-axis poles at the zeros $s = s_\nu$ of $n(s)$ with residues equal to

$$k_\nu = \left[\frac{m(s)}{n'(s)}\right]_{s=s_\nu} \tag{41}$$

in which

$$n'(s) = \frac{dn}{ds} = \frac{dn}{jd\omega} = \frac{d}{d\omega}\left(\frac{n}{j}\right) \tag{42}$$

From Fig. 4 we see (with reference to the zeros of n) that where m is positive the slope of n/j is also positive, and where m is negative there the function n/j has negative slope. It follows that the residues given in Eq. 41 are all real and positive, a result that is a direct consequence of the alternation of the zeros of m and n. By an entirely similar argument it follows that the rational function n/m (the reciprocal of Eq. 40) likewise has all simple poles restricted to the j-axis and positive real residues. In fact, the function $\psi(s)$ and its reciprocal have the limiting form of a p.r. function whose poles all lie on the j-axis. Such a function, we shall see later on, is the driving-point impedance or admittance of a lossless network—one containing only inductances and capacitances (no resistances).

Of importance in this connection is the fact that if a given rational function has this character then its reciprocal has the same character, and the pertinent polynomials $m(s)$ and $n(s)$ must have simple zeros alternating on the j-axis. The simplicity and the alternance of the zeros of $m(s)$ and $n(s)$ are *necessary* conditions for the positiveness of all residues of $\psi(s)$ or of its reciprocal. That is why the converse statement is likewise true, namely, that if the poles of $\psi(s)$ are all simple and the residues positive, then the zeros of $m(s)$ and $n(s)$ must be simple and must alternate —no other arrangement of zeros can yield consistently positive residues. In this case the reciprocal of $\psi(s)$ must obviously have the same property, and the sum $m(s) + n(s) = Q(s)$ must be a Hurwitz polynomial.

Based upon these results, it is an easy matter to formulate a simple testing procedure for determining whether a given polynomial has Hurwitz character. Suppose, merely to be specific, that the given polynomial is of even degree. Then $\psi(s) = m/n$ has a simple pole at infinity, and its residue there (the ratio of the coefficients of the highest power terms in m and n) is positive for the reason that all coefficients in the given polynomial must be positive. Let α_1 denote this residue, and form

$$\begin{aligned} \psi'(s) &= \psi(s) - \alpha_1 s \\ &= \frac{m}{n} - \alpha_1 s = \frac{m'}{n} \end{aligned} \tag{43}$$

which amounts to removing the pole at infinity from $\psi(s)$. The remainder function $\psi'(s)$ has all the other poles of $\psi(s)$ and, of course, with the same residues. If the given polynomial is Hurwitz, these poles are simple and the residues positive, and unless this is the case the given polynomial cannot be Hurwitz. If it is, then the reciprocal function $1/\psi'$ again has all simple poles with positive residues, and if this is not the case then ψ' cannot have this character. In other words, it is just as appropriate to test $1/\psi'$ as ψ' for the necessary character, and we choose at this point to test $1/\psi'$, for the reason that it again has a pole at infinity and can be treated in the same manner as $\psi(s)$ is in the first step. Thus in Eq. 43 it is clear that the polynomial m' is two degrees lower than m because the subtraction removes the highest power term, and the terms in m or in n involve only alternate powers of s. Hence ψ' is zero for $s = \infty$, and its reciprocal has a simple pole there. Unless the residue in this pole is again positive, $1/\psi'$ and hence ψ' cannot have the necessary character to establish that the given polynomial is Hurwitz. As in Eq. 43 we form

$$\begin{aligned} \psi''(s) &= \frac{1}{\psi'(s)} - \alpha_2 s \\ &= \frac{n}{m'} - \alpha_2 s = \frac{n'}{m'} \end{aligned} \tag{44}$$

in which α_2 must be positive, and n' turns out to be two degrees lower than n and hence one degree lower than m'. Hence the function $1/\psi''$ again has a simple pole at infinity with a residue that must be positive.

The pattern of the test is thus established. If continued in the same manner it leads to a continued-fraction expansion of the rational function $\psi(s)$ of the form

$$\psi(s) = \alpha_1 s + \cfrac{1}{\alpha_2 s + \cfrac{1}{\alpha_3 s + \cfrac{1}{\ddots + \cfrac{1}{\alpha_n s}}}} \tag{45}$$

in which n is the degree of the original polynomial, and all coefficients $\alpha_1 \cdots \alpha_n$ must be positive if this polynomial is to have Hurwitz character.

This procedure may be varied if desired by removing poles at $s = 0$ in a continuing sequence from the original function $\psi(s) = m/n$ and from its subsequent inverted remainders. Such a variation yields a continued-fraction development like Eq. 45 in which the terms are constants times $1/s$. Another way of considering this variation in the procedure is to say: In the given polynomial $Q(s)$ let us replace s by $1/s$ and then test the resulting polynomial in the same manner as described originally. Since the transformation corresponding to replacing s by $1/s$ maps the left half-plane upon itself, it is obvious that if $Q(1/s)$ is Hurwitz, then $Q(s)$ must be. It is clear that the coefficients in the polynomial $Q(1/s)$, if arranged in the order of increasing powers of s, are just the coefficients in $Q(s)$ with their order reversed. Hence any Hurwitz criteria (equivalent to Routh's criteria*) expressed directly in terms of these coefficients remain in effect unchanged if one reverses the order of all coefficients.

The discussion so far as this testing procedure is concerned has tacitly assumed that the given polynomial has no zeros on the j-axis. If it does, then the process of continued-fraction development will not continue for n terms but will terminate prematurely, and the j-axis factors will be placed in evidence at the point of termination. The reason for this behavior of the process is due to the fact that j-axis factors, as shown in set 39, have the form $(s^2 + b^2)$, which is an even function of s and hence if $Q(s) = m(s) + n(s)$ contains such a factor then it must separately be contained in both $m(s)$ and $n(s)$. In the rational function $\psi(s) = m/n$

* See, for example, E. J. Routh's *Dynamics of a System of Rigid Bodies*, 3rd ed., The Macmillan Co., London, 1877. See also Vol. 2 of 4th ed.

such a common factor (or factors) virtually cancels, and hence the continued division and inversion process will terminate sooner than it normally would.

This process may be made to seem a little more familiar to the reader if we remind him at this point of an analogous situation in arithmetic, namely, of the process of finding the highest common factor (HCF) of two given integers, say 32 and 56. The process of reasoning upon which the method of determining the HCF is based is known as Euclid's algorithm.* The actual procedure follows the same continued division and remainder inversion process as that discussed above in connection with the function $\psi(s)$. For the two integers given above, the process may be set down compactly as follows

$$\begin{array}{rrrr} & \underline{1} & & \\ 32| & 56 & & \\ & \underline{32|} & \underline{1} & \\ & 24| & 32 & \\ & & \underline{24|} & \underline{3} \\ & & 8| & 24 \\ & & & \underline{24} \\ & & & 0 \end{array} \tag{46}$$

The immediately preceding divisor after the process terminates is revealed as the HCF, in this case 8. The quotients are of no interest so far as determination of the HCF is concerned. In the analogous procedure performed with the rational function $\psi(s)$ these quotients are the terms in the development 45. If in this case the process terminates prematurely, the immediately preceding divisor is revealed as being a common factor of the polynomials $m(s)$ and $n(s)$. Such a common factor, as mentioned above, must be an even polynomial in s. If it is of second degree it will have the form $(s^2 \pm b^2)$. If the sign is plus, it represents a pair of j-axis factors; if it is minus, a pair of real axis zeros symmetrically located about the origin is implied. In the latter case, the polynomial under test would, of course, be discarded.

When the common factor revealed by this process is of fourth degree (second degree in s^2) one can apply the quadratic formula to find the s^2 zeros. Only if these are negative real are the zeros of the common factor restricted to the j-axis. In general, the common factor must have only simple negative real s^2 zeros in order that the polynomial under test be acceptable because only simple j-axis poles are permitted in the im-

---

* *Introduction to Higher Algebra*, M. Bocher, The Macmillan Co., New York, 1927, p. 189.

pedance or admittance whose denominator polynomial is being examined. We shall discuss presently a method for determining whether a given polynomial has any or all simple negative real zeros. In the meantime we wish further to illustrate the present item under discussion.

Suppose we consider the following polynomial

$$Q(s) = s^6 + 2s^5 + 14s^4 + 26s^3 + 49s^2 + 72s + 36 \tag{47}$$

The continued division and inversion process analogous to that shown in 46 reads

$$\begin{array}{r|l} & \quad s/2 \\ \cline{2-2} 2s^5 + 26s^3 + 72s & s^6 + 14s^4 + 49s^2 + 36 \\ & s^6 + 13s^4 + 36s^2 \qquad |2s \\ \cline{2-2} & 0 + \quad s^4 + 13s^2 + 36 \,|\, 2s^5 + 26s^3 + 72s \\ & \qquad\qquad\qquad\quad 2s^5 + 26s^3 + 72s \\ \cline{2-2} & \qquad\qquad\qquad\quad 0 \;+\; 0 \;+\; 0 \end{array} \tag{48}$$

The process terminates prematurely and the immediately preceding divisor

$$s^4 + 13s^2 + 36 = (s^2 + 4)(s^2 + 9) \tag{49}$$

is thus revealed to be a factor of $Q(s)$. Dividing it into $Q(s)$ leaves

$$s^2 + 2s + 1 = (s + 1)^2 \tag{50}$$

In this simple example the method has yielded the complete factorization of the given polynomial which, incidentally, is Hurwitz. This fact is deduced from the positiveness of the two quotients in 48 and from the nature of the common factor 49 which could be factored by inspection and thus be shown to consist of j-axis factors alone.

It is rather obvious from plots of the even and odd parts like those in Fig. 4 that the derivative of a Hurwitz polynomial is again a Hurwitz polynomial, since it is graphically clear that differentiation does not destroy the simplicity and alternation of the zeros of these parts, although the even one becomes odd and the odd one even.* The same reasoning shows that $(m + dm/ds)$ and $(n + dn/ds)$ are Hurwitz polynomials if $(m + n)$ is such a one.

These thoughts yield a simple procedure (not requiring factorization) for determining whether a given even polynomial like 49 has only simple j-axis zeros, for if it has only such zeros then its derivative likewise has only such simple zeros which alternate with those of the given polynomial. The latter divided by its derivative must then be a ψ function like 45 and must yield a continued fraction with all positive coefficients. If this

* A rigorous proof of this property is given by the method suggested in Prob. 45 at the end of this chapter.

test fails, we may conclude that the even polynomial in question does not have all simple j-axis zeros.

When this test is applied to an even polynomial having multiple-order j-axis zeros, then these will reveal themselves just as j-axis factors do in the Hurwitz test described and illustrated above.

By this process of Hurwitz testing we have taken care of item A in criteria 36 and have established whether item C will need to be considered. In this regard we have already taken into account the part of C that requires j-axis poles to be simple. We shall still need to investigate the residues in these poles. The discussion of this item will be easier if we postpone it until the method of checking part B has been taken care of.

The problem here is to study the real part of the given function to see whether it remains positive along the entire j-axis, or more precisely, to see that it is nonnegative along this axis since it may become zero at a finite number of points. To facilitate formulation of the real part, we write the function in the form

$$Z(s) = \frac{P(s)}{Q(s)} = \frac{m_1 + n_1}{m_2 + n_2} \tag{51}$$

in which the m's and n's denote respectively the even and odd parts of the polynomials. For $s = j\omega$, the even parts are real and the odd parts are imaginary. Hence the process of rationalizing the denominator is indicated by

$$Z(s) = \frac{P(s) \times Q(-s)}{Q(s) \times Q(-s)} = \frac{(m_1 + n_1)(m_2 - n_2)}{(m_2 + n_2)(m_2 - n_2)} \tag{52}$$

Multiplying out the numerator and separating it into its even and odd parts gives

$$Z(s) = \frac{m_1 m_2 - n_1 n_2}{m_2^2 - n_2^2} + \frac{m_2 n_1 - m_1 n_2}{m_2^2 - n_2^2} \tag{53}$$

For $s = j\omega$, the common denominator in these two terms becomes the square of an absolute value [note that $n_2(j\omega)$ is a pure imaginary and hence $-n_2^2(j\omega)$ is positive real for all ω as is m_2^2 for $s = j\omega$]. The first term therefore becomes the real part and the second term the imaginary part of $Z(s)$. Hence we have

$$\text{Re}\,[Z(j\omega)] = \left(\frac{m_1 m_2 - n_1 n_2}{m_2^2 - n_2^2}\right)_{s=j\omega} \tag{54}$$

It is significant in this connection to note that the two terms in Eq. 53 become respectively the real and imaginary part of $Z(s)$ *only* for $s = j\omega$. For any complex $s = \sigma + j\omega$ these terms are not respectively real and imaginary.

Our immediate problem is to establish conditions under which expression 54 remains positive (or is nonnegative) for all real ω. Since the denominator is the square of an absolute value and hence is obviously positive, it is sufficient (as well as necessary) to demand that

$$(m_1m_2 - n_1n_2)_{s=j\omega} \geq 0 \text{ for all real } \omega \quad (55)$$

The expression in parentheses, which is the even part of $P(s) \times Q(-s)$, is an even polynomial in s. Since for $s = j\omega$, $s^2 = -\omega^2$, we prefer to denote it as

$$A(-s^2) = m_1m_2 - n_1n_2 \quad (56)$$

and represent it in the factored form

$$A(-s^2) = A_n(s_1^2 - s^2)(s_2^2 - s^2) \cdots (s_n^2 - s^2) \quad (57)$$

where n is the common degree of the polynomials P and Q. [If their degrees are not the same, then the degree of $A(-s^2)$ as a polynomial in s^2 is lower, but the form of 57 is the same.] For $s = j\omega$ we then have

$$A(\omega^2) = A_n(s_1^2 + \omega^2)(s_2^2 + \omega^2) \cdots (s_n^2 + \omega^2) \quad (58)$$

and condition 55 demands that this expression remain positive (or at least not become negative) for all real ω.

An obvious condition is that $A_n > 0$; otherwise $A(\omega^2)$ would become negative for large ω. As for the remaining factors we shall consider separately the possible kinds that can occur. Thus the s^2 zeros of the polynomial $A(-s^2)$, Eq. 56, may be: (*a*) positive real; (*b*) in the form of conjugate complex pairs; and (*c*) negative real. Factors $(s_k^2 + \omega^2)$ in which s_k^2 is a positive real number cannot become negative for any real ω values. A pair of factors $(s_k^2 + \omega^2)(\bar{s}_k^2 + \omega^2)$ involving conjugate complex s_k^2 values are themselves conjugate complex and hence yield a product that is the square of an absolute value. Such factors cannot become negative for any real ω values either. However, factors involving negative real s_k^2 values will surely become negative for some ω values, and this negativeness can be canceled only if there is another identical factor in expression 58. In such a case the polynomial $A(-s^2)$ has a negative real s^2 zero of multiplicity 2, or a double-order j-axis zero.

j-axis zeros and poles of $Z(s)$ are evidently zeros of the polynomial 56, but the converse is not necessarily true, since zeros of this polynomial need not be zeros of $P(s)$ or $Q(s)$ in Eq. 51. At a double-order j-axis zero of 56, the second term in Eq. 53 is a valid representation not only for $Z(s)$ but for the evaluation of its derivative as well; and since $m_2n_1 - m_1n_2 = (n_1/m_2)(m_2^2 - n_2^2)$ at this point, we see that

$$Z(s) = (n_1/m_2) = (m_1/n_2) \quad (59)$$

and

$$\frac{dZ}{ds} = \frac{d}{ds}\left(\frac{n_1}{m_2}\right) = \frac{d}{ds}\left(\frac{m_1}{n_2}\right) \quad (60)$$

wherever the j-axis real part of a p.r. impedance has a zero.

The result, Eq. 59, shows that the impedance is not necessarily zero where its real part is zero. We will see later on (Eq. 26, Ch. 2) that $Z(s)$, according to Eq. 59, behaves like a reactance function (impedance of a lossless network) in the vicinity of such a point, and that its derivative there, as given by Eq. 60, always has a positive real value.

Note further that although Re $[Z(j\omega)]$ must be zero at a j-axis zero of $Z(s)$, it does not have a j-axis pole where $Z(s)$ has one. This fact follows because j-axis zeros of the polynomial 56 must be of even order so that a j-axis factor $(s_k^2 + \omega^2)^2$ contained in the denominator of expression 54 must also be contained in the numerator and hence cancels out. We will see later on (Art. 2, Ch. 8) that Re $[Z(j\omega)]$ has impulses where $Z(s)$ has j-axis poles, but this type of singular behavior is not revealed by the conventional algebraic methods of real-part evaluation. Except for such singular behavior, the j-axis real part of a p.r. impedance is a bounded function.

An impedance having only j-axis poles may readily be seen (by conventional algebraic methods) to have an identically zero real part along the j-axis. Since the residues in these poles must be positive real, such an impedance is given by a sum of functions of the form

$$\frac{1}{s - j\omega_0} + \frac{1}{s + j\omega_0} = \frac{2s}{s^2 + \omega_0^2} \tag{61}$$

which, for $s = j\omega$, clearly have a pure imaginary value.

Any impedance containing j-axis poles may be expressed (through its partial-fraction expansion) as the sum of two components of which one has no j-axis poles while the other has only such poles. The j-axis real part of the total impedance is thus seen to be identically equal to the real part of that component having no j-axis poles. This view of the situation shows that the j-axis real part of an impedance not only remains regular in the vicinity of a j-axis pole of that impedance but that its behavior in this vicinity is not even influenced in the slightest way by the presence of such a pole.

If the polynomial 58 is plotted versus ω, a simple zero is a point where the curve intersects the zero axis at a finite nonzero slope, whereas at a double-order zero the curve touches this axis but does not intersect it. At a third-order zero, the curve again intersects the axis, although at zero slope. In the vicinity of a fourth-order zero the function remains positive, as for the double-order zero, but has smaller values, and so forth. Odd multiplicity zeros of $A(-s^2)$ on the j-axis of the s-plane are thus seen to be points where the real part 54 will surely change sign, while even multiplicity j-axis zeros, or zeros anywhere else in the s-plane (whatever their order), cannot bring about a change of sign of the real part along the j-axis.

It thus becomes clear that expression 58 remains positive for all real ω values unless the polynomial $A(-s^2)$, Eq. 56, has negative real s^2 zeros of *odd* multiplicity. If it does not have such zeros, then the real part 54 is assuredly nonnegative for all real values of ω as required by condition B in 36. The test for this condition is thus reduced to the test of a polynomial to discover whether it has any negative real zeros of odd multiplicity. If we consider the polynomial 58 and let $\omega^2 = x$, then, for $-\infty < \omega < \infty$, x lies in the range $0 < x < \infty$. We then have a polynomial in x which is required to have no positive real zeros of odd multiplicity.

Sturm's theorem* provides a method of testing for the presence of real zeros within a given interval. We shall first present the method and then give a justification for it. Let the polynomial be written

$$f_0(x) = a_0 + a_1x + a_2x^2 + \cdot\cdot\cdot + a_nx^n \tag{62}$$

and its first derivative

$$f_1(x) = a_1 + 2a_2x + \cdot\cdot\cdot + na_nx^{n-1} \tag{63}$$

In terms of these, which are the first two of a set of so-called Sturm functions, we form the remaining ones through a process of continued division similar to that used in the Hurwitz test, with the difference that each division is carried out for a two-term quotient. The successive remainders, differing by unity in degree, with their algebraic signs changed, are the rest of the functions in the set.

More specifically, we begin by dividing $f_1(x)$ into $f_0(x)$ according to the familiar long-division process. The first term in the quotient is a constant times x, and the second term is a constant. The remaining dividend is then one degree lower than $f_1(x)$. This remaining dividend with its algebraic sign changed is denoted as $f_2(x)$—the next of the Sturm functions in the set. The identical process is repeated by dividing $f_2(x)$ into $f_1(x)$ for a two-term quotient, leaving a remaining dividend that is one degree lower than $f_2(x)$. With its algebraic sign reversed, it is denoted as $f_3(x)$. The process finally ends up with the last function $f_n(x) =$ constant, n being the degree of the original polynomial 62.

If we denote the two-term quotients (which are incidentally of no interest) by $q_1, q_2, \cdot\cdot\cdot$, we may compactly summarize the process just described by writing

$$\begin{aligned} f_0 &= q_1f_1 - f_2 \\ f_1 &= q_2f_2 - f_3 \\ f_2 &= q_3f_3 - f_4 \\ &\cdots\cdots\cdots \end{aligned} \tag{64}$$

---

* See, for example, *First Course in the Theory of Equations*, L. E. Dickson, John Wiley & Sons, New York, 1921, p. 75.

The method of establishing whether $f_0(x)$ has any zeros within an interval $a < x < b$ is done in a manner which is best presented in the following tabular form

	f_0	f_1	f_2	$\cdots$	f_n	Variations in Sign
$x = a$	$+$	$-$	$-$	$\cdots$	$+$	v_a
$x = b$	$-$	$+$	$+$	$\cdots$	$+$	v_b

(65)

The entries in the columns headed $f_0, f_1, \cdots$ are merely the algebraic signs of the values of these functions computed for $x = a$ and $x = b$ respectively. The quantities in the last column denoted by v_a and v_b are integers representing the number of sign changes in the respective rows. The absolute value of the difference $v_a - v_b$ equals the number of zeros of $f_0(x)$ within the stated interval, assuming that all of them are simple.

If the polynomial $f_0(x)$ has repeated zeros, then its derivative $f_1(x)$ contains these zeros also. That is to say, $f_0(x)$ and $f_1(x)$ have a common factor. According to Euclid's algorithm, the process of formation indicated in Eqs. 64 will then terminate prematurely and place in evidence such common factors. If these are sufficiently simple so that we can see by inspection what the repeated zeros in $f_0(x)$ are and how many times they are repeated, then we can tell immediately whether the remainder after these factors are divided out of $f_0(x)$ need be further investigated. If the common factor is too complicated to analyze by inspection, we may have to apply the Sturm theorem process to it separately. In any case, if this study reveals any odd multiplicity zeros in $f_0(x)$, then there is no need to go on with the test, for it has already indicated that the given polynomial is unacceptable.

We shall now show why the procedure summarized in tabular form, 65, yields the desired information. To this end we shall first observe some simple principles concerning the number of variations in sign in a given set of quantities. These principles are placed in graphic evidence by the diagrammatic representation in Fig. 5. A set of quantities $f_0 \cdots f_{10}$ are assumed to have a random distribution of algebraic signs. The positive ones are listed in the column to the left of the vertical zero axis and the negative ones are on the right. The broken line connects the various order numbers.

From this diagram it is clear that the number of sign variations among the set of functions equals the number of zero crossings of the broken line. Consider a function like number 2, 3, or 6 for which the immediately adjacent ones are alike in sign. If such a function should have its algebraic sign reversed, then the number of zero crossings changes by *two*. However, when a function like number 1, 4, 5, 7, 8, or 9 (for which the immediately adjacent ones have opposite signs) is shifted from one side of the zero axis to the other, there is no change in the number of zero crossings. Note also that if function number 0 is shifted from one side of the zero axis to the other, the number of zero crossings always changes by *one*.

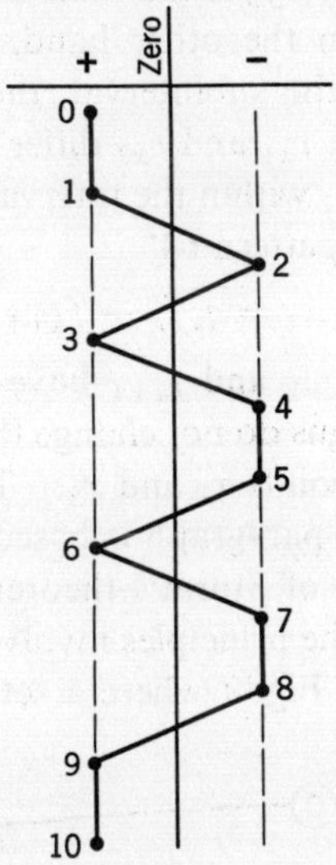

Fig. 5. Graphical portrayal of the sign changes among a set of Sturm functions.

Within the interval $a < x < b$, the function f_0 changes sign at each of its zeros. Several of the other functions $f_1, f_2, \cdots$ may also have zeros within this interval. Suppose all such zeros are denoted by $x = r_1 < r_2 < r_3 \cdots$, and let the quantities $x_1, x_2, x_3, \cdots$ be values of x that separate the zeros. Figure 6 illustrates this arrangement of values along the

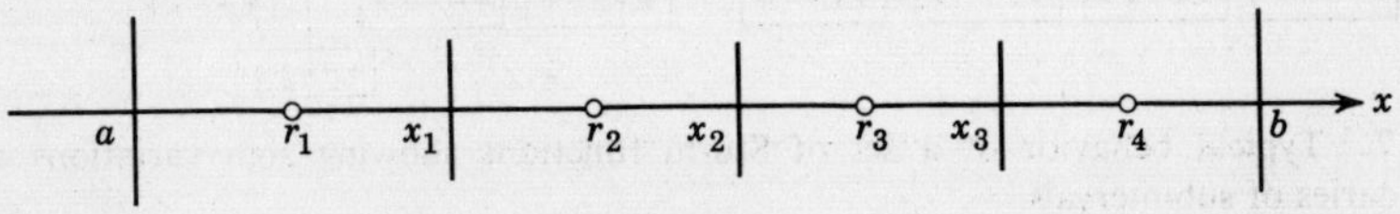

Fig. 6. Zeros of a set of Sturm functions within an interval $a < x < b$.

x-axis for the interval $a < x < b$. It is clear that

$$v_a - v_b = (v_a - v_{x1}) + (v_{x1} - v_{x2}) + (v_{x2} - v_{x3}) + (v_{x3} - v_b) \quad (66)$$

where $v_{x1}, v_{x2}, \cdots$ are the number of variations in sign among the $f_0 \cdots f_n$ at the intermediate points $x_1, x_2, \cdots$.

MICHAEL G. REKOFF, JR.

Throughout any subinterval like $x_1 < x < x_2$, only one of the functions in the set $f_0 \cdots f_n$ passes through zero. This one has opposite signs at x_1 and x_2; the rest have the same signs at x_1 as they do at x_2. It may now be shown that if the function that is zero within the subinterval is any one except f_0, then the ones immediately adjacent to it have opposite signs throughout the subinterval so that, according to Fig. 5, the number of zero crossings (not to be confused with zeros on the x-axis as depicted in Fig. 6) among the set $f_0 \cdots f_n$ is the same at x_1 as it is at x_2; that is to say, $(v_{x1} - v_{x2})$ is zero. On the other hand, if f_0 is the function that passes through zero within the subinterval, then the previous argument concerning Fig. 5 shows that v_{x1} and v_{x2} differ by one.

Suppose f_k is zero at $x = r_2$ within the interval $x_1 < x < x_2$, and $k \neq 0$. According to the formation pattern 64

$$f_{k-1} = q_k f_k - f_{k+1} \tag{67}$$

and hence it is clear that f_{k-1} and f_{k+1} have opposite signs at $x = r_2$ where $f_k = 0$. Since their signs do not change throughout the subinterval, they have opposite signs at points x_1 and x_2. The condition upon which the argument in the previous paragraph is based is therefore fulfilled, and the basis for the justification of Sturm's theorem is established.

In order to clarify further the principles involved in the above reasoning, it may be helpful to refer to Fig. 7 where a set of functions $f_0, f_1, f_2, f_3,$

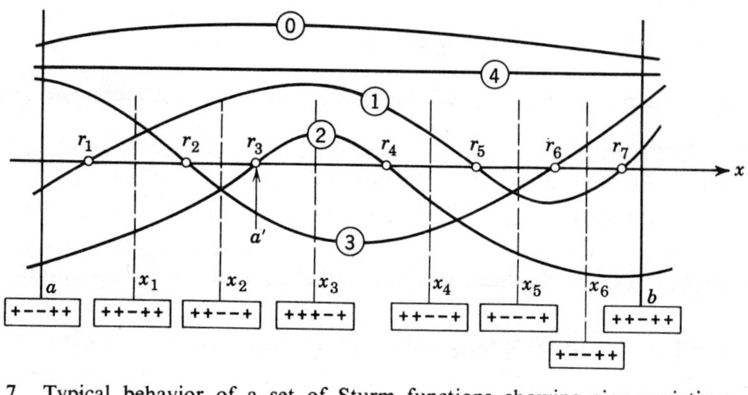

Fig. 7. Typical behavior of a set of Sturm functions showing sign variations at boundaries of subintervals.

and f_4 are shown plotted versus x over an interval $a < x < b$. The last function, in this case f_4, is always a constant and hence plots simply as a horizontal line. Function number 0 (the given polynomial) is assumed to have no zeros in the stated interval. Consequently the number of variations in sign at the boundaries of the various subintervals, as may be seen from the sign sequences indicated at each boundary $a, x_1, x_2, \cdots, b$, are all the same. If boundary a happens to fall at a' where $f_2 = 0$, one can call the sign of f_2 either $+$ or $-$, as though the boundary were slightly to the left or to the right of the point where $f_2 = 0$; the variations in sign are the same either way. Or one can disregard f_2 at this boundary so far as sign variations go; but one does not disregard f_2 altogether, for it is not trivial at the other boundaries where $f_2 \neq 0$.

At this point it may be helpful to give an example of a p.r. test. Suppose the following impedance is specified

$$Z(s) = \frac{s^4 + 3s^3 + 4s^2 + 3s + 1}{s^4 + 7s^3 + 6s^2 + 4s + 1} \tag{68}$$

First we make a Hurwitz test for the denominator polynomial as follows

$$\begin{array}{rl}
 & 0.143s \\
7s^3 + 4s \,\big| & s^4 + \quad 6s^2 + 1 \\
 & s^4 + 0.57s^2 \qquad \big| 1.29s \\
 & \qquad 5.43s^2 + 1 \;\big|\, 7s^3 + 4s \\
 & \qquad\qquad 7s^3 + 1.29s \,\big|\, 2.0s \\
 & \qquad\qquad\qquad 2.71s \,\big|\, 5.43s^2 + 1 \\
 & \qquad\qquad\qquad\qquad 5.43s^2 \quad \big| 2.71s \\
 & \qquad\qquad\qquad\qquad\qquad 1 \,\big|\, 2.71s
\end{array} \tag{69}$$

Since all quotients are positive and the process does not terminate prematurely, the impedance 68 is shown to be analytic in the right half-plane and on the j-axis.

We next form

$$\begin{aligned}(m_1m_2 - n_1n_2) &= (s^4 + 4s^2 + 1)(s^4 + 6s^2 + 1) - 3s(s^2 + 1)s(7s^2 + 4) \\ &= s^8 - 11s^6 - 7s^4 - 2s^2 + 1\end{aligned} \tag{70}$$

or

$$(m_1m_2 - n_1n_2)_{s=j\omega} = \omega^8 + 11\omega^6 - 7\omega^4 + 2\omega^2 + 1 \tag{71}$$

Letting $\omega^2 = x$ we have

$$\begin{aligned} f_0 &= x^4 + 11x^3 - 7x^2 + 2x + 1 \\ f_1 &= \qquad 4x^3 + 33x^2 - 14x + 2 \end{aligned} \tag{72}$$

The next Sturm function is formed as follows

$$\begin{array}{rl}
 & 0.25x + 0.688 \\
4x^3 + 33x^2 - 14x + 2 \,\big| & x^4 + 11x^3 - 7x^2 + 2x + 1 \\
 & x^4 + 8.25x^3 + 3.5x^2 + 0.5x \\
 & \qquad 2.75x^3 - 3.5x^2 + 1.5x + 1 \\
 & \qquad 2.75x^3 + 22.7x^2 - 9.63x + 1.375 \\
 & \qquad\qquad - 26.2x^2 + 11.13x - 0.375
\end{array}$$

Hence

$$f_2 = 26.2x^2 - 11.13x + 0.375 \tag{73}$$

In an analogous manner we get

$$f_3 = -0.68x - 1.503$$

and (74)

$$f_4 = -153$$

Since the process does not terminate prematurely, f_0 has no repeated zeros. The interval of interest is $0 < x < \infty$. The following tabular form is now easily constructed

	f_0	f_1	f_2	f_3	f_4	Variations in Sign
$x = 0$	+	+	+	−	−	1
$x = \infty$	+	+	+	−	−	1

(75)

The number of variations in sign is the same for $x = 0$ and $x = \infty$. Hence the polynomial $f_0(x)$ has no zeros throughout this infinite interval. The impedance 68 is proved to be a p.r. function.

In order to illustrate the use of Sturm's theorem in a more elaborate situation, consider the polynomial

$$f_0(x) = x^7 - 10x^6 + 45x^5 - 120x^4 + 204x^3 - 216x^2 + 128x - 32 \quad (76)$$

having the derivative

$$f_1(x) = 7x^6 - 60x^5 + 225x^4 - 480x^3 + 612x^2 - 432x + 128 \quad (77)$$

Dividing f_1 into f_0, we get a quotient $\frac{1}{7}x - \frac{10}{49}$ and a remainder

$$\frac{30}{49}x^5 - \frac{270}{49}x^4 + \frac{912}{49}x^3 - \frac{1440}{49}x^2 + \frac{1056}{49}x - \frac{228}{49}$$

This polynomial with its sign reversed is $f_2(x)$. We observe, however, that any function may be multiplied by a positive constant without affecting the results of the method, and so we multiply in this case by $\frac{49}{6}$ before changing the algebraic sign and denoting this polynomial as

$$f_2(x) = -5x^5 + 45x^4 - 152x^3 + 240x^2 - 176x + 48 \quad (78)$$

Division of f_2 into f_1 next yields the quotient $-\frac{7}{5}x - \frac{3}{5}$ and a remainder

$$\frac{196}{5}x^4 - \frac{1176}{5}x^3 + \frac{2548}{5}x^2 - \frac{2352}{5}x + \frac{784}{5}$$

Here we multiply by $\frac{5}{196}$, change the algebraic sign, and have

$$f_3(x) = -x^4 + 6x^3 - 13x^2 + 12x - 4 \quad (79)$$

Next we divide f_3 into f_2, get the quotient $5x - 15$, and the remainder

$$3x^3 - 15x^2 + 24x - 12$$

Multiplication by $\frac{1}{3}$ and change of algebraic sign gives

$$f_4(x) = -x^3 + 5x^2 - 8x + 4 \tag{80}$$

The next cycle turns out to be the final one in this sequence, for it yields an identically zero remainder, showing that the polynomial $f_4(x)$ is a common factor of f_0 and f_1. Since this common factor cannot readily be analyzed by inspection, we shall treat it by means of Sturm's theorem. To this end we let the polynomial 80 be denoted as

$$h_0(x) = x^3 - 5x^2 + 8x - 4 \tag{81}$$

and its first derivative by

$$h_1(x) = 3x^2 - 10x + 8 \tag{82}$$

Dividing h_1 into h_0 yields a quotient $\frac{1}{3}x - \frac{5}{9}$ and a remainder

$$-\tfrac{2}{9}x + \tfrac{4}{9}$$

After multiplication by $-\frac{9}{2}$ it becomes

$$h_2(x) = x - 2 \tag{83}$$

Dividing it into h_1, we again find an identically zero remainder, indicating that $(x - 2)$ is a common factor of $h_0(x)$ and $h_1(x)$. This fact tells us that $h_0(x)$ contains the factor $(x - 2)^2$, and it is then a simple matter to find the remaining factor in h_0, whereupon we have

$$h_0(x) = (x - 1)(x - 2)^2 \tag{84}$$

This now is recognized as being a common factor of the original polynomial $f_0(x)$ in Eq. 76 and its derivative 77, from which it is clear that $f_0(x)$ contains the factors $(x - 1)^2(x - 2)^3$. Since $(x - 2)^3$ is a factor of odd multiplicity, the function $(m_1m_2 - n_1n_2)$ from which f_0 is derived cannot remain positive for all real values of ω.

Before concluding this discussion about Sturm's theorem, it may be recalled that the Hurwitz test yields an even function of s when $m(s)$ and $n(s)$ have a common factor, and in this event we must analyze this factor to see whether its zeros are on the j-axis and to determine their multiplicities. Writing $-\omega^2$ for s^2 and then letting $\omega^2 = x$, we get a polynomial $f_0(x)$ whose positive real zeros correspond to pairs of conjugate j-axis zeros in the original polynomial in s (the common factor of m and n). Sturm's theorem is again a means whereby this situation can be handled, although the method already given is computationally simpler.

If this part of the Hurwitz test reveals multiple-order j-axis zeros, then the given impedance or admittance function is, of course, immediately rejected since it cannot possibly be p.r. However, if such j-axis zeros (and hence the corresponding j-axis poles of the impedance) are simple, then we must evaluate the residues of the impedance in these poles to see whether they are real and positive. The computational work involved in these calculations may become rather tedious as the following considerations show.

Given the impedance in the form of Eq. 51, the straightforward evaluation of the residue in a simple j-axis pole at $s_\nu = j\omega_\nu$ reads

$$k_\nu = \left[\frac{P(s)}{Q'(s)}\right]_{s=s_\nu} \tag{85}$$

where $Q'(s)$ is the derivative of $Q(s)$ with respect to s. The polynomials $P(s)$ and $Q'(s)$ evaluated for $s = s_\nu$ are always complex, even if k_ν turns out to be real. It is possible to avoid complex numbers entirely in these computations through taking advantage of the following properties.

When $Z(s)$, Eq. 51, has a simple j-axis pole, either of two distinctly different detailed situations may arise. In one of these, m_2 and n_2 contain the simple factor $(s^2 + \omega_\nu^2)$ as mentioned above. The j-axis real part 54 does not exhibit singular behavior at this point since the polynomials $(m_1m_2 - n_1n_2)$ and $(m_2^2 - n_2^2)$ both contain the squared factor $(s^2 + \omega_\nu^2)^2$. If we write

$$m_2 = m^*_2(s^2 + \omega_\nu^2) \qquad \text{and} \qquad n_2 = n^*_2(s^2 + \omega_\nu^2) \tag{86}$$

then this conclusion leads to the result

$$\left(\frac{m_1}{n_1}\right)_\nu = \left(\frac{n^*_2}{m^*_2}\right)_\nu = \left(\frac{n'_2}{m'_2}\right)_\nu \tag{87}$$

where the prime denotes differentiation with respect to s, and the subscript ν indicates that the pertinent ratio is evaluated for $s = j\omega_\nu$. The residue k_ν, according to Eq. 85, is thus seen to be expressible as

$$k_\nu = \left(\frac{m_1 + n_1}{m'_2 + n'_2}\right)_\nu = \left(\frac{m_1}{n'_2}\right)_\nu = \left(\frac{n_1}{m'_2}\right)_\nu \tag{88}$$

It is alternately possible for the polynomial $(m_1m_2 - n_1n_2)$ to contain the factor $(s^2 + \omega_\nu^2)^2$ because of both m_1 and m_2 containing the simple factor and n_2 containing the squared factor. In this case we can write in place of Eq. 86

$$m_1 = m^*_1(s^2 + \omega_\nu^2), \qquad m_2 = m^*_2(s^2 + \omega_\nu^2), \qquad n_2 = n^*_2(s^2 + \omega_\nu^2)^2 \tag{89}$$

and Eq. 85 is seen to yield

$$k_\nu = \left(\frac{n_1}{m'_2}\right)_\nu \tag{90}$$

The alternate ratio $(m_1/n'_2)_\nu$ appearing in Eq. 88 now becomes both indeterminate and invalid since no relation analogous to that expressed in Eq. 87 applies.

The additional possibility that n_1 and n_2 contain the factor $(s^2 + \omega_\nu^2)$ simply and m_2 contains it squared is merely a variation of this same situation. The results are the same as before with the roles of m and n interchanged. Thus, instead of 89, we have

$$n_1 = n^*_1(s^2 + \omega_\nu^2), \qquad n_2 = n^*_2(s^2 + \omega_\nu^2), \qquad m_2 = m^*_2(s^2 + \omega_\nu^2)^2 \quad (91)$$

and the residue may be calculated from

$$k_\nu = \left(\frac{m_1}{n'_2}\right)_\nu \quad (92)$$

the other ratio in 88 now becoming indeterminate and invalid. There are no other detailed possibilities for which $Z(s)$ has a simple j-axis pole. In any case, therefore, the pertinent residue may be calculated from a ratio of even polynomials evaluated for $s^2 = -\omega_\nu^2$, so that complex numbers are completely avoided.

Since such a ratio must obviously always be real, we see that the non-negativeness of Re $[Z(j\omega)]$, when it has zeros, guarantees the realness of the residues in the corresponding j-axis poles of $Z(s)$. However, it does not guarantee the positiveness of these residues, and so it is still necessary to compute them in order to check condition C in the set 36.

Before concluding this discussion regarding the testing of rational functions to determine their possible p.r. character, a few collaterally interesting (and sometimes useful) items may be pointed out. According to the procedure described in this article, one does not bother to test the numerator polynomial for Hurwitz character, although it must certainly be such a polynomial if it turns out that the given rational function is p.r. However, when the Hurwitz test of the denominator polynomial reveals simple j-axis poles so that it becomes necessary to evaluate residues, it may be advantageous to switch to a test of the inverted rational function, for its denominator polynomial (which is the original numerator polynomial) may have no j-axis zeros. In this case it will be simpler to test the inverted function since no residue evaluations will be called for. In other words, the ABC's of p.r. testing reduce to parts A and B alone when no j-axis poles are present; and since a given rational function may have j-axis poles but no j-axis zeros (or vice versa), a particular orientation of the function (right side up or upside down) can be most advantageous.

In any case, the importance of looking for obvious defects should not be overlooked in making p.r. tests. The polynomials must have positive real coefficients, they can have no missing terms, and their highest as well as their lowest powers cannot differ by more than unity. If any of these characteristics, discernible by inspection, are not in order, then there is obviously no need to proceed with any more elaborate means. Likewise there may be situations where simpler measures than the formal Hurwitz and Sturm tests are adequate. These are matters, however, that need not be further discussed here.

In this connection it may be significant to point out a necessary property of the polynomial $(m_1m_2 - n_1n_2)$, the knowledge of which can in certain instances be collaterally helpful. Clearly, the algebraic sign of m_1m_2 or of $-n_1n_2$ changes at a zero of m_1 or m_2 or of n_1 or n_2. It is also clear that if the sign of $(m_1m_2 - n_1n_2)$ is to remain unchanged, then the signs of m_1m_2 and $-n_1n_2$ cannot change in succession. If m_1m_2 changes sign at some frequency, then $-n_1n_2$ cannot next change sign, otherwise the entire factor $(m_1m_2 - n_1n_2)$ will change sign. Rather, the next sign change must be that of m_1m_2 again; and since the zeros of m_1 and n_1 as well as those of m_2 and n_2 alternate, it follows that the product $-n_1n_2$ must next undergo two sign changes. Thereupon m_1m_2 has two successive zeros; then $-n_1n_2$ has two, and so forth. This double alternance of the zeros of m_1m_2 and $-n_1n_2$ is a necessary though evidently not a sufficient condition to insure the positiveness of $(m_1m_2 - n_1n_2)$ for $s = j\omega$. In situations where the zeros of m_1, m_2, n_1, and n_2 are either known or easily found, one may quickly see whether it is necessary to proceed with a Sturm test.

As Reza has shown,* one may compactly summarize the conditions for p.r. character by plotting the ratio $(m_1m_2)/(-n_1n_2)$ for $s = j\omega$ as a function of ω^2 and noting that it must have the appearance shown in Fig. 8.

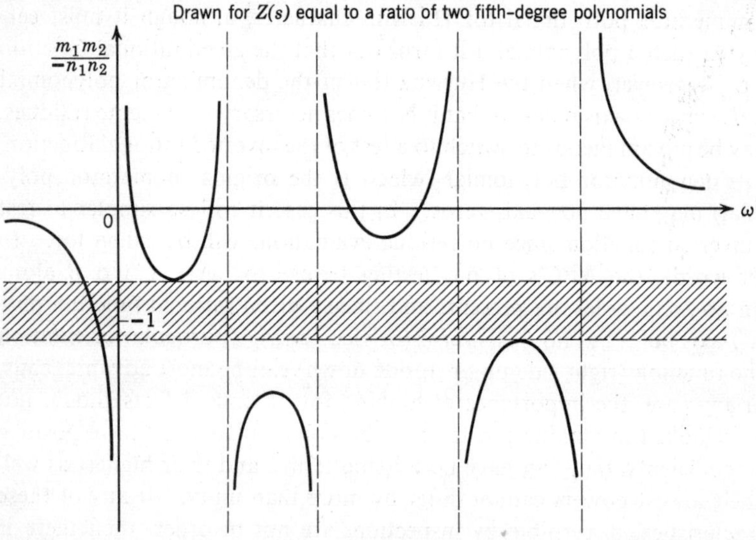

Fig. 8. Illustrating a collaterally useful property of a p.r. function.

* F. M. Reza, *Quart. Prog. Rept.*, Mass. Inst. Technol., Research Lab. Electronics, Apr. 15, 1954, p. 88.

Important is the fact that the double alternance of zeros of m_1m_2 and $-n_1n_2$ alone assures that the smallest minimum of this function cannot be less than its largest maximum for $\omega^2 > 0$, as is emphasized in the figure by the crosshatched strip. Proof of this statement follows from a consideration that the equation $m_1m_2 - kn_1n_2 = 0$ or $(m_1m_2)/(-n_1n_2) = -k$ could otherwise have more positive real ω^2 roots for an appropriate value of k than its degree in ω^2 allows. (For the function of Fig. 8, this number is at most 5.) The pertinent impedance $Z(s)$ is p.r. if (and only if) the point -1 lies within the crosshatched interval.

A closely related artifice that frequently is useful as a means for quickly establishing the p.r. or non-p.r. character of a given rational function, is based upon the manipulation

$$(m_1m_2 - n_1n_2) = m_1n_2\left(\frac{m_2}{n_2} - \frac{n_1}{m_1}\right) = m_1n_2(\psi_2 - \psi_1) \tag{93}$$

As pointed out in connection with the testing of polynomials for Hurwitz character, the functions ψ_1 and ψ_2 have simple zeros and poles alternating along the j-axis. Since their difference changes sign at each pole of ψ_1 or ψ_2, and these are zeros of the factor m_1n_2, the total expression 93 does not change sign at such points. The whole factor 93 can change sign along the j-axis only if the difference $(\psi_2 - \psi_1)$ experiences such a sign change at points where ψ_1 and ψ_2 are both finite.

In terms of a plot of the functions ψ_1 and ψ_2 as shown in Fig. 9, one

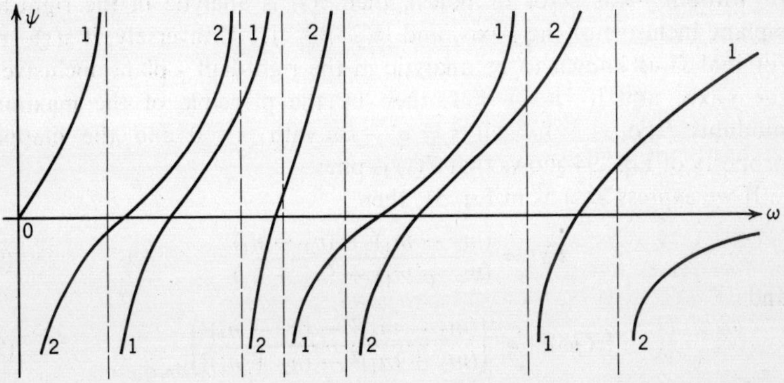

Fig. 9. Behavior of the functions defined by Eq. 93 when the pertinent impedance is p.r.

may say that expression 93 changes sign along the j-axis only if the curves for these two functions intersect. If only their zero and pole locations are known, one can without further computation make a rough sketch which will quickly reveal whether the pertinent rational function is obviously p.r. or whether more detailed computations (such as are involved in the Sturm test) are needed to obtain a decisive result. Rather approximate values for the pole and zero locations (which are computationally not difficult to obtain, even for relatively complex rational functions) are often sufficient to establish a clear-cut answer.

5 Collaterally Useful Properties of P.R. Functions

Suppose we consider the frequently occurring transformation

$$z(s) = \frac{1 - Z(s)}{1 + Z(s)}, \qquad Z(s) = \frac{1 - z(s)}{1 + z(s)} \tag{94}$$

which maps the right half of the Z-plane upon the interior of the unit circle of the z-plane and vice versa. If $Z(s)$ is p.r. then $\text{Re}\,[Z(s)] \geq 0$ for $\text{Re}\,(s) \geq 0$. According to the mapping property of the function 94, it then follows that $|z(s)| \leq 1$ for $\text{Re}\,(s) \geq 0$. Conversely, if $|z(s)| \leq 1$ for $\text{Re}\,(s) \geq 0$, then $Z(s)$ must surely be p.r. (taking for granted that it is real for real s).

This mutual relationship between $z(s)$ and $Z(s)$ can be stated in another way that is found to be practically more useful. If $Z(s)$ is p.r. then $1 + Z(s)$ is surely p.r. and hence has no right half-plane zeros and no j-axis zeros, for $1 + Z(j\omega)$ cannot be zero if $\text{Re}\,[Z(j\omega)] \geq 0$. Hence $z(s)$ is analytic within the right half of the s-plane and on the j-axis. According to the principle of the maximum modulus (see *MCA*, p. 327) it follows that the maximum absolute value of $z(s)$ for all values of $s = \sigma + j\omega$ with $\sigma \geq 0$ must occur for $\sigma = 0$. Hence we can say that if $Z(s)$ is p.r. (with or without j-axis zeros or poles), then $z(s)$ is analytic in the right-half s-plane inclusive of the j-axis, and $|z(j\omega)| \leq 1$. Conversely, if $z(s)$—real for real s—is known to be analytic in the right-half s-plane inclusive of the j-axis, and if $|z(j\omega)| \leq 1$, then by the principle of the maximum modulus $|z(s)| \leq 1$ for all $s = \sigma + j\omega$ with $\sigma \geq 0$ and the mapping property of Eqs. 94 shows that $Z(s)$ is p.r.

If we express $Z(s)$ as in Eq. 51, then

$$z(s) = \frac{(m_2 - m_1) + (n_2 - n_1)}{(m_2 + m_1) + (n_2 + n_1)} \tag{95}$$

and

$$|z(j\omega)|^2 = \left[\frac{(m_2 - m_1)^2 - (n_2 - n_1)^2}{(m_2 + m_1)^2 - (n_2 + n_1)^2}\right]_{s=j\omega} \tag{96}$$

From these expressions we note that an interchange of m_1 and m_2 or of n_1 and n_2 affects neither the denominator of $z(s)$ nor the absolute value of $z(s)$ for $s = j\omega$. Hence if $Z(s)$ is p.r. so that $z(s)$ is analytic in the right half-plane and $|z(j\omega)| \leq 1$ (and conversely), then this situation remains true if one interchanges m_1 and m_2 or n_1 and n_2. That is to say, if

$$Z(s) = \frac{P(s)}{Q(s)} = \frac{m_1 + n_1}{m_2 + n_2} \tag{97}$$

is p.r., then so is

$$Z^*(s) = \frac{m_2 + n_1}{m_1 + n_2} = \frac{P^*(s)}{Q^*(s)} \tag{98}$$

or its reciprocal. The even or odd parts of the polynomials $P(s)$ and $Q(s)$ can be interchanged without affecting the p.r. character of $Z(s)$.

A corollary of this result is the statement that if $Z(s)$ is p.r. then not only are $P(s)$ and $Q(s)$ Hurwitz polynomials but $P^*(s)$ and $Q^*(s)$ resulting from an interchange of even or odd parts are likewise Hurwitz polynomials.

6 Additional Remarks About Transfer Impedances

Returning to the quadratic form 17 and the fact that it must be a p.r. function for all real values of $x_1 \cdots x_p$, let us write the $z_{ik}(s)$ in the rectangular form

$$z_{ik}(s) = r_{ik}(\sigma, \omega) + jx_{ik}(\sigma, \omega) \tag{99}$$

Then, since

$$\text{Re} \sum_{i,k=1}^{p} z_{ik}x_ix_k = \sum_{i,k=1}^{p} r_{ik}x_ix_k \tag{100}$$

application of the previous discussion regarding p.r. functions to the impedance $Z(s)$, Eq. 17, shows that if $Z(s)$ is to be p.r., all the driving-point and transfer impedances $z_{ik}(s)$ must be real for real s, they must be analytic in the right half of the s-plane, and

$$\sum_{i,k=1}^{p} r_{ik}x_ix_k \geq 0 \qquad \text{for } s = j\omega \quad (\sigma = 0) \tag{101}$$

In the event that any $z_{ik}(s)$ has j-axis poles, these must be simple and the residue of $Z(s)$ in such poles must be real and positive. The latter statement follows from the obvious fact that any poles of the $z_{ik}(s)$ are poles of $Z(s)$. If the residues of $z_{ik}(s)$ in their j-axis poles are denoted by k_{ik} and those of $Z(s)$ by k, then it is clear that

$$k = \sum_{i,k=1}^{p} k_{ik}x_ix_k \text{ must be positive real} \tag{102}$$

We see that the necessary p.r. character of $Z(s)$ requires the quadratic forms 101 and 102 to be positive definite. Recalling a familiar way of expressing the necessary and sufficient conditions that a given real quadratic form be positive definite (see *MCA*, p. 150) we may say that, in addition to $r_{ii} \geq 0$ (for $s = j\omega$) and $k_{ii} \geq 0$, the following determinants

$$\begin{vmatrix} r_{11} & \cdots & r_{1p} \\ \cdots & \cdots & \cdots \\ r_{p1} & \cdots & r_{pp} \end{vmatrix} \tag{103}$$

$$\begin{vmatrix} k_{11} & \cdots & k_{1p} \\ \cdot & \cdots & \cdot \\ k_{p1} & \cdots & k_{pp} \end{vmatrix} \tag{104}$$

and all of their principal minors must be positive.

For a two terminal-pair network this result means that for $s = j\omega$ we must have

$$r_{11} \geq 0, \qquad r_{22} \geq 0, \qquad r_{11}r_{22} - r_{12}^2 \geq 0 \tag{105}$$

and for any j-axis poles of the $z_{ik}(s)$, which must be simple, the residues must be real and satisfy the conditions

$$k_{11} \geq 0, \qquad k_{22} \geq 0, \qquad k_{11}k_{22} - k_{12}^2 \geq 0 \tag{106}$$

An important aspect of these results is the fact that r_{12} and k_{12} do not have to be positive. It is this circumstance that allows the zeros of the transfer impedance $z_{12} = z_{21}$ to have any multiplicity and to lie anywhere in the complex frequency plane. It also permits the degree of the denominator polynomial to be more than one degree higher than that of the numerator polynomial. If it is, then clearly the angle of z_{12} changes by more than 90° throughout the frequency interval $0 < \omega < \infty$ and hence the real part r_{12} is certainly negative for some frequencies.

In view of this result we observe that if expression 54 represents the real part of a *transfer* impedance 51, then the polynomial 56 may have simple j-axis factors. An exception occurs if the pertinent frequency is a pole of this impedance, for then the real part 54 will likewise have a pole unless the numerator of this expression has a double-order zero at the same point. Since the j-axis real part of a driving-point impedance cannot have a pole, conditions 105 do not allow the corresponding real part of a transfer impedance to have a pole either. Hence when a transfer impedance has a j-axis pole, its polynomial $(m_1m_2 - n_1n_2)$, like that of a p.r. function, must there have a double-order zero.

For a symmetrical two terminal-pair network $r_{11} = r_{22}$, and conditions 105 read

$$r_{11} \geq 0, \qquad r_{11}^2 - r_{12}^2 = (r_{11} + r_{12})(r_{11} - r_{12}) \geq 0 \tag{107}$$

If we consider this network in the form of a lattice (Fig. 10), we have

$$z_{11} = \tfrac{1}{2}(z_b + z_a), \qquad z_{12} = \tfrac{1}{2}(z_b - z_a) \tag{108}$$

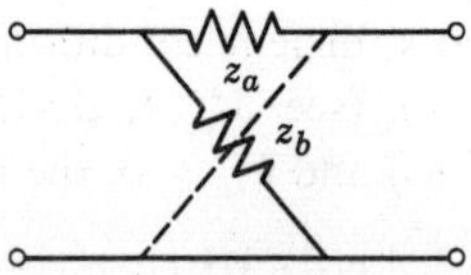

Fig. 10. Symmetrical lattice network. The impedance z_a is repeated in the lower series branch, and z_b in the alternate cross arm (shown dashed).

or

$$z_a = z_{11} - z_{12}, \qquad z_b = z_{11} + z_{12} \tag{109}$$

so that

$$r_a = r_{11} - r_{12}, \qquad r_b = r_{11} + r_{12} \tag{110}$$

Conditions 107 become

$$(r_a + r_b) \geq 0 \quad \text{and} \quad r_a r_b \geq 0 \tag{111}$$

or

$$r_a \geq 0 \quad \text{and} \quad r_b \geq 0$$

Similar relations hold for the residue conditions 106. We see that the realizability conditions for the symmetrical two terminal-pair network become those for the lattice impedances z_a and z_b. Hence, if a symmetrical two terminal-pair network is realizable at all, it is realizable as a lattice network. Stated in another way, the lattice structure is the most general form of symmetrical network. This result is practically very useful, for it tells us that if the lattice is used as a point of departure in the design of a symmetrical network, the design procedure will be as general as possible. The same cannot be said if one assumes the network in the form of a T or Π-network or even a bridged-T. The lattice plays an important part in synthesis procedures for this reason.

A practically important property of a transfer function is the fact that it is essentially the same for any given network and its reciprocal. To appreciate the truth of this statement, consider the network N with its associated source and load as shown in Fig. 11 and the completely

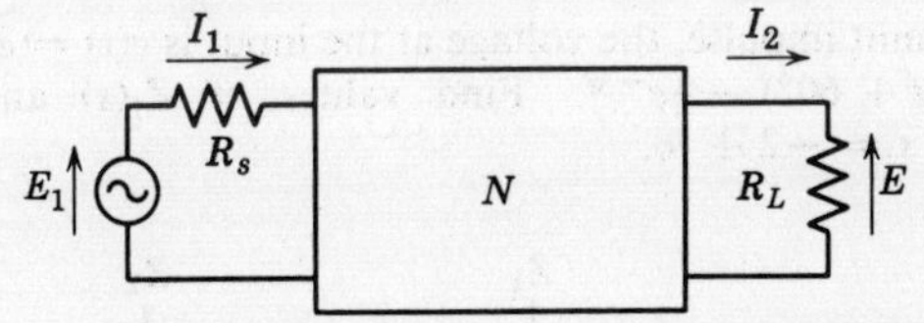

Fig. 11. Passive two terminal-pair with associated source and load.

reciprocal or dual situation given in Fig. 12. Thus the two terminal-pair

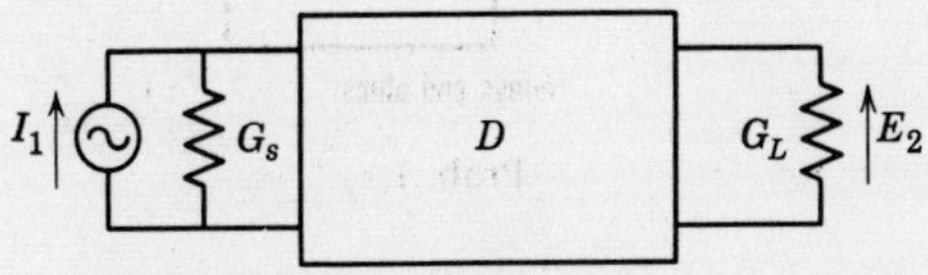

Fig. 12. Dual of situation shown in Fig. 11.

networks N and D are duals or reciprocals and so are the sources and loads. As for the source and load resistances, the following numerical relations hold

$$G_s \doteq R_s \qquad \text{and} \qquad G_L \doteq R_L \tag{112}$$

Since the transfer admittance for the situation in Fig. 11 is numerically equal to the transfer impedance in Fig. 12, we have

$$\left(\frac{I_2}{E_1}\right)_N \doteq \left(\frac{E_2}{I_1}\right)_D \tag{113}$$

However, in Fig. 11

$$\left(\frac{E_2}{E_1}\right)_N = \left(\frac{I_2}{E_1}\right)_N \cdot R_L \tag{114}$$

and if in Fig. 12 we replace the current source in parallel with G_s by a voltage source $E_1 = I_1/G_s \doteq I_1/R_s$ in series with R_s, we have

$$\left(\frac{E_2}{E_1}\right)_D \doteq \left(\frac{E_2}{I_1}\right)_D \cdot R_s \tag{115}$$

Hence

$$\left(\frac{E_2}{E_1}\right)_D \equiv \left(\frac{E_2}{E_1}\right)_N \cdot \frac{R_s}{R_L} \tag{116}$$

If R_s equals R_L, as it frequently does, then the transfer voltage ratio is identical for the network N and its dual or reciprocal. When the source and load impedances have any purely resistive values, the transfer ratios differ at most by a constant multiplier.

PROBLEMS

1. For i_s = unit impulse, the voltage at the input is $e(t) = e^{-t}\sin(10t - 30^0) + 2e^{-2t}\cos(6t + 60^\circ) - \frac{1}{2}e^{-5t}$. Find values of $Z_1(s)$ and $Z_2(s)$ for $s = -1 - j10$ and $s = -2 + j6$.

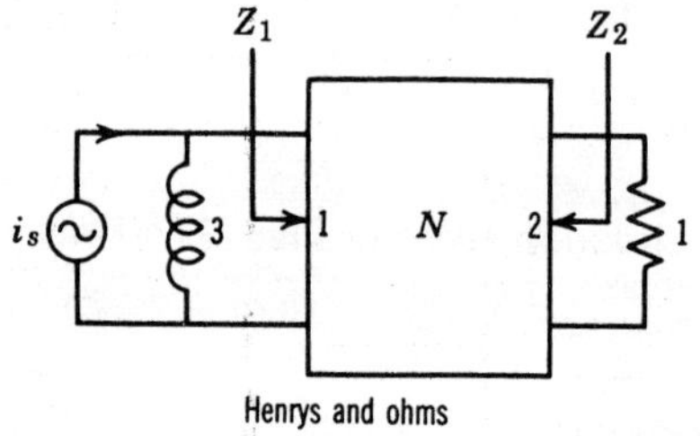

Prob. 1.

2. At a certain frequency the impedance of network (*a*) between nodes *bd* is the conjugate of the impedance of network (*b*) between nodes *eh*. If the two networks are connected by joining separately nodes *b* and *e* and also nodes *d* and *h*, what will be the value of the impedance of the combination at the same frequency for a driving point created through attaching leads at the node pair *a-c*? Again,

for leads attached to the node pair g-h? Also for a driving point created through cutting the branch joining nodes a and c at m?

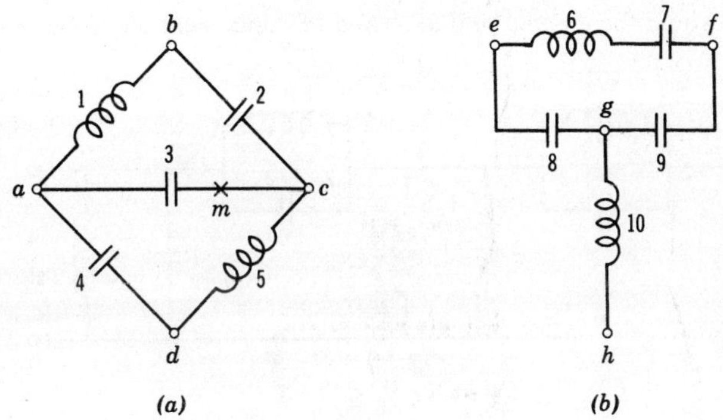

Prob. 2.

3. N_1 and N_2 in the accompanying figure are linear, passive, lossless networks. With a-a' open, one finds at some particular frequency ω_0 that $Z_b = -Z_c$. If the terminals b'-c' and b-c are joined, what is the value of $Z_a(\omega_0)$ and why?

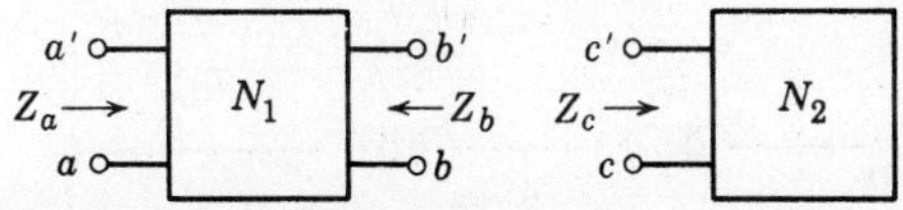

Prob. 3.

4. The accompanying sketch represents a linear, passive, finite, lumped, bilateral network with two accessible terminal pairs labeled 1 and 2. The network is nondegenerate in the sense that all of its open- and short-circuit natural frequencies are excitable from both terminal pairs. The following data are given:

If $i_1(t)$ = a unit step and $i_2 \equiv 0$, then

$$e_1(t) = 1 + Ae^{-t} \sin (t + \phi) \text{ for } t > 0$$

If $i_2(t)$ = a unit impulse and $e_1 \equiv 0$, then

$$e_2(t) = Be^{-t} \cos (2t + \psi) \text{ for } t > 0$$

(a) Determine the driving-point impedance $Z_1(s)$ at terminal pair 1 with terminal pair 2 open, as a rational function of the complex frequency variable s.

(b) Find the values of A and ϕ in the above expression for e_1.

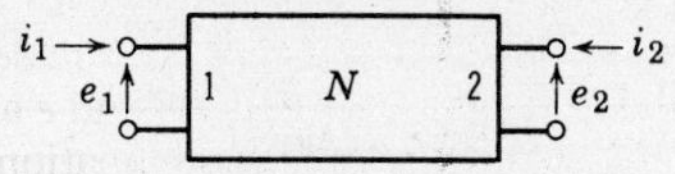

Prob. 4.

5. (*a*) With reference to the circuit shown, find $e_2(t)$ for $t > 0$ if $e_1(t)$ equals a unit impulse.

(*b*) What would the answer be if all L and $(1/C)$ values were multiplied by 100?

(*c*) What would the answer be if all L and C values were multiplied by $\frac{1}{100}$?

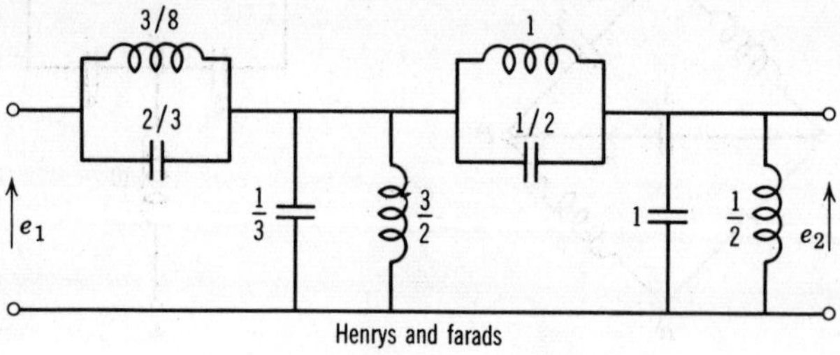

Prob. 5.

6. For the circuit shown and without writing circuit equations, determine the impedances z_{11} and z_{22} at the indicated terminal pairs (with the 1-ohm resistor attached) as functions of s.

With the 1-ohm resistor removed, what is the value of z_{22} for $s = -\frac{1}{2} + j\frac{1}{2}\sqrt{3}$?

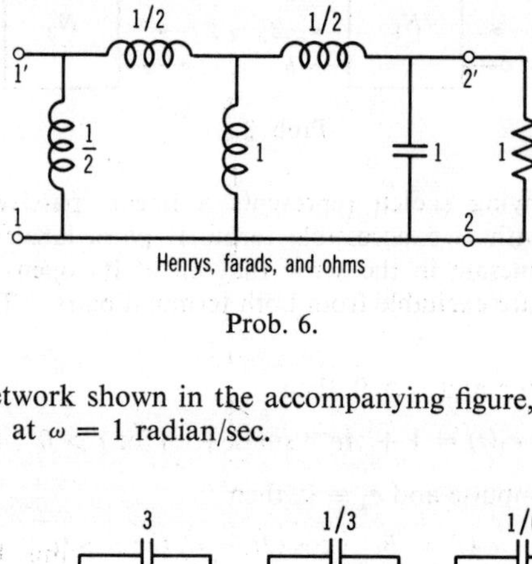

Prob. 6.

7. Given the network shown in the accompanying figure, by inspection find the value of E_2/E_1 at $\omega = 1$ radian/sec.

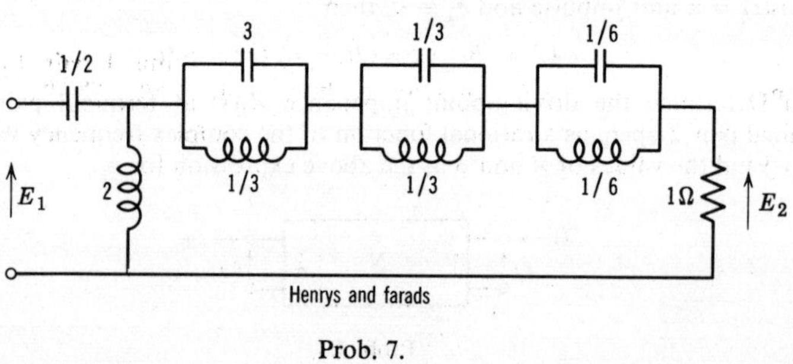

Prob. 7.

8. For the circuit shown in the adjacent sketch, find the impedance $Z(s)$. The resistance is 1 ohm, and the branch inductance matrix corresponding to the indicated reference arrows and branch numbering is

$$[l] = \begin{bmatrix} 1 & -1 & 1 \\ -1 & 2 & -1 \\ 1 & -1 & 2 \end{bmatrix}$$

9. With reference to the network in the accompanying figure, what are the impedance values between the following terminal pairs at a frequency $\omega = 1$ radian/sec? 1-2, 1-3, 1-4, 2-3, 2-4, 3-4.

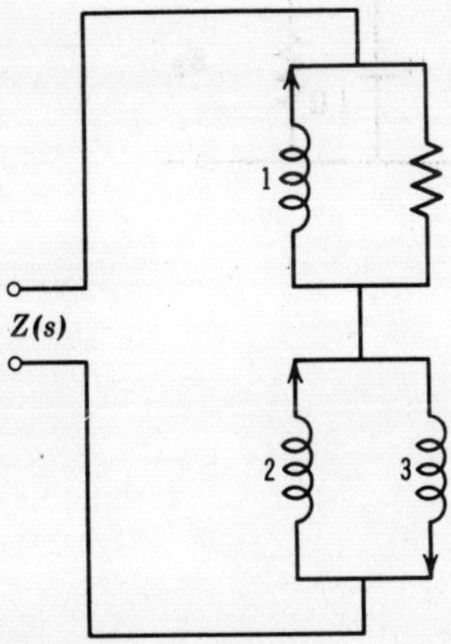

Prob. 8.

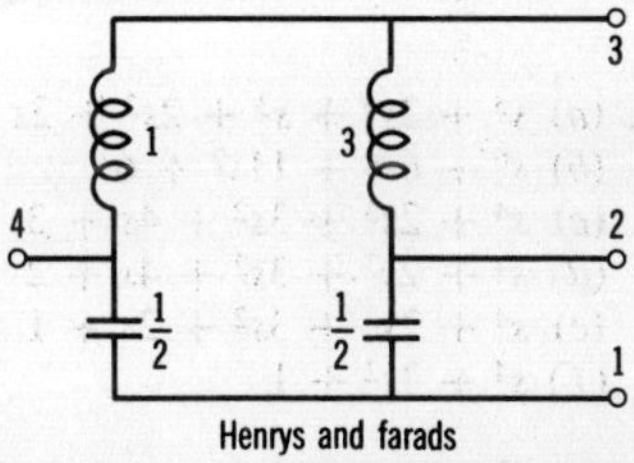

Prob. 9.

10. A student spends several hours finding the voltage transfer ratio of network (*a*), page 44, to be

$$\frac{E_2}{E_1} = \frac{s-1}{s^3 + 2s^2 + 2s + 1}$$

He then discovers that he should have found the voltage transfer ratio of network (*b*). By a few simple manipulations can you find it for him?

In the following five problems, test the polynomials and state which are and which are not Hurwitz, including in this category any that have simple j-axis zeros.

11. (*a*) $s^8 + 3s^7 + 8s^6 + 10s^5 + 17s^4 + 10s^3 + 12s^2 + 3s + 2$
(*b*) $s^8 + 3s^7 + 11s^6 + 30s^5 + 34s^4 + 72s^3 + 45s^2 + 63s + 21$
(*c*) $s^8 + 2s^7 + 8s^6 + 14s^5 + 12s^4 + 10s^3 + 13s^2 + 16s + 8$

12. (*a*) $s^5 + 8s^4 + 26s^3 + 52s^2 + 53s + 20$
(*b*) $s^5 + 6s^4 + 12s^3 + 14s^2 - 13s - 20$
(*c*) $4s^4 + 4s^3 + 29s^2 + 14s + 85$
(*d*) $5s^5 + 4s^4 + 3s^3 + 2s^2 + s + 1$
(*e*) $s^6 - 5s^5 + 4s^4 - 3s^3 + 2s^2 + s - 1$
(*f*) $s^8 + 1$

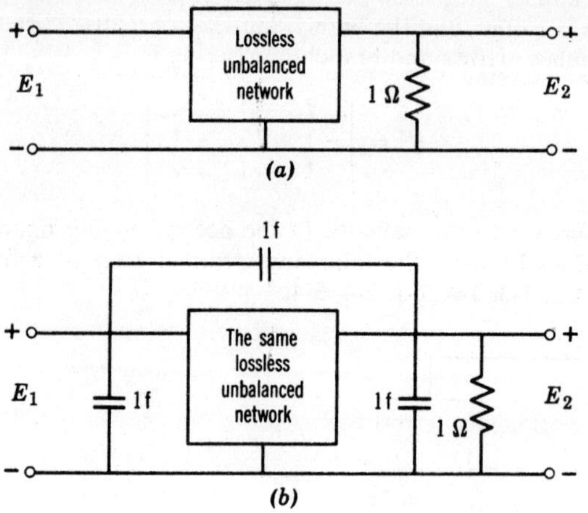

Prob. 10.

13. (*a*) $s^5 + 2s^4 + s^3 + 2s^2 + 2s + 4$
(*b*) $s^6 + 6s^4 + 11s^2 + 6$
(*c*) $s^4 + 2s^3 + 3s^2 + 4s + 3$
(*d*) $s^4 + 2s^3 + 3s^2 + 4s + 2$
(*e*) $s^4 + 2s^3 + 3s^2 + 2s + 1$
(*f*) $s^4 + 3s^2 + 1$

14. (*a*) $s^7 + 2s^6 + 3s^4 + 5s^3 + s^2 + 2$
(*b*) $s^5 + 3s^4 + 2s^3 + 6s^2 + 2s + 6$
(*c*) $s^5 + 4s^3 + 3s$
(*d*) $s^5 + s^4 + 4s^3 + 6s^2 + 3s + 8$
(*e*) $s^5 + s^4 + 6s^3 + 4s^2 + 8s + 3$

15. (*a*) $s^5 + 8s^4 + 26s^3 + 52s^2 + 53s + 20$
(*b*) $s^7 + 6s^5 + 11s^3 + 6s$
(*c*) $s^7 + 4s^6 + 8s^5 + 17s^4 + 13s^3 + 19s^2 + 5s + 4$
(*d*) $s^5 + 6s^4 + 12s^3 + 12s^2 + 11s + 6$
(*e*) $s^7 + 4s^6 + 8s^5 + 17s^4 + 13s^3 + 19s^2 + 5s + 4$
(*f*) $s^5 + 7s^4 + 18s^3 + 23s^2 + 17s + 6$

16. Let

$$P_1(s) = s^n + a_{n-1}s^{n-1} + \cdots + a_1 s + a_0$$
$$P_2(s) = s^{2n} + a_{n-1}s^{2(n-1)} + \cdots + a_1 s^2 + a_0$$

(*a*) If P_1 is Hurwitz can P_2 be Hurwitz?
(*b*) If P_2 is Hurwitz can P_1 be Hurwitz?
(*c*) If P_1 is not Hurwitz can P_2 be Hurwitz?
(*d*) If P_2 is not Hurwitz can P_1 be Hurwitz?

17. The accompanying *s*-plane sketch shows the distribution of zeros of an eighth-degree polynomial $P(s)$. It is desired to represent this polynomial as the

difference of two other polynomials as expressed by

$$P(s) = p(s) - q(s)$$

in which $p(s)$ and $q(s)$ have only even-order zeros restricted to the j-axis. Find the polynomials $p(s)$ and $q(s)$, and determine the locations of their zeros.

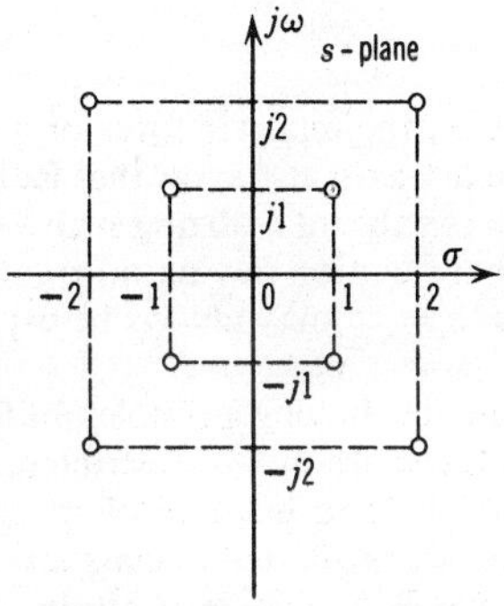

Prob. 17.

18. Which of the following rational functions are p.r.?

$(a)\ \dfrac{s^2 + s + 1}{s^2 + s + 2}$ $(b)\ \dfrac{s^2 + s + 1}{s^2 + s + 4}$ $(c)\ \dfrac{s^2 + s + 1}{s^2 + s + 9}$ $(d)\ \dfrac{5s^2 + s + 1}{s^2 + 4s + 2}$

$(e)\ \dfrac{s^3 + s^2 + 2s + 1}{s^2 + s + 1}$ $(f)\ \dfrac{s^4 + 3s^3 + 4s^2 + 3s + 1}{s^4 + 7s^3 + 6s^2 + 4s + 1}$

$(g)\ \dfrac{s^3 + 3s^2 + 4s + 12}{s^3 + 4s^2 + 9s + 10}$

19. Determine whether each of the following functions is realizable as the driving-point admittance of a passive network.

$(a)\ \dfrac{s^3 + 5s}{s^4 + 8s^2 + 12}$ $(b)\ \dfrac{(s + 1)(s + 3)(s + 4)}{s(s + 2)}$

$(c)\ \dfrac{(s + 1)(s^2 + 2s + 3)(s + 4)}{(s + 2)(s + 3)(s + 5)}$ $(d)\ \dfrac{2(s + 2)^2}{s^2 + 4}$

$(e)\ \dfrac{s^3 + 12s^2 + 44s + 48}{s^3 + 8s^2 + 15s}$ $(f)\ \dfrac{4s^4 + 2s^3 + 5s^2 + 14s + 8}{6s^4 + 7s^3 + 14s^2 + 14s + 4}$

$(g)\ \dfrac{54s^4 + 60s^3 + 36s^2 + 8s + 2}{27s^3 + 21s^2 + 15s + 1}$ $(h)\ \dfrac{s^5 + 7s^3 + 12s}{s^4 + 3s^2 + 2}$

20. If the function $(s + 1)^2/(s + a)$ is to be p.r., what restriction must be placed upon the positive real constant a?

21. Which of the following $Z(s)$ functions are p.r.?

$(a)\ \dfrac{s^2 + 2s + 2}{s^2 + 2s + 1}$ $(b)\ \dfrac{(s + 2)^2}{(s + 4)^2}$ $(c)\ \dfrac{s(s^2 + 3)(s^2 + 5)}{(s^2 + 1)(s^2 + 4)}$

(*d*) $\dfrac{s(s+3)(s+5)}{(s+1)(s+4)}$ (*e*) $\dfrac{(s^2+2)^2}{(s^2+4)^2}$ (*f*) $\dfrac{2s^2+4s+3}{s^3+2s^2+2s+1}$

(*g*) $\dfrac{s^2+1}{(s+1)^2}$ (*h*) $\dfrac{(s+3)}{(s^2+1)(s+2)}$ (*i*) $\dfrac{s^4+7s^3+4s^2+4s+1}{s^4+3s^3+6s^2+4s+1}$

(*j*) $\dfrac{2s^2+s+1}{(s+1)(s^2+1)}$

22. State the restrictions on the analytic form of a p.r. $Z(s)$ having *all* of its zeros restricted to the imaginary axis, and show that Re $[Z(j\omega)]$ has equal maxima. Describe a possible network capable of realizing such a function.

23. Show that any rational function having no right half-plane or j-axis poles and at most a simple pole at $s = \infty$ may always be expressed as the difference of two p.r. functions.

24. $Z(s)$ is a rational function having no poles in the right half-plane or on the imaginary axis, but its zeros are not so restricted. Show that the function $Z + K$ may nevertheless be made to have all of its zeros in the left half-plane through a suitable choice of the positive real constant K.

25. A rational function has a double zero at the point $s = -a$, and poles at the conjugate complex frequencies $s = -a \pm j1$. At infinity the function is regular. Can this function be p.r., and if so, what values of the positive real constant a are permissible?

26. Show in a simple effortless manner whether or not the function

$$z(s) = \left(\frac{s+10}{s+11}\right)^{10}$$

is p.r.

27. A p.r. function has the form

$$Z(s) = \frac{s(s^2+3)}{Q(s)}$$

in which the denominator polynomial $Q(s)$ has all nonzero negative real zeros. Find the simplest possible $Q(s)$.

28. A given servo has an open-loop transfer function $KG(s)$ in which the positive real constant K is a gain factor and the pole-zero pattern for $G(s)$ is as shown. The closed-loop function describing the over-all behavior of the servo is

$$H(s) = \frac{KG}{1+KG}$$

Show whether or not this function will be stable for all values of the gain K.

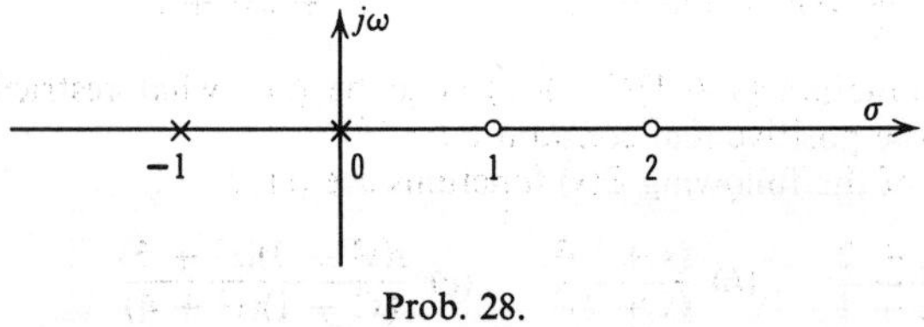

Prob. 28.

29. A p.r. driving-point impedance function has zeros at $s = -1, -2, -3, \cdots$, and its real part for $s = j\omega$ has zeros only at the points $\omega = 0$ and $\omega = \pm 1$. If the corresponding admittance is denoted by $Y(s) = 1/Z(s)$, construct the appropriate rational function

$$\text{Re}\,[Y(j\omega)] = \frac{A(\omega^2)}{B(\omega^2)}$$

of lowest degree.

30. A given rational function (real for real s) has a finite number of simple zeros and poles alternating along a line parallel to, and to the left of, the j-axis. Determine whether or not it is p.r.

31. (*a*) Is the product of two p.r. functions necessarily p.r.?

(*b*) Prove that the quotient of any pair of Hurwitz polynomials can be made up of a product of several p.r. functions.

(*c*) Can a function $Z(s)$ with the pole-zero distribution shown be made up of a product of the impedances of any of the networks shown? If so, which?

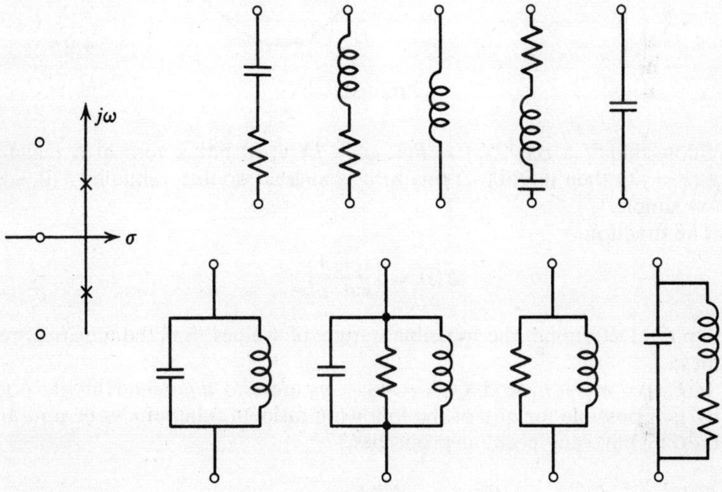

Prob. 31.

32. If the rational function $Z(s)$ is analytic in the right-half s-plane but not p.r., show that $Z(s) \times f(s)$ can be p.r. where $f(s)$ is a rational function having j-axis critical frequencies only.

Suppose the angle of $Z(s)$ for $s = j\omega$ is given in the plot shown on page 48. Construct the appropriate $f(s)$.

33. A rational function has simple poles at the points $s = -1$, $s = -2$, $s = -1 + j2$, $s = -1 - j2$ with residues respectively equal to $k = 1$, $k = 2$, $k = 1 + j1$, $k = 1 - j1$. At $s = \infty$ the function equals unity. Construct this function in the form of a quotient of polynomials. Is it p.r.?

34. (*a*) Given the function

$$f(s) = \frac{g(s)}{h(s)}$$

in which $h(s)$ is a Hurwitz polynomial, and $|g(j\omega)| \leq |h(j\omega)|$ for all real ω.

Making use of any established properties of impedance functions, show that the polynomials $p(s) = h(s) - g(s)$ and $q(s) = h(s) + g(s)$ are Hurwitz.

(*b*) Given a Hurwitz polynomial $m(s) + n(s)$, show that the function

$$t(s) = \frac{n(s + \alpha) - m(s + \alpha)}{n(s + \alpha) + m(s + \alpha)}$$

fulfills the condition $|t(j\omega)| \leq 1$ for any positive real nonzero α.

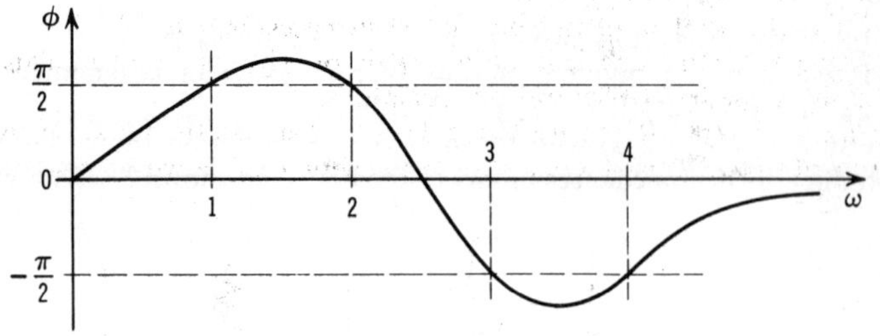

Prob. 32.

35. Show that if a p.r. $Z(s) = R(\sigma, \omega) + jX(\sigma, \omega)$ has a zero at a real frequency ($s = j\omega$) then in $R(0, \omega)$ this zero is at least double, while in $X(0, \omega)$ it is always simple.

36. The function

$$Z(s) = \left(\frac{s + 1}{s + a}\right)^3$$

is to be p.r. Determine the possible range of values for the positive real constant a.

37. If $P(s) = m_1 + n_1$ and $Q(s) = m_2 + n_2$ are two *different* Hurwitz polynomials, is it possible for any of the following rational functions to be p.r., and if so, do they have any peculiar properties?

(*a*) $\dfrac{m_1}{Q}$ (*b*) $\dfrac{n_1}{Q}$ (*c*) $\dfrac{m_2}{Q}$ (*d*) $\dfrac{n_2}{Q}$

Compare with the results of Prob. 22.

38. Given

$$Z_1(s) = \frac{s^3 + \frac{4}{3}s^2 + \frac{2}{3}s}{s^3 + 2s^2 + 2s + 1}$$

(*a*) A second impedance $Z_2(s)$ is to be found such that $Z_1 + Z_2 = \frac{1}{2}$. Is $Z_2(s)$ p.r.?

(*b*) If the specification reads $Z_1 + Z_2 = R$, what is the smallest value of R (a real positive constant) for which $Z_2(s)$ is p.r.?

39. It would be convenient, for synthesis purposes, to be able to split any p.r. impedance

$$Z = \frac{m_1 + n_1}{m_2 + n_2}$$

(where m and n refer to the even and odd parts) into the sum of two impedances

$$Z = Z_1 + Z_2 = \frac{m_1}{m_2 + n_2} + \frac{n_1}{m_2 + n_2}$$

where both Z_1 and Z_2 are p.r. What are the conditions on the impedance Z for which this is possible? (Use results of Prob. 37.)

40. Which of the following polynomials remains positive for all real values of ω?

(a) $\omega^{10} + 5\omega^8 + 15\omega^6 + 25\omega^4 + 24\omega^2 + 10$
(b) $\omega^{10} + \omega^8 + 3\omega^6 - 3\omega^4 + 4\omega^2 + 10$
(c) $\omega^{10} - 3\omega^8 + 3\omega^6 + \omega^4 - 4\omega^2 + 2$
(d) $\omega^{10} - 3\omega^8 + 4\omega^6 - 4\omega^2 + 4$

41. (a) Suppose $Z(s)$ is a p.r. function of s and $W(Z)$ is a p.r. function of Z. Is W a p.r. function of s?

(b) $Z(s)$ is a function with real coefficients which is analytic in the right half-plane and on the imaginary axis. Suppose the minimum real part on the imaginary axis is $-N$. Is the function $Z + N$ p.r.?

42. If the polynomial $P(s)$ is to be Hurwitz, then certain conditions must be met by $\psi(s)$, where

$$\psi(s) = \frac{P(s) + P(-s)}{P(s) - P(-s)}$$

or the reciprocal of this expression. Call the $\psi(s)$ corresponding to a Hurwitz polynomial an allowable $\psi(s)$.

(a) Is the product of two allowable ψ functions necessarily allowable?

(b) Is the sum of two allowable ψ functions necessarily allowable?

(c) When, if ever, could the difference of two allowable ψ functions be allowable?

(d) When, if ever, could the quotient of two allowable ψ functions be allowable?

43. (a) Show that replacing s by $s + \alpha$ in the expression for an impedance [that is, forming $Z(s + \alpha)$] effectively replaces each inductor in the network with an inductor in series with a resistor and replaces each capacitor with a capacitor in parallel with a resistor—the time constant (L/R or $1/RC$) being $1/\alpha$ in each case.

(b) Knowing that the real part of any physically realizable impedance must be positive on the j-axis, use the results of part (a) to show that the real part must also be positive in the whole right half-plane.

44. N (by itself) is an RLC network containing one open-circuit j-axis natural frequency which is a pole of the associated impedances z_{11}, z_{22}, z_{12} with the residues k_{11}, k_{22}, k_{12} respectively. It is desired to connect ideal transformers as shown (page 50) so that the impedance Z does not contain the j-axis pole in question. Determine:

(a) Pertinent values for the ideal transformer ratios n_1 and n_2.

(b) Necessary and sufficient conditions among the residues for which these values exist.

Hint: See *MCA*, Ch. IV, Art. 6.

45. If $m_1 + n_1$ is known to be a Hurwitz polynomial (where m_1 and n_1 are the even and odd parts respectively),

(a) Show that $m_1 + dm_1/ds$ and $n_1 + dn_1/ds$ are both Hurwitz. This is best done graphically.

(*b*) Show that

$$Z_1 = \frac{m_1 + dm_1/ds}{dn_1/ds + n_1}$$

is a p.r. impedance.

(*c*) Show that

$$\frac{dm_1}{ds} + \frac{dn_1}{ds} = \frac{d}{ds}(m_1 + n_1)$$

is Hurwitz and

$$Z_2 = \frac{(d/ds)(m_1 + n_1)}{m_1 + n_1}$$

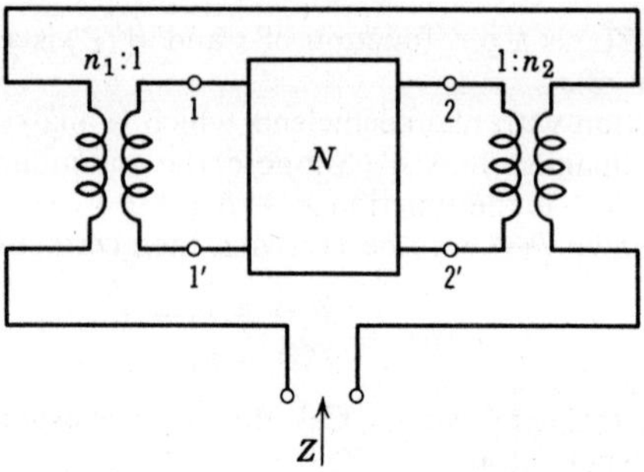

Prob. 44.

is a p.r. impedance. What is the residue of Z_2 in each of its simple poles?

46. $P(s)$ is a polynomial having all real coefficients and all of its zeros on the j-axis.

(*a*) Show that $P(s)$ is either an even or an odd polynomial.

(*b*) Using the results of Prob. 45 show that $P(s)\bigg/\dfrac{dP}{ds}$ is a ψ function.

47. A driving-point impedance function has the pole-zero pattern shown. What are the restrictions on the dimensions a, b, and c for which this impedance is p.r.?

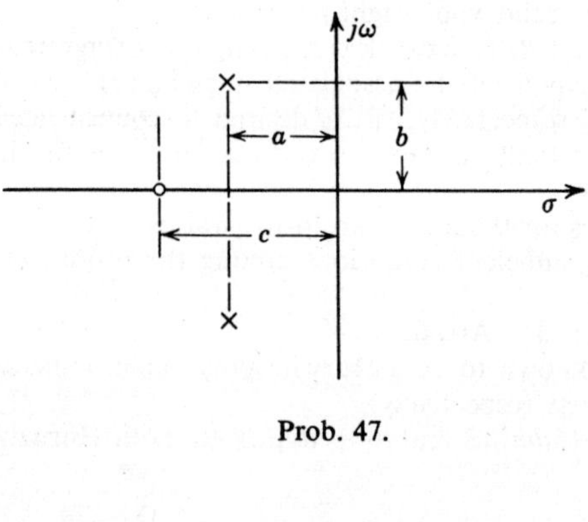

Prob. 47.

C H A P T E R T W O

Driving Point and Transfer Functions of Two-Element-Kind Networks

In this chapter we shall study the detailed characteristics of driving-point and transfer impedances and admittances associated with networks in which only two of the three possible kinds of elements R, L, and C are involved. In other words, we shall determine the various special behavior patterns of those p.r. functions which are the impedances or admittances of so-called two-element-kind networks. In the following chapters we shall see that their relative simplicity, compared with corresponding functions of networks involving all three kinds of elements, makes possible the formulation of consistently simple and straightforward synthesis procedures that not only are practically useful in their own right but also provide an essential background for the development of the more elaborate methods of approach needed in the synthesis of *RLC* networks.

1 *LC* Network Functions

Impedances of *LC* networks are the simplest of the three possible types of two-element-kind network functions since, for real frequencies (as we refer to those complex frequencies $s = j\omega$ for which the associated sinusoid has a steady amplitude—the ones that are most commonly realized in the laboratory), they have only an imaginary part instead of having both a real and an imaginary part. Since it is logical to begin our considerations with the simplest situation, we shall start the present discussion of two-element-kind network functions with the *LC* case.

We begin by considering relation 4 of Ch. 1 for $F_0 \equiv 0$, and $p = 1$, giving

$$z_{11}(s) = Z(s) = \left(sT_0 + \frac{V_0}{s}\right)_{I_1=1} \tag{1}$$

It is clear at once (because of the positive real character of T_0 and V_0) that $Z(j\omega)$ is purely imaginary, as we expect from past experience and from the physical consideration that a network containing no resistances cannot consume any average power in the sinusoidal steady state. Such a network, for which the real part of any impedance or admittance is identically zero for $s = j\omega$, is also referred to as being *lossless* since there can be no average loss of energy associated with it.

If in Eq. 1 we take the stipulation $I_1 = 1$ for granted and equate this expression for $Z(s)$ to zero, we find

$$s = \sqrt{-\frac{V_0}{T_0}} = \pm j\sqrt{\frac{V_0}{T_0}} \tag{2}$$

which is not an explicit relation for s since T_0 and V_0 are implicity functions of s, but the positive realness of T_0 and V_0 nevertheless shows that all the zeros of $Z(s)$ implied by Eq. 2 must have purely imaginary values. All the poles of $Z(s)$, which are the zeros of $Y(s) = 1/Z(s)$ given by an expression analogous to 1, are thus likewise located on the j-axis of the s-plane. Hence we see that the driving-point impedance or admittance of a lossless network is a limiting form of p.r. function for which all zeros and poles lie on the j-axis of the complex frequency plane. Since we already know that such zeros and poles of a p.r. function must be simple, we may say that the driving-point impedance or admittance of a lossless (LC) network has only simple zeros and poles restricted to the j-axis.

The discussion in the previous chapter also showed that residues in j-axis poles must be real and positive and that a function having only j-axis poles with positive real residues [like $\psi(s) = m/n$ formed from the even and odd parts of a Hurwitz polynomial] can have only simple zeros and poles that *alternate*. The driving-point impedance of a lossless network (also called a *reactance* function) can have only simple zeros and poles that alternate on the j-axis. Such a function is identical in its properties with the ratio of the even and odd parts of a Hurwitz polynomial.

We may prove in an alternate manner the simplicity and the alternance of j-axis zeros and poles in the LC case. Consider the function 1 for complex s-values; thus

$$Z(s) = R(\sigma, \omega) + jX(\sigma, \omega) = (\sigma + j\omega)T_0 + \frac{V_0}{\sigma + j\omega} \tag{3}$$

giving

$$R(\sigma, \omega) = \sigma T_0 + \frac{\sigma V_0}{\sigma^2 + \omega^2} \tag{4}$$

and

$$X(\sigma, \omega) = \omega T_0 - \frac{\omega V_0}{\sigma^2 + \omega^2} \tag{5}$$

From Eq. 4 we have

$$\frac{\partial R}{\partial \sigma} = \sigma \frac{\partial T_0}{\partial \sigma} + T_0 + \frac{\sigma}{\sigma^2 + \omega^2} \frac{\partial V_0}{\partial \sigma} + \frac{\omega^2 - \sigma^2}{(\sigma^2 + \omega^2)^2} V_0 \tag{6}$$

or (making use also of Eq. 243, p. 535 of *ICT*)

$$\left(\frac{\partial R}{\partial \sigma}\right)_{\sigma=0} = T_0 + \frac{V_0}{\omega^2} = 4(T_{av} + V_{av}) \tag{7}$$

Since this expression is positive for all real ω (and nonzero since $(T_{av} + V_{av})$ for $I_1 = 1$ can at no j-axis point be zero in any nontrivial situation), the Cauchy-Riemann equation (see *MCA*, p. 256)

$$\frac{\partial R}{\partial \sigma} = \frac{\partial X}{\partial \omega} \tag{8}$$

gives

$$\left(\frac{\partial X}{\partial \omega}\right)_{\sigma=0} > 0 \tag{9}$$

showing that a reactance function has a positive nonzero slope at all j-axis frequencies.

This result shows first of all that zeros must be simple, for the slope at a zero of higher order is zero and changes sign at any zero of even multiplicity. Since the corresponding analysis on an admittance basis shows the slope of $Y(j\omega) = jB(\omega)$—a *susceptance* function—to be positive and nonzero, it follows that the zeros of $Y(s)$ which are poles of $Z(s)$ likewise are simple. Secondly, the positive slope condition 9 evidently requires that poles and zeros must alternate, for it permits a reactance (or susceptance) to have only the general appearance shown plotted in Fig. 1.

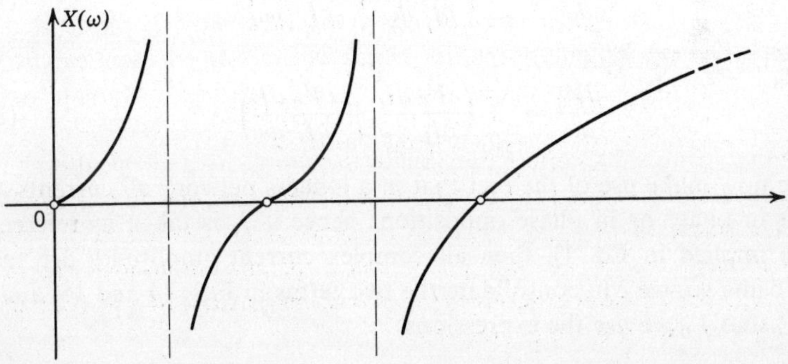

Fig. 1. The general appearance of a reactance (lossless impedance) as a function of frequency.

Since the function can only increase with increasing frequency, it is clear that two zeros cannot follow in succession. Since poles are zeros of the reciprocal function which must have the same character, two poles cannot follow in succession either. If we make use of the customary interpretation of the complex plane for $Z(s)$ according to which it is the surface of a huge sphere whose j-axis is a great circle passing through the south pole at $Z = 0$ and through the north pole at $Z = \infty$, then passage through $X = \infty$ yields the picture in Fig. 1 where, after reaching $+\infty$, the function emerges out of $-\infty$ without violating the condition that it must continuously increase.

The positive slope condition may be made even more specific by means of the following reasoning. From Eq. 3 we get

$$Z(j\omega) = jX(\omega) = j\left(T_0\omega - \frac{V_0}{\omega}\right) \tag{10}$$

and hence for the reactance itself

$$X(\omega) = T_0\omega - \frac{V_0}{\omega} \tag{11}$$

Differentiation, noting that T_0 and V_0 are implicit functions of ω, gives

$$\frac{dX}{d\omega} = T_0 + \frac{V_0}{\omega^2} + \omega\frac{dT_0}{d\omega} - \frac{1}{\omega}\frac{dV_0}{d\omega} \tag{12}$$

In the last two terms we must regard T_0 and V_0 to be functions of all the complex current amplitudes and of their conjugates (see Eqs. 5, 6, 7, Ch. 1) as indicated by writing

$$\begin{aligned} T_0 &= T_0(I_1 \cdots I_l, \bar{I}_1 \cdots \bar{I}_l) \\ V_0 &= V_0(I_1 \cdots I_l, \bar{I}_1 \cdots \bar{I}_l) \end{aligned} \tag{13}$$

Hence the derivatives of these functions become

$$\frac{dT_0}{d\omega} = \sum_{i=1}^{l}\left(\frac{\partial T_0}{\partial I_i}\frac{dI_i}{d\omega} + \frac{\partial T_0}{\partial \bar{I}_i}\frac{d\bar{I}_i}{d\omega}\right) \tag{14}$$

and

$$\frac{dV_0}{d\omega} = \sum_{i=1}^{l}\left(\frac{\partial V_0}{\partial I_i}\frac{dI_i}{d\omega} + \frac{\partial V_0}{\partial \bar{I}_i}\frac{d\bar{I}_i}{d\omega}\right) \tag{15}$$

We now make use of the fact that in a lossless network all currents are either in phase or in phase opposition; hence if I_1 is taken as reference (as is implied in Eq. 1), then all complex current amplitudes are real. This being so, we can consolidate the two terms in Eqs. 14 and 15, and if for T_0 and V_0 we use the expressions

$$T_0 = \sum_{i,\,k=1}^{l} L_{ik}\bar{I}_i I_k, \qquad V_0 = \sum_{i,\,k=1}^{l} S_{ik}\bar{I}_i I_k \tag{16}$$

so that

$$\frac{\partial T_0}{\partial \bar{I}_i} = \sum_{k=1}^{l} L_{ik}I_k \quad \text{and} \quad \frac{\partial V_0}{\partial \bar{I}_i} = \sum_{k=1}^{l} S_{ik}I_k \tag{17}$$

then we find

$$\omega\frac{dT_0}{d\omega} - \frac{1}{\omega}\frac{dV_0}{d\omega} = \frac{2}{j}\sum_{i=1}^{l}\left[\sum_{k=1}^{l} j\left(L_{ik}\omega - \frac{S_{ik}}{\omega}\right)I_k\right]\frac{d\bar{I}_i}{d\omega} \tag{18}$$

The quantity

$$\sum_{k=1}^{l} j\left(L_{ik}\omega - \frac{S_{ik}}{\omega}\right)I_k = E_i \tag{19}$$

is the source voltage in mesh number i. Since all sources except the one in mesh number 1 are dead, because we are assuming that the network has only one point of access, Eq. 18 becomes

$$\omega\frac{dT_0}{d\omega} - \frac{1}{\omega}\frac{dV_0}{d\omega} = \frac{2}{j}E_1\frac{d\bar{I}_1}{d\omega} = 2X(\omega)I_1\frac{dI_1}{d\omega} = 0 \tag{20}$$

because by hypothesis I_1 = constant = 1. Equation 12 therefore yields

$$\frac{dX}{d\omega} = T_0 + \frac{V_0}{\omega^2} > 0 \tag{21}$$

and use of Eq. 11 then gives

$$\frac{dX}{d\omega} - \frac{X}{\omega} = \frac{2V_0}{\omega^2} > 0 \tag{22}$$

Not only does a reactance function have positive slope but this slope at any point on the plotted curve for this reactance must be larger than the slope of a straight line drawn from the origin to that point. Figure 2

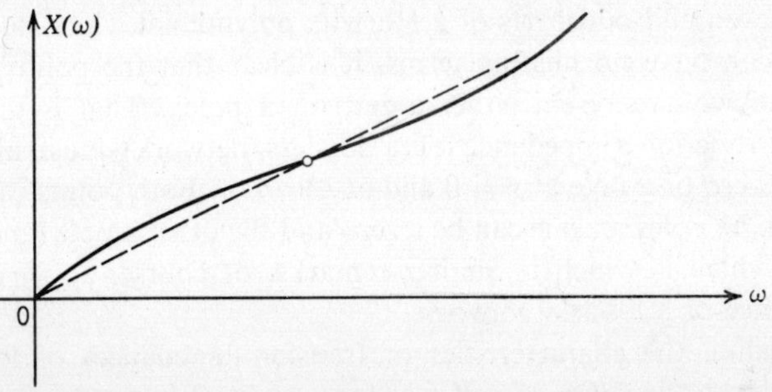

Fig. 2. This reactance is nonphysical because its slope is not everywhere greater than that of a straight line through the origin.

illustrates a situation for which condition 22 is not fulfilled although the slope is everywhere positive. Condition 22 is evidently stronger than the positive slope condition alone. For any graphically given curve it may easily be checked by placing a straight edge upon the drawing and, keeping one of its ends fixed at the origin, letting it sweep across the plotted curve for $X(\omega)$. The slope of the latter wherever it intersects the straight edge must exceed the slope of the straight edge.

The expression for a reactance function as a quotient of polynomials may be obtained from the general expression

$$Z(s) = \frac{P(s)}{Q(s)} = \frac{m_1 + n_1}{m_2 + n_2} \tag{23}$$

by considering the condition that

$$\text{Re}\,[Z(j\omega)] = \left(\frac{m_1 m_2 - n_1 n_2}{m_2^2 - n_2^2}\right)_{s=j\omega} \equiv 0 \tag{24}$$

which yields

$$(m_1 m_2 - n_1 n_2 \equiv 0 \tag{25}$$

since a function that vanishes along a line segment must vanish identically.* Two nontrivial cases can occur for which Eq. 25 is fulfilled. Either $m_1 \equiv n_2 \equiv 0$ or $n_1 \equiv m_2 \equiv 0$. Thus we can have either

$$Z(s) = \frac{n_1}{m_2} \quad \text{or} \quad Z(s) = \frac{m_1}{n_2} \tag{26}$$

Since, as is shown in Art. 5 of Ch. 1, the even and odd parts m_1 and n_2 as well as m_2 and n_1 are respectively the even and odd parts of a Hurwitz polynomial, we see again that a reactance function is identical in its characteristics with the function $\psi(s) = m/n$ (or $1/\psi = n/m$) discussed in connection with the testing of polynomials for Hurwitz character. Of importance is the recognition that a reactance function must be the ratio of two polynomials of which one is odd and the other even. Since these are the even and odd parts of a Hurwitz polynomial which, among other things, can have no missing terms, it is clear that the points $s = 0$ and $s = \infty$ always represent either a zero or a pole. That is to say, a reactance function (impedance of a lossless network) must always have either a zero or a pole at $s = 0$ and at $s = \infty$. Both points can be zeros, both can be poles, or one can be a zero and the other a pole (and it doesn't matter which is which). Similar remarks, of course, apply also to the admittance of a lossless network.

Regarding the characteristics of transfer impedances of lossless networks, we consider Eq. 17 of Ch. 1 for $p = 2$. We are thus led to recognize that the function

$$Z(s) = z_{11}x_1^2 + 2z_{12}x_1x_2 + z_{22}x_2^2 \tag{27}$$

must have all the detailed properties of the lossless driving-point impedance just described for all real values of the variables x_1 and x_2. We see at

* According to an alternate interpretation, which does not invalidate the present conventional analysis, the j-axis real part of a reactance function is seen to have impulses at the poles of that function, and hence is not identically zero. This more general attitude (which is presented in Art. 2, Ch. 8) is, however, not essential to the discussions in this chapter.

once that the transfer impedance z_{12}, as well as the driving-point impedances z_{11} and z_{22}, can have only simple poles restricted to the j-axis. Since the residue

$$k = k_{11}x_1^2 + 2k_{12}x_1x_2 + k_{22}x_2^2 \tag{28}$$

of $Z(s)$ in such poles must be real and positive, we see that all three residues k_{11}, k_{22}, k_{12} of the impedances z_{11}, z_{22}, z_{12} respectively must be real. Thus the residues of the transfer impedance, like those of the driving-point impedances z_{11} and z_{22} must be real. They do not, however, need to be positive, for the positiveness of k in Eq. 28 for all real values of x_1 and x_2 requires in addition to $k_{11} \geq 0$ and $k_{22} \geq 0$ only that the so-called *residue condition*

$$k_{11}k_{22} - k_{12}^2 \geq 0 \tag{29}$$

be fulfilled. It is clear that this condition may be fulfilled with a negative k_{12} value.

The partial-fraction expansion of either the driving-point or transfer impedances has the form

$$z_{ik}(s) = \frac{k_{ik}^{(0)}}{s} + \frac{k_{ik}^{(2)}}{s - s_2} + \frac{k_{ik}^{(2)}}{s + s_2} + \cdots + \frac{k_{ik}^{(2n)}}{s - s_{2n}} + \frac{k_{ik}^{(2n)}}{s + s_{2n}} + k_{ik}^{(2n+2)}s \tag{30}$$

where $s = 0, \pm s_2, \pm s_4, \cdots \infty$ are the j-axis poles, of which conjugates are simply opposite in algebraic sign, and the residues, which in conjugate poles are normally conjugate in their values (in order to assure that the function have real values for all real s), are in the present instance the same in conjugate poles because of their required realness. Combining terms with conjugate poles, Eq. 30 reads

$$z_{ik}(s) = \frac{k_{ik}^{(0)}}{s} + \frac{2k_{ik}^{(2)}s}{s^2 - s_2^2} + \frac{2k_{ik}^{(4)}s}{s^2 - s_4^2} + \cdots + \frac{2k_{ik}^{(2n)}s}{s^2 - s_{2n}^2} + k_{ik}^{(2n+2)}s \tag{31}$$

If either the origin or infinity or both are regular points, then the first or last or both the first and last terms in this expression are absent.

In any case, if this expression is put into the form of a quotient of polynomials, it is clear that one of these polynomials will be even and the other one odd, as was pointed out above to be the case with driving-point reactance functions. Specifically $z_{ik}(s)$ is an even-over-odd quotient if the pole at $s = 0$ is present, and it is an odd-over-even quotient if this pole is absent. In lossless driving-point and transfer impedances alike, the points $s = 0$ and $s = \infty$ are always critical (either a zero or a pole in any combination).

If Eq. 31 represents a transfer impedance ($i \neq k$), so that the residues may have mixed positive and negative values, then the numerator polynomial of $z_{12}(s)$ may have mixed positive and negative coefficients. Unlike the even or odd parts of a Hurwitz polynomial, this numerator polynomial need not have simple zeros restricted to the j-axis. It may have zeros of any multiplicity located anywhere in the finite s-plane. The pattern of these zeros, however, will always have complete symmetry about the j-axis as well as about the real axis, as may be seen by noting that the zeros of an even or odd polynomial are not changed through replacing s by $-s$.

It is also significant to observe that the possible mixed positive and negative values of the residues can result in the numerator being a polynomial of lower degree than the denominator by more than unity, although it cannot be more than one degree higher. In fact, the numerator polynomial may reduce to a constant, but only if the denominator polynomial is odd, for the requirement that the two polynomials whose quotient is z_{12} be respectively even and odd must in any case be met.

Returning our attention to the residue condition 29, we can by inspection make some interesting observations with regard to pole distributions among the three impedances z_{11}, z_{22}, z_{12} characterizing a lossless two terminal-pair network. Thus it is clear that either or both of the driving-point impedances can have poles where the other driving-point impedance or the transfer impedance does not, but the reverse of this situation is not permitted. Wherever the transfer impedance has a pole, there *both* driving-point impedances must likewise have poles (with sufficient residues). If we visualize the network in the form of a T structure, this result is rather obvious since the series branches of the T affect only the driving-point impedance whereas the shunt branch, which equals z_{12}, is an additive component of both driving-point impedances. Although the T structure is not the most general physically realizable form for a two terminal-pair network, it does place in visual evidence this particular property implied by condition 29.

2 *RL* Network Functions

Here the energy function V_0 or V^*_0 is identically zero, and so on an impedance basis Eq. 4 of Ch. 1 (with $I_1 = 1$ understood) yields

$$Z(s) = sT_0 + F_0 = R(\sigma, \omega) + jX(\sigma, \omega) \tag{32}$$

and on an admittance basis Eq. 8 of Ch. 1 (with $E_1 = 1$ understood) gives

$$Y(s) = F^*_0 + \frac{T^*_0}{s} = G(\sigma, \omega) + jB(\sigma, \omega) \tag{33}$$

The zeros of $Z(s)$ must fulfill the condition

$$s = \frac{-F_0}{T_0} \tag{34}$$

and the poles (zeros of Y) must be consistent with the relation

$$s = \frac{-T^*_0}{F^*_0} \tag{35}$$

As in the analogous situation involving the zeros and poles of LC network functions, these expressions are practically useless for the computation of zeros and poles, but they do show—because of the positive realness of the energy functions—that the zeros and poles of RL driving-point impedances and admittances are restricted to lie on the negative real axis of the complex frequency plane.

For $s = \sigma + j\omega$ we get from Eqs. 32 and 33

$$R(\sigma, \omega) = \sigma T_0 + F_0, \qquad X(\sigma, \omega) = \omega T_0 \tag{36}$$

and

$$G(\sigma, \omega) = F^*_0 + \frac{\sigma T^*_0}{\sigma^2 + \omega^2}, \qquad B(\sigma, \omega) = \frac{-\omega T^*_0}{\sigma^2 + \omega^2} \tag{37}$$

Worth mention is the fact that X and B are respectively positive and negative everywhere in the upper half of the complex frequency plane, as is obvious by inspection of the above forms. For the positive j-axis this result checks the familiar fact that the impedance of an RL network has a positive imaginary part; therefore the susceptive part of the corresponding admittance is negative.

From Eqs. 36 we see further that

$$\left(\frac{\partial X}{\partial \omega}\right)_{\omega=0} = T_0 > 0 \tag{38}$$

Hence the Cauchy-Riemann Eq. 8 gives

$$\left(\frac{\partial R}{\partial \sigma}\right)_{\omega=0} = T_0 > 0 \tag{39}$$

For $\omega = 0$ the impedance $Z(s)$ is purely real. The result, Eq. 39, therefore tells us that if we plot an RL driving-point impedance as a function of σ for $\omega = 0$ (that is to say, plot the behavior of $Z(\sigma) = R(\sigma, 0)$ along the real axis, just as in the LC case we plot in Fig. 1 the reactance function along the j-axis), then this curve will everywhere have a positive nonzero slope and in fact will look much like the plot in Fig. 1 since the zeros and poles for the RL driving-point impedance lie along the real axis. Since the slope must have a nonzero value (because T_0 for $I_1 = 1$ in any nontrivial case cannot equal zero), we see that the zeros must be simple and that two zeros cannot follow in succession along the negative real axis.

From Eqs. 37 we obtain

$$\left(\frac{\partial B}{\partial \omega}\right)_{\omega=0} = \frac{-T^*_0}{\sigma^2} < 0 \tag{40}$$

and the Cauchy-Riemann Eq. 8 yields

$$\left(\frac{\partial G}{\partial \sigma}\right)_{\omega=0} = \frac{-T^*_0}{\sigma^2} < 0 \tag{41}$$

An *RL* driving-point admittance plotted as a function of σ therefore has a consistently nonzero negative slope. It follows that zeros of this admittance, which are poles of the *RL* driving-point impedance, must be simple and that two of these cannot follow in succession along the negative real axis.

As a result of these considerations we may say that the zeros and poles of an *RL* driving-point impedance or admittance must be simple and alternate along the negative real axis. The impedance plotted versus σ has a consistently positive slope while the corresponding plot for the admittance has a consistently negative slope. As the sketches of these situations in Fig. 3 indicate, such plots are not unlike the corresponding plots along the *j*-axis for *LC* networks except that there the impedance

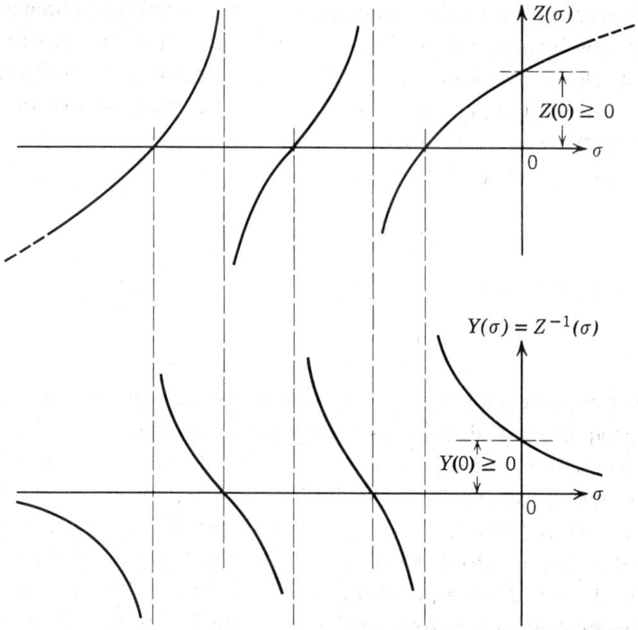

Fig. 3. *RL* impedance and admittance as functions of real values of the complex frequency variable.

and admittance functions both have positive slope, whereas here the slope of one is positive and that of the other is negative. This circumstance can alternately be appreciated from the consideration that the reciprocal of a pure imaginary quantity like jX is $1/jX = -j/X$ and hence has the opposite algebraic sign, while the reciprocal of a real value like $Z(\sigma)$ for an *RL* network has the same algebraic sign. This fact, plus the thought that where the absolute value of Z increases that of Y decreases and vice versa, shows at once that $Z(\sigma)$ and $Y(\sigma)$ in the *RL* case must have opposite slopes as shown in Fig. 3 whereas $Z(j\omega)$ and $Y(j\omega)$ in the *LC* case both have positive slopes.

Aside from these considerations, the *RL* driving-point functions differ in another important respect, namely, they are not necessarily critical at the points $s = 0$ and $s = \infty$, and if they are, these points cannot be either a pole or a zero. This fact is easily appreciated if we remind ourselves that at zero frequency an inductance becomes identical with a short circuit whereas at infinite frequency it becomes an open circuit. Hence it is clear that an *RL* driving-point impedance cannot have a pole at $s = 0$ or a zero at $s = \infty$. The origin of the s-plane is either a zero or a point where $Z(s)$ has a finite value; at infinity this function has either a pole or a finite nonzero value.

We may completely describe an *RL* driving-point impedance as a rational function having simple zeros and poles alternating on the negative real axis, whose lowest critical frequency is a zero. This description allows the numerator polynomial to have a degree that is equal to or greater by unity than the degree of the denominator polynomial, and excludes the possibility of the denominator being of higher degree than the numerator. Hence it adequately meets the restrictions discussed in the previous paragraph. An *RL* driving-point admittance is described in the same way except that the lowest critical frequency is restricted to being a pole. It is taken for granted that the polynomials have positive (not negative) coefficients, so that $Z(0) \geq 0$ as it obviously must be.

Since the residues of $Z(s)$ in its finite poles are equal to the reciprocal values of the slopes of $Y(s)$ at its zeros (refer to Eq. 198, p. 305 of *MCA*), we see that these residues have negative real values. For the same reason, the residues of $Y(s)$ in its poles are positive real. The partial-fraction expansion of an *RL* driving-point impedance reads

$$Z(s) = \frac{k_2}{s - s_2} + \frac{k_4}{s - s_4} + \cdots + \frac{k_{2n}}{s - s_{2n}} + k_{2n+2}s + k_\infty \tag{42}$$

in which the real coefficients k_ν are all negative except k_∞ and k_{2n+2}. Thus while the residues in all finite poles are negative, the residue k_{2n+2} for the possible pole at infinity must, of course, be positive. If we write

$s_k = -\sigma_k$ for the negative real poles, then Eq. 42 is alternately given by

$$Z(s) = \frac{k_2}{s + \sigma_2} + \frac{k_4}{s + \sigma_4} + \cdots + \frac{k_{2n}}{s + \sigma_{2n}} + k_{2n+2}s + k_\infty \tag{43}$$

From this form we get the following expression for the zero-frequency value of the RL driving-point impedance

$$Z(0) = k_\infty + \left(\frac{k_2}{\sigma_2} + \frac{k_4}{\sigma_4} + \cdots + \frac{k_{2n}}{\sigma_{2n}}\right) \geq 0 \tag{44}$$

which, as indicated, must be nonnegative in spite of the negative values of $k_2 \cdots k_{2n}$. When $s = \infty$ is a regular point ($k_{2n+2} = 0$) then $Z(\infty) = k_\infty > 0$. Here the equal sign is not included for reasons already discussed. It is clear from Eq. 44 that $Z(0) \leq Z(\infty)$ in any such situation.

Since

$$\operatorname{Re}\left(\frac{k_\nu}{j\omega + \sigma_\nu}\right) = \frac{k_\nu \sigma_\nu}{\sigma_\nu^2 + \omega^2} \tag{45}$$

is continuously decreasing in absolute value, it is clear from Eq. 43 that

for an RL network, Re $[Z(j\omega)]$ *is a continuously increasing function of* ω (46)

This statement, together with the previously established fact that the imaginary part of $Z(j\omega)$ is positive for all positive real ω, is a useful characterization of the behavior of an RL impedance at real frequencies ($s = j\omega$).

The partial-fraction expansion of an RL driving-point admittance may be written*

$$Y(s) = \frac{h_1}{s - s_1} + \frac{h_3}{s - s_3} + \cdots + \frac{h_{2n-1}}{s - s_{2n-1}} + h_{2n+1} \tag{47}$$

in which the coefficients h_ν have positive values. If we let $s_k = -\sigma_k$ again, this expression reads

$$Y(s) = \frac{h_1}{s + \sigma_1} + \frac{h_3}{s + \sigma_3} + \cdots + \frac{h_{2n-1}}{s + \sigma_{2n-1}} + h_{2n+1} \tag{48}$$

Here the above reasoning shows that

for an RL network, Re $[Y(j\omega)]$ *is a continuously decreasing function of* ω (49)

The imaginary part, as shown in Eq. 37, is negative for all positive real ω values.

---

* Like expression 42 for $Z(s)$, this one for $Y(s)$ represents the most general form that this RL function can have. It is not to be thought that the $Z(s)$ in 42 and the $Y(s)$ in 47 are reciprocals. Thus if representation 47 for $Y(s)$ were to correspond to the $Z(s)$ representation in 42, then $h_{2n+1} = 0$; conversely, if 42 were to correspond to the admittance 47 then k_{2n+2} would have to be zero.

When the *RL* driving-point impedance function is not critical either at $s = 0$ or at $s = \infty$, then its zero-frequency value is always less than its asymptotic value for $s \to \infty$; and for the corresponding *RL* driving-point admittance the zero-frequency value is always greater than the asymptotic value for $s \to \infty$.

In order to determine the characteristics of an *RL* transfer impedance, we apply all of the preceding discussion regarding *RL* driving-point impedances to the expression for $Z(s)$ given by Eq. 17 of Ch. 1 which, for $p = 2$, is Eq. 27 used above in the discussion of lossless networks. Here z_{11} and z_{22} are driving-point impedances of *RL* networks, and z_{12} is a transfer impedance for such a network. We see that z_{12}, like z_{11} and z_{22}, must have simple poles restricted to the negative real axis. Since the residue of $Z(s)$ in terms of the residues of the z_{ik} is given by expression 28, we see by means of reasoning analogous to that used in the *LC* case that the residues of z_{12}, although real, do not have to be negative like those of z_{11} and z_{22} in their finite poles. The residue given by Eq. 28 must (for *RL* networks) be a *negative* definite quadratic form in the variables x_1 and x_2 at all finite poles, and a positive form at the possible pole at infinity. At the finite poles this condition is met by requiring, in addition to $k_{11} \leq 0$ and $k_{22} \leq 0$, that the residue condition 29 (same in form as for the *LC* case and with the same orientation of the inequality sign) be fulfilled.

Although z_{11} or z_{22} may have poles not also contained in either of the other two impedances or both z_{11} and z_{22} may have poles not contained in z_{12}, the residue condition (as in the *LC* case) does not permit poles in z_{12} that are not also contained in both z_{11} and z_{22}.

The partial-fraction expansion of the z_{ik} has the form

$$z_{ik} = \frac{k_{ik}^{(2)}}{s - s_2} + \frac{k_{ik}^{(4)}}{s - s_4} + \cdots + \frac{k_{ik}^{(2n)}}{s - s_{2n}} + k_{ik}^{(2n+2)}s + k_{ik}^{(\infty)} \tag{50}$$

Although $k_{ik}^{(\infty)}$ is not a residue but the asymptotic value of the real part of z_{ik} for $s = j\omega$, the relations expressed by Eqs. 105 and 106 of Ch. 1 show that residue condition 29 applies to these coefficients as well as to the residues in Eq. 50. For the transfer impedance z_{12} the coefficients $k_{ik}^{(\nu)}$ in this partial-fraction expansion may have any positive or negative real values. Hence if this expression for z_{12} is put into the form of a quotient of polynomials, the numerator is a polynomial that can have zeros of any multiplicity located anywhere in the s-plane, and its degree can be anything from zero to one more than the degree of the denominator.

It is interesting to observe that fulfillment of the residue condition, plus the requirement $Z(0) \geq 0$, takes care of the real part conditions mentioned in Ch. 1 as being necessary for the realization of general *RLC* impedances. Thus the property of *RL* driving-point impedances expressed in statement 46, together with a knowledge that $Z(0) \geq 0$, assures fulfillment of the real-part conditions. Residue condition 29 assures that $Z(s)$ as given by Eq. 27 has the property expressed in statement 46.

For a study of the properties of *RL* transfer admittance functions we proceed from the analogous relation to Eq. 27 (see Eq. 26 of Ch. 1 for $p = 2$) which reads

$$Y(s) = y_{11}x_1^2 + 2y_{12}x_1x_2 + y_{22}x_2^2 \tag{51}$$

and require that y_{11}, y_{22}, and $Y(s)$ have the properties of *RL* driving-point admittances discussed above. Thus we see again that y_{12} must have simple poles restricted to the negative real axis with real residues that do not have to be positive like those of the driving-point functions. However, residue condition 29 must be fulfilled at all poles in addition to $Y(\infty) \geq 0$. The transfer admittance y_{12} is seen to have properties exactly like those of z_{12} discussed above (except the possibility of having a pole at infinity).

3 *RC* Network Functions

For the consideration of these functions we set the magnetic energy function T_0 or T^*_0 equal to zero. On an admittance basis, Eq. 8 of Ch. 1 (for $E_1 = 1$) gives

$$Y(s) = sV^*_0 + F^*_0 = G(\sigma, \omega) + jB(\sigma, \omega) \tag{52}$$

while Eq. 4 of Ch. 1 pertinent to an impedance basis yields

$$Z(s) = F_0 + \frac{V_0}{s} = R(\sigma, \omega) + jX(\sigma, \omega) \tag{53}$$

Comparison of Eqs. 52 and 53 with Eqs. 32 and 33 shows that the admittance function $Y(s)$ for an *RC* network is alike in form with the impedance function $Z(s)$ for an *RL* network, and vice versa. We may, therefore, carry over all of the results of the previous article with the identities of Y and Z interchanged.

Thus we may say that *RC* driving-point impedances and admittances have simple zeros and poles alternating on the negative real axis of the s-plane. For the impedance the lowest critical frequency is a pole and for the admittance it is a zero. When plotted versus σ the impedance has negative slope while the admittance has positive slope. The residues of $Z(s)$ in its poles are positive, and those of $Y(s)$ in its finite poles are negative. The point $s = 0$, when it is critical, is a pole of $Z(s)$ or a zero of $Y(s)$; and the point $s = \infty$, when it is critical, is a zero of $Z(s)$ or a pole of $Y(s)$. The partial-fraction expansion of a driving-point impedance has the form

$$Z(s) = \frac{k_1}{s - s_1} + \frac{k_3}{s - s_3} + \cdots + \frac{k_{2n-1}}{s - s_{2n-1}} + k_{2n+1} \tag{54}$$

like Eq. 47 for an *RL* admittance, whereas the partial-fraction expansion of a driving-point admittance has the form*

$$Y(s) = \frac{h_2}{s - s_2} + \frac{h_4}{s - s_4} + \cdots + \frac{h_{2n}}{s - s_{2n}} + h_{2n+2}s + h_\infty \tag{55}$$

like Eq. 42 for an *RL* impedance. In Eq. 54 all k_ν are real and positive, but in Eq. 55, $h_2 \cdots h_{2n}$ are negative real and h_∞ and h_{2n+2} are positive real.

Analogous to statements 46 and 49 we obtain the following:

For an RC network, Re $[Y(j\omega)]$ *is a continuously increasing function of* ω *whereas* Re $[Z(j\omega)]$ is a *continuously decreasing function* (56)

The imaginary part of an *RC* driving-point impedance is negative for all positive real values of ω, and the imaginary part of the corresponding admittance is consistently positive.

To study the properties of transfer impedances and the mutual relations between driving-point and transfer impedances for a two terminal-pair network we simultaneously consider Eqs. 27 and 54 (on an admittance basis, 51 and 55). The z_{ik} for an *RC* network are thus seen to have the partial-fraction expansion

$$z_{ik} = \frac{k_{ik}^{(1)}}{s - s_1} + \frac{k_{ik}^{(3)}}{s - s_3} + \cdots + \frac{k_{ik}^{(2n-1)}}{s - s_{2n-1}} + k_{ik}^{(2n+1)} \tag{57}$$

in which, besides $k_{11} \geq 0$ and $k_{22} \geq 0$, the k_{ik} must fulfill residue condition 29 in order to insure that the residue k of $Z(s)$ given by 28 is a positive definite quadratic form in the real variables x_1 and x_2.

In Eq. 54 the requirement that $k_{2n+1} = Z(\infty)$ be positive insures (because of 56) that the real part of $Z(j\omega)$ be positive for all real values of ω; and in Eq. 55 the requirement that

$$h_\infty - \left(\frac{h_2}{s_2} + \frac{h_4}{s_4} + \cdots + \frac{h_{2n}}{s_{2n}}\right) = Y(0) \geq 0 \tag{58}$$

insures that the real part of $Y(j\omega)$ be positive. Hence it is unnecessary to add any real-part conditions. Those regarding the simplicity and alternation of zeros and poles or the simplicity of poles and the residue requirements, plus $Z(\infty) \geq 0$ or $Y(0) \geq 0$, are sufficient (as with *RL* network functions) to assure the realizability of such impedances or

* As mentioned in connection with expansions 42 and 47, these two for $Z(s)$ and $Y(s)$ are not to be regarded as corresponding to the same network. In order for 55 to correspond to 54, the term with h_{2n+2} would have to be absent; and if 54 were to correspond to 55, k_{2n+1} would be zero.

Table 1. Summary of Properties of Driving-Point and Transfer Impedances

(1) $Z(s)$ real for real s applies to all functions

<table>
<tr><th colspan="4">Driving-Point Impedances</th></tr>
<tr><th>RL</th><th>RC</th><th>LC</th><th>RLC</th></tr>
<tr><td colspan="3">(2) Positive multiplier—simple poles and zeros alternating</td><td rowspan="11">Poles and zeros restricted to left half-plane or j-axis; latter must be simple but former may have any multiplicity and relative positions so long as j-axis real part is nonnegative. Residues in j-axis poles must be positive real; those in other poles may have complex values</td></tr>
<tr><td colspan="2">on negative real axis</td><td>on j-axis</td></tr>
<tr><td colspan="2">(3) Critical frequency nearest $s = 0$ is</td><td rowspan="3">(3) Function is a quotient of respectively even and odd polynomials. $s = 0$ and $s = \infty$ always critical. j-axis real part $\equiv 0$</td></tr>
<tr><td>a zero</td><td>a pole</td></tr>
<tr><td colspan="2">(4) j-axis real part is a monotonically</td></tr>
<tr><td>increasing</td><td>decreasing</td><td rowspan="2">(4) Simple poles on j-axis with positive real residues</td></tr>
<tr><td colspan="2">function of ω</td></tr>
<tr><td>(5) Re $[Z(0)] = Z(0) \geq 0$</td><td>Re $[Z(\infty)] = Z(\infty) \geq 0$</td><td rowspan="2">(5) Slope of reactance function $X(\omega)$ positive and greater than X/ω</td></tr>
<tr><td>(6) Residues in finite poles are negative real</td><td>Residues in all poles are positive real</td></tr>
<tr><td colspan="2">Properties (1), (2), and (3) are sufficient to characterize</td><td>(1) and (2) equivalent to (1) and (4). Either pair is sufficient to characterize</td></tr>
<tr><td colspan="3"></td></tr>
<tr><th colspan="4">Transfer Impedances</th></tr>
<tr><td colspan="3">Simple poles</td><td rowspan="7">Poles restricted to left half-plane or j-axis; latter must be simple and yield real residues; there $k_{11}k_{22} - k_{12}^2 \geq 0$. j-axis real part bounded and must fulfill condition $r_{11}r_{22} - r_{12}^2 \geq 0$</td></tr>
<tr><td colspan="2">on negative real axis</td><td>on j-axis</td></tr>
<tr><td colspan="2">Zeros arbitrary</td><td>Zeros with quadrantal symmetry Polynomials respectively even and odd</td></tr>
<tr><td colspan="3">j-axis real part</td></tr>
<tr><td colspan="2">bounded and $r_{12}^2 \leq r_{11}r_{22}$</td><td rowspan="2">identically zero</td></tr>
<tr><td>at $s = 0$</td><td>at $s = \infty$</td></tr>
<tr><td colspan="3">Residues must be real and fulfill condition $k_{11}k_{22} - k_{12}^2 \geq 0$ at all poles</td></tr>
</table>

admittances, as is demonstrated in the following chapters dealing with the methods of synthesizing two-element-kind networks.

Table 1 is a summary of the properties of driving-point and transfer impedances.

PROBLEMS

1. Show that a simple way of stating the necessary and sufficient conditions that a rational function $(m_1 + n_1)/(m_2 + n_2)$ be p.r. is to require (m_2/n_2) and (n_1/m_1) for $s = j\omega$ to be a pair of *nonintersecting* reactance functions. Thus show (by means of graphical sketches if you like) that if $(m_2 + n_1)$ and $(m_1 + n_2)$ as well as $(m_1 + n_1)$ and $(m_2 + n_2)$ are Hurwitz polynomials differing in degree by no more than unity, then the rational function $(m_1 + n_1)/(m_2 + n_2)$ can always be rendered p.r. through assigning to (n_1/m_1) an appropriate positive real multiplier. The function is assumed to have no j-axis poles.

2. Which of the following functions are relizable impedance functions? Of those that are realizable, which are realizable as two-element-kind networks (*LC*, *RC*, *RL*)?

(*a*) $\dfrac{(s^2 + 2)(s^2 + 4)}{(s^2 + 1)(s^2 + 3)}$ (*b*) $\dfrac{(s + 2)(s + 4)}{(s + 1)(s + 3)}$ (*c*) $\dfrac{(s + 3)}{(s + 1)(s + 2)}$

(*d*) $\dfrac{s(s + 5)}{(s + 3)}$ (*e*) $\dfrac{s^3 + 4s}{(s^2 + 3)(s^2 + 5)}$ (*f*) $\dfrac{s^2 + 1}{(s + 10)(s + \frac{1}{10})}$

(*g*) $\dfrac{s^3 + 8s^2 + 15s}{s^2 + 6s + 8}$ (*h*) $\dfrac{1}{s + 1} + 1 - \dfrac{1}{s + 2}$ (*i*) $\dfrac{s(s^2 + 3)(s^2 + 4)}{(s^2 + 1)(s^2 + 2)}$

3. Of the following impedance functions pick out the realizable ones and state whether the corresponding networks are *RL*, *RC*, *LC*, or *RLC*. Do not find networks.

(*a*) $\dfrac{s^4 + 4s^2 + 3}{s^3 + 2s}$ (*b*) $\dfrac{s^2 + 2s}{s^3 + 3s^2 + 3s + 1}$ (*c*) $\dfrac{s^3 + 2s^2 + 5s}{s^2 + 2s + 1}$

(d) $\dfrac{s^2 + s}{s^2 + 5s + 6}$ (*e*) $\dfrac{s + 1}{s^2 + 5s + 6}$ (*f*) $\dfrac{s^3 + s^2}{s^2 + 5s + 6}$

4. Determine the condition upon a, b, and c for which the function $(s + a)/[(s + b)(s + c)]$ is p.r. for any positive real values of a, b, c. What special condition among these three constants must hold if the function is to represent the impedance or admittance of an *RL* or *RC* network?

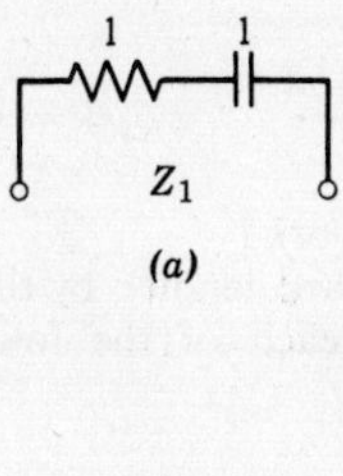

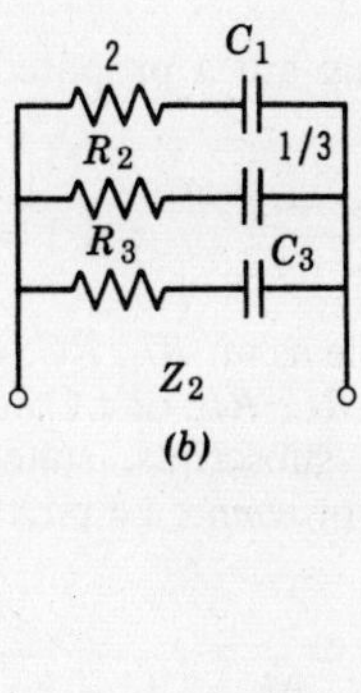

Prob. 5.

5. In the networks shown on page 67, the element values are in ohms and farads. If $Z_1 = Z_2$, what must be the values of R_2, R_3, C_1, C_3? Could the two given parameter values in network (*b*) be completely arbitrary?

6. Which of the following functions are realizable open-circuit transfer impedances z_{12}? State whether they are potentially suitable for *RL*, *RC*, *LC*, or only *RLC* networks.

(*a*) $\dfrac{s^7 - 5s^5}{s^6 + 6s^4 + 11s^2 + 6}$ (*b*) $\dfrac{s^6 + 6s^4 + 11s^2 + 6}{s^7 - 5s^5}$ c) $\dfrac{8s^4 + 5s^2 - 5}{s^3 + 7s}$

(*d*) $\dfrac{s^5 - 8s^4 + 5s^2 - 5}{s^4 + 3s^2 + 2}$ (*e*) $\dfrac{4s^5 + 3s^4 - 8}{s^3 + 6s^2 + 11s + 6}$ (*f*) $\dfrac{s^4 + s + 1}{s^2 + 2s + 5}$

(*g*) $\dfrac{-4s^2 + s - 3}{s^3 + 6s^2 + 11s + 6}$

7. The open-circuit transfer impedance of an *RC* network is to be the function

$$z_{12} = \frac{h(s^2 + 4)}{(s + 1)(s + 4)}$$

Of the following z_{11} and z_{22} functions, which pair may be associated with this z_{12} function, and what is the largest permissible value of the constant multiplier h?

(*a*) $z_{11} = \dfrac{s + 2}{(s + 1)(s + 4)}$ (*b*) $z_{11} = \dfrac{s + 2}{(s + 1)(s + 3)}$

(*c*) $z_{22} = \dfrac{s + 3}{(s + 1)(s + 4)}$ (*d*) $z_{22} = \dfrac{(s + 0.5)(s + 2)}{(s + 1)(s + 4)}$

8. Which of the following y_{12} functions can be realized as the short-circuit transfer admittances of *RL* networks? Give reasons.

(i) $\dfrac{1}{(s + a)(s + b)(s + c)}$

(ii) $\dfrac{s^2 + \alpha s + \beta}{(s + a)(s + b)(s + c)}$

(iii) $\dfrac{(s^2 + \alpha s + \beta)}{(s + a)}$

where a, b, c, α, and β are real positive constants [applies to (i)–(iii)]

(iv) $\dfrac{s^2 + \alpha s + \beta}{s(s + a)}$ where only a is positive but a, α, and β are all real

9. The following are a proposed set of impedances for a two terminal-pair network:

$$z_{11}(s) = \frac{s^2 + 6s + 6}{s^2 + 3s + 2} \qquad z_{22}(s) = \frac{s^2 + 7s + 4}{s^2 + s} \qquad z_{12}(s) = \frac{s}{s + 1}$$

Is the set realizable as an *RL*, *RC*, or *LC* network?

10. If any two *RC*, *RL*, or *LC* impedances are denoted by the letters z and Z with appropriate subscripts, state whether each of the following functions *must* be, *may* be, or *cannot* be p.r.:

(*a*) $\dfrac{z_{RC}}{Z_{RC}}$ (*b*) $\dfrac{z_{RC}}{Z_{RL}}$ (*c*) $z_{RC}Z_{RL}$ (*d*) $1 - Z_{RC}$ (*e*) $Z_{LC} - z_{LC}$

(*f*) $z_{RC} - Z_{LC}$

Give reason for each answer.

11. Are the following statements true or false? Justify your answers.

(*a*) The imaginary part of an *RL* impedance is positive along the entire positive half of the *j*-axis and negative along the negative half.

(*b*) The residue of an *RL* impedance is negative in any finite pole, but in a pole at $s = \infty$ it must be positive.

(*c*) The residue in a pole of an *RLC* admittance, when the pole is not on the *j*-axis, may have any complex value.

(*d*) If the denominator of a rational function $F(s) = p(s)/q(s)$ has all of its zeros in the left half-plane, exclusive of the *j*-axis, and if its real part on the *j*-axis is nonnegative, then $p(s)$ must be a Hurwitz polynomial.

(*e*) The complement of a p.r. function must always be p.r. also, the complementary relationship being $Z_1 + Z_2 = 1$.

(*f*) A p.r. function must have simple poles on the *j*-axis of the *s* plane, but it may have multiple-order zeros there.

(*g*) A rational function that has multiple-order poles and/or zeros in the left-half of the *s*-plane may still be p.r.

(*h*) A rational function having a positive multiplier and only simple zeros and poles in alternating order on the negative real axis is always p.r. and realizable as an *RL* or *RC* network.

12. Indicate whether each of the following statements is true or false and give a reasonable justification for your answer.

(*a*) Sufficient conditions for a rational function to be realizable as the driving-point impedance of a passive network are that it have no poles in the right half-plane and that the real part of the function be nonnegative at all points on the $j\omega$ axis.

(*b*) The real part of a function representing the admittance of a lossless (*LC*) network may not be positive at any point in the left half-plane.

(*c*) The derivative of a p.r. function may never equal zero at zeros of the function.

(*d*) The sum of two functions may be p.r. even if neither of the functions is p.r.

(*e*) If each of two functions is realizable as the impedance of an *RL* network, the sum of the two functions is also realizable as the impedance of an *RL* network.

(*f*) The product of two Hurwitz polynomials is necessarily Hurwitz.

13. Two impedances are said to be complementary (with respect to *R*) if they satisfy the relationship

$$Z_a + Z_b = R$$

(*a*) What is the smallest value of *R* for which the following impedance has a complement

$$Z = \frac{(s+2)(s+4)}{(s+1)(s+3)(s+5)}$$

(*b*) Show that an *LC* impedance (or any impedance having *j*-axis poles) can have no complement.

14. Network *A* has certain natural frequencies for the open-circuit condition of the terminal pair as shown. If a conductance *G* is shunted across the ter-

minals, the natural frequencies shift from their original locations where $Y(s) = 0$ to new locations where $Y(s) = -G$, these new locations always remaining in the left half-plane.

(*a*) Give an appropriate expression for the shift in natural frequencies for very small values of shunted conductance G.

(*b*) Using the result of (*a*), state restrictions upon the angle of dY/ds at zeros of the admittance $Y(s)$ which fall on $s = j\omega$.

(*c*) Repeat (*a*) and (*b*) for finite nonzero natural frequencies on the j-axis and small values of shunted capacitance.

(*d*) Repeat (*c*) for large values of shunted inductance.

(*e*) On the basis of (*b*), (*c*), and (*d*), show that the angle of dY/ds must be zero at zeros of $Y(s)$ lying on the j-axis.

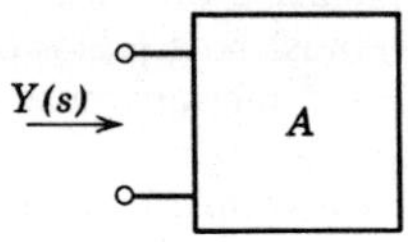

Prob. 14.

15. $Z(s)$ and $h(\lambda)$ are rational impedance functions having the properties stated in (i) through (v) below. Consider the impedance function, $F(\lambda) = Z[h(\lambda)]$, and state:

(*a*) whether it is p.r.

(*b*) how it is realizable (*RC*, *RL*, *LC*, or *RLC*) if:

(i) $Z(s)$ and $h(\lambda)$ are both *RL* realizable.

(ii) $Z(s)$ and $h(\lambda)$ are both *RC* realizable.

(iii) $Z(s)$ is *RC* realizable and $h(\lambda)$ is *RL* realizable.

(iv) $Z(s)$ is *LC* realizable and $h(\lambda)$ is *RL* realizable.

(v) $Z(s)$ is *RL* realizable and $h(\lambda)$ is *LC* realizable.

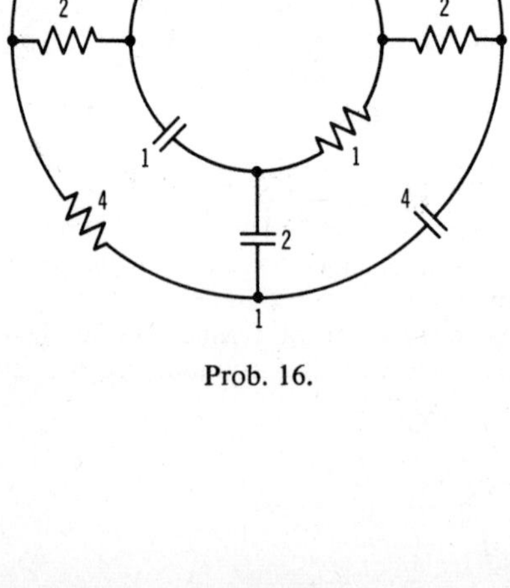

Prob. 16.

16. The parameter values in the RC network on page 70 are in ohms and farads. Consider the node pair 1-2 as the driving point and denote the impedance at this point by $Z(\lambda)$, the corresponding admittance by $Y(\lambda)$. Determine:

(*a*) The largest value of R for which $Z(\lambda) - R$ remains p.r.

(*b*) The largest value of G for which $Y(\lambda) - G$ remains p.r.

17. The following questions relate to the network shown:

(*a*) It is desired to synthesize an admittance $Y'(s)$ equal to the given admittance minus a positive conductance

$$Y'(s) = Y(s) - G_1$$

For what values of G_1 is this possible?

(*b*) It is desired to synthesize an admittance $Y''(s)$ equal to

$$Y''(s) = G_2 - Y(s)$$

For what values of G_2 is this possible?

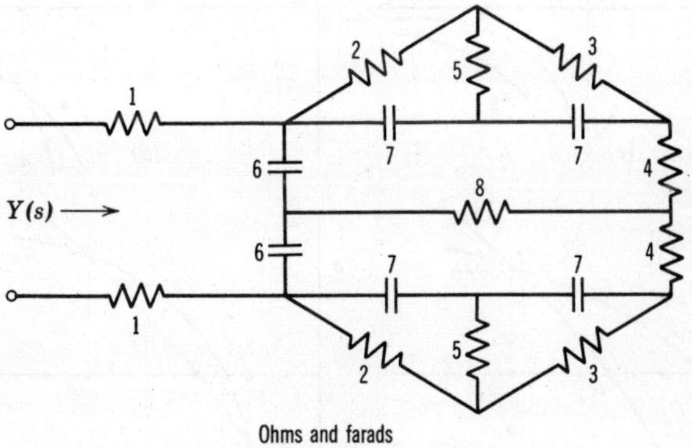

Prob. 17.

18. Sketches (*a*) through (*f*) on page 72 represent frequency functions which are to be realized by pairs of physical reactance functions z_a and z_b. The exact forms of these curves are not so much of interest as are certain mutual characteristics such as intersections at prescribed frequencies, coincidence of prescribed critical frequencies, relative ordinates over certain frequency ranges, and so forth. The functions have no other critical frequencies beyond the portion of the frequency spectrum shown except the obvious one at $\omega = \infty$.

Determine for each case whether or not the reactance functions are physically realizable in view of the particular mutual characteristics called for.

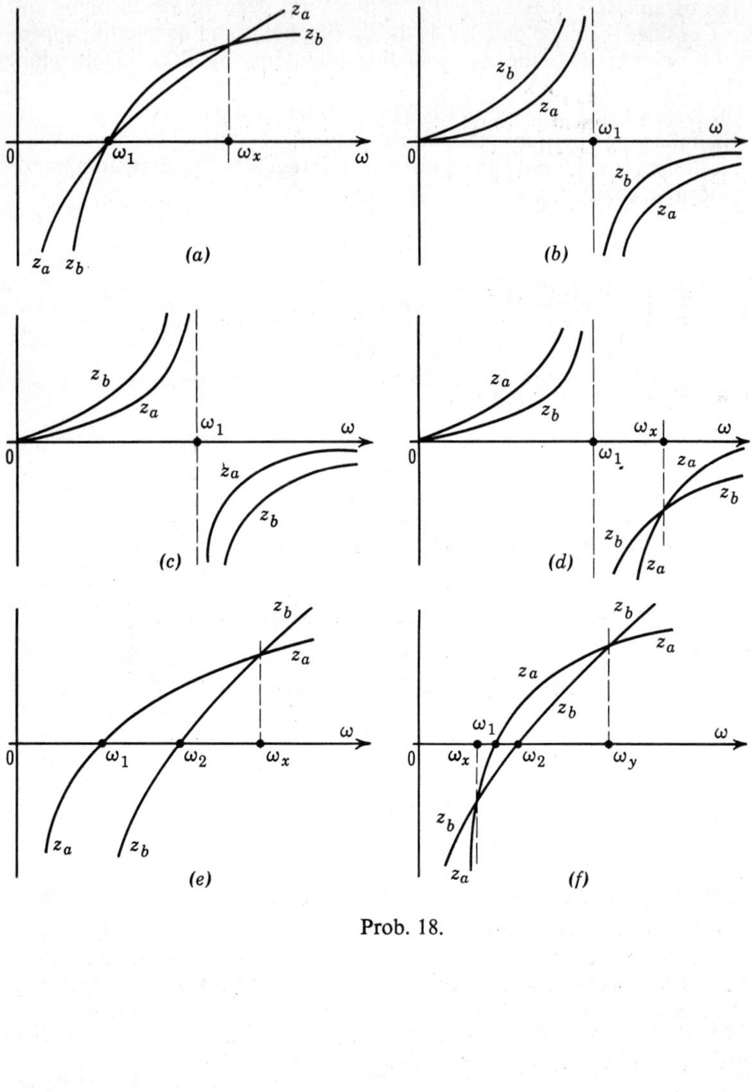

Prob. 18.

CHAPTER THREE

Synthesis of *LC* Driving-Point Impedances

1 General Remarks About Synthesis Procedures

In network theory we are concerned with three things: an excitation, a response, and a network. When any two of these are given, the third can be found. If the network is one of the two given things, then we speak of the problem as one involving network *analysis*. When the network is the unknown quantity, the process of solution is commonly called network *synthesis*. While this definition of the term synthesis is adequate from a broader point of view, it is not sufficiently specific to describe the methods to which the discussions in this text apply.

The term network synthesis when correctly used does not apply to any and all procedures whereby a circuit is designed to meet stated requirements, nor is it to be regarded simply as a more elegant substitute for the term network design. Thus it is quite common in practice to design a network by making a judicious guess (based upon past experience) as to its general configuration, and then through a continued cut-and-try process, supplemented by possible minor revisions in circuit structure, to succeed ultimately in coming acceptably close to a specified behavior pattern. This sort of thing is not network synthesis, the essence of which resides in the fact that the structure of the network and all of its element values are produced by the process itself and that all questions regarding tolerances and the acceptability of behavior patterns are settled at the outset while a rational response function (quotient of polynomials in the complex frequency variable) is constructed to meet the desired behavior pattern on the one hand and the pertinent realizability conditions on the other. This initial step, which is, to be sure, an important part of the entire synthesis process, is referred to as dealing with the pertinent *approximation problem*. Once its solution is accomplished and an appropriate rational function is found, there is no more cutting and trying or guessing involved. The synthesis procedure yields a family of so-called *equivalent networks* any one of which exactly realizes the stated rational function.

As the reader may have observed from various things that have been said so far, the synthesis problem differs from the analysis problem in several important respects. The first of these is concerned with the existence of solutions. When a network and an excitation are given, there is little question as to the existence of a response; and if instead we specify the response and ask for the excitation that will yield it, the same problem is involved with merely the labels of "known" and "unknown" interchanged. It is significant to observe in this connection that the response function does not become inverted, for this would imply an interchange of the labels "excitation" and "response."

However, if we specify an excitation and a response, and ask for the network, then a question as to the existence of a solution very definitely does arise. How we deal with this question depends on whether the given functions are specified in the time or in the frequency domain. If in the latter, their ratio is the desired response function which must be p.r. if a solution for a driving-point function is to exist, and rational if a finite network is to be expected. If the given data are in the form of time functions, then conversion to the frequency domain is first called for, and there is in this situation not even an assurance that the resulting response function formed from a ratio of the pertinent frequency domain functions fulfills the necessary stability requirement to be realized by a passive network.

The second important difference between the analysis and synthesis problems lies in the fact that when the conditions for the existence of a solution are fulfilled, then that solution is not unique. A theoretically infinite number of solutions (known as equivalent networks) exist. These matters will be elaborated upon in the discussions to follow.

The problem with which we will concern ourselves initially is that of finding network realizations for driving-point impedances or admittances in the form of p.r. rational functions. Given such a function, the problem is to find ways of constructing a network which, at a stated point of entry, exactly realizes this function. Although we shall in this chapter and in the following one consider only the special forms of p.r. functions that characterize two-element-kind networks, the philosophy upon which their synthesis procedures is based will be found to establish a basic pattern of thought that continues to be useful throughout the development of all the more elaborate schemes presented later on in connection with the generalization of our problem to include all three kinds of circuit elements and more accessible terminal pairs.

As we contemplate the problem of relating a physical network structure to a given rational function, our first thought is that this process is rather easy if the given function is simple enough. For example, if the impedance function is simply a constant times the variable s or a constant divided by s, then obviously the solution is given by a single inductance or by a single capacitance respectively. The rational function that characterizes a series or parallel combination of these two elements is likewise simple enough, so that with a little practice we can easily recognize the appropriate physical realization by inspection. The question is: What do we do when the given function is rather complicated and the process of recognizing the answer by inspection is no longer possible?

Here the thought rather logically suggests itself that perhaps in such a situation we could find it possible (or at least see if it is possible) to split off from the given function a simple part that is recognizable as a physical component. If the remainder is simpler and (perhaps after some manipulation) permits splitting off another simple part, continued repetitions of this sort should ultimately yield complete development of the original function into a network realization. This scheme is based upon the thought that if the given function is too elaborate to permit synthesis by inspection, perhaps we can wear it away by splitting off a little at a time.

Simple as it may seem, this basic idea is found to yield a successful attack upon many aspects of the synthesis problem, but only if it is possible to conduct each step (or series of steps) so as to fulfill three conditions, namely:

(*a*) The part that we split off from a given function must be realizable and sufficiently simple so that its realization can be recognized by inspection.

(*b*) The remainder must still be p.r. That is to say, the splitting-off process must not destroy the p.r. character of the given function.

(*c*) The remainder must be of lower degree than the given function.

Unless all three of these conditions are met, it is obvious that the process cannot be successful. If they are met by the contemplated cycle of manipulations, then its repeated application will surely yield a complete realization of the given function.

In some instances it may be possible to shorten this process of complete development by breaking down the given function at once into additive components that are realizable and recognizable. We shall see that such a simplified procedure is successful in all two-element-kind cases and sometimes even in the general *RLC* case. The key to carrying out a method of this sort lies in finding a mathematical process in terms of which any rational function can be decomposed into simple (and realizable) component terms. The method of partial-fraction expansion offers such a process of decomposition. Sometimes it can be applied directly to the given function and other times only after some judicious preliminary manipulation. Insight and ingenuity are helpful and often indispensable in carrying through such schemes successfully. The pertinent details of this synthesis method are discussed with reference to *LC* driving-point impedance and admittance functions in the immediately following articles of this chapter.

2 Synthesis of Reactance and Susceptance Functions Through Their Partial-Fraction Expansions

Suppose a reactance function has simple zeros at the points $s = \pm s_1, \pm s_3, \pm s_5, \cdots$, and simple poles at $s = 0, \pm s_2, \pm s_4, \cdots$ As a quotient of factored polynomials it has the form

$$Z(s) = \frac{H(s^2 - s_1^2)(s^2 - s_3^2)(s^2 - s_5^2) \cdots}{s(s^2 - s_2^2)(s^2 - s_4^2)(s^2 - s_6^2) \cdots} \tag{1}$$

in which each factor represents a pair of conjugate imaginary critical frequencies. Thus for $s_\nu = j\omega_\nu$ the factor $(s^2 - s_\nu^2) = (s^2 + \omega_\nu^2) = (s + s_\nu)(s - s_\nu) = (s + j\omega_\nu)(s - j\omega_\nu)$ clearly places in evidence the j-axis critical frequencies at the conjugate points $s = +j\omega_\nu$ and $s = -j\omega_\nu$. If the last factor is in the numerator, then its degree is evidently larger by one than the degree of the denominator, and the function $Z(s)$ has a simple pole at $s = \infty$. If the last factor is in the denominator, then the denominator degree exceeds that of the numerator by one, and $Z(s)$ has a zero at $s = \infty$. The pole at $s = 0$ results from the factor s in the denominator, which incidentally is an odd polynomial. If instead, $Z(s)$ is an odd-over-even ratio, then the point $s = 0$ becomes a zero instead of a pole. Whether the point $s = 0$ is a zero or a pole is independent of whether the point $s = \infty$ is a zero or a pole.

It is important to note that, because of the alternation property of the zeros and poles as expressed by

$$0 < \omega_1 < \omega_2 < \omega_3 < \omega_4 < \cdots \tag{2}$$

a specification of the finite nonzero critical frequencies (zeros and poles) of $Z(s)$ fixes the nature of the points $s = 0$ and $s = \infty$ automatically. Thus if the lowest nonzero critical frequency is a zero (as in Eq. 1), then $s = 0$ must be a pole. If the lowest nonzero critical frequency is a pole, then $s = 0$ must be a zero. Similarly, if the highest finite critical frequency is a zero, then $s = \infty$ must be a pole, and vice versa. Having specified the finite nonzero critical frequencies, it is redundant, therefore, to specify also the nature of the points $s = 0$ and $s = \infty$. Such additional data are irrelevant. Except for the constant multiplier H (also called an *impedance level factor*), the analytic expression for a reactance function (as in Eq. 1) is uniquely characterized by its finite nonzero critical frequencies alone.

A useful graphical way of specifying a reactance function (except for a

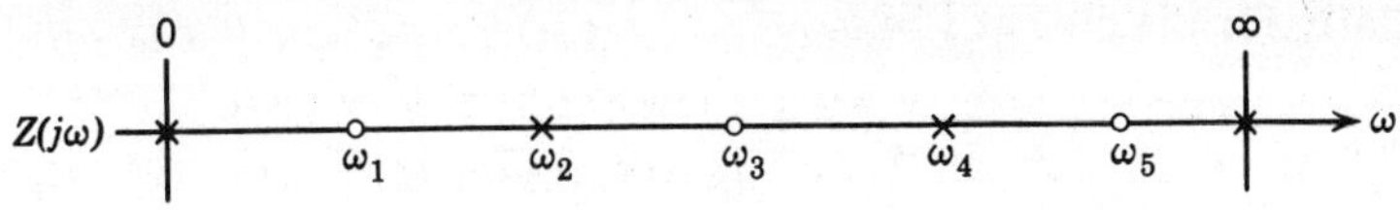

Fig. 1. Frequency pattern for the impedance given by Eq. 3.

constant multiplier) is shown in Fig. 1, which is referred to as its *frequency pattern.* It consists merely of a frequency axis (labeled ω) on which the points $\omega = \infty$ and $\omega = 0$ are marked off as finite points, and the finite nonzero zeros and poles are indicated by circles and crosses respectively. These are denoted as the frequencies ω_1, ω_2, $\cdots$. In a numerical example one may replace these symbols by their respective numerical values. No attempt is made to portray quantitatively the values of any of the critical frequencies. The frequency pattern is thus merely a qualitative representation for the reactance function, but it does fix (except for a level factor) the analytic expression for this function. Thus, setting the level factor for the function in Fig. 1 equal to unity, one can write immediately

$$Z(s) = \frac{(s^2 + \omega_1^2)(s^2 + \omega_3^2)(s^2 + \omega_5^2)}{s(s^2 + \omega_2^2)(s^2 + \omega_4^2)} \tag{3}$$

This analytic expression for $Z(s)$ and the frequency pattern in Fig. 1 are entirely equivalent in the sense that we can construct either one when the other is given.

Recognition of a network which realizes the reactance function is effected through its partial-fraction expansion after pairs of terms representing conjugate poles are combined as in the step from Eq. 30 to Eq. 31 in the previous chapter. Writing the expansion for $Z(s)$ as given by Eq. 3 in this resultant form immediately we have, for example

$$Z(s) = \frac{k_0}{s} + \frac{2k_2 s}{s^2 + \omega_2^2} + \frac{2k_4 s}{s^2 + \omega_4^2} + k_6 s \tag{4}$$

in which k_0, k_2, k_4, and k_6 are the residues of $Z(s)$ in its poles at $s = 0$, $s = \pm j\omega_2$, $s = \pm j\omega_4$, and $s = \infty$ respectively. Their evaluation may be accomplished in several ways. For example, we may use the well-known formula (see *MCA*, p. 306, Eq. 204)

$$k_\nu = [(s - s_\nu)Z(s)]_{s=s_\nu} \tag{5}$$

In the present example for $s_\nu = 0$

$$(s - s_\nu)Z(s) = \frac{(s^2 + \omega_1^2)(s^2 + \omega_3^2)(s^2 + \omega_5^2)}{(s^2 + \omega_2^2)(s^2 + \omega_4^2)} \tag{6}$$

whence Eq. 5 yields

$$k_0 = \frac{\omega_1^2\omega_3^2\omega_5^2}{\omega_2^2\omega_4^2} \tag{7}$$

Similarly for the pole at $s = j\omega_2$

$$(s - s_\nu)Z(s) = \frac{(s^2 + \omega_1{}^2)(s^2 + \omega_3{}^2)(s^2 + \omega_5{}^2)}{s(s + j\omega_2)(s^2 + \omega_4{}^2)} \tag{8}$$

Since $s = j\omega_2$ implies $s^2 = -\omega_2{}^2$, Eq. 5 gives

$$k_2 = \frac{(\omega_1{}^2 - \omega_2{}^2)(\omega_3{}^2 - \omega_2{}^2)(\omega_5{}^2 - \omega_2{}^2)}{-2\omega_2{}^2(\omega_4{}^2 - \omega_2{}^2)} \tag{9}$$

which, incidentally, is seen to be positive as one expects in view of the discussion in the previous chapter.

A direct method for the evaluation of these residues is to consider the immediate vicinity of any one pole and recognize that in the partial-fraction expansion 4 the dominant term alone gives an adequate representation for $Z(s)$ in that vicinity. For example, for $s \to j\omega_2$ or $s^2 \to -\omega_2{}^2$, Eqs. 3 and 4 yield

$$Z(s) = \frac{(s^2 + \omega_1{}^2)(s^2 + \omega_3{}^2)(s^2 + \omega_5{}^2)}{s(s^2 + \omega_2{}^2)(s^2 + \omega_4{}^2)} \to \frac{2k_2 s}{s^2 + \omega_2{}^2} \tag{10}$$

After cancellation of the factor $(s^2 + \omega_2{}^2)$ this result gives

$$k_2 \to \left[\frac{(s^2 + \omega_1{}^2)(s^2 + \omega_3{}^2)(s^2 + \omega_5{}^2)}{2s^2(s^2 + \omega_4{}^2)}\right] \text{ for } s^2 \to -\omega_2{}^2 \tag{11}$$

which is seen to agree with Eq. 9.

Finally, the formula (*MCA*, p. 305, Eq. 198)

$$k_\nu = \left[\frac{dY(s)}{ds}\right]^{-1}_{s=s_\nu} \tag{12}$$

may be used, where $Y(s) = 1/Z(s)$ is the corresponding susceptance function having zeros where $Z(s)$ has poles. It is easy to show that the results agree with those given above. For example, considering the function given by Eq. 3, we have

$$Y(s) = \frac{s(s^2 + \omega_2{}^2)(s^2 + \omega_4{}^2)}{(s^2 + \omega_1{}^2)(s^2 + \omega_3{}^2)(s^2 + \omega_5{}^2)} = \frac{Q(s)}{P(s)} \tag{13}$$

If differentiation is indicated by means of a prime, we have

$$\left(\frac{dY}{ds}\right)_{s=s_2} = \left[\frac{P(s)Q'(s) - Q(s)P'(s)}{P^2(s)}\right]_{s=s_2} = \frac{Q'(s_2)}{P(s_2)} \tag{14}$$

as is clear from the fact that $Q(s_2) = 0$. If we now write

$$Q(s) = (s^2 + \omega_2{}^2)q(s) \tag{15}$$

then

$$Q'(s_2) = [(s^2 + \omega_2{}^2)q'(s) + 2sq(s)]_{s=s_2} = 2s_2 q(s_2) \tag{16}$$

Substitution of this value into Eq. 14 and the result into Eq. 12 yields

$$k_2 = \frac{P(s_2)}{2s_2q(s_2)} \tag{17}$$

Noting the expressions for $P(s)$ and $q(s)$ from Eqs. 13 and 15, we get

$$k_2 = \left[\frac{(s^2 + \omega_1^2)(s^2 + \omega_3^2)(s^2 + \omega_5^2)}{2s^2(s^2 + \omega_4^2)}\right]_{s^2=-\omega_2^2} \tag{18}$$

which again is in agreement with Eq. 9.

The residue in the simple pole at $s = \infty$ is found from noting the behavior of $Z(s)$ for large s. Thus when $Z(s)$ has a pole at infinity, it behaves like a constant times s for $s \to \infty$. The constant is the residue in this pole. Formally we can write

$$k_6 = \left[\frac{Z(s)}{s}\right] \text{ for } s \to \infty \tag{19}$$

which yields unity for the function 3 since we assumed a unit constant multiplier.

Comparing the partial-fraction expansion of $Z(s)$ in Eq. 4 with the frequency pattern shown in Fig. 1, we see that to each pole in the pattern there corresponds a term in the expansion. The first and last terms represent the poles at $s = 0$ and at $s = \infty$ respectively, while the remaining ones correspond to the finite nonzero poles. Once the analytic form for each of these three kinds of terms is recognized, it is a simple matter to write down the partial-fraction expansion for $Z(s)$ directly from its frequency pattern, or from the expansion to sketch the corresponding frequency pattern.

Regarding the pertinent network realization we note that the first and last terms are respectively the reactance of a capacitance and that of an inductance, while the term for a finite nonzero pole is the reactance of the

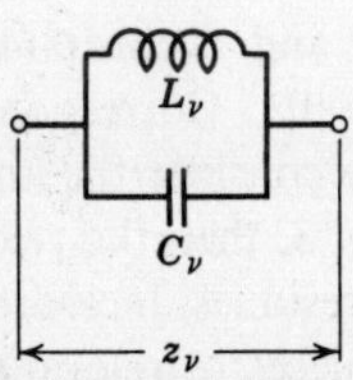

Fig. 2. Realization for a typical term in the partial-fraction expansion of a lossless impedance.

parallel LC combination in Fig. 2 for which

$$z_\nu = \frac{1}{C_\nu s + (1/L_\nu s)} = \frac{L_\nu s}{1 + L_\nu C_\nu s^2} = \frac{(1/C_\nu)s}{s^2 + (1/L_\nu C_\nu)} \tag{20}$$

Comparison with an appropriate term in the partial-fraction expansion shows that

$$\omega_\nu^2 = \frac{1}{L_\nu C_\nu}$$
$$2k_\nu = \frac{1}{C_\nu} \tag{21}$$

from which the values of L_ν and C_ν are readily calculated.

The total network for the function given in Eq. 4, therefore, is recognized to have the form shown in Fig. 3 where L_2, C_2, L_4, C_4 are determined from relations 21, and

$$C_0 = \frac{1}{k_0} \qquad L_6 = k_6 \tag{22}$$

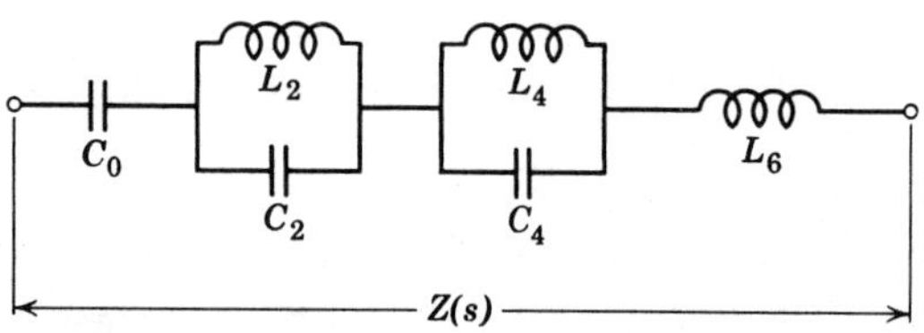

Fig. 3. Realization of the impedance, Eq. 3, according to its partial-fraction expansion in Eq. 4.

Two things are of special interest about this result. First, it should be appreciated that the positive realness of the residues guarantees the positive realness of all circuit element values, and hence we see that the p.r. character of $Z(s)$ is sufficient to assure the physical realizability of the driving-point impedance in this LC case. Second, the number of elements in the realization exactly equals the number of pieces of relevant data determining $Z(s)$. Thus function 3 is determined by five finite nonzero critical frequencies and a constant multiplier (equal to unity), making six pieces of data in all. Correspondingly, the network in Fig. 3 involves six elements. If we consider the analytic representation for $Z(s)$ given in Eq. 4 instead of Eq. 3, then the relevant data are the finite nonzero pole locations and the residues, including those in the poles at $s = 0$ and at $s = \infty$. Again six pieces of relevant data are involved. For six given numbers characterizing the function $Z(s)$ in either of the forms given by Eq. 3 or 4, we obtain six element values in the realizing network. It is logically clear that this method of synthesis yields a realization involving a minimum number of elements. For this reason it is referred to as a *canonic* form for the realization of the given $Z(s)$.

Precisely the same method of synthesis may be applied to the susceptance

13. It has poles only at the finite nonzero critical frequencies $\pm j\omega_1$, $\pm j\omega_3$, $\pm j\omega_5$, the points $s = 0$ and $s = \infty$ being zeros. Hence the pertinent partial-fraction expansion reads

$$Y(s) = \frac{2k_1 s}{s^2 + \omega_1{}^2} + \frac{2k_3 s}{s^2 + \omega_3{}^2} + \frac{2k_5 s}{s^2 + \omega_5{}^2} \tag{23}$$

Evaluation of residues follows any one of the three methods discussed for the reactance function.

Each term here is recognized as the susceptance of a series LC combination as in Fig. 4, for which

$$y_\nu = \frac{1}{L_\nu s + (1/C_\nu s)} = \frac{C_\nu s}{1 + L_\nu C_\nu s^2} = \frac{(1/L_\nu)s}{s^2 + (1/L_\nu C_\nu)} \tag{24}$$

Comparison with a typical term in expansion 23 shows that

$$\begin{aligned} \omega_\nu{}^2 &= \frac{1}{L_\nu C_\nu} \\ 2k_\nu &= \frac{1}{L_\nu} \end{aligned} \tag{25}$$

The complete network for the realization of function 23 is thus found to be that in Fig. 5 in which the element values are found from Eqs. 25 in terms of the appropriate residues and pole frequencies.

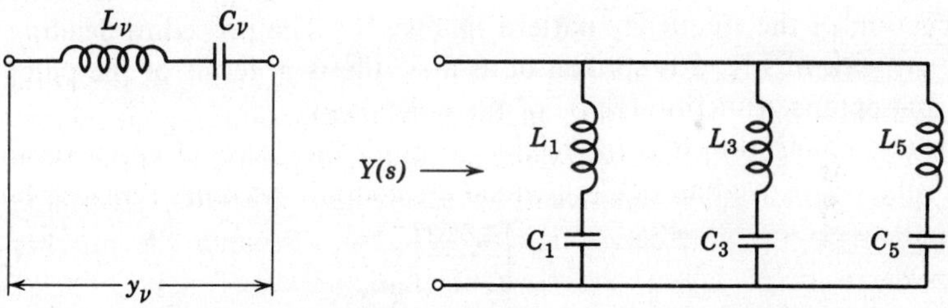

Fig. 4 (left). Realization of a typical term in expansion 23.
Fig. 5 (right). Realization of impedance, Eq. 3, according to the partial-fraction expansion of the corresponding admittance function as given by Eq. 23. This circuit and that in Fig. 4 are equivalent Foster forms.

It is evident that this procedure (not the end result) is completely dual to that on the impedance basis leading to the network in Fig. 3. Thus the partial-fraction expansions 4 and 23 are basically the same in form, while the component networks in Figs. 2 and 4 as well as the resultant networks in Figs. 3 and 5 are dual in form, and the specific relations for the elements given by Eqs. 21 and 25 are the same except for an interchange in the dual elements L and C. Comments relative to the sufficiency of the p.r. character of $Y(s)$ in guaranteeing its realization as well as to the canonic character* of the resulting network in that it effects a realization in terms of the minimum number of elements are the same as for the impedance basis.

3 A Useful Property of These Network Realizations

If we compare the network of Fig. 3 with the frequency pattern of Fig. 1 it is apparent that each pole of the function $Z(s)$ is responsible for one of the four components in the complete series combination. Thus the pole at $s = 0$ is provided by the capacitance C_0, the poles at the points $s = s_2$ and $s = s_4$ are placed in evidence by the two parallel LC circuits resonant respectively at these frequencies, and the pole at $s = \infty$ is provided by the final series inductance. In connection with this interpretation it is helpful to regard the capacitance C_0 and the inductance L_6 as degenerate forms of parallel LC networks that are resonant respectively at $s = 0$ and at $s = \infty$. We speak of the network of Fig. 3 as placing the poles of the reactance function in evidence. The synthesis method leading to this result is correspondingly referred to as a synthesis in terms of the poles of the given reactance.

Analogously do the series-resonant component LC networks in the realization shown in Fig. 5 place in evidence the poles of the corresponding susceptance function $Y(s)$ or the zeros of the reactance $Z(s)$. This network as well as that in Fig. 3 can be sketched in form directly by inspection of the frequency pattern in Fig. 1. The procedure leading to the network of Fig. 5 is spoken of as a synthesis in terms of the poles of the susceptance function (zeros of the reactance).

In this connection it is interesting to point out a useful application of formula 12 for residues of $Z(s)$, which incidentally becomes replaced by

$$k_\nu = \left[\frac{dZ(s)}{ds}\right]^{-1}_{s=s_\nu} \tag{26}$$

when the residues of $Y(s)$ are to be calculated. If, with reference to the network of Fig. 3, Eqs. 12 and 21 are combined, we have for the element values

$$C_\nu = \frac{1}{2}\left[\frac{dY(s)}{ds}\right]_{s=s_\nu}$$

and

$$L_\nu = \frac{1}{C_\nu \omega^2_\nu}, \quad \text{for } \nu = 2, 4, \cdots \tag{27}$$

* The network forms shown in Figs. 3 and 5, obtained from the partial-fraction expansions of pertinent impedance and admittance functions, are also referred to as *Foster* forms since this synthesis procedure for reactance functions was first pointed out by R. M. Foster in his classic paper "A Reactance Theorem," Bell System Tech. J., Vol. 3, 1924, p. 259. This paper, incidentally, is the first to present the exact synthesis of any p.r. rational function. It marks the beginning, therefore, of our modern theory of network synthesis.

Here the $s_\nu = j\omega_\nu$ are poles of $Z(s)$ or zeros of $Y(s)$. The network elements 27 (relative to the parallel LC combinations in Fig. 3) are expressed in terms of the zeros of $Y(s)$ and the slopes of $Y(s)$ at these zeros.

Similarly, if we combine Eqs. 25 and 26 with reference to the network of Fig. 5, we find for its element values the expressions

$$L_\nu = \frac{1}{2}\left[\frac{dZ(s)}{ds}\right]_{s=s_\nu}$$

$$C_\nu = \frac{1}{L_\nu \omega^2{}_\nu}, \quad \text{for } \nu = 1, 3, 5 \cdots \tag{28}$$

in which the $s_\nu = j\omega_\nu$ are the poles of $Y(s)$ or zeros of $Z(s)$. The network elements 28 (relative to the series LC combinations in Fig. 5) are expressed in terms of the zeros of $Z(s)$ and the slopes of $Z(s)$ at these zeros.

When zeros and slopes of a reactance or susceptance function are given or readily computable, their network realizations in the form of either of the circuits shown in Fig. 3 or 5 are particularly easy to determine. A specific example will serve best to illustrate the usefulness of this method. Figure 6 illustrates schematically a uniform lossless transmission line of

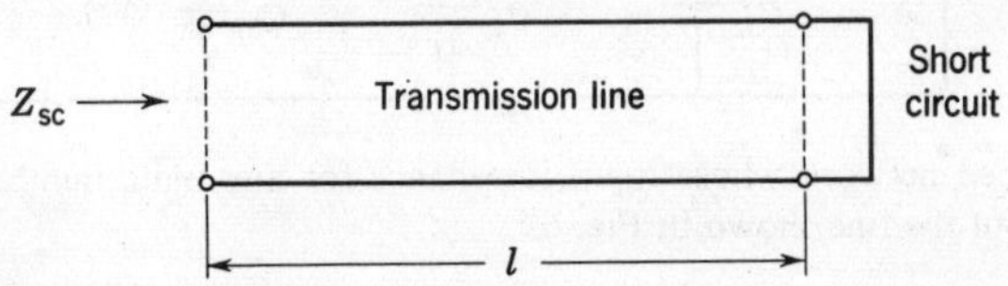

Fig. 6. Short-circuited lossless transmission line whose input impedance is a reactance function with uniformly spaced zeros and poles, unlimited in number.

length l, short-circuited at the far end. If L and C are the distributed inductance and capacitance parameters per unit length, then the input impedance is given by the well-known expression

$$Z_{sc} = j\sqrt{\frac{L}{C}} \tan \omega l\sqrt{LC} \tag{29}$$

and the corresponding admittance by

$$Y_{sc} = -j\sqrt{\frac{C}{L}} \cot \omega l\sqrt{LC} \tag{30}$$

These are transcendental meromorphic functions and hence have an infinite number of ordinary poles in the finite s-plane. Therefore, they can at best be only approximated by a finite network. However, since the poles are all simple and lie on the j-axis, and the residues are real and positive, it is reasonable that a lumped, finite network can be constructed which simulates the impedance Z_{sc} over a finite range of frequencies. Such a simulating network has various practical uses, and so it is of some interest to discuss a possible way of synthesizing it.

From Eq. 29 we see that the zeros of Z_{sc} are located at the points $s = j\omega_\nu$ where

$$\omega_\nu l\sqrt{LC} = \theta_\nu = \frac{\nu\pi}{2} \quad \text{or} \quad \omega_\nu = \frac{\nu\pi}{2l\sqrt{LC}} \quad \text{for } \nu = 0, 2, 4, \cdots \tag{31}$$

and that the slopes of Z_{sc} at these points are given by

$$\left(\frac{dZ_{sc}}{j\,d\omega}\right)_{\omega=\omega_\nu} = \sqrt{\frac{L}{C}}\frac{l\sqrt{LC}}{\cos^2\theta_\nu} = Ll = L_0 \tag{32}$$

where L_0 is an abbreviation for the total inductance of the line. According to relations 28 and the network structure in Fig. 5 to which they pertain, we see at once that a possible simulating network is given by the configuration shown in Fig. 7 where the first shunt branch is an inductance

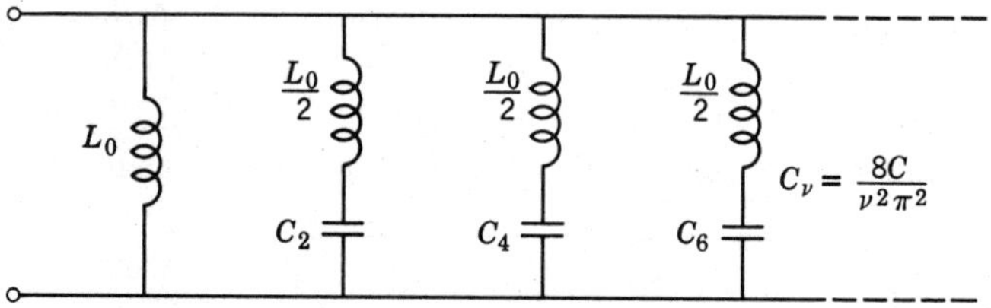

Fig. 7. Lumped network whose input impedance for any finite number of elements simulates that of the line shown in Fig. 6.

twice as large as the others in order that the slope condition may be met also at $s = 0$.

Alternately we find from Eq. 30 that the zeros of Y_{sc} are located at the points $s = j\omega_\nu$, where the ω_ν are given by the same expression 31 for ν equal to odd integer values, and that

$$\left(\frac{dY_{sc}}{j\,d\omega}\right)_{\omega=\omega_\nu} = \sqrt{\frac{C}{L}}\frac{l\sqrt{LC}}{\sin^2\theta_\nu} = Cl = C_0 \tag{33}$$

where C_0 is an abbreviation for the total capacitance of the line. According to relations 27 pertinent to the network of Fig. 3 we obtain the simulating network shown in Fig. 8.

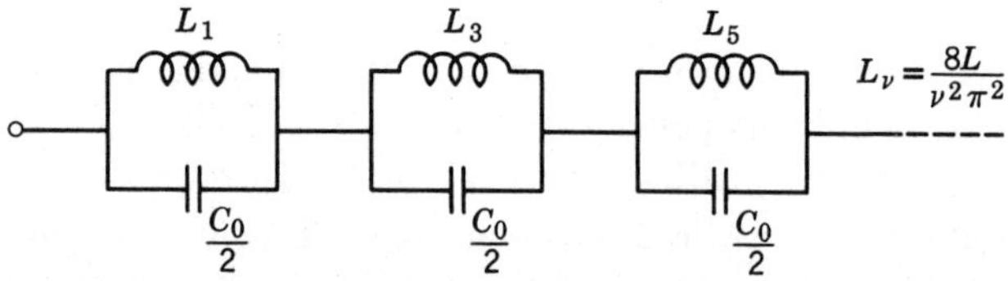

Fig. 8. Simulating network alternate to that shown in Fig. 7.

In terms of the known pole and zero positions, one can write down an expression for Z_{sc} in the form of Eq. 1 and from its partial-fraction expansion (for a finite number of poles) obtain also a simulating network of the form shown in Fig. 8; or from a partial-fraction expansion of the corresponding Y_{sc} we may obtain one like the network of Fig. 7. The element values in these networks will not check exactly with those given in Figs. 7 and 8, however, since the latter are designed so as to yield Y_{sc} and Z_{sc} functions respectively that have the correct poles and residues of the pertinent transcendental functions instead of having the correct poles and zeros. Questions regarding the exact simulating properties which these various approximating networks have are more appropriately gone into at some later time.

4 Alternate Canonic Network Realizations

In terms of the ideas involved in the realization procedures discussed in the preceding articles it is not difficult to construct other network forms that likewise represent the same given reactance function. Thus the Foster forms shown in Figs. 3 and 5 are arrived at by removing at once all the poles of the given function [$Z(s)$ or $Y(s)$ respectively]. Instead we can remove from $Z(s)$ only one or two of its poles, leaving a remainder function that contains the rest of the poles. This remainder is surely a p.r. reactance function and its reciprocal is surely a p.r. susceptance function which we can realize in the Foster manner as series LC components in parallel. A series connection of the network portion representing the poles removed from the given $Z(s)$ and that realizing the inverted remainder in the alternate Foster form is evidently a valid synthesis of the given function.

It is clear that more elaborate ways of combining the two Foster procedures may easily be carried out. For example, one can begin by removing just one pole from $Z(s)$, then inverting the remainder and removing one pole, and from the inverted second remainder again remove one pole, and so forth. The first pole removed from $Z(s)$ yields a series branch in the resultant network, whereas the pole removed from the inverted first remainder function yields a shunt branch because this function is an admittance. The second inverted remainder function is again an impedance, and so a pole removed from it yields a series branch, since component impedances add in series but component admittances add in parallel. It is not difficult to see that this procedure leads to an unbalanced ladder structure of the general form shown in Fig. 9 in which it is tacitly assumed that all of the poles removed are at finite nonzero frequencies so that no degenerate forms of the series- or parallel-resonant LC circuits occur.

This process can be varied by occasionally removing more than one pole from an inverted remainder function. There can be no question as to the possibility of carrying through any of these combination schemes because the removal of one or more poles from a p.r. *LC* driving-point function must surely leave a remainder that is p.r. and *LC* realizable either right side up or upside down. It is also not difficult to appreciate that all these forms of realization will involve the same minimum number of elements.

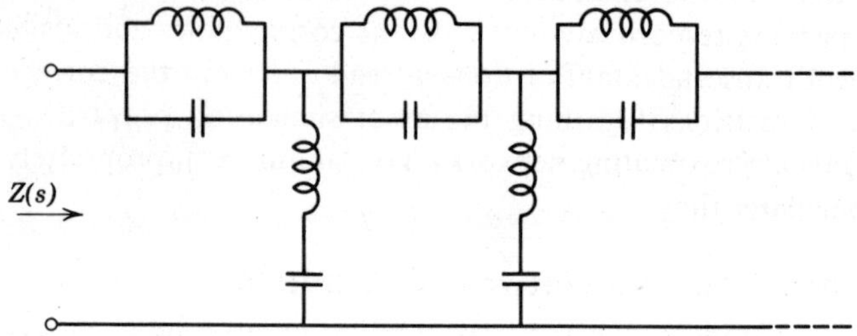

Fig. 9. A possible canonic realization for a reactance function.

An infinity of variations in this mixed procedure becomes possible if we realize that at any stage we do not have to remove a pole completely. When we remove a pole, we may regard this process as the subtraction of a term from the partial-fraction expansion of the pertinent impedance or admittance function. This term has the form $2ks/(s^2 + \omega_\nu^2)$, where k is the residue in the particular pole. If, instead of subtracting this whole term, we subtract $2k's/(s^2 + \omega_\nu^2)$, where $k' < k$, then the pertinent pole is only partially removed from the function in question. Such partial pole removals can be repeated as often as we wish, with the result that a truly infinite variety of network realizations may be constructed for any given function. These networks will obviously no longer involve the minimum number of elements, but they offer an opportunity for revising element value distributions and for this reason they are of considerable practical value.

Two particular network forms of this sort are of special theoretical interest. One of these is obtained by consistently removing only the pole at $s = \infty$ from the given function and its subsequent inverted remainders, the other one by consistently removing only the pole at $s = 0$. Thus, suppose the given $Z(s)$ has a pole at $s = \infty$. Removal of this pole in the form of a series inductance leaves a remainder that is surely p.r. and *LC* realizable and one that surely has a zero at $s = \infty$ (because a reactance function must have a zero or a pole at $s = \infty$, and the pole has just been removed). The inverted remainder (a susceptance function) therefore has a pole at $s = \infty$, and its removal yields a shunt capacitance. The remaining susceptance again is p.r. and *LC* realizable, and has a zero at $s = \infty$, so that its reciprocal has a pole there. At this stage the original state of affairs again obtains, namely, we have a p.r. and LC realizable reactance function with a pole at $s = \infty$. The same cycle of manipulations as just carried out yields another series inductance followed by a shunt capacitance and another reactance function with a pole at $s = \infty$. After each cycle the remaining reactance function has one less finite nonzero pole and zero (two finite nonzero critical frequencies less than at the beginning of the cycle). When all poles and zeros are exhausted, the process ends, and the given function is developed into an unbalanced ladder network in which all series branches are inductances and all shunt branches are capacitances, as illustrated in Fig. 10.

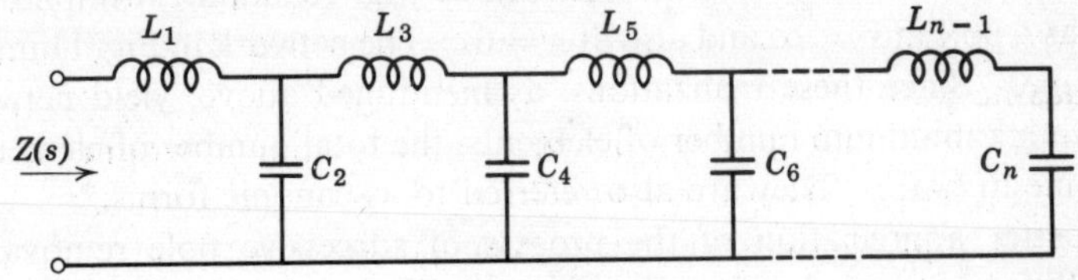

Fig. 10. Canonic realization known as a "Cauer form," obtainable from a continued-fraction development of the pertinent impedance function, and corresponding to consistent pole removals at $s = \infty$.

The procedure, alternate to this one, in which only poles at $s = 0$ are removed from the given reactance and its subsequent inverted remainders is entirely similar in character and in the proof of its general applicability. Here we may suppose that the given $Z(s)$ has a pole at $s = 0$. Its removal in the form of a series-capacitance branch yields a remaining reactance that is surely p.r. and LC realizable, having a zero at $s = 0$. The inverted remainder, a susceptance, has a pole at $s = 0$, and its removal in the form of a shunt inductance leaves a remainder which, turned over, has the same character as the original reactance function. Repeated applications of this alternate cycle lead finally to a complete development in the form of an unbalanced ladder with series capacitances and shunt inductances as illustrated in Fig. 11.

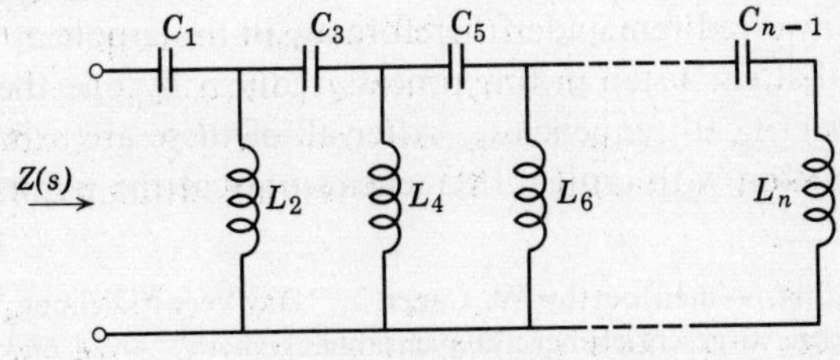

Fig. 11. Alternate Cauer form to the one shown in Fig. 10. Here poles are consistently removed at $s = 0$.

It is useful to note that the first and last elements in each of these ladder networks determines the behavior of the function at $s = 0$ and at $s = \infty$. Specifically in the network of Fig. 10, if $Z(s)$ has a pole at $s = \infty$, then $L_1 \neq 0$, and if $s = \infty$ is a zero, then evidently $L_1 = 0$. If $s = 0$ is a pole, then $C_n \neq \infty$, and if it is a zero, $C_n = \infty$ (a short circuit). When the given $Z(s)$ does not have a pole at $s = \infty$, therefore, the first step is to turn it over and consider $Y(s) = 1/Z(s)$ which has a pole at $s = \infty$, yielding the shunt capacitance C_2.

Similarly the network of Fig. 11 will lack the series capacitance C_1 if $Z(s)$ has no pole at $s = 0$, and it will lack the final inductance L_n if $Z(s)$ has no pole at $s = \infty$. As the network of Fig. 10 stands, it implies that $Z(s)$ has a pole at $s = \infty$ and also at $s = 0$. The network in Fig. 11 implies the same. Since these realizations, as mentioned above, yield networks involving a minimum number of elements, the total number of elements is the same in both. They are also referred to as canonic forms.*

A better appreciation of the process of successive pole removals at $s = 0$ or at $s = \infty$ is had by observing graphically the details involved in a typical step of this sort as illustrated in Fig. 12. In the top part of this figure is a plot of a reactance $X_1(\omega)$ having a pole at $s = \infty$ which is to be removed as a series inductance L_1. Shown dashed is the asymptote of the function $X_1(\omega)$, a straight line with a slope equal to the residue in the pole at $s = \infty$. This slope therefore equals the value of the series inductance L_1. Removal of the pole at $s = \infty$ from $X_1(\omega)$ amounts to the subtraction of its linear asymptote. This process leaves the positions of the remaining poles unchanged but shifts the zeros at a and b to the positions a' and b' located vertically below the points of intersection between the $X_1(\omega)$ curve and the linear asymptote. For the zero at c this intersection lies at infinity, and so this zero is shifted to the point $s = \infty$.

The inverted remainder function $B_2(\omega)$ shown below the curve for $X_1(\omega)$ in Fig. 12 thus clearly has a pole at $s = \infty$, and the residue in this pole is again equal to the slope of the linear asymptote, which is the value C_2 of the capacitance in the next shunt branch of the ladder development. Removal of this pole from $B_2(\omega)$ shifts its highest zero to the point $s = \infty$, and the next inverted remainder therefore again has a pole at infinity. It is thus clear that each step in this process reduces by one the number of finite nonzero critical frequencies. After all of these are exhausted there remains a function with critical frequencies only at the points $s = 0$ and

* They were first pointed out by W. Cauer in "Die Verwirklichung von Wechselstromwiderständen vorgeschriebener Frequenzabhängigkeit," *Arch. Elektrotech.*, Vol. 17, 1927, p. 355, and are commonly referred to as the *Cauer* forms in contrast to the Foster forms discussed previously.

$s = \infty$. This function represents simply an inductance or a capacitance and so we see that the total number of elements in the complete ladder development is equal to the number of finite nonzero critical frequencies plus one, which is just equal to the number of pieces of relevant data characterizing the given reactance or susceptance function as pointed out earlier. We thus have an independent verification of the fact that this method of realization involves a minimum number of elements.

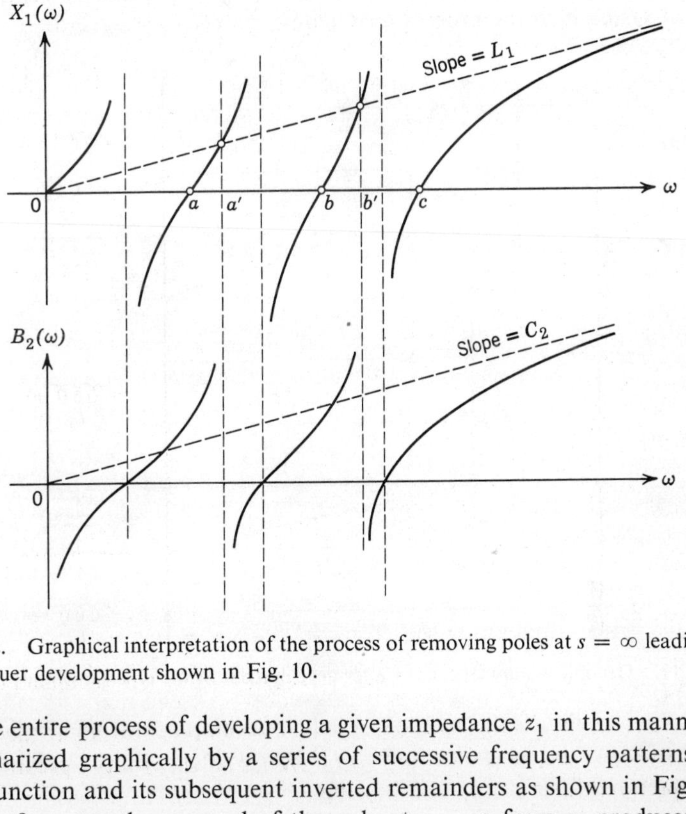

Fig. 12. Graphical interpretation of the process of removing poles at $s = \infty$ leading to the Cauer development shown in Fig. 10.

The entire process of developing a given impedance z_1 in this manner is summarized graphically by a series of successive frequency patterns for this function and its subsequent inverted remainders as shown in Fig. 13. In the first step the removal of the pole at $s = \infty$ from z_1 produces the series inductance L_1 and leaves a remainder z_2 whose finite nonzero poles are the same as those of z_1 but whose zeros are shifted from their original positions; in particular, the highest zero in z_1 shifts to the point $s = \infty$. On the third line of Fig. 13 is next shown the frequency pattern for the inverted z_2 function, the admittance $y_2 = 1/z_2$. It has poles where z_2 has zeros and vice versa. Removal of the pole at $s = \infty$ produces the shunt capacitance C_2 and again shifts the highest zero to the point at infinity, leaving a remainder y_3. The next impedance $z_3 = 1/y_3$ has the same character as z_1, so that the same cycle of events can be reapplied yielding successively the series branch L_3 and a following shunt branch C_4. At this point all of the original finite nonzero zeros and poles are exhausted, and the remainder y_5 or its reciprocal the impedance z_5 has only the critical frequencies at $s = 0$ and $s = \infty$. As indicated, this remainder yields a final series inductance L_5. Such a graphical portrayal of the sequence of operations involved in the complete ladder development provides a compact way of observing the step-by-step effects of the method and of seeing how the process terminates.

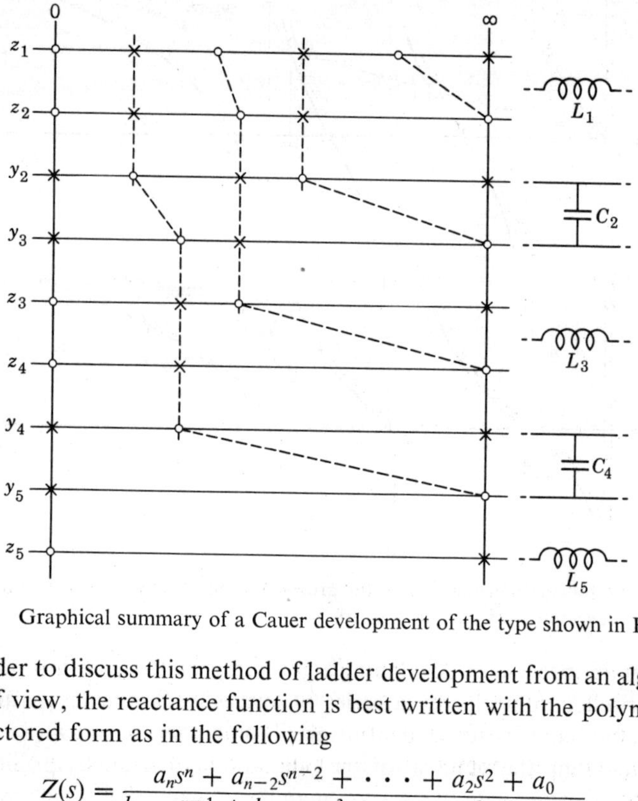

Fig. 13. Graphical summary of a Cauer development of the type shown in Fig. 10.

In order to discuss this method of ladder development from an algebraic point of view, the reactance function is best written with the polynomials in unfactored form as in the following

$$Z(s) = \frac{a_n s^n + a_{n-2} s^{n-2} + \cdots + a_2 s^2 + a_0}{b_{n-1} s^{n-1} + b_{n-3} s^{n-3} + \cdots + b_3 s^3 + b_1 s} \tag{34}$$

Removal of the pole at $s = \infty$ is done through dividing the denominator polynomial into the numerator polynomial so as to obtain a quotient $(a_n s/b_{n-1})$ and a remainder function which is the ratio of an $(n - 2)$ to an $(n - 1)$ degree polynomial, as indicated below

$$Z(s) = L_1 s + \frac{a_{n-2}' s^{n-2} + a_{n-4}' s^{n-4} + \cdots + a_2' s^2 + a_0'}{b_{n-1} s^{n-1} + b_{n-3} s^{n-3} + \cdots + b_3 s^3 + b_1 s} \tag{35}$$

where $L_1 = a_n/b_{n-1}$ is the value of the first series inductance. Inverting the remainder and repeating the process yields

$$Z(s) = L_1 s + \cfrac{1}{C_2 s + \cfrac{b_{n-3}'s^{n-3} + \cdots + b_1's}{a_{n-2}'s^{n-2} + \cdots + a_2's^2 + a_0'}} \tag{36}$$

in which C_2 is the value of the succeeding shunt capacitance, and the second remainder function is the ratio of an $(n - 3)$ to an $(n - 2)$ degree polynomial. Continuation of this process evidently yields the following continued-fraction development of the given driving-point impedance

$$Z(s) = L_1 s + \cfrac{1}{C_2 s + \cfrac{1}{L_3 s + \cfrac{1}{C_4 s + \cdot \, \cdot \, \cdot + \cfrac{1}{L_{n-1}s + \cfrac{1}{C_n s}}}}} \tag{37}$$

It should be observed that expression 34 for $Z(s)$ tacitly assumes that n is even. If it is odd, then $Z(s)$ is an odd-over-even rational function and the final term is an inductive reactance $L_n s$. This process of successive division and inversion is evidently identical with that used in making a Hurwitz test. As has been pointed out, the reactance function is the same kind of limiting form of a p.r. function as the ratio of odd-over-even or even-over-odd parts of a Hurwitz polynomial, for which the continued-fraction development (Eq. 45, Ch. 1), yields all positive real coefficients. The physical realizability of the network obtained by this process is thus independently verified.

The alternate ladder development in which poles are consistently removed at $s = 0$ corresponds to a continued-fraction expansion in terms of $(1/s)$. To this end, the polynomials in $Z(s)$ are written in ascending powers of s as follows

$$Z(s) = \frac{a_0 + a_2 s^2 + a_4 s^4 + \cdots + a_n s^n}{b_1 s + b_3 s^3 + b_5 s^5 + \cdots + b_{n-1}s^{n-1}} \tag{38}$$

Division of the denominator polynomial into the numerator polynomial yields a quotient $(a_0/b_1 s)$ and a remainder function as shown in

$$Z(s) = \frac{1}{C_1 s} + \frac{a_2's^2 + a_4's^4 + \cdots + a_n's^n}{b_1 s + b_3 s^3 + \cdots + b_{n-1}s^{n-1}} \tag{39}$$

where $C_1 = (b_1/a_0)$ is the value of the first series capicitance (Fig. 11). The inverted remainder function again has a pole at $s = 0$ whose removal is indicated in

$$Z(s) = \frac{1}{C_1 s} + \cfrac{1}{\cfrac{1}{L_2 s} + \cfrac{b_3' s^3 + \cdots + b_{n-1}' s^{n-1}}{a_2' s^2 + \cdots + a_n' s^n}} \tag{40}$$

in which $L_2 = (a_2'/b_1)$ is the value of the next shunt inductance. Continuation yields the following continued-fraction development

$$Z(s) = \frac{1}{C_1 s} + \cfrac{1}{\cfrac{1}{L_2 s} + \cfrac{1}{\cfrac{1}{C_3 s} + \cfrac{1}{\cfrac{1}{L_4 s} + \cdots + \cfrac{1}{\cfrac{1}{C_{n-1} s} + \cfrac{1}{\cfrac{1}{L_n s}}}}}} \tag{41}$$

The successive terms here are $1/C_1 s$, $1/L_2 s$, $1/C_3 s$, and so forth, and hence clearly represent poles at $s = 0$. The process is like the test of a Hurwitz polynomial $P(s)$ where one first makes the change of variable $s \to 1/s$, as discussed in Art. 3 of Ch. 1. The positiveness of all the L's and C's is thus assured by the p.r. and LC character of the reactance function $Z(s)$ which is identical with a ratio of the even and odd parts of a Hurwitz polynomial.

In carrying out these Cauer developments for a numerical example, it is essential that one does not merely try to follow a continued division and inversion pattern without paying attention to the physical implications of at least the initial steps in the process, for it is possible otherwise to become badly confused. As we have shown, the process may have to be begun with the given function $Z(s)$ right side up or upside down depending on whether this reactance has poles at $s = 0$ and $s = \infty$ or not. Since one of the ladder developments is done with the polynomials arranged in the order of ascending powers and the other one with these polynomials turned end for end, it is clear that a particular development contemplated calls for one of *four* possible initial orientations of the $Z(s)$ function, which can be turned upside down or end for end in four distinct combinations. Two of these initial orientations in any given situation yield the two Cauer forms; the other two yield nonsense. The best way to avoid the nonsense is to interpret physically the significance of a contemplated manipulation according to the discussions given above. In other words one must be alert and not merely try to follow a pattern.

5 Numerical Examples

Consider a reactance function with the frequency pattern shown in Fig. 14. The analytic expression for it as a quotient of frequency factors and its partial-fraction expansion are seen to be

$$Z(s) = \frac{s(s^2 + 4)}{(s^2 + 1)(s^2 + 9)} = \frac{\frac{3}{8}s}{s^2 + 1} + \frac{\frac{5}{8}s}{s^2 + 9} \tag{42}$$

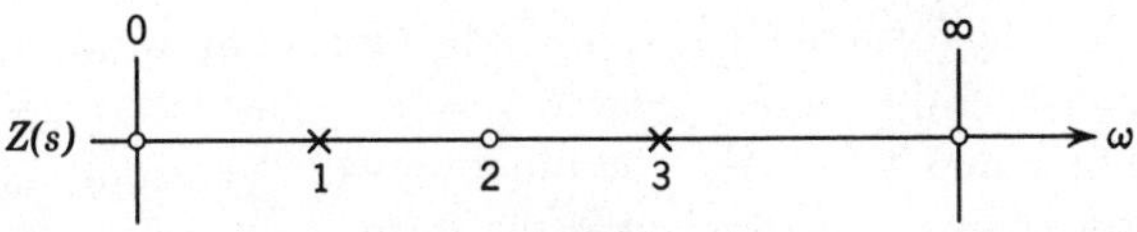

Fig. 14. Frequency pattern for the example given by Eq. 42.

The corresponding susceptance function and its partial-fraction expansion reads

$$Y(s) = \frac{(s^2 + 1)(s^2 + 9)}{s(s^2 + 4)} = \frac{\frac{9}{4}}{s} + \frac{\frac{15}{4}s}{s^2 + 4} + s \tag{43}$$

The two Foster forms associated with these expansions are shown in Fig. 15*a* and *b*.

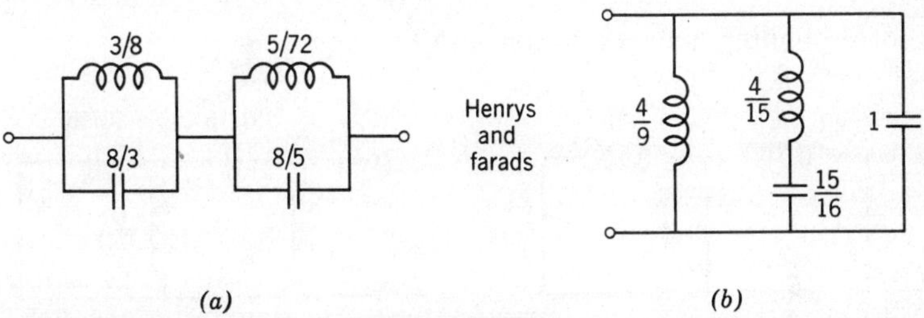

Fig. 15. The two Foster networks realizing impedance 42 whose frequency pattern is shown in Fig. 14.

It is possible to make some partial checks on numerical accuracy by noting whether these circuits yield the same results for very low and very high frequencies. Thus at $s = 0$, the circuit of Fig. 15*a* reduces to an inductance of the value $\frac{3}{8} + \frac{5}{72} = \frac{4}{9}$, which checks the behavior of the circuit in Fig. 15*b* at low frequencies where the capacitances are open circuits. At high frequencies the circuit of part (*a*) degenerates into an elastance of the value $\frac{3}{8} + \frac{5}{8} = 1$, thus agreeing with the corresponding asymptotic behavior of the circuit of part (*b*). We can make still another check. If we short-circuit either network the natural frequency should be the zero of $Z(s)$ at $s^2 = -4$ or $s = \pm j2$. In the circuit of part (*b*) the short circuit leaves the central series LC component closed upon itself for which $\omega_0^2 = 1/LC$ gives $\omega_0^2 = 4$ or $\omega_0 = 2$. In the circuit of part (*a*) the short circuit puts the capacitances and the inductances in parallel. The resultant capacitance is $\frac{8}{3}$ plus $\frac{8}{5}$ or $\frac{64}{15}$; the resultant inductance is the reciprocal of $\frac{8}{3}$ plus $\frac{72}{5}$ or $\frac{15}{32}$; the *LC* product therefore is $4 = 1/\omega_0^2$ and so $\omega_0 = 2$.

To get the Cauer forms we must put $Z(s)$ in the form

$$Z(s) = \frac{s^3 + 4s}{s^4 + 10s^2 + 9} \tag{44}$$

If we want first the form that has series inductances and shunt capacitances we should consistently remove poles at $s = \infty$. Since $Z(s)$ has no pole at $s = \infty$, the first step is to invert the function 44. The continued division and inversion process therefore takes the form

$$\begin{array}{l}
\qquad\qquad\qquad s \\
s^3 + 4s \,|\, s^4 + 10s^2 + 9 \\
\qquad\quad\; s^4 + 4s^2 \qquad |\, s/6 \\
\qquad\qquad\quad 6s^2 + 9 \,|\, s^3 + 4s \\
\qquad\qquad\qquad\qquad s^3 + \frac{3}{2}s \,|\, 12s/5 \\
\qquad\qquad\qquad\qquad\quad \frac{5}{2}s \,|\, 6s^2 + 9 \\
\qquad\qquad\qquad\qquad\qquad 6s^2 \qquad |\, 5s/18 \\
\qquad\qquad\qquad\qquad\qquad\quad + 9 \,|\, \frac{5}{2}s
\end{array} \tag{45}$$

The corresponding network is shown in Fig. 16*a*.

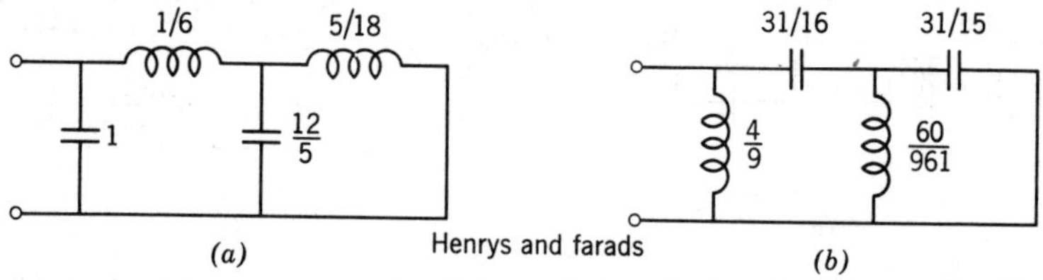

Fig. 16. The two Cauer networks realizing impedance 42 with the frequency pattern of Fig. 14.

Since $Z(s)$ has no pole at $s = 0$ either, the division and inversion process leading to the alternate Cauer form likewise begins with an inversion. The computations are given by

$$\begin{array}{l}
\qquad\qquad\quad 9/4s \\
4s + s^3 \,|\, 9 + 10s^2 + s^4 \\
\qquad\qquad 9 + \frac{9}{4}s^2 \qquad |\, 16/31s \\
\qquad\qquad\quad \frac{31}{4}s^2 + s^4 \,|\, 4s + s^3 \\
\qquad\qquad\qquad\qquad 4s + \frac{16}{31}s^3 \,|\, 961/60s \\
\qquad\qquad\qquad\qquad\quad \frac{15}{31}s^3 \,|\, \frac{31}{4}s^2 + s^4 \\
\qquad\qquad\qquad\qquad\qquad \frac{31}{4}s^2 \qquad |\, 15/31s \\
\qquad\qquad\qquad\qquad\qquad\quad s^4 \,|\, \frac{15}{31}s^3
\end{array} \tag{46}$$

The network is shown in Fig. 16*b*. Both networks degenerate into an inductance of $\frac{4}{9}$ henry at $s = 0$ and a capacitance of 1 farad at $s = \infty$. With their input terminals short-circuited, both networks have the natural frequency $\omega_0 = 2$ as they should.

6 Equivalent and Reciprocal Networks

The fact that the synthesis problem does not have a unique solution is amply demonstrated by the discussion in this chapter. Thus, for each p.r. reactance function we have found not only two Foster forms and two Cauer forms for its realization but have also seen that innumerable additional networks may be constructed that likewise have the same input impedance at a stated terminal pair. All of these circuits are referred to as being *equivalent* with respect to this particular driving point. In this connection it is significant that the term "equivalent" implies not only that the networks have certain configurations but that their elements have the proper values to render the driving-point impedances identical as functions of the complex frequency variable. Thus the four structures in Figs. 15 and 16 without the designation of specific element values would be referred to merely as *potentially* equivalent. By this term it is meant that the networks will become equivalent if proper values are assigned to the elements.

Since a given p.r. rational function may be regarded as an impedance whether it is right side up or upside down, it is clear that the synthesis procedures given above are also the means for constructing a circuit whose driving-point reactance function is the reciprocal of that of some given network. In other words, we are able by these methods to construct reciprocal networks with respect to a single terminal pair. In a situation where the dual can be constructed (see *ICT*, pp. 42, 535), a desired reciprocal can, of course, be found without recourse to these methods; but when a nonmappable graph is involved or if mutual inductances as well as self-inductances prevail, then the reciprocal network can still be found by first computing the driving-point impedance in the standard manner, turning it upside down, and then synthesizing it in any of the forms discussed above. It is clear, therefore, that a reciprocal network, unlike the dual one, always exists and can be constructed in a straightforward manner.

The reciprocals of any of the canonic forms shown in Figs. 15 and 16 can readily be constructed by the use of the simple duality principle that an inductance and a capacitance of the same numerical value are duals and that a parallel connection of elements or combinations of elements is the dual of a series connection of their respective duals. We thus see without difficulty that the circuits of Fig. 17 are respectively reciprocal to the Foster forms of Fig. 15, and those in Fig. 18 are reciprocals of the Cauer forms in Fig. 16.

Since the dual of a Foster or a Cauer form is structurally similar to the corresponding equivalent circuit, it is necessary to distinguish these types rather carefully to avoid confusing them. As with equivalent networks, it is also necessary to distinguish reciprocal and potentially reciprocal networks. Since element value designations are usually absent in sketches of potentially equivalent or potentially reciprocal networks, their distinction requires even greater care.

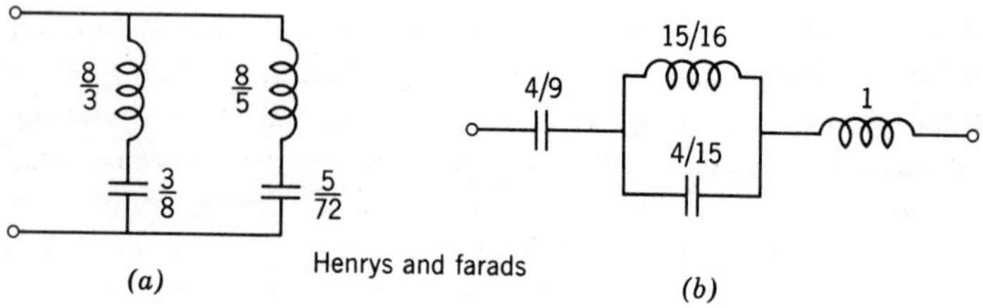

Fig. 17. Reciprocals of the Foster forms shown in Fig. 15.

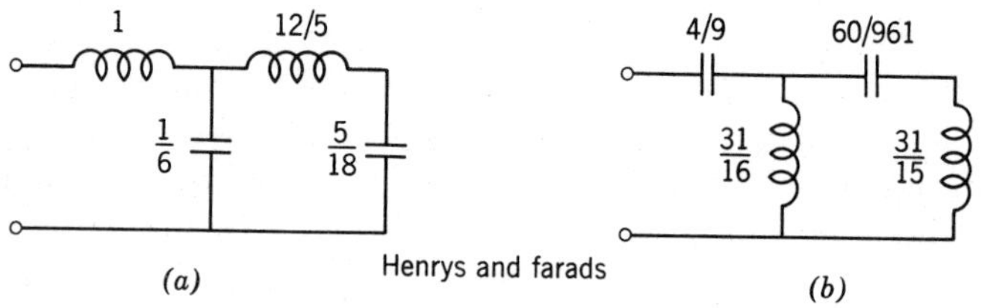

Fig. 18. Reciprocals of the Cauer forms shown in Fig. 16.

7 Relation to Normal Co-ordinates

Of all the equivalent networks that may be constructed by any of the methods discussed above, the Foster forms have the unique distinction of placing in evidence the normal co-ordinates common to all of these networks. Not only is it of collateral theoretical interest to discuss the mathematical and physical aspects of this circumstance but we gain thereby an alternate useful method of constructing reciprocal networks and a better understanding of the relationship between the analysis and the synthesis aspects of the general network problem.

Let T and V be respectively the instantaneous magnetic and electrical energy functions associated with any physical LC network. Expressed in terms of a set of loop currents $i_1 \cdots i_l$ and their indefinite integrals, the loop charges $q_1 \cdots q_l$, these are given by the positive definite quadratic forms (see *ICT*, p. 511)

$$T = \tfrac{1}{2} \sum_{i,\,k=1}^{l} L_{ik} i_i i_k \tag{47}$$

and

$$V = \frac{1}{2} \sum_{i,\, k=1}^{l} S_{ik} q_i q_k \tag{48}$$

in which L_{ik} and S_{ik} are the inductance and elastance parameters on the loop basis. These quadratic forms may be written in the equivalent matrix notation (see *MCA*, p. 132)

$$2T = \underline{i} \times [L] \times i] \tag{49}$$

$$2V = \underline{q} \times [S] \times q] \tag{50}$$

in which $[L]$ and $[S]$ are the loop inductance and elastance matrices (*ICT*, Ch. 10, Eq. 76, p. 501) and

$$i] = \begin{bmatrix} i_1 \\ i_2 \\ \cdot \\ \cdot \\ \cdot \\ i_l \end{bmatrix} \qquad q] = \begin{bmatrix} q_1 \\ q_2 \\ \cdot \\ \cdot \\ \cdot \\ q_l \end{bmatrix} \tag{51}$$

are column matrices for the loop currents and charges.

According to established methods (*MCA*, p. 156) we determine a real nonsingular transformation of the variables

$$i] = [A] \times i'], \qquad q] = [A] \times q'] \tag{52}$$

with the matrix $[A]$ which simultaneously carries T and V over into a sum of squares so that Eqs. 49 and 50 become

$$2T = \underline{i'} \times [A]_t \times [L] \times [A] \times i'] \tag{53}$$

$$2V = \underline{q'} \times [A]_t \times [S] \times [A] \times q'] \tag{54}$$

in which $[A]_t$ is the transpose of $[A]$ and

$$[A]_t \times [L] \times [A] = [L'] \tag{55}$$

$$[A]_t \times [S] \times [A] = [S'] \tag{56}$$

are diagonal matrices. The characteristic equation of the matrix: $[L']s + [S']s^{-1} = 0$ therefore appears in factored form and places the normal modes (natural frequencies) in evidence. The co-ordinates to which the variables i'_k and q'_k relate have the property that they isolate the normal modes and hence are called the *normal co-ordinates* of the original network.

Ordinarily we regard these co-ordinates as fictitious things that are interesting only because of their mathematical implication. We shall see, however, that we can construct an actual network, equivalent to the original one with respect to a driving point located in mesh 1, in which these normal co-ordinates appear as actual loops.

To this end we make the additional transformation

$$i'] = [D] \times i''] \tag{57}$$

with the diagonal matrix

$$[D] = \begin{bmatrix} a_{11}^{-1} & 0 & 0 & \cdots & 0 \\ 0 & a_{12}^{-1} & 0 & \cdots & 0 \\ \cdots & \cdots & \cdots & \cdots & \cdots \\ 0 & 0 \cdot & \cdots & \cdots & \cdot a_{1l}^{-1} \end{bmatrix} \tag{58}$$

in which $a_{11}, a_{12}, \cdots, a_{1l}$ are the coefficients in the first row of the matrix $[A]$. The result of this further transformation is simply that we now have

$$i] = [A] \times [D] \times i''] = [B] \times i''] \tag{59}$$

where

$$[B] = \begin{bmatrix} 1 & 1 & \cdots & 1 \\ b_{21} & b_{22} & \cdots & b_{2l} \\ \cdots & \cdots & \cdots & \cdots \\ b_{l1} & b_{l2} & \cdots & b_{ll} \end{bmatrix} \tag{60}$$

as indicated, has a first row in which all the coefficients are unity, so that the first of the transformation Eqs. 59 reads

$$i_1 = i_1'' + i_2'' + i_3'' + \cdots + i_l'' \tag{61}$$

while the parameter matrices pertinent to the variables i_k'' (and q_k'') are still diagonal and are given by

$$[L''] = [B]_t \times [L] \times [B] \tag{62}$$

$$[S''] = [B]_t \times [S] \times [B] \tag{63}$$

Actually the additional transformation amounts merely to changing the impedance levels of the various normal co-ordinates.

Each normal co-ordinate consists of a series *LC* branch with parameters given by matrices 62 and 63; and the sum of the currents in these branches, according to Eq. 61, is numerically equal to the input current of the original network for a given applied voltage excitation. The input impedance for the network implied by this situation is therefore the same as for the given network with the parameter matrices $[L]$ and $[S]$. In other words, transformation 59 keeps the input impedance invariant while transforming to the normal form in which the natural frequencies are placed in evidence. From the schematic in Fig. 19 which represents the end result, we see that this form is identical with one of the Foster forms obtained from the partial-fraction expansion of the input admittance.

The other Foster form is obtainable in an exactly parallel manner by carrying out the same transformation on the node basis; and so we see that *the two canonic forms which we designate as Foster networks are actually those forms in which the normal co-ordinates of the system become a reality.*

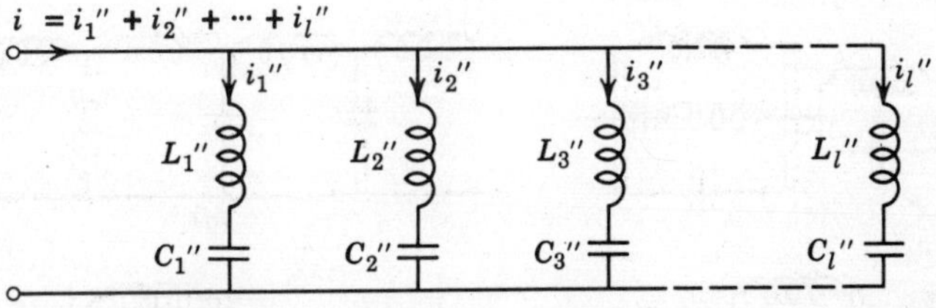

Fig. 19. Transformation to normal co-ordinates on the loop basis leads to this network configuration recognizable as a Foster form. The other Foster form is similarly obtainable on the node basis.

Although the original network with the matrices $[L]$ and $[S]$ may be nonmappable and may involve arbitrary mutual inductive couplings, we have in this way obtained a direct method of transforming to any of the equivalent (and hence also reciprocal) canonic forms.

PROBLEMS

1. Which of the following functions represent the reactance of a physical network?

$$(a)\ Z_{11} = \frac{5s^2 + 20}{s^4 + 10s^2 + 9} \qquad (b)\ Z_{11} = \frac{s^4 + 13s^2 + 36}{s^3 + 6.25s}$$

$$(c)\ Z_{11} = \frac{4s^3 + 3s}{4s^2 + 1} \qquad (d)\ Z_{11} = \frac{s^3 + 4s}{s^4 + 2s^2 + 1}$$

$$(e)\ Z_{11} = \frac{4s^4 + 5s^2 + 1}{4s^3 + 3s} \qquad (f)\ Z_{11} = \frac{s^5 + 4s^3 + 3s}{s^5 + 6s^3 + 8s}$$

For the realizable ones carry out a partial-fraction expansion, and determine a corresponding network. Sketch the three fundamental equivalent networks in each case.

2. From the following rational impedance functions pick out those that are p.r., and for those that are realizable by *LC* networks find the network realizations in any *one* of several possible equivalent forms.

$$(a)\ \frac{s^2 + 2s + 2}{s^3 + 4s} \qquad (b)\ \frac{s^3 + 4s}{s^2 + 1} \qquad (c)\ \frac{s^4 + 20s^2 + 64}{s^5 + 10s^3 + 9s}$$

$$(d)\ \frac{s^5 + 20s^3 + 64s}{s^4 + 10s^2 + 9} \qquad (e)\ \frac{s^3 + 3s^2 + 7s + 5}{s^2 + 2s + 2} \qquad (f)\ \frac{s^3 + 2s^2 + 5s}{s^2 + 2s + 2}$$

3. From the following rational functions pick out those that are p.r., and point out the ones that can be realized by *LC* networks.

$$(a)\ \frac{s^2 + s + 1}{s^2 + s + 9} \qquad (b)\ \frac{s^2 + s + 1}{s^2 + s + 2} \qquad (c)\ \frac{5s^2 + s + 1}{s^2 + 4s + 2}$$

(*d*) $\dfrac{s^4 + 2s^2 + 1}{s^5 + 3s^3 + 2s}$ (*e*) $\dfrac{s^4 + 4s^2 + 3}{s^5 + 6s^3 + 8s}$ (*f*) $\dfrac{s^5 + 4s^3 + 3s}{s^4 + 6s^2 + 8}$

4. Of the following networks, which are potentially equivalent and which are potentially inverse? Point out any superfluous elements.

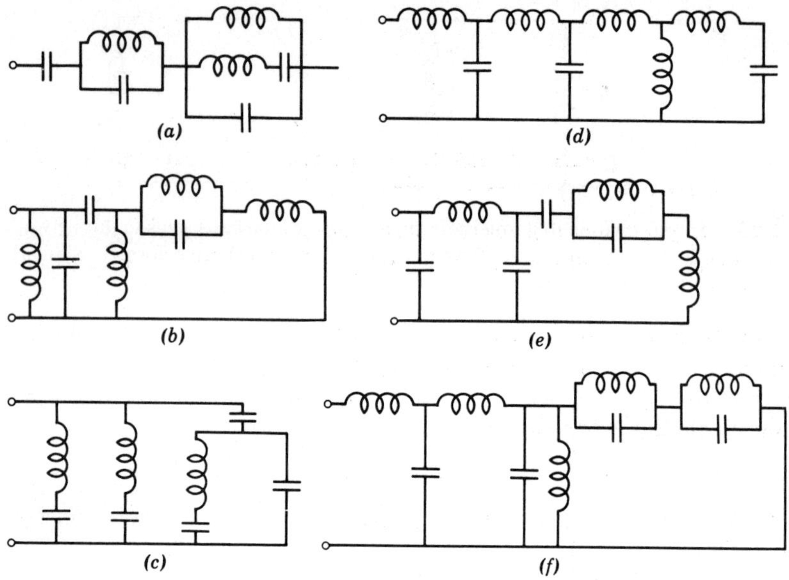

Prob. 4.

5. Determine whether the following networks are realizable in terms of positive elements only, and if so, convert to an equivalent Foster form. Numerical values are in henrys and darafs.

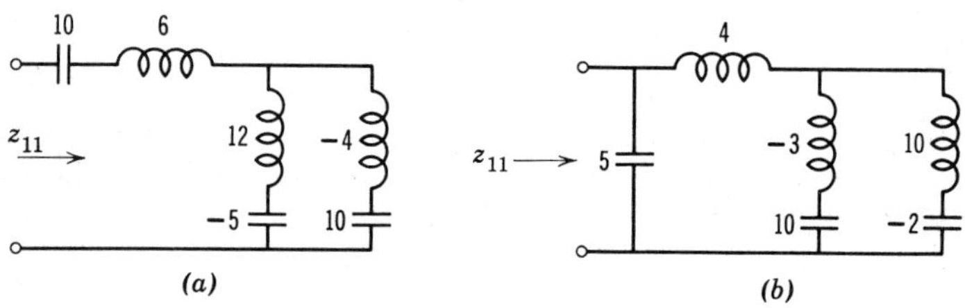

Prob. 5.

6. The frequency patterns for three impedances are given in sketch (*a*). Draw the frequency patterns for the three connections shown in sketch (*b*).

7. Find the element values in the following network such that the impedance at 1-1′ with 2-2′ open has poles at $\omega = 1, 3, 5$, and with 2-2′ short-circuited it has poles at $\omega = 2, 4$. The capacitance C across 2-2′ is 1 farad.

8. If jx_a and jx_b are reactance functions (impedances of lossless networks),

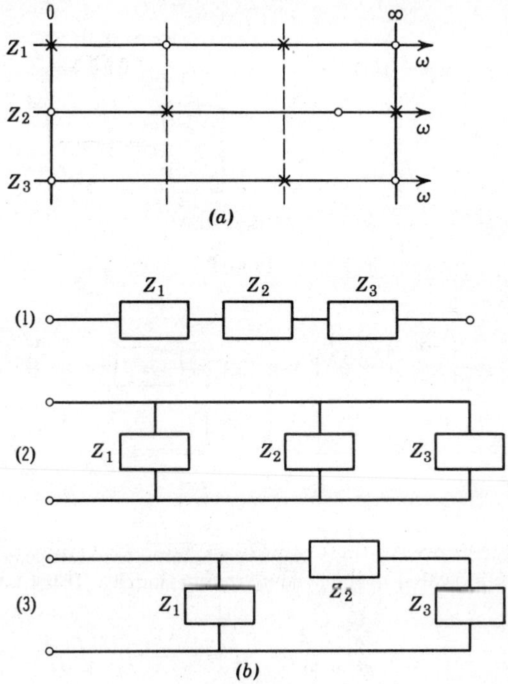

Prob. 6.

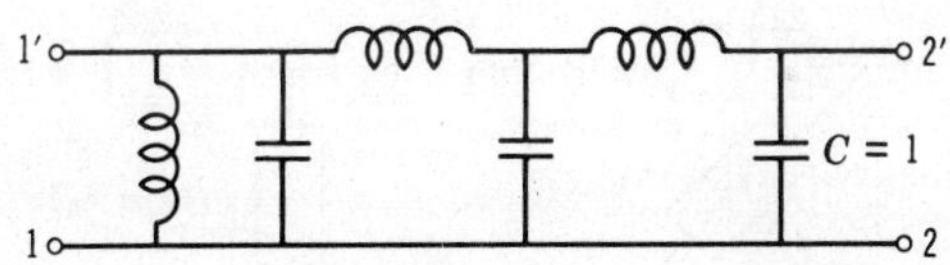

Prob. 7.

state the necessary and sufficient conditions for $j(x_a - x_b)$ to be a physically realizable reactance function.

9. For what largest positive real m value is the function $Z = (Z_b - mZ_a)$ still realizable? See figure. Determine a Foster network for this limiting form of Z.

10. Given that

$$z_a = \frac{33(s^9 + 26s^7 + 231s^5 + 806s^3 + 880s)}{14(s^2 + 1)(s^2 + 4)(s^2 + 7)(s^2 + 10)}$$

$$z_b = \frac{s^7 + 21s^5 + 126s^3 + 216s}{s^6 + 15s^4 + 54s^2 + 40}$$

state whether $z_a - z_b$ is realizable, and if so, find a network.

11. With reference to the accompanying sketches, determine qualitatively the relative locations of the poles and zeros of the functions jx, z', and z''.

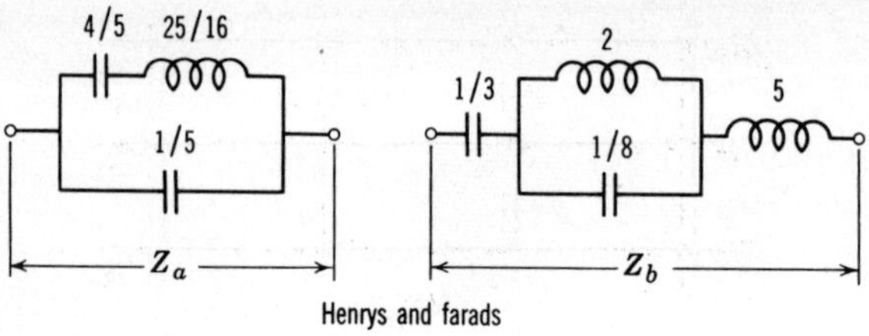

Prob. 9.

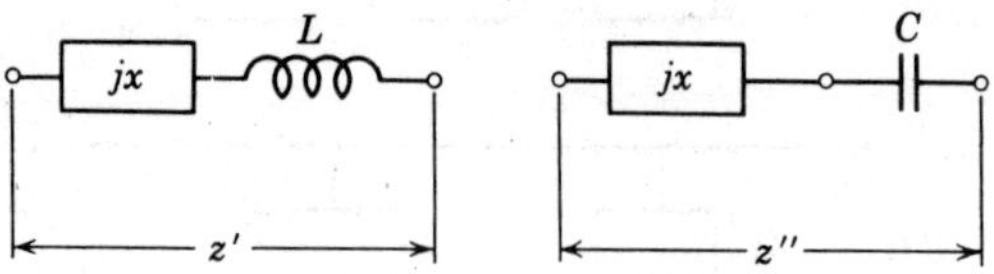

Prob. 11.

12. The equivalence relations between a symmetrical lattice and a symmetrical T network are indicated in the accompanying sketch. If the lattice impedances are

$$z_a = \frac{2s^2 + 4}{s^3 + 4s} \qquad z_b = \frac{4s^4 + 11s^2 + 4}{s^5 + 5s^3 + 4s}$$

determine whether the equivalent T exists, and if it does, find the structure and element values of its branches.

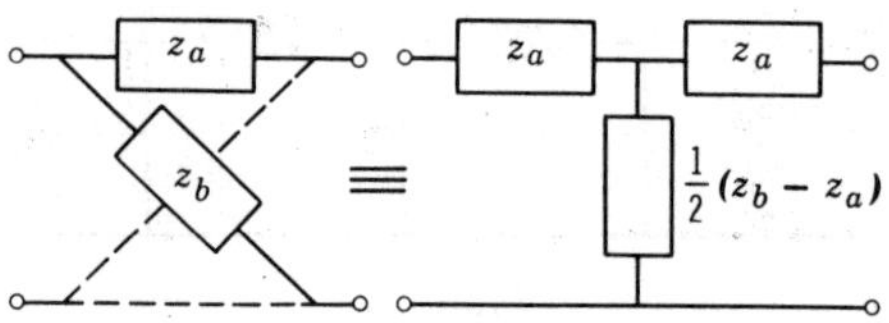

Prob. 12.

13. The following reactive networks are the components of a symmetrical lattice low-pass filter. The parameters have the values

$$L_2 = \frac{10^{-4}}{2} \qquad C_2 = 10^{-6} \qquad L_4 = 10^{-4} \qquad L_1 = 3 \times 10^{-4} \qquad C_1 = \frac{10^{-6}}{9}$$

Determine the possibility of converting this lattice into an equivalent symmetrical Π network.

14. In the following network the numerical values are in henrys and darafs. Determine the structure and parameter values in the equivalent Foster and Cauer networks.

15. A reactance function has poles at $\omega = 0$, $\omega = 2$, $\omega = 4$. At these frequencies the corresponding susceptance has a slope equal to unity. Determine the zeros of the reactance and find the four canonic networks which realize it.

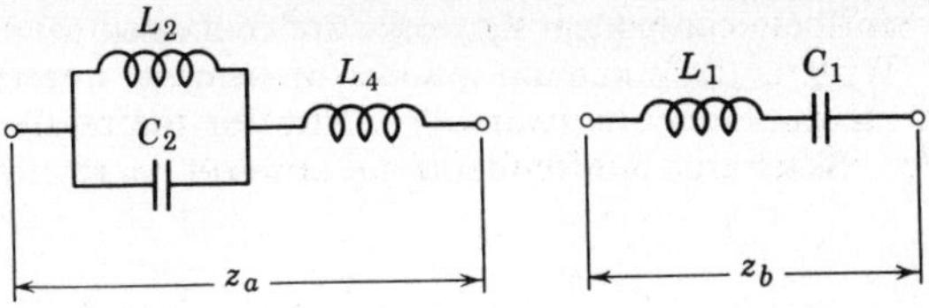

Prob. 13.

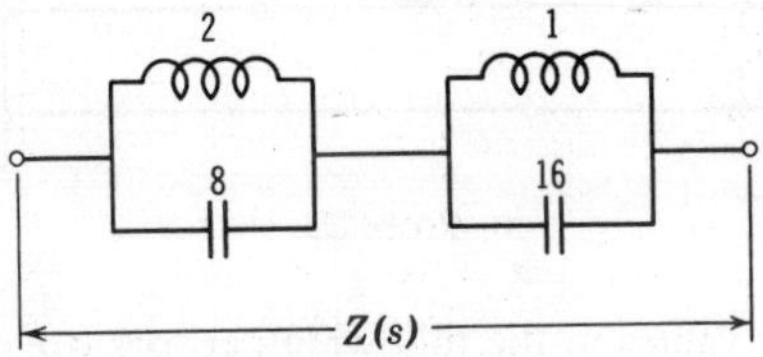

Prob. 14.

16. A reactance function has poles at the frequencies $\omega_1 = 1000$, $\omega_3 = 3000$, and a zero at $\omega_2 = 2000$. At this zero the function has a slope equal to $\frac{8}{15}$ ohm/radian/sec. Find a network which realizes this function.

17. A reactive (lossless) network in parallel with a 1-ohm resistance is to have an impedance varying in magnitude between 0.707 and 1.0 in the frequency ranges $0.98 < \omega < 1.02$ and $2.99 < \omega < 3.01$, exhibiting elsewhere a smaller magnitude. Find an appropriate network.

18. Two reactance functions $Z_1(s)$ and $Z_2(s)$ have zeros at $s = 0, \pm j2 \pm j4$ only. At these frequencies $dZ_1/ds = 1, 2, 3$ respectively and $dZ_2/ds = 1, 4, 3$ respectively.

If $Y_1 = 1/Z_1$ and $Y_2 = 1/Z_2$ are the corresponding susceptance functions, state whether $Y = Y_1 - Y_2$ is realizable, and if so, find a network for it.

19. When 1 ohm is placed across the input terminals to a lossless network, it has four natural modes, all coincident at $s = -1$. Construct the driving-point reactance function for this network and synthesize it in both of its Cauer forms.

20. A reactance function $Z(s)$ has poles at $s = 0$, $s = \pm j2$, $s = \pm j4$. The residues of $Z(s)$ in these poles are all equal. The corresponding admittance, $Y(s) = 1/Z(s)$, has a real part $G(\sigma, \omega)$, for which

$$\left(\frac{\partial G}{\partial \sigma}\right)_{s=0} = \frac{1}{2}$$

(*a*) Find $Z(s)$.

(*b*) Synthesize a Foster network for $Z(s)$.

21. $Z(s)$ is a reactance function with the property that $(d^n/ds^n)[Z(s) + 1]$ evaluated at $s = 1$ is zero for $n = 0, 1, 2,$ and 3. Construct the reactance function $[m(s)]/[n(s)]$ of lowest degree which has this property, and synthesize it in both Cauer forms.

22. Frequency patterns for the reactances z_1, z_2, and z_3 are shown in the accompanying sketch. Determine the frequency pattern for the reactance

which results when these component networks are connected (*a*) in series, and (*b*) in parallel. What is the minimum number of elements in terms of which each component network may be realized? Ditto for the resultant networks under (*a*) and (*b*). Sketch the four fundamental forms of the resultant networks in (*a*) and (*b*).

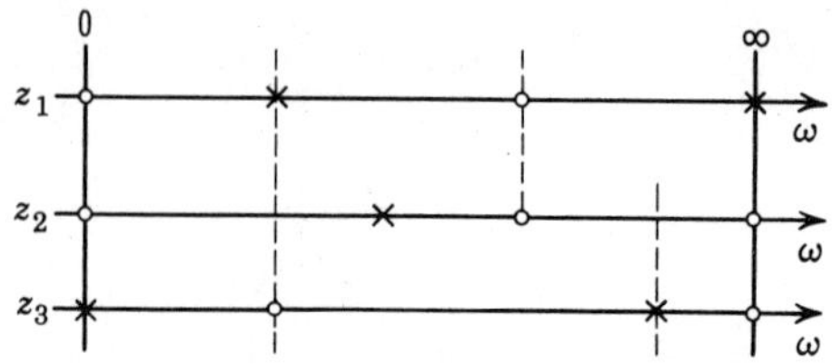

Prob. 22.

23. The parameter values in the following network are in henrys and darafs. The mutual inductance is positive for the indicated polarity marks, and the coupling coefficient is unity. Determine an equivalent network in the form of antiresonant components in series.

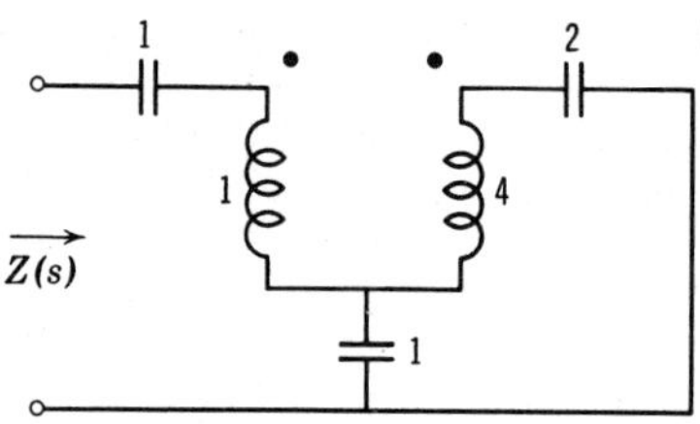

Prob. 23.

24. In network (*a*) shown here, the parameter values are given in henrys and farads. The mutual inductances are numerically positive for the indicated polarity markings on the coils. Determine the parameter values in the equivalent network (*b*) in which no mutual inductances appear.

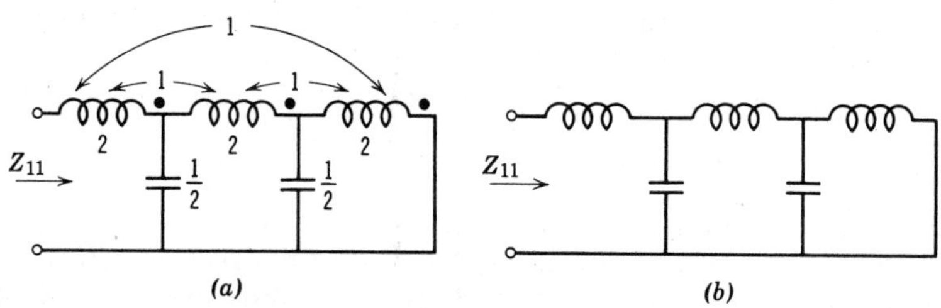

Prob. 24.

25. A driving-point impedance function has simple poles at $s = -1 \pm j1$ and at $s = -1 \pm j2$. The residues in these poles are all equal to 2, and the function

is zero for $s = \infty$. Determine a corresponding network. Show that the function must be p.r. so long as the residues are positive real, regardless of their values.

26. Construct a p.r. impedance having all of its zeros restricted to the j-axis and poles at $s = -1 \pm j1$ and at $s = -1 \pm j2$, such that the maximum value of its real part for $s = j\omega$ equals unity. If the solution is not unique find all possible solutions and determine corresponding networks. Find the zeros of each impedance and the frequencies at which its real part is a maximum.

27. Show that the function

$$Z(s) = \frac{s^4 + 10s^2 + 9}{s^4 + s^3 + 10s^2 + 4s + 9}$$

is p.r. and find its network realization.

28. Given the following networks for z_a and z_b, state whether $z_b - z_a$ is a realizable reactance function, and if it is, find its realization.

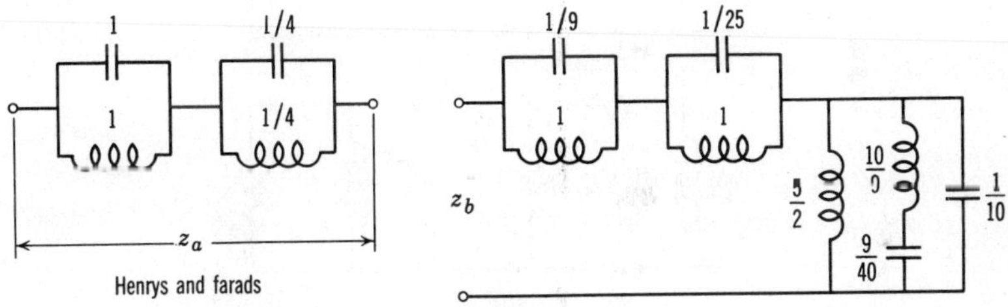

Prob. 28.

29. Test the following p.r. driving-point impedance functions for possible poles on the j-axis. If present, evaluate residues there and remove by Foster's method; then reciprocate remainder and again test for j-axis poles. Attempt complete synthesis by continuing this process.

$$(a)\quad \frac{s^7 + 4s^6 + 8s^5 + 17s^4 + 13s^3 + 19s^2 + 5s + 4}{s^7 + 2s^6 + 6s^5 + 7s^4 + 11s^3 + 7s^2 + 6s + 2}$$

$$(b)\quad \frac{s^8 + 3s^7 + 8s^6 + 10s^5 + 17s^4 + 10s^3 + 12s^2 + 3s + 2}{2s^7 + 6s^6 + 13s^5 + 12s^4 + 17s^3 + 5s^2 + 4s}$$

30. If an LC impedance has a pole at both zero and infinite frequency, a ladder development is possible consisting of the following steps:

(a) The poles at zero and infinity are removed leaving zeros in both places.

(b) The remainder function is inverted and the process is repeated until the function is completely synthesized.

Synthesize the following admittance function by this method. Notice that the resulting network is particularly simple because the pole-zero constellation is geometrically symmetrical about the point $\omega = 1$.

$$Y = \frac{(s^2 + \frac{1}{4})(s^2 + 1)(s^2 + 4)}{s(s^2 + \frac{1}{2})(s^2 + 2)}$$

31. (*a*) Synthesize the impedance

$$Z = \frac{2s^5 + 5s^3 + 2s}{(s^2 + 1)(s^4 + 3s^2 + 1)}$$

in the form shown in sketch (*a*).

(*b*) Synthesize the impedance

$$Z = \frac{s(2s^2 + 7s + 2)}{s^3 + 6s^2 + 9s + 2}$$

in the form indicated in sketch (*b*).

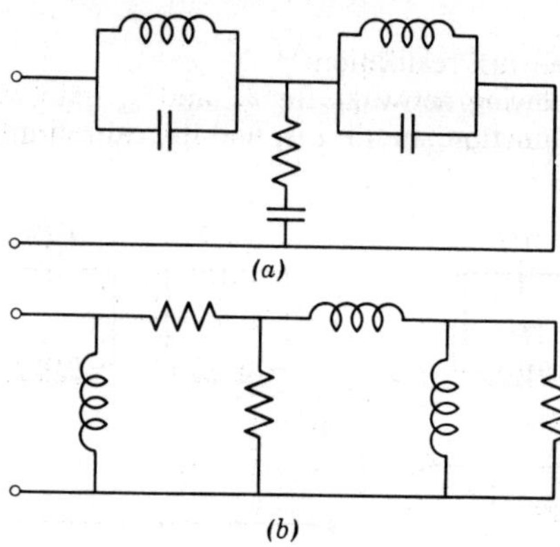

Prob. 31.

C H A P T E R F O U R

Synthesis of *RC* and *RL* Driving-Point Impedances

1 The Foster-like Realizations for *RC* Networks

In Ch. 2 it was shown that if a rational function has simple poles restricted to the finite negative real axis where it has positive real residues, and if its value for $s = \infty$ is nonnegative and finite, then it satisfies the necessary conditions for being the driving-point impedance of an *RC* network. We may readily see that these conditions are also sufficient by observing that a rational function fulfilling only these requirements is always realizable. The pattern that we follow in order to achieve the synthesis of this function is exactly parallel to that used in the previous chapter for the realization of *LC* driving-point functions.

The partial-fraction expansion of a rational function of this sort reads

$$Z(s) = \frac{k_1}{s - s_1} + \frac{k_3}{s - s_3} + \cdots + \frac{k_{2n-1}}{s - s_{2n-1}} + k_{2n+1} \tag{1}$$

with

$$-\infty < s_{2n-1} < \cdots < s_3 < s_1 \leq 0 \tag{2}$$

or writing σ_ν for $-s_\nu$ we have

$$Z(s) = \frac{k_1}{s + \sigma_1} + \frac{k_3}{s + \sigma_3} + \cdots + \frac{k_{2n-1}}{s + \sigma_{2n-1}} + k_{2n+1} \tag{3}$$

in which the k's are positive real numbers and $k_{2n+1} \neq 0$ unless $Z(s)$ has a zero at $s = \infty$. If we observe that the expression for the impedance of the parallel *RC* circuit in Fig. 1 is given by

$$z_\nu = \frac{1}{(1/R_\nu) + C_\nu s} = \frac{R_\nu}{1 + R_\nu C_\nu s} = \frac{(1/C_\nu)}{s + (1/R_\nu C_\nu)} \tag{4}$$

we may identify the terms in Eq. 3 with the z_ν's in Eq. 4 where

$$k_\nu = \frac{1}{C_\nu} \tag{5}$$

and

$$\sigma_\nu = \frac{1}{R_\nu C_\nu} \quad (\text{except } \nu = 2n + 1) \tag{6}$$

Since k_ν and σ_ν are guaranteed positive real by the stated properties of

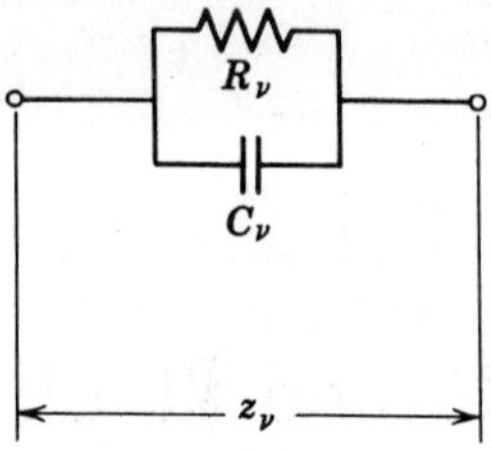

Fig. 1. Realization of a typical term in the partial-fraction expansion of an *RC* impedance.

$Z(s)$, we see that the values of R_ν and C_ν are assuredly real and positive. The function $Z(s)$ is, therefore, always realizable in the form shown in Fig. 2 in which the element values are given by

$$R_\nu = k_\nu/\sigma_\nu, \qquad C_\nu = 1/k_\nu \quad \text{for } \nu = 1, 3, \cdots 2n-1$$

and

$$R_{2n+1} = k_{2n+1} = Z(\infty) \qquad \text{(ohms and farads)} \tag{7}$$

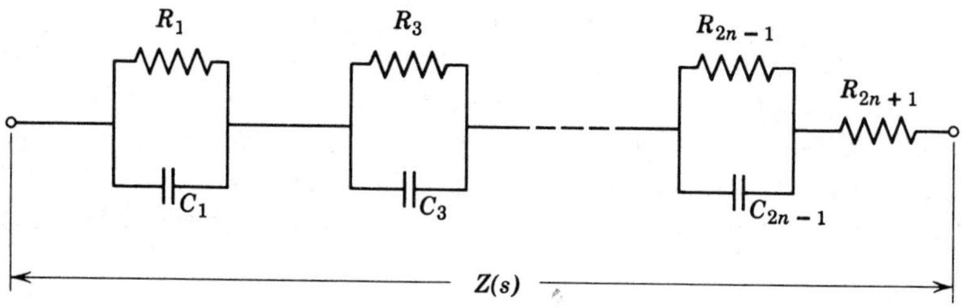

Fig. 2. Foster-like realization of the *RC* impedance, Eq. 1, from its partial-fraction expansion.

If the lowest pole in $Z(s)$ occurs at $s = 0$, then $\sigma_1 = 0$, $R_1 = \infty$, and the first parallel *RC* circuit in the network of Fig. 2 degenerates into the capacitance C_1 alone.

Because of its similarity to the Foster form for the realization of an *LC* driving-point impedance from its partial-fraction expansion, the network of Fig. 2 is referred to as a *Foster-like* form.* It involves the minimum number of elements necessary for realization of the given $Z(s)$, as is evident from the fact that this number exactly equals the number of constants $k_1 \cdots k_{2n+1}$ and $\sigma_1 \cdots \sigma_{2n-1}$ characterizing the function in Eq. 3. Like the analogous *LC* network, it is referred to as a canonic form.

It is logical to expect at this point that an alternate Foster-like network is obtainable from the partial-fraction expansion of the corresponding

* This and the other forms of *RC* and *RL* network realizations discussed in this chapter were first pointed out by W. Cauer. See footnote, p. 88, Ch. 3.

admittance function $Y(s) = 1/Z(s)$. Before attempting this approach it is helpful to look somewhat closer at the detailed properties of *RC* driving-point impedance and admittance functions and thus become familiar with the necessary and sufficient conditions for their realization from several additional points of view. First let us consider the alternate statement of these conditions in terms of zeros and poles instead of poles and residues. Thus (as pointed out in Ch. 2) a rational function is p.r. and *RC* realizable (as a driving-point impedance) if, in addition to $0 \leq Z(\infty) < \infty$, its zeros and poles are simple and interlace on the negative real axis, the lowest critical frequency being a pole.

Forbidding critical frequencies to lie in the right half-plane is merely the usual stability requirement, the significant points in this statement about necessary conditions being the restriction of poles and zeros to the real axis which is responsible for the realness of the residues, and their simplicity and alteration which is a necessary condition for the positiveness of the residues. We may obtain an understanding of these relationships through an interesting graphical interpretation of formula 5 of Ch. 3 for the residues, which we repeat here for convenience

$$k_\nu = [(s - s_\nu)Z(s)]_{s=s_\nu} \tag{8}$$

If we think of $Z(s)$ as a quotient of frequency factors, then the quantity within brackets is this function with the νth denominator factor (pole) missing. For $s = s_\nu$ the other factors may be regarded as vectors emanating from the remaining zeros and poles, and all terminating upon the νth pole position. In Fig. 3, which shows a typical pole-zero pattern along the

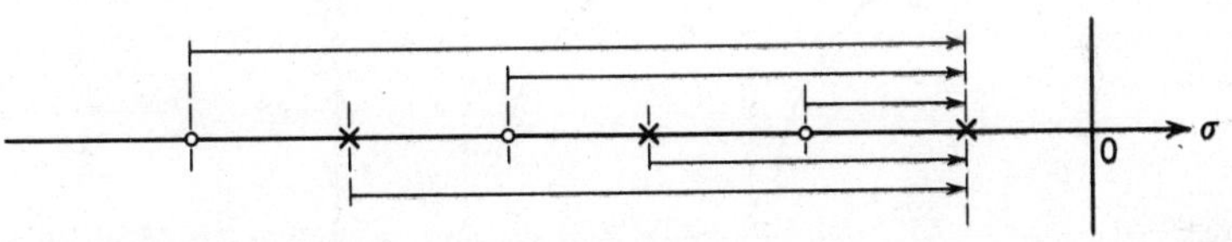

Fig. 3. Frequency factors in the residue evaluation at the pole nearest the origin.

negative σ-axis of the s-plane ($s = \sigma + j\omega$), such a group of frequency factors are drawn as vectors for the evaluation of the residue at the pole nearest the origin. Zero factors are drawn above the σ-axis and pole factors below it. Since all vectors are positive real, the ratio of the product of zero factors to the product of pole factors is evidently positive real.

In Fig. 4 another set of factors are drawn as they appear from the evaluation of the residue at the next pole. Because of the alternation of zeros and poles we see that one zero factor and one pole factor have changed sign, and so the residue is again positive real. In terms of this graphical portrayal of the process of residue evaluation, it is not difficult to see that the alternation of zeros and poles (starting with a pole nearest the origin) is both necessary and sufficient to assure the positiveness of all the residues (their restriction to the real axis is necessary and sufficient to render all factors real and hence assure the realness of the residues*).

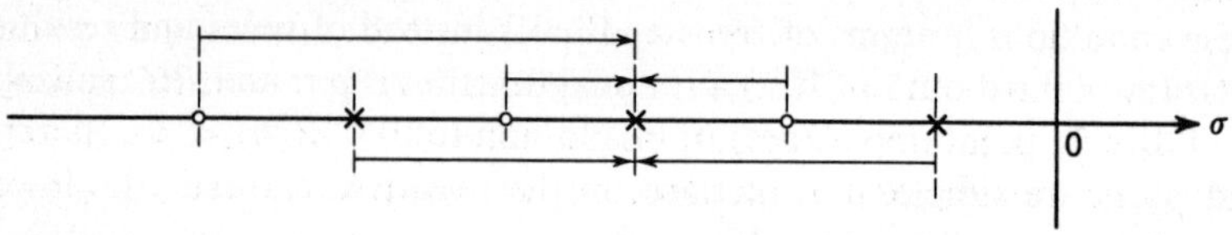

Fig. 4. Frequency factors in the residue evaluation at the next-lowest pole.

We see, therefore, that the statement about necessary conditions in terms of zeros and poles is entirely equivalent to the one in terms of poles and residues whose sufficiency is demonstrated above. It follows that the statement in terms of zeros and poles is likewise sufficient.

A simple rational function $Z(s)$ fulfilling these conditions is shown plotted in Fig. 5 for values of the variable s along the real axis (σ-axis).

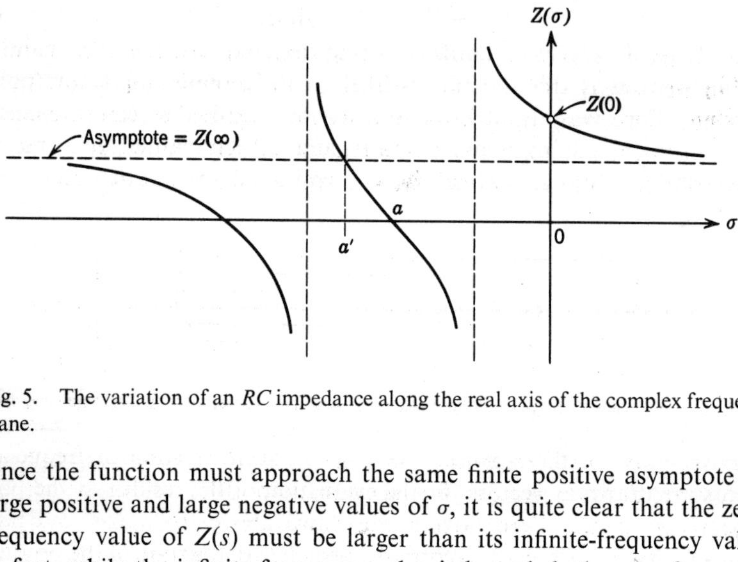

Fig. 5. The variation of an *RC* impedance along the real axis of the complex frequency plane.

Since the function must approach the same finite positive asymptote for large positive and large negative values of σ, it is quite clear that the zero-frequency value of $Z(s)$ must be larger than its infinite-frequency value. In fact, while the infinite-frequency value is bounded, the zero-frequency value is unbounded, for the lowest pole† may coincide with the point

* The realness of the residues actually requires only that the poles be located on the real axis since complex zeros must occur in conjugate pairs; but their positiveness requires that the zeros lie on the real axis and that poles and zeros be simple and alternate.

† The terms "low" and "high" are here used to indicate values respectively close to, or far from, the value $s = 0$, regardless of their algebraic signs.

$s = 0$. It is also clear from this figure that the special properties of $Z(s)$ as described above are not damaged by subtracting a positive real constant so long as its value is less than $Z(\infty)$. Subtraction of such a constant merely shifts the zeros of $Z(s)$ to the left, this shift becoming infinite for the highest zero when the subtracted constant has the value $Z(\infty)$. If a larger constant is subtracted, then the resulting curve intersects the *positive* real axis which is not permitted since no p.r. function may have zeros in the right half-plane.

The fact that subtraction of a positive constant less than $Z(\infty)$ leaves none of the properties of $Z(s)$ impaired proves that the minimum value of the real part of $Z(j\omega)$ for all real values of ω occurs at the point $s = \infty$, as is implied by statement 56 in Ch. 2. For if the minimum value of Re $[Z(j\omega)]$ occurred at a finite ω-value, then subtraction of the value $Z(\infty)$, presumably larger than this minimum value, would surely ruin the p.r. character of $Z(s)$. In connection with this reasoning it is essential to regard infinity as a point which one may approach by proceeding along either the real or imaginary axes; and since $Z(s)$ is rational, its limiting value becomes the same irrespective of the path along which one proceeds.

The result of subtracting $Z(\infty)$ from $Z(s)$ in Fig. 5 is shown in the top part of Fig. 6 where the remainder function is denoted as $Z'(s)$ and is again plotted along the σ-axis. The bottom part of this same figure shows the corresponding admittance $Y'(s)$ whose poles are the zeros of $Z'(s)$ and vice versa. Thus since $Z'(s)$ becomes zero for $s \to \infty$, $Y'(s)$ has a pole there as may be recognized from its indicated asymptotic behavior. Although these figures are drawn for rather simple functions, their general character evidently remains unaltered if we insert additional zeros and poles to the left of those shown, so that we may safely assume that the following as well as the preceding conclusions hold true generally.

Since the slope of the admittance curves is consistently positive while that of the impedance curves is negative, it is readily appreciated that the residues of $Z(s)$ in its poles [equal to the reciprocal slopes of $Y(s)$ in its zeros] are positive while the residues of $Y(s)$ in its finite poles [equal to the reciprocal slopes of $Z(s)$ in its finite zeros] are negative. The residue of an admittance $Y(s)$ in its pole at $s = \infty$ [as for the function $Y'(s)$ in Fig. 6], however, is positive since it is equal to the slope of its linear asymptote. The equation of this linear asymptote reads

$$y = C(s - s_0) = Cs + g_0 \tag{9}$$

where C is the slope of the asymptote and g_0 is its zero intercept. The degenerate behavior of the admittance for $s \to \infty$ is like that of a capacitance C in parallel with the conductance g_0.

More generally, an *RC* driving-point admittance having a pole at $s = \infty$

may be written

$$Y(s) = \frac{(s + \sigma_1)(\sigma + \sigma_3) \cdots (s + \sigma_n)}{(s + \sigma_2)(s + \sigma_4) \cdots (s + \sigma_{n-1})} \quad \text{for } n \text{ odd} \tag{10}$$

The asymptotic behavior for large s is found by dividing the denominator into the numerator for a two-term quotient of the form

$$y = s + g_0 \tag{11}$$

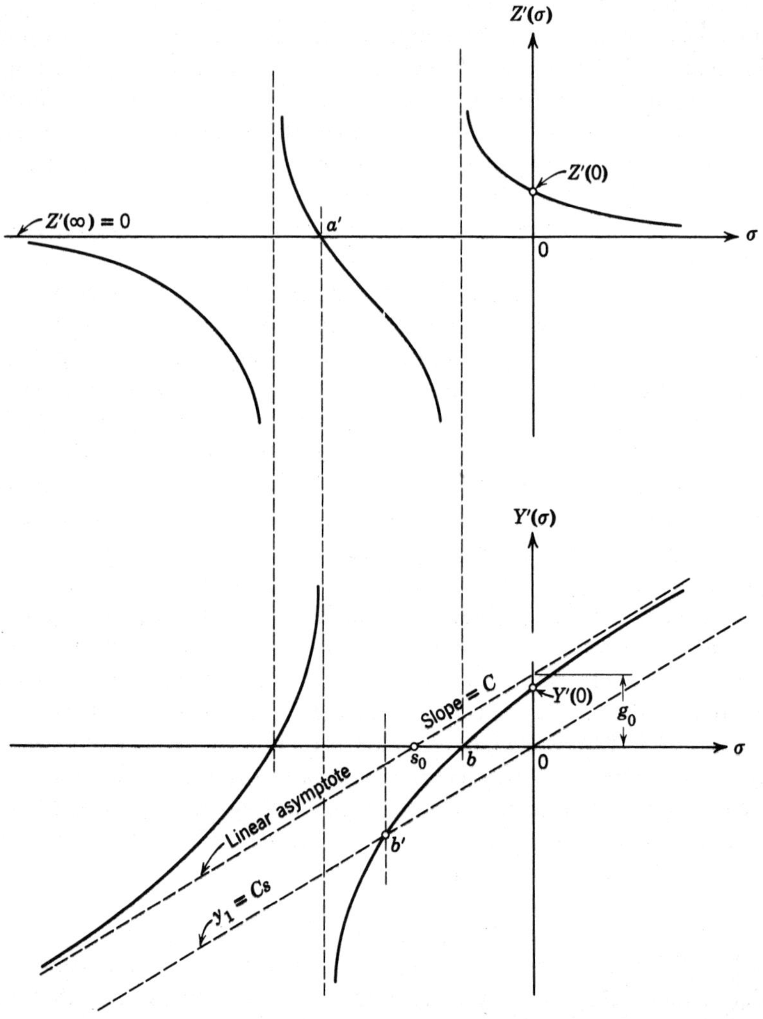

Fig. 6. The impedance of Fig. 5 after subtraction of its asymptotic value (top) and the corresponding admittance (bottom).

in which one readily finds for the zero intercept

$$g_0 = (\sigma_1 + \sigma_3 + \cdots + \sigma_n) - (\sigma_2 + \sigma_4 + \cdots + \sigma_{n-1}) \qquad (12)$$

If with reference to Fig. 6 we denote the admittance $Y'(s)$ minus the straight line Cs through the origin by $Y''(s)$, then, as Fig. 7 shows, the asymptote of $Y''(\sigma)$ is a constant equal in value to the intercept g_0 in Eq. 9. The zero-frequency value of Y', or of Y'' which equals it, is clearly less than the value of this constant asymptote. That is to say, $Y''(0) < Y''(\infty)$, a property of *RC* driving-point admittances that evidently (from a sketch like the one in Fig. 7) is seen to be generally true (as is also shown in Ch. 2).

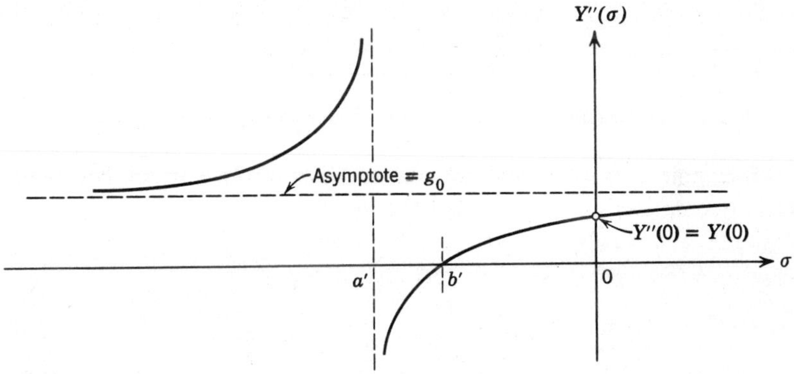

Fig. 7. The admittance of Fig. 6 after subtraction of the component $y_1 = Cs$.

It is likewise clear from the character of these plots that the special properties of an *RC* driving-point admittance are not impaired by the subtraction of a positive constant so long as its value is less than the zero-frequency value, for such a subtraction merely shifts the zeros closer to the origin, the closest one becoming coincident with the origin if the subtracted constant equals $Y(0)$, the zero-frequency value of the pertinent admittance. Clearly, the subtraction of a larger constant value than this would shift a zero into the right half-plane and, at once, ruin the p.r. character of the function. That is why one could not, for example, subtract the whole of the asymptotic behavior as given by Eq. 9 or Eq. 11 and expect to have an undamaged remainder.

From these same plots one may furthermore see that the minimum value of the real part of $Y(j\omega)$ for all real values of ω occurs at the point $s = 0$, as is implied by statement 56 in Ch. 2. For if the minimum value of Re $[Y(j\omega)]$ occurred at a finite ω value, the subtraction of the value $Y(0)$, presumably larger than this minimum value, would surely ruin the p.r. character of $Y(s)$.

Suppose we now consider the realization of $Y(s)$ from its partial-fraction expansion in a Foster-like manner similar to the procedure applied above to the impedance $Z(s)$, leading to the network of Fig. 2. Immediately the negativeness of the residues in all finite poles appears to be an obstacle, since a breakdown of $Y(s)$ into additive terms with minus signs does not look very promising. What we intuitively expect to get in the way of a final network is a parallel connection of component admittances like the series *RC* circuit shown in Fig. 8 (with appropriate degenerate forms for a

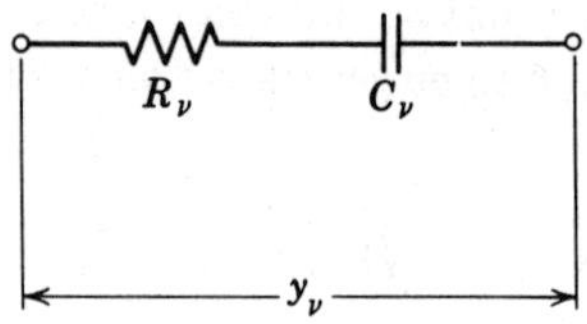

Fig. 8. Realization of a typical term in expansion 16 of an *RC* admittance.

possible pole at $s = \infty$ and for the additive zero-frequency constant). Such a component admittance has the form

$$y_\nu = \frac{1}{R_\nu + (1/C_\nu s)} = \frac{C_\nu s}{1 + R_\nu C_\nu s}$$
$$= \frac{(1/R_\nu)s}{s + (1/R_\nu C_\nu)} = \frac{k_\nu s}{s - s_\nu} = \frac{k_\nu s}{s + \sigma_\nu} \tag{13}$$

where

$$k_\nu = \frac{1}{R_\nu}$$
$$-s_\nu = \sigma_\nu = \frac{1}{R_\nu C_\nu} \tag{14}$$

Here another apparent difficulty presents itself in that the typical term in a partial-fraction expansion which reads $k_\nu/(s - s_\nu)$ evidently lacks the factor s appearing in the numerator of expression 13 for y_ν. A little reflection upon these two items, however, soon reveals a simple cure for both of them. Thus we say to ourselves that the missing factor s can be supplied if we apply the normal partial-fraction expansion procedure, not to $Y(s)$, but to the function $Y(s)/s$, and afterwards multiply the expansion by s. This altered procedure will give us a term-by-term breakdown for $Y(s)$ in which each term at least has the necessary form as shown in Eq. 13. All that we need now is for the terms to have positive real coefficients k_ν. But since these are the residues of $Y(s)/s$, not of $Y(s)$, they *are* positive for the very reason that those of $Y(s)$ are negative! This fact is obvious if we remind ourselves that a residue of $Y(s)/s$ in a pole at $s = s_\nu$ is the corresponding residue of $Y(s)$ divided by s_ν, which is a negative real number. The function $Y(s)/s$ will, of course, not have a pole at $s = \infty$ and will instead have one (in general) at $s = 0$, but these minor matters will readily straighten themselves out as we proceed with the general plan, which appears now to look rather encouraging.

Accordingly we consider the partial-fraction expansion

$$\frac{Y(s)}{s} = \frac{k_0}{s} + \frac{k_2}{s - s_2} + \frac{k_4}{s - s_4} + \cdots + \frac{k_{2n-2}}{s - s_{2n-2}} + k_{2n} \tag{15}$$

and through multiplication by s obtain the following representation for the admittance

$$Y(s) = k_0 + \frac{k_2 s}{s - s_2} + \frac{k_4 s}{s - s_4} + \cdots + \frac{k_{2n-2} s}{s - s_{2n-2}} + k_{2n} s \tag{16}$$

in which all coefficients k_ν are assuredly positive, including $k_0 = Y(0)$ and k_{2n} which is the residue of $Y(s)$ in its pole at $s = \infty$. A complete network realization for this expression is recognized to be given by the schematic shown in Fig. 9 in which, according to Eq. 14,

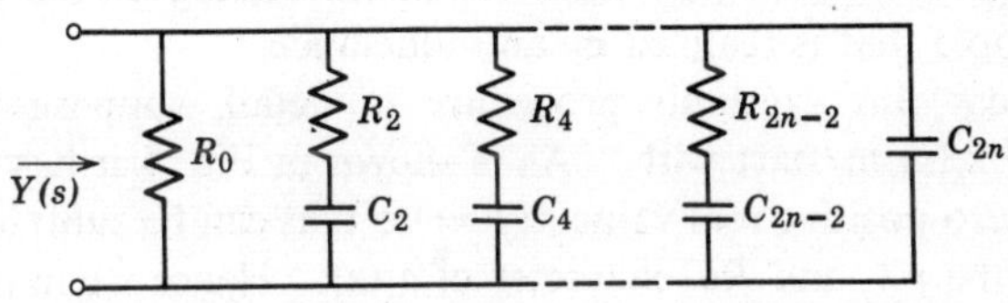

Fig. 9. Realization of an *RC* admittance according to expansion 16.

$$R_\nu = 1/k_\nu, \qquad C_\nu = k_\nu/\sigma_\nu \quad \text{for } \nu = 0, 2, 4, \cdots 2n - 2$$

and (17)

$$C_{2n} = k_{2n} \qquad \text{(ohms and farads)}$$

As in the alternate Foster-like form shown in Fig. 2, this one involves a minimum number of elements since their number exactly equals the number of determining constants in the expression for the admittance 16. The networks of Figs. 2 and 9 are, of course, not pertinent to the same driving-point impedance; the one for the network of Fig. 2 has a finite nonzero value at $s = \infty$ whereas that for the network of Fig. 9 is zero at $s = \infty$, the corresponding admittance having a pole at this point.

2 Other Canonic Forms for the Realization of *RC* Impedances

In terms of these two basic procedures for the realization of *RC* driving-point impedances, we can now find many other equivalent networks by carrying out various combination schemes as we did with *LC* networks. Two of these are analogous to the procedures leading to the Cauer forms or *LC* ladder networks discussed in the preceding chapter. In one of these, it will be recalled, we consistently removed poles at $s = \infty$ from the original function and its successive inverted remainders; in the other one (dual in character) we consistently removed poles at $s = 0$. The essential difference that we encounter when an analogous procedure is attempted with *RC* network functions is that we must alternately remove a pole and a constant value at $s = \infty$ or at $s = 0$ instead of consistently removing poles. The reason for this difference is readily clarified if we note that only an *RC* driving-point impedance can have a pole at $s = 0$, and only an admittance can have a pole at $s = \infty$. At $s = \infty$ an impedance has a (positive real) finite value, and so does an admittance at $s = 0$ (see Figs. 5 and 7).

This difference, incidentally, fits in with the results shown in Figs. 2 and 9 in which a comparison with the analogous *LC* networks reveals that resistances take the place of inductances. For an impedance the inductance represents a pole at $s = \infty$; for an admittance it represents a pole at $s = 0$. By sheer parallelism, therefore, we would expect in the *RC* case to remove a constant value wherever in the analogous *LC* situation we remove a pole that is realized by an inductance.

To discuss this synthesis procedure in detail, suppose we have an impedance $z_1(s)$ to start with. As is shown in Fig. 5 it has in general a finite nonzero positive real value at $s = \infty$ that can be subtracted without damaging the p.r. and *RC* character of $z_1(s)$. Hence we may remove an initial series resistance having this value. The remaining impedance (let us call it z_2) is now zero at $s = \infty$, as shown in the top part of Fig. 6, and so the reciprocal $y_2 = 1/z_2$, shown in the bottom part of the same figure, has a pole at $s = \infty$. Removal of this pole produces a shunt capacitance equal in value to the slope of the linear asymptote shown in Fig. 6. The remainder function (we will here call it y_3) is plotted in Fig. 7. It has a positive real nonzero value at $s = \infty$ because this value must be larger than the nonnegative zero-frequency value. Hence the reciprocal $z_3 = 1/y_3$ is again like z_1 in that it has a finite nonzero positive real value that can be subtracted to begin a second cycle in the ladder development exactly like the one just completed.

Figure 10 summarizes the complete process of ladder development in terms of a series of successive frequency patterns (like the ones drawn for reactance functions except that the horizontal axes here represent the negative real axis of the s-plane). Thus removal of the first series resistance $R_1 = z_1(\infty)$ shifts the lowest zero to the point $s = -\infty$ so that the inverted remainder, the function y_2, has a pole there. This pole is next removed as a succeeding shunt capacitance C_2 in the ladder development. Observe that this step removes not only the pole at $s = \infty$ from y_2 but also the next adjacent zero. Thus y_3 has two less critical frequencies than y_2 whereas z_2 or y_2 has the same total number of critical frequencies as z_1.

However, it would be incorrect to say that z_2 or y_2 is no simpler than z_1, for both do involve one less parameter as may be seen if we write these functions as quotients of frequency factors. Clearly, z_2 has one less numerator factor than does z_1.

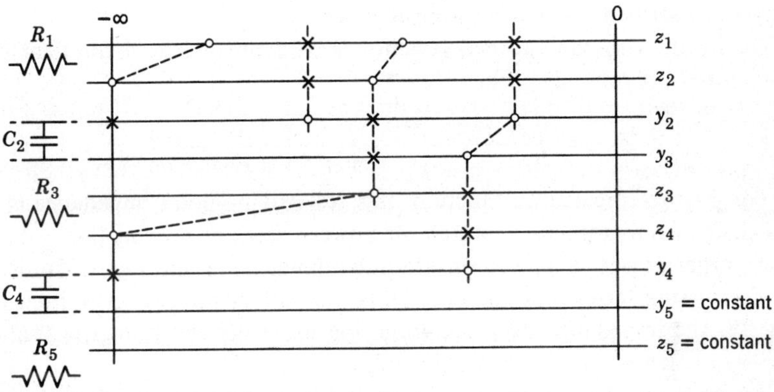

Fig. 10. Summary showing successive frequency patterns in the ladder (Cauer) development of an *RC* impedance by consistent subtraction of components at $s = \infty$.

The inverted remainder, $z_3 = 1/y_3$, is like z_1 and permits the subtraction of $R_3 = z_3(\infty)$ as in the first step, and so forth. Note that after we extract the capacitance C_4 from y_4, all poles and zeros are exhausted, but we still have a constant left which yields the final branch R_5. The complete ladder development is shown in Fig. 11. It evidently realizes the given

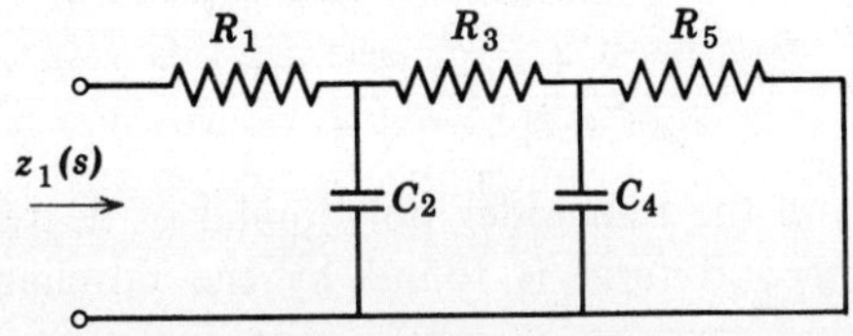

Fig. 11. Network obtained from the development summarized in Fig. 10.

function with a minimum number of elements since this number equals the number of finite nonzero critical frequencies plus one, which equals the number of parameters characterizing the given impedance. From the nature of the process just described we see that this situation is generally true since it yields an element per finite nonzero critical frequency until all of these are exhausted and then it yields one more for the remaining constant.

The separate elements removed are shown at appropriate places in the left-hand column of Fig. 10, and the various undeveloped remainder functions to which the frequency patterns pertain are indicated in a right-hand column of this figure which compactly contains all relevant information about the complete development except numerical element values. These must be gotten from the corresponding algebraic procedure which may be described in the following manner.

We begin with the rational function $Z(s)$ expressed as a quotient of unfactored polynomials, thus

$$Z(s) = \frac{a_n s^n + a_{n-1} s^{n-1} + \cdots + a_1 s + a_0}{b_n s^n + b_{n-1} s^{n-1} + \cdots + b_1 s + b_0} \tag{18}$$

In contrast to a reactance function, this RC driving-point impedance is a quotient of polynomials in which all powers of s are present (instead of every other power as in even or odd polynomials). There can, of course, be no missing terms and all coefficients must be positive. As written in Eq. 18, the impedance $Z(s)$ has finite nonzero values at both $s = 0$ and $s = \infty$, which is its most general form.

If a function of this form were given and there were no assurance that it was p.r. and RC realizable, we could, of course, settle this question through solving the polynomials for their zeros, possibly checking first their zero distributions through use of Sturm's theorem. It is much faster, however, to apply the algebraic process about to be discussed. If it is successful, the function is obviously p.r. and RC realizable.

The ladder development process, to which the series of manipulations portrayed in Fig. 10 refers, involves the alternate removal of a constant value and a pole at the point $s = \infty$, beginning with the constant value $Z(\infty)$. Accordingly, the initial algebraic step is given by

$$Z(s) = R_1 + \frac{a_{n-1}' s^{n-1} + a_{n-2}' s^{n-2} + \cdots + a_1' s + a_0'}{b_n s^n + b_{n-1} s^{n-1} + \cdots + b_1 s + b_0} \tag{19}$$

where $R_1 = a_n/b_n$ and the remainder polynomial of degree $n - 1$ in the numerator of the second term is found by the familiar long-division process. This second term is evidently zero for $s = \infty$. Hence the reciprocal of this remainder, an admittance, has a pole there. Its removal constitutes the second step in the expansion process and is again carried out by the method of long division. We now have

$$Z(s) = R_1 + \cfrac{1}{C_2 s + \cfrac{b_{n-1}' s^{n-1} + \cdots + b_0'}{a_{n-1}' s^{n-1} + \cdots + a_0'}} \tag{20}$$

where $C_2 = b_n/a_{n-1}'$. The reciprocal of the rational function remainder now is an impedance with the same properties as $Z(s)$ since neither of the two steps involved in the removal of R_1 and C_2 can (according to previous discussion) damage the p.r. and *RC* character of the function. Hence we can repeat the cycle of manipulations just completed until the remainder is reduced to a constant.

Observe that each cycle yields two circuit elements and reduces by two the number of polynomial coefficients. Hence we must get a minimal form or canonic form. The continued algebraic procedure is the familiar division and inversion process used in the Cauer development of *LC* impedances and in Hurwitz testing. It yields the continued-fraction representation

$$Z(s) = R_1 + \cfrac{1}{C_2 s + \cfrac{1}{R_3 + \cfrac{1}{C_4 s + \cdot \; \cdot \; \cdot + \cfrac{1}{R_{n-1} + \cfrac{1}{C_n s + \cfrac{1}{R_{n+1}}}}}}} \tag{21}$$

Although $n + 2$ coefficients appear in expression 18 for $Z(s)$, only $n + 1$ of them are independent, since the constant multiplier is the ratio a_n/b_n. Hence the development procedure will yield $n + 1$ elements which, as pointed out before, equal the number of finite nonzero critical frequencies plus one. The resulting network is that shown in Fig. 12.

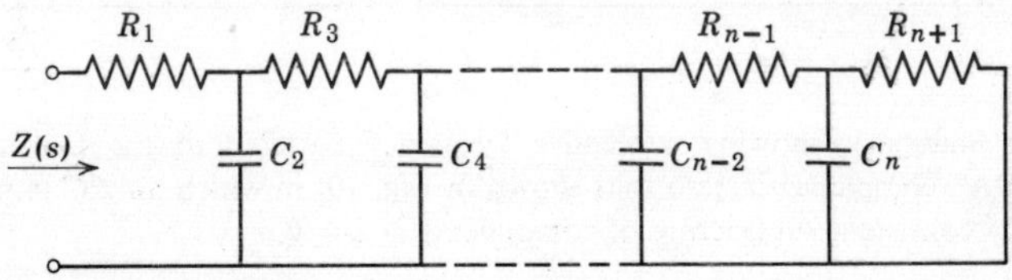

Fig. 12. General Cauer development for an *RC* impedance of the sort to which the summary in Fig. 10 is relevant.

The other ladder development analogous to the second Cauer form for *LC* impedances involves the alternate removal of a constant value and a pole at $s = 0$. Since a constant value at $s = 0$ is subtractable from an *RC* admittance (see Fig. 7 and the associated discussion), we see that the pole at $s = 0$ is removed from an impedance and the constant value from its inverted remainder in a continuing sequence that yields a ladder network with series capacitances (because series branches are removed from impedances) and shunt resistances. Figure 13 shows graphically a summary of such a sequence of operations analogous to those illustrated in Fig. 10 for the complementary Cauer development of the same driving-point impedance $z_1(s)$.

Since the given impedance does not have a pole at $s = 0$, one must first maneuver the critical frequencies so as to produce a pole in that position. This operation requires two steps because poles cannot be moved but zeros can. Hence the first step is to invert z_1, changing poles into zeros and zeros into poles. The lowest zero of y_1 may then be shifted to the origin by the subtraction of a conductance $G_1 = y_1(0)$ which becomes the first shunt branch in the desired ladder development. The remainder function at this point is y_2. The corresponding impedance $z_2 = 1/y_2$ now has the desired pole at $s = 0$, which is subtracted as the series-capacitive branch

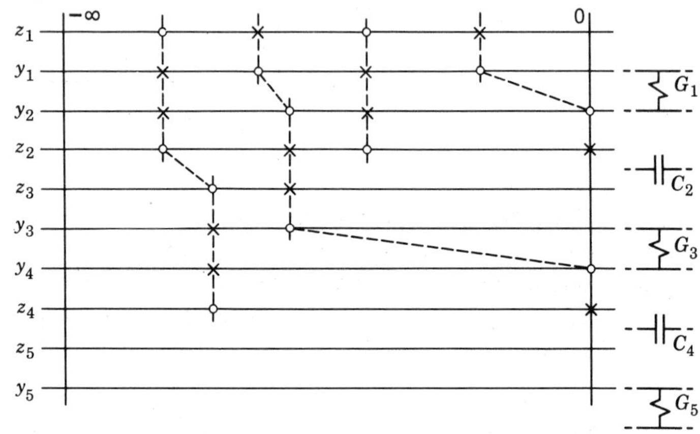

Fig. 13. Summary showing successive frequency patterns in the ladder (Cauer) development (complementary to that shown in Fig. 10) in which an *RC* impedance is realized by consistent subtraction of components at $s = 0$.

C_2, leaving a remainder z_3 having the same character as z_1. The cycle is repeated, yielding branches G_3 and C_4, after which only a constant remains. The complete ladder network is shown in Fig. 14.

Again we observe that the p.r. and *RC* character of the given function suffices to guarantee the positiveness of all elements, and that their

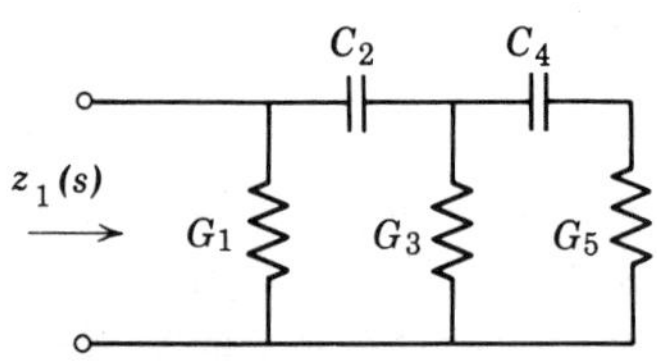

Fig. 14. Network obtained from the development summarized in Fig. 13.

number equals the number of finite nonzero critical frequencies plus one, the same as in all the other canonic forms.

For a discussion of the corresponding algebraic process we turn our attention to the expression for $Z(s)$ given in Eq. 18. Since we are to remove constant values and poles at $s = 0$, the polynomials must be oriented in ascending powers of s, and so we turn the function in Eq. 18 end for end. The first step is to invert the expression because the constant value at $s = 0$ must be removed from an admittance, not from an impedance. The first step in the continued long-division and inversion process, therefore, reads

$$Z(s) = \cfrac{1}{G_1 + \cfrac{b_1's + b_2's^2 + \cdots + b_n's^n}{a_0 + a_1s + \cdots + a_ns^n}} \tag{22}$$

with $G_1 = b_0/a_0$. From here on the process follows the now familiar pattern, yielding alternately terms in $1/s$ and constants, as shown by

$$Z(s) = \cfrac{1}{G_1 + \cfrac{1}{(1/C_2s) + \cfrac{1}{G_3 + \cfrac{1}{(1/C_4s) + \cdots + \cfrac{1}{G_{n-1} + \cfrac{1}{(1/C_ns) + \cfrac{1}{G_{n+1}}}}}}}} \tag{23}$$

The complete development involves $n + 1$ terms which, as before, equal the number of finite nonzero critical frequencies plus one. The corresponding canonic network is shown in Fig. 15.

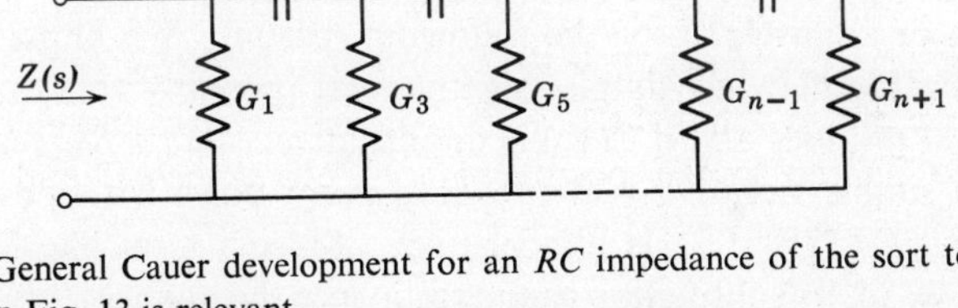

Fig. 15. General Cauer development for an RC impedance of the sort to which the summary in Fig. 13 is relevant.

As in the LC ladder networks, the first and last elements determine the behavior at $s = 0$ and at $s = \infty$. Thus the network of Fig. 15 will lack the shunt conductance G_1 if $Z(s)$ has a pole at $s = 0$, and it will lack the final shunt conductance G_{n+1} if $Z(s)$ has a zero at $s = \infty$. Similarly, in the network of Fig. 12 the first resistance R_1 is missing if $Z(s)$ has a zero at $s = \infty$, and the last resistance R_{n+1} is missing if $Z(s)$ has a pole at $s = 0$.

In contemplating the continued-fraction development like Eq. 21 or 23 so as to obtain a particular one of the ladder forms shown in Figs. 12 and 15, one must be careful to start off the process of continued division and inversion with the polynomials properly oriented as to ascending or descending powers and with the proper identification as to which one is the dividend and which one the divisor. Only one out of four possible combinations is the right one for obtaining the anticipated network, and two possible arrangements will lead to an impossible remainder function at some point in the division and inversion process. These spurious arrangements, moreover, do not display their fallacious character at the outset, and so the unwary operator who attempts to follow a pattern instead of using common sense and reason can easily be taken in and "bog down" in the middle of the process. Since it is also possible to "bog down" merely because of numerical error or insufficient accuracy, it is quite essential that one be sure of one's ground before embarking upon one of these ladder development procedures. The physical and mathematical reasoning given above is an indispensable guide in carrying through any of the synthesis methods.

3 More General "Mixed" Procedures for *RC* Synthesis

Additional ladder developments can evidently be obtained by combining the Foster-like procedures in a way that is more general than considering merely the removal of a component representing asymptotic behavior at $s = 0$ and at $s = \infty$. One may remove one or more or part of one finite nonzero pole as well. In this regard it must be remembered, however, that finite nonzero poles of an *RC* function may be removed either from $Z(s)$ or from $Y(s)/s$, not from $Y(s)$.

Subtracting a pole, algebraically a term like $k_\nu/(s - s_\nu)$, from such a function cannot damage the p.r. and *RC* character of the driving-point impedance or admittance for the following reason. We know that $Z(s)$ is *RC* realizable if it has simple negative real (possibly zero) poles with positive real residues and if $Z(\infty)$ is nonnegative; $Y(s)$ is *RC* realizable if $Y(s)/s$ has simple negative real (possibly zero) poles with positive real residues and if $[Y(s)/s]_{s=\infty}$ is nonnegative. Since a term like $k_\nu/(s - s_\nu)$ is zero at $s = \infty$, its subtraction cannot influence the nonnegative character of $Z(s)$ or $Y(s)/s$ at this point, and since its subtraction obviously cannot influence the similar terms representing the remaining poles, a mixed procedure involving any arbitrary sequence of pole removals of this sort must always lead to a successful realization for the given function. It will, moreover, be a realization involving a minimum number of elements if poles are always totally removed. When partial pole removals are involved, then superfluous elements are certain to occur.

4 Examples of *RC* Driving-Point Synthesis

Suppose we consider the *RC* driving-point impedance

$$Z(s) = \frac{(s+2)(s+4)}{(s+1)(s+3)(s+5)} = \frac{\frac{3}{8}}{s+1} + \frac{\frac{1}{4}}{s+3} + \frac{\frac{3}{8}}{s+5} \tag{24}$$

for which the residues are easily evaluated by simply dropping out the pertinent pole factor on the left and evaluating the remaining expression at the point where this factor is zero. For example, for the residue at $s = -3$ we have

$$k_3 = \left[\frac{(s+2)(s+4)}{(s+1)(s+5)}\right]_{s=-3} = \frac{(-1)(1)}{(-2)(2)} = \frac{1}{4} \tag{25}$$

Since each term in the partial-fraction expansion 24 represents a parallel *RC* circuit, the total network realizing this impedance is recognized to be that shown in Fig. 16.

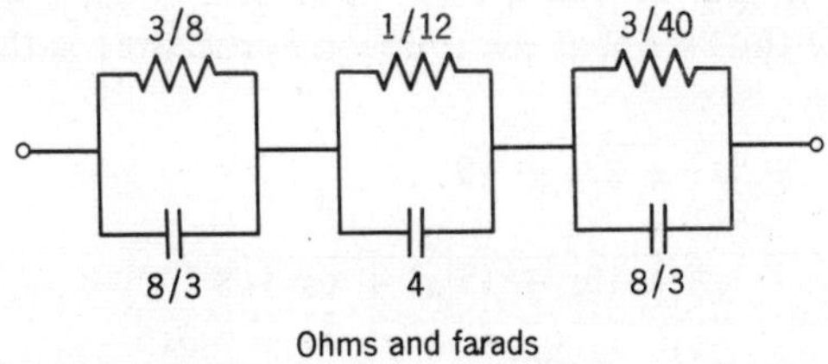

Fig. 16. Realization in Foster-like form of the *RC* impedance given by Eq. 24.

To obtain the second Foster-like form, we expand the corresponding $Y(s)/s$ in partial fractions, thus

$$\frac{Y(s)}{s} = \frac{(s+1)(s+3)(s+5)}{s(s+2)(s+4)} = \frac{15}{8s} + \frac{\frac{3}{4}}{s+2} + \frac{\frac{3}{8}}{s+4} + 1 \tag{26}$$

The constant term +1 is needed in order to take care of the behavior of the function for $s \to \infty$ and represents what in partial-fraction expansion parlance is called the rational entire remainder (see *MCA*, pp. 307–310). The residues are evaluated as easily as in the example above.

Multiplication of expansion 26 by s yields

$$Y(s) = \frac{15}{8} + \frac{\frac{3}{4}s}{s+2} + \frac{\frac{3}{8}s}{s+4} + s \tag{27}$$

which is recognized by inspection to be the admittance of the network in Fig. 17.

In order to obtain the ladder networks, we first put the impedance 24 in the unfactored form

$$Z(s) = \frac{s^2 + 6s + 8}{s^3 + 9s^2 + 23s + 15} \tag{28}$$

Suppose we want the ladder network involving resistances in its series branches and capacitances in its shunt branches. This type of development will require the subtraction of constant terms from an impedance

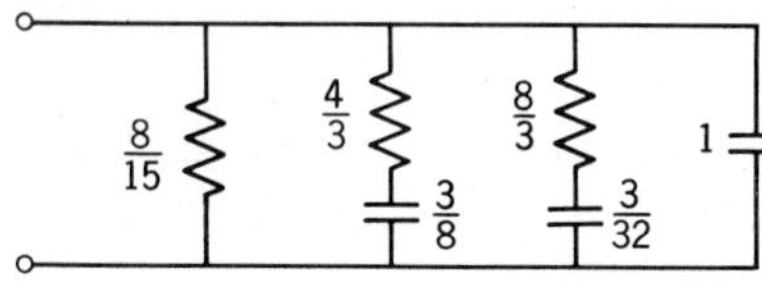

Fig. 17. Alternate Foster-like realization of the impedance in Eq. 24.

and a capacitive susceptance from an admittance and hence will involve the behavior of the original and subsequent remainder functions at $s = \infty$. Accordingly, the polynomials should be arranged in descending powers of s. Since $Z(s)$ in Eq. 28 has a zero value at $s = \infty$, the first step is an inversion. We thus see that the following procedure is the correct one

$$\begin{array}{l}
s^2 + 6s + 8\,\overline{\big)\, s^3 + 9s^2 + 23s + 15}\quad s \\
\qquad\qquad\;\; s^3 + 6s^2 + 8s \\
\qquad\qquad\;\; \overline{3s^2 + 15s + 15\,\big)\, s^2 + 6s + 8}\quad \tfrac{1}{3} \\
\qquad\qquad\qquad\qquad\qquad\;\; s^2 + 5s + 5 \\
\qquad\qquad\qquad\qquad\qquad\;\; \overline{s + 3\,\big)\, 3s^2 + 15s + 15}\quad 3s \\
\qquad\qquad\qquad\qquad\qquad\qquad\quad 3s^2 + 9s \\
\qquad\qquad\qquad\qquad\qquad\qquad\quad \overline{6s + 15\,\big)\, s + 3}\quad \tfrac{1}{6} \\
\qquad\qquad\qquad\qquad\qquad\qquad\qquad\qquad s + \tfrac{5}{2} \\
\qquad\qquad\qquad\qquad\qquad\qquad\qquad\qquad \overline{\tfrac{1}{2}\,\big)\, 6s + 15}\quad 12s \\
\qquad\qquad\qquad\qquad\qquad\qquad\qquad\qquad\quad 6s \\
\qquad\qquad\qquad\qquad\qquad\qquad\qquad\qquad\quad \overline{15\,\big)\, \tfrac{1}{2}}\quad \tfrac{1}{30}
\end{array} \tag{29}$$

The numerical work is easily accommodated on a normal-sized sheet by moving the remainder polynomials in the continued process of division to the left and below the previous line of computations each time the right-hand side of the page is reached. The successive quotients, s, $\frac{1}{3}$, $3s$, $\frac{1}{6}$, $12s$, $\frac{1}{30}$, are alternately admittances and impedances respectively of the ladder branches. Hence one recognizes easily that the desired realization is the network shown in Fig. 18.

To obtain the alternate ladder in which series branches are capacitances and shunt branches are resistances, we must remove poles at $s = 0$ from impedances and constant values at $s = 0$ from admittance functions. The polynomials should therefore be arranged in ascending powers, and so we turn expression 28 for $Z(s)$ end for end. We also turn it upside down, for we observe that this particular impedance has no pole at $s = 0$. The proper algebraic procedure is thus seen to take the form

$$
\begin{array}{l}
\qquad\qquad\qquad\quad |\tfrac{15}{8} \\
8 + 6s + s^2 \,|\, 15 + 23s + 9s^2 + s^3 \\
\qquad\qquad\qquad\quad 15 + \tfrac{45}{4}s + \tfrac{15}{8}s^2 \qquad |32/47s \\
\qquad\qquad\qquad\qquad \tfrac{47}{4}s + \tfrac{57}{8}s^2 + s^3 \,|\, 8 + 6s + s^2 \\
\qquad\qquad\qquad\qquad\qquad\qquad\qquad 8 + \tfrac{228}{47}s + \tfrac{32}{47}s^2 \,|\, \tfrac{2209}{216} \\
\qquad\qquad\qquad\qquad\qquad\qquad\qquad\quad \tfrac{54}{47}s + \tfrac{15}{47}s^2 \,|\, \tfrac{47}{4}s + \tfrac{57}{8}s^2 + s^3 \\
\qquad\qquad\qquad\qquad\qquad\qquad\qquad\qquad\qquad\quad \tfrac{47}{4}s + \tfrac{235}{72}s^2 \\
\qquad\qquad\qquad\qquad\qquad\qquad\qquad\qquad\qquad\quad\ \tfrac{139}{36}s^2 + s^3 \\
\qquad\qquad\quad |1944/6533s \\
\tfrac{139}{36}s^2 + s^3 \,|\, \tfrac{54}{47}s + \tfrac{15}{47}s^2 \\
\qquad\qquad\quad \tfrac{54}{47}s + \tfrac{1944}{6533}s^2 \,|\, \tfrac{19321}{108} \\
\qquad\qquad\qquad\qquad \tfrac{3}{139}s^2 \,|\, \tfrac{139}{36}s^2 + s^3 \\
\qquad\qquad\qquad\qquad\quad \tfrac{139}{36}s^2 \qquad |3/139s \\
\qquad\qquad\qquad\qquad\qquad\quad s^3 \,|\, \tfrac{3}{139}s^2
\end{array}
\tag{30}
$$

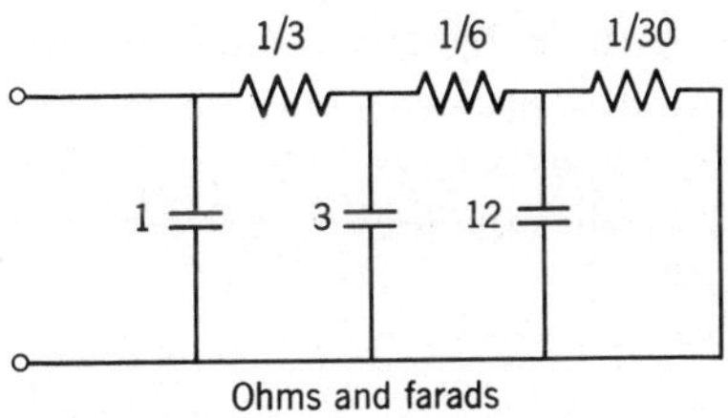

Fig. 18. Cauer development of impedance 24 through consistent removal of components at $s = \infty$.

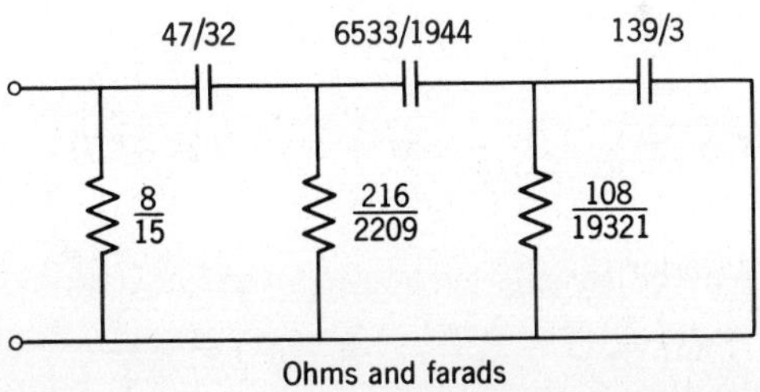

Fig. 19. Cauer development of impedance 24 through consistent removal of components at $s = 0$.

The successive quotients again are alternately admittances and impedances respectively, and their interpretation is readily recognized to yield the circuit shown in Fig. 19. In more elaborate situations than the present example it will be advantageous to use decimal fractions rather than actual fractions as is done in the computations 30.

Finally, we shall develop this same impedance into a ladder network by removing finite nonzero poles alternately from $Z(s)$ and $Y(s)/s$. Returning our attention, therefore, to Eq. 24, suppose we start by removing from $Z(s)$ the pole at $s = -3$. Denoting the remainder function by $Z_1(s)$, we compute it in the following way

$$Z_1(s) = \frac{(s+2)(s+4)}{(s+1)(s+3)(s+5)} - \frac{\frac{1}{4}}{s+3}$$

$$Z_1(s) = \left[\frac{s^2+6s+8}{s^2+6s+5} - \frac{1}{4}\right]\frac{1}{s+3} = \frac{\frac{3}{4}s^2 + \frac{9}{2}s + \frac{27}{4}}{(s^2+6s+5)(s+3)} = \frac{\frac{3}{4}s + \frac{9}{4}}{s^2+6s+5} \tag{31}$$

Observe that the subtraction in the square bracket leads to a numerator polynomial that must be divisible by the factor $(s + 3)$ since the remainder $Z_1(s)$ can no longer contain the pole at $s = -3$. The factor $(s + 3)$ must at this point be divided out, and the fact that this division must leave no remainder is often useful as a check on the numerical accuracy.

We now form $Y_1(s)/s$ where $Y_1(s) = 1/Z_1(s)$ and note that its only finite nonzero pole again (by sheer accident) lies at $s = -3$, for Eq. 31 yields

$$\frac{Y_1(s)}{s} = \frac{\frac{4}{3}(s+1)(s+5)}{s(s+3)} \tag{32}$$

The residue of this function at $s = -3$ is

$$k = \left[\frac{\frac{4}{3}(s+1)(s+5)}{s}\right]_{s=-3} = \frac{\frac{4}{3}(-2)(2)}{-3} = \frac{16}{9} \tag{33}$$

and the subtraction of this pole involves the computations

$$\frac{\frac{4}{3}(s+1)(s+5)}{s(s+3)} - \frac{\frac{16}{9}}{s+3} = \left[\frac{\frac{4}{3}s^2 + 8s + \frac{20}{3}}{s} - \frac{16}{9}\right]\frac{1}{s+3}$$

$$= \frac{\frac{4}{3}s^2 + \frac{56}{9}s + \frac{20}{3}}{s(s+3)} = \frac{12s^2 + 56s + 60}{9s(s+3)} = \frac{12s+20}{9s} = \frac{Y_2(s)}{s} \tag{34}$$

so the second remainder is

$$Y_2(s) = \tfrac{4}{3}s + \tfrac{20}{9}, \qquad \text{or } Z_2(s) = \frac{\frac{3}{4}}{s + \frac{5}{3}} \tag{35}$$

The impedance removed in the first series branch, as indicated in the subtraction 31, is evidently

$$z_1 = \frac{\frac{1}{4}}{s+3} \tag{36}$$

The admittance of the next shunt branch, according to the computations 34, is given by

$$y_2 = \frac{\frac{16}{9}s}{s+3} \tag{37}$$

and the final series branch is the impedance 35. The corresponding ladder network is shown in Fig. 20.

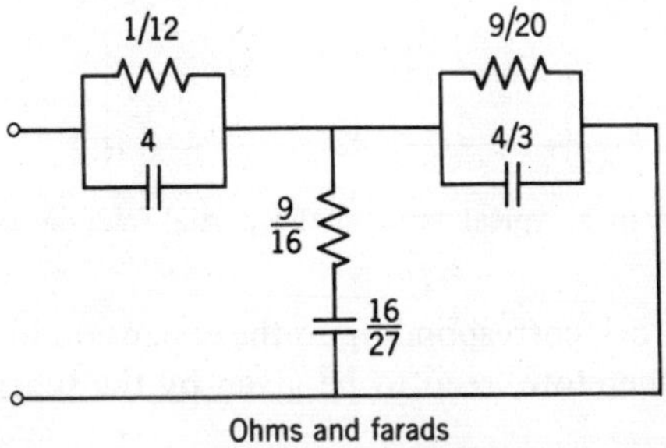

Fig. 20. Alternate realization of the *RC* impedance 24 in the manner of a Cauer development except that poles are removed at finite nonzero frequencies.

5 Analogous Procedures for *RL* Networks

As is evident from the discussions in Ch. 2, the *RL* driving-point admittance function is identical in its form and in its detailed properties with the *RC* driving-point impedance function and vice versa. Thus the driving-point admittance $Y(s)$ of an *RL* network has simple poles restricted to the negative real axis where its residues are positive real, and $0 \leq Y(\infty) < \infty$. We may replace that part of this statement involving poles and residues by saying that the zeros and poles are simple and interlace on the negative real axis, the lowest critical frequency being a pole.

If we write down a quotient of frequency factors like

$$\frac{(s+2)(s+4)(s+6)\cdots}{(s+1)(s+3)(s+5)\cdots} \tag{38}$$

being mindful only that numerator and denominator factors interlace on the negative real axis, then the finiteness and nonnegativeness for $s = \infty$ is automatically fulfilled regardless of whether the last factor is in the numerator or in the denominator. Such an expression is therefore either an *RC* driving-point impedance or an *RL* driving-point admittance. The lowest frequency factor $(s + 1)$ may become simply the factor s, in which case the function has a pole at the origin.

Realization of an *RL* $Y(s)$ in the Foster-like manner proceeds from its partial-fraction expansion identical with that given by Eqs. 1 and 3 for an *RC* $Z(s)$. Each term in this expansion is now recognized as having the form of the admittance of a typical series *RL* circuit as shown in Fig. 21, for which

$$y_\nu = \frac{1}{R_\nu + L_\nu s} = \frac{1/L}{s + (R_\nu/L_\nu)} \tag{39}$$

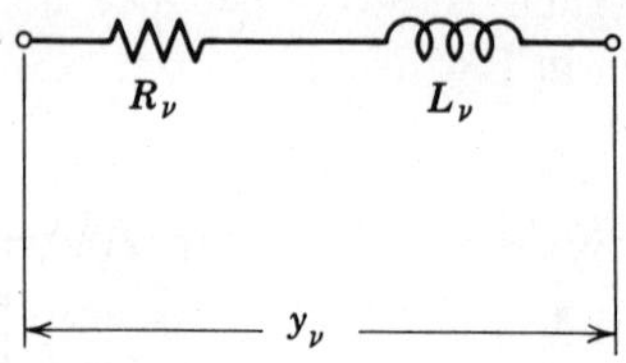

Fig. 21. Realization of a typical term in the partial-fraction expansion of an *RL* admittance.

The complete network corresponding to the expansion in Eq. 3 for the *RL* $Y(s)$ function is, therefore, seen to be given by the schematic in Fig. 22, in which

$$R_\nu = \sigma_\nu/k_\nu, \qquad L_\nu = 1/k_\nu, \quad \text{for } \nu = 1, 3, \cdots 2n-1 \quad \text{(ohms and henrys)}$$

and

$$1/R_{2n+1} = k_{2n+1} = Y(\infty) \tag{40}$$

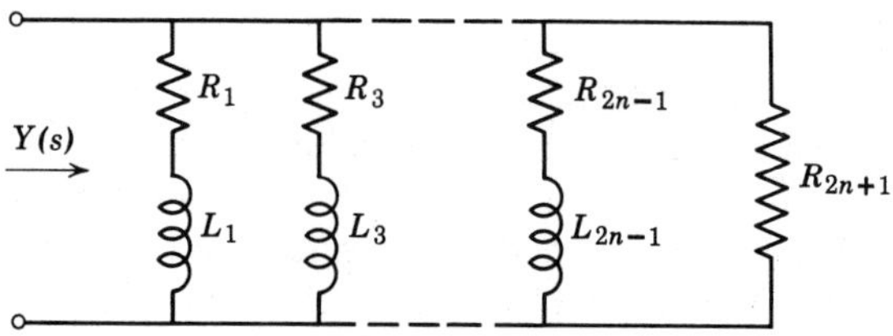

Fig. 22. Realization of an *RL* admittance in the Foster-like manner.

The second Foster-like form is obtained from a partial-fraction expansion of the function $Z(s)/s$ like the one for the *RC* $Y(s)/s$ function in Eq. 15. Thus a representation for $Z(s)$ is obtained which is identical in form with the right-hand side of Eq. 16. The typical term in this expression is recognized as being the impedance of a parallel *RL* circuit as shown in Fig. 23 for which

$$z_\nu = \frac{1}{(1/R_\nu) + (1/L_\nu s)} = \frac{R_\nu s}{s + (R_\nu/L_\nu)} \tag{41}$$

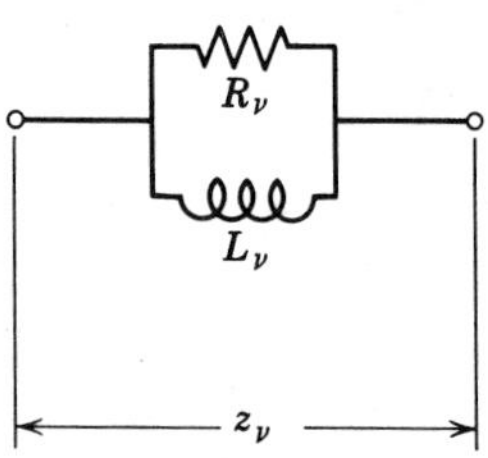

Fig. 23. Realization of a typical term in the expansion of an *RL* impedance.

The complete network is thus seen to be given by the circuit in Fig. 24, in which

$$R_\nu = k_\nu,\ L_\nu = k_\nu/\sigma_\nu, \text{ for } \nu = 0, 2, 4, \cdots, 2n-2, \text{ and } L_{2n} = k_{2n} \tag{42}$$

The positive realness of the residues plus the finite nonnegativeness of $Y(s)$ or $Z(s)/s$ for $s = \infty$ are seen to guarantee the realizability of these *RL* Foster-like networks. It is evident, moreover, that they involve a minimum number of elements.

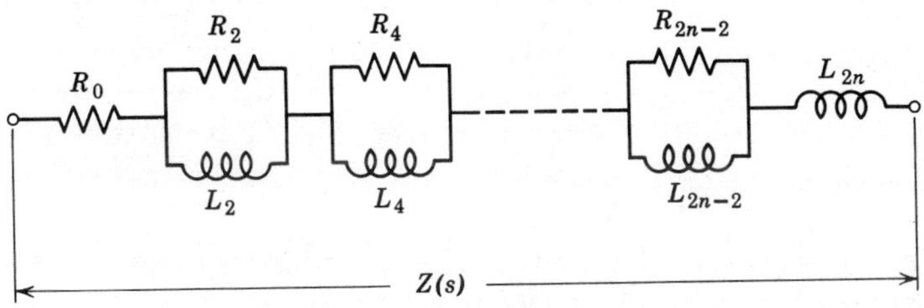

Fig. 24. Realization of an *RL* impedance in the Foster-like manner.

The ladder development procedures are likewise identical in form with those for *RC* networks, with the identities of $Y(s)$ and $Z(s)$ interchanged. For an *RL* network the value $Z(0)$ is always subtractable from $Z(s)$ or the value $Y(\infty)$ from $Y(s)$ since these values are (according to statements 46 and 49 of Ch. 2) equal to the minimum real part of the pertinent function along the entire j-axis. We may therefore obtain a ladder with series resistances and shunt inductances by subtracting alternately a constant value and a pole at $s = 0$ from the impedance $Z(s)$ and its subsequent inverted remainders; or we may obtain a ladder with series inductances and shunt resistances by subtracting alternately a pole and a constant value at $s = \infty$ from the impedance $Z(s)$ and its subsequent inverted remainders.

Thus, starting from the expression for $Z(s)$ given by Eq. 18 (which in analytic form may equally well represent an *RL* or an *RC* impedance), we orient the polynomials in ascending powers for the first of these two ladders, and in descending powers for the second. The familiar process of continued division and inversion will then yield on the one hand the continued-fraction development

$$Z(s) = R_1 + \cfrac{1}{(1/L_2s) + \cfrac{1}{R_3 + \cfrac{1}{(1/L_4s) + \cdots + \cfrac{1}{R_{n-1} + \cfrac{1}{(1/L_ns) + \cfrac{1}{R_{n+1}}}}}}} \tag{43}$$

and on the other

$$Z(s) = \cfrac{1}{G_1 + \cfrac{1}{L_2 s + \cfrac{1}{G_3 + \cfrac{1}{L_4 s + \cdot \, \cdot \, \cdot + \cfrac{1}{G_{n-1} + \cfrac{1}{L_n s + \cfrac{1}{G_{n+1}}}}}}}} \tag{44}$$

The resulting networks are shown in Figs. 25 and 26 respectively. Presence or absence of the first and last elements depends upon the specific nature of the function $Z(s)$ at $s = 0$ and $s = \infty$ in a manner that is entirely analogous to the corresponding situation in the RC ladder networks.

Finally, one may produce an unlimited number of additional ladder networks through removal in whole or in part of finite nonzero poles alternately from the functions $Z(s)/s$ and $Y(s)$. The details involved in this process are so analogous to those already discussed for RC networks that further elaboration should be unnecessary.

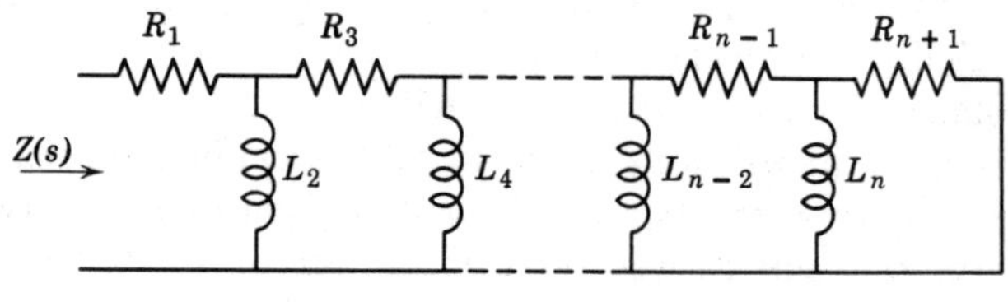

Fig. 25. Ladder development of an RL impedance through consistent removal of components at $s = 0$.

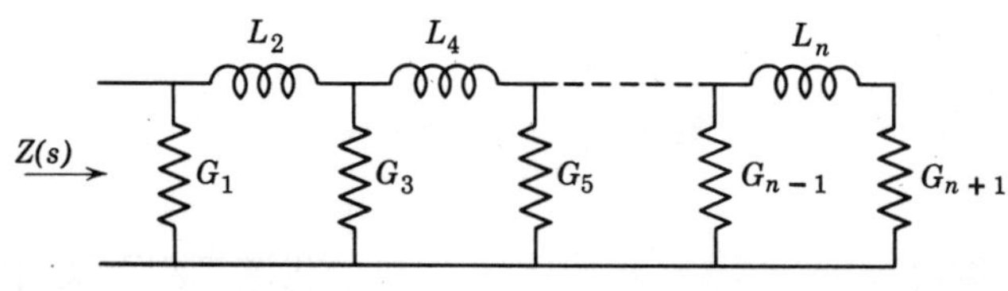

Fig. 26. Ladder development of an RL impedance through consistent removal of components at $s = \infty$.

6 Equivalent, Reciprocal, and Complementary Networks

For the sake of emphasis it is appropriate to call attention to the equivalence of the various *RC* or *RL* networks found for a given driving-point impedance or admittance function by the different synthesis methods discussed above. Again, as in the analogous situation discussed in Ch. 3 for *LC* networks, it is in general necessary to distinguish between equivalence and potential equivalence, and to call attention to the fact that all of these networks are equivalent only with respect to the one driving point to which the given impedance $Z(s)$ pertains.

In regard to *RL* or *RC* networks whose driving-point impedances are reciprocal, the situation is somewhat different from that pertaining to *LC* networks. Thus the reciprocal of an *LC* network is again an *LC* network, but the reciprocal of an *RC* network is an *RL* network and vice versa. This result should be obvious from the fact that the *RL* synthesis methods are identical in form with the *RC* methods, except that the identities of the $Z(s)$ and $Y(s)$ functions are interchanged. That is to say the reciprocal of an *RC* impedance, if again regarded as an impedance, fulfills the realizability conditions of an *RL* impedance, and vice versa. Solution to the problem of constructing a network whose driving-point impedance is the reciprocal of the impedance of a given *RC* or *RL* network is, therefore, straightforward; and it is also obvious that it can be carried out whether or not the given network has a mappable graph and/or mutual inductive couplings. The use of normal co-ordinate transformations as discussed in this connection for *LC* circuits in Art. 7 of Ch. 3 is also applicable here in unaltered form. One applies to the quadratic forms T and F or V and F the process of simultaneous reduction to sums of squares that in the analogous *LC* discussion is applied to the forms T and V.

Besides the mutual relationship of being equivalent or reciprocal with respect to their specific driving points, a pair of networks may in appropriate circumstances display another important mutual property, namely, that of being complementary. If $Z_1(s)$ and $Z_2(s)$ denote the pertinent driving-point impedances, then we can summarize these various mutual properties in a conformal manner by stating that equivalent, reciprocal, and complementary networks are characterized respectively by the relationships

$$Z_1(s)/Z_2(s) = 1 \tag{45}$$

$$Z_1(s) \times Z_2(s) = 1 \tag{46}$$

$$Z_1(s) + Z_2(s) = 1 \tag{47}$$

The constant involved in these expressions need not necessarily be unity, but it is always possible to so normalize the given impedances (by choice of their constant multipliers) as to render this constant unity, and thus, in most cases, appreciably simplify the numerical work associated with the process of constructing one network when the other is given.

For example, if, for a given network with a mappable graph, we construct the reciprocal by first finding the dual, then the value unity for the constant in relation 46 is tacitly implied. In this case inductance values in one network become capacitance values in the other as do resistance and conductance values also, with the result that no numerical calculations whatever are called for.

With regard to the complementary relationship expressed by Eq. 47, it may be pointed out first of all that this mutual property cannot exist for lossless networks whose impedances have j-axis poles, because the residues in such poles must be real and positive in both Z_1 and Z_2 and hence cannot possibly cancel so as to yield a sum having no poles at all. On the other hand, we have seen that *RC* driving-point impedances have positive residues and those for *RL* impedances are negative. Here a network having the complementary impedance is evidently constructible, and we see that the complement of an *RC* network is an *RL* network and vice versa.

In this connection, however, one important point must be observed. Denoting *RL* and *RC* impedances with these letters as subscripts, we have found, among other things, that realizability requires

$$Z_{RC}(\infty) \geq 0, \qquad Y_{RC}(0) \geq 0 \tag{48}$$

$$Z_{RL}(0) \geq 0, \qquad Y_{RL}(\infty) \geq 0 \tag{49}$$

If Z_{RC} is given and we are to construct a network for $Z_{RL} = 1 - Z_{RC}$, then condition 49 will require $Z_{RC}(0) \leq 1$. Of if Y_{RC} is given, the construction of a network for $Y_{RL} = 1 - Y_{RC}$ will require $Y_{RC}(\infty) \leq 1$. Conditions 48 impose similar requirements upon a given *RL* impedance or admittance if its complementary network is to exist. Since these conditions can always be met, even if they are not originally fulfilled, by introducing appropriate scale factors for the given functions (and correspondingly modifying the pertinent networks), they do not present a serious obstacle in most practical situations.

As an illustrative example suppose we consider the *RC* impedance

$$Z(s) = \frac{(s+2)(s+4)}{(s+1)(s+3)} = \frac{s^2+6s+8}{s^2+4s+3} \tag{50}$$

A ladder network consisting of series resistances and shunt capacitances is found from the division and inversion process

$$
\begin{array}{l}
\qquad\qquad\qquad\quad |1 \\
s^2 + 4s + 3|s^2 + 6s + 8 \\
\qquad\qquad\qquad s^2 + 4s + 3|s/2 \\
\qquad\qquad\qquad\qquad 2s + 5|s^2 + 4s + 3 \\
\qquad\qquad\qquad\qquad\qquad s^2 + \tfrac{5}{2}s \qquad |\tfrac{4}{3} \\
\qquad\qquad\qquad\qquad\qquad \tfrac{3}{2}s + 3|2s + 5 \\
\qquad\qquad\qquad\qquad\qquad\qquad\quad 2s + 4|3s/2 \\
\qquad\qquad\qquad\qquad\qquad\qquad\qquad\quad 1|\tfrac{3}{2}s + 3 \\
\qquad\qquad\qquad\qquad\qquad\qquad\qquad\qquad \tfrac{3}{2}s \qquad |\tfrac{1}{3} \\
\qquad\qquad\qquad\qquad\qquad\qquad\qquad\qquad\qquad 3|1
\end{array}
\tag{51}
$$

This network is shown in Fig. 27.

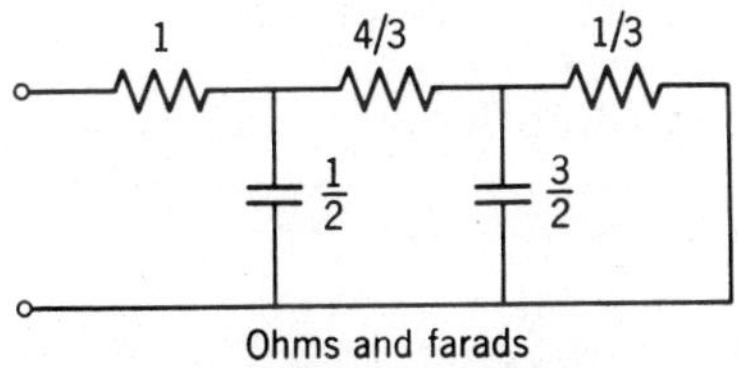

Fig. 27. Realization of the *RC* impedance 50.

Suppose we want to find a second network that we can connect in parallel with this one so as to have a resultant input impedance which is a pure resistance of 1 ohm. To do this we must construct a network whose admittance is complementary to the admittance $Y(s) = 1/Z(s)$ in Eq. 50. Since this *RC* admittance has the value unity for $s = \infty$, the necessary condition for the existence of its complement is fulfilled and it is given by

$$
Y_{RL} = 1 - \frac{s^2 + 4s + 3}{s^2 + 6s + 8} = \frac{2s + 5}{s^2 + 6s + 8} \tag{52}
$$

To synthesize it in the form of a ladder with series L's and shunt R's, the following computations are pertinent

$$
\begin{array}{l}
\qquad\qquad |s/2 \\
2s + 5|s^2 + 6s + 8 \\
\qquad\quad s^2 + \tfrac{5}{2}s \qquad |\tfrac{4}{7} \\
\qquad\qquad\quad \tfrac{7}{2}s + 8|2s + 5 \\
\qquad\qquad\qquad\qquad 2s + \tfrac{32}{7}|49s/6 \\
\qquad\qquad\qquad\qquad\qquad\quad \tfrac{3}{7}|\tfrac{7}{2}s + 8 \\
\qquad\qquad\qquad\qquad\qquad\qquad \tfrac{7}{2}s \qquad |\tfrac{3}{56} \\
\qquad\qquad\qquad\qquad\qquad\qquad\qquad 8|\tfrac{3}{7}
\end{array}
\tag{53}
$$

The corresponding network is shown in Fig. 28. If this one and that given in Fig. 27 are connected in parallel at their input terminals, the resultant admittance or impedance becomes indistinguishable from a pure resistance of 1 ohm. Practical applications of networks of this sort will be discussed later on.

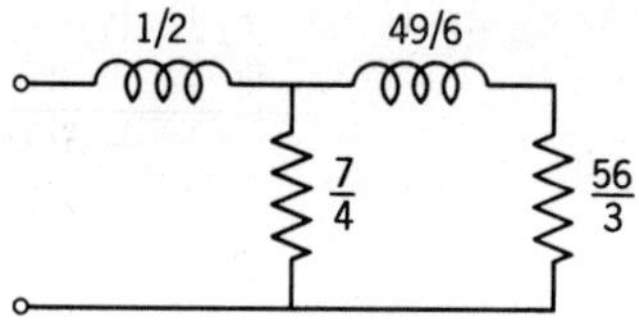

Fig. 28. Realization of the *RL* admittance 52, complementary to the admittance of the network shown in Fig. 27.

PROBLEMS

1. Of the following impedance functions, pick those that are p.r., and find one network each for those that are two-element realizable.

$(a)\ \dfrac{s(s+1)(s+4)}{(s+2)(s+3}$ $\quad (b)\ \dfrac{(s+1)(s+2)}{s(s+3)(s+4)}$ $\quad (c)\ \dfrac{(s^2+1)(s^2+3)}{s^2(s^2+2)(s^2+4)}$

$(d)\ \dfrac{(s^2+1)(s^2+3)}{s(s^2+2)(s^2+4)}$ $\quad (e)\ \dfrac{(s^2+1)(s^2+2)}{s(s^2+3)(s^2+4)}$ $\quad (f)\ \dfrac{(s+1)(s+3)}{(s+2)(s+4)}$

$(g)\ \dfrac{(s+1)(s+3)}{s(s+2)(s+4)}$ $\quad (h)\ \dfrac{(s+2)(s^2+2)}{s(s+1)(s^2+1)}$ $\quad (i)\ \dfrac{(s^2+1)}{(s+1)(2s^2+s+3)}$

2. If realizable, find networks for the following impedance functions:

$(a)\ \dfrac{s^4+5s^2+4}{s^3+9s}$ $\quad (b)\ \dfrac{s^2+3s+2}{s^2+3s}$ $\quad (c)\ \dfrac{s^2+2s+1}{s^2+4s+4}$

$(d)\ \dfrac{s^4+10s^2+9}{s^5+20s^3+64s}$ $\quad (e)\ \dfrac{s^2+4s+3}{s^2+6s^2+8s}$

$(f)\ \dfrac{s^5+20s^3+64s}{2s^5+s^4+40s^3+10s^2+128s+9}$ $\quad (g)\ \dfrac{s^5+10s^3+9s}{s^4+20s^2+64}$

$(h)\ \dfrac{s^3+4s^2+3s}{s^2+6s+8}$ $\quad (i)\ \dfrac{s^2+3s+2}{s+3}$

Note that these functions may represent networks involving any combination or all of the three kinds of parameters *R*, *L*, and *C*.

3. From the following list of driving-point impedances, pick out the realizable ones and find a network for each.

$(a)\ \dfrac{(s+1)(s^2+2s+2)}{s(s^2+1)}$ $\quad (b)\ s+2-\dfrac{3}{s+3}$ $\quad (c)\ \dfrac{s(s+3)}{(3s+2)(s+4)}$

$(d)\ \dfrac{s^3+4s^2+6s+2}{s^4+2s^3+3s^2+s+1}$ $\quad (e)\ \dfrac{s^3+s}{2s^4+5s^2+2}$

$(f)\ \dfrac{6s^2+s+1}{24s^3+10s^2+4s+1}$

4. When a given network is excited at its driving point by a voltage $e(t)$ equal to the unit step $u_{-1}(t)$, the resulting current at this point is given by

$$i(t) = 6e^{-t} - 15e^{-2t} + 10e^{-3t}$$

Find the driving-point impedance $Z(s)$ as a quotient of polynomials and synthesize the network.

If this same network is excited by a current step, what is the resulting voltage?

5. e and i are the voltage and current at the terminals of a network, as shown in the figure. $u_{-1}(t)$ is the unit step function which is 0 for $t < 0$ and 1 for $t > 0$. If a current source $i_s(t) = u_{-1}(t)$ is applied to the network, the resulting voltage is

$$e(t) = 1 + e^{-t} \sin t \qquad t > 0$$

If a voltage source, $e(t) = u_{-1}(t)$, is applied to the network, the resulting current is

$$i(t) = 1 - e^{-t} + e^{-2t} \qquad t > 0$$

(*a*) What is the impedance of the network?
(*b*) Synthesize the network.

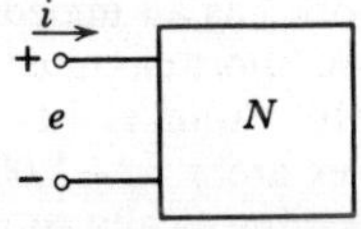

Prob. 5.

6. With reference to the *RC* network shown in the sketch: Poles of Z_1 with terminals 2-2′ open are at $s = -1, -3, -5$. Poles of Z_1 with terminals 2-2′ short-circuited are at $s = -2, -4$. The shunt resistance directly across terminals 1-1′ is $\frac{2}{5}$ ohm. Construct analytic expressions for both driving-point impedances Z_1 and Z_2, and determine all other circuit elements.

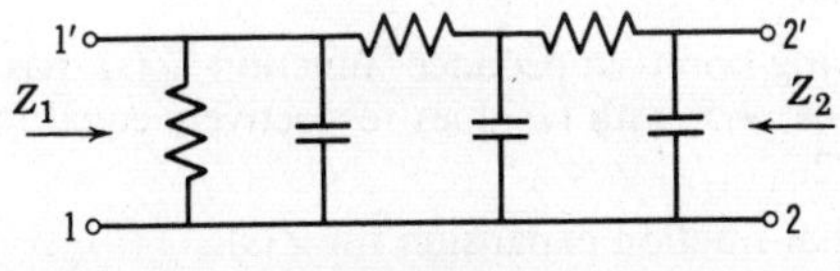

Prob. 6.

7. The impedance $Z(s)$ of the network shown has poles at $s = -1$ and $s = -3$ and zeros at $s = -2$ and $s = -4$. The resistance R equals $\frac{7}{3}$ ohm. Determine the other elements of the network.

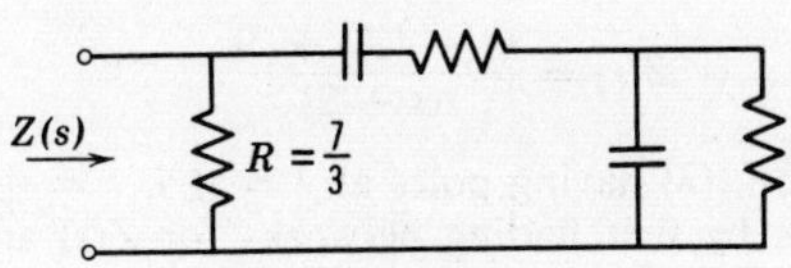

Prob. 7.

8. Each of the functions given below is a realizable impedance function. For those that are realizable as two-element-kind networks, find all four canonic

forms. For those that are not two-element-kind realizable you should be able to proceed in a straightforward manner, using the methods of Chs. 3 and 4 to find one network.

(*a*) $\dfrac{(s^2+2)(s^2+4)}{s(s^2+3)}$ (*b*) $\dfrac{(s+2)(s+3)}{s+5}$ (*c*) $\dfrac{s^5+10s^3+5s}{(s+1)^5}$

(*d*) $\dfrac{(s+1)(s+4)}{s(s+3)}$ (*e*) $\dfrac{s^4+2s^3+7s^2+3s+5}{(s+1)(s^2+1)}$ (*f*) $\dfrac{(s+1)(s+4)}{s+3}$

(*g*) $s+1-\dfrac{1}{s+1}$ (*h*) $\dfrac{s}{s^2+s+1}$ (*i*) $\dfrac{s+1}{s^2+s+1}$ (*j*) $\dfrac{s(s+1)}{s^2+s+1}$

(*k*) $\dfrac{s^4+2s^3+3s^2+3s+1}{s^3+2s^2+2s+1}$

9. Check the following function for p.r. character and realize

$$Z(s)=\frac{s^5+8s^4+17s^3+13s^2+12s+3}{s^5+4s^4+4s^3+4s^2+3s}$$

10. A two-element-kind network has an impedance $Z(s)$ at its input terminal pair. When these terminals are short-circuited, the natural frequencies are $s=-1$ and $s=-2$. When a 10^{-3} ohm resistance is placed across these same terminals, the natural frequencies are $s=-1.001$ and $s=-2.002$. At very high frequencies $Z(s)$ becomes asymptotically equal to 10 ohms. Construct the function $Z(s)$ and synthesize the network.

11. A two-element-kind impedance $Z(s)$ has poles at $s=-1, -2$, and -3 only. When a 5000-ohm resistance is placed in parallel with the input terminals of the network, the poles shift to the points $-1.002, -2.002, -3.002$. $Z(\infty)=0$. Making reasonable approximations:

(*a*) Construct $Z(s)$.
(*b*) Find its zeros.
(*c*) Synthesize a network.

12. A given driving-point impedance function $Z(s)$ has simple poles at $s=-1, -2, -3, -4, -5$, with residues respectively equal to $1, -2, 3, -4, 5$. For $s\to\infty$, $Z(s)\to 2$.

(*a*) Write the partial-fraction expansion for $Z(s)$.
(*b*) Find a network realization.

13. Show that a p.r. function of a p.r. function must be p.r. That is to say, if $Z(s)$ is p.r. and if $z(\lambda)$ is also p.r., then show that $Z(z)=F(\lambda)$ must be a p.r. function.

Illustrate this result for the function

$$Z(s)=\frac{(s+1)(s+3)}{s(s+2)}$$

and a reactance function $z(\lambda)$ having poles at $\lambda=\pm j$, $\lambda=\pm j2$, with residues equal to $+1$. Do this by first finding networks for $Z(s)$ and $z(\lambda)$, and then constructing the network for the impedance $F(\lambda)$ without further computation or synthesis.

14. A given p.r. impedance has poles at $s=\pm j2$ and at $s=-2$. It has one pair of conjugate complex zeros with a real part equal to $-\frac{1}{2}$ as shown in the

s-plane sketch below. $Z(0) = \frac{1}{4}$. Find the analytic expression for this impedance and its network realization.

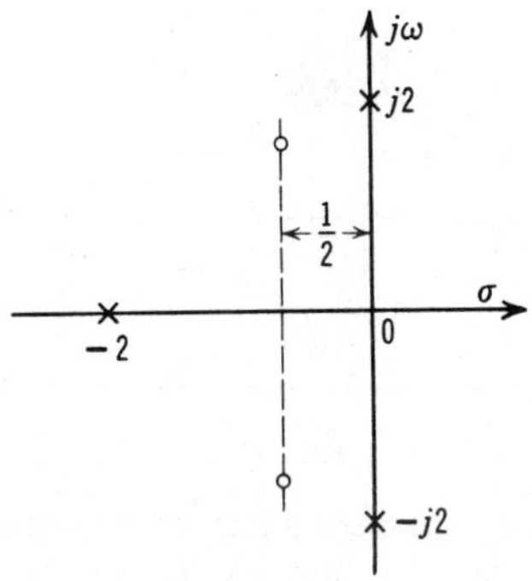

Prob. 14.

15. A given impedance has poles only on the negative real axis at the points $s = -1, -2, -3, -4$. At $s = -1$ and at $s = -3$ the residues are equal to $+1$; at $s = -2$ and at $s = -4$ they equal -1. This impedance may presumably be realized as a series combination of an RC and an RL network. Tell whether such a possibility exists if.

(a) $Z(\infty) = 1$
(b) $Z(\infty) = \frac{1}{2}$

and in the event that it does, find a network realization of this form.

16. A given impedance has poles at $s = -1$ and at $s = -1 \pm j1$. The residues in these poles are positive real and equal. If $Z(\infty) = 0$:

(a) Find the locations of the zeros.
(b) Find the common value of the residues.
(c) Show that the impedance must be p.r.
(d) Find a network realization.

17. Find the largest multiplier K for which $Z_2 = 1 - KZ_1$ is realizable and determine a ladder network for Z_2. See figure. Combine this network with an appropriate modification of the given network so as to obtain a resultant network having a constant resistance input impedance equal to 1 ohm.

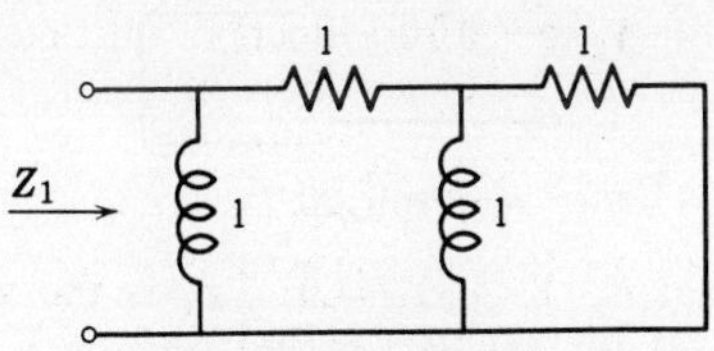

Prob. 17.

18. The RC network shown on the following page has the input admittance

$$y(s) = \frac{(s+1)(s+3)}{(s+2)}$$

(a) If it is specified that $R_2C_2 = \frac{1}{4}$, find values of all the elements.

(*b*) Write down by inspection the open-circuit transfer impedance $z_{12}(s)$ for this network including the value of its constant multiplier.

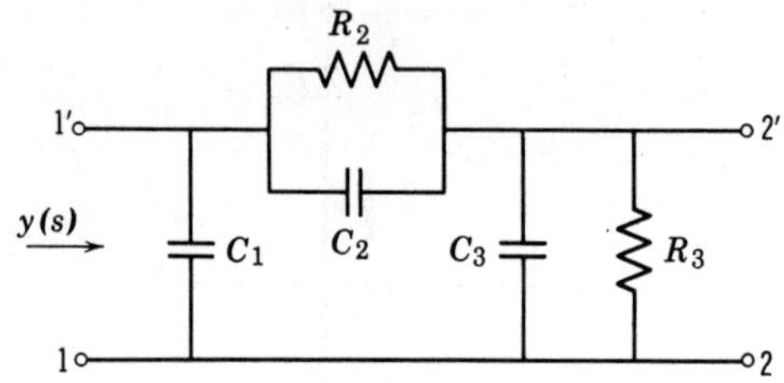

Prob. 18.

19. A given impedance function $Z(s)$ has three simple poles and three simple zeros distributed in the s-plane as shown in the sketch. For $s \to \infty$, $Z(s) \to 1$. If p.r., find a network realization for this impedance.

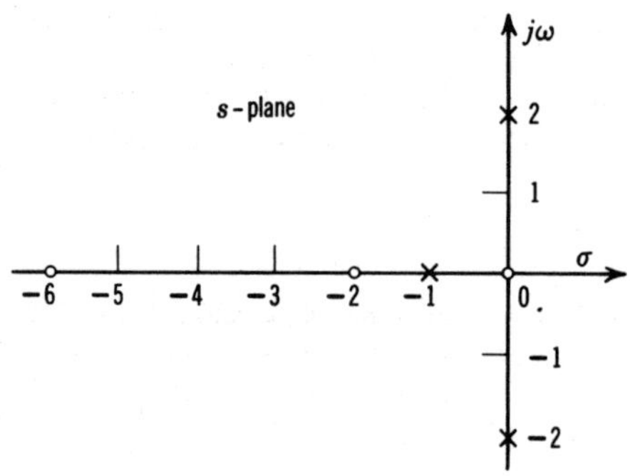

Prob. 19.

20. The sketch below shows two networks whose driving-point admittances

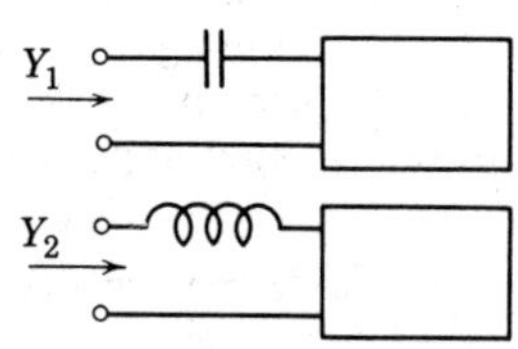

Prob. 20.

are complementary; that is, $Y_1 + Y_2 = 1$. Y_1 is the admittance of an RC network, having poles at the points $s = -1, -2, -3$. Its residues have a common value. Construct both admittances and synthesize a network for Y_2.

21. Synthesize

$$Z = \frac{m}{(s+1)(s+2)(s+3)(s+4)}$$

where m is an even polynomial (which you may choose to suit your convenience) and $Z(0) = 1$.

22. Find a network of the form indicated in the accompanying sketch having the impedance

$$Z = \frac{s^2 + 13s + 32}{3s^2 + 27s + 44}$$

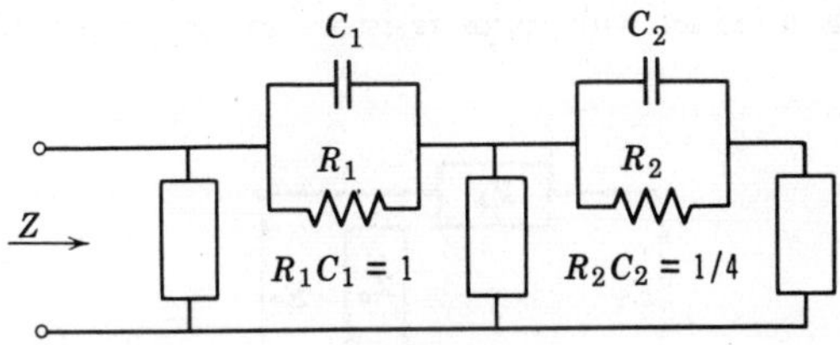

Prob. 22.

23. It would be convenient to be able to realize an RLC driving-point impedance as a RC network in series with an RL network. Show that if $Z(s)$ is a p.r. impedance having all simple poles on the negative real axis and zeros anywhere in the left half-plane, $Z(s) + R$ can always be realized as an RC impedance in series with an RL impedance for some real positive values of R. Under some conditions R can be made equal to zero.

Find the minimum value of R for which each of the following functions can be realized in the manner described above.

(a) $$Z(s) = \frac{s^2 + 2}{(s+1)(s+2)}$$

(b) $$Z(s) = \frac{(s+2)^4}{s(s+1)(s+3)(s+4)}$$

(c) $$Z(s) = \frac{(s+2)^2(s^2+s+1)}{s(s+1)(s+3)(s+4)}$$

24. (a) Show that a rational function having all simple poles on the negative real axis (and no poles at infinity) can always be realized as the transfer impedance of either an RC or an RL lattice (that is, both RC and RL lattices are always possible).

(b) Synthesize an RC lattice and an RL lattice, each of which possess the transfer impedance

$$Z_{12} = \frac{s^2 - 1}{(s+2)(s+3)}$$

(c) Show that any rational function, which is an even polynomial over an odd polynomial or an odd polynomial over an even polynomial, having all simple poles on the j-axis can always be realized as the transfer impedance of an LC lattice. Synthesize an LC lattice which possesses the transfer impedance $(s^2 - 1)/[s(s^2 + 1)]$.

25. A possible procedure for synthesis of voltage transfer functions E_2/E_1 is indicated in the following sketch. If $Z_a + Z_b = R$, then $E_2/E_1 = KZ_a$.

(a) What are the restrictions on the given voltage transfer function E_2/E_1 for which this procedure is possible?

(*b*) The following transfer function E_2/E_1 is to be synthesized

$$\frac{E_2}{E_1} = K\frac{(s + 3)(s + 7)}{(s + 1)(s + 5)}$$

(i) What is the largest possible K, using this method of synthesis?

(ii) Employing this value of K, synthesize the given E_2/E_1 using this procedure. Synthesize Z_a as a ladder with series resistors and Z_b as a ladder with shunt resistors.

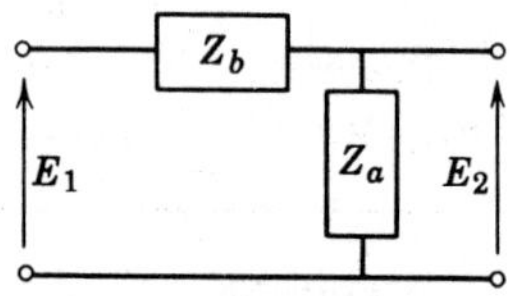

Prob. 25.

CHAPTER FIVE

More about Equivalent and Reciprocal Networks

1 Equivalent Networks by Linear Transformation of Variables

The subject of network equivalence to which we are introduced by the realization procedures discussed in the two preceding chapters is one of the most interesting and useful aspects of the network synthesis problem. It is also one of the most subtle and least well-understood aspects of this problem to date. Additional discussion of this topic by itself is useful not only because of the independent means which it provides for carrying out this important part of the solution to a synthesis problem but also for its value in orienting the student with regard to the present status of this collateral topic and in suggesting some methods for its further investigation.

A possible approach to the solution of the problem of network equivalence, and one that is fundamental in character as well as apparently powerful in its potential capabilities, makes use of the idea of subjecting the network variables to a linear transformation restricted only by the desired invariance of specified driving-point and transfer impedances.* For example, if only one driving-point impedance is to remain unchanged, and if the pertinent driving-point is located in loop 1 of the network, then a real nonsingular transformation of the loop currents, which is arbitrary except for the restriction that the loop current i_1 shall not change, should lead to networks involving a number of variations in structure and element values, while presenting the same input impedance at each one's single driving point.

Except in a very restricted sense, this apparently powerful approach has not led to fruitful results. In order to better appreciate the useful portion of this method and to understand the reason for its failure, so far at least, in realizing the potential accomplishments which its original

* This method was first suggested by W. Cauer in a paper entitled "Vierpole," *Elek. Nachr. Tech.*, Vol. 6, No. 7, 1929. It was later evaluated by N. Howitt in "Group Theory and the Electric Circuit," *Phys. Rev.*, Vol. 37, No. 12, 1931, pp. 1583–1595.

proponents had visualized as being achievable, a brief discussion of the pertinent mathematical background seems worthwhile.

The method may be developed equally well for the loop or node methods of characterization of the network in terms of sets of parameters and independent variables. We arbitrarily choose the loop basis in the immediately following discussion. The parameter matrices $[L]$, $[R]$, and $[S]$, together with the loop current variables $i_1 \cdots i_l$ and their indefinite integrals the loop charges $q_1 \cdots q_l$, determine the three quadratic forms

$$T = \tfrac{1}{2} \sum_{i,\,k=1}^{l} L_{ik} i_i i_k \tag{1}$$

$$F = \tfrac{1}{2} \sum_{i,\,k=1}^{l} R_{ik} i_i i_k \tag{2}$$

$$V = \tfrac{1}{2} \sum_{i,\,k=1}^{l} S_{ik} q_i q_k \tag{3}$$

representing the stored energy and loss functions for the network. If the variables are subjected to the linear transformation

$$\begin{aligned} i_1 &= a_{11}i'_1 + a_{12}i'_2 + \cdots + a_{1l}i'_l \\ i_2 &= a_{21}i'_1 + a_{22}i'_2 + \cdots + a_{2l}i'_l \\ &\cdots\cdots\cdots\cdots\cdots\cdots\cdots \\ i_l &= a_{l1}i'_1 + a_{l2}i'_2 + \cdots + a_{ll}i'_l \end{aligned} \tag{4}$$

with the matrix

$$[A] = \begin{bmatrix} a_{11} & \cdots & a_{1l} \\ \cdots & \cdots & \cdots \\ a_{l1} & \cdots & a_{ll} \end{bmatrix} \tag{5}$$

then the parameter matrices in the quadratic forms 1, 2, 3 become replaced by (see *MCA*, p. 135)

$$\begin{aligned} [L'] &= [A]_t \times [L] \times [A] \\ [R'] &= [A]_t \times [R] \times [A] \\ [S'] &= [A]_t \times [S] \times [A] \end{aligned} \tag{6}$$

The quadratic forms T, F, and V do not lose their positive definite character so long as matrix 5 is real and nonsingular. It will be recalled that it is this positive definite character that assures the p.r. character of the driving-point impedances associated with the network. We have also seen (for the two-element-kind networks at least) that the positive definite character of the associated quadratic forms assures the success of a normal co-ordinate transformation leading to the realization of equivalent canonic network configurations and hence this property of the forms T, F, and V may be regarded as the fundamental necessary and sufficient realizability criterion. Therefore, if the matrices $[L]$, $[R]$, and $[S]$ are derived from a given physical network, then the transformed version of this network to which matrices 6 pertain must be realizable, although the manner in which this realization is to be effected is not clear except in some special situations to be described presently.

If the given network has a single point of access located, let us say, in loop 1, then the invariance of the current in this loop (for a given excitation voltage located there) is assured by choosing the matrix $[A]$ such that

$$a_{11} = 1, \quad \text{and} \quad a_{1k} = 0 \text{ for } k \neq 1 \tag{7}$$

Although the remaining real elements in $[A]$ are arbitrary (except that this matrix must be nonsingular), the transformed network with matrices 6 will have the same driving-point impedance as the given network. If the latter has two points of access (say in loops 1 and 2) and we wish to have the driving-point and transfer impedances z_{11}, z_{22}, z_{12} remain invariant, then, in addition to conditions 7, we must specify

$$a_{22} = 1, \quad \text{and} \quad a_{2k} = 0 \text{ for } k \neq 2 \tag{8}$$

If we continue in this way and finally assume a point of access in every loop and demand that all driving-point and transfer impedances remain invariant, then $[A]$ reduces to the identity matrix and the transformation becomes trivial.

The reason for the restricted usefulness of this method is that there is no one-to-one correspondence between the parameter matrices and the associated network unless we restrict the choice of loops and loop-current variables in a specific way, and even then the correspondence is difficult to interpret in terms of graphs and element values unless this choice is a very simple one.

To simplify the process of interpretation it is helpful to consider the transformation matrix $[A]$ as a product of elementary transformation matrices (see *MCA*, p. 53) and correspondingly consider the transformation of the original network to be carried out in simple stages. The three elementary transformations of a matrix are recalled to involve:

(*a*) The interchange of a pair of rows or columns.

(*b*) The addition of the elements of one row (or column) to the respective ones of another row (or column).

(*c*) The multiplication of a row or column by a constant factor.

Because of the symmetrical property of the transformations indicated in 6, it is necessary always to carry out on corresponding columns the same transformation that is applied to chosen rows. The same transformation must, of course, be applied to all relevant parameter matrices; and only such transformations may be considered that do not violate the desired invariance conditions like 7 or 8.

In this connection it turns out that the transformation (*a*) is trivial, for it amounts merely to a renumbering of the meshes; (*b*) is either difficult to interpret physically or is trivial also; and all that is left, practically speaking, of this very general and elegant mathematical method for the production of equivalent networks is that fragment relating to results achievable by the elementary transformation (*c*). In spite of this rather disappointing result, let us not be hasty in condemning the method entirely, for the practical usefulness of this single nontrivial aspect easily justifies the effort needed in acquiring an understanding of its mathematical and physical significance.

Although the above discussion leads one to conclude that all equivalent networks derivable by these methods must have the same number of geometrically independent loops, or on an analogous node basis that they must have the same number of geometrically independent node pairs, there are special devices (to be illustrated presently) whereby the number of loops on a loop basis or the number of nodes on a node basis may be changed in spite of their apparent invariance under the pertinent transformation. These artifices together with the flexibility afforded by switching from a loop to a node basis, or vice versa, provide opportunity to accomplish a number of interesting and useful results with this method.

To illustrate these ideas, consider first the single 10-ohm resistance in Fig. 1*a* and its redrawn version as a two-mesh circuit in Fig. 1*b*. Although

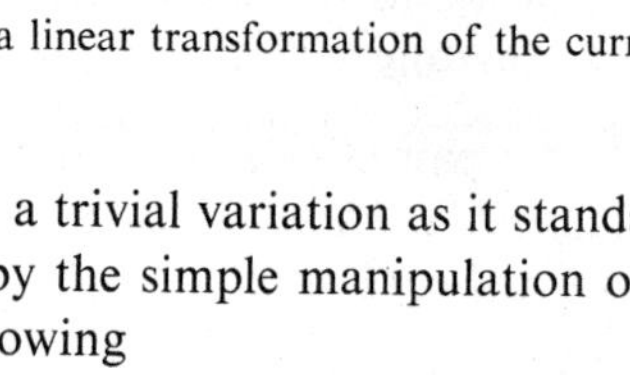

Fig. 1. Circuit to which a linear transformation of the current variables with matrix 10 is applied.

this redrawn version is a trivial variation as it stands, it may be converted into a nontrivial one by the simple manipulation of its resistance matrix as indicated in the following

$$[R] = \begin{bmatrix} 15 & 5 \\ 5 & 5 \end{bmatrix}, \qquad [R'] = \begin{bmatrix} 15 & 5 \\ 5 & 5 \end{bmatrix} \begin{matrix} \\ \leftarrow 2 \end{matrix} \longrightarrow \begin{bmatrix} 15 & 10 \\ 10 & 20 \end{bmatrix} \tag{9}$$
$$\qquad\qquad\qquad\qquad \uparrow 2$$

where we have multiplied the second row and column by the factor 2, an operation that amounts to the choice of a transformation matrix 5 given by

$$[A] = \begin{bmatrix} 1 & 0 \\ 0 & 2 \end{bmatrix} \tag{10}$$

and hence clearly meets the invariance condition 7.

The circuit of Fig. 2 is readily seen to correspond, on a simple mesh

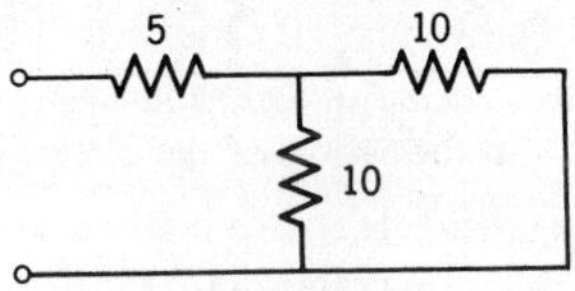

Fig. 2. Transformed version of the circuit of Fig. 1 having the same input impedance.

basis, to the resistance matrix $[R']$. Its input impedance is still 10 ohms. This process may be continued in the same manner. Thus in the three-mesh circuit of Fig. 3 we have introduced in mesh 2 the additional 5-ohm

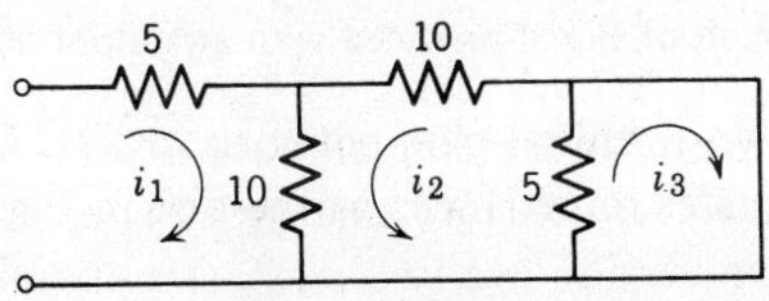

Fig. 3. Redrawn version of the circuit of Fig. 2 to which a further transformation is applied.

resistance by-passed by a short circuit. Here we carry out the manipulations indicated by

$$\begin{bmatrix} 15 & 10 & 0 \\ 10 & 25 & 5 \\ 0 & 5 & 5 \end{bmatrix} \begin{matrix} \\ \\ \leftarrow 2 \end{matrix} \longrightarrow \begin{bmatrix} 15 & 10 & 0 \\ 10 & 25 & 10 \\ 0 & 10 & 20 \end{bmatrix} \tag{11}$$
$$\uparrow 2$$

and obtain the network of Fig. 4 which is a nontrivial three-mesh circuit having the same 10-ohm input impedance.

If we introduce a second terminal pair in mesh 3 as shown in Fig. 5 and manipulate the matrix in the following manner

$$\frac{3}{2}\rightarrow \begin{bmatrix} 15 & 10 & 0 \\ 10 & 25 & 10 \\ 0 & 10 & 20 \end{bmatrix} \longrightarrow \begin{bmatrix} 15 & 15 & 0 \\ 15 & 56.25 & 15 \\ 0 & 15 & 20 \end{bmatrix} \tag{12}$$
$$\uparrow \frac{3}{2}$$

which corresponds to a transformation matrix

$$[A] = \begin{bmatrix} 1 & 0 & 0 \\ 0 & \frac{3}{2} & 0 \\ 0 & 0 & 1 \end{bmatrix} \tag{13}$$

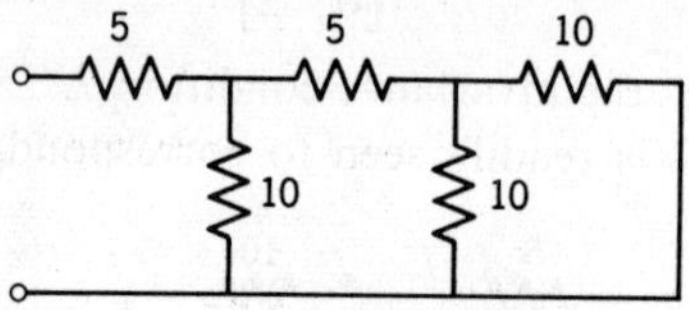

Fig. 4. Result of the further transformation of the circuit of Fig. 3 as indicated by matrix manipulation 11.

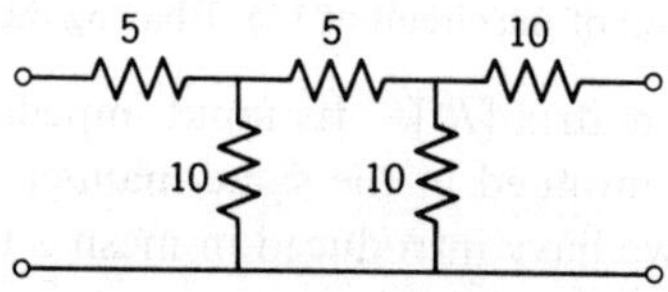

Fig. 5. The circuit of Fig. 4 provided with an output terminal-pair.

then we obtain the two terminal-pair network in Fig. 6 having the same driving-point and transfer impedances as the one in Fig. 5.

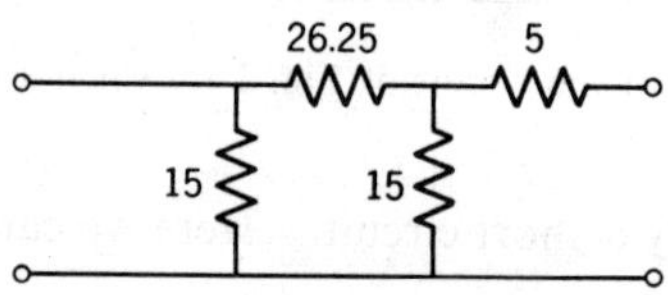

Fig. 6. The circuit of Fig. 5 with the impedance level of its central mesh raised by a factor $\frac{9}{4}$. Its driving point and transfer impedances are unchanged.

We can, of course, also use this method if we wish to change an input impedance by a desired constant factor. Thus, if we multiply the first row and column in matrix 11, representing the circuit of Fig. 4, by the factor 2 as indicated in

$$\begin{matrix} 2\rightarrow \\ \\ \\ \end{matrix}\begin{bmatrix} 15 & 10 & 0 \\ 10 & 25 & 10 \\ 0 & 10 & 20 \end{bmatrix} \longrightarrow \begin{bmatrix} 60 & 20 & 0 \\ 20 & 25 & 10 \\ 0 & 10 & 20 \end{bmatrix} \tag{14}$$
$$\uparrow 2$$

then we get the circuit shown in Fig. 7 for which the input impedance is larger by the square of this factor (yielding $Z_{\text{in}} = 40$ ohms) because the corresponding transformation 4 makes $i_1 = 2i'_1$, $i_2 = i'_2 \cdots i_l = i_l'$, while the quadratic form F, Eq. 2, for the total input power remains invariant.* Physically the effect of this transformation is the same as though we had placed an ideal transformer at the input terminals of the circuit of Fig. 4 as shown in Fig. 8 where the new input current $i_1' = i_1/2$ and the net power dissipation for a given i_1 is the same as in the circuit of Fig. 4. The voltage on the input side of the ideal transformer must be double in order to have the same i_1, and so the input impedance is multiplied by the square of the ideal transformer ratio, as is well known. We speak of the process of transformation from the circuit of Fig. 4 to that of Fig. 7 as "raising the impedance level" of mesh 1 by the factor 4.

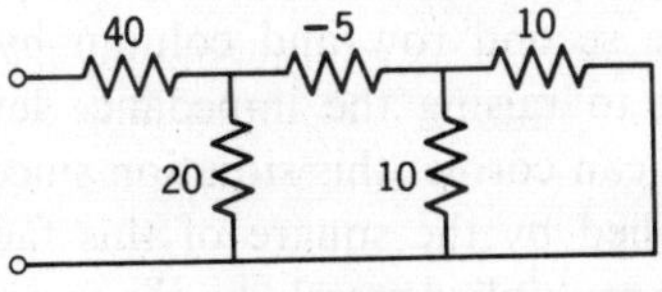

Fig. 7. The circuit of Fig. 4 with the impedance level of its input mesh raised by a factor 4 according to matrix manipulation 14.

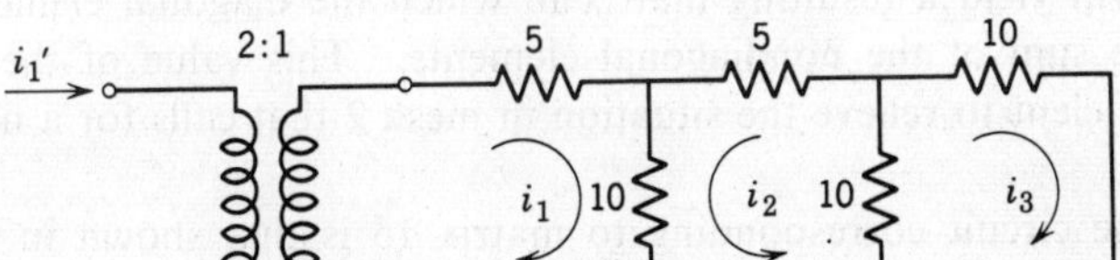

Fig. 8. The circuit of Fig. 4 with the impedance level of its input mesh raised by a factor 4 in a more conventional manner than that shown in Fig. 7.

A disturbing feature about the circuit of Fig. 7 is the appearance of a negative resistance in it. This fact, incidentally, does not mean that the circuit is no longer passive, for the loss function F is still positive definite since this property is invariant to the linear congruent transformation with a real nonsingular matrix $[A]$. The circuit of Fig. 7 is practically difficult to construct because of the −5-ohm resistor, and so we proceed to get rid of it by raising the impedance level of mesh 2 just sufficiently to relieve this situation, as is indicated in the following manipulation of

* For given values of $i_1 \cdots i_l$, the function F has a certain value. Substitution of transformation 4 into Eq. 2 for F merely expresses this same value in terms of a new set of currents $i_1' \cdots i_l'$, and Eqs. 4 enable us to compute the appropriate values of the new currents for which this condition obtains.

matrix 14

$$\frac{6}{5}\rightarrow\begin{bmatrix}60 & 20 & 0\\ 20 & 25 & 10\\ 0 & 10 & 20\end{bmatrix} \longrightarrow \begin{bmatrix}60 & 24 & 0\\ 24 & 36 & 12\\ 0 & 12 & 20\end{bmatrix} \qquad (15)$$
$$\uparrow \frac{6}{5}$$

In order to understand how the factor $\frac{6}{5}$ is arrived at, observe first of all that the reason for the appearance of the -5-ohm resistance in the circuit of Fig. 7 is that the diagonal elements in matrix 14 (which represent total loop resistance values) are not consistently larger than the sums of nondiagonal elements in the same row or column. In the second row the diagonal element is 25 and the sum of the nondiagonal elements is 30. Multiplication of the second row and column by a factor larger than unity (corresponding to raising the impedance level of mesh 2 by the square of this factor) can correct this situation since the diagonal element thus becomes multiplied by the square of this factor whereas the nondiagonal elements are multiplied merely by the factor itself. It is a matter of simple algebra to see that a factor equal to the sum of the nondiagonal elements divided by the diagonal element [in the above example $(20 + 10)/25 = \frac{6}{5}$] will yield a resulting matrix in which the diagonal element just equals the sum of the nondiagonal elements. This value of the factor is just sufficient to relieve the situation in mesh 2 that calls for a negative resistance.

Thus the circuit corresponding to matrix 15 is that shown in Fig. 9,

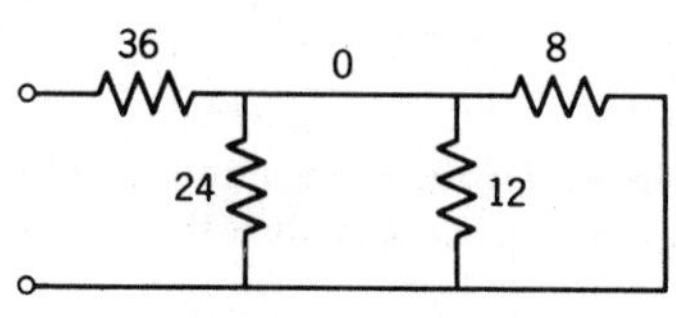

Fig. 9. The circuit of Fig. 7 with the impedance level of its central mesh raised by a factor just sufficient to relieve the appearance of a negative resistance. The pertinent matrix manipulation is 15.

which could be contracted into a two-mesh circuit by combining the 24- and 12-ohm branches in parallel. Observe that if the factor in the matrix operation 15 is chosen larger than 2, then a negative resistance will appear in mesh 3. The factor $\frac{6}{5}$ needed to relieve the negativeness in mesh 2 is fortunately not large enough to cause trouble in mesh 3. We may say that mesh 3 has enough positive resistance to absorb the negativeness which operation 15 shifts from mesh 2 to mesh 3.

The following question may be raised at this point: Is it always possible to get completely rid of negative resistance trouble by this process of "passing it along," so to speak? The answer is definitely yes, for the most unfavorable circumstances cannot call for more than the necessity of having to multiply *all* rows and columns in the given matrix by the *same* factor, since this operation leads to the familiar result of getting a circuit with a times as much impedance by increasing all resistance values by a factor a. In any less unfavorable situation, only the resistances near the driving point are affected, and the effect disappears gradually as we proceed into the network. Certainly, the process of absorbing negative element values by passing them on to the surrounding meshes of a network, thereby leaving the input impedance invariant, is a result of the linear transformation technique that alone makes this process of obtaining equivalent networks worth the effort it takes for its consideration.

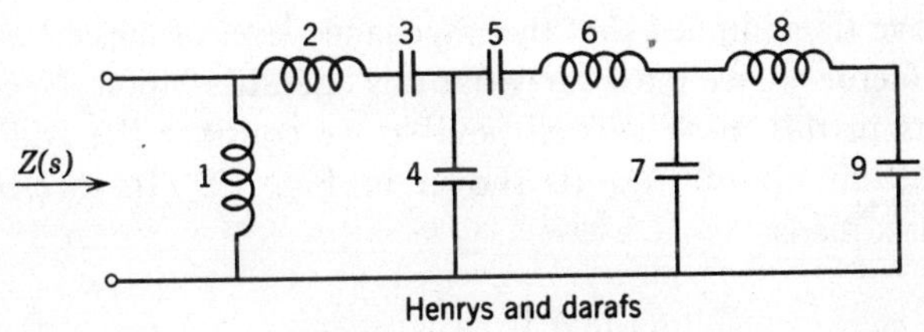

Fig. 10. Network whose input impedance is to be multiplied by the factor 25.

In order to show how it may best be used in somewhat more elaborate situations, let us consider the *LC* circuit of Fig. 10 and require that we modify it so as to produce an input impedance 25 times as large. We could, of course, write down the complete loop inductance and elastance matrices and operate in an identical manner upon their rows and columns, as required by transformations 6. In a ladder network like the one in Fig. 10, however, it is not necessary to proceed in such a cumbersome fashion. It is much easier to work with a section of this circuit at a time, choosing each section so that only one kind of circuit element is involved. For example, if we split off the first two inductance branches as shown in Fig. 11, then we have initially only to consider the following

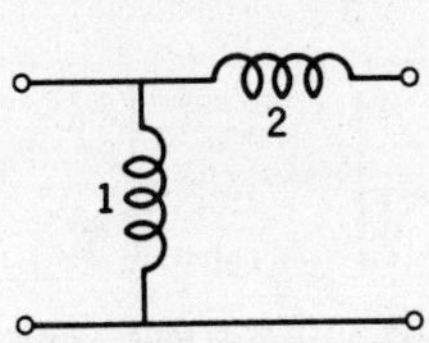

Fig. 11. Input portion of the circuit of Fig. 10 to which the transformation indicated in manipulation 16 is applied.

transformation of a simple two-by-two inductance matrix

$$[L] = \begin{bmatrix} 1 & 1 \\ 1 & 3 \end{bmatrix}, \qquad [L'] = \begin{matrix} \overset{5}{\rightarrow} \\ \\ \end{matrix}\!\!\underset{\uparrow 5\quad}{\begin{bmatrix} 1 & 1 \\ 1 & 3 \end{bmatrix}} \longrightarrow \begin{bmatrix} 25 & 5 \\ 5 & 3 \end{bmatrix} \tag{16}$$

From the form of the resulting matrix we see that we will have some negative inductance in mesh 2 unless we raise its impedance level also. The operation that just relieves the difficulty is the following

$$\underset{\quad\uparrow \frac{5}{3}}{\begin{bmatrix} 25 & 5 \\ 5 & 3 \end{bmatrix}}\!\!\begin{matrix} \\ \leftarrow \frac{5}{3} \end{matrix} \longrightarrow \begin{bmatrix} 25 & \frac{25}{3} \\ \frac{25}{3} & \frac{25}{3} \end{bmatrix} \tag{17}$$

Since we have thus implied that the impedance level of mesh 2 is to be raised by the factor $\frac{25}{9}$, we must carry out this operation upon the capacitance elements in this mesh as well, and so we consider the additional portion of the circuit of Fig. 10 shown in Fig. 12. Its two-by-two elastance matrix reads

$$[S] = \begin{bmatrix} 7 & 4 \\ 4 & 9 \end{bmatrix} \tag{18}$$

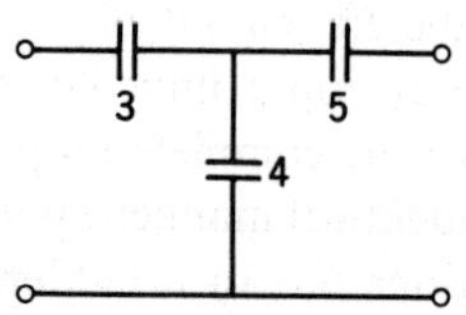

Fig. 12. Following capacitive portion of the circuit of Fig. 10 to which some additional impedance leveling must be applied to meet the changes called for in the portion shown in Fig. 11.

Here row and column number one are pertinent to mesh 2 of the original network. Hence the transformation consistent with 17 is

$$\begin{matrix} \frac{5}{3}\rightarrow \\ \\ \end{matrix}\!\!\overset{\downarrow \frac{5}{3}\quad}{\begin{bmatrix} 7 & 4 \\ 4 & 9 \end{bmatrix}} \longrightarrow \begin{bmatrix} \frac{175}{9} & \frac{20}{3} \\ \frac{20}{3} & 9 \end{bmatrix} \tag{19}$$

No further negative element difficulty arises, and so the negative inductance trouble resulting from the original transformation 16 is taken care of. It is interesting to observe that here we have effectively absorbed some negative inductance by some positive capacitance (actually it is more correct to say that the negative inductance is absorbed by an appropriate impedance level increase in mesh 2, and that this increase can be tolerated by the capacitive aspects of that mesh).

The inductive circuit pertinent to matrix 17 cascaded with the capacitive circuit pertinent to matrix 19, with the latter followed by the unaltered remainder of the original circuit in Fig. 10, as shown in Fig. 13, constitutes

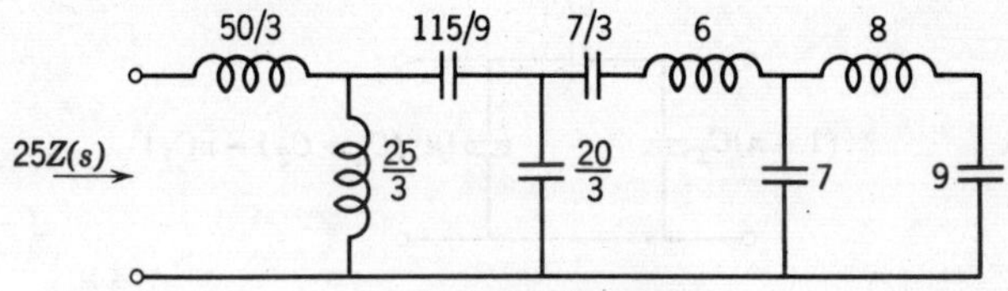

Fig. 13. Result of multiplying the input impedance of the circuit of Fig. 10 by a factor 25.

the solution to the given problem. It is interesting to make the observation that transformation 16 alone accomplishes the desired end so far as multiplication of the input impedance by the factor 25 is concerned. The subsequent manipulations, which we refer to as *internal* impedance level transformations, leave the driving-point impedance invariant.

Fundamental to such internal impedance level transformations in a ladder network is the transformation of the simple "L" circuit (actually a reoriented L) shown in Fig. 14 in terms of capacitances, although the

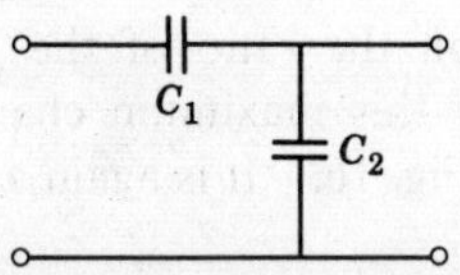

Fig. 14. Pertinent to the transformation of a capacitive L circuit.

basic principle has nothing to do with the kind of circuit element involved so long as both branches of the L involve the same kind of element. For the sake of variety we will treat this circuit on the node basis. Considering the upper left- and right-hand nodes as nodes 1 and 2 with respect to the base line as a datum, the capacitance matrix becomes

$$[C] = \begin{bmatrix} C_1 & -C_1 \\ -C_1 & (C_1 + C_2) \end{bmatrix} \tag{20}$$

Now let us multiply the admittance level of node 2 by a factor n^2, giving the transformed capacitance matrix

$$[C'] = \begin{bmatrix} C_1 & -C_1 \\ -C_1 & (C_1 + C_2) \end{bmatrix} \begin{matrix} \\ \leftarrow n \end{matrix} \longrightarrow \begin{bmatrix} C_1 & -nC_1 \\ -nC_1 & n^2(C_1 + C_2) \end{bmatrix} \quad (21)$$
$$\uparrow n$$

to which the Π circuit of Fig. 15 corresponds. In order that all element

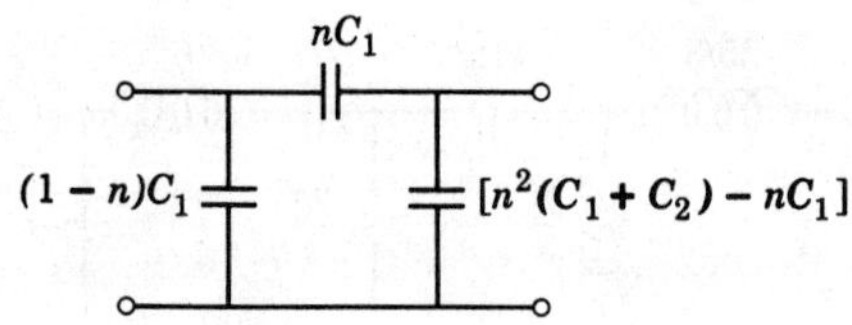

Fig. 15. The circuit of Fig. 14 with its output admittance level multiplied by n^2.

values remain positive, the factor n is restricted to a range of values defined by

$$\frac{C_1}{C_1 + C_2} \le n \le 1 \quad (22)$$

from which it is clear that, for the orientation in Fig. 14, the admittance level can only be lowered (the impedance level raised) on the right-hand side. The maximum amount of admittance level change is obtained for

$$n = \frac{C_1}{C_1 + C_2} = \frac{1}{1 + (C_2/C_1)} \quad (23)$$

which depends solely upon the ratio of the given capacitance values. The resulting network for this maximum change in admittance or impedance level is shown in Fig. 16. It is again an L circuit, but oppositely

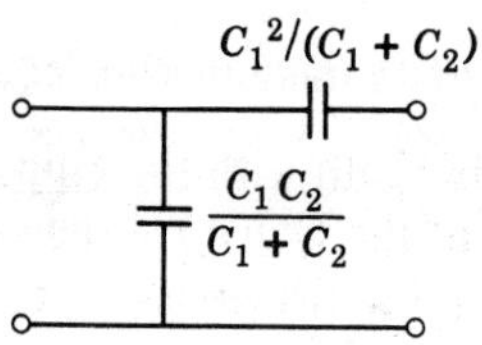

Fig. 16. The completely transformed L circuit of Fig. 14 obtained by choosing n in the circuit of Fig. 15 equal to its limiting value 23.

oriented to the one in Fig. 14. We speak of it as the completely transformed version of the given L circuit, the intermediate versions being represented by the circuit of Fig. 15 for an n value within the range 22.

A practical application of these results is found in the design of band-pass filters. Thus a common type of such a filter has the schematic shown in Fig. 17*a* in which $L_1C_1 = L_2C_2$, but in many cases the ratio $L_1/L_2 = C_2/C_1$ turns out to be extremely large, 10^4 or more. Such a spread of element values is difficult to accommodate, especially in the inductances. By scaling the impedance level of the entire network we can only change both inductance values by the same factor, so that one is either excessively large or excessively small when the other is reasonable in size.

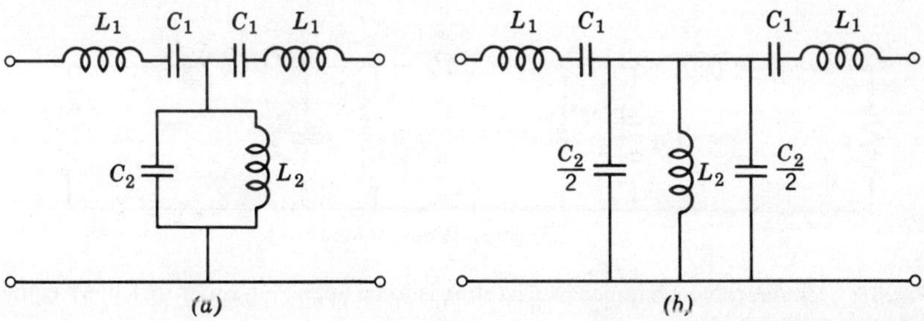

Fig. 17. A familiar band-pass filter schematic for which internal impedance leveling may be applied to obtain more nearly equal parameter values.

We can alleviate this situation by making an internal impedance level transformation. To this end we first redraw the circuit as shown in Fig. 17*b*. The shunt inductance is now flanked on the left and on the right by L circuits in symmetrical orientations. In terms of the discussion pertinent to the circuits of Figs. 14, 15, and 16, we see that the impedance environment of the inductance L_2 can be raised by a factor that may be chosen larger than adequate to make L_2 and L_1 alike in value. Some capacitance value ratios will still be fairly substantial, but these can be accommodated more easily in a practical circuit.

There are several additional ways in which the circuit of Fig. 17*a* may be modified to render it more reasonable for practical construction, one of these making use of mutual inductive coupling in which the looseness of coupling reflects the relative smallness of the shunt inductance. These variations of the present theme we shall leave as exercises for the reader, and turn our attention to a more elaborate impedance leveling procedure for which the circuit in Fig. 18 forms the starting point. The problem is so to manipulate the reactive portion of this network as to make all inductance values equal. Since we can subsequently scale all inductances to any desired uniform value, we shall initially attempt to make them all equal to unity.

The first step toward achieving the desired end is to redraw the network as shown in Fig. 19 where each inductance is flanked by capacitive L circuits and the amount by which each one's impedance environment is to be changed, either up or down, is indicated by an arrow and a numerical factor. The literal capacitance values must fulfill the conditions

$$C_1 + C_2 = 10 \tag{24}$$

$$\frac{1}{C_3} + \frac{1}{C_4} = 5 \tag{25}$$

$$C_5 + C_6 = 3 \tag{26}$$

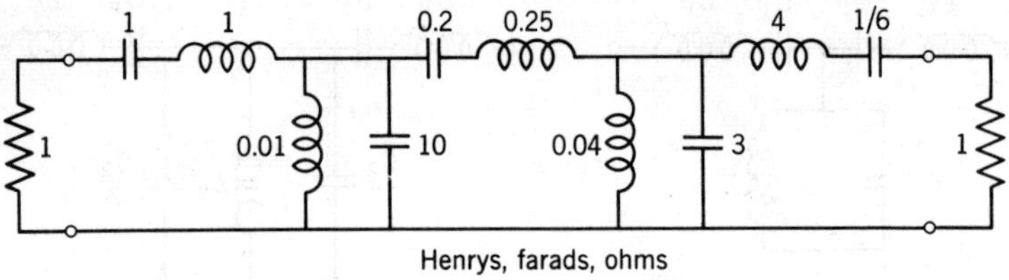

Fig. 18. Circuit to which impedance leveling is to be applied in order to achieve equal values for the inductances.

The next objective is to determine values for the capacitances $C_1 \cdots C_6$ that will satisfy Eqs. 24, 25, and 26 on the one hand and the contemplated impedance level changes on the other. This process is, of course, not unique because the transformations of the capacitive L circuits can yield an infinity of Π circuits for a given level change depending upon the

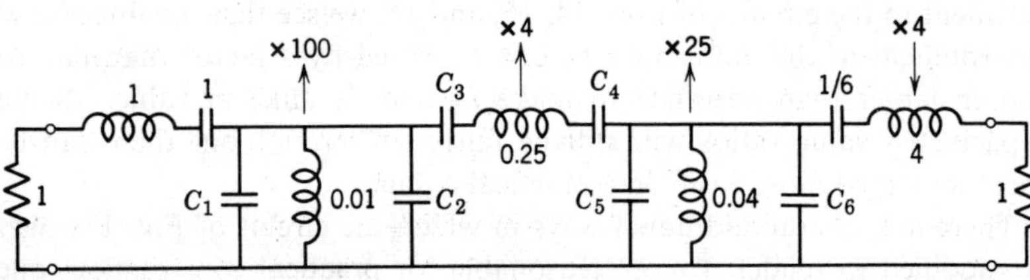

Fig. 19. The circuit of Fig. 18 redrawn to accommodate the contemplated impedance leveling process.

distribution of capacitance values between C_1 and C_2, C_3 and C_4, C_5 and C_6. It will become unique, however, if we demand that each L shall become completely transformed into its opposite orientation, a condition that may not be consistently fulfillable in view of the given element values but in any event can be adhered to so far as possible.

Accordingly we begin with the first L circuit on the left and apply the desired transformation to its capacitance matrix, as follows

$$\begin{bmatrix} 1 & -1 \\ -1 & 1+C_1 \end{bmatrix} \leftarrow \tfrac{1}{10} \quad \uparrow \tfrac{1}{10} \longrightarrow \begin{bmatrix} 1 & -\dfrac{1}{10} \\ -\dfrac{1}{10} & \dfrac{1+C_1}{100} \end{bmatrix} \tag{27}$$

If this resultant matrix is to represent a completely transformed L, then we must have

$$\frac{1}{10} = \frac{1+C_1}{100} \quad \text{or} \quad C_1 = 9 \tag{28}$$

which can be met by Eq. 24 with

$$C_2 = 1 \tag{29}$$

Next we write down the capacitance matrix for the adjacent L circuit consisting of the capacitances C_2 and C_3 and note that this one must be transformed so as to accommodate an admittance factor of $\frac{1}{100}$ on the left and an admittance factor of $\frac{1}{4}$ on the right. The corresponding matrix operations are

$$\tfrac{1}{10} \rightarrow \overset{\tfrac{1}{10}\downarrow}{\begin{bmatrix} 1+C_3 & -C_3 \\ -C_3 & C_3 \end{bmatrix}} \leftarrow \tfrac{1}{2} \quad \uparrow \tfrac{1}{2} \longrightarrow \begin{bmatrix} \dfrac{1+C_3}{100} & -\dfrac{C_3}{20} \\ -\dfrac{C_3}{20} & \dfrac{C_3}{4} \end{bmatrix} \tag{30}$$

Here the condition for complete conversion reads

$$\frac{1+C_3}{100} = \frac{C_3}{20} \quad \text{or} \quad C_3 = \tfrac{1}{4} \tag{31}$$

whence Eq. 25 yields

$$C_4 = 1 \tag{32}$$

so that again the complete conversion is possible.

The L circuit involving C_4 and C_5 must have its admittance level multiplied by $\frac{1}{4}$ on the left and by $\frac{1}{25}$ on the right. The pertinent capacitance matrix is, therefore, transformed as follows

$$\begin{array}{ccccc} & \tfrac{1}{2} & & & \\ & \downarrow & & & \\ \tfrac{1}{2}\rightarrow & \left[\begin{matrix} 1 \\ \\ -1 \end{matrix}\right. & \left.\begin{matrix} -1 \\ \\ 1+C_5 \end{matrix}\right] & \leftarrow \tfrac{1}{5} & \\ & & \uparrow & & \\ & & \tfrac{1}{5} & & \end{array} \longrightarrow \begin{bmatrix} \dfrac{1}{4} & \dfrac{-1}{10} \\ \\ \dfrac{-1}{10} & \dfrac{1+C_5}{25} \end{bmatrix} \tag{33}$$

and the condition for complete conversion requires that

$$\frac{1}{10} = \frac{1+C_5}{25} \quad \text{or} \quad C_5 = \tfrac{3}{2} \tag{34}$$

from which Eq. 26 gives

$$C_6 = \tfrac{3}{2} \tag{35}$$

Finally, the last L circuit calls for transformation according to the following manipulation of its capacitance matrix

$$\begin{array}{ccccc} & \tfrac{1}{5} & & & \\ & \downarrow & & & \\ \tfrac{1}{5}\rightarrow & \left[\begin{matrix} \tfrac{5}{3} \\ \\ -\tfrac{1}{6} \end{matrix}\right. & \left.\begin{matrix} -\tfrac{1}{6} \\ \\ \tfrac{1}{6} \end{matrix}\right] & \leftarrow 2 & \\ & & \uparrow & & \\ & & 2 & & \end{array} \longrightarrow \begin{bmatrix} \tfrac{1}{15} & \tfrac{-1}{15} \\ \\ \tfrac{-1}{15} & \tfrac{2}{3} \end{bmatrix} \tag{36}$$

which evidently yields a completely transformed section. Matrices 27, 30, and 33, with the values of C_1, C_3, and C_5 substituted from Eqs. 28, 31, and 34 respectively, are

$$\begin{bmatrix} 1 & \tfrac{-1}{10} \\ \\ \tfrac{-1}{10} & \tfrac{1}{10} \end{bmatrix} \quad \begin{bmatrix} \tfrac{1}{80} & \tfrac{-1}{80} \\ \\ \tfrac{-1}{80} & \tfrac{1}{16} \end{bmatrix} \quad \begin{bmatrix} \tfrac{1}{4} & \tfrac{-1}{10} \\ \\ \tfrac{-1}{10} & \tfrac{1}{10} \end{bmatrix} \tag{37}$$

These together, with matrix 36 fix all of the resultant capacitance values in the completely transformed network which is shown in Fig. 20. Consistent with dropping the impedance level of the output mesh by the factor 4, the load resistance becomes $\tfrac{1}{4}$ ohm. Compared with the given network in Fig. 18, the transformed one has the same input impedance z_{11}, an impedance z_{22} one-fourth as large, and a transfer impedance z_{12} half as large. The frequency dependence of these functions is unchanged by the transformations.

It is of some interest to note that the resultant network involves three more capacitances than the original one. From a practical point of view, one is quite willing to pay this price for the privilege of being able to readjust certain element values as is done in the present example.

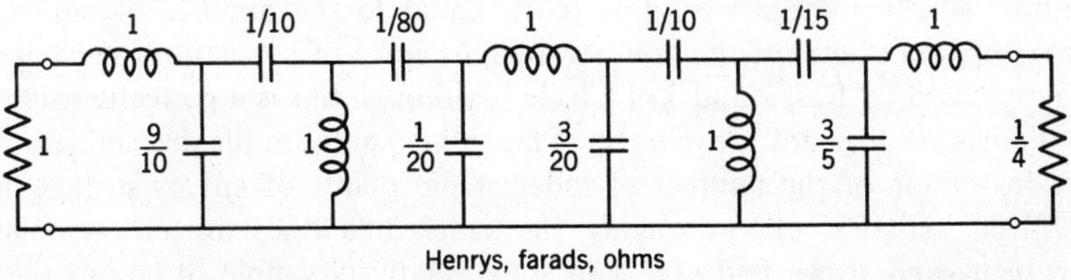

Fig. 20. The circuit of Fig. 18 after impedance leveling has rendered its inductance values equal.

Although this example was constructed so as to make complete conversion of each transformed L section possible, it would not be fair to say that similar impedance leveling procedures would seldom be possible in random practical situations. To begin with, practical situations are never random. Designing networks is not like tossing dice; there is a lot of order involved in the process as well as opportunity to influence the character of even a tentative result. It is, of course, not always possible to make things come out so nicely as in the above example, but even if some of the transformed L sections remain Π sections, the price is only a few more capacitances. In a rather unfavorable situation we might not be able to make all inductance values equal, but the method is still very useful if it enables one to reduce considerably an initial excessive spread in these element values.

2 A Method of Determining Potentially Equivalent Networks

Since the open- and short-circuit natural frequencies of a network for a given driving point are the poles and zeros of the pertinent impedance function, the problem of establishing the potential equivalence of different network configurations is essentially the problem of being able to determine by inspection the number of natural frequencies of a network from its geometrical structure and element-kind distribution alone. This latter problem, namely, that of being able to predict the degree of the characteristic equation by inspection of a given network, is one that has been of interest in connection with the analysis aspects of network theory for a long time. Its consideration is therefore of collateral as well as of direct interest to us in our present studies.

An early approach to the problem of establishing the degree of the characteristic equation by inspection was based upon the principle that the value of this integer must equal the number of independent points of energy storage (inductances and capacitances) within the network because the latter must equal the number of independently specifiable initial conditions capable of determining an equal number of integration constants, which number we know to be equal to the number of natural frequencies because of the familiar analytic form of the natural behavior as expressed by Eqs. 2 and 3 of Ch. 1. Although this is a perfectly sound argument, it does not prove to be of much help because the determination by inspection of the number of independent points of energy storage in a given network offers roughly the same difficulties as our original problem, and so we find except in some relatively simple situations that we have merely traded one set of difficulties for another similar set.

A more effective approach makes use of the concept of normal co-ordinates. Each normal co-ordinate places in evidence one natural frequency, or a pair of conjugate natural frequencies when these are complex, and so our problem may alternately be regarded as that of finding a method of predicting the number of normal co-ordinates associated with a given dynamic system. Through this approach, one important aspect of our problem becomes immediately apparent, namely, that there cannot be a clear-cut solution when all three kinds of network elements are involved, because normal co-ordinates in their usual sense are not definable for such a system since it is in general not possible simultaneously to reduce three quadratic forms (the forms T, F, and V) to sums of squares (see *MCA*, p. 164). All of the following discussion relative to this problem is therefore pertinent only to networks involving two kinds of elements.

The way to determine the number of normal co-ordinates of a dynamic system is to establish its degrees of freedom, or the number of dynamically independent variables needed to describe completely the state of the system at any moment. The important word in this last sentence is the adjective "dynamically." In deriving equilibrium equations on loop and node bases, we have talked about numbers of independent variables from a different point of view than the one that we must now adopt; and it is important also that we clearly understand the meaning of the term "degrees of freedom" which is often erroneously and loosely used as an alternative for the number of variables involved in expressing the dynamic equilibrium of a system, like the number of loop currents on a loop basis or the number of node-pair voltages on a node basis. Since these numbers are in general not alike, the error in identifying either with the number of degrees of freedom should be evident since the latter number must be fixed by the inherent physical character of the dynamic system and therefore cannot depend upon the peculiarities of some special algebraic procedure for establishing a set of equilibrium equations.

To state the same idea in a different way, we may say that the number of dynamically independent variables among the loop currents or among the node-pair voltages must be the same, and this number can at most be equal to the smaller of the two numbers l and n which are respectively the number of geometrically independent loops and node pairs associated with the network graph. Thus, in discussing the question of independent variables in connection with loop and node methods of setting up equilibrium equations, we were concerned only with obtaining as many independent equations as the number of variables in terms of which they are expressed. Consideration of this question involves only the geometry of the network graph, and all matters pertaining to the independence of the variables involved are relevant only with regard to their topological or geometrical independence, which has no relation to the kinds of circuit elements involved and hence has no relation to questions of dynamic independence.

To make the distinction between geometrical and dynamic independence strikingly clear, we need merely call attention to the fact that while a purely resistive network may involve l geometrically independent loop-current variables, only one of these has any independence in a dynamic sense. The whole system has only one degree of freedom, for if at a given instant the value of one current is fixed then the values of all the others are rigidly determined. There is only one independent adjustment that can be made amongst all the currents; fixing that one fixes all others at any given instant. From a dynamic point of view we regard the l independent equations relating the loop currents as a set of *constraints*.

Equations of constraint are ordinary algebraic equations relating several variables, as distinct from a set of *differential* equations which involve derivative terms of several different orders. As soon as derivatives of various orders are involved in a set of equations, then it is no longer possible to say that fixing the value of one of the variables at a given instant fixes the values of all the others. It takes more than the mere value of one variable to determine the values of the others at a given moment; one must additionally know the values of one or more successive derivatives of this variable at the same instant, or one must know the values of not just one but of several variables in order to be able to ascertain the values of all of them. The necessary number of variables that one must know, or the necessary number of pieces of relevant information that one must have whether it be about one variable or about several of them, equals the number of dynamically independent variables or the number of degrees of freedom of the system.

Constraint equations are like clamps; they take away a certain amount of freedom from a given system—freedom in a dynamic sense. In a purely resistive circuit, all the equilibrium equations are clamps. The whole system is rigidly clamped together. That is why a knowledge of one variable suffices to find all the rest. Except for certain canonic forms, dynamic systems ordinarily have incorporated in them a number of clamps or "freedom snatchers" expressible in terms of linear algebraic constraint equations amongst the variables. The number of dynamically independent variables in such a situation is equal to the total number of variables in terms of which the equilibrium is uniquely expressed, diminished by the number of independent constraint equations existing among these variables. We will now show how this number of constraint equations can always be found in a straightforward and unambiguous manner.

The first step in this direction is to determine the kind of physical situation that gives rise to a constraint relation. On a loop basis it is rather easy to see that a constraint relation exists whenever it is possible to find embedded within the network a closed path on which only one kind of circuit element is encountered. Figure 21 illustrates this situation

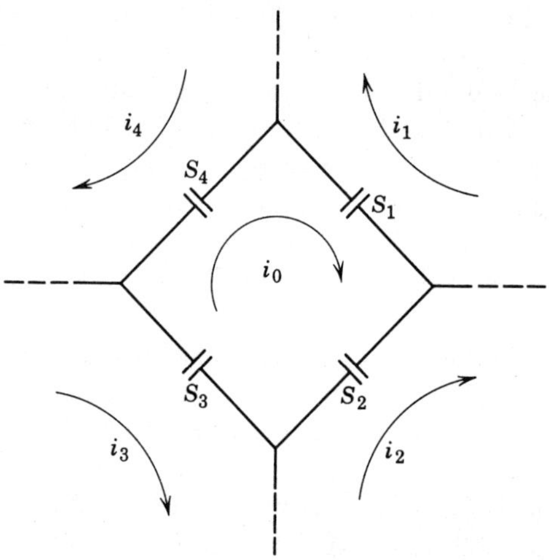

Fig. 21. A capacitance mesh, giving rise to the constraint relation 39.

for a simple mesh, the kind of element being elastance (or capacitance). A Kirchhoff voltage-law equation written for this mesh reads

$$\int [S_1(i_1 - i_0) + S_2(i_2 - i_0) + S_3(i_3 - i_0) + S_4(i_4 - i_0)]dt = 0 \quad (38)$$

in which $(i_1 - i_0)$, $(i_2 - i_0)$, $\cdot\cdot\cdot$ are the currents in the separate capacitances. Differentiation and rearrangement of terms yields

$$S_1i_1 + S_2i_2 + S_3i_3 + S_4i_4 = (S_1 + S_2 + S_3 + S_4)i_0 \quad (39)$$

which clearly is a constraint relation involving the current i_0 in the so-called capacitance loop and the adjacent loop currents.

It is obvious that one obtains the same kind of constraint relation when the capacitance loop is not a simple mesh but some more devious closed path. It is also clear that the same kind of a situation obtains when inductance is the only kind of circuit element encountered, and that the presence or absence of mutual inductive coupling is immaterial. Thus suppose branches 1, 2, · · ·, τ are any subset of branches among the inductive branches 1, 2, · · ·, λ which collectively form some devious closed path. For the voltage drops in these branches we may write the voltage-law equation

$$v_1 + v_2 + \cdots + v_\tau = 0 \tag{40}$$

Now

$$v_i = \sum_{k=1}^{\lambda} l_{ik} \frac{dj_k}{dt} \tag{41}$$

where $j_1 \cdots j_\lambda$ are currents in all of the inductive branches and the l_{ik} are the pertinent self- and mutual inductance coefficients. These branch currents are related to the chosen set of loop currents by linear equations of the form

$$j_i = \sum_{k=1}^{l} \beta_{ki} i_k \tag{42}$$

in which, for the ordinary garden variety of loop currents, the β's are ± 1 or zero. Substitution of Eq. 42 into 41 and the resulting v's into 40 yields an equation of the form

$$L_1 \frac{di_1}{dt} + L_2 \frac{di_2}{dt} + \cdots + L_l \frac{di_l}{dt} = 0 \tag{43}$$

in which the $L_1 \cdots L_l$ are some sort of resultant inductance coefficients (their interpretation is of no interest at the moment). Integration of Eq. 43 gives

$$L_1 i_1 + L_2 i_2 + \cdots + L_l i_l = \text{constant} \tag{44}$$

which is the desired constraint equation.

A similar situation obtains on the node basis. Here there is a condition of linear constraint among the node-pair voltages if there exists within the network a cut set for which all branches involve the same kind of element. In the case of an inductance cut set, the presence or absence of arbitrary mutual inductive coupling is immaterial. Figure 22 illustrates a situation of this sort for a cut set consisting of the branches converging toward a single node, the element kind being capacitance. A Kirchhoff current-law equation written for this node reads

$$\frac{d}{dt}[C_1(e_1 - e_0) + C_2(e_2 - e_0) + C_3(e_3 - e_0) + C_4(e_4 - e_0)] = 0 \quad (45)$$

Integrating and rearranging terms gives

$$C_1e_1 + C_2e_2 + C_3e_3 + C_4e_4 = (C_1 + C_2 + C_3 + C_4)e_0 + \text{constant} \quad (46)$$

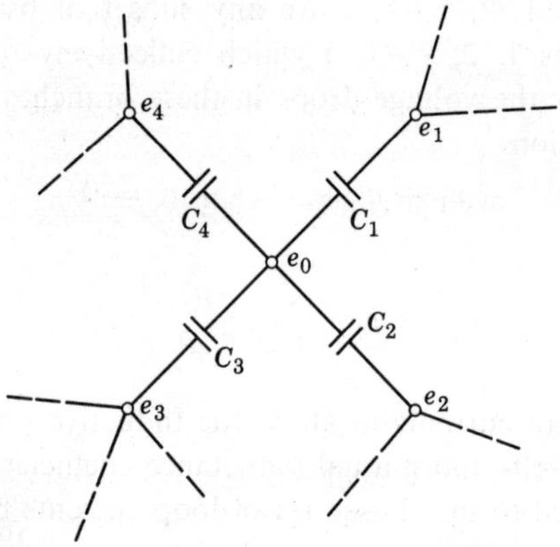

Fig. 22. A capacitance node, giving rise to the constraint relation 46.

Situations of this sort are called resistance, inductance, or capacitance nodes or cut sets, depending upon the kind of element involved. On the loop basis we refer to the analogous degeneracies causing linear constraint relations as resistance, inductance, or capacitance meshes or loops. The problem now is to devise means for unambiguously determining the number of *independent* nodes, cut sets, meshes, or loops of this sort in any given random network. The fact that we must be sure about the independence of these things suggests that we not depend upon spotting them in any haphazard fashion, for, just as in any network graph there are many more distinct closed contours than there are geometrically independent loops, so the number of distinct capacitance loops (for example) that we can discover within a given network is apt to be larger than the number of independent ones of this variety.

The key to the solution of this collateral problem is found by drawing for the particular two-element-kind network not only the complete graph but also a subgraph for each of the two kinds of elements. For example, in the case of an *LC* network, we draw, in addition to the complete graph, a subgraph consisting only of those branches that are inductances and another consisting only of those branches that are capacitances. Or we may say that the so-called *inductance subgraph* is obtained by removing from the total network graph those branches which represent capacitances, whereas the *capacitance subgraph* is formed from the total graph by removing those branches representing inductances.

The subgraphs render the determination of the number of inductance or capacitance loops straightforward. Thus the number of geometrically independent loops in each subgraph is respectively equal to the number of independent loops of that element kind. Since the number of branches b, the total number of nodes n_t, and the number of separate parts s are easily and unambiguously determined for each subgraph, therefore the respective number of independent loops $l = b - n_t + s$ may as surely and simply be computed. Thus in a network in which all of the loop currents are dynamically independent, the subgraphs will separately exhibit no closed paths. If they do exhibit closed paths, then the total number of such independent loops for both subgraphs equals the number of dynamically superfluous loop-current variables. That is to say, the number of geometrically independent loop currents for the total network, diminished by the number of independent loops for both subgraphs, equals the number of dynamically independent loop currents, and hence equals the number of normal co-ordinates or normal modes of the system.

We may alternately obtain the number of dynamically independent variables on the node basis by subtracting from the number of geometrically independent node pairs for the total network the number of independent inductance and capacitance cut sets which may likewise be determined from the subgraphs in a straightforward manner. These are, however, not simply the number of independent cut sets for both subgraphs as might at first be expected. To obtain the correct number of inductance or capacitance cut sets we must consider separately those cut sets that are formed by branches converging toward single nodes (inductance nodes or capacitance nodes), and those that converge toward a group of nodes whose associated branches form a smaller subgraph within the subgraph under consideration.

With regard to inductance or capacitance nodes we observe that an inductance node, for example, is absent in the capacitance subgraph because all the branches converging upon it are inductances. A capacitance node likewise is absent in the inductance subgraph. Therefore the number of independent inductance nodes equals the total number of nodes in the graph for the whole network (the original graph) minus the total number of nodes in the capacitance subgraph; and the number of independent capacitance nodes equals the total number of nodes in the original graph minus the total number in the inductance subgraph. If n_t denotes the total number of nodes in the original graph and n_{Lt} and

n_{Ct} denote the same quantities for the inductance and capacitance subgraphs, then

$$\begin{aligned} \text{the number of independent inductance nodes} &= n_t - n_{Ct} \\ \text{the number of independent capacitance nodes} &= n_t - n_{Lt} \end{aligned} \tag{47}$$

Next we observe that removal of the branches in a cut set converging toward a group of nodes separates this group with its associated branches (the smaller subgraph referred to above) from the rest of the network. If this cut set is formed by inductance branches, then the capacitance subgraph will contain an additional separate part as compared with the original graph. If s denotes the number of separate parts in the original graph, and if s_L and s_C denote the same quantities for the inductance and capacitance subgraphs, then this reasoning shows that

$$\begin{aligned} \text{the number of independent inductance cut sets} &= s_C - s \\ \text{the number of independent capacitance cut sets} &= s_L - s \end{aligned} \tag{48}$$

Actually these relations will include also the inductance or capacitance cut sets that converge toward single nodes if, in the removal of branches to form the subgraphs, we do not consider the nodes as removable with the branches and if we regard a single isolated node as constituting a subgraph. It is not customary to do this, however; and so we consider the inductance nodes and capacitance nodes determined by relations 47 as distinct from the cut sets determined by 48.

If we denote the total number of independent inductance and capacitance nodes and cut sets by n_{Ls} and n_{Cs} respectively, then relations 47 and 48 show that

$$n_{Ls} = (n_t - n_{Ct}) + (s_C - s) \tag{49}$$

$$n_{Cs} = (n_t - n_{Lt}) + (s_L - s) \tag{50}$$

Their sum

$$n_s = n_{Ls} + n_{Cs} \tag{51}$$

equals the total number of independent conditions of linear constraint that exist among the node-pair voltage variables (the total number of dynamically superfluous variables).

If, in addition to the notation introduced above, the letters b, n, and l are used to denote numbers of branches, geometrically independent node pairs, and geometrically independent loops for the original graph, and also, with appropriate subscripts, are used to denote these respective quantities for the subgraphs, then according to established theory

$$n_t = n + s, \qquad n_{Ct} = n_C + s_C, \qquad n_{Lt} = n_L + s_L \tag{52}$$

and

$$b = n + l, \qquad b_C = n_C + l_C, \qquad b_L = n_L + l_L \tag{53}$$

The quantities l_L and l_C, according to the above discussion, are the numbers of independent inductance and capacitance loops, and their sum

$$l_s = l_L + l_C \tag{54}$$

equals the number of independent conditions of linear constraint among the loop currents, or the number of dynamically superfluous loop-current variables, just as n_s in Eq. 51 represents the number of dynamically superfluous node-pair voltage variables. By using Eqs. 49 and 50, this number is expressible as

$$n_s = 2(n_t - s) - (n_{Lt} - s_L) - (n_{Ct} - s_C) \tag{55}$$

or by Eqs. 52 we have

$$n_s = 2n - n_L - n_C \tag{56}$$

This result, noting Eqs. 53, now yields

$$n - n_s = n_L + n_C - n = (b_L - l_L) + (b_C - l_C) + (l - b)$$

or

$$n - n_s = l - l_L - l_C = l - l_s \tag{57}$$

where the obvious relation

$$b = b_L + b_C \tag{58}$$

in addition to Eq. 54 is made use of.

The result expressed by Eq. 57 is one that we could have predicted heuristically on the basis of physical reasoning. Thus $n - n_s$ equals the number of dynamically independent variables on a node basis and $l - l_s$ equals the number of dynamically independent variables on a loop basis. That these numbers should be equal is a necessary consequence of the fact that the number of dynamically independent variables characterizing a given physical system (equal to the number of normal modes or normal co-ordinates) is dependent solely upon the physical character of the system and cannot vary according to the particular algebraic process used in formulating the equilibrium equations.*

Although discussed specifically with reference to *LC* networks, the above derivations, with appropriate modification in the notation, obviously apply also to *RL* and *RC* networks. For reasons already pointed out, these results are not straightforwardly extendable to networks involving all three kinds of elements.

We shall now discuss some of the things that can be done with these results. Suppose we consider the *LC* network of Fig. 23 and inquire initially as to the number of its finite nonzero natural frequencies. To this end we construct the three graphs shown in Fig. 24 and determine

* This result (based upon heuristic reasoning), together with the following applications, was first pointed out by the writer at the Round Table Conference on Network Analysis and Synthesis during the Winter Convention of the AIEE in Jan. 1939. The above topological proof was contributed at the same time by R. M. Foster.

in a straightforward manner the quantities n, n_L, n_C, l, l_L, l_C as indicated there. From Eqs. 54 and 56 we find

$$l_s = 2 + 0 = 2, \qquad \text{and } n_s = 18 - 8 - 5 = 5 \tag{59}$$

which tells us that on a loop basis two variables are dynamically superfluous and on a node basis five are superfluous. On either basis the equilibrium can be expressed in terms of

$$l - l_s = n - n_s = 4 \tag{60}$$

dynamically independent variables. The circuit has four finite nonzero natural frequencies and possesses a normal form involving four normal co-ordinates.

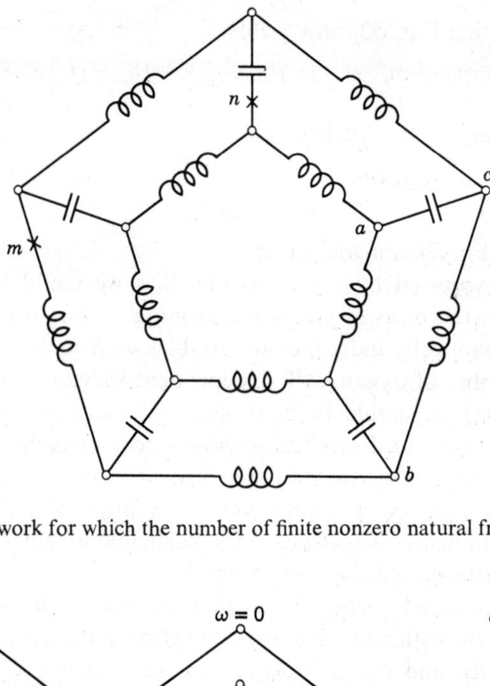

Fig. 23. A network for which the number of finite nonzero natural frequencies is to be determined.

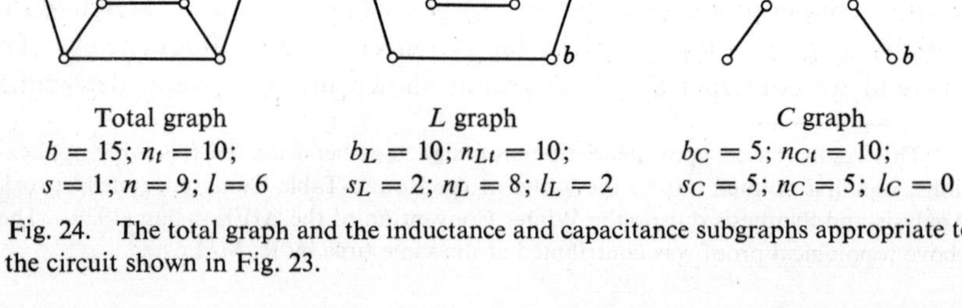

Total graph $b = 15$; $n_t = 10$; $s = 1$; $n = 9$; $l = 6$

L graph $b_L = 10$; $n_{Lt} = 10$; $s_L = 2$; $n_L = 8$; $l_L = 2$

C graph $b_C = 5$; $n_{Ct} = 10$; $s_C = 5$; $n_C = 5$; $l_C = 0$

Fig. 24. The total graph and the inductance and capacitance subgraphs appropriate to the circuit shown in Fig. 23.

Now we will show how, with this information, we can construct the frequency pattern for a driving-point reactance function corresponding to any chosen type of entry, and thus sketch at once the pertinent canonic or other equivalent networks. In this regard we distinguish between the soldering-iron and the pliers type of entries (see *ICT*, p. 92) and make use of the principle that all impedances at terminal pairs created by soldering leads to node pairs must have the same poles, while all impedances at terminal pairs created by cutting branches must have the same zeros, and that these zeros are the same as the poles of the impedances at soldering-iron–type entries (see *ICT*, pp. 430–431). All such sets of critical frequencies are simply the finite nonzero natural frequencies of the given network without entries of any kind.

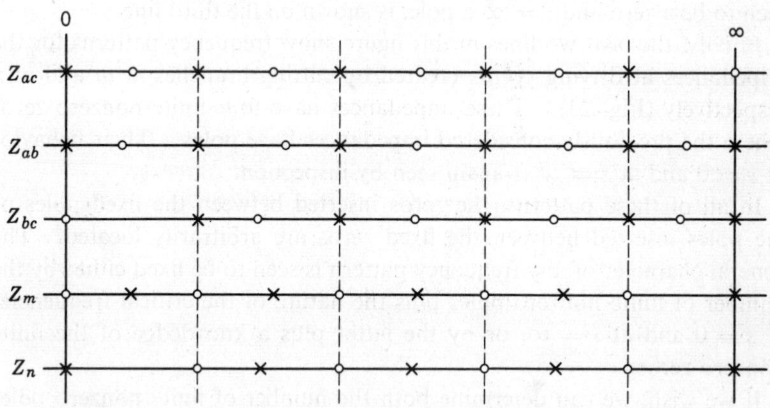

Fig. 25. Frequency patterns for soldering-iron and pliers-type impedances pertinent to the circuit of Fig. 23.

On the basis of these statements it should not be inferred that all impedances for entries made by soldering leads to node pairs have the same frequency patterns or that all impedances for pliers-type entries have the same frequency patterns. The appropriate frequency pattern is dependent upon the particular entry in question because the behavior of the impedance function at $s = 0$ and at $s = \infty$ is determined by the conditions that obtain at the particular entry considered. Thus suppose we choose a driving point across the node pair *a-c* in Fig. 23. At $s = 0$ the pertinent driving-point impedance is clearly infinite whereas at $s = \infty$ it is zero. Hence we know that this impedance Z_{ac} has four finite nonzero poles (Eq. 60) besides having a pole at $s = 0$ and a zero at $s = \infty$. The frequency pattern is shown on the top line of Fig. 25.

If we choose the node pair *a-b* as a driving point, the impedance has the same finite nonzero poles, but now both points $s = 0$ and $s = \infty$ are poles. With reference to the subgraphs in Fig. 24 it is helpful to note that, since capacitances are open circuits at $s = 0$ whereas inductances are open circuits at $s = \infty$, we can regard the inductance subgraph as a zero-frequency graph and the capacitance subgraph as an infinite-frequency graph. Moreover, since inductances and capacitances are short circuits at $s = 0$ and $s = \infty$ respectively, one may regard the branches in these subgraphs as short circuits when considering them as representing the original network for the limiting frequencies $s = 0$ and $s = \infty$. Thus for a certain driving point like the node pair *a-b*, one can tell at a glance whether the critical frequency at $s = 0$ or at $s = \infty$ is a zero or a pole for the pertinent impedance function. The frequency pattern for Z_{ab} is shown in the second line of Fig. 25, and that for Z_{bc}, for which $s = 0$ is seen to be a zero and $s = \infty$ a pole, is shown on the third line.

Finally, the last two lines in this figure show frequency patterns for the impedances at driving points created by cutting branches at *m* and at *n* respectively (Fig. 23). These impedances have four finite nonzero zeros where the previously considered impedances have poles. Their behavior at $s = 0$ and at $s = \infty$ is again seen by inspection.

In all of these patterns, the zeros inserted between the fixed poles or the poles inserted between the fixed zeros are arbitrarily located. The general character of any frequency pattern is seen to be fixed either by the number of finite nonzero poles plus the nature of the critical frequencies at $s = 0$ and at $s = \infty$, or by the latter plus a knowledge of the finite nonzero zeros.

If we wish, we can determine both the number of finite nonzero poles and the number of finite nonzero zeros. If the driving point is across a node pair then we superimpose these nodes to form a modified graph; if the driving point is created by cutting a branch then we remove the branch to form a modified graph. The number of natural frequencies determined for the modified graph are finite nonzero zeros of an impedance pertinent to a node pair and they are finite nonzero poles of an impedance pertinent to an entry made by cutting a branch. If the numbers of finite nonzero poles and zeros thus found for a given driving point are alike (they can at most differ by one) then we must still determine the nature of the critical frequencies at either $s = 0$ or $s = \infty$ in order to fix the frequency pattern. Except as a means for checking the results gotten by the simpler process discussed above, there is no point in getting the frequency pattern by this second method which requires drawing two sets of graphs and subgraphs (one for the poles and one for the zeros).

In practical situations we sometimes encounter circuits in which considerable symmetry or uniformity is involved. We might, for example, have a circuit like the one in Fig. 23 in which all inductances and capacitances are alike. The approximate lumped-circuit equivalent for the anode structure of a strapped magnetron is very similar to such a uniform network. It is not out of place, therefore, to call attention to the fact that the resulting degeneracy of such a system leads to a considerable reduction in the number of natural frequencies that it would otherwise have.

Suppose we reconsider the impedance Z_{ac} for the network of Fig. 23 assuming that the element values are uniform instead of random. If we imagine a current source applied to the node pair *a-c*, then the symmetry at once tells us that currents in the diametrically opposite branches are zero. So far as the resultant voltage at the terminal pair *a-c* is concerned, we can remove these branches; and since the remaining network portions to the left and right of the driving point are identical, it will suffice to consider only one of them as shown in Fig. 26. By inspection we see that its frequency pattern is that drawn in Fig. 27 which exhibits only two finite nonzero poles instead of four as in the random circuit.

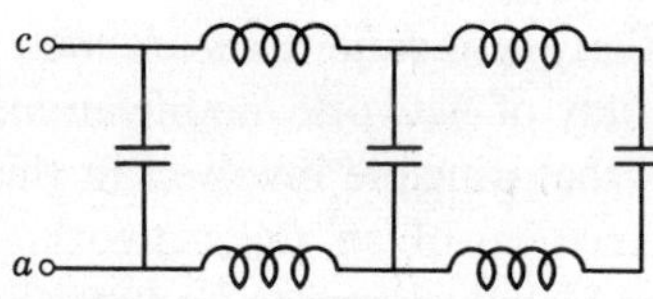

Fig. 26. Potentially equivalent network for the impedance Z_{ac} of the circuit of Fig. 23 if the elements have uniform values.

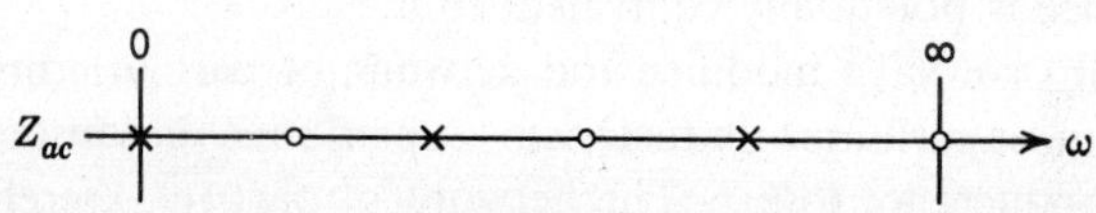

Fig. 27. Frequency pattern for the input impedance of the network in Fig. 26.

A further interesting and useful application of these ideas is to the rapid construction of potentially equivalent network schematics. Often in the practical application of networks one likes to know whether adding a capacitance here or inserting an inductance there and removing one elsewhere will upset the potential equivalence of the network. That is to say, one contemplates certain changes in the network configuration (sometimes dictated by unavoidable parasitics), and is desirous of knowing whether the same network response can be retained after the changes have been incorporated. The answer, of course, is that they can be if the contemplated changes in configuration lead to a potentially equivalent structure, for then an appropriate revision in element values will restore the original network response.

The method whereby we can rapidly indicate circuit modifications which will keep invariant the potential equivalence of a driving-point impedance is based upon the fact, established by the preceding studies, that the frequency pattern of this impedance remains unchanged so long as the number of natural modes plus the behavior at $s = 0$ and at $s = \infty$ remain the same. If we introduce into a network a circuit element that can be shown to be superfluous, then its insertion will not change the number of dynamically independent variables, and hence it will not change the number of natural frequencies. By the same token, an element can be removed from the network if it is a superfluous element in the sense that its reinsertion does not affect the number of dynamically independent variables.

The element in question is superfluous if its insertion into the network creates an inductance or capacitance loop or cut set, or if its removal from the network destroys an inductance or capacitance loop or cut set. Its insertion or removal must not, of course, influence the behavior of the pertinent impedance function at $s = 0$ or at $s = \infty$. In many cases one can see by inspection that these conditions are met, and thus it is possible to consider a large variety of network modifications rapidly.

Figure 28 illustrates the principle involved in this procedure. In part (*a*) is a familiar Foster form of an *LC* network. In part (*b*) we have inserted a capacitance which we know to be superfluous since its insertion has created a capacitance loop. The input impedance for the network of part (*b*) has the same frequency pattern as that for the network of part (*a*) and hence is potentially equivalent to it.

In part (*c*) we have modified the network of part (*a*) through the insertion of a superfluous inductance—superfluous because its insertion creates an inductance loop. The network of part (*c*), therefore, likewise has a potentially equivalent driving-point impedance function.

In part (*d*) we have inserted three superfluous inductances and three superfluous capacitances into the network of part (*a*), and so (*a*) and (*d*) are still potentially equivalent with respect to their driving points.

Now we shall remove superfluous elements one by one so as to destroy by each such removal an inductance or capacitance loop; observe that we do not have to remove the same elements that were inserted in going from network (*a*) to network (*d*). Elements that may be removed in this way (one at a time or altogether) are shown dashed in part (*e*) of this same figure. It is interesting to note that when all dashed elements are removed we have left the alternate Foster form with respect to the one in part (*a*) with which we began this series of modifications.

In passing it may be remarked that when one removes superfluous elements it is well to be wary that additional separate parts are thereby

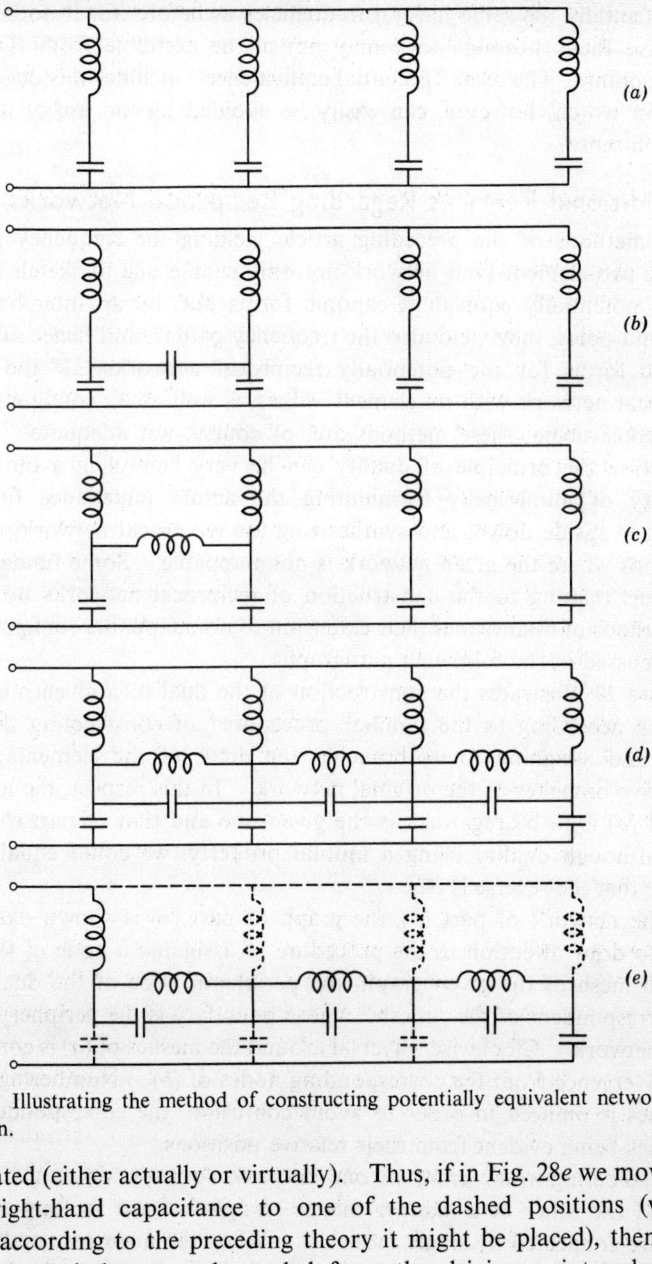

Fig. 28. Illustrating the method of constructing potentially equivalent networks by inspection.

not created (either actually or virtually). Thus, if in Fig. 28*e* we move the lower right-hand capacitance to one of the dashed positions (where strictly according to the preceding theory it might be placed), then part of the network becomes decoupled from the driving point unless we assume that the inductances are mutually coupled (which, of course, they may be). This result does not contradict the theory that the system still has potentially the same natural frequencies as before, for it does indeed still have them although they may not all be excitable from the same driving point. The term "potential equivalence" includes this degenerate situation which, however, can easily be avoided by the use of a bit of common sense.

3 Additional Remarks Regarding Reciprocal Networks

The methods of the preceding article yielding the frequency pattern for any two-element-kind network not only enable one to sketch at once all the potentially equivalent canonic forms but, by an interchange of zeros and poles, they yield also the frequency pattern and hence all of the canonic forms for the potentially reciprocal networks. If the actual reciprocal network with its element values as well as its configuration is to be constructed, these methods are, of course, not adequate. In this connection the principle of duality can be very helpful in avoiding the necessity of numerically formulating the actual impedance function, turning it upside down, and synthesizing the reciprocal network, even in situations where the given network is nonmappable. Some fundamental questions relating to the construction of reciprocal networks by use of the methods of duality and their extension to nonmappable configurations are discussed in the following paragraphs.

Figure 29 illustrates the construction of the dual to a given mappable network according to the familiar procedure* of constructing the dual graph and assigning to its branches the duals of the elements in the respective branches of the original network. In this respect, the network of part (*a*) is to be regarded as the given one and that of part (*b*) as its dual, although duality being a mutual property, we could equally well assume that the reverse is true.

In the network of part (*b*), the graph of part (*a*) is shown dashed in order to draw attention to the procedure of assigning a node of the dual to each mesh of the given graph, the peripheral node of the dual being the correspondent of the "mesh" whose boundary is the periphery of the given network. Clockwise traversal around the meshes of (*a*) is correlated with divergence from the corresponding nodes of (*b*). Numbering of the branches is omitted in order to avoid confusion, the correspondence of branches being evident from their relative positions.

A "soldering-iron" entry in one network corresponds to a "pliers" entry in the other in a manner that is straightforward so long as node pairs are connected by single branches. When the node pair under consideration is not directly connected by a branch, the method of finding

* For a discussion of these matters see *ICT*, pp. 42–51 and 535–539.

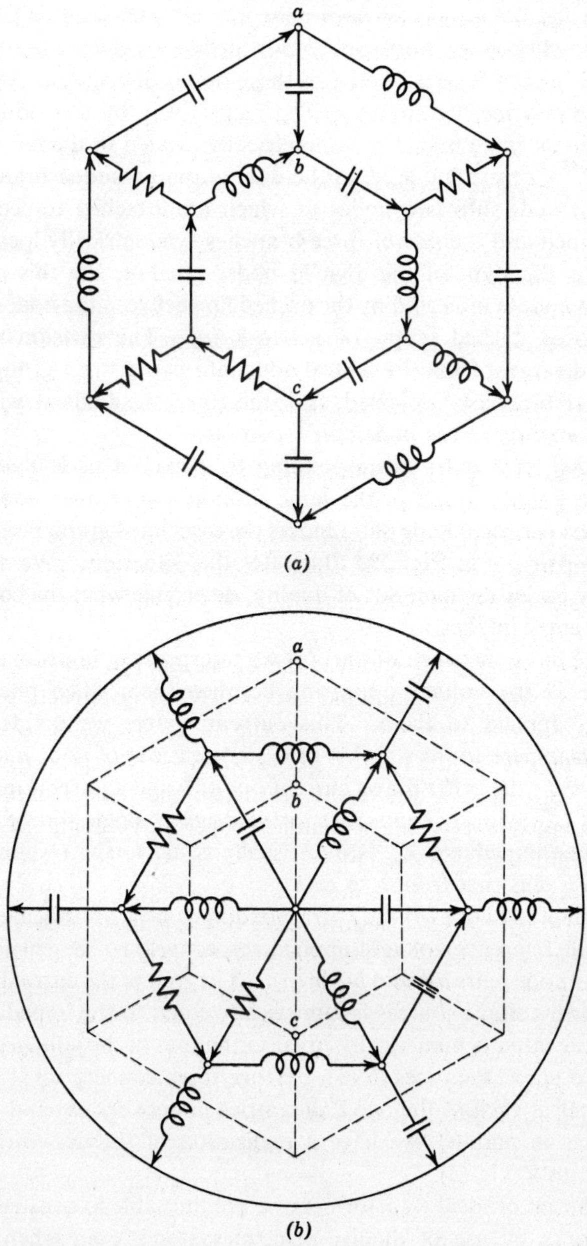

Fig. 29. A given network and its dual, illustrating a direct topological method for constructing the latter.

the corresponding entry in the dual network is only slightly more complicated when the graphs of both networks are indicated as in Fig. 29*b*. Thus if we choose the node pair *b-c* in network (*a*) we observe that an additional branch inserted between these nodes divides the central mesh in (*a*) into two meshes circumscribed respectively by this added branch and the sets of three branches symmetrically located to the left and to the right of it. Correspondingly, in the dual graph an added branch divides the central node into two nodes to which are attached respectively this added branch and the sets of three branches symmetrically located to the left and to the right of the divided node. In Fig. 29*b* this division of the central node is indicated by the dashed line between the nodes *b-c* in the superimposed dashed graph of network (*a*). The division of the six branches divergent from the central node into two distinct groups of three is thus unambiguously indicated, as is, therefore, the dual entry in network (*b*) corresponding to the node pair *b-c* in (*a*).

Any other dual entry corresponding to a chosen node pair in either network is readily found in the same fashion except if an added branch between the pertinent node pair renders the associated graph nonmappable. The node pair *a-c* in Fig. 29*a* illustrates this situation. We can, nevertheless, by use of the methods of duality, determine what the construction of a dual entry implies.

Suppose in the network of part (*a*) we interpret the impedance between nodes *a-c* as the voltage appearing between them when one reference ampere is applied to them. This current source we divide into two separate 1-ampere sources. One draws current out of node *a* and squirts it into node *b*; the other draws current out of *b* and squirts it into *c*. The end result is obviously equivalent to the original single source between *a* and *c*. The impedance Z_{ac} is numerically equal to the resultant voltage from *a* to *b* plus that from *b* to *c*.

In the dual network (*b*) the corresponding source arrangement consists of two unit reference voltages applied respectively to the entries that are dual to the node pairs *a-b* and *b-c* in (*a*). The sum of the currents delivered by these two voltage sources is numerically equal to the impedance Z_{ac} in (*a*) since the latter is numerically equal to the sum of the voltages produced by the two current sources in (*a*). Hence if we connect a 1:1 ratio ideal transformer at each of the two dual entries in network (*b*) and place their secondaries in parallel, we have a realization of the desired impedance reciprocal to Z_{ac} in (*a*).

With the aid of ideal transformers we are thus able to construct reciprocal networks by use of duality principles alone, even when the given problem involves a nonmappable graph. Although ideal transformers are not very practical devices, this method of constructing reciprocal networks is definitely of theoretical interest and may be useful practically as well in situations where the ideal transformer is associated with inductances in such a way that the combination may be replaced by a pair of coupled coils.

An entry in the network of Fig. 29*b*, dual to the node pair *a-c* in network (*a*), constructed with the help of ideal transformers according to these ideas, is shown in Fig. 30 where the points *a'-c'* represent the appropriate

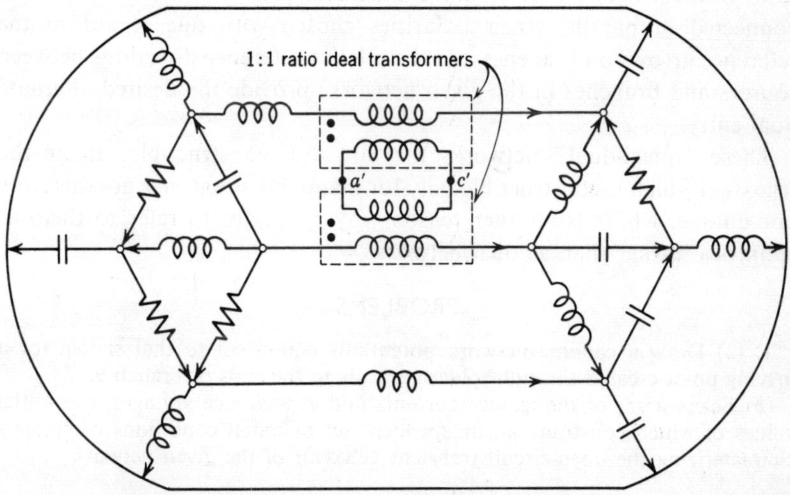

Fig. 30. Construction of an entry in the network of Fig. 29*b*, dual to the entry *a-c* in the network of Fig. 29*a*, with the help of ideal transformers.

terminal pair at which the impedance function is the reciprocal of Z_{ac}. Except for the appearance of the ideal transformers, this network configuration with a voltage source E_s across *a'-c'* is in every sense the dual of the network in Fig. 29*a* with a current source I_s (numerically equal to E_s) bridged across the node pair *a-c*. If we remove the current source then $I_s = 0$, and so we must short-circuit the points *a'-c'*. If we place any circuit element across the node pair *a-c* in the network of Fig. 29*a*, and the dual element across node pair *a'-c'* in the network of Fig. 30, then the two networks are completely dual so far as we are willing to disregard the extent to which the presence of the ideal transformers mars the otherwise mutual characteristic of this relationship.

From this example we see that it is possible in this same restricted sense to construct a dual network for any given nonmappable one by dealing with appropriate branches in the manner in which a branch across *a-c* in the network of Fig. 29*a* is dealt with above. It should be clear that this latter process is not unique, and that the branches chosen to be handled in this manner can be varied in several different ways. A variant in the above example is to replace a current source across the node pair *a-c* in the network of Fig. 29*a* by a series of identical current sources across any set of confluent branches leading from node *a* to node *c*, their reference directions being such that no net current enters or leaves any node except the pair in question. In the dual network of Fig. 29*b* we imagine identical voltage sources in series with the respective dual branches. Ideal 1:1-ratio transformers placed in these dual branches with their secondaries connected in parallel (their polarities chosen with due regard to the reference arrows on branches and the relative reference directions between sources and branches in the given network) provide the desired alternate dual entry.

These "quasi-dual" networks that are thus constructible, unlike the true dual which is constructible only for mappable situations, are therefore not unique, which is another reason why we prefer to refer to them as reciprocal rather than as dual networks.

PROBLEMS

1. (*a*) Draw a canonic network potentially equivalent to that shown for a driving point created through soldering leads to the ends of branch 9.

(*b*) Select a set of inductance currents and capacitance voltages, the initial values of which constitute an independent set of initial conditions completely characterizing the open-circuit transient behavior of the given network.

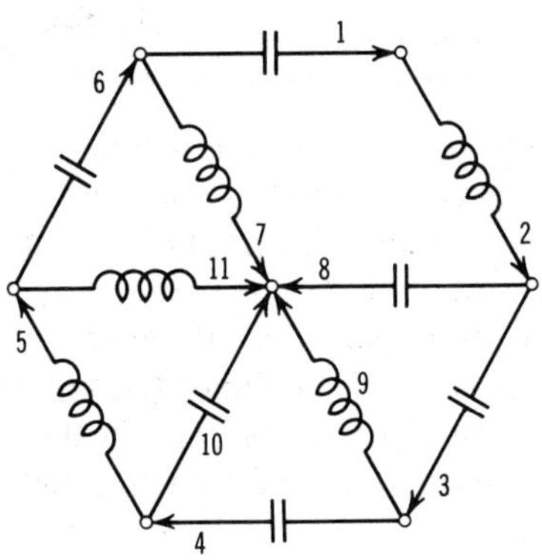

Prob. 1.

2. (*a*) Determine the pole-zero pattern of the impedance $Z(s)$, and sketch the four canonic networks which are potentially equivalent to the one shown here.

(*b*) Repeat part (*a*), letting the inductances be replaced by capacitances.

3. For the *RC* network shown, determine equivalent canonic networks for driving-point impedances obtained by

(*a*) Attaching leads to nodes 1 and 2.
(*b*) Attaching leads to nodes 1 and 3.
(*c*) Making a cut at *m*.

Write the analytic expression for the impedance in each case.

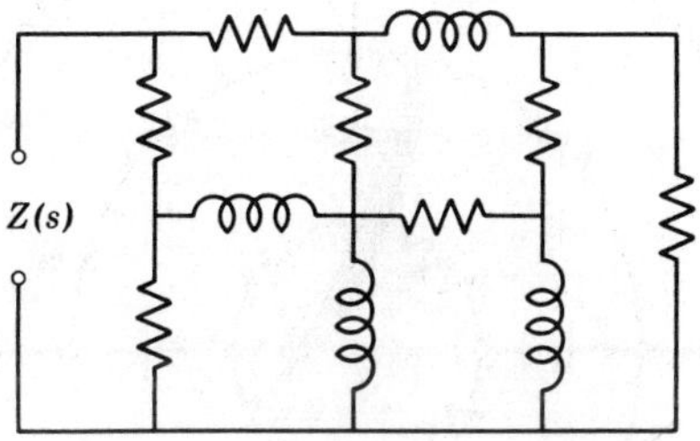

Prob. 2.

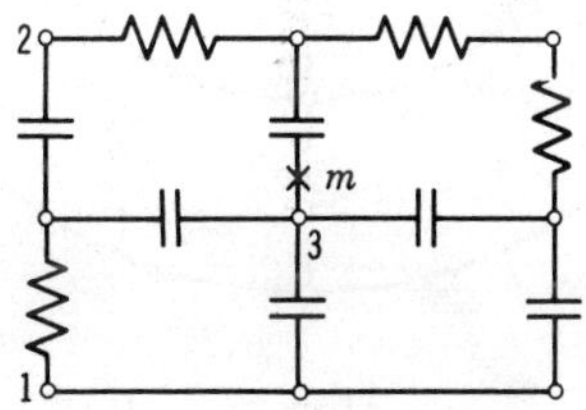

Prob. 3.

4. Considering the *RC* network shown in the sketch on page 178, find canonic forms (only one in each case) for potentially equivalent networks when the driving points in the given network are:

(*a*) Nodes 1 and 2.
(*b*) Nodes 1 and 3.
(*c*) A cut at *m*.

5. It is desired that networks be found that have driving-point impedances which are reciprocal to those involved in Prob. 4. Point out for each impedance the method that yields the desired result with a minimum of computational effort. Give the configuration of the appropriate network in each case.

6. If the *R*'s in the network of Prob. 4 are replaced by *L*'s, write down analytic expressions for reactance functions at the stated driving points, assuming that the critical frequencies lie at $\omega = 1, 2, 3, 4, \ldots$ radians/sec. and that the impedance level is unity.

7. For the network shown in the sketch on page 178, determine frequency patterns for reactance functions obtained by

(*a*) Attaching leads at node pair 1-2.
(*b*) Attaching leads at node pair 3-4.
(*c*) Making a cut at *m*.
(*d*) Making a cut at *n*.

Sketch an equivalent ladder network for each case. How many of the loop currents or node-pair voltages are dynamically independent?

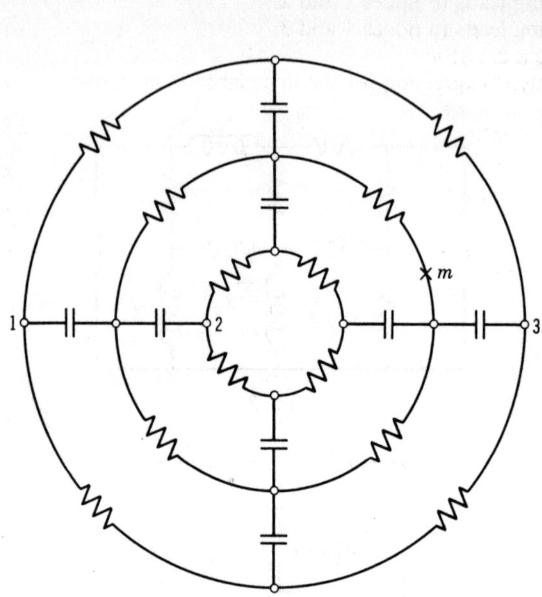

Prob. 4.

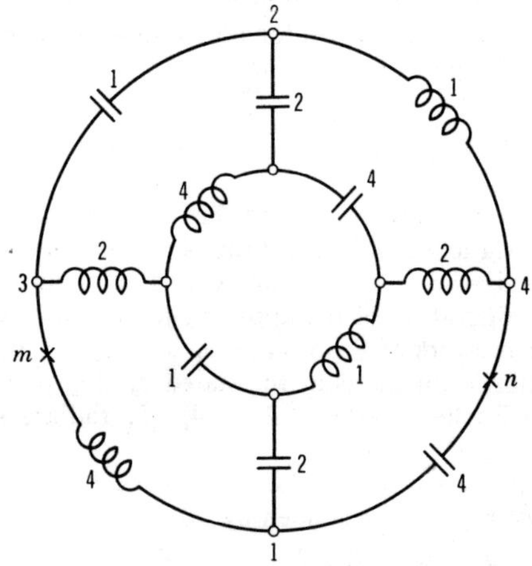

Prob. 7.

8. Find the dual of the network configuration in Prob. 7 indicating all element values. The given values are in henrys and farads. Indicate methods of creating driving points in the dual network such that the impedances at these

points will be the reciprocals of those stated in the preceding problem under (*a*), (*b*), (*c*), and (*d*) respectively.

9. With reference to the network in the adjacent sketch, determine the number of dynamically independent loop currents or node-pair voltages needed for its characterization.

Draw the frequency patterns for driving-point impedances corresponding to

(*a*) A cut at *m*.

(*b*) A node pair *a*-*b*. Sketch Cauer forms with series *L*'s and shunt *R*'s for each case.

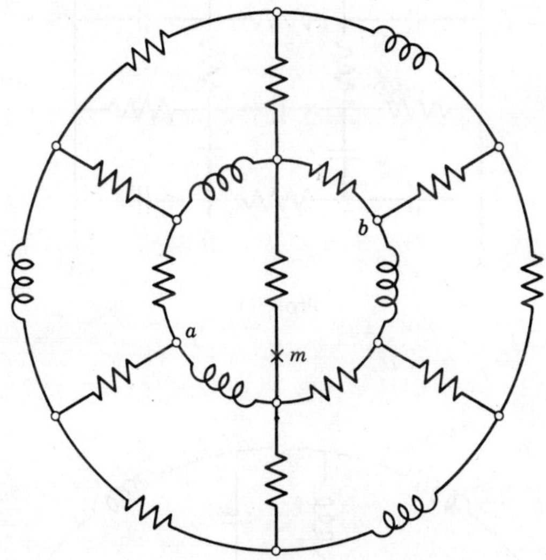

Prob. 9.

10. Construct the dual of the network in Prob. 9. Indicate points of access in this dual network corresponding to the entries specified, making use of ideal transformers if necessary.

11. (*a*) Determine the number of dynamically independent loop currents and node-pair voltages for the network illustrated on page 180.

(*b*) Sketch the frequency patterns for the driving-point impedances corresponding to (i) node pair *A*-*A*; (ii) node pair *B*-*B*; (iii) a cut at *m*.

(*c*) Construct the dual of this network. Indicate the points of access dual to those above, using ideal transformers if necessary.

12. The anode structure of a strapped magnetron has an approximate equivalent circuit like that shown on page 180, in which all parallel *LC* circuits are identical and the inductances joining these (the strap inductances) are all equal.

(*a*) Determine the frequency pattern for the reactance at the terminals *a*-*b*, and thus determine the number of natural modes of this equivalent anode structure.

(*b*) Sketch the four canonic forms for this reactance.

(*c*) How many superfluous inductances and capacitances does the given anode circuit involve?

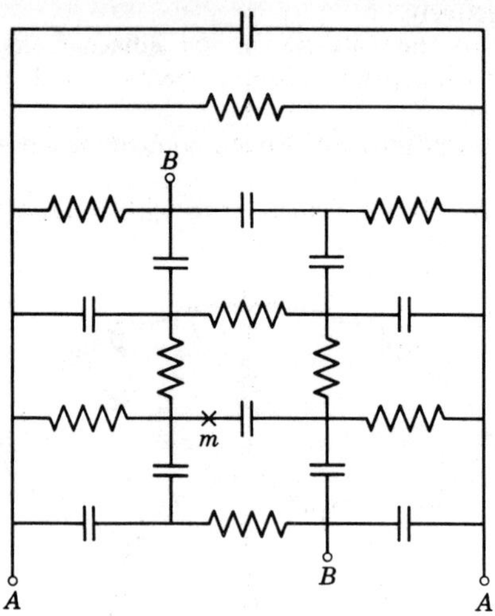

Prob. 11.

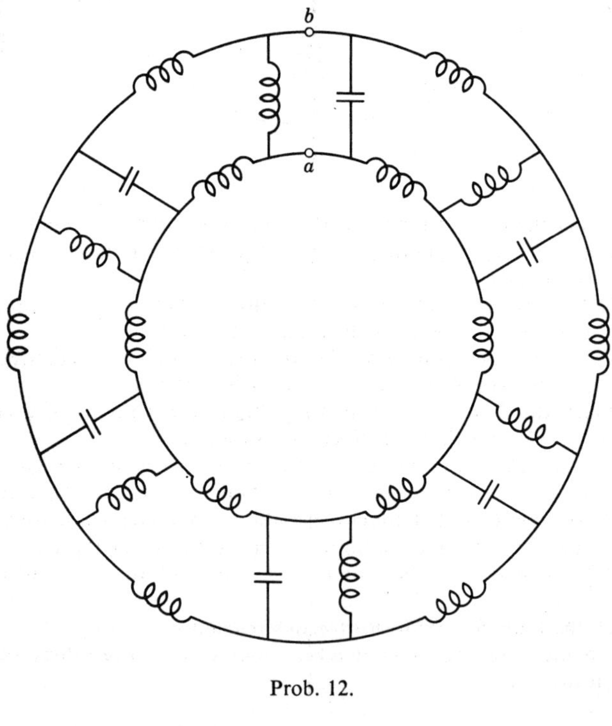

Prob. 12.

13. In the circuit shown here, assume the values $L = C = 1$, $C_1 = 0.01$. Determine the numerical values of all the finite nonzero natural frequencies of this network.

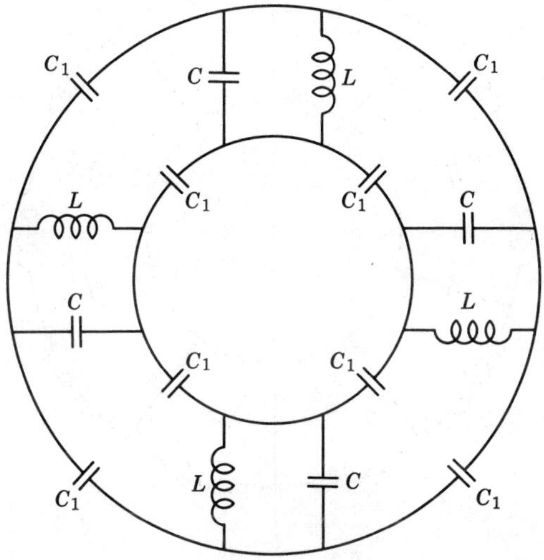

Prob. 13.

14. The network shown in the sketch on page 182 has symmetry about lines *a-a* and *b-b* so far as its element values are concerned. These are to be determined such that the natural modes of the structure are $\omega = 1, \sqrt{2}, \sqrt{3}, \sqrt{4}, \sqrt{5}$. The impedance level is fixed by the statement that an impedance formed by cutting the outside ring should reduce to $64s$ for $s \to 0$.

15. A reactance function $z(s)$ has the frequency pattern shown on the next page, and $[(dz)/(ds)]_{s=0} = 1$.

(*a*) Construct the analytic expression for $z(s)$.

(*b*) Determine the Cauer form obtained by removing poles at $s = 0$ only.

(*c*) Through an impedance level transformation, modify this network so that all inductance values become equal (keeping the input impedance the same).

16. Given the impedance function

$$Z(s) = \frac{2s(s^2 + 1)}{s^4 + 4s^2 + 2}$$

realize this impedance in the form of the network shown on the next page, in which each inductance is 1 henry.

17. The circuit for a given band-pass filter is shown on page 182 in which the element values are in henrys and farads. Make an internal impedance level transformation so that all three inductance values become 10^{-5} henry, and give the structure and element values for the final network. Approximations involving numerical accuracy within a few per cent are sufficient.

18. Through impedance level transformations, convert this figure into a

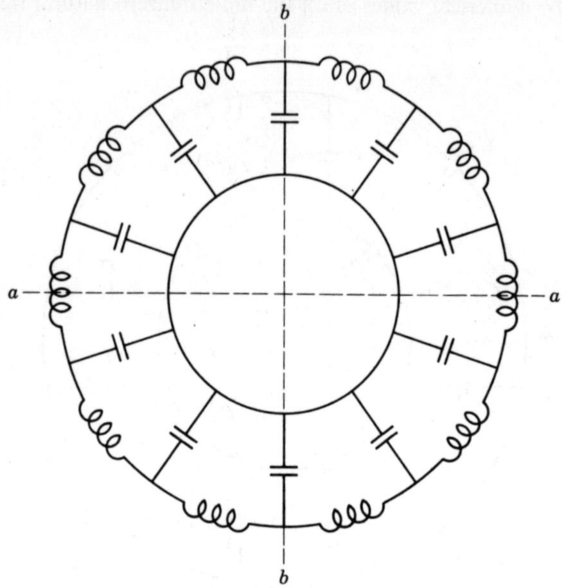

Prob. 14.

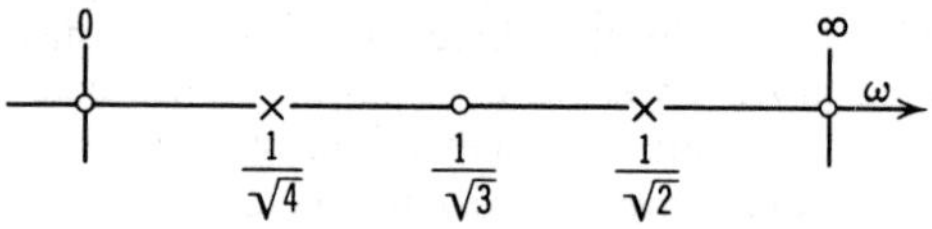

Prob. 15.

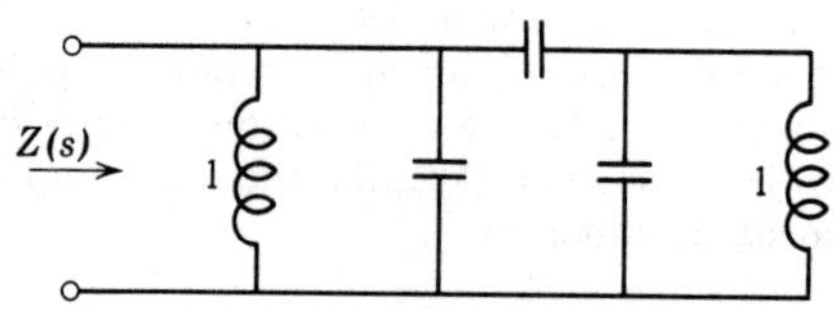

Prob. 16.

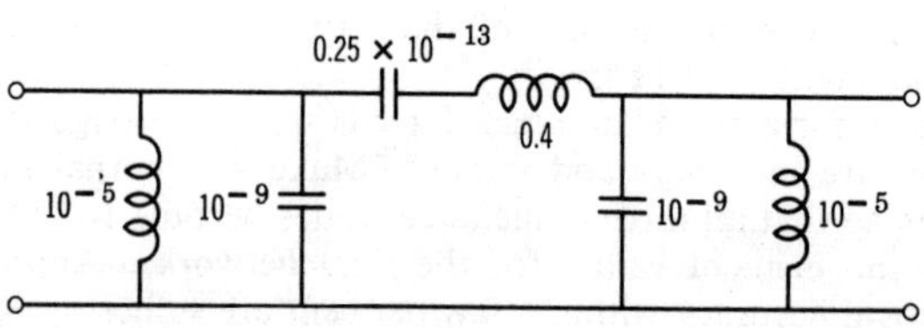

Prob. 17.

network which has a voltage transfer ratio

$$\frac{E_2'}{E_1} = K\frac{E_2}{E_1}$$

where K is as large as possible with the impedance faced by the source remaining constant.

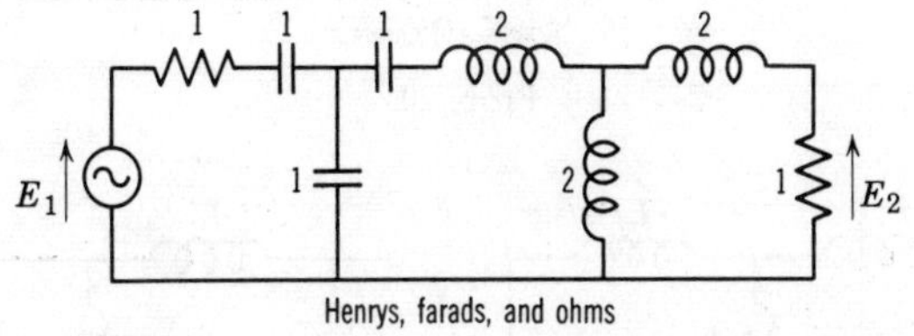

Prob. 18.

19. Through impedance level transformations, convert sketch (a) to the form shown in sketch (b).

Determine the values of C_1, C_2, C_3, R, and E_2'/E_2 for a given I_1.

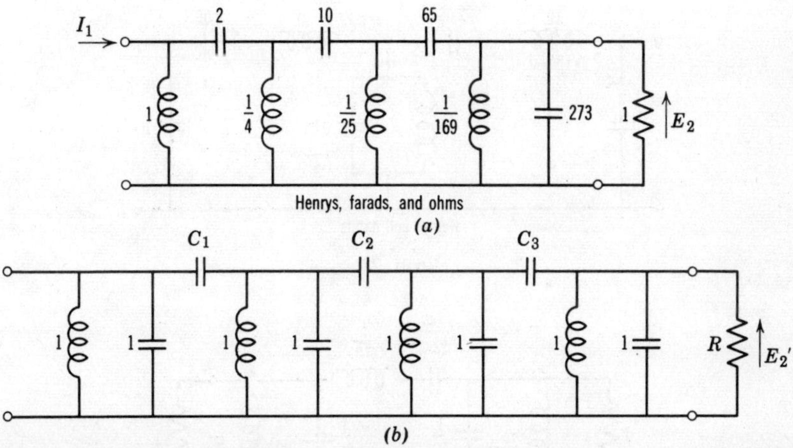

Prob. 19.

20. Given the ladder network shown on page 184, modify this network so that its input impedance becomes $25Z(s)$ with a minimum increase in the impedance levels of its internal meshes.

21. By means of matrix manipulation methods, show that the networks on the next page are realizable in terms of positive elements without mutual inductance, and determine such equivalent forms.

22. To the reactive two terminal-pair network shown on the next page, apply an internal impedance leveling process which makes all inductances equal to 1 henry and leaves the conditions at the terminals unchanged.

23. Sketch (a) shows the circuit of a conventional band-pass filter with its impedance level normalized at 1 ohm for the element values indicated, where

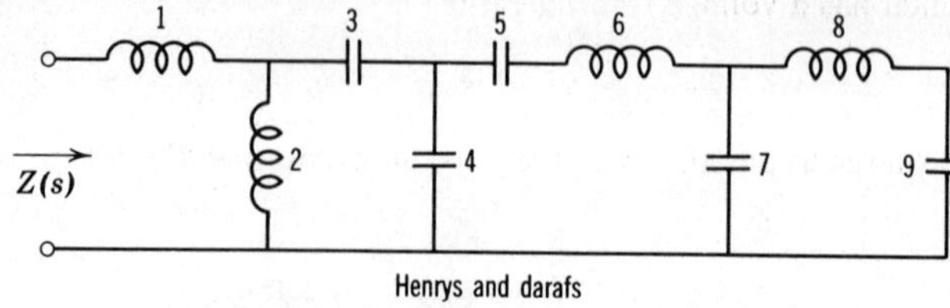

Prob. 20.

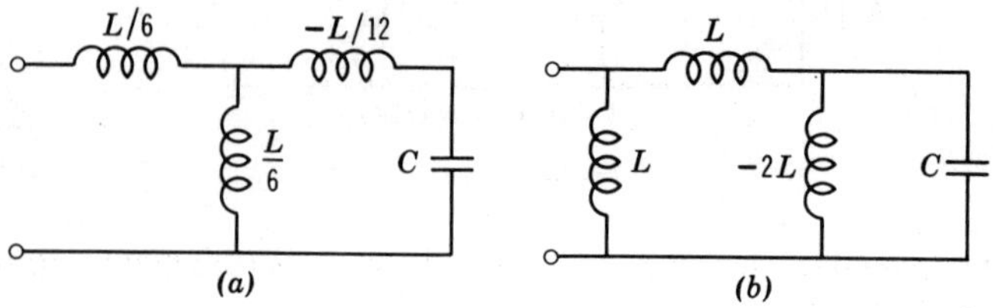

Prob. 21.

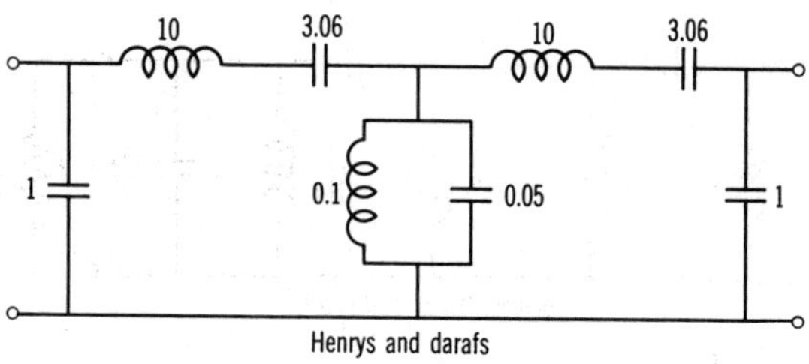

Prob. 22.

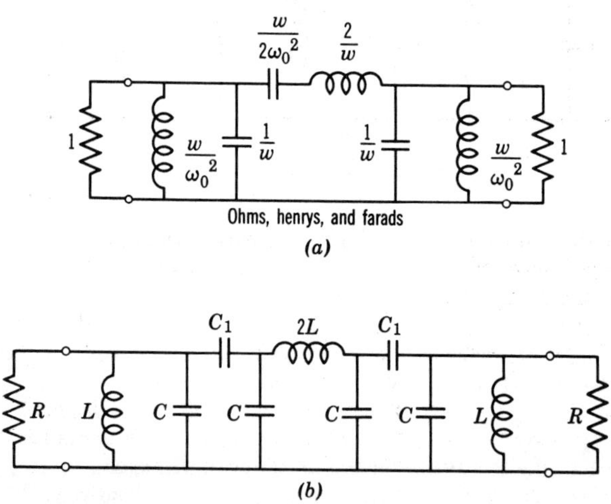

Prob. 23.

$\omega_0 = \sqrt{\omega_a \omega_b}$, $w = \omega_b - \omega_a$, and ω_a, ω_b are the limits of the pass band; that is, they are the so-called cutoff frequencies. If this filter is to be designed for $\omega_0 = 2\pi 10^6$, $w = 2\pi 10^4$, it becomes necessary to make internal impedance level transformations in order to prevent an excessive spread of element values. Show that the result given in sketch (*b*) is achievable in this way and determine the values of L, C, C_1 in microhenrys and micromicrofarads appropriate to source and load resistances $R = 10^5$ ohms.

24. The following network represents a band-pass filter. The parameters L and S are in henrys and darafs, $R = 1$ ohm, and a, b, and c are numerical factors. k_1, k_2, and k_3 are coupling coefficients.

Four specific cases are of interest:

(i) $a = b = c = 1$.
(ii) $a = b = 1$; $c = 10^3$.
(iii) $a = 10^2$; $b = 10^4$; $c = 10^3$.
(iv) $a = 10^2$; $b = 10^4$; $c = 1$.

For all of these, $k_1 = k_2 = k_3 = k$.

Compare the general characteristics of the ratio E_2/E_1 as a function of frequency, that is, point out the essential differences (if any) in this ratio as a frequency function in the above four cases, demonstrating the validity of any significant conclusions.

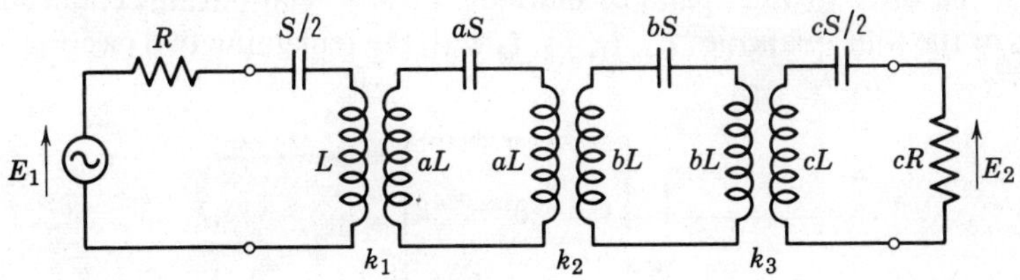

Prob. 24.

C H A P T E R S I X

Properties of Two Terminal-Pair Networks

1 Parameters Used to Characterize Two Terminal-Pair Networks*

In the synthesis procedures relating to *RLC* driving-point and transfer impedances, extensive use is made of various properties of two terminal-pair networks and the interrelationship of different structural forms for their realization. It is essential, therefore, that we discuss these collateral topics before continuing with the main theme of this text.

Of the three distinct pairs of mutually inverse relationships connecting two of the four quantities E_1, I_1, E_2, I_2 with the remaining two (see Fig. 1),

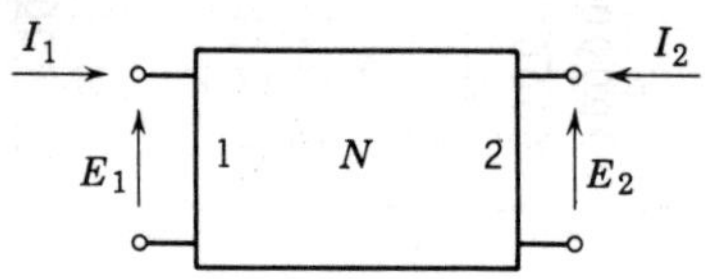

Fig. 1. Two terminal-pair network showing voltage and current references.

the ones that interest us most are the familiar y and z systems.

$$\begin{aligned} y_{11}E_1 + y_{12}E_2 &= I_1 \\ y_{21}E_1 + y_{22}E_2 &= I_2 \end{aligned} \tag{1}$$

and

$$\begin{aligned} z_{11}I_1 + z_{12}I_2 &= E_1 \\ z_{21}I_1 + z_{22}I_2 &= E_2 \end{aligned} \tag{2}$$

and the so-called general circuit parameter or *ABCD* equations

$$\begin{aligned} AE_2 - BI_2 &= E_1 \\ CE_2 - DI_2 &= I_1 \end{aligned} \tag{3}$$

and

$$\begin{aligned} DE_1 - BI_1 &= E_2 \\ CE_1 - AI_1 &= I_2 \end{aligned} \tag{4}$$

* It is assumed that the reader has a reasonable acquaintance with this topic from his introductory circuit studies. The treatment in this article is intended only as a brief review and summary.

In terms of the determinants of Eqs. 1 and 2

$$|y| = y_{11}y_{22} - y_{12}^2 \tag{5}$$

$$|z| = z_{11}z_{22} - z_{12}^2 \tag{6}$$

the interrelation of the coefficients in these inverse equations reads

$$y_{11} = z_{22}/|z|, \qquad y_{22} = z_{11}/|z|, \qquad y_{12} = -\,z_{12}/|z| \tag{7}$$

and

$$z_{11} = y_{22}/|y|, \qquad z_{22} = y_{11}/|y|, \qquad z_{12} = -\,y_{12}/|y| \tag{8}$$

By inspection of Eqs. 1 through 4, we see that

$$A = \left(\frac{E_1}{E_2}\right)_{I_2=0} = -\frac{y_{22}}{y_{12}} = \frac{z_{11}}{z_{12}} = -\left(\frac{I_2}{I_1}\right)_{E_1=0} \tag{9}$$

$$B = -\left(\frac{E_1}{I_2}\right)_{E_2=0} = \frac{-1}{y_{12}} = \frac{|z|}{z_{12}} = -\left(\frac{E_2}{I_1}\right)_{E_1=0} \tag{10}$$

$$C = \left(\frac{I_1}{E_2}\right)_{I_2=0} = \frac{1}{z_{12}} = -\frac{|y|}{y_{12}} = \left(\frac{I_2}{E_1}\right)_{I_1=0} \tag{11}$$

$$D = \left(\frac{I_1}{I_2}\right)_{E_2=0} - \frac{z_{22}}{z_{12}} = -\frac{y_{11}}{y_{12}} = \left(\frac{E_2}{E_1}\right)_{I_1=0} \tag{12}$$

From these we get

$$z_{12} = \frac{1}{C}, \qquad z_{11} = \frac{A}{C}, \qquad z_{22} = \frac{D}{C} \tag{13}$$

$$y_{12} = \frac{-1}{B}, \qquad y_{11} = \frac{D}{B}, \qquad y_{22} = \frac{A}{B} \tag{14}$$

and

$$AD - BC = 1 \tag{15}$$

Also we see that electrical symmetry ($z_{11} = z_{22}$ or $y_{11} = y_{22}$) implies in the *ABCD* system that

$$A = D \tag{16}$$

A symmetrical two terminal-pair network is characterized by *two* parameters, an unsymmetrical one by *three*.

The third pair of mutually inverse relations linking the voltages and currents in a two terminal-pair network are less commonly used but sometimes needed. They are the mixed systems

$$\begin{aligned} g_{11}E_1 + g_{12}I_2 &= I_1 \\ g_{21}E_1 + g_{22}I_2 &= E_2 \end{aligned} \tag{17}$$

and

$$\begin{aligned} h_{11}I_1 + h_{12}E_2 &= E_1 \\ h_{21}I_1 + h_{22}E_2 &= I_2 \end{aligned} \tag{18}$$

One readily finds

$$g_{11} = \left(\frac{I_1}{E_1}\right)_{I_2=0} = \frac{|y|}{y_{22}} = \frac{1}{z_{11}} \tag{19}$$

$$g_{22} = \left(\frac{E_2}{I_2}\right)_{E_1=0} = \frac{1}{y_{22}} = \frac{|z|}{z_{11}} \tag{20}$$

$$g_{12} = \left(\frac{I_1}{I_2}\right)_{E_1=0} = \frac{y_{12}}{y_{22}} = \frac{-z_{12}}{z_{11}} \tag{21}$$

$$g_{21} = \left(\frac{E_2}{E_1}\right)_{I_2=0} = \frac{-y_{12}}{y_{22}} = \frac{z_{12}}{z_{11}} \tag{22}$$

and the corresponding relations for the coefficients in the inverse equations

$$h_{11} = \left(\frac{E_1}{I_1}\right)_{E_2=0} = \frac{1}{y_{11}} = \frac{|z|}{z_{22}} \tag{23}$$

$$h_{22} = \left(\frac{I_2}{E_2}\right)_{I_1=0} = \frac{|y|}{y_{11}} = \frac{1}{z_{22}} \tag{24}$$

$$h_{12} = \left(\frac{E_1}{E_2}\right)_{I_1=0} = \frac{-y_{12}}{y_{11}} = \frac{-z_{12}}{z_{22}} \tag{25}$$

$$h_{21} = \left(\frac{I_2}{I_1}\right)_{E_2=0} = \frac{y_{12}}{y_{11}} = \frac{-z_{12}}{z_{22}} \tag{26}$$

It is interesting to note that these mixed Eqs. 17 and 18 are skew symmetrical. Their determinants

$$|g| = g_{11}g_{22} - g_{12}g_{21} = \frac{z_{22}}{z_{11}} = \frac{y_{11}}{y_{22}} \tag{27}$$

and

$$|h| = h_{11}h_{22} - h_{12}h_{21} = |g|^{-1} \tag{28}$$

are impedance ratios and hence are dimensionless.

Another set of parameters that is found to be particularly useful in conventional filter design methods involves the so-called *image impedances* and the *image propagation function*. When the network is terminated in its appropriate image impedances at both ends, then a condition of impedance match obtains at both terminal pairs. In terms of this basic definition of the image impedances it is a simple matter to derive expressions for them in terms of the above parameter sets. With reference to the situation in Fig. 2, noting that $Z_{I1} = E_1/I_1$, $Z_{I2} = E_2/(-I_2)$, we get

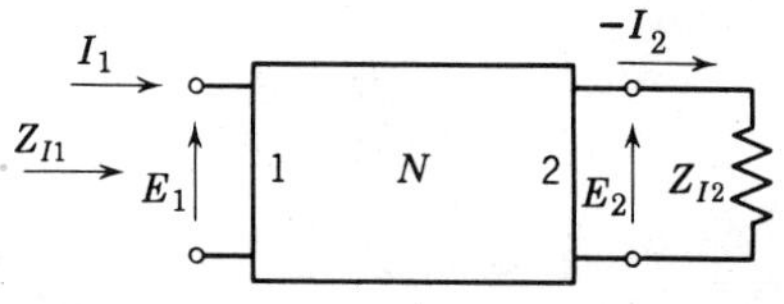

Fig. 2. Appropriate to computation of the image impedance Z_{I1}, Eq. 29.

through division of the first of Eqs. 3 by the second

$$Z_{I1} = \frac{AZ_{I2} + B}{CZ_{I2} + D} \tag{29}$$

If we reverse the orientation of things as shown in Fig. 3 and note that

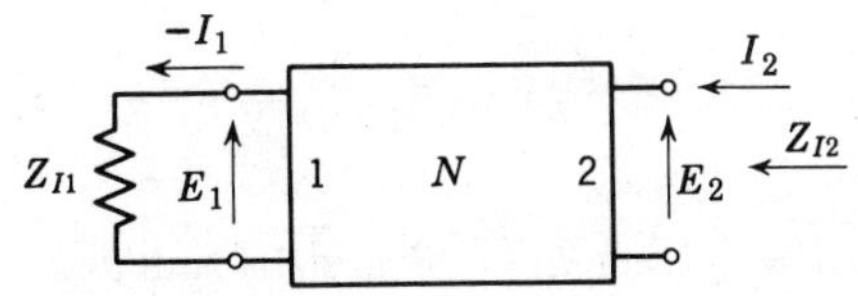

Fig. 3. Appropriate to computation of the image impedance Z_{I2}, Eq. 30.

now $Z_{I2} = E_2/I_2$, $Z_{I1} = E_1/(-I_1)$, we get through division of the first Eq. 4 by the second

$$Z_{I2} = \frac{DZ_{I1} + B}{CZ_{I1} + A} \tag{30}$$

Simultaneous solution of Eqs. 29 and 30 yields

$$Z_{I1} = \sqrt{AB/CD} = Z_{I2} \times A/D \tag{31}$$

In terms of Eqs. 9 through 12 the image impedances may alternately be expressed in terms of the y_{ik} and z_{ik} parameters, thus

$$Z_{I1} = \sqrt{z_{11}/y_{11}}, \quad Z_{I2} = \sqrt{z_{22}/y_{22}} \tag{32}$$

It is interesting to note that each image impedance is equal to the geometric mean of the open- and short-circuit impedances pertaining to the same terminal pair of the network (the open- or short-circuit constraint being applied at the opposite terminal pair).

With the network terminated in its image impedance as in Fig. 2, we get from Eqs. 3 the voltage ratio

$$\frac{E_1}{E_2} = \sqrt{\frac{A}{D}}(\sqrt{AD} + \sqrt{BC}) \tag{33}$$

and the current ratio

$$\frac{I_1}{-I_2} = \sqrt{\frac{D}{A}}(\sqrt{AD} + \sqrt{BC}) \tag{34}$$

whereas for the opposite orientation as in Fig. 3, Eqs. 4 yield the same relations with subscripts 1 and 2, and the letters A and D interchanged.

The so-called image propagation function θ is defined by

$$e^{\theta} = \sqrt{\frac{E_1 I_1}{-E_2 I_2}} = \sqrt{AD} + \sqrt{BC} \tag{35}$$

Through use of Eq. 15 we see that

$$e^{-\theta} = \sqrt{AD} - \sqrt{BC} \tag{36}$$

hence

$$\cosh \theta = \sqrt{AD}, \qquad \sinh \theta = \sqrt{BC} \tag{37}$$

If we let

$$\sqrt{A/D} = \sqrt{Z_{I1}/Z_{I2}} = n \tag{38}$$

then we find

$$\begin{aligned} A &= n \cosh \theta, \qquad B = nZ_{I2} \sinh \theta \\ C &= \frac{1}{n} \frac{\sinh \theta}{Z_{I2}}, \qquad D = \frac{\cosh \theta}{n} \end{aligned} \tag{39}$$

With these values, Eqs. 3 take the form

$$\begin{aligned} E_1 &= n(E_2 \cosh \theta - I_2 Z_{I2} \sinh \theta) \\ I_1 &= \frac{1}{n}\left(E_2 \frac{\sinh \theta}{Z_{I2}} - I_2 \cosh \theta\right) \end{aligned} \tag{40}$$

which look like the equations for a transmission line with characteristic impedance Z_{I2}, propagation function θ, and an n:1 ratio transformer at the input end.

Characterization of the lumped network in terms of image parameters therefore leads naturally to an interpretation of its transmission properties in terms of wave propagation concepts which are particularly useful in the analysis of uniform recurrent structures (identical two terminal-pair networks in cascade) or of a cascade of dissimilar networks for which adjacent image impedances are alike (or nearly alike). In numerous other practical situations where the concepts of wave transmission and reflection provide the basis for a quasi-physical interpretation of circuit behavior, there the image parameter (or analogous) characterization of the pertinent networks is relevant.

In connection with Eqs. 40 it is useful to observe that they remain valid if we interchange the subscripts 1 and 2, and reciprocate the ratio factor n. They then have the same form as the equations for a transmission line having the characteristic impedance Z_{I1}, a propagation function θ, and an n:1 ratio transformer at the output end.

The propagation function may be expressed in terms of admittances and impedances through noting first that Eqs. 35 and 36 enable one to write

$$e^{2\theta} = \frac{\sqrt{AD/BC} + 1}{\sqrt{AD/BC} - 1} \tag{41}$$

whereupon use of Eqs. 9 through 12 yields

$$e^{2\theta} = \frac{\sqrt{y_{11}z_{11}}+1}{\sqrt{y_{11}z_{11}}-1} = \frac{\sqrt{y_{22}z_{22}}+1}{\sqrt{y_{22}z_{22}}-1} \tag{42}$$

2 Interconnection of Two Terminal-Pair Networks

When several two terminal-pair networks are connected in series, in parallel, in series-parallel, or in cascade, it is useful in analysis problems to be able to express the parameters for the resultant network in terms of those characterizing the individual component networks. Conversely, in synthesis we may wish to use these same relationships as a means for breaking down an over-all response function in a way that will render the realization and interconnection of its components evident by inspection. The following discussions are aimed at providing the essential principles involved in either process (of combination or decomposition).

The series, parallel, and series-parallel interconnections of component networks are shown respectively in parts (*a*), (*b*), and (*c*) of Fig. 4. The

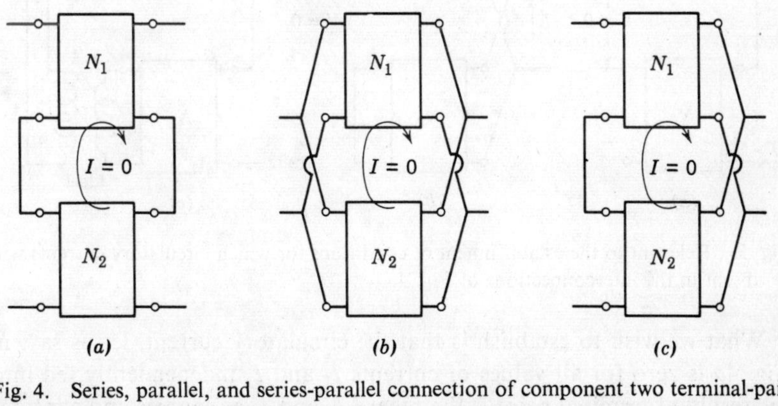

Fig. 4. Series, parallel, and series-parallel connection of component two terminal-pair networks.

parameters for the resultant two terminal-pair network may easily be found in terms of those characterizing the components provided that there is no current circulating between the component networks as is indicated in Fig. 4. The reason for this restriction is appreciated by noting that the normal situation at a terminal pair is to have the same current enter at one terminal that leaves at the other one, and that this situation is upset by the presence of a circulating current. When the current entering and leaving at each terminal pair is not the same, then the equations relating the voltages and currents for the component networks are no longer those that characterize this network when it is by itself. In other words, the equations which correctly express the relations between voltages and currents at the terminal pairs for each separate network become invalid when the networks are interconnected, unless the act of interconnection does not cause a circulatory current between them.

When two component networks N_1 and N_2 are given, and their interconnection is contemplated, one must be able to devise some method of determining whether a circulatory current will exist when the interconnection is carried out and arbitrary currents or voltages are applied to the resultant terminal pairs. A simple test for making such a determination is shown for the series and parallel connections in parts (*a*) and (*b*) of Fig. 5. Use is here made of Thévenin's theorem and of the superposition of responses due to separate sources.

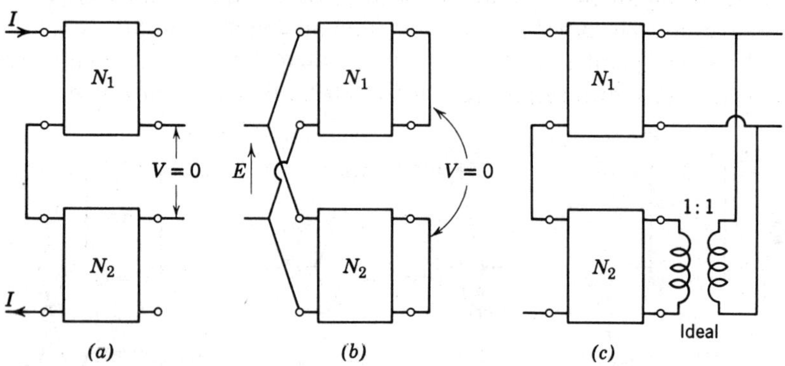

Fig. 5. Relevant to the establishment of conditions for which circulatory currents will be absent in the interconnections of Fig. 4.

What we wish to establish is that the circulatory current, let us say, in Fig. 4*a* is zero for all values of currents I_1 and I_2 independently fed into the resultant terminal pairs. We regard I_1 and I_2 as sources and consider them one at a time. With $I_1 = I$ and $I_2 = 0$ we open the loop around which current can circulate and see if this open-circuit voltage (the voltage V in Fig. 5*a*) is zero. Next we make a similar test with $I_2 = I$ and $I_1 = 0$. If $V = 0$ both times, then no current will circulate for any values of I_1 and I_2 acting simultaneously.

The analogous pair of tests for a contemplated parallel connection are illustrated in Fig. 5*b*. In the parallel connection of Fig. 4*b* we must be sure that the circulatory current is zero for all values of voltages E_1 and E_2 independently applied to the resultant terminal pairs. We regard E_1 and E_2 as sources and consider them one at a time. With $E_1 = E$ and $E_2 = 0$ (both output terminal pairs short-circuited) we open the loop around which current can circulate and see if this open-circuit voltage (the voltage V in Fig. 5*b*) is zero. Then we make a similar test with $E_2 = E$ and $E_1 = 0$ (both input terminal pairs separately short-circuited). If $V = 0$ both times, then no current will circulate for any values of E_1 and E_2 acting simultaneously.

The tests in the contemplated series-parallel connection amount to an obvious combination of the ones just described for the series and for the parallel connections. Thus the series-connected side is dealt with like the two sides in the series connection and the parallel-connected side like the two sides in the parallel connection. Two tests must again yield $V = 0$ in order to assure no circulatory current in the resultant interconnection (Fig. 4*c*).

In any situation where the tests reveal that a circulatory current will exist, one can force it to be zero by placing an ideal 1:1 ratio transformer at one of the terminal pairs as is shown for the series-parallel connection in Fig. 5*c*. In synthesis procedures where one is attempting to construct a resultant response function by an appropriate interconnection of components, this expedient may frequently be resorted to.

The appropriate method of expressing the resultant in terms of component behavior when no circulatory current exists is easily established for each contemplated interconnection. For the series connection the input currents and the output currents are the same for both component networks and are equal respectively to the resultant input and output currents, and the resultant voltages are equal to the sums of input and output voltages of the components. Hence one obtains the z system, Eqs. 2, for the series combination by adding the corresponding equations for the separate components. Factoring out the common currents thus shows that the z-matrix for the resultant equations is simply the sum of the z-matrices for the components.

In an analogous manner one sees that the parallel connection calls for addition of the component y-matrices; the series-parallel combination involves addition of the component h-matrices while for a parallel-series connection the addition of g-matrices is appropriate (see Eqs. 17 and 18).

The cascade connection shown in Fig. 6 presents no difficulties. Here the output of the first network is the input to the next one. Hence if their individual behaviors are characterized by $ABCD$ equations like Eqs. 3, then the over-all behavior is obtained through substitution of one of these equations into the other. If they individually are written in the matrix form

$$\begin{bmatrix} A & B \\ C & D \end{bmatrix} \times \begin{bmatrix} E_2 \\ -I_2 \end{bmatrix} = \begin{bmatrix} E_1 \\ I_1 \end{bmatrix} \tag{43}$$

in which the reference directions for both I_1 and $-I_2$ are from left to right, then one readily sees that the over-all *ABCD* matrix is simply the product of the individual *ABCD* matrices in the same left-to-right sequence in which the networks are arranged. With reference to the arrangement in Fig. 6, the *ABCD* matrix of N_1 is premultiplied into the *ABCD* matrix of N_2.

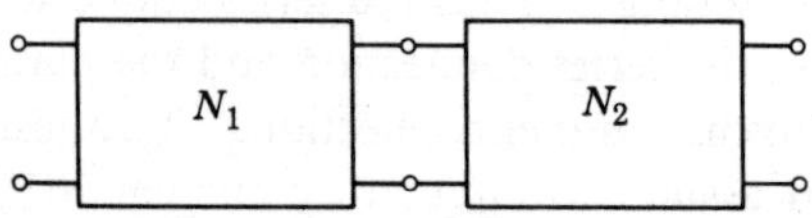

Fig. 6. Cascade (or tandem) connection of component networks.

When more than two networks are involved in the series or parallel connections, then all of their z- or y-matrices are added, provided, of course, that the conditions yielding no circulating currents are fulfilled for the first two components and again as each succeeding component is added to the existing combination. Again, these conditions can be forced by means of ideal transformers attached to either the input or output sides of all but one of the component networks.

When more than two networks are involved in a cascade connection, the over-all *ABCD* matrix is given by the multiple product of all component *ABCD* matrices arranged in the same sequence as the respective networks to which they pertain.

3 The Symmetrical Lattice Network

Since symmetrical networks require only two parameters for their characterization, they are simpler to work with in many synthesis problems where the restriction upon achievable results imposed by the symmetry condition is not serious. A collateral advantage of working with a symmetrical network is due to the fact, pointed out in Art. 6 of Ch. 1, that the symmetrical lattice is the most general symmetrical two terminal-pair network, a result that makes it possible to gain the advantage of using a specific network configuration as a point of departure without thereby subjecting the design procedure to any restrictions beyond those imposed by symmetry.

The two parameters characterizing the symmetrical lattice, Fig. 7, are its series and crossarm impedances z_a and z_b. It is helpful noting at the outset that twisting one pair of terminals is equivalent to an interchange of the impedances z_a and z_b. Such an interchange therefore merely introduces a minus sign into the expression for the transfer impedance and has no effect upon the driving-point impedance, as may be seen from the relations

$$\begin{aligned} z_{11} &= \tfrac{1}{2}(z_b + z_a) \\ z_{12} &= \tfrac{1}{2}(z_b - z_a) \end{aligned} \tag{44}$$

The z-determinant for the lattice is

$$|z| = z_{11}^2 - z_{12}^2 = \tfrac{1}{4}(z_b + z_a)^2 - \tfrac{1}{4}(z_b - z_a)^2 = z_a z_b \tag{45}$$

and hence the short-circuit admittances, according to Eqs. 7, are

$$y_{11} = \frac{z_b + z_a}{2z_a z_b}, \qquad y_{12} = \frac{z_b - z_a}{2z_a z_b} \tag{46}$$

or in terms of the lattice admittances

$$y_a = 1/z_a, \qquad y_b = 1/z_b \tag{47}$$

we have

$$\begin{aligned} y_{11} &= \tfrac{1}{2}(y_b + y_a) \\ y_{12} &= \tfrac{1}{2}(y_b - y_a) \end{aligned} \tag{48}$$

which are identical with relations 44 except for a replacement of the letter z by the letter y. The lattice is its own dual.

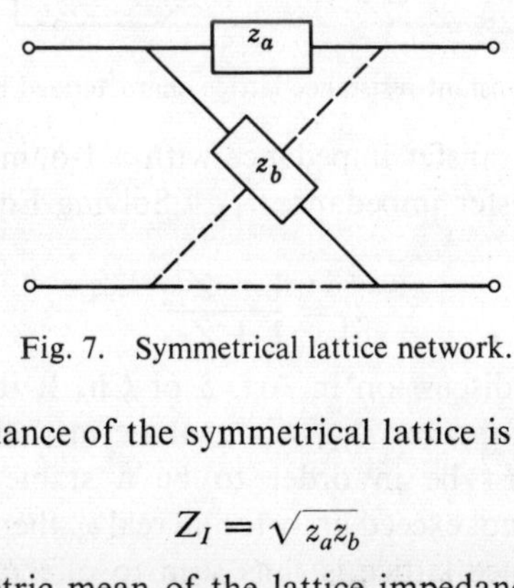

Fig. 7. Symmetrical lattice network.

The image impedance of the symmetrical lattice is found from Eqs. 32, 44, and 46 to be

$$Z_I = \sqrt{z_a z_b} \tag{49}$$

which is the geometric mean of the lattice impedances. Substitution of Eqs. 44 and 48 into Eq. 42 yields for the propagation function θ the relation

$$e^{2\theta} = \frac{z_b + z_a + 2\sqrt{z_a z_b}}{z_b + z_a - 2\sqrt{z_a z_b}} = \left(\frac{\sqrt{z_b} + \sqrt{z_a}}{\sqrt{z_b} - \sqrt{z_a}}\right)^2 \tag{50}$$

from which

$$e^{\theta} = \frac{1 + \sqrt{z_a/z_b}}{1 - \sqrt{z_a/z_b}} \tag{51}$$

or

$$\theta = \ln\left(\frac{1 + \sqrt{z_a/z_b}}{1 - \sqrt{z_a/z_b}}\right) \tag{52}$$

If the lattice impedances are reciprocal so that

$$z_a z_b = 1 \tag{53}$$

then the image impedance, Eq. 49, becomes a pure resistance of 1 ohm. When terminated in a 1-ohm resistance, such a lattice evidently has an input impedance equal to a resistance of 1 ohm. A symmetrical lattice fulfilling condition 53 is therefore called a *constant-resistance* lattice. Since function 51 is the voltage ratio E_1/E_2 for a terminal impedance equal to Z_I, we see that the constant-resistance lattice terminated in 1 ohm (Fig. 8) has the transfer function

$$\frac{E_2}{E_1} = \frac{E_2}{I_1} = Z_{12} = \frac{1 - z_a}{1 + z_a} \tag{54}$$

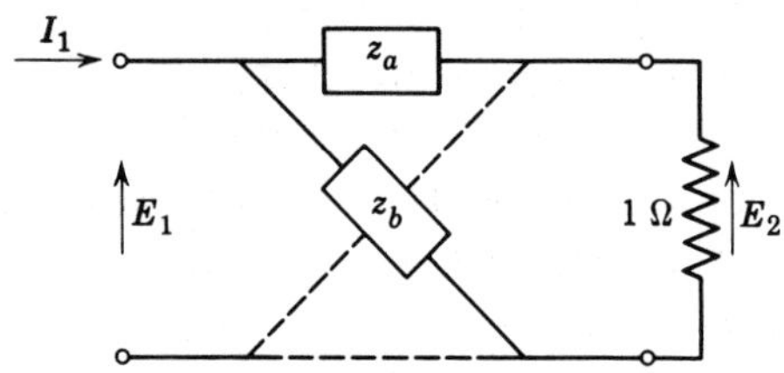

Fig. 8. Terminated constant-resistance lattice characterized by relations 53, 54, 55.

in which Z_{12} is the transfer impedance with a 1-ohm load in contrast to the open-circuit transfer impedance z_{12}. Solving Eq. 54 for z_a in terms of Z_{12} we have

$$z_a = \frac{1 - Z_{12}}{1 + Z_{12}} \tag{55}$$

According to the discussion in Art. 5 of Ch. 1, it follows that if the transfer impedance function $Z_{12}(s)$ is analytic in the right half of the s-plane (which it must be in order to be a stable transfer function) and if $|Z_{12}(j\omega)|$ does not exceed unity for all real ω, then z_a is a p.r. function. The constant-resistance lattice is thus seen to offer a possible approach to the synthesis of transfer functions that we shall exploit in our discussions of this topic later on.

4 Bartlett's Bisection Theorem

Although the symmetrical lattice is analytically convenient, it is in many practical situations undesirable because of its balanced character about a horizontal center line. For this and other reasons (economy of elements for one thing) one is usually interested in converting a lattice design into some other equivalent form. In this regard it is useful to have simple and effective means for expressing equivalence relations between the lattice and other symmetrical structures. The following general property of physically and electrically symmetrical networks is particularly useful in establishing such relations with a minimum of computational effort.

Figure 9*a* represents a two terminal-pair network that is both physically and electrically symmetrical about a vertical center line. It may be balanced or unbalanced so far as symmetry about a horizontal center line is concerned. Its electrical behavior, according to Eqs. 2, is expressed by

$$\begin{aligned} z_{11}I_1 + z_{12}I_2 &= E_1 \\ z_{12}I_1 + z_{11}I_2 &= E_2 \end{aligned} \tag{56}$$

In Fig. 9*b* this network is shown as it appears after the box containing it is bisected and the two halves are pulled slightly apart so as to expose the connecting wires.

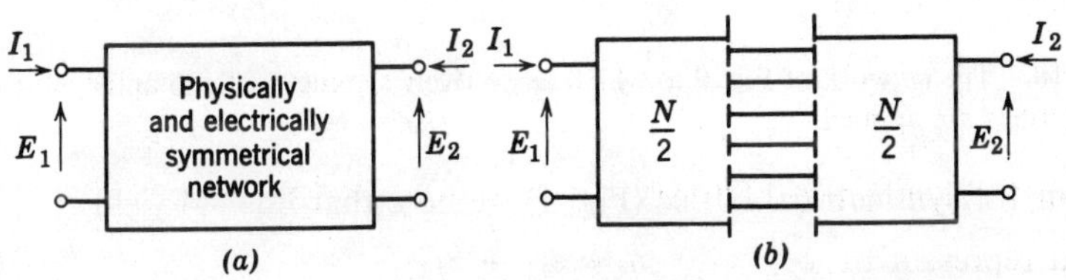

Fig. 9. Pertinent to the discussion of Bartlett's bisection theorem.

We now consider two special forms of excitation to be applied to the terminal pairs, namely: (*a*) symmetrical excitation, for which $E_1 = E_2 = E$; and (*b*) antisymmetrical excitation, for which $E_1 = E$, $E_2 = -E$. Because of the symmetry of the network it follows that $I_1 = I_2 = I$ for the symmetrical excitation, and that $I_1 = I$, $I_2 = -I$ for the antisymmetrical excitation. For the symmetrical excitation we see, moreover, that the currents in all the connecting wires between network halves are zero, and hence that the terminal conditions are unaffected if these wires are cut as shown in Fig. 10*a*. For the antisymmetrical excitation, on the other hand, the network symmetry shows that the potentials of all the connecting wires are alike and that they may be joined into a common node, as shown by the short-circuiting bar *B-B'* in Fig. 10*b*, without affecting the terminal conditions. The truth of this statement is seen by noting that the currents indicated in Fig. 10*b* by the circulatory arrows on either side of the bar *B-B'* are equal in adjacent pairs, and hence that no current exists in the bar itself.

For these special conditions of excitation, we see that the terminal conditions are the same as they are for the situations shown in Fig. 11 where only one-half of the original network is involved. Equations 56 evaluated respectively for the conditions $E_1 = E_2 = E$, $I_1 = I_2 = I$, and

$E_1 = -E_2 = E$, $I_1 = -I_2 = I$, are thus seen to yield

$$z_{och} = z_{11} + z_{12} \tag{57}$$

and

$$z_{sch} = z_{11} - z_{12} \tag{58}$$

where the subscripts are intended to indicate that these are respectively the open- and short-circuit impedances of half of the original symmetrical network.

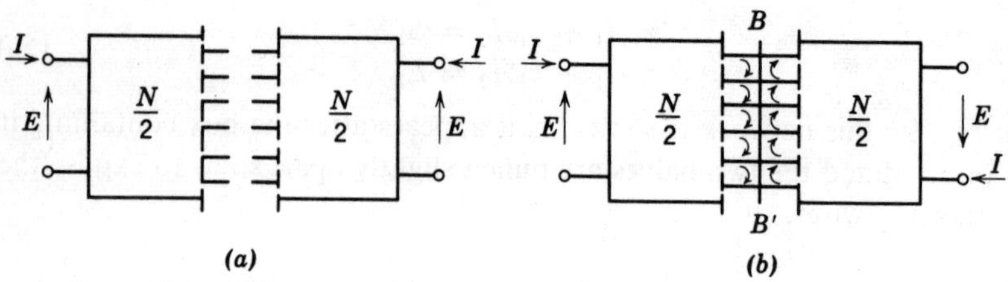

Fig. 10. The network of Fig. 9 to which respectively symmetrical and antisymmetrical excitations are applied.

For the symmetrical lattice (Fig. 7) we note that Eqs. 44 yield

$$z_b = z_{11} + z_{12} \tag{59}$$

and

$$z_a = z_{11} - z_{12} \tag{60}$$

from which we see that

$$z_a = z_{sch}, \qquad \text{and} \qquad z_b = z_{och} \tag{61}$$

which represent in very general terms the equivalence relations between the impedances of a symmetrical lattice and any other form of symmetrical structure into which the lattice may be converted.*

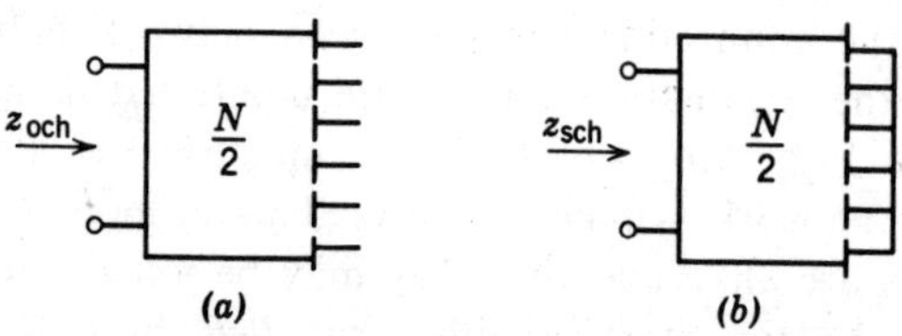

Fig. 11. Input impedances respectively equal to z_b and z_a in the lattice of Fig. 7 equivalent to the symmetrical network of Fig. 9.

The above derivation of the result expressed by Eqs. 61 is subject to an important restriction, namely, that the connecting wires between the

* This result is known as Bartlett's bisection theorem, first published by A. C. Bartlett in *Phil. Mag.*, Vol. 4, Nov. 1927, p. 902. The above derivation, which is far simpler than the algebraic one given in the original paper, is due to O. Brune: "Note on Bartlett's Bisection Theorem," *Phil. Mag.*, Vol. 14, Ser. 7, Nov. 1932, p. 806.

network halves, as shown in Fig. 9*b*, are none of them crossed, as one pair would be if the lattice itself were represented in this manner. Networks for which this condition is fulfilled were referred to by Bartlett as being "bisectable." The lattice itself was regarded by him as being not bisectable. Brune pointed out a simple way in which this restriction may, in certain situations at least, be obviated.

As shown in Fig. 12*a*, the lattice may be redrawn with a $-1:1$ ratio ideal transformer taking the place of the crossover in the z_b branches. The equivalence of this redrawn version and the normal lattice is readily shown by regarding it as a parallel combination of component networks with the y-matrices

$$\begin{bmatrix} \frac{y_a}{2} & \frac{-y_a}{2} \\ \frac{-y_a}{2} & \frac{y_a}{2} \end{bmatrix} \quad \text{and} \quad \begin{bmatrix} \frac{y_b}{2} & \frac{y_b}{2} \\ \frac{y_b}{2} & \frac{y_b}{2} \end{bmatrix} \tag{62}$$

the sum of which evidently yields the correct y-matrix for the lattice.

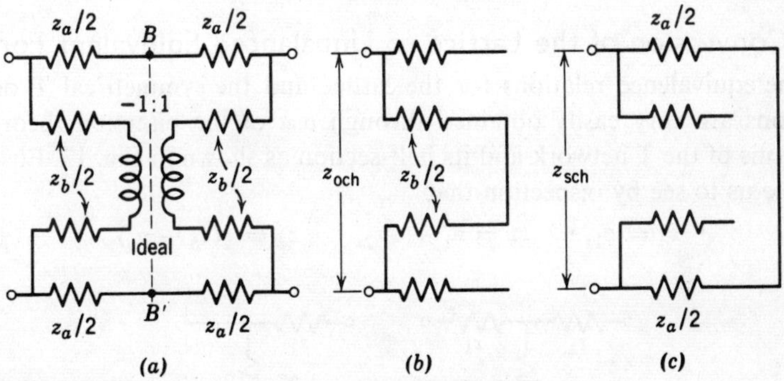

Fig. 12. Relevant to the application of Bartlett's bisection theorem to the lattice of Fig. 7.

When symmetrical excitation is applied to this redrawn version of the lattice, it is rather easy to see that there is no current in the z_a branches while the ideal transformer acts like a short circuit for the z_b branches because a zero voltage at the transformer is the only value that can simultaneously satisfy the demand of the symmetrical excitation and the voltage constraint of the transformer. Figure 12*b* represents this condition.

When antisymmetrical excitation is applied to the redrawn version of the lattice, then the ideal transformer acts like an open circuit because a zero current is the only value that can simultaneously satisfy the demand of the antisymmetrical excitation and the current constraint of the transformer, while the points B-B' are at the same potential and can be connected by a short-circuiting bar as in the situation shown in Fig. 10*b*. Figure 12*c* is now pertinent.

From Fig. 12*b* and *c* we see at once that relations 61 follow. We may conclude that if the given symmetrical network exhibits crossed pairs of wires when its halves are separated as in Fig. 9*b*, the results given by Eqs. 61 may still be applied if in the sketch of Fig. 11*a* the crossed pairs are short-circuited whereas in the sketch of Fig. 11*b* they are open-circuited. However, this modified aspect of Bartlett's bisection theorem is permitted only if replacement of the crossed wires in the original network by a $-1:1$ ratio ideal transformer leaves its behavior as a two terminal-pair network unchanged, as in the case of the lattice itself.* As a counterexample consider an ordinary symmetrical lattice with one of its z_a branches removed. It is still physically and electrically symmetrical, but the crossed wires can now not be replaced by a $-1:1$ ratio ideal transformer. This mutilated form of the lattice structure is not bisectable in the sense in which Bartlett used the term, even though we revise Fig. 11 as described above.

5 Conversion of the Lattice to Unbalanced Equivalent Forms

The equivalence relations for the lattice and the symmetrical T or Π sections are very easily obtained through use of the bisection theorem. In terms of the T network and its half-section as shown in Fig. 13, Eqs. 61 enable us to see by inspection that

$$z_a = z_1, \qquad z_b = z_1 + 2z_2, \qquad z_2 = \tfrac{1}{2}(z_b - z_a) \tag{63}$$

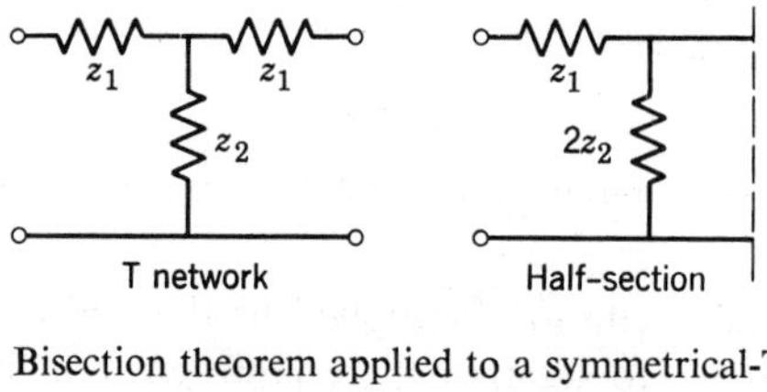

Fig. 13. Bisection theorem applied to a symmetrical-T network.

Since the expression for z_2 involves a subtraction, it is clear that the equivalent T is not always realizable.

For the Π network of Fig. 14 we have analogously (on an admittance basis)

$$y_b = y_2, \qquad y_a = y_2 + 2y_1, \qquad y_1 = \tfrac{1}{2}(y_a - y_b) \tag{64}$$

* It is readily appreciated that this condition is fulfilled for any network that has symmetry about a horizontal center line.

Here a subtraction is involved in obtaining the series branch of the equivalent Π, and again the conversion may not be possible.

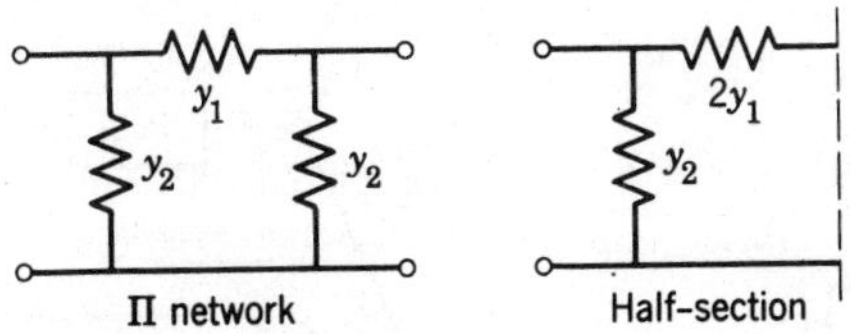

Fig. 14. Bisection theorem applied to a symmetrical-Π network.

Figure 15 illustrates an application of the bisection theorem to the conversion of a bridged-T network to an equivalent lattice. Since the lattice is analytically so much easier to work with, it is practically often expedient to make such a transformation. The equivalence relations made evident in these sketches enable one also to convert a lattice having the indicated form into an equivalent bridged T, although the following methods are more generally helpful in this regard.

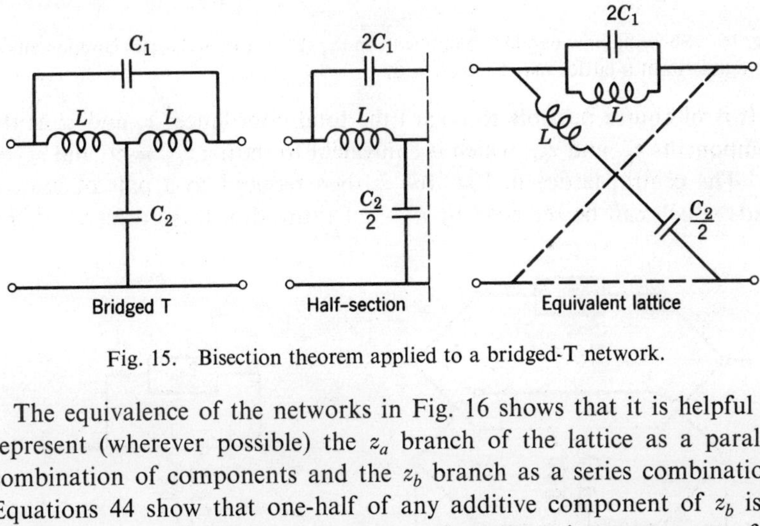

Fig. 15. Bisection theorem applied to a bridged-T network.

The equivalence of the networks in Fig. 16 shows that it is helpful to represent (wherever possible) the z_a branch of the lattice as a parallel combination of components and the z_b branch as a series combination. Equations 44 show that one-half of any additive component of z_b is a common term in z_{11} and z_{22} and hence that any series component of z_b may be removed from the lattice and accounted for by means of a separate two terminal-pair network connected in series as shown at the bottom of Fig. 16*b*. The over-all z-matrix remains unchanged by this maneuver which, however, calls for the insertion of an ideal transformer for reasons discussed above. Removal of the parallel component z_{a1} in the z_a branch, and its reassociation in the form of a separate two terminal-pair network connected in parallel, as shown at the top of Fig. 16*b*, also calls for the insertion of an ideal transformer because the combination of the two z_{a1} components (one in each of the two z_a branches) into a single impedance $2z_{a1}$ unbalances the pertinent two terminal-pair network.

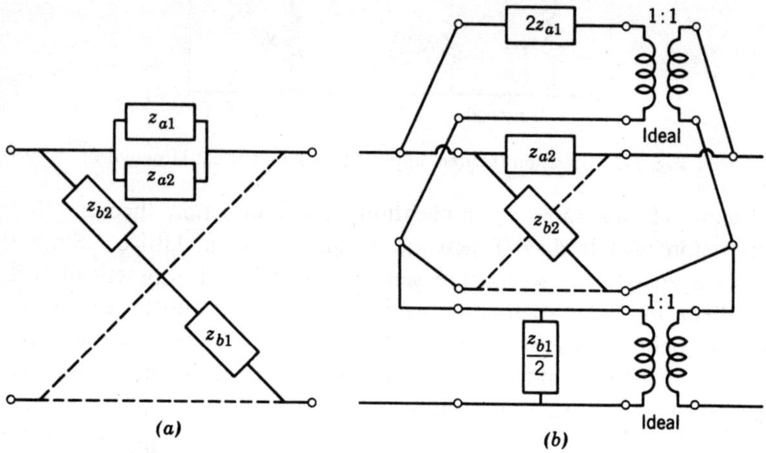

Fig. 16. Showing how parallel components in the series arm and series components in the crossarm of a lattice may be dealt with.

It is of course possible to regard the total impedances z_a and z_b as the components z_{a1} and z_{b1}, which is equivalent to setting $z_{a2} = \infty$ and $z_{b2} = 0$. The central lattice in Fig. 16*b* is then reduced to a pair of crossed leads which can be replaced by a $-1:1$ ratio ideal transformer. If this

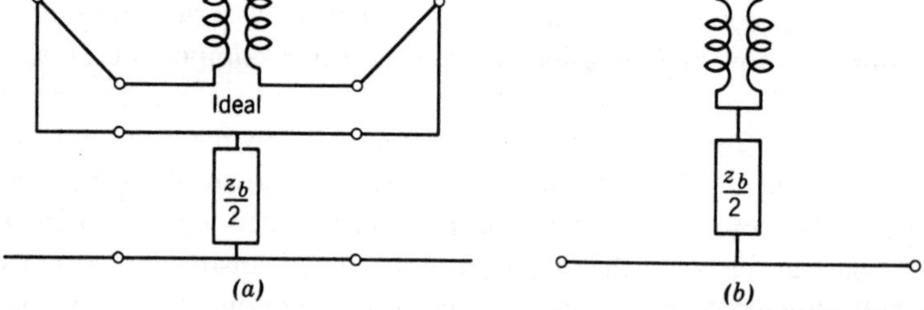

Fig. 17. Resulting form of the lattice transformation shown in Fig. 16 when $z_{a2} = \infty$ and $z_{b2} = 0$.

is done, then the other two ideal transformers are not needed, and we have the network in Fig. 17*a*, which can be redrawn as shown in Fig. 17*b*.

If the central lattice in Fig. 16*b* can be transformed to an equivalent unbalanced form, as, for example, the lattice in Fig. 15 can be, then no ideal transformers are needed at all, and we have succeeded in transforming the original lattice of Fig. 16*a* into an unbalanced form that is practically constructible. Since in most situations the given lattice will permit several different decompositions of z_a and z_b into respectively parallel and series components, it is quite likely that for one of these at least, such a complete transformation to a practical unbalanced form will be successful.

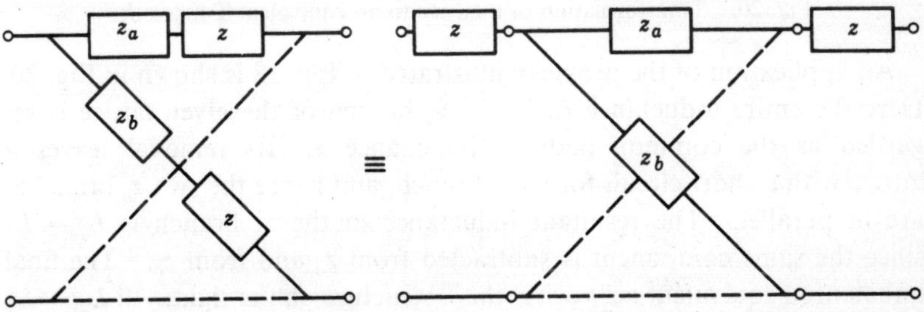

Fig. 18. Possible transformation when the lattice arms have a common series component.

Another property of the lattice that is useful when attempting its transformation to an equivalent unbalanced form is illustrated in Figs. 18 and 19. Thus Eqs. 61 show that any additive component that is common to z_a and z_b is also common to z_{sch} and z_{och}, and hence may alternately be considered in series with both pairs of terminals. In Fig. 18 the impedance z is such a common additive component.

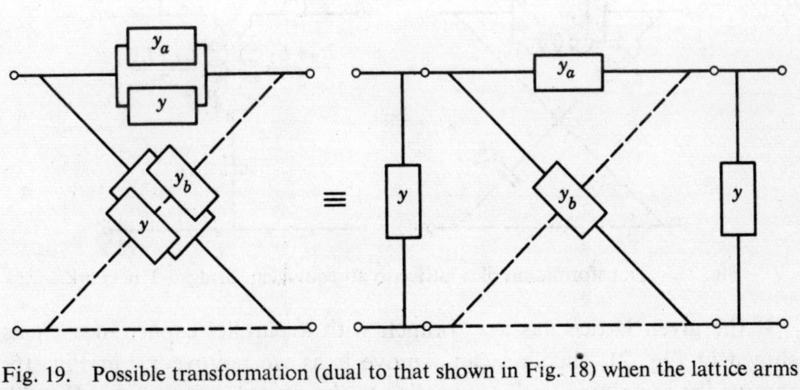

Fig. 19. Possible transformation (dual to that shown in Fig. 18) when the lattice arms have a common parallel component.

The dual situation results if we rewrite Eqs. 61 in terms of admittances. The common additive component (an admittance y) is then physically in parallel with each lattice branch and may alternately be regarded as being in parallel with each terminal pair, as shown in Fig. 19.

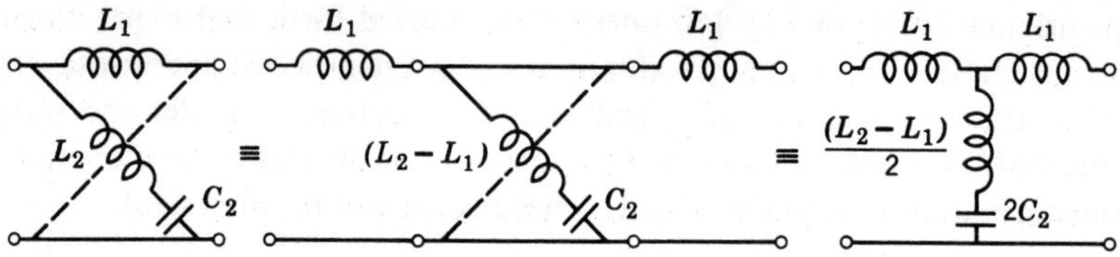

Fig. 20. Transformation of a lattice to an equivalent-T network.

An application of the property illustrated in Fig. 18 is shown in Fig. 20. Here the entire inductance L_1 in the z_a branch of the given lattice is regarded as the common additive impedance z. Its removal leaves a lattice with a short circuit for its z_a branch, and hence the two z_b branches are in parallel. The resultant inductance in the z_b branch is $L_2 - L_1$ since the same component is subtracted from z_a and from z_b. The final unbalanced equivalent network is the T structure on the right. If $L_2 - L_1$ is negative, one can replace the T arrangement of inductances by a pair of mutually coupled inductances. The coupling coefficient is seen to be

$$k = \left|\frac{L_2 - L_1}{L_2 + L_1}\right| < 1 \tag{65}$$

and hence this arrangement is always realizable.

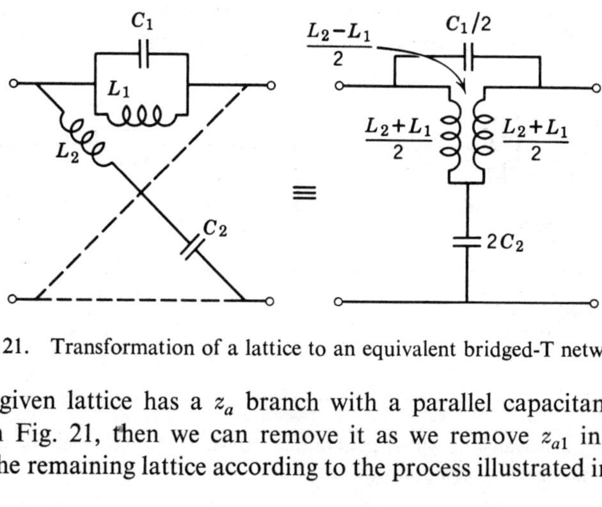

Fig. 21. Transformation of a lattice to an equivalent bridged-T network.

If the given lattice has a z_a branch with a parallel capacitance C_1 as shown in Fig. 21, then we can remove it as we remove z_{a1} in Fig. 16, convert the remaining lattice according to the process illustrated in Fig. 20, and then reassociate this parallel capacitance which must be an impedance twice as large and hence has one-half of its original value. The result is a bridged-T structure. The indicated mutual inductive coupling is, of course, needed only if $L_1 > L_2$.

It is interesting to observe that alternate use of the properties illustrated in Figs. 18 and 19 may in some cases yield a complete ladder development of a given lattice. Suppose the z_a and z_b branches of the lattice have the forms shown in Fig. 22*a* and *b*. After removing the first common series inductance L_1 according to the principle illustrated in Fig. 18, we can remove C_1 as a common shunt component by the principle shown in Fig. 19, and so forth. The completely developed lattice is the ladder network in Fig. 22*c*. Application of Bartlett's bisection theorem to this resultant ladder yields again the networks for z_a and z_b.

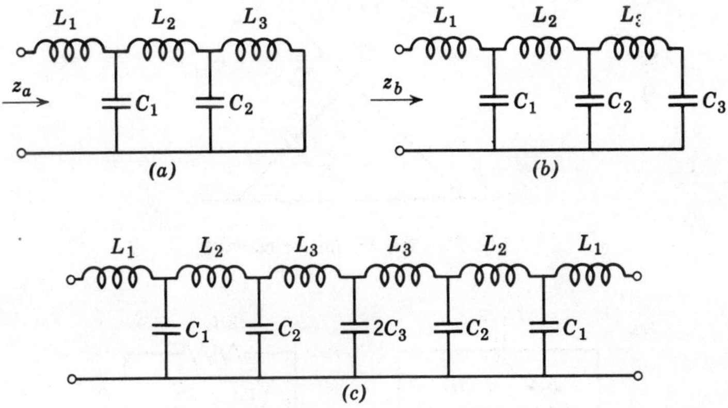

Fig. 22. Lattice impedances (*a*) and (*b*) leading to a complete ladder development (*c*).

In dealing with *RC* networks the possibility of using mutual inductive coupling, as is done in some of the above examples, does not exist; hence additional techniques are needed to accomplish the conversion of a lattice to an equivalent unbalanced structure when such a conversion is at all possible. To illustrate some of these techniques, let us consider the *RC* lattice shown in Fig. 23.

To begin with we can attempt to convert this lattice into one or the other of the two bridged-T sections shown in Fig. 24. The one in part (*a*) is obtained by removing the parallel capacitance C_1 in the z_a branch according to the principle illustrated in Fig. 16 and then converting the remaining lattice into an equivalent-T section as shown in Fig. 20. The one in part (*b*) is similarly obtained with the roles of R_1 and C_1 interchanged. The conditions under which either of these conversions is possible are stated under the respective schematics.

When neither of these conditions is fulfilled, we may still be able to convert the lattice of Fig. 23 to an equivalent unbalanced form by splitting the z_b branch into parallel components and correspondingly representing the lattice as a parallel combination of two lattices, as shown in Fig. 25 in which α is a positive numeric less than unity. The upper lattice is convertible into an equivalent T if

$$R_1 \leq R_2/\alpha \tag{66}$$

and the lower one if

$$C_1 \geq (1-\alpha)C_2 \tag{67}$$

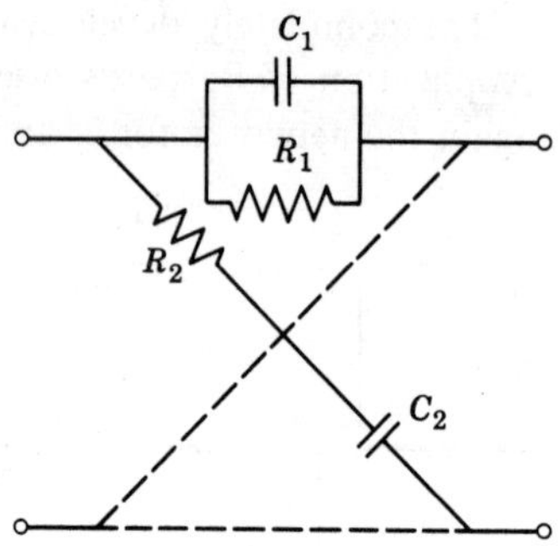

Fig. 23. An *RC* lattice network.

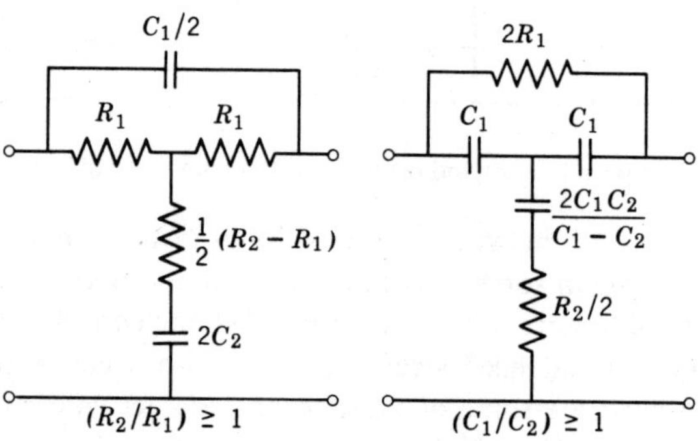

Fig. 24. Bridged-T equivalents of the lattice shown in Fig. 23.

The numerical factor α may thus have any value within the range

$$1 - \frac{C_1}{C_2} \leq \alpha \leq \frac{R_2}{R_1} \tag{68}$$

In order for this range to exist, it is necessary that

$$\frac{R_2}{R_1} + \frac{C_1}{C_2} \geq 1 \tag{69}$$

which becomes the condition for which the present method of lattice conversion is possible. We see that it is more lenient than either of the conditions given in Fig. 24 for conversion of the lattice to a bridged-T section.

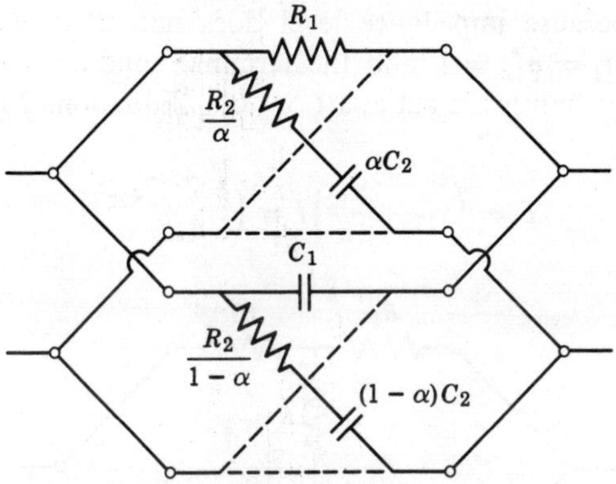

Fig. 25. Redrawn version of the lattice in Fig. 23 in which the crossarm is split into parallel components.

When condition 69 is fulfilled with the inequality sign, then we have theoretically an infinite possible variety of unbalanced equivalents. If we choose α equal to its lower limiting value according to the range 68, then the equivalent T for the lower lattice in Fig. 25 involves two instead of three capacitances. Hence this choice will ordinarily be preferred. The resultant unbalanced equivalent for the lattice of Fig. 23 is then the twin-T circuit shown in Fig. 26 in which

$$R = \frac{R_1 C_2}{2}\left[\frac{(R_2/R_1) + (C_1/C_2) - 1}{C_2 - C_1}\right] \tag{70}$$

Evidently all element values are positive unless $C_2 < C_1$, in which case the bridged-T circuit of Fig. 24*b* is possible and also involves not more than three capacitances.

6 Dissymmetrical Networks Derivable From the Symmetrical Lattice

In order to retain the advantages of analytic simplicity and manipulative flexibility of the symmetrical lattice in situations where unequal impedances at the input and output terminal pairs must be accommodated, it is possible to expand its field of usefulness through the following extension of Bartlett's bisection theorem.

Figure 27 represents a symmetrical network for which the impedance level of the right-hand half is multiplied by the real factor $a > 0$. In a manner analogous to Brune's proof of Bartlett's bisection theorem we say:

(*a*) Let $E_2 = E_1 = E$. If, under these circumstances, the wires between the halves were cut, the voltages between them would be the same for both halves because impedance level does not affect a voltage ratio. Clearly then, $I_1 = aI_2 = I$, and the terminal conditions are unchanged whether the two halves are cut apart or not. Equations 2 for the over-all network yield

$$E = \left(z_{11} + \frac{z_{12}}{a}\right)I = \left(z_{12} + \frac{z_{22}}{a}\right)I \tag{71}$$

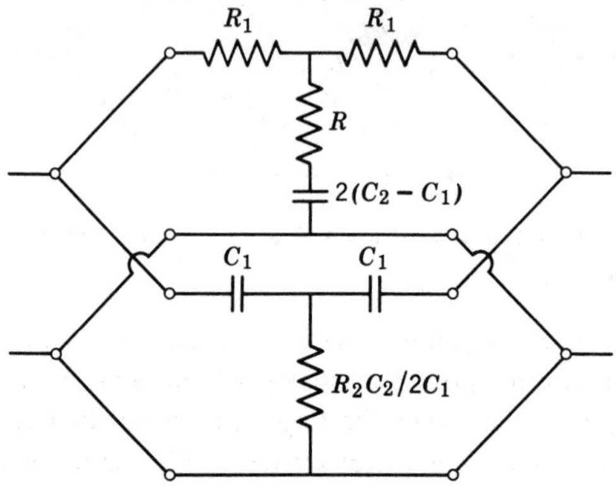

Fig. 26. Twin-T equivalent of the lattice shown in Fig. 23, possible if condition 69 is fulfilled.

(*b*) Let $E_2/a = -E_1 = -E$. If, for this excitation, the connecting wires between the halves were connected by a short-circuiting bar, the currents on the two sides of this bar would be equal in their magnitudes but circulating in opposite directions, so that no net currents would traverse the short circuit. Under these circumstances $I_2 = -I_1 = -I$, and it is clear that the terminal conditions are unchanged whether the two halves are isolated by the short circuit or not. Equations 2 now yield

$$E = (z_{11} - z_{12})I = (z_{22} - z_{12})I/a \tag{72}$$

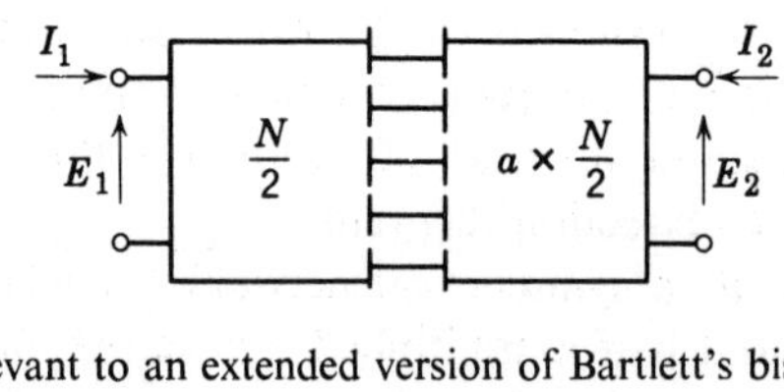

Fig. 27. Relevant to an extended version of Bartlett's bisection theorem.

With reference to Fig. 28, in which $N/2$ represents the left-hand half of the network of Fig. 27, Eqs. 71 and 72 enable us to write

$$z_{\text{och}} = z_{11} + \frac{z_{12}}{a} = z_{12} + \frac{z_{22}}{a} = z_b \tag{73}$$

and

$$z_{\text{sch}} = z_{11} - z_{12} = \frac{z_{22} - z_{12}}{a} = z_a \tag{74}$$

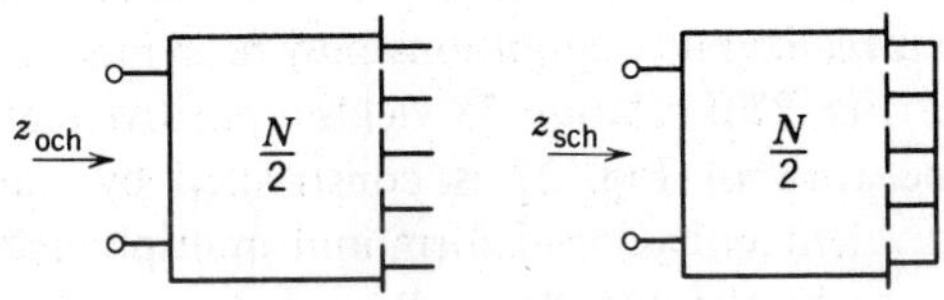

Fig. 28. The left-hand half of the network of Fig. 27.

in which z_a and z_b are the impedances of a symmetrical lattice from which the dissymmetrical network in Fig. 27 is derived. From these equations we obtain

$$z_{11} = \frac{az_b + z_a}{a + 1} \tag{75}$$

$$\frac{z_{22}}{a} = \frac{z_b + az_a}{a + 1} \tag{76}$$

$$z_{12} = \frac{a(z_b - z_a)}{a + 1} \tag{77}$$

$$a = \frac{z_{22} - z_{12}}{z_{11} - z_{12}} \tag{78}$$

It is interesting to see that the transfer impedance, except for a constant multiplier, is the same as for the symmetrical lattice. Although the poles of the driving-point impedances are the same as those of the driving-point impedance of the symmetrical lattice, their zeros are not, nor are the zeros the same for both z_{11} and z_{22}. In this respect the network of Fig. 27 is more general than the symmetrical lattice cascaded with an ideal transformer.

In a synthesis problem in which z_{11}, z_{22}, and z_{12} are given, one can compute z_a and z_b from Eqs. 73 and 74, and the level factor a from Eq. 78. Besides fulfilling the usual realizability conditions (analyticity in the right half-plane, plus the conditions expressed by Eqs. 105 and 106 of Ch. 1), the given impedances must yield a positive real number for the factor a. The usual realizability conditions, as in the symmetrical case, assure the realizability of the lattice impedances z_a and z_b. Thus if r_{11}, r_{22}, r_{12} denote the respective real parts of the impedances z_{11}, z_{22}, z_{12} and

k_{11}, k_{22}, k_{12} their residues in j-axis poles, one has by Eqs. 75, 76, and 77

$$r_{11} = \frac{ar_b + r_a}{a + 1}, \qquad r_{22} = \frac{a(r_b + ar_a)}{a + 1} \tag{79}$$

$$r_{11}r_{22} - r_{12}^2 = ar_ar_b$$

and similar expressions for the k_{ik}. It follows that the conditions expressed by Eqs. 105 and 106 of Ch. 1 guarantee the p.r. character of z_a and z_b. Hence we may say that any set of impedances z_{11}, z_{22}, z_{12} fulfilling the usual realizability conditions may be synthesized in the form of the network of Fig. 27 if relation 78 yields a positive real value for a.

The desired network of Fig. 27 is constructed by transforming the lattice to an equivalent unbalanced form and multiplying the right-hand half of this network by the impedance level factor a. To carry out this latter operation it may be necessary first to bisect the equivalent unbalanced network, multiply one half by the factor a, and then rejoin the two halves.

Since it is sometimes necessary or advantageous to work on an admittance rather than on an impedance basis, it is essential that we be familiar with the dual form of the preceding results. In this connection we observe from the fundamental dual character of the T and Π structures of Figs. 13 and 14, together with their lattice equivalence relations as expressed by Eqs. 63 and 64, that whereas structurally the lattice is its own dual the analytic relationships expressing the transition from a given lattice to its dual are given by

$$z_a \rightarrow y_b, \qquad z_b \rightarrow y_a$$

and

$$z_{12} \rightarrow -y_{12} \tag{80}$$

The relations respectively dual to Eqs. 73 through 78 are thus seen to read

$$y_{\text{och}} = y_{11} - ay_{12} = -y_{12} + ay_{22} = y_a \tag{81}$$

$$y_{\text{sch}} = y_{11} + y_{12} = a(y_{22} + y_{12}) = y_b \tag{82}$$

$$y_{11} = \frac{ay_b + y_a}{a + 1} \tag{83}$$

$$ay_{22} = \frac{ay_a + y_b}{a + 1} \tag{84}$$

$$y_{12} = \frac{y_b - y_a}{a + 1} \tag{85}$$

$$a = \frac{y_{11} + y_{12}}{y_{22} + y_{12}} \tag{86}$$

in which the factor a is still an impedance level factor.

PROBLEMS

1. Two identical symmetrical lattice networks in cascade may be considered equivalent to a single resultant symmetrical lattice, as indicated in the following sketch. Determine the networks for the resultant components z_{a0} and z_{b0} in terms of z_a and z_b.

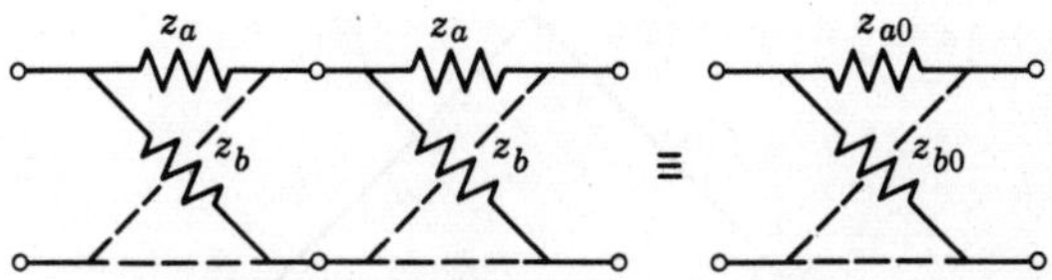

Prob. 1.

2. For the cascade of two identical lattices shown in the adjacent sketch, find an equivalent unbalanced network not involving ideal transformers or negative elements or closely coupled inductances. Determine the analytic expression for the over-all transfer impedance $Z_{12}(s)$ when this network is terminated in a 1-ohm resistance. Assume $L = C = 1$.

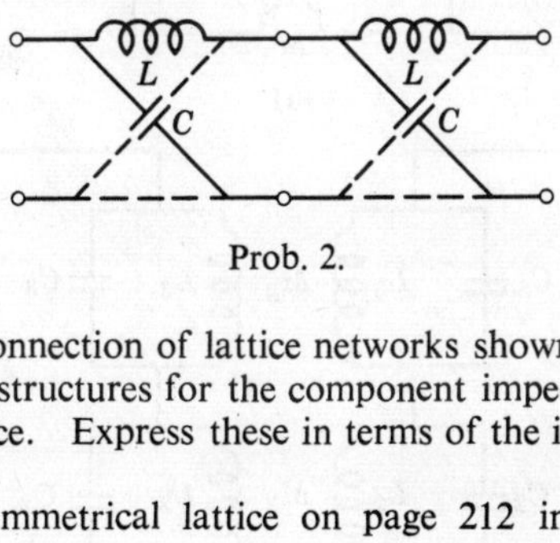

Prob. 2.

3. For the series connection of lattice networks shown in the accompanying figure, determine the structures for the component impedances z_{a0} and z_{b0} of a single equivalent lattice. Express these in terms of the impedances z_{a1}, z_{a2}, z_{b1}, and z_{b2}.

4. Develop the symmetrical lattice on page 212 into an equivalent unbalanced ladder form.

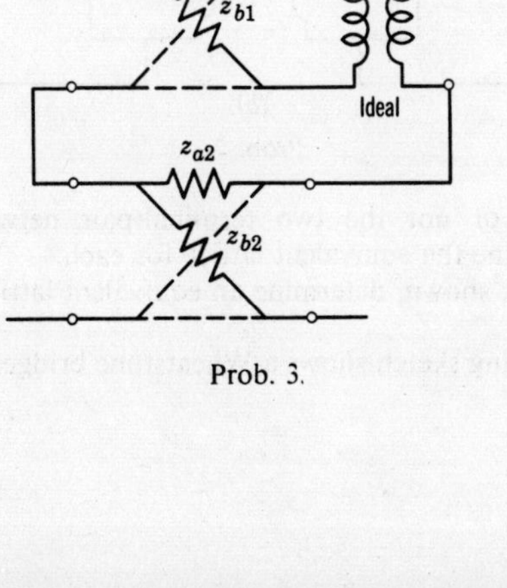

Prob. 3.

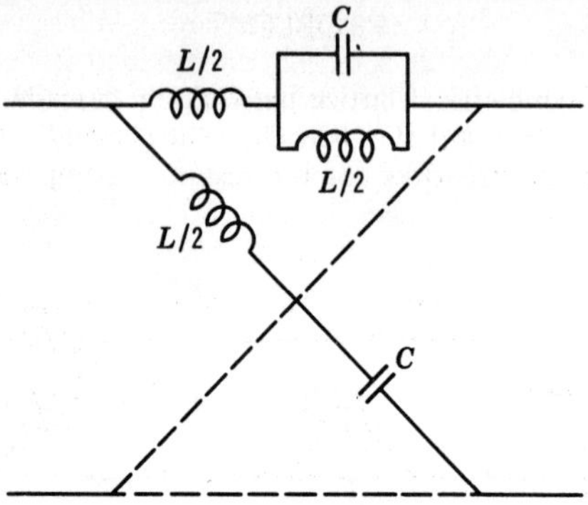

Prob. 4.

5. Determine the lattice equivalent for the network shown in sketch (*a*). Using this result, and that of Prob. 3, determine the lattice equivalent of the network shown in sketch (*b*).

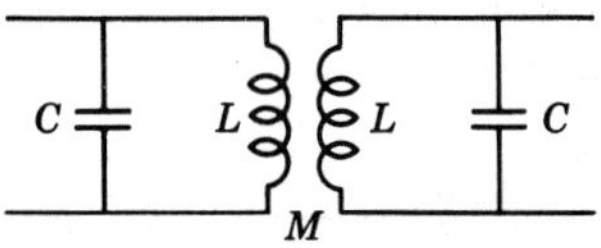

(*a*)

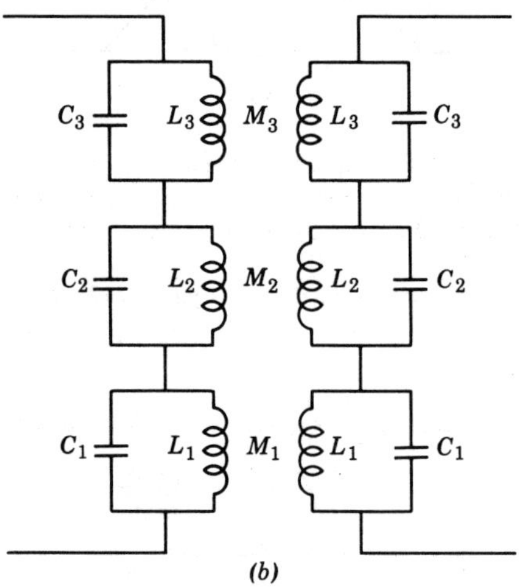

(*b*)

Prob. 5.

6. Show whether or not the two terminal-pair networks illustrated are equivalent. Determine the equivalent lattice for each.

7. For the network shown, determine an equivalent lattice and an equivalent unbalanced network.

8. The accompanying sketch shows a Wheatstone bridge circuit which is used

in practice to substantially prevent any alternating-voltage component present in the source E from producing a current in the load impedance Z_L. This is accomplished by making the blocking capacitors C sufficiently large so that they present practically no impedance to the alternating-voltage components, so that for these the bridge remains balanced.

By regarding this bridge as a symmetrical lattice network, determine an equivalent unbalanced network which accomplishes the same end as this Wheatstone bridge circuit.

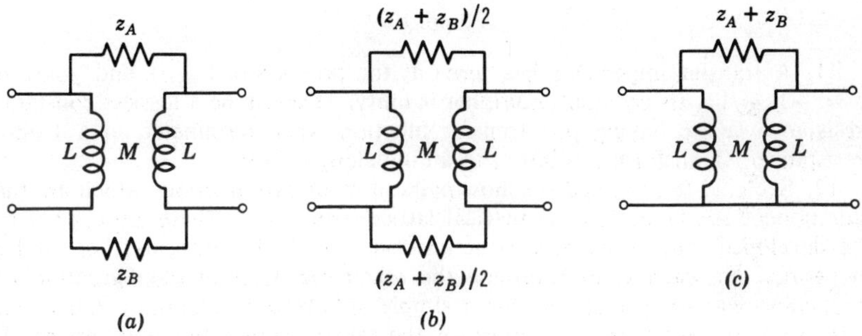

Prob. 6.

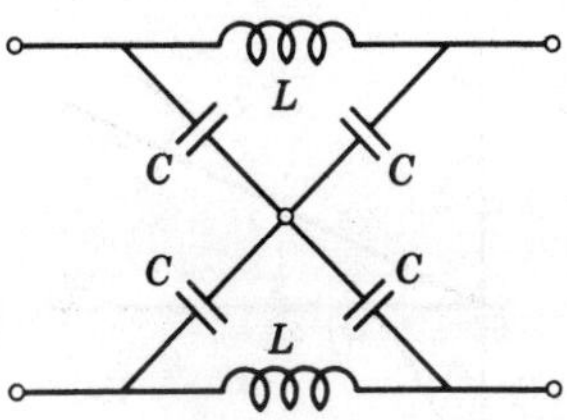

Prob. 7.

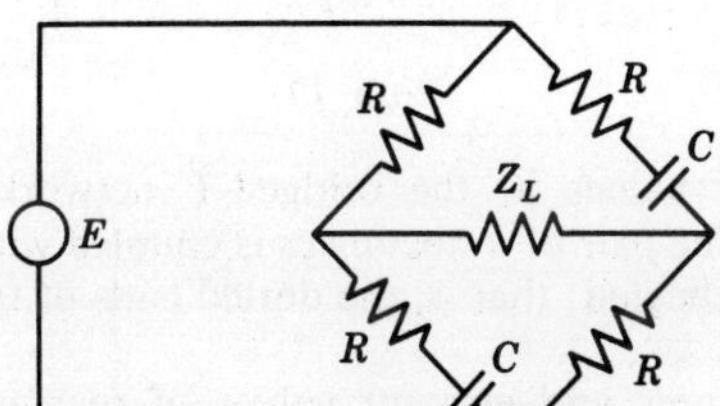

Prob. 8.

9. Under what circumstances, if any, can a constant-resistance lattice be a two-element-kind network? Show your reasoning. What practical use could be made of the result?

10. Determine a constant-resistance lattice having the same transfer impedance $Z_{12} = 2E_2/I_1$ as that of the network shown in the accompanying sketch. What is the magnitude of this transfer impedance at real frequencies?

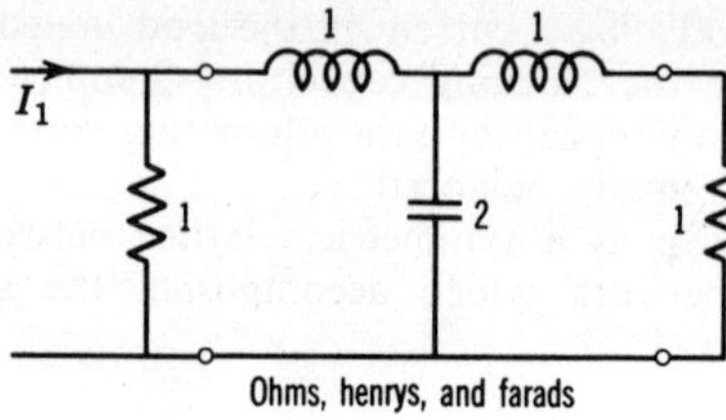

Prob. 10.

11. A transfer impedance has zeros at the points $s = 1 \pm j1$ and poles at $s = -1 \pm j1$. Its constant multiplier is unity. Determine a lossless constant-resistance lattice having this transfer function when terminated in a 1-ohm resistance. Transform this lattice to an unbalanced form.

12. Sketches (*a*), (*b*), and (*c*) show pairs of reactance functions which are the component reactances of symmetrical lattice structures. These lattices are to be developed into unbalanced structures such as T, Π, bridged-T, or ladder networks. In each case determine the schematic diagram (configuration of inductances and capacitances) for a simple unbalanced potentially equivalent network. In sketch (*a*) the values of the two reactance functions approach each other for very large frequencies.

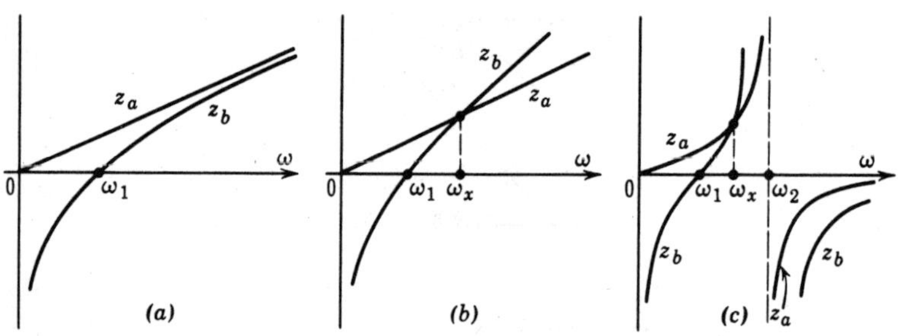

Prob. 12.

13. The parameter values in the bridged-T network shown here are in henrys and farads. The pair of inductances is coupled with a mutual of 1 henry having the polarity indicated; that is, the dotted ends of these coils have corresponding polarities.

(*a*) Find the structures and element values of reactances z_a and z_b of an equivalent symmetrical lattice network.

(*b*) Find the structures and element values of reactances z_{a0} and z_{b0} of a symmetrical lattice equivalent to two bridged-T networks in cascade.

(*c*) Determine the zeros and poles of z_{a0} and z_{b0}, and sketch their frequency patterns.

14. Show that the lattice in the adjacent sketch is transformable to an equivalent unbalanced network without mutual inductive coupling if

$$\frac{R_b}{R_a} + \frac{L_b}{L_a} + \frac{C_a}{C_b} \geq 1$$

and describe the method of transformation giving the structure and element values in the resultant network.

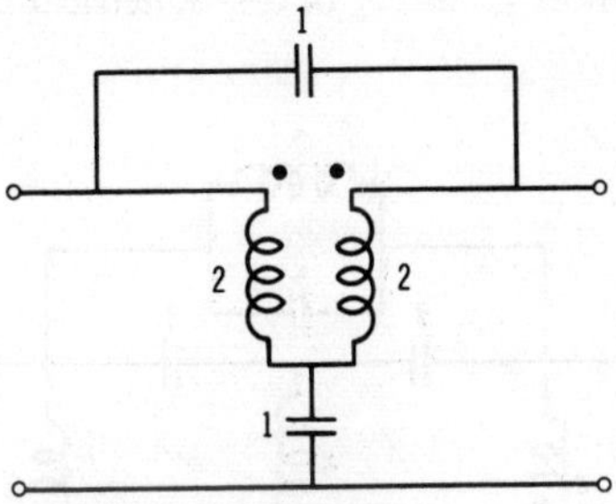

Prob. 13.

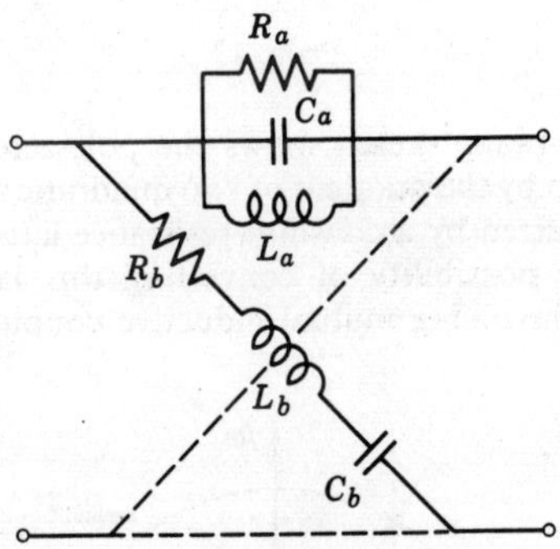

Prob. 14.

15. For the mutilated lattice shown in the adjacent sketch find an equivalent-T circuit and absorb any negative inductance by mutual inductive coupling. Find an equivalent lattice. Now consider two of these networks in cascade and find an equivalent lattice for the resultant combination. Develop this lattice into an unbalanced ladder network.

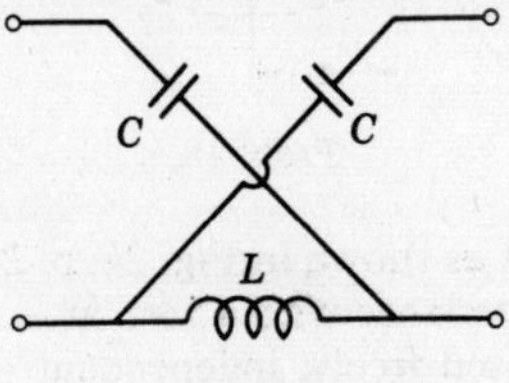

Prob. 15.

16. Consider the *RC* lattice shown in Fig. 23, p. 206, and determine the restrictions upon the pole-zero pattern of its open-circuit transfer impedance

imposed by stipulating that it shall be transformable to an equivalent unbalanced network according to the methods discussed in this chapter.

17. Determine the structures (reduced to simplest form) and element values of the component impedances z_a and z_b of a symmetrical lattice equivalent to the network shown here.

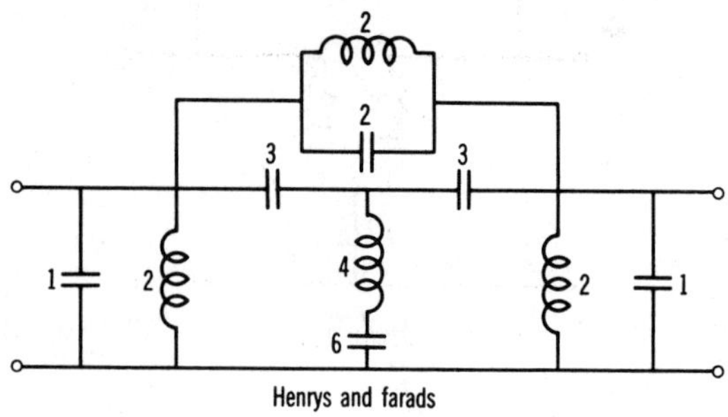

Prob. 17.

18. The following s-plane sketch shows the pole-zero pattern of a transfer impedance $Z_{12}(s)$ given by the quotient of two quadratic factors. If this transfer impedance is to be realized by a constant-resistance lattice, determine the latter and comment on the possibility of converting this lattice to an equivalent unbalanced form not involving mutual inductive coupling.

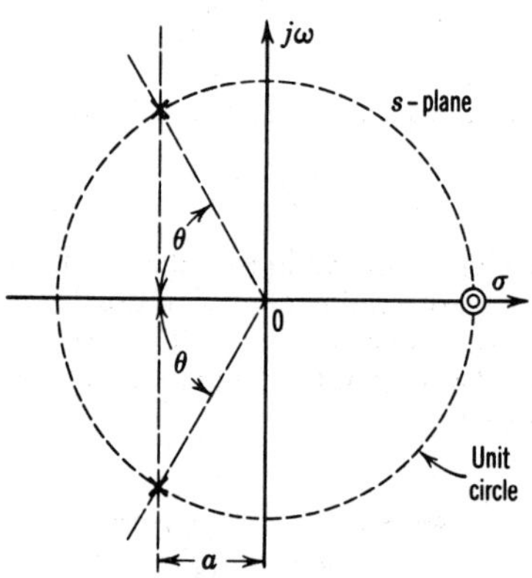

Prob. 18.

19. An RC twin-T circuit as shown in Fig. 26, p. 208, is to be designed having an open-circuit transfer impedance with a zero at $\omega = \omega_0$. Show that the time constant R_1C_1 may be chosen freely, independent of the value ω_0. Choosing unity for the value of this time constant, determine the expression for the transfer impedance z_{12} with an arbitrary constant multiplier and express all element values in the twin-T in terms of this multiplier and the critical frequency ω_0.

20. A symmetrical lattice has the impedances

$$(a)\ z_a = s, \qquad z_b = (s^2 + 1)/s$$

Determine the image impedance z_0 and propagation function θ for this lattice as functions of $s = j\omega$. Defining the attenuation function α by writing $\theta = \alpha + j\beta$, show that this lattice is a low-pass filter by sketching the functions z_0 and α versus ω.

Now revise the impedances z_a and z_b by writing

$$(b)\ z_a = ms, \qquad z_b = (s^2 + 1)/ms$$

in which m is a positive numeric. Again determine and sketch z_0 and α versus ω, choosing $m = 0.6$.

If the two lattices for the respective impedances (a) and (b) are connected in cascade, sketch the resultant z_0 and α functions versus ω.

CHAPTER SEVEN

Synthesis of Lossless Two Terminal-Pair Networks

1 An Extension of Foster's Reactance Theorem

The purpose of the following discussions is to extend the methods described in Ch. 3 for the synthesis of driving-point reactance functions to situations involving two pairs of terminals.* As pointed out in Ch. 2, such reactive two terminal-pair networks are characterized by the set of driving-point and transfer impedances z_{11}, z_{22}, z_{12} having only simple poles on the j-axis, in which z_{11} and z_{22} have positive real residues k_{11} and k_{22} respectively, and the residues k_{12} of z_{12}, although real, need not be positive but must fulfill the residue condition

$$k_{11}k_{22} - k_{12}^2 \geq 0 \tag{1}$$

The requirement that the impedances be real for real values of the complex frequency variable $s = \sigma + j\omega$ results, of course, in the j-axis poles occurring in conjugate pairs. Functions of this character (as has also been seen in Ch. 2) are expressed as quotients of polynomials of which one is even and the other odd, so that the point $s = 0$ and likewise the point $s = \infty$ is either a zero or a pole. Zeros and poles of the driving-point functions are both simple and restricted to conjugate points on the j-axis where they alternate with each other, whereas the zeros of the transfer impedance may have any multiplicity and may lie anywhere in the complex frequency plane although their distribution pattern (because of the even or odd character of the numerator polynomial of z_{12}) must exhibit complete symmetry with respect to both the imaginary as well as the real axis.

We shall show that these conditions concerning the properties of z_{11}, z_{22}, and z_{12}, which the discussions in Ch. 2 point out to be necessary for the realizability of lossless networks, are also sufficient. We do this by observing that we can successfully carry through a synthesis proce-

* The general extension to n terminal-pair networks, which follows the same pattern, is given in a paper by W. Cauer entitled "Ein Reaktanztheorem," *Sitzber. preuss. Akad. Wiss. Physik. math. Kl.*, Vol. 30, 1931, pp. 3–11.

dure for a set of impedance functions fulfilling only these necessary conditions.

In devising a synthesis procedure of this sort it is natural that one should begin by logically extending the methods that are found to be successful with driving-point functions. Thus the Foster procedure which makes use of partial-fraction expansions is again found to yield a possible solution, although we must now consider simultaneously the expansions of all three of the given impedances. Expressed in the form in which terms involving conjugate poles are combined (see Eq. 31, Ch. 2), these read

$$
\begin{aligned}
z_{11} &= \frac{k_{11}^{(0)}}{s} + \frac{2k_{11}^{(2)}s}{s^2 - s_2^2} + \cdots + \frac{2k_{11}^{(2n-2)}s}{s^2 - s_{2n-2}^2} + k_{11}^{(2n)}s \\
z_{22} &= \frac{k_{22}^{(0)}}{s} + \frac{2k_{22}^{(2)}s}{s^2 - s_2^2} + \cdots + \frac{2k_{22}^{(2n-2)}s}{s^2 - s_{2n-2}^2} + k_{22}^{(2n)}s \\
z_{12} &= \frac{k_{12}^{(0)}}{s} + \frac{2k_{12}^{(2)}s}{s^2 - s_2^2} + \cdots + \frac{2k_{12}^{(2n-2)}s}{s^2 - s_{2n-2}^2} + k_{12}^{(2n)}s
\end{aligned}
\tag{2}
$$

We may regard these three expansions as one expansion of the matrix

$$
[z] = \begin{bmatrix} z_{11} & z_{12} \\ z_{21} & z_{22} \end{bmatrix} \tag{3}
$$

in the form

$$
[z] = [z]_0 + [z]_2 + \cdots + [z]_{2n-2} + [z]_{2n} \tag{4}
$$

in which a typical term for a finite nonzero pole is given by

$$
[z]_\nu = \begin{bmatrix} \dfrac{2k_{11}^{(\nu)}s}{s^2 - s_\nu^2} & \dfrac{2k_{12}^{(\nu)}s}{s^2 - s_\nu^2} \\[2ex] \dfrac{2k_{21}^{(\nu)}s}{s^2 - s_\nu^2} & \dfrac{2k_{22}^{(\nu)}s}{s^2 - s_\nu^2} \end{bmatrix} \tag{5}
$$

If we construct a reactive two terminal-pair network with a z-matrix equal to this typical one, then the discussion in Art. 2 of Ch. 6 shows that the series interconnection of such networks for the various terms in the expansion 4 will yield the desired realization of the total matrix 3 (provided, of course, that the pertinent conditions for this method of expressing over-all behavior as pointed out in Art. 2, Ch. 6, are fulfilled).

Construction of a component network realizing the typical matrix 5 can always be accomplished in the form shown in Fig. 1, as the following discussion demonstrates. In order to include the first and last as well as any intermediate term in expansions 2, we will accordingly let

$$
f(s) = \frac{1}{s} \tag{6}
$$

or

$$f(s) = \frac{2s}{s^2 - s_v^2} \tag{7}$$

or

$$f(s) = s \tag{8}$$

and rewrite the typical impedance matrix in the simpler form

$$\begin{bmatrix} k_{11}f(s) & k_{12}f(s) \\ k_{21}f(s) & k_{22}f(s) \end{bmatrix} \tag{9}$$

in which we have dropped the superscripts on the k_{ik} since they are not needed in the following detailed discussion.

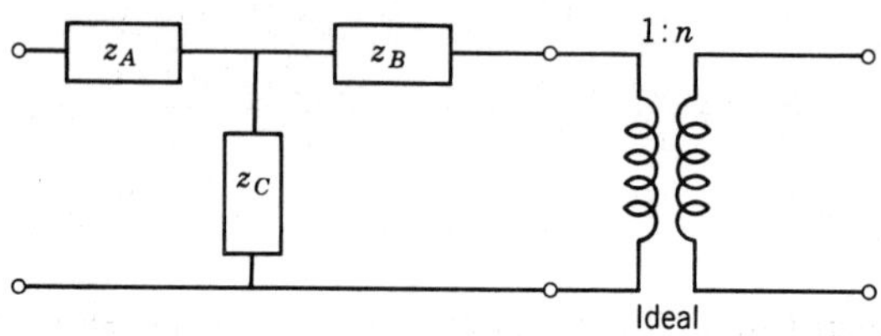

Fig. 1. Component two terminal-pair network for the realization of the impedance matrix 5.

The function $f(s)$, regarded as an impedance, is evidently represented by a capacitance, or by a parallel LC circuit, or by an inductance, as shown in Fig. 2, according to its analytic form given by Eq. 6, 7, or 8 respectively. Hence a positive real constant times $f(s)$ is easily realized by inspection.

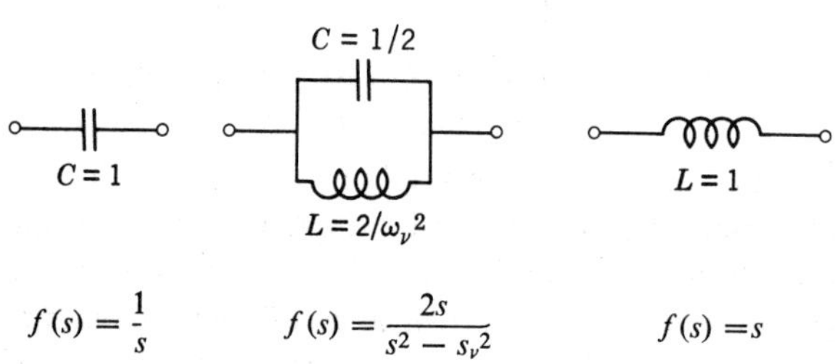

Fig. 2. Realization of the impedance function $f(s)$ according to any one of the three forms given by Eq. 6, 7, or 8.

If for the impedances in the T circuit of Fig. 1 we write

$$z_A = af(s), \qquad z_B = bf(s), \qquad z_C = cf(s) \tag{10}$$

in which a, b, and c are positive real numerics, then this T circuit is certainly physically realizable and readily constructible as soon as the numerical factors are known. For the complete circuit of Fig. 1 we have by inspection

$$\begin{aligned} z_{11} &= z_A + z_C \\ z_{22} &= n^2(z_B + z_C) \\ z_{12} &= nz_C \end{aligned} \tag{11}$$

and since these impedances are to be identified with the respective elements of matrix 9, we get

$$k_{11} = a + c, \qquad k_{22} = n^2(b + c), \qquad k_{12} = nc \tag{12}$$

Solving for the numerics a, b, and c, we find

$$a = k_{11} - \frac{k_{12}}{n}, \qquad b = \frac{k_{22}}{n^2} - \frac{k_{12}}{n}, \qquad c = \frac{k_{12}}{n} \tag{13}$$

Since these must have positive values, we observe first of all that the ideal transformer ratio n must have the same sign as the residue k_{12}. Having fixed this much, we see that the following conditions suffice.

$$k_{11} \geq \frac{|k_{12}|}{|n|} \qquad \text{and } k_{22} \geq |n| \cdot |k_{12}| \tag{14}$$

or more simply expressed

$$\frac{|k_{12}|}{k_{11}} \leq |n| \leq \frac{k_{22}}{|k_{12}|} \tag{15}$$

which yields a range of possible values for the ideal transformer ratio. In order for this range to exist, however, it is necessary and sufficient that

$$\frac{|k_{12}|}{k_{11}} \leq \frac{k_{22}}{|k_{12}|} \qquad \text{or } k_{12}^2 \leq k_{11}k_{22} \tag{16}$$

which is guaranteed fulfilled by residue condition 1.

The typical matrix 9 is, therefore, always realizable as a component network of the form shown in Fig. 1, and a series combination of these realizes the given set of impedances 2. The impedances in the T circuit of Fig. 1 are given by one of the three components in Fig. 2 multiplied by appropriate level factors as expressed by relations 10. Formally, at least, a procedure for the synthesis of reactive two terminal-pair networks is thus established and the sufficiency of the realizability conditions stated in Ch. 2 is demonstrated.

When residue condition 1 is fulfilled with the inequality sign, then theoretically an infinite number of choices for the magnitude of the ideal transformer ratio, according to Eq. 15, are possible. The two limiting values in this range yield more economical results since the T circuit of Fig. 1 degenerates into a reoriented L. Thus for

$$|n| = |k_{12}|/k_{11}, \qquad a = 0, \qquad \text{and } z_A = 0 \tag{17}$$

and for

$$|n| = k_{22}/|k_{12}|, \qquad b = 0, \qquad \text{and } z_B = 0 \tag{18}$$

If the residue condition 1 is fulfilled with an equal sign, then both z_A and

z_B are zero and the component two terminal-pair network realizing the typical matrix 9 degenerates into the shunt impedance z_C followed by the ideal transformer.

The ideal transformer, incidentally, may not always be needed so far as the realization of the typical circuit of Fig. 1 is concerned. Thus if the residue of k_{12} in the pertinent pole is positive, and if

$$k_{11} \geq k_{12} \geq 0 \qquad \text{and } k_{22} \geq k_{12} \tag{19}$$

then the range 15 permits the choice $n = +1$, and the ideal transformer can be dispensed with. If this condition is consistently met at all poles, then the complete synthesis assumes the form of a T circuit as shown in Fig. 3. This situation is not of particular practical interest, however,

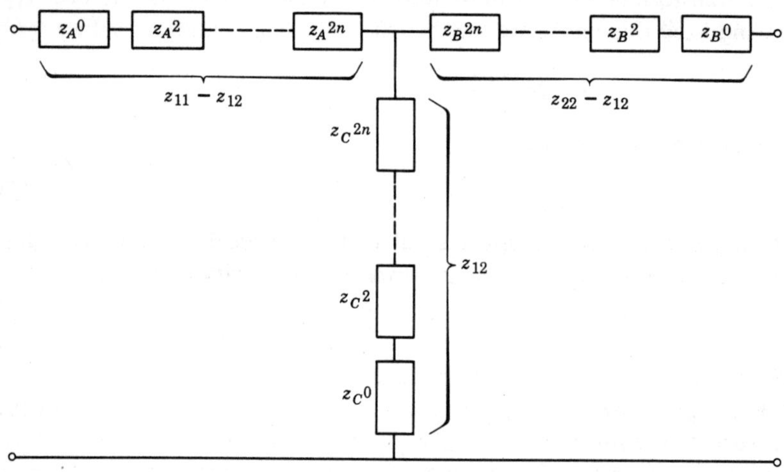

Fig. 3. Physical realization for the impedances 2 if the special conditions 19 are fulfilled at all poles.

since conditions 19 are so restrictive as not to allow very useful response characteristics. As indicated in the figure, conditions 19 are equivalent to stating that $z_{11} - z_{12}$ and $z_{22} - z_{12}$ as well as z_{12} are realizable driving-point reactance functions. In other words, the transfer impedance, which is capable of having any distribution of zeros in the complex plane, reverts to the drab and uninteresting behavior of a driving-point reactance. In most practical examples where reactive networks occur, the transfer impedance is not apt to have the same properties as a driving-point reactance.

It is nevertheless possible in such cases to circumvent the necessity of using ideal transformers in another way. Consider the circuit of Fig. 1 for an $f(s)$ given by Eq. 7 and the choice 18 for the ideal transformer ratio yielding $z_B = 0$. In this circuit, which is shown in Fig. 4*a*, the ideal transformer adjacent to the shunt inductance L_C may be replaced by a pair of mutually coupled inductances as shown in Fig. 4*b*. The ideal transformer may similarly be dealt with in the circuit for which $f(s)$ is an inductance but not in the purely capacitive circuit corresponding to an $f(s)$ function given by Eq. 6. To avoid the ideal transformer here we observe that in most practical problems one can usually tolerate an impedance level change at, say, the output terminals of the total network, which amounts to multiplying z_{12} by a factor m and z_{22} by m^2. If this operation is carried out on Eqs. 2, we can choose m so as to make

$$k_{11}^{(0)} \geq m^2 k_{22}^{(0)} = m k_{12}^{(0)} > 0 \tag{20}$$

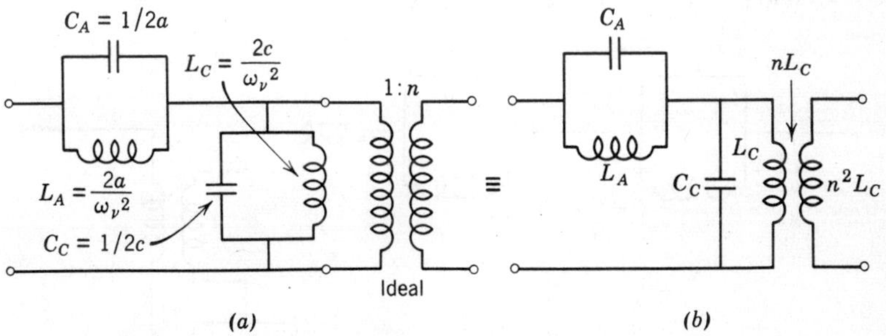

Fig. 4. Special form of the circuit of Fig. 1 with $f(s)$ given by Eq. 7, that results if the ideal transformer ratio is chosen according to Eq. 18.

The corresponding capacitive circuit of Fig. 1 is then simply an L structure involving no ideal transformer. If for the remaining terms in expansions 2 we avoid the ideal transformer requirement according to the equivalence illustrated in Fig. 4, then the total realization for the given z_{ik} (except for the output impedance level change) takes the form shown in Fig. 5.

Although no ideal transformers are involved in this resultant circuit, the pairs of inductances in the shunt branch are closely coupled (the coupling coefficients are all unity). However, this situation is easily remedied by means of the inductance in the series branch which may just as well be added on to the first shunt inductance. This step relieves the close coupling in the topmost pair of coils, a condition that we may equally well attribute to the fact that the right-hand coil in this pair has an excess inductance over the close coupling value. This excess may be distributed among all of the right-hand coils to relieve the close coupling in all pairs. Unless the given z_{ik} have a pole at $s = \infty$ at which the residue condition is fulfilled with the inequality sign, however, this procedure for the relief of the close coupling requirement cannot be carried out.

The analogy to the Foster form of realization for a driving-point reactance through its partial-fraction expansion is very evident in the circuit structure of Fig. 5. The process of obtaining the alternate Foster form by partial-fraction expansion of the corresponding susceptance function may be extended to two terminal-pair networks in an exactly similar fashion. The driving-point and transfer admittances y_{11}, y_{22}, and y_{12}, having precisely the same properties as the impedances z_{11}, z_{22}, and z_{12}, are expanded in partial fractions as are the impedance functions in Eqs. 2. All of the steps in the synthesis are exactly dual to those described for the impedance basis, the typical component network (dual to that in Fig. 1) being a Π circuit in cascade with an ideal transformer; and the total realization takes the form of a parallel combination of these components.

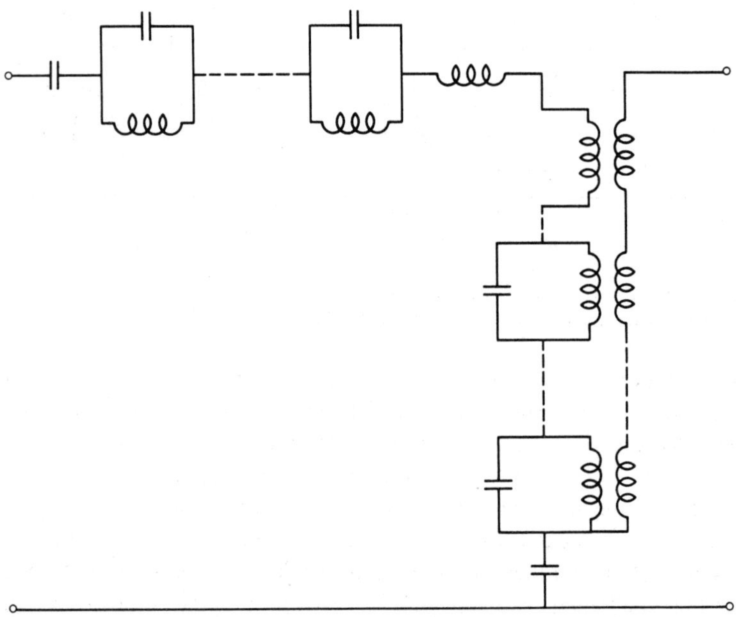

Fig. 5. Total realization for the impedances 2 (involving only mutual inductive coupling instead of ideal transformers) that is always possible if one can tolerate an appropriate impedance level change at one of the terminal pairs.

Figure 6 illustrates the form that the final network has on an admittance basis if the admittance level at one terminal pair is modified so as to eliminate the ideal transformer for the component network representing a possible pole at $s = \infty$, and if the ideal transformer ratios for the rest of the component networks are chosen so as to yield these in the form of L structures as shown. The shunt inductance in the L circuit for the pole at $s = 0$, if transferred to the right-hand side of its adjacent ideal transformer, appears in parallel with all ideal transformers and hence may be distributed between them. Each ideal transformer with its associated series and shunt inductance may then be converted into a pair of mutually coupled inductances involving less than unity coupling coefficients. A redrawn version of the resulting network in which these changes are incorporated is shown in Fig. 7, which is comparable with the network of Fig. 5 obtained on an impedance basis. Again the analogy to the Foster form for one terminal-pair is evident.

A numerical example will serve to clarify these synthesis procedures. We choose the following set of impedances

$$\begin{aligned} z_{11} &= \frac{s(s^2+2)(s^2+4)}{(s^2+1)(s^2+3)} = \frac{\frac{3}{2}s}{s^2+1} + \frac{\frac{1}{2}s}{s^2+3} + s \\ z_{12} &= \frac{s^5}{(s^2+1)(s^2+3)} = \frac{\frac{1}{2}s}{s^2+1} - \frac{\frac{9}{2}s}{s^2+3} + s \\ z_{22} &= \frac{s(3s^4+134s^2+132)}{3(s^2+1)(s^2+3)} = \frac{\frac{1}{6}s}{s^2+1} + \frac{\frac{81}{2}s}{s^2+3} + s \end{aligned} \tag{21}$$

Here z_{11} is written down for some arbitrarily chosen zero and pole factors, whereas the numerator in z_{12} is selected because it is simple and yet illustrates a typical departure from the familiar interlaced pole-zero pattern which is required for the driving-point impedances of lossless networks. The impedance z_{22} in this example is associated with the other two, after these are expanded in partial fractions, by choosing its residues with such values that the residue condition is fulfilled with the equal sign at each pole.

From these partial-fraction expansions, the resultant network, according to the pattern of Fig. 5, is readily constructed as shown in Fig. 8. The series branch of the structure of Fig. 5 is absent in this example because the residue condition is fulfilled with the equal sign at all poles, including the one at $s = \infty$. The pairs of inductances are therefore all closely coupled.

Now let us realize the same set of impedances on an admittance basis. In order to compute the corresponding y_{ik}, we must first form the z-determinant. In this connection it is important to note that fulfillment of the residue condition with an equal sign at all poles results in the z-determinant having only simple poles in spite of its quadratic character as a function of the z_{ik}. Thus if we visualize evaluating the expression $z_{11}z_{22} - z_{12}^2$ by substitution of the partial-fraction expansions 21, all terms involving denominator factors like $(s^2 + 1)^2$ or $(s^2 + 3)^2$ cancel, and only cross-product terms involving the simple factors $(s^2 + 1)$ or $(s^2 + 3)$ remain. We may say, therefore, that fulfillment of the residue condition with an equal sign at all poles results in the z-determinant having only simple poles; conversely, if a z-determinant has only simple poles, then the residue condition must be satisfied with an equal sign at all poles.

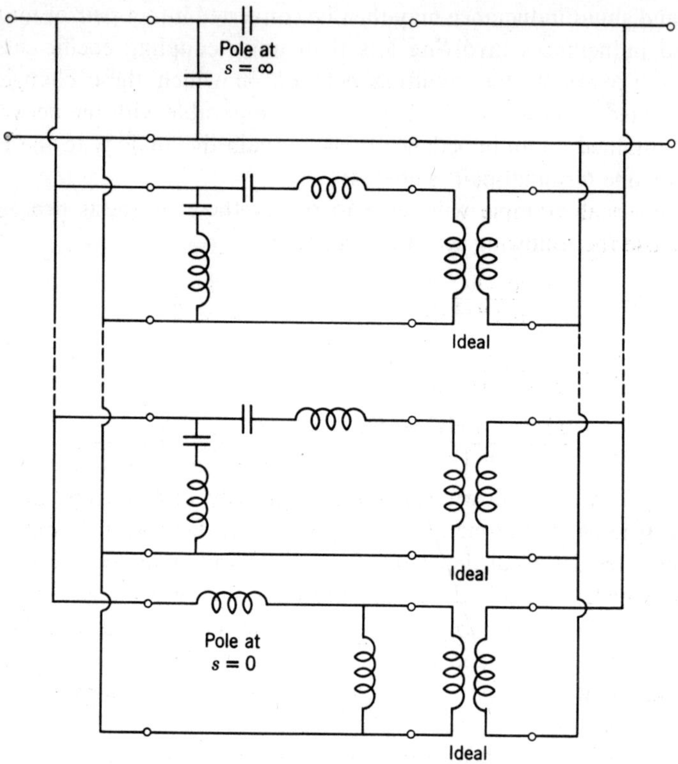

Fig. 6. Form of realization for a reactive two terminal-pair, dual to that obtained from a series combination of components like that in Fig. 4*a*.

The product of the numerator polynomials in z_{11} and z_{22} minus the square of the numerator of z_{12} must be divisible by the factors $(s^2 + 1)$ and $(s^2 + 3)$. We thus find

$$z_{11}z_{22} - z_{12}^2 = \frac{152s^2(s^2 + \frac{44}{19})}{3(s^2 + 1)(s^2 + 3)} \tag{22}$$

and hence

$$\begin{aligned}
y_{11} &= \frac{3s^4 + 134s^2 + 132}{152s(s^2 + \frac{44}{19})} = \frac{132}{352s} + \frac{0.461s}{s^2 + \frac{44}{19}} + \frac{3s}{152} \\
y_{22} &= \frac{3(s^2 + 2)(s^2 + 4)}{152s(s^2 + \frac{44}{19})} = \frac{24}{352s} + \frac{0.00445s}{s^2 + \frac{44}{19}} + \frac{3s}{152} \\
y_{12} &= \frac{-3s^3}{152(s^2 + \frac{44}{19})} = 0 + \frac{0.0453s}{s^2 + \frac{44}{19}} - \frac{3s}{152}
\end{aligned} \tag{23}$$

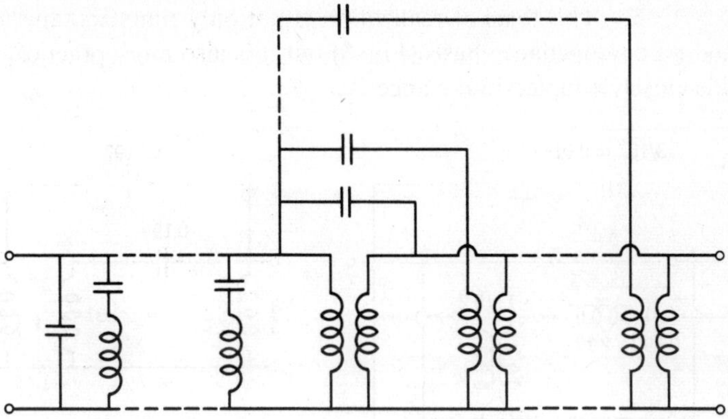

Fig. 7. Form of realization on an admittance basis that is dual to that shown for the impedance basis in Fig. 5.

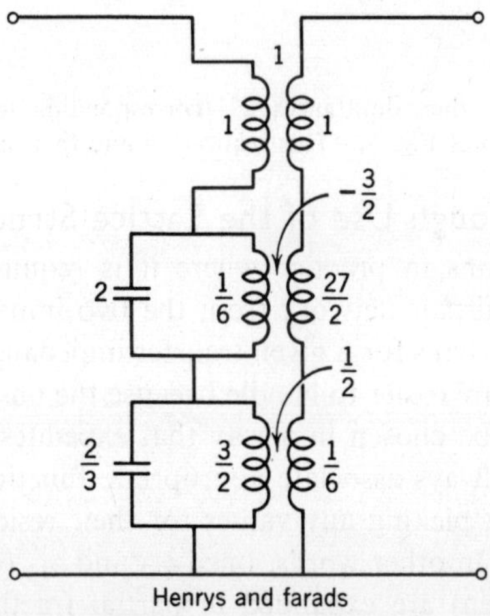

Fig. 8. Realization of the impedances 21 according to the procedure leading to the form shown in Fig. 5.

Realization according to the pattern of Fig. 6 takes the form of the network in Fig. 9*a*. The first two component networks have no shunt branches because of the residue condition being fulfilled with the equal sign. The third component network for the pole at $s = 0$ has no series branch because the transfer admittance has no pole at $s = 0$. The ideal transformer with its associated series and shunt inductances is readily converted into a pair of mutually coupled inductances as shown by the redrawn version of this circuit in Fig. 9*b*. The coupling coefficient turns out to be 0.25. This form of realization is not only simpler than the one obtained on an impedance basis (Fig. 8) but it is also more practical since it avoids closely coupled inductances.

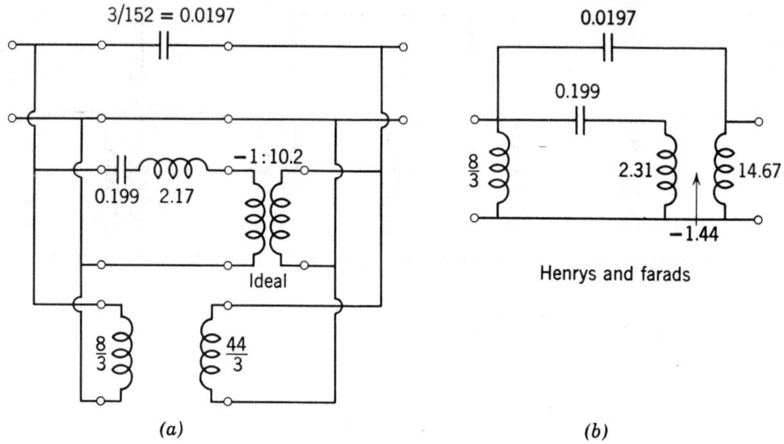

Fig. 9. Realization of the admittances 23 (corresponding to the impedances 21) according to the pattern of Fig. 6. This realization and that in Fig. 8 are equivalent.

2 Synthesis Through Use of the Lattice Structure

There are situations in practice where it is required to synthesize a lossless two terminal-pair network from the two impedances z_{11} and z_{12} or z_{22} and z_{12}, or perhaps for a given transfer impedance z_{12} alone. Such synthesis problems are easier to handle because the unspecified impedance or impedances can be chosen in a way that expedites the synthesis procedure. One can always associate appropriate functions for the missing impedances through picking any values for their residues that fulfill the residue condition. In other words, once z_{12} and z_{11} (or z_{12} alone if only this function is given) are expanded in partial fractions, it is a simple matter to write down a corresponding partial-fraction expansion for z_{22} (or for z_{11} and z_{22}) by assuming any values for the residues that fulfill the residue condition, and one has, of course, the option of fulfilling it with the equal sign or not, according to whichever choice leads to the most acceptable final network.

In problems of this sort it may be possible to use the symmetrical lattice as a design basis since there is no reason why we should not choose $z_{11} = z_{22}$ so long as the given data permit this added specification. Thus if z_{11} and z_{12} are given and if at all poles $k_{11} \geq |k_{12}|$, then the choice $z_{22} = z_{11}$ will evidently not violate any of the residue conditions. Note, however, that if the condition $k_{11} \geq |k_{12}|$ is not fulfilled at *all* poles, then the choice $z_{22} = z_{11}$ cannot be made, and a realization in the form of a symmetrical network is impossible. The most economical symmetrical design results if at all poles $k_{11} = |k_{12}|$.

Let us illustrate this method by assuming that only z_{12} in the set 21 is specified. It is not even necessary to write down the corresponding expansions for $z_{11} = z_{22}$ letting $k_{11} = k_{22} = |k_{12}|$ at the various poles. We simply make use of the lattice relation (Eqs. 44, Ch. 6).

$$z_{12} = \tfrac{1}{2}(z_b - z_a) \tag{24}$$

and assign the positive terms in the partial-fraction expansion of z_{12} to $z_b/2$, the negative ones to $z_a/2$. For z_{12} in the set 21, this process gives

$$z_b = \frac{s}{s^2 + 1} + 2s \tag{25}$$

$$z_a = \frac{9s}{s^2 + 3} \tag{26}$$

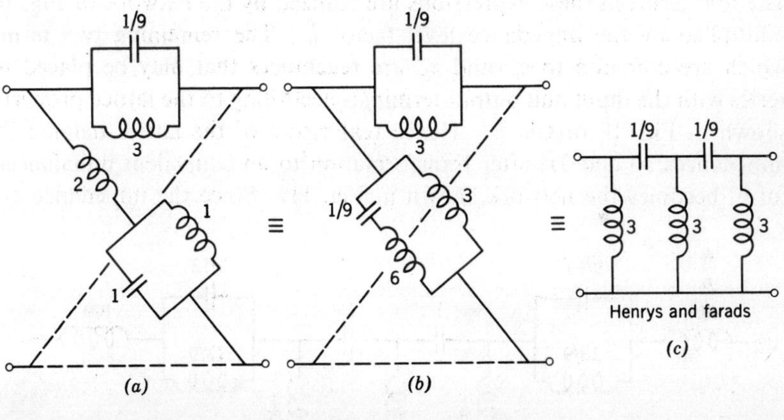

Fig. 10. Symmetrical lattice realizing the transfer impedance of z_{12} in the set 21 and its development into an equivalent ladder form.

The corresponding lattice is shown in Fig. 10*a*. If we replace the z branch by its alternate Foster form as in part (*b*) of this figure, then the development of this lattice into the equivalent unbalanced ladder of part (*c*) follows without difficulty according to the techniques discussed in Art. 5 of Ch. 6.

If the output impedance level is of no consequence, then the condition $k_{11} \geq |k_{12}|$ at all poles can always be fulfilled through multiplying z_{12} by an appropriate constant (equivalent to connecting an ideal transformer at the output terminals of the network). Thus in Eqs. 21, if we multiply z_{12}

by the factor $\frac{1}{9}$, then we have

$$z_{11} = \tfrac{1}{2}(z_b' + z_a') = \frac{\frac{3}{2}s}{s^2+1} + \frac{\frac{1}{2}s}{s^2+3} + s$$
$$z_{12} = \tfrac{1}{2}(z_b' - z_a') = \frac{\frac{1}{18}s}{s^2+1} - \frac{\frac{1}{2}s}{s^2+3} + \tfrac{1}{9}s \tag{27}$$

from which

$$z_b' = \frac{\frac{14}{9}s}{s^2+1} + \tfrac{10}{9}s \tag{28}$$

and

$$z_a' = \frac{\frac{13}{9}s}{s^2+1} + \frac{s}{s^2+3} + \tfrac{8}{9}s \tag{29}$$

Comparison with Eqs. 25 and 26 shows that

$$z_b' = \frac{z_b}{9} + \frac{\frac{13}{9}s}{s^2+1} + \tfrac{8}{9}s \tag{30}$$

$$z_a' = \frac{z_a}{9} + \frac{\frac{13}{9}s}{s^2+1} + \tfrac{8}{9}s \tag{31}$$

The first terms in these expressions are realized by the network of Fig. 10 multiplied by the impedance level factor $\frac{1}{9}$. The remaining two terms which are common to z_a' and z_b' are reactances that may be placed in series with the input and output terminals according to the lattice property shown in Fig. 18 of Ch. 6. Hence realization of the lattice defined by impedances 30 and 31, after transformation to an equivalent unbalanced form, becomes the network shown in Fig. 11. Since the impedance z_{22}

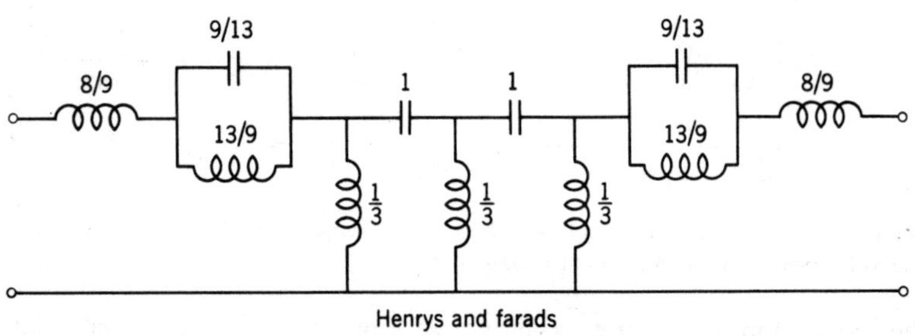

Fig. 11. Realization of the impedances given in Eq. 27.

presumably is of no interest (since only z_{11} and z_{12} are assumed to be specified in the present problem), the extra series branch on the right can be dropped.

As it stands, the network of Fig. 11 realizes the impedances 27 with the tacit assumption that an associated $z_{22} = z_{11}$. For this set of driving-point and transfer impedances the residue condition is fulfilled with the equal sign only for the pair of conjugate poles at $s = \pm j\sqrt{3}$. For the other two poles the residue condition is fulfilled with the inequality sign; that is to say, with a "margin of safety," so to speak. This margin of safety yields the extra series arms on both sides of the network of Fig. 11. Unless there is a specific reason for padding the residue condition in this way, the practice of allowing a margin of safety for this condition is not to be recommended since it merely yields extra circuit elements and hence leads to an uneconomical design. Unlike the practice followed by civil engineers in building bridges, the best design in electrical network synthesis usually results when the necessary and sufficient conditions for the network's realizability are fulfilled with *nothing* to spare.

3 Synthesis Through Ladder Development of One Driving-Point Impedance

Another synthesis procedure is possible when only one driving-point impedance and the transfer impedance are specified. This method, which leads directly to an unbalanced ladder network, is based upon the principle that the development of a driving-point impedance can be accomplished in such a way that when an output terminal pair is attached to the opposite end of the ladder the resulting transfer impedance will have the desired zeros, its poles automatically being the same as those of the given driving-point impedance (assuming that degeneracies tending to interfere with this condition have not been permitted to occur). We speak of the process as a development of the driving-point impedance in such a way as to control the zeros of the transfer impedance.

This control of the zeros of the transfer impedance in a ladder structure is based upon the initial logical observation that such a zero (also called a *zero of transmission*) results either when a series branch of the ladder is equivalent to an open circuit or when a shunt branch is equivalent to a short circuit. We shall see presently that this condition, although necessary for the production of a zero of transmission, is not always sufficient. The transfer impedance of a ladder structure is not necessarily zero where the impedance of a series branch is infinite or where the impedance of a shunt branch is zero, for if this condition obtains at some frequency at which the transfer impedance of the network portion to the right or to the left of this branch has a pole, then the over-all transfer impedance will there have a finite nonzero value.

In Art. 4 of Ch. 3 the development of a driving-point reactance into a ladder structure is shown to be possible in a large variety of ways by removing poles from the given impedance and from its subsequently inverted remainders in a random manner, and it is also shown that this process can be made to yield an even greater variety of distinct network realizations if poles are at times only partially instead of totally removed. When a pole is removed from an impedance so as to form a series branch in the pertinent ladder development, the impedance of that series branch is infinite at the frequency of that pole; when a pole is removed from an admittance in the form of a shunt branch, its impedance is zero at the respective pole frequency. However, such branches at their pertinent critical frequencies do not produce transmission zeros for the over-all ladder network if their formation is due to a *partial* pole removal, for then the remainder function still contains that pole and, therefore, so does the transfer impedance or admittance of the rest of the ladder which the continued development of this remainder function yields. At the critical frequency in question the pertinent series or shunt branch is, to be sure, an infinite impedance or admittance, but the voltage or current acting upon it is likewise infinite, and both infinities (or poles) are simple. Hence the over-all transfer function is not critical at this frequency.

This circumstance gives us the means for controlling the production of transmission zeros in the ladder development of a driving-point reactance function, for if we want a step in this process to produce a transmission zero then the pertinent pole is totally removed, whereas if no such zero is wanted at the particular pole frequency, then we simply leave part of this pole in the remainder function.

We have also seen in the discussion of this ladder development process that pole removals, either total or partial, are the means for shifting zeros, which become the poles of the inverted remainder function. Since we can thus shift a pole into any desired position (by two or more successive shifting steps if necessary), it becomes clear that the development of a given driving-point function can be carried out so as to produce any desired distribution of transmission zeros.

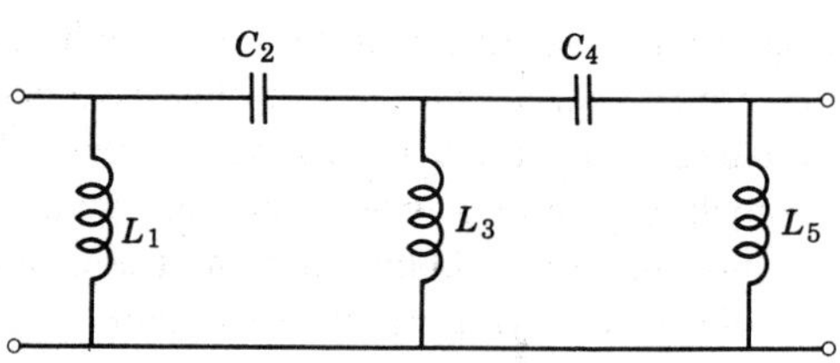

$L_1 = \frac{8}{3}$ $\quad C_2 = \frac{7}{32}$

$L_3 = \frac{88}{49}$ $\quad C_4 = \frac{21}{968}$

$L_5 = \frac{44}{3}$

Fig. 12. Realization of the impedances in the set 21 through the ladder development of z_{11}. This network is equivalent to either of the two shown in Figs. 8 and 9.

To illustrate this process let us begin with the functions z_{11} and z_{12} in the set 21. The zeros of this transfer impedance are all at $s = 0$. Hence the desired transmission zeros are produced by totally removing poles at

$s = 0$ from z_{11} and/or its subsequent inverted remainder functions in a continuing sequence for five steps, since the zero of z_{12} at $s = 0$ is of fifth order. This process is recognized as being identical with one of the canonic Cauer developments. No zero-shifting steps are needed in this simple example. The computations read

$$\begin{array}{l}
\qquad\qquad\quad 3/8s \\
8s + 6s^3 + s^5 \,\overline{)\,3 + 4s^2 + s^4} \\
\qquad\qquad\quad 3 + \tfrac{9}{4}s^2 + \tfrac{3}{8}s^4 \,|\, 32/7s \\
\qquad\qquad\qquad \overline{\tfrac{7}{4}s^2 + \tfrac{5}{8}s^4} \,|\, 8s + 6s^3 + s^5 \\
\qquad\qquad\qquad\qquad\qquad 8s + \tfrac{20}{7}s^3 \qquad |\, 49/88s \\
\qquad\qquad\qquad\qquad\qquad \overline{\tfrac{22}{7}s^3 + s^5} \,|\, \tfrac{7}{4}s^2 + \tfrac{5}{8}s^4 \\
\qquad\qquad\qquad\qquad\qquad\qquad\quad \tfrac{7}{4}s^2 + \tfrac{49}{88}s^4 \,|\, 968/21s \\
\qquad\qquad\qquad\qquad\qquad\qquad\qquad\quad \overline{\tfrac{3}{44}s^4} \,|\, \tfrac{22}{7}s^3 + s^5 \\
\qquad\qquad\qquad\qquad\qquad\qquad\qquad\qquad\quad \tfrac{22}{7}s^3 \qquad |\, 3/44s \\
\qquad\qquad\qquad\qquad\qquad\qquad\qquad\qquad\qquad \overline{s^5} \,|\, \tfrac{3}{44}s^4
\end{array} \tag{32}$$

and the resulting network is shown in Fig. 12.

Since this method only assures that z_{12} will have the proper poles and zeros, it exercises no control over the constant multiplier associated with this function. From the final network it is, however, a simple matter to compute the constant multiplier without going to the trouble of computing the entire function z_{12}. Thus, with reference to Fig. 12, we see that for very high frequencies the circuit degenerates into a single shunt inductance having the reciprocal value

$$\frac{1}{L_1} + \frac{1}{L_3} + \frac{1}{L_5} = \frac{3}{8} + \frac{49}{88} + \frac{3}{44} = 1 \tag{33}$$

Hence for $s \to \infty$, $z_{12} \to s$, which is precisely the same as for the transfer impedance in the set 21. The network of Fig. 12 therefore happens to yield the correct constant multiplier for z_{12}.

We can alternately check this multiplier by computing the asymptotic behavior of z_{12} for $s \to 0$ in the following way. Assume 1 volt at the output. The current through L_5 is $1/L_5s$ and across C_4 it produces a voltage $(1/L_5s) \times (1/C_4s)$ which very nearly is the voltage across L_3. Hence the current through L_3 is $(1/L_5s) \times (1/C_4s) \times (1/L_3s)$ which very nearly is the current through C_2 because the impedance of the L_3 branch is much smaller than the impedance of the C_4 branch. Continuing this simple argument, we see that

$$\text{for } s \to 0,\ (I_1/E_2) = (1/z_{12}) \to 1/L_1C_2L_3C_4L_5s^5 \tag{34}$$

According to the element values in Fig. 12, this gives

$$\text{for } s \to 0,\ z_{12} \to \frac{8}{3} \times \frac{9}{32} \times \frac{88}{49} \times \frac{21}{968} \times \frac{44}{3}s^5 = \frac{s^5}{3} \tag{35}$$

Reference to Eqs. 21 shows again that this result yields the same constant multiplier.

In this simple example we also produce the same z_{22} function as that given in the set 21. The easiest way to establish this fact is to develop function z_{22}, as given in the set 21, into a ladder network precisely as z_{11} is developed above. The result is the network of Fig. 12, as the reader may check for himself. Hence this two terminal-pair network is entirely equivalent to the networks given in Figs. 8 and 9, which is an interesting result that could not have been anticipated at the outset unless we are aware of properties of the ladder development process elaborated upon below. It is partly for reasons such as this that we should have several different synthesis procedures for the same type of problem, because then we can obtain a variety of solutions from which to pick out the one most suitable from practical or any other considerations.

As a variation of this problem, suppose we take the same z_{11} function as in 21 but change z_{12} so as to have only a third-order zero at $s = 0$, thus

$$z_{12} = \frac{s^3}{(s^2 + 1)(s^2 + 3)} \tag{36}$$

Observe, incidentally, that the numerator in this function could not be chosen equal to s^4 or s^2 because it must be odd since the denominator is even. This modified z_{12} function has three zeros at $s = 0$ and one at $s = \infty$.

We can begin the expansion of z_{11} as in the previous example, but the process must be discontinued after three steps. The remainder must be expanded so as to produce a transmission zero at $s = \infty$. According to development 32, this remainder is the impedance

$$z' = \frac{s^2 + \frac{22}{7}}{\frac{3}{44}s} = \frac{44s}{3} + \frac{968}{21s} \tag{37}$$

The already developed portion of the network consists of the branches L_1, C_2, and L_3 of the network in Fig. 12. Since the transfer impedance of this portion has a pole at $s = \infty$ (as we can see by inspection), the removal of the pole in the remainder z' at $s = \infty$ in the form of a series branch produces no transmission zero at this frequency. However, removal of the pole at $s = \infty$ from the next inverted remainder (the admittance $21s/968$) in the form of a shunt branch does produce the desired transmission zero at $s = \infty$ and completes the development.

The end result is shown in Fig. 13 in which the element values are the same as in Fig. 12. We see that the last two elements C_4 and L_5 have merely interchanged places. By considering the asymptotic behavior for $s \to 0$ or $s \to \infty$ we find that the network of Fig. 13 realizes the transfer impedance 36 with a multiplier of $\frac{22}{7}$.

Alternately we may approach this same synthesis problem by beginning the development of z_{11} so as to produce the transmission zero at $s = \infty$. Since z_{11} has a pole at $s = \infty$, we can begin its development by removing this pole as a series branch. Again this step will produce no zero in the transfer impedance, however, since the input is tacitly a current source. We must therefore continue the same expansion process for a second step before switching to the removal of poles at $s = 0$. Hence the computations appear in the form

$$
\begin{array}{rl}
 & s \\
s^4 + 4s^2 + 3 | & s^5 + 6s^3 + 8s \\
 & s^5 + 4s^3 + 3s \,|\, s/2 \\
 & 2s^3 + 5s \,|\, s^4 + 4s^2 + 3 \\
 & \qquad s^4 + \tfrac{5}{2}s^2 \\
 & \qquad \tfrac{3}{2}s^2 + 3
\end{array}
\tag{38}
$$

$$
\begin{array}{rl}
 & 3/5s \\
5s + 2s^3 | & 3 + \tfrac{3}{2}s^2 \\
 & 3 + \tfrac{6}{5}s^2 \,|\, 50/3s \\
 & \tfrac{3}{10}s^2 \,|\, 5s + 2s^3 \\
 & \qquad 5s \qquad |\, 3/20s \\
 & \qquad 2s^3 \,|\, \tfrac{3}{10}s^2
\end{array}
\tag{39}
$$

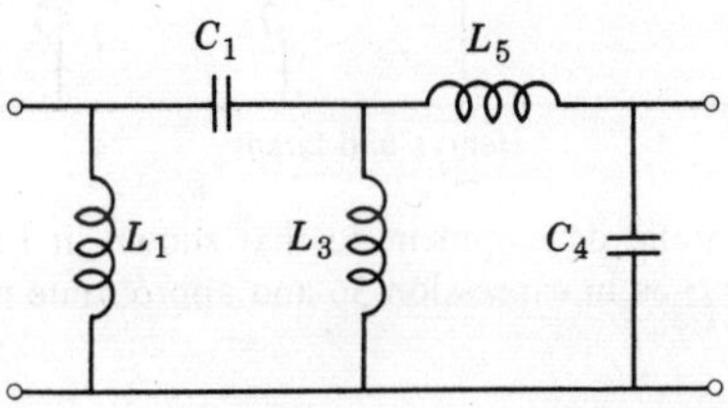

Fig. 13. Modification of the realization in Fig. 12 appropriate to replacing z_{12} in the set 21 by the expression 36 (and introducing appropriate constant multipliers in z_{12} and z_{22}).

The computations 38 relate to the first two steps in the development yielding the transmission zero at $s = \infty$; the ones in 39 pertain to the last three steps in which we remove poles at $s = 0$ to produce the three required zeros of z_{12} at this point.

The resulting network for this development takes the form shown in Fig. 14. Consideration of the asymptotic behavior of this network for $s \to 0$ or $s \to \infty$ shows that the transfer impedance 36 is realized in this case with a constant multiplier of 2.

As a further illustration of this method of ladder development let us consider the impedances

$$z_{11} = \frac{(s^2 + 1)(s^2 + 9)}{s(s^2 + 4)(s^2 + 16)} \tag{40}$$

$$z_{12} = \frac{(s^2 + 25)(s^2 + 36)}{s(s^2 + 4)(s^2 + 16)} \tag{41}$$

in which the transfer impedance involves some finite nonzero zeros in addition to one at $s = \infty$. Various developments are possible according to the order in which we produce the zeros of transmission. We shall begin by producing the transmission zero at $s = \infty$ by removal of a shunt capacitance which is possible since the admittance $y_1 = 1/z_{11}$ has a pole at $s = \infty$. By inspection, the residue of this admittance at $s = \infty$ is unity, and so the first branch in our ladder development is a shunt capacitance

$$c_1 = 1 \tag{42}$$

The remainder is

$$y_2 = y_1 - s = \frac{10s(s^2 + 5.5)}{(s^2 + 1)(s^2 + 9)} \tag{43}$$

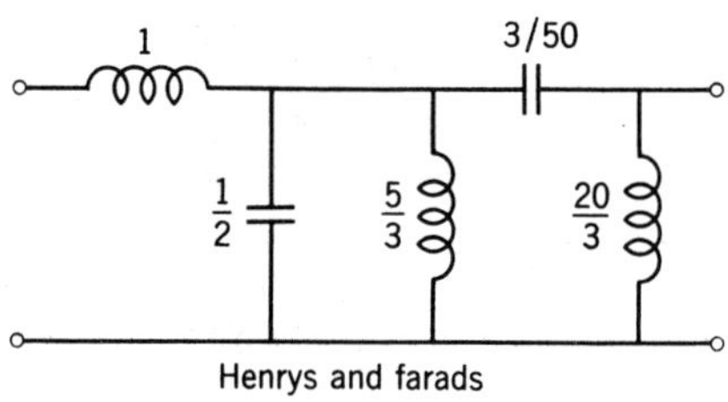

Fig. 14. Alternate canonic development to that shown in Fig. 13. Both networks realize the set 21 with z_{12} as in expression 36 and appropriate multipliers provided for z_{12} and z_{22}.

Parts (*a*), (*b*), and (*c*) of Fig. 15 illustrate the graphical aspect of this initial step. Specifically, part (*a*) shows the given reactance function z_{11}. Part (*b*) shows its reciprocal, the susceptance $y_1 = 1/z_{11}$ having the pole at $s = \infty$. The residue in this pole is equal to the slope of the asymptote, and this slope equals the value of the capacitance c_1. The remainder y_2 after subtracting the asymptote from y_1 is shown in part (*c*), and its reciprocal, the reactance $z_2 = 1/y_2$, is sketched in part (*d*).

We propose next to produce a transmission zero at the frequency $s = \pm j5$, and so we must shift a zero of z_2 into this position so that the inverted remainder will have a pole there. As part (*d*) shows, this zero-shifting can be done by removing an appropriate part of the pole at $s = \infty$, this appropriate part being equal to the slope of a straight line passing through the origin and a point on the z_2 curve corresponding to $\omega = 5$.

Through use of expression 43, we compute

$$z_2(j5) = \frac{(-25+1)(-25+9)}{j50(-25+5.5)} = j0.394 \tag{44}$$

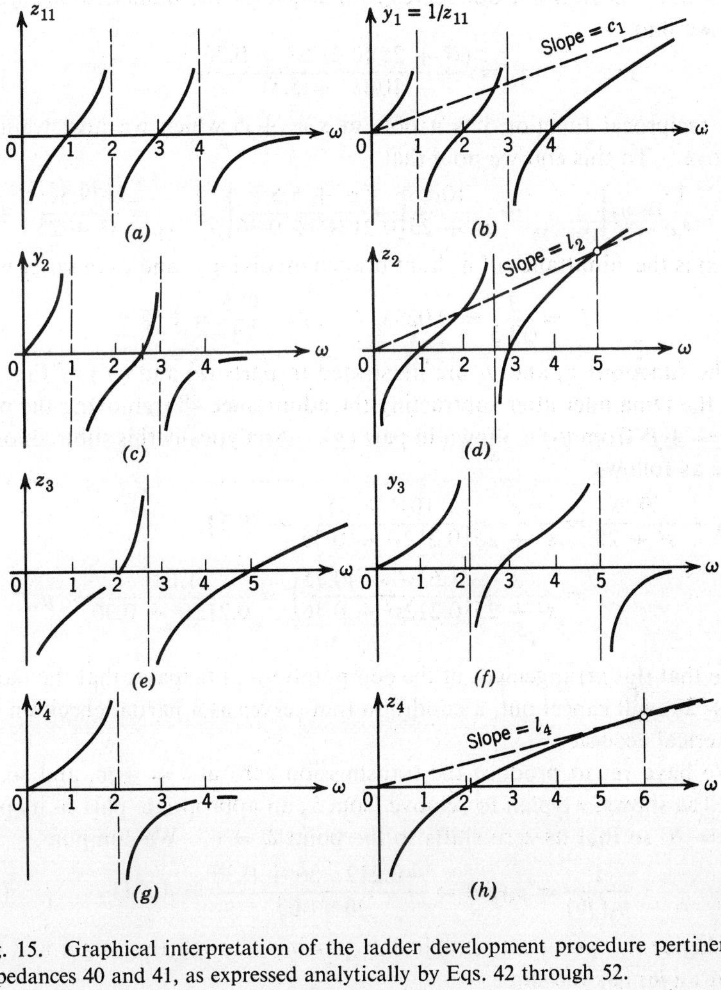

Fig. 15. Graphical interpretation of the ladder development procedure pertinent to impedances 40 and 41, as expressed analytically by Eqs. 42 through 52.

The slope of the pertinent straight line, equal to the inductance l_2 of a series branch in the ladder development, equals

$$l_2 = \frac{0.394}{5} = 0.0788 \tag{45}$$

The remainder function after subtracting the corresponding inductive reactance from z_2 is

$$z_3 = z_2 - l_2 s = \frac{0.212s^4 + 5.66s^2 + 9}{10s(s^2 + 5.5)} \tag{46}$$

The numerator polynomial in this expression must contain the factor $(s^2 + 25)$. Dividing it out serves as a check on the numerical accuracy, and we find

$$z_3 = \frac{(s^2 + 25)(0.212s^2 + 0.36)}{10s(s^2 + 5.5)} \tag{47}$$

The reciprocal function has a pole at $s = \pm j5$ which we now want to remove. To this end we note that

$$\left.\frac{1}{z_3} = y_3\right]_{s^2 \to -25} \rightarrow \frac{10s}{s^2 + 25}\left[\frac{s^2 + 5.5}{0.212s^2 + 0.36}\right]_{s^2 = -25} = \frac{39.5s}{s^2 + 25} \tag{48}$$

which is the admittance of a shunt branch involving l_3 and c_3 in series with

$$l_3 = \frac{1}{39.5} = 0.0253, \qquad c_3 = \frac{39.5}{25} = 1.58 \tag{49}$$

The functions z_3 and y_3 are illustrated in parts (e) and (f) of Fig. 15, and the remainder after subtracting the admittance 48 (removing the pole at $s = \pm j5$ from y_3) is shown in part (g). Analytically this subtraction is done as follows

$$\begin{aligned} y_3 - \frac{39.5s}{s^2 + 25} &= \frac{s}{s^2 + 25}\left(\frac{10s^2 + 55}{0.212s^2 + 0.36} - 39.5\right) \\ &= \frac{s}{s^2 + 25}\left(\frac{1.63s^2 + 40.75}{0.212s^2 + 0.36}\right) = \frac{1.63s}{0.212s^2 + 0.36} = y_4 \end{aligned} \tag{50}$$

Note that this arrangement of the computations anticipates that the factor $(s^2 + 25)$ will cancel out, a condition that serves as a partial check on the numerical accuracy.

We have yet to produce the transmission zero at $s = \pm j6$, and so, as Fig. 15h shows, we plan to remove from z_4 an appropriate part of its pole at $s = \infty$ so that its zero shifts to the point $\omega = 6$. We compute

$$\frac{1}{y_4(j6)} = z_4(j6) = \frac{-0.212 \times 36 + 0.36}{j6 \times 1.63} = j0.743 \tag{51}$$

and hence the pertinent straight line passing through the origin and the point $z_4(j6)$ has the slope

$$l_4 = \frac{0.743}{6} = 0.124 \tag{52}$$

which is the value of the series inductance forming the next branch in the ladder development. Subtracting this inductive reactance from z_4 leaves

$$z_5 = z_4 - l_4 s = \frac{s^2 + 36}{163s} = \frac{1}{y_5} \tag{53}$$

The admittance y_5 forms the final shunt branch in the ladder development. It consist of l_5 and c_5 in series with

$$l_5 = \frac{1}{163} = 0.00613, \qquad c_5 = \frac{163}{36} = 4.47 \tag{54}$$

The resulting network involving the element values 42, 45, 49, 52, and 54 is shown in Fig. 16. In this circuit the shunt branches are responsible for producing the transmission zeros at $s = \infty$, $s = \pm j5$ and $s = \pm j6$. The series branches represent steps in the development process which are necessary in order to make the transmission zeros fall at the desired frequencies. We refer to these steps as "zero-shifting" steps. Those producing the shunt branches of the ladder network in Fig. 16 are referred to as "zero-producing" steps. The ladder development process thus consists alternately of zero-shifting and zero-producing steps.

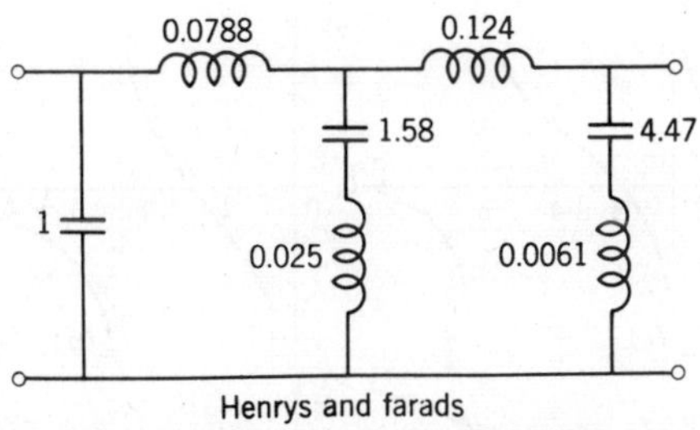

Fig. 16. Realization of impedances 40 and 41 according to the ladder development to which the graphical interpretations in Fig. 15 are relevant.

In the synthesis just completed, we have used series branches in the ladder to do the zero-shifting and shunt branches to do the zero-producing. In Fig. 17 are summarized graphically the steps in an alternate ladder development realizing the same functions 40 and 41 in which (except for for the last branch) we have done the zero-shifting with shunt branches and zero-producing with series branches.

Figure 17*a* again shows $y_1 = 1/z_{11}$ as in Fig. 15*b*; but now we do not contemplate removing the pole at $s = \infty$ completely. Rather, we remove just the right amount of this pole to shift the zero of y_1 at $\omega = 4$ to the point $\omega = 5$. This zero-shifting step produces a shunt capacitance branch with the capacitance value c_1 equal to the slope of the dashed line passing through the point $y_1(j5)$ at $\omega = 5$. The remainder function y_2, having a zero at $\omega = 5$, is shown in part (*b*) of Fig. 17, and its reciprocal z_2 appears in part (*c*).

The next step in the development process is the total removal of the pole of z_2 at $\omega = 5$. This zero-producing step yields a series branch consisting of l_2 and c_2 in parallel, resonant at $\omega = 5$. The remainder function z_3 after this step is sketched in part (*d*) of Fig. 17 and its reciprocal y_3 is shown in part (*e*). At this point in the development process we have completed one cycle and are again at the stage where we must shift a zero into the proper position, as was done with the initial function y_1 in part (*a*).

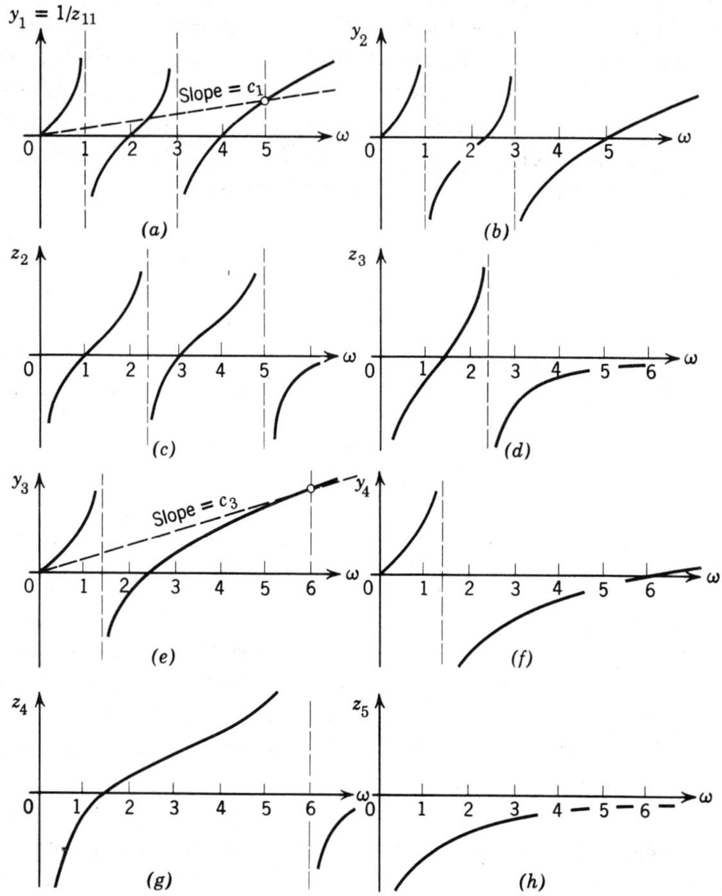

Fig. 17. Sequence of steps in an alternate ladder development realizing impedances 40 and 41.

This time we shift a zero to the point $\omega = 6$ (the second required finite nonzero zero of z_{12}) by removing part of the pole at $s = \infty$. This step yields a second shunt capacitance c_3 and a remainder function z_4 (Fig. 17*f*) having a zero at $\omega = 6$. The reciprocal function z_4 (Fig. 17*g*) therefore has a pole at $\omega = 6$ which we remove as a series branch consisting of l_4 and c_4 in parallel, resonant at $\omega = 6$.

The remainder function z_5 now has only the pole at $s = 0$; its reciprocal is an admittance having only one pole at $s = \infty$. Total removal of this pole in the form of a final shunt capacitance c_5 produces the required zero of transmission at $s = \infty$ and completes the ladder development process. The structure of the resulting network is shown in Fig. 18.

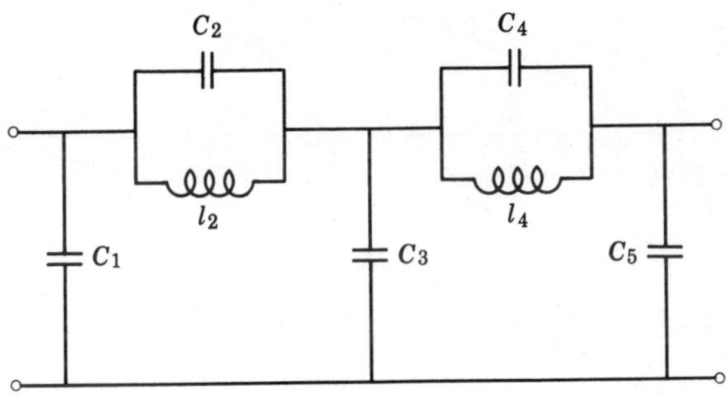

Fig. 18. Resulting structure obtained from the ladder development to which the sketches in Fig. 17 are relevant. This network and that in Fig. 16 have the same z_{11} (Eq. 40) and the same z_{12} (Eq. 41) except for a constant multiplier.

It is worth noting that we have produced this network structure without making a single calculation. The ability to carry through the structural part of a synthesis merely by drawing sketches of the relevant frequency functions as is done in Fig. 17 is practically useful because it saves computational labor until we have selected the structure that appears to best suit the needs of the problem at hand.

To illustrate further the flexibility of this method of synthesis, suppose we return to the point in the procedure just finished, at which we remove the second shunt capacitance c_3 in order to shift a zero to the frequency $\omega = 6$ (Fig. 17*e*). The admittance y_3 pertinent to this stage in the development process is shown in Fig. 19*a*, but now we propose to remove completely the pole at $s = \infty$ by making c_3 equal to the slope of the asymptote. This step produces the required zero of transmission at $s = \infty$, and leaves the remainder function y_4 (Fig. 19*b*) or its reciprocal z_4 (Fig. 19*c*).

We now shift a zero to the point $\omega = 6$ by removing an appropriate series inductance l_4 whose value equals the slope of the dashed line in Fig. 19*c* passing through the point $z_4(j6)$. After this step the remainder z_5 (Fig. 19*d*) is a series combination of l_5 and c_5 resonant at $\omega = 6$, which becomes the final shunt branch in the ladder development and accounts for the required transmission zero at $\omega = 6$.

The structure of the complete network thus found is shown in Fig. 20. Its unique feature is that the zero-shifting process is shared by the shunt and series branches instead of being done consistently by one or the other as in the previous two realizations for this same set of driving-point and transfer impedances. Thus the branches c_1 and l_4 represent zero-shifting processes; the rest account for the required zeros of the transfer impedance.

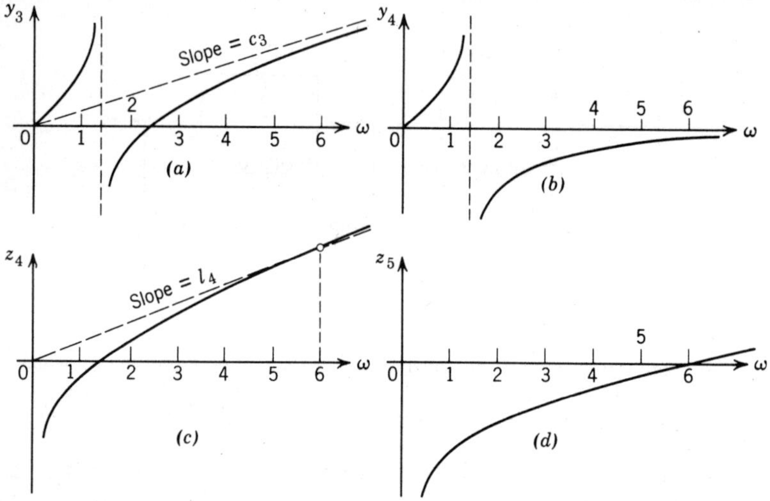

Fig. 19. Steps alternate to those in parts e, f, g, h in Fig. 17, leading to a modified ladder development of impedances 40 and 41.

In these examples we have done all the required zero-shifting by the partial removal of a pole at $s = \infty$. It may in some situations become necessary to remove part of a finite nonzero pole or of a pole at $s = 0$. The principle involved in any such process is that the partial removal of a pole causes all zeros on both sides of this pole to move toward it. They are attracted toward the pole by its partial removal. The immediately adjacent zeros are affected more strongly than are those farther removed from the pertinent pole, and so the zero-shifting ought usually be done by the nearest pole in order that the desired amount of shifting can be accomplished without having to remove too large a part of the entire pole.*

It is significant also to observe that zeros at $s = 0$ and at $s = \infty$ cannot be shifted by the partial removal of the next adjacent pole, but this

* The question involved here has to do with the corrupting influence of parasitic loss which is unavoidable in a physical structure. If only a small fraction of a pole remains, incidental loss can easily obliterate it entirely. It is better to do the required shifting in several easy stages even though more elements will be needed.

situation is easily met by reciprocating the driving-point function in hand. Since either series or shunt branches in the ladder network can be used as the means for zero-shifting, the given function or its reciprocal are equally appropriate to a particular zero-shifting step. Thus we may say that the required zero-shifting can always be done, subject only to the restriction that the required zeros of the given transfer impedance or admittance lie upon the j-axis. When these are located in the complex frequency plane, then the present method of synthesis is not applicable.

At this point the question may arise as to how one can be sure that the development of the driving-point impedance z_{11} lasts just long enough to produce the required number of zeros in z_{12}. In other words, can it happen that the development of z_{11} is completed before all the zeros in z_{12} are taken care of; or, after all the zeros in z_{12} are accounted for, can one still have an undeveloped part of z_{11} to realize?

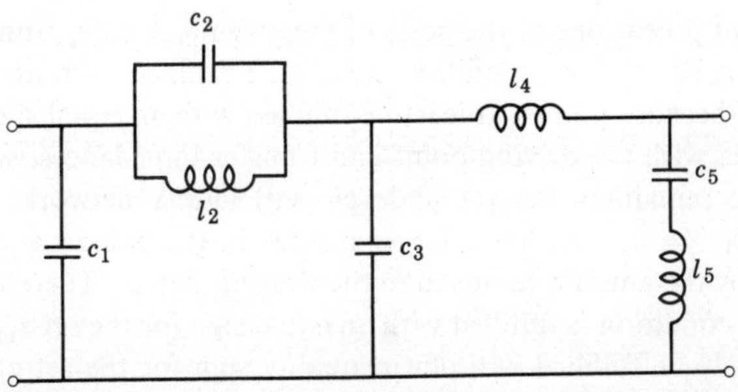

Fig. 20. Modified form of the realization of impedances 40 and 41 in Fig. 18 obtained by substituting the steps in Fig. 19 for the last four of Fig. 17.

The answers to both of these questions is, of course, "no"; and a careful study of the preceding examples reveals the reason for this conclusion. If the numerator of z_{12} is a constant, so that all transmission zeros lie at $s = \infty$, then the pertinent development of z_{11} yields a canonic Cauer network and the number of zero-producing steps equals the degree of the denominator polynomial in either z_{11} or z_{12}. If, instead, z_{12} has some or all of its zeros at $s = 0$, the development again yields a canonic network, and again the total number of elements and the number of transmission zeros must agree. If a transmission zero is to be produced at a finite nonzero frequency, one or more zero-shifting steps may be necessary, but *these do not reduce the degree of the driving-point impedance which is being developed.* Only the zero-producing steps reduce the degree of this function, and always by exactly the number of transmission zeros produced, just as in the canonic developments.

It has been pointed out that the partial removal of a pole in a zero-shifting step does not produce a zero of transmission in spite of the fact that the pertinent series or shunt branch is an infinite impedance or admittance at the pole frequency in question, because the remainder function contains this pole and hence the transfer function of the undeveloped portion of the ladder likewise contains this pole. When the synthesis problem is approached from given z_{11} and z_{12} functions, this situation has a bearing upon the nature of the z_{22} function that is produced.

Although we initially approached this synthesis method with the attitude that we are not interested in controlling z_{22}, it is helpful to be able to predict its general character and to account for certain of its peculiarities. For example, this method in general will not yield a z_{22} function such that the residue condition is fulfilled with an equal sign at all poles. The driving-point impedance z_{22} may even contain poles* not present in z_{11} or z_{12}. It is useful to know how such results come about.

If the zero-shifting step is done at the outset with the given z_{11}, then the pertinent pole is one of the poles of the given z_{11} and z_{12} functions, but observe that the residue condition cannot be fulfilled with an equal sign at this pole because it must at least be fulfilled with an equal sign when z_{22} is associated with the driving-point and transfer impedances, say z_{11}' and z_{12}', of the remaining (as yet undeveloped) ladder network. While at this stage $z_{12}' = z_{12}$, z_{11} has a larger residue in the pole in question than does z_{11}' (by the amount removed in the shifting step). Therefore, even if the residue condition is fulfilled with an equal sign for the set z_{11}', z_{12}', z_{22}, it must surely be fulfilled with an inequality sign for the set z_{11}, z_{12}, z_{22}.

If the zero-shifting step, done with an impedance, involves some remainder function (in contrast with z_{11}), then the pole in question will not necessarily be a pole of z_{11} because poles of the remainder functions in general bear no relation to those of the original function (for example, in Fig. 15 compare z_3 with z_{11}, or in Fig. 17 compare y_3 with y_1), yet the impedance z_{22} for the over-all network contains this pole because z_{22}' for the undeveloped portion of the ladder contains it and the already developed portion is virtually isolated from the undeveloped portion at the pertinent pole frequency by the series branch produced in this zero-shifting step.

An exception to this situation occurs only if, in some subsequent step in the development, this same pole is totally removed so as to produce a transmission zero at that frequency, for then it is no longer excitable

* At such a pole the residue condition is fulfilled with an equal sign in a degenerate way since $k_{11} = k_{12} = 0$, while $k_{22} \neq 0$. In order to distinguish this situation from the *normal* one in which the residue condition is fulfilled with an equal sign, we will refer to the latter as one in which the z_{ik} are *compact* (a term first suggested by B. J. Dasher, *Tech. Rept.* 215, Mass. Inst. Technol., Research Lab. Electronics, Nov. 20, 1951).

from either end of the total network although it is a natural frequency of an internal portion.

It becomes clear, therefore, that zero-shifting done with an impedance (yielding a series branch in the ladder development) will cause z_{22} for the over-all network to have an excess residue over that needed to fulfill the residue condition with an equal sign; and if done with some remainder function, this process will in general introduce poles into z_{22} not present in either of the given impedances z_{11} or z_{12}.

When zero-shifting is done with an admittance (yielding a shunt branch in the ladder development), the same effect is produced upon the resulting z_{22} function, although the reason for it is not so evident as in the situation just discussed. The reason may, however, be seen from the following argument. Consider the network N in Fig. 21a and suppose that its open-circuit input impedance z_{11} has a zero at the frequency $s = s_k$. This frequency then is one of the natural frequencies when the terminals 1-1′ for the network N are short-circuited, the open-circuit natural frequencies being the poles of z_{11} and also those of z_{22} for the same network, which we will assume to be obtained through a canonic ladder development of z_{11}.

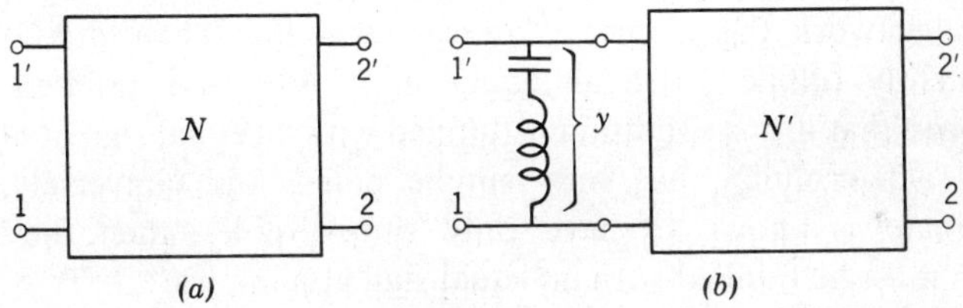

Fig. 21. Pertinent to the argument that the partial removal of a pole in the driving-point admittance at any stage in the ladder development introduces a spurious pole in the resulting z_{22} function.

Now suppose we form $y_1 = 1/z_{11}$ and remove from it a part of the pole at $s = s_k$ in the form of a shunt branch y as shown in Fig. 21b in which N' is obtained from a canonic ladder development of the remainder function $z_{11}' = (y_1 - y)^{-1}$. This impedance has the same zeros as z_{11} but not the same poles because these have been shifted in the process of removing the branch y. Since s_k is a zero of z_{11}' it is a natural frequency when the terminals 1-1′ in the network of Fig. 21b are short-circuited. But at the frequency $s = s_k$, the branch y is virtually a short circuit, and so s_k is also a natural frequency of this network when its terminals 1-1′ are open-circuited. Although not a pole of z_{11} because of the degeneracy of the situation at the terminals 1-1′, the frequency s_k is a pole of the open-circuit impedance z_{22}' of the network of part (b). Thus the impedance

z_{22}' contains this pole in addition to the poles of z_{11} or z_{22} of the network of part (*a*); and we see that zero-shifting with an admittance can also create excess poles in the resulting z_{22} function.

An exception occurs if the zero at s_k in z_{11} falls at the point $s = \infty$ so that z_{12} is likewise zero at $s = \infty$ (because the alternative of z_{12} having a pole at $s = \infty$ contradicts the assumption that z_{11} equals zero there). In this case the branch y in Fig. 21*b* reduces to a capacitance; and if the zero in z_{12} at $s = \infty$ is produced by the removal of a final shunt capacitance, then the associated z_{22} function also becomes zero at $s = \infty$. The situation is an example of the above-mentioned exception in which the same pole that is partially removed in a zero-shifting step is eventually totally removed to produce a transmission zero.

Hence we see that if the given z_{sk} have zeros at $s = \infty$, and in the development of z_{11} to produce the required zeros of z_{12} we zero-shift solely by shunt capacitance branches, then the resulting network has z_{sk} that are compact. Analogously, if the given z_{sk} have zeros at $s = 0$, and in the development of z_{11} we zero-shift solely by shunt inductances, then the resulting network likewise has z_{sk} that are compact; and the duals of these situations provide corresponding results in terms of admittances.

The following reasoning is collaterally useful in determining whether, for a given network, the z_{sk} form a compact set so that the residue condition is consistently fulfilled with an equal sign. We have pointed out in Art. 1 above that if this condition is fulfilled with an equal sign at all poles, then the z-determinant has only simple poles; and conversely, if the z-determinant is known to have only simple poles, then the residue condition must be fulfilled with an equal sign at all poles.

Now let us consider a lossless two terminal-pair network terminated at its output end in a resistance R. The input impedance under these circumstances is given by

$$Z_1 = \frac{z_{11}z_{22} - z_{12}^2 + z_{11}R}{z_{22} + R} \tag{55}$$

as may readily be computed. From this expression it may be seen that if the z-determinant $|z| = z_{11}z_{22} - z_{12}^2$ has only simple poles then the input impedance is regular at these j-axis points (except in the degenerate case in which z_{11} has a pole that is not also contained in z_{12} or z_{22}, a situation which we shall for the moment exclude from our present considerations). Since the denominator $z_{22} + R$ cannot be zero on the j-axis, it follows that

> *Z_1 can have no j-axis poles if $|z|$ has only simple poles present in all three z_{sk}; conversely, if Z_1 has no j-axis poles, then $|z|$ can have only simple poles* (56)

By using the previous result regarding the residue condition, this conclusion may be restated (with the exception of degenerate cases) as follows

Z_1 can have no j-axis poles if the z_{sk} characterizing the lossless two terminal-pair network are compact; conversely, if Z_1 has no j-axis poles then the z_{sk} must form a compact set (57)

Since the poles of Z_1 are natural frequencies of the network with its input terminals open-circuited, and since the preceding argument applies equally well with the lossless network turned end for end, we may say that if a lossless two-terminal-pair network with a resistance across one terminal pair has no j-axis natural frequencies, then (excepting degeneracies *) the z_{sk} form a compact set, and it does not matter which terminal pair the resistance is connected to.

Exceptions because of degeneracies affect only the converse portion in statement 57. Thus, Z_1 does not have a j-axis pole that is contained in z_{22} but not in z_{11} or z_{12}, and hence the absence of this pole in Z_1 does not assure fulfillment of the compact condition at this frequency. Although, in this situation, the resistance across terminal pair 2 is virtually isolated at the pertinent pole frequency, the latter is not a natural frequency excitable from the input end. It is, however, excitable from terminal pair 2 with the resistance across terminal pair 1.

In the light of these results one may see that any development requiring zero-shifting cannot, in general, yield z_{ik} fulfilling the compact condition at all poles. With exception of the above-mentioned special cases, the compact condition will fail to be fulfilled at those poles involved in the zero-shifting steps. The examples in Figs. 16 and 20, involving partial pole removals at $s = \infty$ from impedance remainder functions, yield z_{ik} fulfilling the compact condition at all poles except the one at $s = \infty$. The development yielding the network of Fig. 18 exemplifies one of the exceptional cases in which zero-shifting does not prevent the resulting z_{ik} from fulfilling the compact condition at *all* poles. In the canonic ladder developments yielding the networks in Figs. 12, 13, and 14, for which no zero-shifting steps are involved, the compact condition is, of course, likewise fulfilled at all poles.

4 Cascade Synthesis

From a practical point of view it is essential that we have generally applicable methods of synthesis for lossless two terminal-pair networks leading to unbalanced structures that may readily be constructed (apart from realizing the lossless character of the elements, of course). Although

* Owing to a pole contained in z_{11} or in z_{22} alone.

the method discussed in the previous article leads to practically useful networks in this sense, it can be applied only in situations where the transmission zeros occur at real frequencies (points on the j-axis of the complex frequency plane). This restriction we wish now to remove, while retaining the otherwise practical features of the structure obtained with that method, namely, (*a*) its unbalanced character, and (*b*) the fact that the zeros of transmission are controlled by independent parts of the network and hence can readily be adjusted to their proper values.

When the zeros of transmission occur at complex frequencies, we find that it is still possible to control independently the zeros of transmission, although no longer by means of single branches as in a ladder structure. What is needed is a cascade of component two terminal-pair networks as shown in Fig. 22, each of which places one set of complex transmission

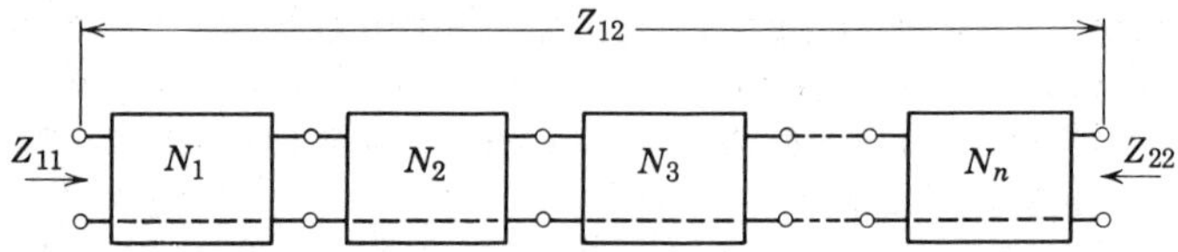

Fig. 22. Form of unbalanced development that is needed when the transmission zeros are complex. The result guarantees realization of one driving-point impedance (not both) and the transfer impedance within a constant multiplier.

zeros in evidence. By a "set" we mean two pairs of conjugate values that are negatives of each other. As has been pointed out before, the zeros of a lossless transfer function (being the zeros of an even or odd polynomial) form a pattern in the complex frequency plane that is symmetrical with respect to both the real and imaginary axes. Hence the complex ones occur in sets of quadruplets.*

Each component network in the structure of Fig. 22 (hereinafter also referred to as a "zero section") produces one complex quadruplet set of transmission zeros. It is always realizable as an unbalanced network (in the form of either a bridged-T or twin-T structure), and in many cases does not require mutual inductive coupling.

We will denote the driving-point and transfer impedances of the total lossless network by Z_{11}, Z_{22}, and Z_{12}, and those of a typical component two terminal pair by z_{11}, z_{22}, and z_{12}. Specifically, we shall show how we can extract from the given driving-point impedance Z_{11}, or from a typical intermediate remainder function which we shall call Z_1, the set of im-

* We will discuss here the method of dealing with these quadruplet zeros. An appropriately modified form suitable for situations involving pairs of zeros on the real axis will then become rather evident.

pedances z_{11}, z_{22}, and z_{12} characterizing a component network (realizable in the desired form) which produces one quadruplet of zeros of the function z_{12}, and leaves an LC realizable remainder function Z_2 having the same detailed character as Z_1.

This typical cycle in the development process, as illustrated in Fig. 23, can then be repeated until there is no remainder. Any j-axis zeros that the initially given Z_{12} function may have had (except possibly a zero at $s = 0$ and/or at $s = \infty$) are assumed to have been produced by a previous ladder development according to methods already discussed, the present network then being a continuation (in appropriately altered form) of that development procedure.

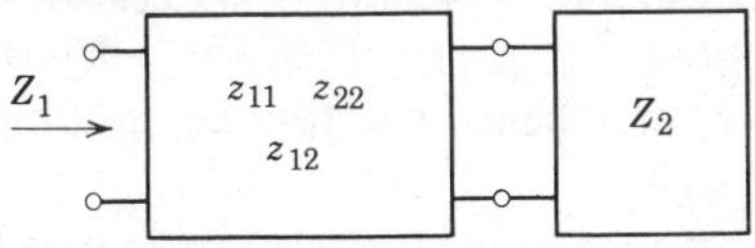

Fig. 23. Schematic representation of a typical cycle in the cascade development process.

Although each component network in the development of Fig. 22 is compact, the over-all network does not have this character. We will see, however, that in most cases it will be possible to have the over-all impedances satisfy the compact condition at all poles except the ones at $s = 0$ and $s = \infty$. Thus if such a resulting network is supplied with a resistance termination at one end, the only finite nonzero frequencies at which this resistance can virtually be isolated are complex. Hence no finite nonzero j-axis frequencies can be natural frequencies of this system.

Considering the typical development step illustrated in Fig. 23, and assuming that the lossless network characterized by z_{11}, z_{22}, and z_{12} produces a complex quadruplet of transmission zeros at the conjugate points $s = \pm s_0$ and $s = \pm \bar{s}_0$ with $s_0 = \sigma_0 + j\omega_0$, the simplest form for the pertinent transfer impedance reads

$$z_{12} = \frac{Kt(s)}{s(s^2 - s_a^2)} \tag{58}$$

K is a positive real constant and the transmission zeros are represented by

$$t(s) = (s^2 - s_0^2)(s^2 - \bar{s}_0^2) = [(s + \sigma_0)^2 + \omega_0^2][(s - \sigma_0)^2 + \omega_0^2] \tag{59}$$

The partial-fraction expansion of this transfer function is given by

$$z_{12} = Ks + \frac{k_0}{s} - \frac{2k_{12}s}{s^2 - s_a^2} \tag{60}$$

in which $-k_{12}$ is its residue in the j-axis pole at $s = s_a$, and the terms for the conjugate pair of these poles are combined in the usual manner. The associated driving-point impedances may be written

$$z_{11} = Ks + \frac{k_0}{s} + \frac{2k_{11}s}{s^2 - s_a^2} \tag{61}$$

and

$$z_{22} = Ks + \frac{k_0}{s} + \frac{2k_{22}s}{s^2 - s_a^2} \tag{62}$$

with the tacit assumption that

$$k_{11}k_{22} - k_{12}^2 = 0 \tag{63}$$

which must automatically be fulfilled zince Z_1 does not contain the pole at $s = s_a$. Thus, if in Eq. 55 we substitute Z_2 for R, it becomes evident that if Z_1 does not contain the pole s_a then the z-determinant must have a simple pole at $s = s_a$ and hence the residue condition must there be fulfilled with an equal sign.

Although all three component impedances are assumed to have the same residue in their pole at $s = \infty$ and at $s = 0$, their residues in the j-axis pole at $s = s_a$ are left arbitrary except for condition 63. It will be seen later on that this degree of arbitrariness in the choice of these impedances is all that can be tolerated if a simple form of unbalanced network is to be had for their realization.

From Eq. 58 we have for the residues of z_{12}

$$k_0 = Kt(0)/(-s_a^2) \tag{64}$$

and

$$k_{12} = Kt(s_a)/(-2s_a^2) \tag{65}$$

The driving-point impedance for the pertinent development cycle we assume in the form

$$\begin{aligned} Z_1 &= A\frac{sp(s)}{q(s)} + \frac{\rho_0}{s} + \rho_\infty s \\ &= A\frac{s(s^2 - s_{p1}^2)(s^2 - s_{p2}^2) \cdots (s^2 - s_{pn-1}^2)}{(s^2 - s_1^2)(s^2 - s_2^2) \cdots (s^2 - s_n^2)} + \frac{\rho_0}{s} + \rho_\infty s \end{aligned} \tag{66}$$

Here A is a positive real constant and the simple j-axis zeros and poles alternate in the usual fashion. The remainder function Z_2 should have the same form involving polynomials that are respectively four degrees lower than $sp(s)$ and $q(s)$. Important about this assumed Z_1 function is the fact that it has poles at the points $s = 0$ and $s = \infty$.

The analytic relationship between the five quantities z_{11}, z_{12}, z_{22}, Z_1 and Z_2 is, for our present purposes, best written in the form

$$(z_{11} - Z_1)(z_{22} + Z_2) = z_{12}^2 = \frac{K^2t^2(s)}{s^2(s^2 - s_a^2)^2} \tag{67}$$

Considering expression 66 for Z_1 and the form of z_{11} in Eq. 61, it is clear that the factor $(z_{11} - Z_1)$ has the denominator $s(s^2 - s_a^2)q$; and from Eq. 62 we see that the denominator of the factor $(z_{22} + Z_2)$ must, among other things, contain the factors $s(s^2 - s_a^2)$. The right-hand side contains $t^2(s)$ in its numerator, and so this factor must appear on the left-hand side of Eq. 67 also. We cannot allot it to $(z_{22} + Z_2)$ which is p.r. and LC, and hence can have only j-axis zeros; it must therefore be associated with the numerator of $(z_{11} - Z_1)$ which can contain this factor because of the minus sign. These considerations lead us to conclude that the left-hand factors in expression 67 are given by

$$(z_{11} - Z_1) = \frac{K^2t^2h}{s(s^2 - s_a^2)q} \tag{68}$$

$$(z_{22} + Z_2) = \frac{q}{s(s^2 - s_a^2)h} \tag{69}$$

Thus $h(s)$ in the numerator of $(z_{11} - Z_1)$ must be an even polynomial of degree $n - 2$ in s^2 since the whole factor must be the ratio of polynomials that are respectively even and odd, and must have a simple pole at $s = \infty$ like z_{11} and Z_1 do. That is to say, since the denominator of this factor has the degree $2n + 3$, the numerator must have the degree $2n + 4$; and $t(s)$ being of fourth degree, $t^2(s)$ is of degree eight, which requires $h(s)$ to have the degree $2n - 4$, or, as an even polynomial, its degree becomes $n - 2$ in s^2. The detailed form of $(z_{22} + Z_2)$ in Eq. 69 is now easily recognized since, in addition to the denominator $s(s^2 - s_a^2)$, it must contain whatever other factors are needed to make the left- and right-hand sides of Eq. 67 equal.

Essentially our problem now is to construct the polynomial $h(s)$, which we will write in the form

$$h(s) = H(s^2 - s_{h1}^2)(s^2 - s_{h2}^2) \cdots (s^2 - s_{hn-2}^2) \tag{70}$$

and determine the finite nonzero pole s_a for the z_{ik}. Once these are known, the z_{ik} for the component network producing the desired transmission zero (the zero section) and the remainder function Z_2 will readily be extractable from the pertinent factors in Eqs. 68 and 69.

To determine $h(s)$ we observe from Eq. 68 that the rational function

$$\frac{h(s)}{q(s)} = \frac{s(s^2 - s_a^2)(z_{11} - Z_1)}{K^2t^2(s)} \tag{71}$$

can be written explicitly in terms of its partial-fraction expansion. A typical pair of terms in this expansion for the conjugate pole-pair at $s = \pm s_\nu$ is given by

$$\frac{-s_\nu(s_\nu^2 - s_a^2)k_\nu}{K^2t^2(s_\nu)}\left(\frac{1}{s - s_\nu} - \frac{1}{s + s_\nu}\right) = \frac{-2s_\nu^2k_\nu}{K^2t^2(s_\nu)} \times \frac{s_\nu^2 - s_a^2}{s^2 - s_\nu^2} \tag{72}$$

in which the residues of $Z_1(s)$ in these finite nonzero poles are denoted by

$$k_\nu = \left[\frac{A(s - s_\nu)sp(s)}{q(s)}\right]_{s=s_\nu} = \left[\frac{A(s^2 - s_\nu^2)p(s)}{2q(s)}\right]_{s=s_\nu} \tag{73}$$

We thus obtain for h/q the representation

$$\frac{h(s)}{q(s)} = \frac{1}{K^2}\sum_{\nu=1}^{n} m_\nu\left(\frac{s_\nu^2 - s_a^2}{s^2 - s_\nu^2}\right) \tag{74}$$

with the abbreviation

$$m_\nu = \frac{-2s_\nu^2 k_\nu}{t^2(s_\nu)} \tag{75}$$

Since the degree of $h(s)$ as a polynomial in s^2 is $n - 2$ and that of $q(s)$ is n, the rational function 74 should behave like $1/s^4$ for large s. To place in evidence the asymptotic behavior of 74 we write

$$\frac{1}{s^2 - s_\nu^2} = \frac{1}{s^2}\left(\frac{1}{1 - s_\nu^2/s^2}\right) = \frac{1}{s^2}\left(1 + \frac{s_\nu^2}{s^2} + \frac{s_\nu^4}{s^4} + \cdots\right) \tag{76}$$

It is clear, therefore, that we must have

$$\sum_{\nu=1}^{n} m_\nu(s_\nu^2 - s_a^2) = 0 \tag{77}$$

and the constant multiplier H in Eq. 70 is given by

$$H = \frac{1}{K^2}\sum m_\nu s_\nu^2(s_\nu^2 - s_a^2) \tag{78}$$

Equation 77 determines the finite nonzero pole-pair of the z_{sk}, giving

$$s_a^2 = \frac{\sum m_\nu s_\nu^2}{\sum m_\nu} \tag{79}$$

Since the m_ν are positive and the s_ν^2 are negative real numbers, the value for s_a^2 comes out negative real as it should.

If we interpret the quantities m_ν as a set of masses located at the points s_ν^2 (on an axis along which the negative real s_ν^2 values are plotted), then s_a^2 may be regarded as the center of gravity of these masses. This interpretation provides a useful physical and geometrical way of compactly visualizing the determination of the pole s_a.

In expression 78 for H we may regard the quantities $m_\nu(s_\nu^2 - s_a^2)$ as a set of forces, negative to the left of s_a^2 and positive to the right of this point whose algebraic sum is zero, as indicated in Fig. 24. Thus we see that multiplication by $s_\nu^2 < 0$, followed by addition, must lead to a positive value for H.

In view of forms 61 and 62 assumed for z_{11} and z_{22}, and the requirement that the remainder Z_2, like Z_1, shall have poles at the points $s = 0$ and $s = \infty$, we must at $s = 0$ fulfill the conditions

$$(sz_{11})_{s=0} \geq k_0 = \frac{Kt(0)}{-s_a^2}, \qquad \text{and } [s(z_{22} + Z_2)]_{s=0} > \frac{Kt(0)}{-s_a^2} \tag{80}$$

and at $s = \infty$ require that

$$\left(\frac{z_{11}}{s}\right)_{s=\infty} \geq K, \qquad \text{and } \left(\frac{z_{22} + Z_2}{s}\right)_{s=\infty} > K \tag{81}$$

By use of Eqs. 68 and 69, conditions 80 become

$$\frac{K^2t^2(0)h(0)}{-s_a^2q(0)} + \rho_0 \geq \frac{Kt(0)}{-s_a^2}, \qquad \text{and } \frac{q(0)}{-s_a^2h(0)} > \frac{Kt(0)}{-s_a^2} \tag{82}$$

while those expressed by 81 read

$$K^2H + \rho_\infty \geq K, \qquad \text{and } \frac{1}{H} > K \tag{83}$$

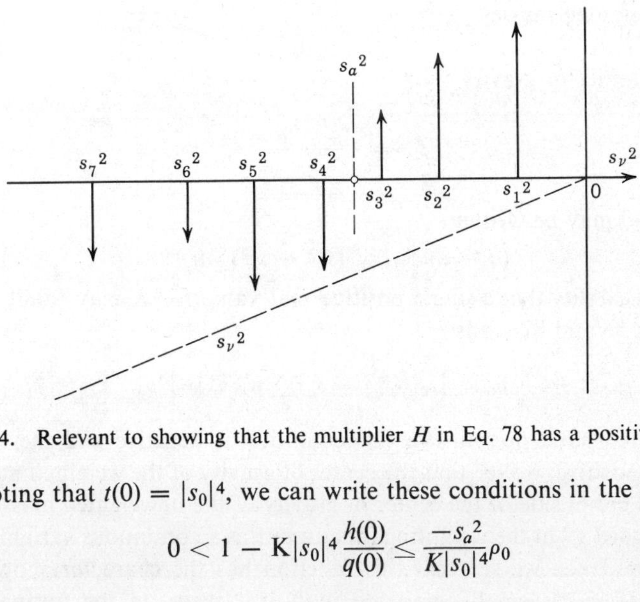

Fig. 24. Relevant to showing that the multiplier H in Eq. 78 has a positive value.

Noting that $t(0) = |s_0|^4$, we can write these conditions in the form

$$0 < 1 - \mathrm{K}|s_0|^4\frac{h(0)}{q(0)} \leq \frac{-s_a^2}{K|s_0|^4}\rho_0 \tag{84}$$

and

$$0 < 1 - HK \leq \frac{\rho_\infty}{K} \tag{85}$$

Using Eq. 74 we obtain in place of 84

$$0 < K - |s_0|^4 \sum \frac{m_\nu}{-s_\nu^2}(s_\nu^2 - s_a^2) \leq \frac{-s_a^2}{|s_0|^4}\rho_0 \tag{86}$$

and substitution of Eq. 78 into 85 gives

$$0 < K - \sum m_\nu s_\nu^2(s_\nu^2 - s_a^2) \leq \rho_\infty \tag{87}$$

The last two conditions have the form

$$\begin{aligned} 0 < K - a \leq b \\ 0 < K - c \leq d \end{aligned} \tag{88}$$

Straightforward study shows that their simultaneous fulfillment with a single K value is possible only if

$$-b < (a - c) < d \tag{89}$$

With reference to Eqs. 86 and 87 (noting that $s_\nu^2 < 0$) we have

$$(a - c) = |s_0|^2 \sum \left(\left|\frac{s_0}{s_\nu}\right|^2 + \left|\frac{s_\nu}{s_0}\right|^2 \right) m_\nu (s_\nu^2 - s_a^2) \tag{90}$$

If we define a weighting function

$$w_\nu = \left|\frac{s_0}{s_\nu}\right|^2 + \left|\frac{s_\nu}{s_0}\right|^2 \tag{91}$$

a set of weighted masses

$$\hat{m}_\nu = w_\nu m_\nu \tag{92}$$

and their center of gravity

$$\hat{s}_a^2 = \frac{\sum \hat{m}_\nu s_\nu^2}{\sum \hat{m}_\nu} \tag{93}$$

then Eq. 90 may be written

$$(a - c) = |s_0|^2 (\hat{s}_a^2 - s_a^2) \sum \hat{m}_\nu \tag{94}$$

and the possibility that a single positive real value for K may fulfill both conditions 86 and 87 reads

$$-\left|\frac{s_a}{s_0}\right|^2 \rho_0 < |s_0|^4 (\hat{s}_a^2 - s_a^2) \sum \hat{m}_\nu < |s_0|^2 \rho_\infty \tag{95}$$

Since the left-hand quantity in this condition is negative and the right-hand one positive, we see that the center of gravity of the weighted masses may lie on either side of the center of gravity of the unweighted masses.

If we regard s_ν^2 in the weighting function 91 as a continuous variable for the moment, then we see that this function has the character shown in Fig. 25. If $|s_0|$ is small compared with $|s_a|$, then, on the normalized abscissa scale of Fig. 25, the points $s_1^2 \cdots s_n^2$ all (or mostly) lie to the left of the point -1, and the effect of the weighting function is to shift the center of gravity $\hat{s}_a^2$ to the left of s_a^2 so that the quantity $(\hat{s}_a^2 - s_a^2)$ becomes negative. Similarly, if $|s_0|$ is large compared with $|s_a|$, the points $s_1^2 \cdots s_n^2$ are to the right of the point -1 in Fig. 25, and the weighting function causes $\hat{s}_a^2$ to shift to the right of s_a^2 so that $(\hat{s}_a^2 - s_a^2)$ becomes positive. The absolute value of $(\hat{s}_a^2 - s_a^2)$ cannot, however, exceed the value of $|s_n^2 - s_a^2|$ for small $|s_0|$ or that of $|s_a^2 - s_1^2|$ for large values of

$|s_0|$; and if the values of $|s_0|$ and $|\hat{s}_a|$ are not appreciably different, then the absolute value of $(\hat{s}_a^2 - s_a^2)$ will be rather small since the weighting function shifts the center of gravity of the masses by only a small amount.

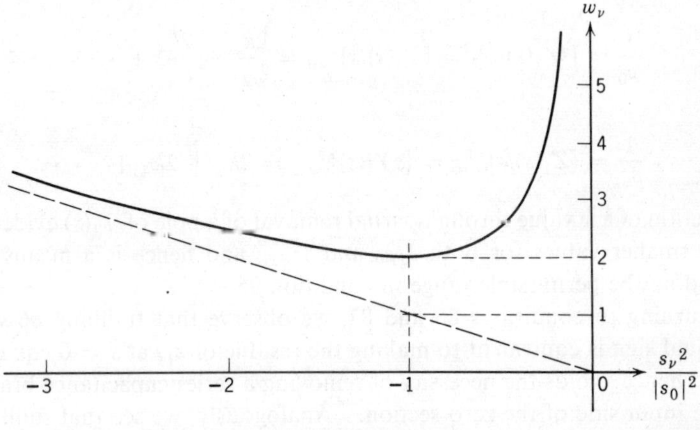

Fig. 25. Behavior of the weighting function 91 versus s_ν^2 as the variable.

In considering the probability of being able to fulfill condition 95, it is significant to observe the dependence of $\hat{m}_\nu$ upon s_0 as expressed by Eqs. 59, 75, 91, and 92. Thus for large values of $|s_0|$, $t^2(s_\nu) \approx |s_0|^8$. Since w_ν varies as $|s_0|^2$ and m_ν inversely as $|s_0|^8$, $\hat{m}_\nu$ is inversely proportional to $|s_0|^6$. The central quantity in condition 95 then varies inversely as $|s_0|^2$; and since it is positive, the right-hand inequality is pertinent. The right-hand quantity, however, is proportional to $|s_0|^2$, and so we see that condition 95 will easily be met. For small values of $|s_0|$, on the other hand, Eq. 59 shows that $t^2(s_\nu) \approx s_\nu^8$. w_ν now varies inversely as $|s_0|^2$, and so does $\hat{m}_\nu$, since m_ν is substantially independent of the value of $|s_0|$. The central quantity in condition 95 then varies directly as $|s_0|^2$; and since it is negative, the left-hand inequality is pertinent. The left-hand quantity being inversely proportional to $|s_0|^2$ is now very large, and so again the condition is easily met. Finally, in an intermediate case for which $|s_0|$ and $|s_a|$ are not appreciably different, condition 95 is apt to be fulfillable since $(\hat{s}_a^2 - s_a^2)$ is small.

In this connection it is significant to observe that the values of both ρ_0 and ρ_∞ can at any stage in the development be increased by the partial removal of a pole of the admittance $Y_1(s) = 1/Z_1(s)$. If, for the moment, we indicate a partial-fraction expansion of $Z_1(s)$ by the expression

$$Z_1(s) = \frac{2sk_1}{s^2 + \omega_1^2} + \frac{2sk_3}{s^2 + \omega_3^2} + \cdots + \frac{\rho_0}{s} + \rho_\infty s \tag{96}$$

and the expansion of the corresponding admittance by

$$Y_1(s) = \frac{2sk_2}{s^2 + \omega_2^2} + \frac{2sk_4}{s^2 + \omega_4^2} + \cdots \tag{97}$$

then we have

$$\frac{1}{\rho_0} = [sZ_1(s)]_{s=0}^{-1} = [Y_1(s)/s]_{s=0} = \frac{2k_2}{\omega_2^2} + \frac{2k_4}{\omega_4^2} + \cdots \tag{98}$$

and

$$\frac{1}{\rho_\infty} = [Z_1(s)/s]_{s=\infty}^{-1} = [sY_1(s)]_{s=\infty} = 2k_2 + 2k_4 + \cdots \tag{99}$$

Reduction of a k value through partial removal of a pole of $Y_1(s)$ evidently yields smaller values for both $1/\rho_0$ and $1/\rho_\infty$, and hence is a means for expanding the permissible range in condition 95.

Returning to conditions 86 and 87, we observe that fulfilling 86 with the equal sign is equivalent to making the residue of z_{11} at $s = 0$ equal to k_0 and hence avoids the necessity of removing a series-capacitance branch on the input side of the zero section. Analogously, we see that fulfilling 87 with the equal sign is equivalent to making the residue of z_{11} at $s = \infty$ equal to unity and avoids the necessity of removing a series inductance on the input side of the zero section. In terms of form 88 of conditions 86 and 87, the possibility of avoiding the series capacitance is found to exist if

$$-b < (a - c) \leq d - b \tag{100}$$

and the series inductance may be avoided if

$$d - b \leq (a - c) < d \tag{101}$$

For small values of $|s_0|$ for which $(a - c)$ is negative, condition 100 is fulfillable whenever 89 is; whereas for large values of $|s_0|$ for which the value of b is small, it is only moderately more restrictive than 89. For large values of $|s_0|$ for which $(a - c)$ is positive, condition 101 is fulfillable whenever 89 is; whereas for small values of $|s_0|$ for which the value of d is small, it is only moderately more restrictive than 89.

Although the possibility of avoiding both a series inductance and a series capacitance is rather remote, the chances of avoiding one or the other are reasonably good. By resorting to the partial removal of a pole of $Y_1(s)$ as discussed above, one may also trade the removal of a series branch for the removal of a shunt branch wherever this switch seems advisable.

Finally it is necessary to show that the remainder function Z_2 is p.r. and LC realizable, which will be so if the factor $(z_{22} + Z_2)$ in Eq. 69 is p.r. and LC realizable because conditions 86 and 87 insure that the residues of $(z_{22} + Z_2)$ in the poles at $s = 0$ and $s = \infty$ are larger than those of z_{22}.

According to Eqs. 68 and 69 we can form the function

$$\frac{1}{z_{22} + Z_2} = \frac{s^2(s^2 - s_a{}^2)^2(z_{11} - Z_1)}{K^2t^2(s)} \tag{102}$$

which has only the finite nonzero poles of $Z_1(s)$ with residues given by the expression

$$\frac{-s_\nu{}^2(s_\nu{}^2 - s_a{}^2)^2k_\nu}{K^2t^2(s_\nu)} \tag{103}$$

Since these values are clearly positive, function 102 and hence $(z_{22} + Z_2)$ must be p.r. and LC realizable. The p.r. and LC character of both z_{22} and Z_2 is, therefore, established.

The positiveness of both K and k_0 in Eq. 62 thus being assured, Eq. 64 substantiates the fact that $-s_a{}^2$ is positive; and a positive real value k_{12} then follows from Eq. 65. The p.r. and LC character of z_{22} further insures the positive realness of k_{22}, whereupon residue condition 63 yields a positive real value for k_{11}.

The computational procedure for finding Z_2 is to expand the factor $(z_{22} + Z_2)$ into partial fractions, after the zeros of $h(s)$ and the value of s_a have first been determined. The terms for the poles at $s = \pm s_a$ plus k_0/s and Ks belong to z_{22}; the rest is Z_2 with all of its residues computed, ready for carrying out the next development cycle.*

We come now to the synthesis of the zero section from its z_{ik}. In order to avoid complicated algebraic formulas, we wish to derive this section from the symmetrical lattice, and in order to accommodate the dissymmetry of the zero section we will make use of the extension of Bartlett's bisection theorem discussed in Art. 6 of Ch. 6. To apply the results of this extension to our present problem we make specific use of Eqs. 73 and 74 for the lattice impedances and Eq. 78 for the impedance level factor, which are given in Ch. 6 and are repeated below for convenience in our present discussions

$$z_a = z_{11} - z_{12} \tag{104}$$

$$z_b = z_{11} + \frac{z_{12}}{a} \tag{105}$$

$$a = \frac{z_{22} - z_{12}}{z_{11} - z_{12}} \tag{106}$$

Now let us use these results to find the lattice, and derive from it the desired zero section. Substitution from Eqs. 60, 61, and 62 into Eq. 106 gives

$$a = \frac{k_{22} + k_{12}}{k_{11} + k_{12}} = \frac{k_{22}}{k_{12}} = \frac{k_{12}}{k_{11}} \tag{107}$$

---

* Any value for K may be chosen that simultaneously fulfills conditions 86 and 87.

where use is also made of condition 63. Since the residues k_{11} and k_{22} are surely positive, and k_{12} as pointed out above is also real and positive, it follows that a positive real value for the impedance level factor a is assured.

Substitution of Eqs. 60, 61, and 62 into Eqs. 104 and 105 yields for the lattice impedances

$$z_a = \frac{2(k_{11} + k_{12})s}{s^2 - s_a^2} \tag{108}$$

and

$$z_b = \left(Ks + \frac{k_0}{s}\right)\frac{k_{11} + k_{12}}{k_{12}} \tag{109}$$

in which k_0 is readily computed from Eq. 64. k_{22}, as mentioned above, is determined from the partial-fraction expansion of $(z_{22} + Z_2)$ given by Eq. 69; k_{12} is obtained from Eq. 65 after s_a^2 is computed from Eq. 79, and k_{11} is then calculated from residue condition 63.

The impedances 108 and 109 lead to the lattice of Fig. 26a with the element values

$$L_a = \frac{2(k_{11} + k_{12})}{-s_a^2}, \qquad C_a = \frac{1}{2(k_{11} + k_{12})} \tag{110}$$

$$L_b = \frac{(k_{11} + k_{12})K}{k_{12}}, \qquad C_b = \frac{k_{12}}{k_0(k_{11} + k_{12})} \tag{111}$$

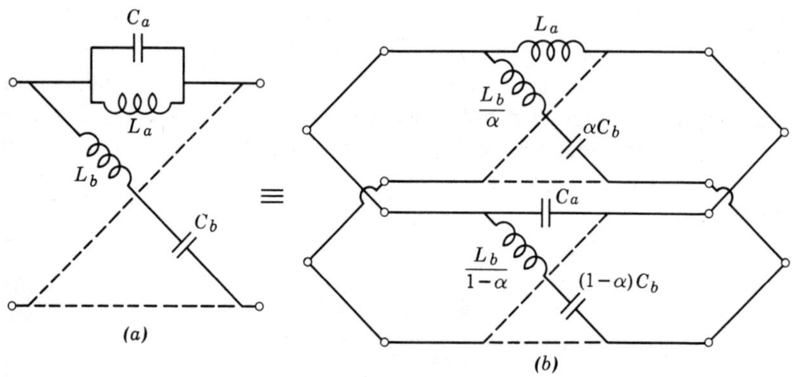

Fig. 26. Realization of the zero section as a lattice with the element values 110 and 111.

From this lattice we can now derive an equivalent unbalanced structure according to the methods discussed in Art. 5, Ch. 6, and illustrated there for an analogous RC circuit by Figs. 23, 24, 25, and 26. Thus we can always convert the lattice into the equivalent bridged-T circuit of Fig. 27a since the condition $L_b/L_a \geq 1$ is relevant only if we wish to avoid mutual inductive coupling. The bridged-T circuit in Fig. 27*b* is realizable if $C_a/C_b \geq 1$ as indicated.

If we are striving to avoid mutual inductive coupling and neither of the conditions noted in Fig. 27 is fulfilled, then the lattice is first decomposed into the parallel combination shown in Fig. 26*b* in which the numeric α (as pointed out in Art. 5, Ch. 6) is most advantageously chosen equal to the value

$$\alpha = 1 - (C_a/C_b) \tag{112}$$

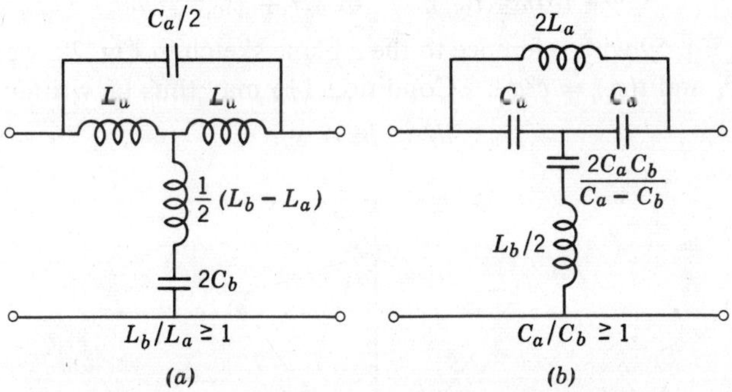

Fig. 27. Bridged-T circuits equivalent to the lattice of Fig. 26*a*.

(which is positive when the bridged-T circuit of Fig. 27*b* does not exist) for then the capacitance in the crossarm of the lower lattice in Fig. 26*b* becomes

$$(1 - \alpha)C_b = C_a \tag{113}$$

the same as the capacitance in the series arm. Conversion of the lattice into the twin-T of Fig. 28 is then straightforward, and (according to Eq. 69,

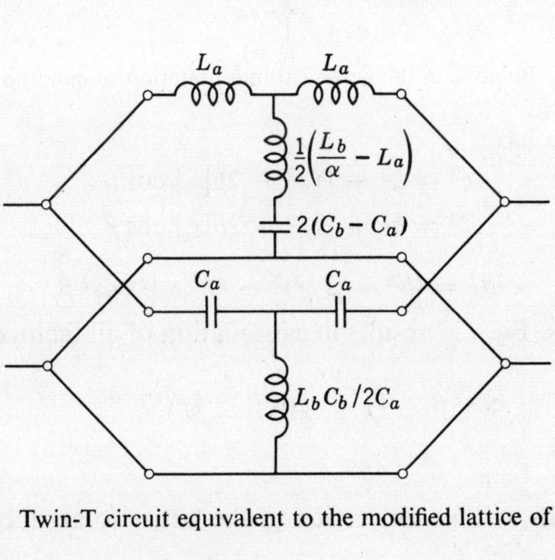

Fig. 28. Twin-T circuit equivalent to the modified lattice of Fig. 26*b*.

Ch. 6) is realizable without mutual inductive coupling if

$$\frac{L_b}{L_a} + \frac{C_a}{C_b} \geq 1 \tag{114}$$

Substitution from Eqs. 110 and 111 converts this condition into the form

$$\frac{-s_a^2K + k_0}{2k_{12}} \geq 1 \tag{115}$$

and through use of Eqs. 64 and 65 we get

$$s_a^4 + t(0) \geq t(s_a), \qquad \text{or } s_a^4 + |s_0|^4 \geq t(s_a) \tag{116}$$

Recalling Eq. 59 with reference to the s-plane sketch in Fig. 29, we see that $|s_0|^4 = b^4$, and $t(s_a) = c^2d^2$. Condition 116 may thus be written

$$c^2d^2 \leq b^4 + s_a^4 \tag{117}$$

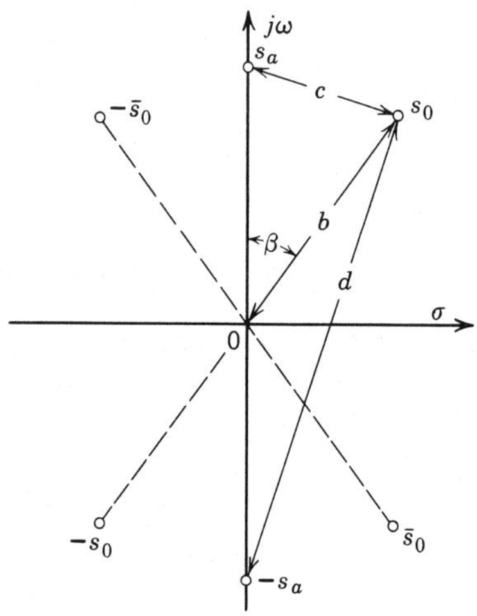

Fig. 29. Relative to the graphical interpretation of condition 116.

By geometry we have

$$c^2 = b^2 + |s_a|^2 - 2b|s_a| \cos\beta$$
$$d^2 = b^2 + |s_a|^2 + 2b|s_a| \cos\beta$$

so that

$$c^2d^2 = (b^2 + |s_a|^2)^2 - 4b^2|s_a|^2 \cos^2\beta$$

Substitution into Eq. 117 results in cancellation of all factors except $\cos\beta$, and we have

$$\cos^2\beta \geq \tfrac{1}{2}, \qquad \cos\beta \geq 1/\sqrt{2} \tag{118}$$

or

$$\beta \leq 45° \tag{119}$$

This result tells us that conversion of the lattice to an unbalanced form without mutual inductive coupling is not possible for all complex zeros of transmission but only for those lying on or between the j-axis and the 45° lines in the complex s-plane.

The less lenient conditions relevant to the bridged-T circuits of Fig. 26 read

$$\frac{L_b}{L_a} = \frac{s_a^4}{t(s_a)} = \frac{s_a^4}{c^2d^2} \geq 1, \qquad \text{or } cd < s_a^2 \tag{120}$$

$$\frac{C_a}{C_b} = \frac{t(0)}{t(s_a)} = \frac{|s_0|^4}{c^2d^2} \geq 1, \qquad \text{or } cd < |s_0|^2 \tag{121}$$

which are interpreted geometrically with the help of Fig. 29. With mutual inductive coupling the bridged-T circuit of Fig. 27a will realize any complex value for the transmission zero s_0, and unless the ratio $L_b/L_a = 0$ (which is practically impossible) the coupling coefficient is less than unity, so that close coupling can be avoided in any case.

The desired zero section is finally formed through bisecting the pertinent unbalanced structure, multiplying the impedance level of the right-hand half by the factor a, Eq. 107, and rejoining the halves.

It remains to point out that a completely analogous procedure may be carried out on an admittance basis for which the development applies to an input admittance $Y_1(s)$ having the same form as $Z_1(s)$ in Eq. 66. Here the component two terminal-pair network (zero section) is characterized in terms of the short-circuit driving-point and transfer admittances

$$-y_{12} = \frac{Kt(s)}{s(s^2 - s_a^2)} = Ks + \frac{k_0}{s} - \frac{2k_{12}s}{s^2 - s_a^2} \tag{122}$$

$$y_{11} = \quad \cdots \quad = Ks + \frac{k_0}{s} + \frac{2k_{11}s}{s^2 - s_a^2} \tag{123}$$

$$y_{22} = \quad \cdots \quad = Ks + \frac{k_0}{s} + \frac{2k_{22}s}{s^2 - s_a^2} \tag{124}$$

and the remainder function is an admittance $Y_2(s)$ having the same detailed character as $Y_1(s)$. Since every step in this development procedure on an admittance basis is the dual of the corresponding step in the process just described (so that all analytic expressions are unaltered except for the trivial changes in notation due to writing the letter y in place of z), there is no need to carry through the details.

In this connection it is pertinent to mention that use of the admittance basis may be made under either of two distinctly different situations.

1. We may be given the short-circuit driving-point and transfer admittances $Y_{11}(s)$ and $Y_{12}(s)$ and be asked to find a corresponding lossless two terminal-pair network, just as on the previously discussed impedance basis our problem was to associate a lossless two terminal-pair network with the open-circuit driving-point and transfer impedances $Z_{11}(s)$ and $Z_{12}(s)$.

2. We may be given the impedances $Z_{11}(s)$ and $Z_{12}(s)$ and be asked to determine the pertinent lossless two terminal-pair network, but $Z_{11}(s)$ has zeros at the points $s = 0$ and $s = \infty$ instead of having poles there.

We then simply reciprocate $Z_{11}(s)$ and proceed on an admittance basis *as though* this reciprocated $Z_{11}(s)$ were the short-circuit driving-point admittance $Y_{11}(s)$ which, of course, it is not. Nevertheless, if we develop this Y_{11} function into a cascade of component two terminal-pair networks characterized by the admittances 122, 123, and 124, choosing the same transmission zeros as those specified for the over-all transfer impedance $Z_{12}(s)$, then we will obtain a resultant development that is just as pertinent to the given $Z_{11}(s)$ and $Z_{12}(s)$ as we would have obtained if these had been developable on an impedance basis.

The truth of this statement follows from the fact that the development of $Y_{11} = 1/Z_{11}$ on an admittance basis must just as surely yield the desired Z_{11} function; and the z_{12} function for any component network must have the same zeros as the y_{12} function for that network because $z_{12} = -y_{12}/(y_{11}y_{22} - y_{12}^2)$ in which the y_{sk} fulfill the residue condition with an equal sign at all finite poles. Hence the development of $Y_{11} = 1/Z_{11}$ on an admittance basis provides

1. the correct input impedance, and
2. the desired over-all transmission zeros,

which is neither more nor less than is provided by a development of $Z_{11}(s)$ on an impedance basis.

These ideas tell us that in conditions 86 and 87 the left-hand inequalities can at any stage in the total development procedure be replaced by equality signs if we wish. By doing so, the remainder function $Z_2(s)$ will not have poles at the points $s = 0$ and $s = \infty$, and hence will have zeros there. We cannot normally proceed further on an impedance basis but we can reciprocate the remainder and proceed on an admittance basis. This ability to switch from one basis to another, perhaps repeatedly during an entire development, provides a degree of flexibility that may prove useful under certain circumstances.

For the synthesis of the zero section on an admittance basis we make use of Eqs. 81, 82, and 86 of Ch. 6 for the determination of the lattice admittances and the impedance level factor. These equations are repeated below

$$y_a = y_{11} - ay_{12} \tag{125}$$

$$y_b = y_{11} + y_{12} \tag{126}$$

$$a = \frac{y_{11} + y_{12}}{y_{22} + y_{12}} \tag{127}$$

Substitution from Eqs. 122, 123, and 124 gives

$$a = \frac{k_{11} + k_{12}}{k_{22} + k_{12}} = \frac{k_{12}}{k_{22}} = \frac{k_{11}}{k_{12}} \tag{128}$$

which formally is the reciprocal of the value 107, and

$$y_a = \left(Ks + \frac{k_0}{s}\right)\left(1 + \frac{k_{11}}{k_{12}}\right) \tag{129}$$

$$y_b = \frac{2(k_{11} + k_{12})s}{s^2 - s_a^2} \tag{130}$$

These again yield the lattice of Fig. 26*a*, but instead of the element values 110 and 111 we now have

$$L_a = \frac{k_{12}}{k_0(k_{11} + k_{12})}, \qquad C_a = \frac{(k_{11} + k_{12})K}{k_{12}} \tag{131}$$

$$L_b = \frac{1}{2(k_{11} + k_{12})}, \qquad C_b = \frac{2(k_{11} + k_{12})}{-s_a^2} \tag{132}$$

The transformation methods for converting the lattice to an equivalent unbalanced structure and the conditions under which this is possible without the need for mutual inductive coupling are the same as before.

An entirely analogous procedure applies when the desired zeros of transmission are a pair of symmetrical points on the real axis of the complex frequency plane. The z_{sk} in this case have poles at the points $s = 0$ and $s = \infty$ only, and the corresponding zero section always involves mutual inductive coupling. Details of the procedure appropriate to this situation are almost identical with those pertinent to the particular development of an *RLC* driving-point impedance to be discussed in Art. 5 of Ch. 9, and its further discussion at this time is therefore omitted except to mention that if two pairs of real-axis transmission zeros are considered simultaneously, then the method pertinent to quadruplet zeros is applicable with Eq. 59 replaced by

$$t(s) = (s^2 - \sigma_1^2)(s^2 - \sigma_2^2) \tag{133}$$

in which $\pm\sigma_1$ and $\pm\sigma_2$ are the two pairs of transmission zeros in question.

We shall now consider a numerical example to illustrate the procedures discussed above. Let us assume for the driving-point and transfer impedances the functions

$$Z_{11}(s) = \frac{(s^2 + 1)(s^2 + 3)}{s(s^2 + 2)(s^2 + 4)} \tag{134}$$

and

$$Z_{12}(s) = \frac{s^4 + 4}{s(s^2 + 2)(s^2 + 4)} \tag{135}$$

Besides the quadruplet of transmission zeros at the points $s = \pm 1 \pm j$, the transfer impedance 135 has one zero at the point $s = \infty$. We will see that this one is produced by the final termination provided by the remainder function. The driving-point function 134 does not have a pole at $s = \infty$, and so $\rho_\infty = 0$. We choose this situation deliberately in order to show that the development method is nevertheless straightforwardly applicable.

From the partial-fraction expansion

$$Z_{11}(s) = \frac{\frac{1}{4}s}{s^2 + 2} + \frac{\frac{3}{8}s}{s^2 + 4} + \frac{\frac{3}{8}}{s} \tag{136}$$

we see that

$$2k_1 = \tfrac{1}{4}, \qquad 2k_2 = \tfrac{3}{8}, \qquad \rho_0 = \tfrac{3}{8} \tag{137}$$

Since $s_1^2 = -2$ and $s_2^2 = -4$, we have

$$t(s_1) = s_1^4 + 4 = 8, \qquad \text{and } t(s_2) = s_2^4 + 4 = 20 \tag{138}$$

From Eq. 75 we compute the masses

$$m_1 = \frac{2 \times \frac{1}{4}}{64} = \frac{1}{128}, \qquad m_2 = \frac{4 \times \frac{3}{8}}{400} = \frac{3}{800} \tag{139}$$

and in terms of these

$$\sum m_\nu = \frac{1}{128} + \frac{3}{800} = \frac{37}{3200} \tag{140}$$

and

$$\sum m_\nu s_\nu^2 = \frac{-2}{128} - \frac{4 \times 3}{800} = \frac{-49}{1600} \tag{141}$$

Equation 79 then gives

$$s_a^2 = -\tfrac{98}{37} = -2.648 \tag{142}$$

Next we compute

$$|s_0|^4 \sum \frac{m_\nu}{-s_\nu^2}(s_\nu^2 - s_a^2) = 4\left[\frac{1}{256}\left(-2 + \frac{98}{37}\right) + \frac{3}{3200}\left(-4 + \frac{98}{37}\right)\right] = \frac{3}{16 \times 37} \tag{143}$$

$$\sum m_\nu s_\nu^2 (s_\nu^2 - s_a^2) = \frac{-2}{128}\left(\frac{24}{37}\right) - \frac{3 \times 4}{800}\left(\frac{-50}{37}\right) = \frac{3}{8 \times 37} \tag{144}$$

and

$$\frac{-s_a^2}{|s_0|^4}\rho_0 = \frac{98 \times 3}{37 \times 32} = \frac{147}{16 \times 37} \tag{145}$$

Conditions 86 and 87 therefore read

$$0 < K - \frac{3}{16 \times 37} \leq \frac{147}{16 \times 37} \tag{146}$$

$$0 < K - \frac{3}{8 \times 37} \leq 0 \tag{147}$$

Condition 89 becomes

$$\frac{-147}{16\times 37} < \frac{-3}{16\times 37} < 0 \tag{148}$$

which is evidently fulfilled. Since $\rho_\infty = 0$, condition 147 leaves no choice for K except the value

$$K = \frac{3}{8\times 37} = \frac{3}{296} = \frac{1}{98.67} \tag{149}$$

and Eqs. 78 and 144 then give

$$H = 98.67 \tag{150}$$

From Eq. 64 we now find

$$k_0 = \frac{4\times 37}{98}\times\frac{3}{8\times 37} = \frac{3}{196} = \frac{1.51}{98.67} \tag{151}$$

and since

$$t(s_a) = s_a^4 + 4 = \left(\frac{98}{37}\right)^2 + 4 \tag{152}$$

Eq. 65 gives

$$k_{12} = \left(\frac{98}{74} + \frac{74}{98}\right)\frac{1}{98.67} = \frac{2.08}{98.67} \tag{153}$$

According to Eq. 69 and the fact that the polynomial $h(s)$ reduces to the multiplier H alone, we have

$$(z_{22} + Z_2) = \frac{1}{98.67}\frac{(s^2+2)(s^2+4)}{s(s^2+\frac{98}{37})} = \left[s + \frac{3.02}{s} + \frac{0.331s}{(s^2+\frac{98}{37})}\right]\frac{1}{98.67} \tag{154}$$

Hence we see that

$$k_{22} = \frac{0.1655}{98.67} \tag{155}$$

and the residue condition 63 then gives

$$k_{11} = \frac{1}{98.67}\times\frac{(2.08)^2}{0.1655} = \frac{26.13}{98.67} \tag{156}$$

We are now ready to determine the parameters of the zero section. From Eq. 107 we find the impedance level multiplier

$$a = \frac{k_{12}}{k_{11}} = \frac{k_{22}}{k_{12}} = \frac{0.1655}{2.08} = 0.0796 \tag{157}$$

Computing the lattice parameters from Eqs. 110 and 111, we find

$$L_a = \frac{2\times 28.21}{2.684\times 98.67} = 0.216, \qquad C_a = \frac{98.67}{2\times 28.21} = 1.748 \tag{158}$$

$$L_b = \frac{28.21}{2.08\times 98.67} = 0.1375, \qquad C_b = \frac{2.08\times 98.67}{1.51\times 28.21} = 4.815 \tag{159}$$

From these we have

$$\frac{L_b}{L_a} = 0.637, \qquad \text{and } \frac{C_a}{C_b} = 0.363 \tag{160}$$

The sum of these two values equals unity, and so condition 114 is satisfied with the equal sign, which is the result of our having assumed transmission zeros lying upon the 45° lines in the s-plane.

From Eq. 154 (noting that $k_0 = 1.51/98.67$) we have

$$Z_2 = \frac{1.51}{98.67s} = \frac{0.0153}{s} \tag{161}$$

Since the right-hand inequality in condition 146 is not fulfilled with the equal sign, z_{11} has a larger residue at $s = 0$ than the value k_0. Specifically, we may conclude directly from 146 that

$$(sz_{11})_{s=0} = \frac{k_0}{K}\left(\frac{147}{16 \times 37} + \frac{3}{16 \times 37}\right) = \frac{150k_0}{16 \times 37 \times K} = 25k_0 \tag{162}$$

The zero section is therefore preceded by a series branch with the elastance value

$$[(sz_{11})_{s=0} - k_0] = 24k_0 = \frac{24 \times 3}{196} = \frac{18}{49} \text{ darafs} \tag{163}$$

Terminating the zero section in the impedance 161 produces the required transmission zero at $s = \infty$; and the resulting twin-T according to Fig. 28 with its associated input and output branches is shown in Fig. 30*a*.

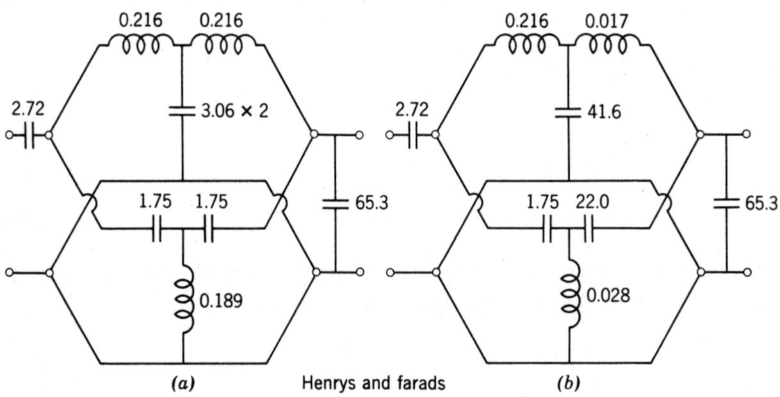

Fig. 30. Network (*b*) realizes the impedances 134 and 135; network (*a*) is the result prior to multiplying its right-hand half (exclusive of the terminal capacitance) by the impedance level factor 157.

Bisection of the twin-T and multiplication of its right-hand half by the impedance level factor 157, followed by recombination of the two halves, leads to the final network shown in Fig. 30*b*. Observe that the terminal capacitance is not part of the twin-T and hence its value is not affected by the impedance leveling process.

Now let us consider the same driving-point impedance 134, but for a transfer impedance with

$$t(s) = s^4 + 16 \tag{164}$$

for which $|s_0|^4 = 16$ instead of 4 as in the example just completed. This switch to a large $|s_0|$ value will be seen to alter the way in which the realization procedure can successfully be carried through.

Equations 136 and 137 remain valid, but the masses 139 now become

$$m_1 = \frac{2 \times \frac{1}{4}}{400} = \frac{1}{800}, \qquad m_2 = \frac{4 \times \frac{3}{8}}{1024} = \frac{3}{2048} \tag{165}$$

We now find

$$\sum m_\nu = \frac{139}{51{,}200}, \qquad \text{and} \sum m_\nu s_\nu^2 = \frac{-107}{12{,}800} \tag{166}$$

so that

$$s_a^2 = -428/139 \tag{167}$$

Also

$$|s_0|^4 \sum \frac{m_\nu}{-s_\nu^2}(s_\nu^2 - s_a^2) = \frac{3}{4 \times 139} \tag{168}$$

and

$$\sum m_\nu s_\nu^2 (s_\nu^2 - s_a^2) = \frac{3}{8 \times 139} \tag{169}$$

The quantity $(a - c)$ in condition 89, which is 168 minus 169, is evidently positive, and since $d = \rho_\infty = 0$, we see that this condition cannot be satisfied. Hence we switch to an admittance basis and proceed from the driving-point function

$$\frac{1}{Z_{11}} = Y_{11} = \frac{s(s^2+2)(s^2+4)}{(s^2+1)(s^2+3)} = \frac{\frac{3}{2}s}{s^2+1} + \frac{\frac{1}{2}s}{s^2+3} + s \tag{170}$$

We now have

$$s_1^2 = -1, \qquad s_2^2 = -3 \tag{171}$$

$$2k_1 = \tfrac{3}{2}, \qquad 2k_2 = \tfrac{1}{2} \tag{172}$$

$\rho_\infty = 1$, and $\rho_0 = 0$. Since Eq. 164 yields

$$t(s_1) = 17, \qquad \text{and } t(s_2) = 25 \tag{173}$$

the masses now become

$$m_1 = \frac{\frac{3}{2}}{289} = \frac{3}{578}, \qquad m_2 = \frac{3 \times \frac{1}{2}}{625} = \frac{3}{1250} \tag{174}$$

In terms of these values we find

$$\sum m_\nu = \frac{3 \times 457}{289 \times 625}, \qquad \text{and} \sum m_\nu s_\nu^2 = \frac{-3 \times 746}{289 \times 625} \tag{175}$$

so that

$$s_a^2 = -746/457 = -1.632 \tag{176}$$

Next we compute

$$|s_0|^2 \sum \frac{m_v}{-s_v^2}(s_v^2 - s_a^2) = \frac{16}{457} \tag{177}$$

and

$$\sum m_v s_v^2 (s_v^2 - s_a^2) = \frac{3}{457} \tag{178}$$

Conditions 86 and 87 are

$$0 < K - \frac{16}{457} \leq 0 \tag{179}$$

and

$$0 < K - \frac{3}{457} \leq 1 \tag{180}$$

Here it is evident that we must choose

$$K = \frac{16}{457} = \frac{1}{28.55} \tag{181}$$

and Eq. 178 then gives

$$H = \frac{3 \times 457}{256} \tag{182}$$

From Eq. 64 we find

$$k_0 = \frac{16 \times 16}{746} = \frac{128}{373} = 9.8/28.55 \tag{183}$$

and with

$$t(s_a) = (746/457)^2 + 16 \tag{184}$$

Eq. 65 gives

$$k_{12} = \frac{2 \times 16}{457}\left(\frac{746}{1828} + \frac{1828}{746}\right) = 5.716/28.55 \tag{185}$$

Now we obtain

$$(y_{22} + Y_2) = \frac{16}{457} \times \frac{(s^2+1)(s^2+3)}{s(s^2+\frac{746}{457}) \times \frac{3}{16}} = \frac{k_0}{s} + \frac{2k_{22}s}{(s^2+\frac{746}{457})} + \frac{256s}{3 \times 457} \tag{186}$$

from which

$$k_{22} = \frac{16}{457} \times \frac{4 \times 289 \times 625}{3 \times 457 \times 373} = 1.412/28.55 \tag{187}$$

and

$$Y_2(s) = (16/457)(13/3)s = 0.1517s \tag{188}$$

From 180 (equivalent to 81) we have

$$\left(\frac{y_{11}}{s}\right)_{s=\infty} = \left(1 + \frac{3}{457}\right) = \frac{460}{457} \tag{189}$$

and hence we see that there will be a shunt capacitance at the input equal to

$$C_1 = \left(\frac{y_{11}}{s}\right)_{s=\infty} - \frac{16}{457} = \frac{444}{457} = 0.972 \text{ farad} \tag{190}$$

The residue condition 63, together with the values 185 and 187, gives

$$k_{11} = \frac{1}{28.55} \frac{(5.716)^2}{1.412} = 23.12/28.55 \tag{191}$$

and the impedance level factor for the zero section according to Eq. 128 is

$$a = \frac{k_{12}}{k_{22}} = \frac{5.716}{1.412} = 4.046 \tag{192}$$

In computing the lattice parameters from Eqs. 131 and 132, we find

$$L_a = \frac{5.716 \times 28.55}{9.8 \times 28.84} = 0.577, \qquad C_a = \frac{28.84}{5.716 \times 28.55} = 0.1766 \tag{193}$$

$$L_b = \frac{28.55}{2 \times 28.84} = 0.495, \qquad C_b = \frac{2 \times 28.84}{1.632 \times 28.55} = 1.237 \tag{194}$$

from which

$$\frac{L_b}{L_a} = 0.857, \quad \text{and } \frac{C_a}{C_b} = 0.143 \tag{195}$$

Again the sum of these ratios equals unity as it should. The terminal capacitance represented by the remainder 188 again provides the required transmission zero at $s = \infty$. The final network (comparable to the one shown in Fig. 30*b*) is given in Fig. 31. Both networks have the same Z_{11}, Eq. 134.

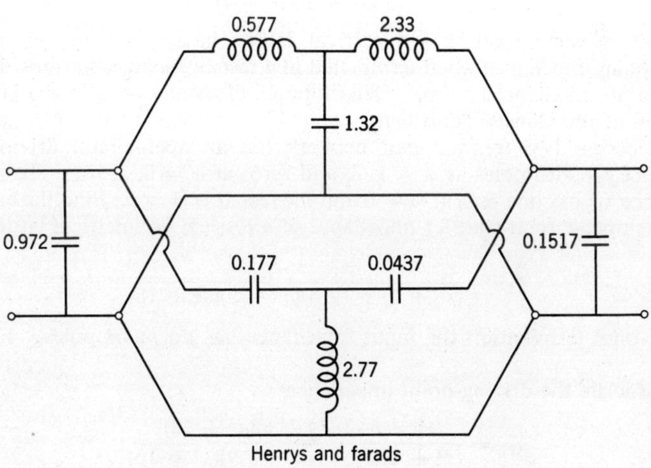

Fig. 31. Network having the same Z_{11} as that in Fig. 30*b*, but for transmission zeros defined by Eq. 164 instead of the numerator in Eq. 135.

PROBLEMS

1. Following are the driving-point and transfer impedances of a lossless two terminal-pair network

$$z_{22} = \frac{s^2 + 1}{s(s^2 + 4)} \qquad z_{12} = \frac{1}{s(s^2 + 4)}$$

If this network is terminated in a resistance of 1 ohm, determine the corresponding minimum-reactive input impedance (having no j-axis poles). Synthesize the network.

2. Develop the short-circuit driving-point admittance

$$y_{11} = \frac{5s^4 + 5s^2 + 1}{3s^3 + 2s}$$

into a ladder network in such a way that this network (when provided with output terminals) has the short-circuit transfer admittance

$$y_{12} = \frac{(s^2 + 1)^2}{3s^3 + 2s}$$

3. A symmetrical lossless two terminal-pair network has an open-circuit transfer impedance

$$z_{12} = \frac{1}{s(s^2 + 1)}$$

Find the associated z_{11} and z_{22} such that the input impedance for a 1-ohm termination becomes minimum reactive. Find the analytic expression for this impedance. Find the short-circuit driving-point and transfer admittances y_{11}, y_{22}, y_{12} for the lossless network and from these synthesize the network.

4. Given the open-circuit transfer impedance of a lossless network

$$z_{12} = \frac{s^5 + 5s^3 + 7s}{(s^2 + 1)(s^2 + 4)}$$

determine a corresponding symmetrical lattice having a minimum-reactive driving-point impedance when terminated in a resistive load, and transform this lattice to an unbalanced form. Determine all element values in the lattice as well as in its unbalanced equivalent.

5. A lossless two terminal-pair network has an open-circuit driving-point impedance z_{11} with poles at $\omega = 1, 3$, and zeros at $\omega = 0, 2, \infty$. The transfer impedance z_{12} has one zero at $\omega = 0$ and the rest at $\omega = \infty$. Find the network.

6. The open-circuit transfer impedance of a lossless symmetrical lattice is

$$z_{12} = \frac{s}{(s^2 + 1)(s^2 + 4)(s^2 + 9)(s^2 + 16)}$$

For a 1-ohm termination the input impedance has no j-axis poles. Find the lattice.

7. Associate the driving-point impedance

$$z_{22} = \frac{s(s^2 + 2)(s^2 + 6)(s^2 + 12)}{(s^2 + 1)(s^2 + 4)(s^2 + 9)(s^2 + 16)}$$

with the transfer impedance of Prob. 6, and find an appropriate reactive two terminal-pair network by developing z_{22} into a ladder.

8. Given the following driving-point and transfer impedances of a lossless two terminal-pair network

$$z_{11} = \frac{2s(s^2 + 5)}{(s^2 + 1)(s^2 + 9)} \qquad z_{22} = \frac{s(s^2 + 4)}{(s^2 + 1)(s^2 + 9)} \qquad z_{12} = \frac{s}{(s^2 + 1)(s^2 + 9)}$$

(*a*) Show that the network is realizable.

(*b*) If terminated in a 1-ohm resistance, does the input impedance $Z_1(s)$ possess a complement?

(*c*) Synthesize the network.

9. The functions

$$z_{22} = \frac{s(s^2 + 4)}{(s^2 + 1)(s^2 + 9)} \qquad z_{12} = \frac{s}{(s^2 + 1)(s^2 + 9)}$$

are the open-circuit driving-point and transfer impedances of a lossless two terminal-pair network. Develop z_{22} into a ladder so as to yield that network.

10. For the impedances given by Eqs. 40 and 41 carry through the development leading to the network in Fig. 18. Compute z_{22}, and verify that the residue condition is fulfilled with an equal sign at all poles.

11. Given

$$z_{11} = \frac{16s^4 + 9s^2 + 1}{26s^5 + 21s^3 + 3s} \qquad z_{12} = \frac{(4s^2 + 1)(s^2 + 1)}{26s^5 + 21s^3 + 3s}$$

synthesize these in geometrical configurations (*a*) and (*b*). For network (*a*), produce the zeros of z_{12} in the order $s^2 = -1$, $s^2 = -\frac{1}{4}$, $s = \infty$. For network (*b*) use the order $s = \infty$, $s^2 = -1$, $s^2 = -\frac{1}{4}$.

For each network point out whether the residue condition is fulfilled with an equal sign at any or all poles.

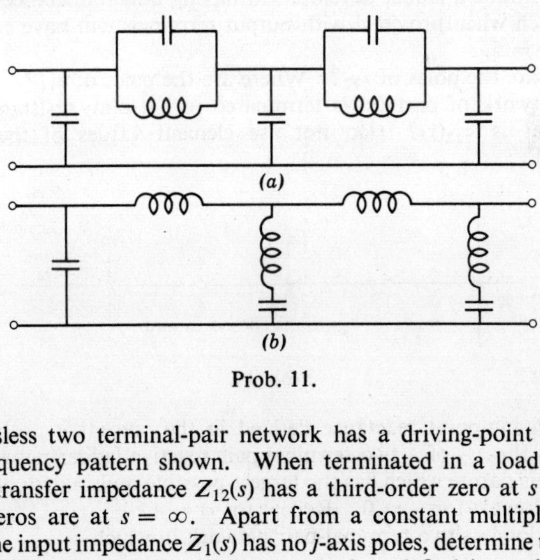

Prob. 11.

12. A lossless two terminal-pair network has a driving-point reactance z_{22} with the frequency pattern shown. When terminated in a load resistance of 1 ohm, the transfer impedance $Z_{12}(s)$ has a third-order zero at $s = 0$, and the rest of its zeros are at $s = \infty$. Apart from a constant multiplier, construct $Z_{12}(s)$. If the input impedance $Z_1(s)$ has no j-axis poles, determine the reactances z_{11}, z_{22}, z_{12} characterizing the lossless network and find its realization. The constant multiplier of z_{22} is unity.

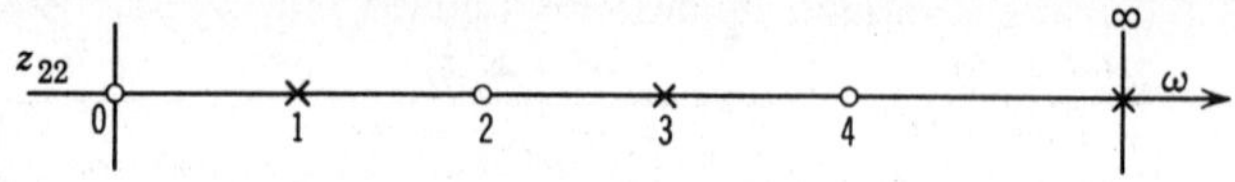

Prob. 12.

13. The open-circuit driving-point impedance z_{11} of a lossless network is illustrated here. The associated z_{12} function has zeros at real finite nonzero values of ω. For appropriately chosen zeros of z_{12}, determine the ladder development of z_{22} that produces the zeros of z_{12} such that the residue condition for z_{11}, z_{22}, z_{12} is fulfilled with an equal sign at all poles.

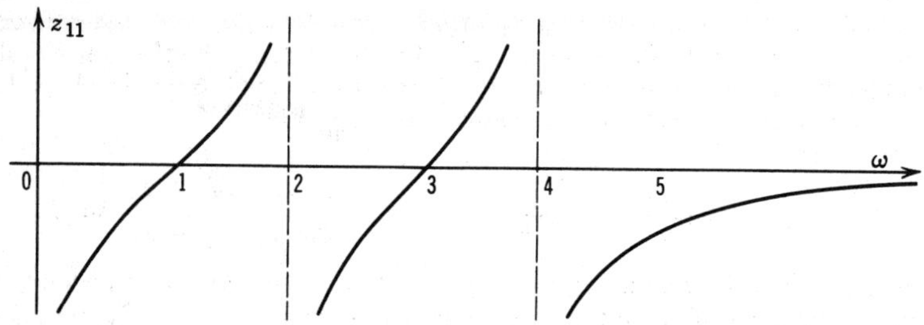

Prob. 13.

14. (*a*) The frequency pattern of a reactance z_{22} is shown in the accompanying sketch. Determine a ladder development having only inductances in the series branches which when provided with output terminals will have zeros of z_{12} at $\omega = 3$ and $\omega = 5$.

(*b*) Where are the poles of z_{12}? Where are the poles of z_{11}?

(*c*) The network of part (*a*) is terminated in a 1-ohm resistance. What is $Z_{12}(s)$? What is $z_{12}(s)$? (Do not use element values of the network to compute z_{12}.)

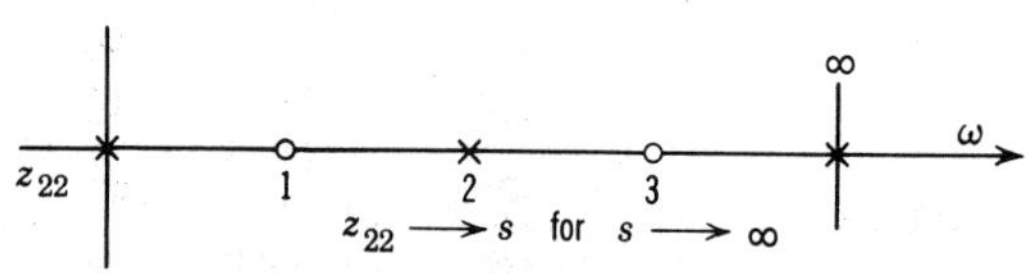

Prob. 14.

15. The driving-point reactance defined in the preceding problem is to be considered as the z_{22} of a two terminal-pair reactive ladder network. Find a network of this nature which has the largest possible multiplicity of zeros of Z_{12} at $\omega = \infty$. Repeat for $\omega = 0$. Repeat for $\omega = 4$.

16. Find a lossless two terminal-pair network in which

$$z_{12} = \frac{s}{s^2 + 1} + \frac{2s}{s^2 + 2} + \frac{5}{s}$$

(*a*) Specifying z_{11} and z_{22} in such a manner that no ideal transformers are required and that the residue condition is fulfilled at each pole with an equal sign.

(*b*) Specifying z_{22} in such a manner that $Z_{12} = (z_{12})/(1 + z_{22})$ (see figure) is $\frac{1}{4}$ at $\omega = 1$, $\frac{1}{2}$ at $\omega = \sqrt{2}$, and $\frac{1}{3}$ at $\omega = 0$. For this case choose z_{11} so that the network involves no ideal transformers.

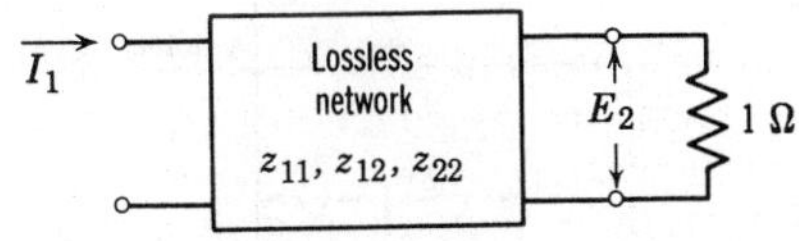

$$\frac{E_2}{I_1} = Z_{12} = \frac{z_{12}}{1 + z_{22}}$$

Prob. 16.

17. The adjacent sketch shows the pole-zero pattern of a transfer impedance $Z_{12}(s)$. Its constant multiplier is determined by the condition $Z_{12} \rightarrow 1$ for $s \rightarrow \infty$. Realize this transfer function by a lossless symmetrical lattice terminated in a 1-ohm resistance, and transform the lattice to an equivalent unbalanced form not involving mutual inductive coupling.

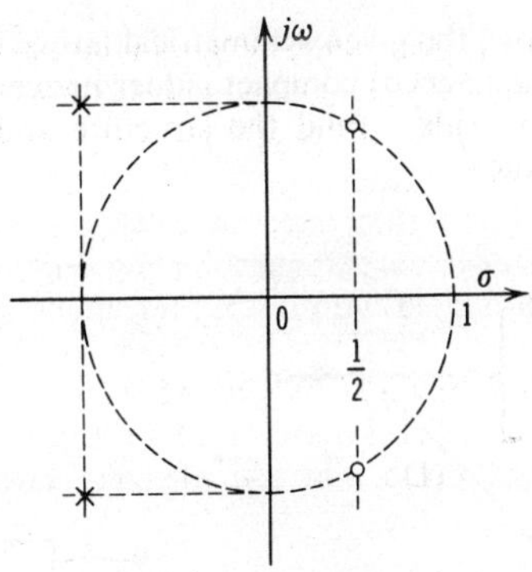

Prob. 17.

18. The transfer impedance Z_{12} of a lossless two terminal-pair network terminated in a 1-ohm load has three poles, the locations of which in the s-plane are shown in the sketch on page 274. The zeros of Z_{12} are all at $s = \infty$, and its zero-frequency value is unity. Determine the open-circuit driving-point and transfer impedances z_{11}, z_{22}, z_{12} of the nondissipative network such that the input impedance for the terminated network is minimum reactive. Synthesize the network.

19. Given

$$Z_{12} = \frac{(s^2 + 1)^2}{(s + 1)^4} = \frac{z_{12}}{1 + z_{22}}$$

find a lossless network terminated in a 1-ohm resistance having this transfer impedance. Synthesize the network through developing z_{22} into a ladder so as to produce the associated z_{12}.

20. Determine simple lossless ladder networks whose transfer impedances when terminated in 1 ohm are respectively

$$Z_{12} = \frac{1}{(s+1)^4} \qquad Z_{12} = \frac{s^4}{(s+1)^4}$$

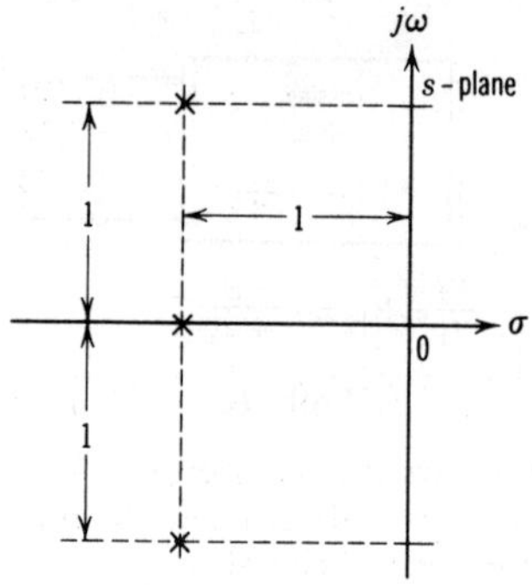

Prob. 18.

21. As indicated below, the given symmetrical lattice is to be converted into a dissymmetrical (and unbalanced) compact ladder network with additional series reactance on the output side. Find the structure and element values of the network and the reactance.

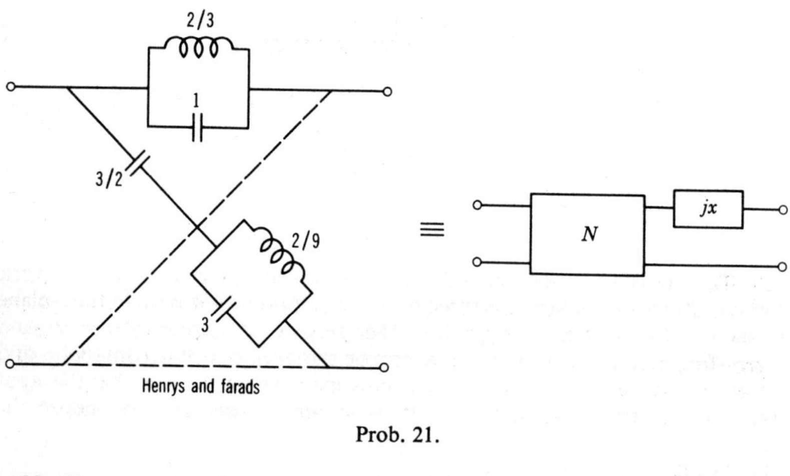

Prob. 21.

22. The accompanying sketch shows a lossless symmetrical lattice terminated in a 1-ohm load resistance (the remaining indicated element values being in henrys and farads). As may readily be verified, the input impedance is not minimum reactive. Find a network having the same transfer function E_2/I_1 (apart from a constant multiplier) and a minimum-reactive input impedance. Put this network into its simplest possible form.

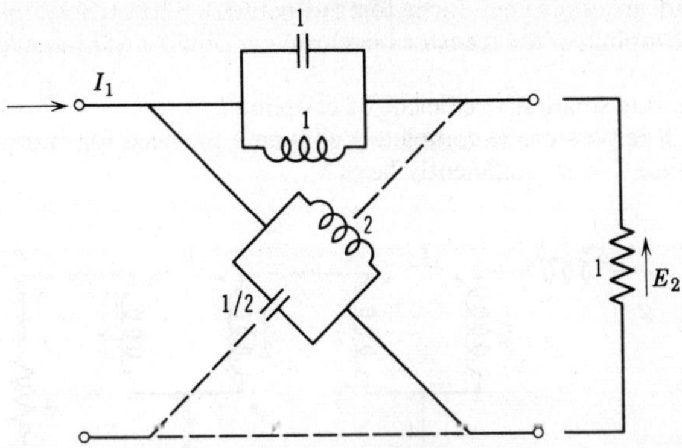

Prob. 22.

23. By the method of cascade synthesis find a lossless unbalanced two terminal-pair network realizing the driving-point and transfer impedances

$$z_{11} = \frac{(s^2+1)(s^2+3)}{s(s^2+2)(s^2+4)} \qquad z_{12} = \frac{(s^2+s+1)(s^2-s+1)}{s(s^2+2)(s^2+4)}$$

24. By means of an unbalanced network, realize the impedances

$$z_{11} = \frac{s^6+7s^4+12s^2+3}{s^7+8s^5+18s^3+11s} \qquad z_{12} = \frac{(s^2+s+1)(s^2-s+1)}{s^7+8s^5+18s^3+11s}$$

25. Realize the set of impedance functions

$$Z_{11} = \frac{\frac{1}{2}(s^2+1)(s^2+4)}{s(s^2+2)}$$

$$Z_{12} = \frac{s^4}{s(s^2+2)}$$

$$Z_{22} = \frac{4(s^2+1)(s^2+6)}{s(s^2+2)}$$

as a two terminal-pair network in which all ideal transformers are replaced by nonunity coupled coils having equal coefficients of coupling.

26. Given the set of open-ciruit driving-point and transfer impedances

$$z_{11} = \frac{4s^5+23s^3+22s}{(s^2+1)(s^2+4)}$$

$$z_{12} = \frac{2s^5+7s^3+8s}{(s^2+1)(s^2+4)}$$

$$z_{22} = \frac{4s^5+29s^3+28s}{(s^2+1)(s^2+4)}$$

synthesize the corresponding lossless two terminal-pair network and convert to a form in which the mutual coupling coefficients are equal and less than unity. What is the value of this common coupling coefficient? (The realization should obviously involve no ideal transformers.)

27. (*a*) Find an equivalent form for the network shown here in which the coefficient of coupling of the transformers are equal and have the smallest possible value.

(*b*) What is this smallest coefficient of coupling?

(*c*) Would it be possible to completely eliminate the need for mutual coupling if the inductance L were sufficiently large?

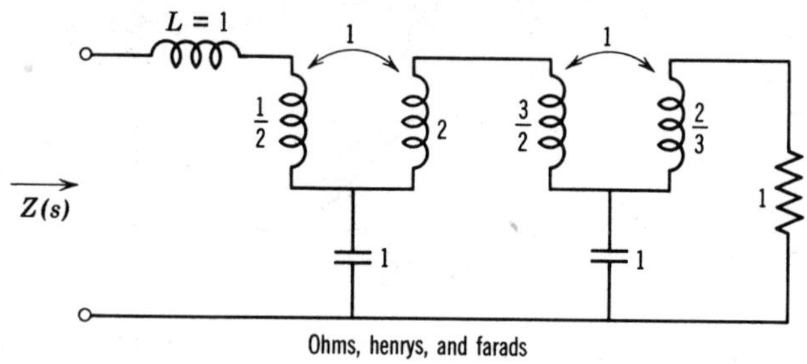

Prob. 27.

28. (*a*) Synthesize a lossless two terminal-pair network which has the following open-circuit impedances, without ideal transformers if possible:

$$z_{11} = \frac{7}{s} + \frac{4s}{s^2 + 1} + s$$

$$z_{12} = \frac{6}{s} - \frac{2s}{s^2 + 1} + s$$

$$z_{22} = \frac{8}{s} + \frac{s}{s^2 + 1} + 6s$$

(*b*) If this network is terminated at its output end in a 1-ohm resistance, will the driving-point impedance at the input end have no j-axis poles? If not, modify the given functions (if possible) so that the input impedance will have no j-axis poles but the transfer impedance $Z_{12}(s)$ of the resistance loaded network will be unchanged.

(*c*) If the network is terminated as in part (*b*), will the transfer impedance $Z_{12}(s)$ have right half-plane zeros?

29. The transfer impedance of a lossless network terminated in a 1-ohm resistance has three poles and two j-axis zeros as shown in the adjacent sketch. $Z_{12}(0) = 1$.

(*a*) Write the expression for $Z_{12}(s)$ and obtain z_{12} and z_{22} for the lossless network.

(*b*) Find a network realization by developing z_{22} into a ladder so as to control the zeros of z_{12}, at the same time achieving a minimum-reactive input impedance.

30. The open-circuit driving-point and transfer impedances of a lossless two terminal-pair network are specified to be

$$z_{11} = z_{22} = \frac{3s^4 + 6s^2 + 2}{s^5 + 3s^3 + 2s} \qquad z_{12} = \frac{K}{s^5 + 3s^3 + 2s}$$

in which K is a positive real number. What is the largest permissible value of

K for which this network is realizable? If the latter is constructed as a lattice with the impedances $z_a = z_{11} - z_{12}$ and $z_b = z_{11} + z_{12}$, synthesize this lattice. If only z_{12} is specified (any $z_{11} = z_{22}$ is acceptable), can a simpler lattice be found? Under these conditions construct the simplest lattice for $K = 1$.

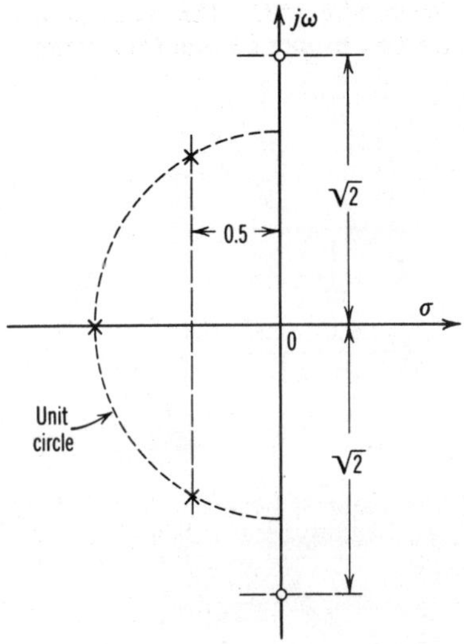

Prob. 29.

31. A given transfer impedance $Z_{12}(s)$ is realized by a lossless symmetrical lattice terminated in 1 ohm. The impedances of the lattice have poles at the frequencies: $s = 0$, $s = \pm j2$, $s = \pm j3$, and the residues of z_{22} in these poles are all equal to unity. If

$$Z_{12}(0) = 1, \qquad Z_{12}(\pm j2) = -1, \qquad Z_{12}(\pm j3) = 1$$

find the lattice network and transform it to an equivalent unbalanced form involving no mutual inductive coupling.

32. The transfer impedance of a lossless symmetrical two terminal-pair network terminated in 1 ohm is given by

$$Z_{12}(s) = \frac{m - n}{m + n}$$

with $m + n = s^3 + 2s^2 + 2s + 1$. Find expressions for the impedances z_{11}, z_{22}, z_{12} of the lossless network, and synthesize this network.

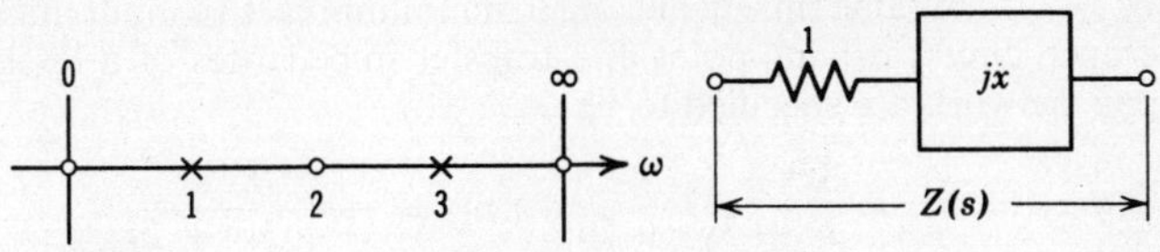

Prob. 33.

33. The accompanying sketch shows the frequency pattern of a reactance function jx. This reactance in series with 1 ohm forms an impedance $Z(s)$ as shown at the right. A second impedance $Z_1(s)$ having no j-axis poles is realized as the input impedance of a lossless two terminal-pair network terminated in 1 ohm. The corresponding transfer impedance $Z_{12}(s)$ has no finite nonzero zeros, and for $s \to \infty$ behaves like $1/s^4$. The poles of $Z_1(s)$ are the zeros of $Z(s)$. Find $Z_1(s)$ inclusive of its proper constant multiplier.

CHAPTER EIGHT

Real-Part Sufficiency and Related Topics

1 Determination of an Impedance From Its Real Part

The property of passive networks, which in the time domain asserts itself in the boundedness of their natural behavior and in the frequency domain results in the right half-plane analyticity of their associated driving-point and transfer impedances or admittances, imposes upon these functions certain constraints whose clarification and exploitation prove invaluable to the further development of synthesis procedures. Foremost among these constraint relations is the property that proclaims the sufficiency of the j-axis real part of an impedance or admittance in uniquely characterizing that function and through it the response of the network in the transient as well as in the steady state. Thus a knowledge of the real part of an impedance $Z(s)$ for $s = j\omega$ completely fixes this function throughout the complex frequency plane in the sense that one is able to construct $Z(s)$ from a knowledge of the values of Re $[Z(j\omega)]$ alone. If the latter is a given rational function of ω, then one can construct the corresponding rational function $Z(s)$ in terms of the complex frequency variable s, and if only the values of Re $[Z(j\omega)]$ are provided, as for example by a graphical plot, then one can (theoretically at least) compute values of $Z(s)$ for any given complex value of s.

In order to meet the uniqueness claim contained in this statement, it is necessary to enlarge our customary interpretation of the properties of real-part functions to include impulses as legitimate components, a step which logically fits in, however, with established practice in other branches of network theory. These and related ideas will engage our attention throughout the discussions of this chapter.

Before embarking upon algebraic and other analytic means for demonstrating this interesting property of impedance functions, let us reflect for a moment upon its underlying philosophy so that we may see clearly why it is that only passive network impedances and admittances, and not all rational functions, are so endowed. As stated in the opening sentence of this chapter, this property is the result of the stability of a passive network, reflecting itself in the right half-plane analyticity of its driving-point and transfer impedances, as is almost immediately evident if we express the real-part function in the following way

$$\text{Re}\,[Z(j\omega)] = \tfrac{1}{2}[Z(s) + Z(-s)]_{s=j\omega} \tag{1}$$

Thus, for $s = j\omega$, $Z(-s)$ is the complex conjugate of $Z(s)$; and the sum of any complex number and its conjugate equals twice the real part of that number. If we are given the real-part function, then by writing s for $j\omega$ or $-s^2$ for ω^2, we have (except for the factor $\frac{1}{2}$) the sum of $Z(s)$ and $Z(-s)$. If we can separate this sum into its constituent parts, then the desired construction of $Z(s)$ from Re $[Z(j\omega)]$ is achieved. This separation is uniquely possible *only* if $Z(s)$ is analytic in either the right or left half-plane, for then we know that the poles in one half-plane belong to $Z(s)$ and those in the other one belong to $Z(-s)$. If it were not for this half-plane analyticity of the function $Z(s)$, then we would have no way of telling which of the poles of the total rational function $Z(s) + Z(-s)$ belong to $Z(s)$ and which belong to $Z(-s)$; we could not uniquely construct $Z(s)$, although we could construct a number of possible $Z(s)$ functions.

In effect what we are doing in this construction process is using the j-axis as a "sorting boundary," for, of all the poles of the given function, we know that the ones belonging to $Z(s)$ lie to the left of this boundary. We also see at once that j-axis poles present a special problem since for these the boundary loses its sorting ability. This difficulty and the fact that the real-part function loses its boundedness at such points combine to produce a situation that requires a more careful study for its disentanglement. But one major issue stands out clearly from this simple interpretation of the problem of constructing $Z(s)$ from its real part, and that is the fact, contrary to a common attitude created by first impressions, that its solution does not require the construction of a whole function when only half of it is given. Thus one is apt to regard the real and imaginary parts of an impedance as separately containing half as much information as the whole complex impedance. As Eq. 1 shows, the fallacy in this attitude lies in assuming that the real part contains less information than (presumably half as much as) the whole impedance $Z(s)$. The information that characterizes a rational function is conveyed by its poles. The real-part function contains not only the left half-plane poles of $Z(s)$ but also their images about the origin. Thus the real part is not half a function but the whole function with a certain redundancy in its characterizing elements. One constructs the whole function by discarding the redundant part of the given "half."

An additional point of view, more directly related to the time-domain aspects of this problem, is useful because of the correlation that it provides with the physical aspects of network behavior. Thus the voltage response to an applied unit current impulse may be expressed in the well-known form

$$e(t) = \frac{1}{2\pi j}\int_C Z(s)e^{st}\,ds \tag{2}$$

in which the contour C may be taken coincident with the j-axis, and the point at infinity on the complex sphere is passed on the left for $t > 0$ and on the right for $t < 0$ along semicircular arcs (see *MCA*, pp. 547 ff.). In the ordinary s-plane these arcs may be regarded as semicircles of sufficient size to enclose all poles of the integrand, which are the poles of $Z(s)$. Since its poles are restricted to the left half-plane, the contour for $t > 0$ encloses poles and the one for $t < 0$ encloses none. By Cauchy's residue theorem the value of the integral is therefore zero for $t < 0$ and nonzero for $t > 0$. The absence of a response before the occurrence of an excitation, which is an obvious stability requirement, is thus seen to be the result of the right half-plane analyticity of the impedance $Z(s)$. Analyticity in the right half-plane on the part of a network response function is a necessary and sufficient condition to assure that a response will not precede the signal which is its cause. A network which responds before the advent of a signal is actually taking off under its own power and hence cannot be passive.

Stability or the inability of the natural behavior to grow without limit, the right half-plane analyticity of the frequency-response function, and the time response being dead before the advent of an excitation are three different manifestations of the same thing: the passivity of the network.

The fact that the real part of $Z(j\omega)$ determines $Z(s)$ may easily be seen from the time-domain formulation given by Eq. 2. The evaluation of $e(t)$ for $t > 0$ may as well be expressed in terms of the following real integration

$$e(t) = \frac{1}{2\pi}\int_{-\infty}^{\infty} Z(j\omega)e^{j\omega t}\,d\omega \tag{3}$$

in which we may write

$$Z(j\omega) = R(\omega) + jX(\omega) \tag{4}$$

Since the real and imaginary parts of $Z(j\omega)$ are respectively even and odd, substitution of Eq. 4 into 3 and consequent separation of the integral into two parts yields

$$e(t) = \frac{1}{\pi}\int_0^{\infty} R(\omega)\cos\omega t\,d\omega - \frac{1}{\pi}\int_0^{\infty} X(\omega)\sin\omega t\,d\omega \tag{5}$$

The first of these two integrals is evidently an even function of time and the second an odd function. Since we know that their sum is identically zero for $t < 0$, it follows that for $t > 0$ they are equal and hence their sum equals two times either function. Hence for $t > 0$ we may write

$$e(t) = \frac{2}{\pi}\int_0^\infty R(\omega)\cos\omega t\, d\omega = \frac{-2}{\pi}\int_0^\infty X(\omega)\sin\omega t\, d\omega \tag{6}$$

which shows that the real part or the imaginary part alone suffices to determine the impulse response of the network. Since the Laplace transform of $e(t)$ is $Z(s)$, we see again that this complex function is determined by its real part; and we see also that the right half-plane analyticity of $Z(s)$ is essential to the possibility of this determination because it implies $e(t) \equiv 0$ for $t < 0$, which in turn is the basis for obtaining expressions 6.

Now let us get down to the detailed algebra involved in the process of constructing $Z(s)$ from a given rational function representing its real part. If we write

$$Z(s) = \frac{a_0 + a_1 s + a_2 s^2 + \cdots + a_n s^n}{b_0 + b_1 s + b_2 s^2 + \cdots + b_n s^n} = \frac{P(s)}{Q(s)} = \frac{m_1 + n_1}{m_2 + n_2} \tag{7}$$

as in Eq. 51 of Ch. 1, then as given there by Eqs. 52, 53, and 54, the real part for $s = j\omega$ is expressed by

$$\begin{aligned}\text{Re}\,[Z(j\omega)] &= \left(\frac{m_1 m_2 - n_1 n_2}{m_2^2 - n_2^2}\right)_{s=j\omega} \\ &= \frac{A_0 + A_1\omega^2 + \cdots + A_n\omega^{2n}}{B_0 + B_1\omega^2 + \cdots + B_n\omega^{2n}} = \frac{A(\omega^2)}{B(\omega^2)}\end{aligned} \tag{8}$$

This quotient of polynomials in ω^2 is positive for all real values of ω only when $Z(s)$ is a driving-point impedance, and it remains bounded only if the impedance has no j-axis poles. For the time being we shall consider our problem within this restricted framework, that is to say, for a p.r. $Z(s)$ having no j-axis poles.

By means of Eq. 1 we are able to write

$$\tfrac{1}{2}[Z(s) + Z(-s)] = \frac{A(-s^2)}{B(-s^2)} = f(s) \tag{9}$$

where $f(s)$ is a function having the left half-plane poles of $Z(s)$ and their images about the origin, yielding altogether a pole pattern having symmetry about both the real and imaginary axes in the complex s-plane. If the rational function 8 for Re $[Z(j\omega)]$ is given, we have

$$f(s) = \frac{A_0 - A_1 s^2 + A_2 s^4 - \cdots + (-1)^n A_n s^{2n}}{B_0 - B_1 s^2 + B_2 s^4 - \cdots + (-1)^n B_n s^{2n}} \tag{10}$$

From the following detailed form of the denominator polynomial

$$B(-s^2) = Q(s)\cdot Q(-s) = (m_2 + n_2)(m_2 - n_2) \tag{11}$$

the character of this pole pattern of the function $f(s)$ is again apparent.

If in a complete partial-fraction expansion of $f(s)$ we strike out all terms involving right half-plane poles, multiply by two the ones for the left half-plane poles, and leave intact the rational entire remainder which is simply the constant

$$R = f(\infty) = \frac{A_n}{B_n} \tag{12}$$

then we have the desired function $Z(s)$. Since the determination of the partial-fraction expansion for a rational function is a well-known procedure, the construction of $Z(s)$ by this method is straightforward.

In the usual situation for which the poles of $Z(s)$ or of $f(s)$ are all simple, and the residues are given by

$$k_\nu = [(s - s_\nu)f(s)]_{s=s_\nu} \tag{13}$$

where $s_1, s_2, \cdots, s_n$ are the zeros of $Q(s)$ or the poles of $Z(s)$, one has

$$Z(s) = \frac{2k_1}{s - s_1} + \frac{2k_2}{s - s_2} + \cdots + \frac{2k_n}{s - s_n} + R \tag{14}$$

in which conjugate poles involve conjugate complex values for the corresponding residues. It is significant in this connection to note that the residues of $f(s)$ in its left half-plane poles are the residues of $Z(s)/2$ in these poles. Except for the additive constant R, the impedance $Z(s)$ is completely characterized by the residues in its simple poles.* Thus $\text{Re}\,[Z(j\omega)]$ which determines $f(s)$, determines $Z(s)$ also.

Let us illustrate this construction process with the example

$$\text{Re}\,[Z_0(j\omega)] = \frac{1}{1 + \omega^6} \tag{15}$$

for which

$$f(s) = \frac{1}{1 - s^6} \tag{16}$$

The poles of this function are given by

$$s_\nu = e^{j\pi\nu/3} \quad \text{for } \nu = 0, 1, \cdots, 5 \tag{17}$$

and the left half-plane poles by

$$s_2 = e^{j(2\pi/3)}, \qquad s_3 = -1, \qquad s_4 = e^{-j(2\pi/3)} \tag{18}$$

For the residue at $s = s_2$ we have

$$k_2 = \frac{-1}{(s_2 - s_1)(s_2 - s_3)(s_2 - s_4)(s_2 - s_5)(s_2 - s_0)} \tag{19}$$

* This method of constructing $Z(s)$ from its j-axis real part is, of course, not restricted to impedances with simple poles, although the expression given by Eq. 14 implies this assumption. When poles of higher multiplicity are involved, the corresponding terms in the partial-fraction expansion of $f(s)$ are appropriately modified according to a well-established procedure (see *MCA*, pp. 307 ff.).

From the sketch of these frequency factors as shown in Fig. 1 and the simple geometry involved we see that the sum of the angles of all the factors equals $\pi + (\pi/3) + (\pi/2) + (2\pi/3) + (5\pi/6) = 4\pi - 2\pi/3$, and the product of their magnitudes equals $1 \cdot 1 \cdot \sqrt{3} \cdot 2 \cdot \sqrt{3} = 6$. Hence we have

$$k_2 = -e^{j(2\pi/3)}/6, \qquad k_3 = \tfrac{1}{6}, \qquad k_4 = -e^{-j(2\pi/3)}/6 \tag{20}$$

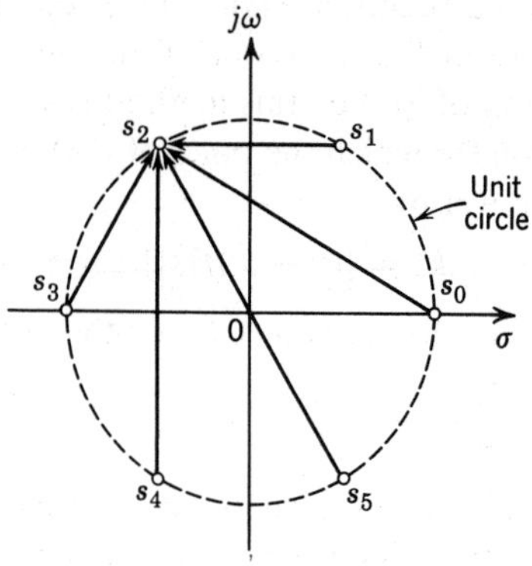

Fig. 1. Frequency factors in the determination of the residue k_2, Eq. 19.

The value of R according to Eq. 12 is in this case zero, so that Eq. 14 yields

$$Z_0(s) = \frac{-1}{3}\left(\frac{e^{j(2\pi/3)}}{s - s_2} - \frac{1}{s+1} + \frac{e^{-j(2\pi/3)}}{s - s_4}\right) \tag{21}$$

and with the values 18, we get

$$Z_0(s) = \tfrac{1}{3}\left(\frac{s+2}{s^2+s+1} + \frac{1}{s+1}\right) = \frac{\frac{2}{3}s^2 + \frac{4}{3}s + 1}{s^3 + 2s^2 + 2s + 1} \tag{22}$$

This method* yields the impedance in its partial-fraction expanded form and hence is best suited in connection with problems where this form is desired for the subsequent use that is to be made of the impedance function. When this use calls for the impedance in the form of a quotient of polynomials as in Eq. 22, it is more expedient to derive the impedance from its real part in the following manner.†

From Eqs. 7 we see that

$$\begin{aligned}(m_1m_2 - n_1n_2) = {} & (a_0 + a_2s^2 + \cdots + a_ns^n)(b_0 + b_2s^2 + \cdots + b_ns^n) \\ & - s^2(a_1 + a_3s^2 + \cdots + a_{n-1}s^{n-1})(b_1 + b_3s^2 + \cdots + b_{n-1}s^{n-1})\end{aligned} \tag{23}$$

* First suggested by H. W. Bode.

† This method is due to C. M. Gewertz who first developed an algebraic procedure for constructing an impedance from its real part as discussed in his book *Network Synthesis*, Williams and Wilkins, Baltimore, 1933, pp. 142–149.

where we have arbitrarily assumed that the degree n is even. Letting $s = j\omega$ or $s^2 = -\omega^2$, we get for the numerator of Eq. 8

$$\begin{aligned} A_0 + A_1\omega^2 + A_2\omega^4 + \cdots + A_n\omega^{2n} \\ = (a_0 - a_2\omega^2 + a_4\omega^4 - + \cdots)(b_0 - b_2\omega^2 + b_4\omega^4 - + \cdots) \\ + \omega^2(a_1 - a_3\omega^2 + a_5\omega^4 - + \cdots)(b_1 - b_3\omega^2 + b_5\omega^4 - + \cdots) \end{aligned} \tag{24}$$

If we multiply out the right-hand side of this equation and then equate coefficients of like powers of ω^2 on both sides, we get the following set of equations

$$\begin{aligned} A_0 &= a_0b_0 \\ A_1 &= -a_0b_2 + a_1b_1 - a_2b_0 \\ A_2 &= a_0b_4 - a_1b_3 + a_2b_2 - a_3b_1 + a_4b_0 \\ A_3 &= -a_0b_6 + a_1b_5 - a_2b_4 + a_3b_3 - a_4b_2 + a_5b_1 - a_6b_0 \\ &\cdots\cdots\cdots\cdots\cdots\cdots \end{aligned} \tag{25}$$

A compact form for these relations is given by

$$A_r = \sum_{s=-r}^{s=r} a_{r+s}b_{r-s}(-1)^s \tag{26}$$

The coefficients $b_0, b_1, b_2, \cdots$ in the polynomial $Q(s)$ of Eq. 7 are found by forming this polynomial from the left half-plane zeros of the known polynomial $B(-s^2)$ in Eqs. 10 and 11. Thus we can assume that

$$Q(s) = (s - s_1)(s - s_2) \cdots (s - s_n) \tag{27}$$

where, as in Eqs. 13 and 14, $s_1, s_2, \cdots, s_n$ are the left half-plane zeros of $B(-s^2)$. Except for a constant multiplier, which in our present problem can just as well be allotted to the numerator polynomial of $Z(s)$, a polynomial is uniquely constructible from its zeros.

In Eqs. 25 we now know the coefficients $A_0, A_1, \cdots, A_n$, and $b_0, b_1, \cdots, b_n$. We may therefore regard them as a set of simultaneous equations for the determination of the coefficients $a_0, a_1, \cdots, a_n$ in the numerator polynomial of $Z(s)$, Eq. 7. It should be observed in connection with Eqs. 25 that any terms involving coefficients a_k or b_k for $k > n$ are zero. The number of equations is thus always equal to the number of unknowns, and so a unique solution exists provided only that the determinant is nonzero. This determinant reads

$$D = \begin{vmatrix} b_0 & 0 & 0 & 0 & 0 & \cdot \\ -b_2 & b_1 & -b_0 & 0 & 0 & \cdot \\ b_4 & -b_3 & b_2 & -b_1 & b_0 & \cdot \\ -b_6 & b_5 & -b_4 & b_3 & -b_2 & \cdot \\ b_8 & -b_7 & b_6 & -b_5 & b_4 & \cdot \\ \cdot & \cdot & \cdot & \cdot & \cdot & \cdot \\ \cdot & \cdot & \cdot & \cdot & \cdot & \cdot \\ \cdot & \cdot & \cdot & \cdot & \cdot & \cdot \end{vmatrix} \tag{28}$$

If we change the algebraic signs of all elements in rows 1, 3, 5, · · · and subsequently change the signs of all elements in the corresponding columns, then the signs of all elements are positive and the value of D is unchanged. Next we observe that

$$D = b_0 \times \begin{vmatrix} b_1 & b_0 & 0 & 0 & \cdot \\ b_3 & b_2 & b_1 & b_0 & \cdot \\ b_5 & b_4 & b_3 & b_2 & \cdot \\ b_7 & b_6 & b_5 & b_4 & \cdot \\ \cdot & \cdot & \cdot & \cdot & \cdot \\ \cdot & \cdot & \cdot & \cdot & \cdot \\ \cdot & \cdot & \cdot & \cdot & \cdot \end{vmatrix} \tag{29}$$

The determinant involved here must be positive and nonzero according to the Hurwitz criteria as expressed by Eqs. 664 and 665 on p. 405 of *MCA*, since the b_k are known to be the coefficients of a Hurwitz polynomial. Therefore, Eqs. 25 are sure to possess unique solutions for the coefficients $a_0, a_1, \cdots, a_n$.

To illustrate with the real part 15, one begins by finding the left half-plane zeros of $B(-s^2)$ as given in 18, and from these forming the polynomial $Q(s)$, Eq. 27, thus

$$\begin{aligned} Q(s) &= (s - e^{j(2\pi/3)})(s - e^{-j(2\pi/3)})(s + 1) \\ &= (s^2 + 2s + 1)(s + 1) = s^3 + 2s^2 + 2s + 1 \end{aligned} \tag{30}$$

Eqs. 25 may then be written. They read

$$\begin{aligned} A_0 &= 1 = a_0 \\ A_1 &= 0 = -2a_0 + 2a_1 - a_2 \\ A_2 &= 0 = \qquad\quad -a_1 + 2a_2 - 2a_3 \\ A_3 &= 0 = \qquad\qquad\qquad\qquad\quad a_3 \end{aligned} \tag{31}$$

Their solution is readily found, and yields

$$a_0 = 1, \qquad a_1 = \tfrac{4}{3}, \qquad a_2 = \tfrac{2}{3}, \qquad a_3 = 0 \tag{32}$$

which is seen to check with the coefficients in the numerator polynomial of $Z_0(s)$ as given in Eq. 22. The denominator polynomial is the same as in Eq. 30.

An interesting alternate method,* which becomes useful in the development of some synthesis procedures to be discussed later on (Ch. 10), is obtained from the following considerations. In connection with the notation given in Eqs. 7 and 8 let us write for the sake of abbreviation

$$M = m_1 m_2 - n_1 n_2 \tag{33}$$

---

* Pointed out by F. Miyata in "A New System of Two-Terminal Synthesis," *J. Inst. Elec. Engrs. (Japan)*, Vol. 35, May 1952, pp. 211–218.

and consider the construction of an auxiliary impedance $Z'(s) = P'(s)/Q(s)$ having a particular numerator polynomial $P'(s) = m_1' + n_1'$ such that

$$m_1'm_2 - n_1'n_2 = 1 \tag{34}$$

It is clear from the preceding discussion that such a polynomial $P'(s)$ can always be found,† in fact the example given by the real part 15 calls for precisely condition 34.

Multiplying Eq. 34 on both sides by M, we get

$$m_1'Mm_2 - n_1'Mn_2 = m_1m_2 - n_1n_2 \tag{35}$$

from which we cannot conclude simply that $m_1'M = m_1$ and $n_1'M = n_1$, but rather that

$$m_1'M = m_1 + m^*_1, \qquad \text{and } n_1'M = n_1 + n^*_1 \tag{36}$$

in which m^*_1 and n^*_1 are, respectively, even and odd polynomials fulfilling the identity.

$$m^*_1m_2 - n^*_1n_2 \equiv 0 \tag{37}$$

The results 36 enable us to write

$$F(s) = \frac{(m_1' + n_1')M}{m_2 + n_2} = \frac{m_1 + n_1}{m_2 + n_2} + \frac{m^*_1 + n^*_1}{m_2 + n_2} \tag{38}$$

and observe from 37 that

$$\frac{m^*_1 + n^*_1}{m_2 + n_2} = \frac{m^*_1}{n_2} = \frac{n^*_1}{m_2} = p(s) = \text{an odd rational function} \tag{39}$$

so that

$$F(s) = Z(s) + p(s) \tag{40}$$

Since by hypothesis $Z(s)$ has no j-axis poles, Eqs. 38 and 40 show that $p(s)$ can have no j-axis poles. Equation 39, on the other hand, demands that $p(s)$ can have no poles other than j-axis poles. The only possible conclusion is that $p(s)$ has no finite nonzero poles at all, and hence must be an odd finite polynomial. If we think of $F(s)$ in Eq. 40 expanded into partial fractions, the rational entire remainder function, except for its constant term, is $p(s)$.

The method of obtaining $Z(s)$ from the auxiliary impedance $Z'(s)$ is thus clear. We first form the improper rational function $MZ'(s) = M(m_1' + n_1')/(m_2 + n_2)$ and, by the familiar process of long division, reduce the degree of the numerator polynomial until it no longer exceeds the degree of the denominator polynomial. To illustrate, let us suppose that we are given

$$\text{Re}\,[Z(j\omega)] = \frac{\omega^2}{1 + \omega^6} \tag{41}$$

† An especially effective method for doing this is discussed in Art. 2, Ch. 11.

which is Eq. 15 multiplied by ω^2 and implies $M = -s^2$. The pertinent function $F(s)$ is the impedance of Eq. 22 multiplied by $-s^2$. The appropriate long-division process takes the form

$$\begin{array}{r|l} & -\tfrac{2}{3}s \\ \hline s^3 + 2s^2 + 2s + 1 & -\tfrac{2}{3}s^4 - \tfrac{4}{3}s^3 - s^2 \\ & -\tfrac{2}{3}s^4 - \tfrac{4}{3}s^3 - \tfrac{4}{3}s^2 - \tfrac{2}{3}s \\ \hline & \tfrac{1}{3}s^2 + \tfrac{2}{3}s \end{array} \tag{42}$$

and the desired impedance is seen to be

$$Z_1(s) = \frac{\tfrac{1}{3}s^2 + \tfrac{2}{3}s}{s^3 + 2s^2 + 2s + 1} \tag{43}$$

If we next want an impedance with the real part

$$\text{Re}\,[Z(j\omega)] = \frac{\omega^4}{1 + \omega^6} \tag{44}$$

we multiply $Z_1(s)$ in Eq. 43 by $-s^2$ and carry out the division

$$\begin{array}{r|l} & -\tfrac{1}{3}s \\ \hline s^3 + 2s^2 + 2s + 1 & -\tfrac{1}{3}s^4 - \tfrac{2}{3}s^3 \\ & -\tfrac{1}{3}s^4 - \tfrac{2}{3}s^3 - \tfrac{2}{3}s^2 - \tfrac{1}{3}s \\ \hline & \tfrac{2}{3}s^2 + \tfrac{1}{3}s \end{array} \tag{45}$$

The corresponding impedance is

$$Z_2(s) = \frac{\tfrac{2}{3}s^2 + \tfrac{1}{3}s}{s^3 + 2s^2 + 2s + 1} \tag{46}$$

Finally, to obtain an impedance with the real part

$$\text{Re}\,[Z(j\omega)] = \frac{\omega^6}{1 + \omega^6} \tag{47}$$

we multiply $Z_2(s)$ in Eq. 46 by $-s^2$ and carry out the division

$$\begin{array}{r|l} & -\tfrac{2}{3}s \\ \hline s^3 + 2s^2 + 2s + 1 & -\tfrac{2}{3}s^4 - \tfrac{1}{3}s^3 \\ & -\tfrac{2}{3}s^4 - \tfrac{4}{3}s^3 - \tfrac{4}{3}s^2 - \tfrac{2}{3}s \\ \hline & s^3 + \tfrac{4}{3}s^2 + \tfrac{2}{3}s \end{array} \tag{48}$$

yielding

$$Z_3(s) = \frac{s^3 + \tfrac{4}{3}s^2 + \tfrac{2}{3}s}{s^3 + 2s^2 + 2s + 1} \tag{49}$$

This result could, of course, also be obtained directly from $Z_0(s)$ in Eq. 22 through multiplication by $-s^6$ followed by three steps of long division. It is interesting in passing to note that the sum of $Z_0(s)$ in Eq. 22 and $Z_3(s)$ in Eq. 49 equals unity, as is to be expected from the fact that the corresponding real parts 15 and 47 add up to unity.

If we now want an impedance with the real part

$$\text{Re}\,[Z(j\omega)] = \frac{A_0 + A_1\omega^2 + A_2\omega^4 + A_3\omega^6}{1 + \omega^6} \tag{50}$$

the results by given Eqs. 22, 43, 46, and 49 may be used to evaluate the desired solution in the form

$$Z(s) = A_0Z_0(s) + A_1Z_1(s) + A_2Z_2(s) + A_3Z_3(s) \tag{51}$$

These methods of constructing an impedance from its real part normally cannot yield a function having j-axis poles since corresponding frequency factors of the form $(s^2 + \omega_0^2)$ will cancel out of the even part of $Z(s)$ given by Eq. 10. Thus if the denominator polynomial $Q(s) = m_2 + n_2$ contains the factor $(s^2 + \omega_0^2)$, then $m_2^2 - n_2^2$ contains this factor squared, and the polynomial $m_1m_2 - n_1n_2$ must likewise contain this factor squared in order to insure the nonnegativeness of the real part for $s = j\omega$.

We may also recognize this restriction on the part of these methods from the consideration that j-axis poles alone constitute a reactance function (impedance of a lossless network) the real part of which is zero for all j-axis frequencies. Hence the addition of a reactance function to any given impedance does not affect the resulting real part, and so the impedance which is constructible from a given real part cannot have j-axis poles. Such a driving-point impedance is spoken of as being *minimum reactive* since its imaginary part or reactive part can have any arbitrary pure reactance added to it and still be appropriate to the given real part.

This argument assumes, of course, that factors of the form $(s^2 + \omega_0^2)^2$ contained in the numerator and denominator of expression 10 are canceled before proceeding with the formal process of setting up Eqs. 25. If expression 8 for the real part is given in polynomial form, common factors of the numerator and denominator are not in evidence, and it is interesting to inquire as to what happens if Eqs. 25 are formed from the given expression containing a pair of cancelable factors. In this event $Q(s)$ will, of course, be found to contain a pair of j-axis zeros, and, according to the methods of testing for its Hurwitz character, the determinant 28 becomes zero. The inhomogeneous Eqs. 25 then have nontrivial solutions only for certain values of the coefficients $A_0, A_1, A_2, \cdots$ (see *MCA*, pp. 100 ff.), but the fact that the numerator polynomial in Eq. 10 also contains the j-axis factors automatically fulfills these special conditions, and the resulting nontrivial solution yields an impedance $Z(s)$ with the j-axis pole.

Thus if cancelable j-axis factors in the real-part expression 8 are not canceled, the present method of constructing $Z(s)$ is nevertheless straightforwardly applicable (provided one is familiar with the pertinent theory of algebraic equations), and the impedance thus found contains the appropriate j-axis poles.

Since factors may arbitrarily be inserted into the given real-part expression, it follows that the construction of an impedance by these methods is not unique. This situation, which in some problems is objectionable, may be remedied by observing that the real part of a reactance function, although zero everywhere on the j-axis in the ordinary sense, possesses impulses at the j-axis poles which the usual algebraic procedure fails to detect. The means for developing this characteristic of the real-part function and thus rendering the determination of an impedance from its real part unique is discussed in the following article.

2 Impulses in the Real Part

In order to see that the real part of a reactance function is not zero along the j-axis as the usual algebraic method for its derivation indicates, it is effective to study carefully the behavior of the real part associated with a function having a single left half-plane pole, as that pole is allowed to move closer and closer to the j-axis. Suppose we consider the function

$$Z(s) = \frac{1}{\pi(s + \alpha)} \tag{52}$$

having a simple pole on the negative real axis, α units to the left of the origin. For $s = j\omega$ we find

$$Z(j\omega) = \frac{\alpha}{\pi(\alpha^2 + \omega^2)} - j\frac{\omega}{\pi(\alpha^2 + \omega^2)} \tag{53}$$

and so we have for the real and imaginary parts according to Eq. 4

$$R(\omega) = \frac{\alpha}{\pi(\alpha^2 + \omega^2)}, \qquad X(\omega) = \frac{-\omega}{\pi(\alpha^2 + \omega^2)} \tag{54}$$

If we set $\alpha = 0$, the pole in the impedance 52 becomes a j-axis pole, and $R(\omega)$ in Eq. 54 appears to become zero. The latter phenomenon, however, is a mirage, as the following more careful study of this function reveals. From the plot in Fig. 2 we see that the peak value of $R(\omega)$ at $\omega = 0$ equals $1/\pi\alpha$, and that its width at the half-value points is 2α. The net area under this function is readily found from the following integration

$$\int_{-\infty}^{\infty} R(\omega)\, d\omega = \frac{1}{\pi}\int_{-\infty}^{\infty} \frac{\alpha\, d\omega}{\alpha^2 + \omega^2} \tag{55}$$

Letting $\omega/\alpha = x$ we have for this integral

$$\int_{-\infty}^{\infty} R(\omega)\, d\omega = \frac{1}{\pi}\int_{-\infty}^{\infty} \frac{dx}{1 + x^2} = \frac{1}{\pi} \tan^{-1} x \Big]_{-\infty}^{\infty} = 1 \tag{56}$$

The important part about this result is the fact that the area is independent of α. Normalization of this area to the value unity accounts for the factor $1/\pi$ in the original expression 52 for $Z(s)$.

As α is allowed to become smaller and smaller, and the pole moves closer to the j-axis, the pulselike function $R(\omega)$ becomes taller and slimmer, but the area it encloses remains equal to unity. From the picture presented by this limit process it becomes clear that $R(\omega)$ approaches uniformly the unit impulse function $u_0(\omega)$ as α is allowed to become zero (see *ICT*, pp. 190–203).

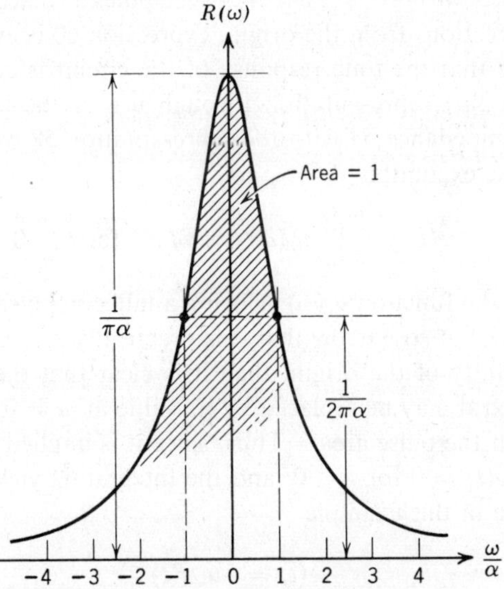

Fig. 2. Showing the behavior of the real part of the impedance 52 for the vicinity of $s = 0$.

On the basis of this result we may say that the reactance function

$$Z(s) = \frac{1}{\pi s} \tag{57}$$

when evaluated for $s = j\omega$ yields

$$Z(j\omega) = u_0(\omega) - j/\pi\omega \tag{58}$$

Similarly, a reactance having simple j-axis poles at $s = \pm j\omega_0$ with positive real residues k_0/π is given by

$$Z(s) = \frac{1}{\pi}\left(\frac{k_0}{s - j\omega_0} + \frac{k_0}{s + j\omega_0}\right) \tag{59}$$

For $s = j\omega$ it has the real part

$$R(\omega) = k_0[u_0(\omega - \omega_0) + u_0(\omega + \omega_0)] \tag{60}$$

and the imaginary part

$$X(\omega) = \frac{2k_0}{\pi} \frac{\omega}{\omega_0^2 - \omega^2} \tag{61}$$

The result 61 is familiar. The one stated by Eq. 60 follows directly from the fact that what is said above about the real part for a pole on the real axis obviously remains intact when this pole is merely displaced parallel to the j-axis. Since we now have a pair of such poles displaced equal distances in opposite directions from the origin, expression 60 is evidently correct.

We may see that the time response of the circuit is correctly obtained from this real-part representation through use of the integral in Eq. 6. Thus for the impedance 57 with the representation 58 we have for a unit current impulse excitation

$$e(t) = \frac{2}{\pi}\int_0^\infty u_0(\omega) \cos \omega t \, d\omega \text{ for } t > 0 \tag{62}$$

If we visualize the function $u_0(\omega)$ as being a tall slim pulse having the form shown in Fig. 2 but so narrow that it is practically zero everywhere in the immediate vicinity of the origin, then it is clear that the function $\cos \omega t$ under the integral may be replaced by its value at $\omega = 0$, and the integral then yields half the pulse area. Thus, since it is implied in the derivation of Eq. 6 that $e(t) = 0$ for $t < 0$, and the integral 62 yields $e(t) = 1/\pi$ for $t > 0$, we have in this example

$$e(t) = \frac{1}{\pi} u_{-1}(t) \tag{63}$$

which is a step function of value $1/\pi$, a result that is seen to agree with the unit current impulse response for the impedance 57 as calculated by elementary circuit analysis methods.

For the real-part 60, Eq. 6 yields for $t > 0$

$$e(t) = \frac{2k_0}{\pi}\int_0^\infty [u_0(\omega - \omega_0) + u_0(\omega + \omega_0)] \cos \omega t \, d\omega \tag{64}$$

Here the integrand is zero everywhere except at the points $\omega = \pm\omega_0$ where $\cos \omega t$ equals $\cos \omega_0 t$, and since the integration extends only over positive ω values the impulse $u_0(\omega + \omega_0)$ contributes nothing. Hence we get for $t > 0$

$$e(t) = \frac{2k_0}{\pi} \cos \omega_0 t \int_0^\infty u_0(\omega - \omega_0) d\omega = \frac{2k_0}{\pi} \cos \omega_0 t \tag{65}$$

which also agrees with the well-known unit current impulse response of a circuit having the impedance 59.

To summarize: (a) the real part associated with an impedance having only j-axis poles is not identically zero for $s = j\omega$ but consists of a sum of impulses located at the pole positions, equal in value to π times the respective residues of the impedance in these poles; (*b*) it may be stated without reservations of any kind that an impedance (or admittance or dimensionless transfer function) is uniquely characterized by and constructible from its real part when the latter is specified, inclusive of any impulses that it may contain; and (*c*) one can carry through in a straightforward manner all the usual analysis procedures in the frequency or time domains with real-part functions consisting of or containing impulses as constituent components. In fact, it is usually easier to work with impulses since the values of associated functions (like $\cos \omega t$ in the integrals 62 and 64) are unimportant except at the frequencies of occurrence of the impulses.

3 Construction of an Arbitrary Real Part From Impulses

A familiar artifice in network analysis is to consider the impulse as a building block for the construction of arbitrary functions. As indicated in Fig. 3, some given function $R(\omega)$ is regarded as consisting of a sum of

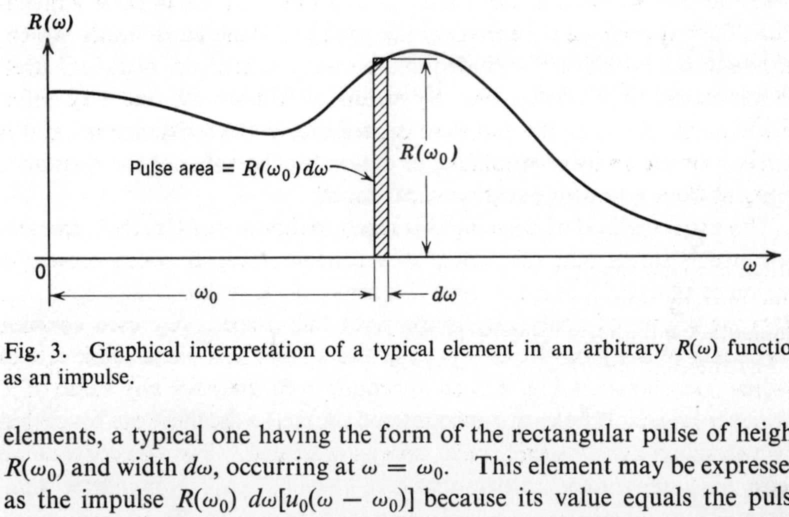

Fig. 3. Graphical interpretation of a typical element in an arbitrary $R(\omega)$ function as an impulse.

elements, a typical one having the form of the rectangular pulse of height $R(\omega_0)$ and width $d\omega$, occurring at $\omega = \omega_0$. This element may be expressed as the impulse $R(\omega_0)\, d\omega[u_0(\omega - \omega_0)]$ because its value equals the pulse area which is $R(\omega_0)\, d\omega$. Since the real-part function $R(\omega)$ is even so that an identical element occurs also at $\omega = -\omega_0$, as in expression 60, the incremental reactance function associated with this element is given, according to Eq. 59, by

$$dZ(s) = \frac{R(\omega_0)\, d\omega}{\pi}\left(\frac{1}{s - j\omega_0} + \frac{1}{s + j\omega_0}\right) = \frac{2sR(\omega_0)\, d\omega}{\pi(s^2 + \omega_0{}^2)} \tag{66}$$

From this result we can deduce a useful property regarding the asymptotic behavior of impedances. Thus for $s \to \infty$ we obtain from

Eq. 66

$$dZ(s) \rightarrow \frac{2R(\omega)\,d\omega}{\pi s} \tag{67}$$

and for the total impedance having the real part $R(\omega)$

$$Z(s) \rightarrow \frac{2}{\pi s}\int_0^\infty R(\omega)\,d\omega \quad \text{for } s \rightarrow \infty \tag{68}$$

where we have dropped the subscript zero on ω since it is no longer needed. The result 68 states that every impedance having a finite non-zero total area under its real-part function must, at very large frequencies, behave like a capacitive reactance $1/Cs$. If a driving-point impedance is involved, we may say that at sufficiently high frequencies the network at its input terminals degenerates into a single capacitance whose value is fixed by the relation

$$\int_0^\infty R(\omega)\,d\omega = \frac{\pi}{2C} \tag{69}$$

This result which is known as the *resistance-integral* theorem* has many practical uses as we shall see in later discussions. In one of these applications the capacitance C represents the parasitic shunt capacitance which, although not specifically included in a network schematic, manifests itself in the actual circuit behavior. By means of relation 69, one may either calculate the value of this parasitic capacitance if the resistance integral is directly or indirectly computable, or determine the value of the resistance integral from a known parasitic capacitance.

The usual method of deriving this result makes use of Cauchy's integral law which states that the value of a contour integral taken around a region of analyticity of the integrand is zero. With regard to an impedance $Z(s)$ the region of analyticity is the right half-plane. A closed contour consisting of the j-axis from $-j\rho$ to $j\rho$ and a right half-plane semicircle of radius ρ as shown in Fig. 4 is an appropriate contour for any value of ρ, however large. If poles are encountered on the j-axis, these are by-passed on vanishingly small semicircular arcs as indicated. For this contour we have

$$\oint Z(s)\,ds = 0 \tag{70}$$

In the evaluation of this integral we consider separately the part contributed by the j-axis and that due to the semicircular arc, the radius of which

* First pointed out and applied usefully by H. W. Bode. See his book *Network Analysis and Feedback Amplifier Design*, D. Van Nostrand Co., New York, 1945, p. 280. The basic property expressed here, however, was anticipated by R. Feldtkeller in his article "Zur Frequenzabhängigkeit der Reaktanzen," *Elek. Nacht.-Tech.*, Vol. 13, No. 12, Dec. 1936, pp. 401–404.

we now assume to be infinitely large. Thus Eq. 70 is written in the form

$$\int_{-\infty}^{\infty} Z(j\omega)\, d(j\omega) + \int_{\text{arc}} Z(s)\, ds = 0 \tag{71}$$

In the first of these integrals we write $Z(j\omega) = R(\omega) + jX(\omega)$ and observe that the integral of $X(\omega)$ over the whole j-axis is zero because $X(\omega)$ is an odd function. In the second integral which pertains to values of s in the neighbourhood of the point at infinity, it is sufficient to substitute the asymptotic behavior for $Z(s)$ which is $1/Cs$. We then have in place of 71

$$j\int_{-\infty}^{\infty} R(\omega)\, d\omega + \frac{1}{C}\int_{\text{arc}} \frac{ds}{s} = 0 \tag{72}$$

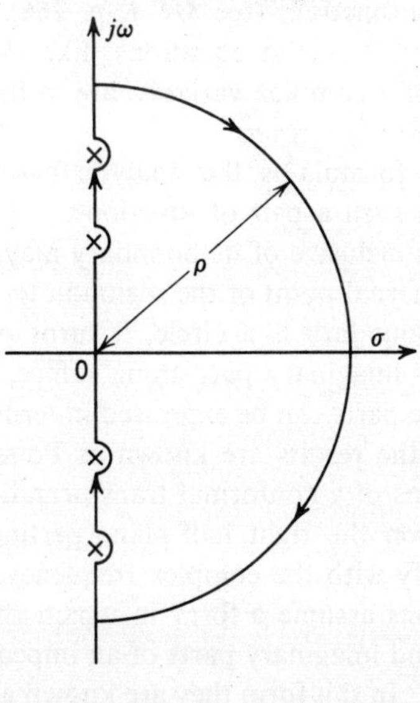

Fig. 4. Illustrating the path of integration for the integral 70.

For values of s on the large semicircular arc we write

$$s = \rho e^{j\theta}, \qquad ds = j\rho e^{j\theta}\, d\theta \tag{73}$$

and hence

$$\frac{ds}{s} = j\, d\theta \tag{74}$$

Integration along the semicircular arc implies that θ varies from $\pi/2$ to $-\pi/2$. Therefore Eq. 72 yields

$$j\int_{-\infty}^{\infty} R(\omega)\, d\omega + \frac{j}{C}\int_{\pi/2}^{-\pi/2} d\theta = 0 \tag{75}$$

or

$$\int_{-\infty}^{\infty} R(\omega)\, d\omega = \frac{1}{C}\int_{-\pi/2}^{\pi/2} d\theta = \frac{\pi}{C} \tag{76}$$

Since $R(\omega)$ is an even function we can evaluate the left-hand integral between the limits 0 and ∞, and multiply by 2. The result then agrees with that given in Eq. 69.

4 Analytic Procedure in Terms of Hilbert Transforms

Analytically the problem of constructing the imaginary part of a complex function from its real part or vice versa (which is an alternate way of looking at the construction of the complex function from either of its real or imaginary parts) has its origin in potential theory. Thus it is well known that the real and imaginary parts of a complex function satisfy Laplace's equation throughout any region of analyticity and hence may be regarded as potential functions (see *MCA*, p. 266). Two such functions satisfying the Cauchy-Riemann equations, like the real and imaginary parts of a function of a complex variable, are in this regard spoken of as being *conjugate potential functions*.

Cauchy's integral formula is the analytic basis for formulating the relationship between such a pair of functions. A function $Z(s)$ analytic over a certain region inclusive of its boundary may, with this formula, be computed for any internal point of the region in terms of its values on the boundary. If the boundary is a circle, it turns out that the boundary values of the real or imaginary part alone suffice, or that the boundary values of one of these parts can be expressed in terms of those of the other part. In this form the results are known as Poisson's integrals (*MCA*, p. 330). If, by means of a conformal transformation, the interior of the circle is mapped upon the right half-plane pertinent to a new variable which we can identify with the complex frequency variable s, then these same integral relations assume a form in which they are appropriate to expressing the real and imaginary parts of an impedance or admittance in terms of each other. In this form they are known as Hilbert transforms.* It is well to be aware of their origin in the Cauchy integral formula, for the conditions under which they are applicable are less likely to be lost track of. These are that the right half-plane inclusive of the j-axis must be the region of analyticity for the complex function whose real and imaginary parts are related by these transforms. These conditions agree as they should with those pointed out at the beginning of this chapter as being essential to the possibility of constructing the whole function from its real part.

It is possible to obtain the Hilbert transforms more directly than through

* First applied in circuit theory by Y. W. Lee, "Synthesis of Electric Networks by Means of Fourier Transforms of Laguerre's Functions," *J. Math. and Phys.*, Vol. 11, 1932, pp. 83–113.

the use of Cauchy's integral formula and Poisson's integrals. To this end we use as our point of departure the argument in Art. 2 showing that the real and imaginary parts of the complex function 58 are a typical pair of conjugate potential functions. In other words, this discussion tells us that the conjugate of a unit impulse located at $\omega = 0$ is $-1/\pi\omega$, or the conjugate of a unit impulse located at $\omega = \omega_0$ is $-1/\pi(\omega - \omega_0)$. Since the impulse is a building block for the construction of arbitrary functions (as indicated in Fig. 3) one can evidently construct the conjugate of anything in terms of the known conjugate of an impulse. The results obtained by carrying out this thought are the Hilbert transforms, as we shall now show.

In order to keep the argument noncommittal as to whether an impedance, admittance, or dimensionless response function is involved, we refer to the latter as a *system function* and write it

$$h(j\omega) = h_1(\omega) + jh_2(\omega) \tag{77}$$

If we visualize the real part $h_1(\omega)$ plotted as $R(\omega)$ is in Fig. 3, and if the symbol ξ instead of ω_0 is used to denote the arbitrary position of a typical element, then its value as an impulse is $h_1(\xi)\,d\xi$. The conjugate of this element, therefore, is $-h_1(\xi)\,d\xi/\pi(\omega - \xi)$, and the conjugate of the entire $h_1(\omega)$ function is given by the integral

$$h_2(\omega) = -\frac{1}{\pi}\int_{-\infty}^{\infty} \frac{h_1(\xi)\,d\xi}{\omega - \xi} \tag{78}$$

where it is to be noted that the integration extends from $-\infty$ to $+\infty$ since the integrand represents the conjugate of only a single element and nothing is as yet implied as to the even or odd character of $h_1(\omega)$, the only implication being that it is the real part of a complex function that is analytic in the right half-plane and on the j-axis.*

If we want the companion integral to 78, namely, the one that expresses the real part in terms of the imaginary part, we have but to observe that the Cauchy-Riemann equations

$$\frac{\partial h_1}{\partial \sigma} = \frac{\partial h_2}{\partial \omega}, \qquad \frac{\partial h_1}{\partial \omega} = -\frac{\partial h_2}{\partial \sigma} \tag{79}$$

which relate these conjugate potential functions remain unchanged if we interchange the functions h_1 and h_2 and change the algebraic sign of one of them. Doing this in Eq. 78 yields the desired complementary relation

$$h_1(\omega) = \frac{1}{\pi}\int_{-\infty}^{\infty} \frac{h_2(\xi)\,d\xi}{\omega - \xi} \tag{80}$$

The integrals 78 and 80 are the so-called Hilbert transforms. A large

---

* It is taken for granted that $h_1(\omega)$ remains bounded for all ω including the point at infinity.

variety of modified forms for these relations are derivable by recognizing the even and odd character of h_1 and h_2 respectively and thus obtaining integrals that extend only over the positive half of the spectrum, or by introducing a logarithmic frequency scale, and by other means. We shall not enter upon such elaborations here, but merely call attention to a few fundamental properties of these integrals that are useful in dealing with various types of problems.

First of all, the linearity of these relations enables us to simplify the consideration of complicated functions by breaking them down into handy pieces, finding their separate conjugates, and adding these together again. We also can differentiate or integrate both functions any number of times; and we can differentiate or integrate the one under the integral sign, then carry out the integration, and finally integrate or differentiate so as to undo the respective initial operation, whatever that was. It is furthermore useful to observe that these integrals express the convolution of h_1 or h_2 with the function $1/\pi\omega$. Like any convolution integral, they may be modified by interchanging the arguments ξ and $\omega - \xi$, or by differentiating one of the two functions being convolved and integrating the other one, and repeating this process as often as we like. If differentiating one function converts it into impulses, we need not integrate at all, for we know that anything convolved with an impulse yields that same thing again, displaced according to the location of the impulse. These and similar devices simplify the use of these integrals so that in most cases little, if any, integration needs to be done at all, especially if we can be content with an approximate result.*

A few specific examples will serve to illustrate the kind of problem for which this analytic approach, in contrast with the algebraic procedures given above, is appropriate. Primarily it is appropriate when the given function is specified graphically or by some nonrational function or by several of these pieced together over confluent frequency intervals. In such situations the preceding methods given in this chapter are obviously useless.

The first example of this sort to be discussed is shown in Fig. 5. The given $h_1(\omega)$ function is a rectangular pulse centered at the origin. For $-\omega_1 < \omega < \omega_1$ it equals unity; for $|\omega| > \omega_1$ it is identically zero. The integral 78 becomes

$$h_2(\omega) = -\frac{1}{\pi}\int_{-\omega_1}^{\omega_1} \frac{d\xi}{\omega - \xi} \tag{81}$$

---

* See "Computational Techniques Which Simplify the Correlation between Steady-State and Transient Response of Filters and Other Networks," *Proc. Nat. Electronics Conf.*, Vol. 9, Feb. 1954, or *Tech. Rept.* 268, Mass. Inst. Technol., Research Lab. Electronics, Sept. 2, 1953. Article IX of this paper is pertinent to the present discussion.

and yields

$$h_2(\omega) = \frac{1}{\pi}\ln(\omega - \xi)\Big]_{-\omega_1}^{\omega_1} = \frac{1}{\pi}\ln\left|\frac{\omega-\omega_1}{\omega+\omega_1}\right| \tag{82}$$

The absolute value signs are needed here to preserve the real character of $h_2(\omega)$. We can alternately see the need of using the absolute value signs here by returning to expressions 52 and 53 whence all our present results stem. If we had not set α equal to zero, the integral 81 would be replaced by

$$h_2(\omega) = -\frac{1}{\pi}\int_{-\omega_1}^{\omega_1}\frac{(\omega-\xi)\,d\xi}{\alpha^2+(\omega-\xi)^2}$$

$$= \frac{1}{2\pi}\int_{\alpha^2+(\omega-\omega_1)^2}^{\alpha^2+(\omega+\omega_1)^2}\frac{du}{u} = \frac{1}{\pi}\ln\sqrt{\frac{\alpha^2+(\omega-\omega_1)^2}{\alpha^2+(\omega+\omega_1)^2}} \tag{83}$$

Setting α equal to zero now clearly yields expression 82.

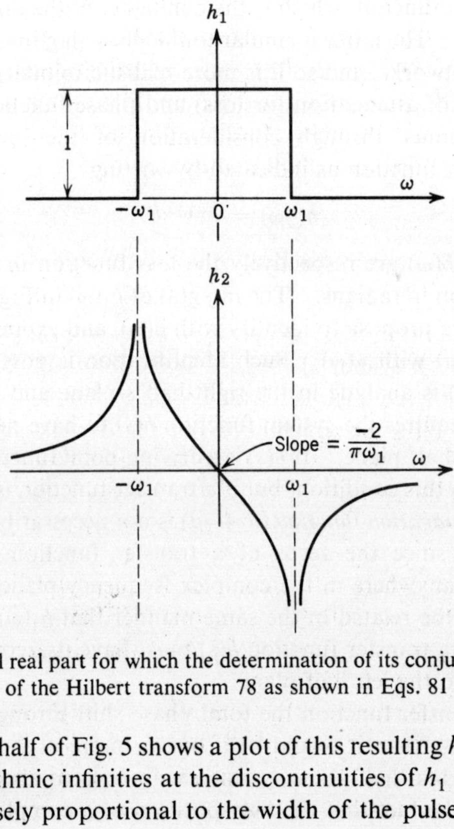

Fig. 5. A typical real part for which the determination of its conjugate mate is accomplished by means of the Hilbert transform 78 as shown in Eqs. 81 and 82.

The bottom half of Fig. 5 shows a plot of this resulting h_2 function which exhibits logarithmic infinities at the discontinuities of h_1 while its slope at $\omega = 0$ is inversely proportional to the width of the pulselike h_1 function.

As ω_1 becomes larger and larger, h_1 degenerates into a constant and h_2 becomes identically zero. The conjugate of a constant is zero.

We can derive a new pair of conjugate potential functions by integrating the ones shown in Fig. 5. Observe in this connection that integration (or differentiation) changes an even function into an odd function (except for a possible constant which, as we have seen, can be dropped) and vice versa. Since the real parts of network functions are always even and the imaginary parts are odd, we must interchange the identities of the functions after integration, which, as pointed out above, necessitates reversal of the algebraic sign of one of them. If we integrate h_1 in Fig. 5 we obtain a function that equals a linear ramp for $-\omega_1 < \omega < \omega_1$ and is otherwise constant. Integration of expression 82 from $-\infty$ to a variable upper limit ω, and reversal of the algebraic sign, yields

$$\frac{\omega_1}{\pi}\left(\ln|x^2-1| - x\ln\left|\frac{x-1}{x+1}\right| - 2\right) \text{ with } x = \frac{\omega}{\omega_1} \tag{84}$$

This is an even function which is the conjugate of the quasi-linear ramp described above. The latter is similar to the phase-lag function associated with low-pass networks, and so it is more realistic to interpret the present results as a pair of attenuation (or loss) and phase functions resulting in the familiar manner through consideration of the logarithm of the reciprocal system function as indicated by writing

$$h(j\omega) = e^{-(\alpha+j\beta)} \tag{85}$$

where $\alpha(\omega)$ and $\beta(\omega)$ are respectively the loss function in nepers and the phase-lag function in radians. The integral of $h_1(\omega)$ in Fig. 5 (without the constant term) we propose to identify with $\beta(\omega)$, and expression 84 (minus its constant term) with $\alpha(\omega)$. Such identification is possible only if the logarithm of $h(s)$ is analytic in the right-half s-plane and on the j-axis, a condition that requires the system function $h(s)$ to have neither zeros nor poles in the right half-plane. If $h(s)$ is a driving-point function, then its p.r. character assures this condition, but if a transfer function is involved, then the so-called *propagation function* $(\alpha + j\beta)$ is not necessarily analytic in the right half-plane since the zeros of a transfer function are in general permitted to lie anywhere in the complex frequency plane. In order for $\alpha(\omega)$ and $\beta(\omega)$ to be related in the same manner that $h_1(\omega)$ and $h_2(\omega)$ are, the corresponding transfer function $h(s)$ must have its zeros as well as its poles restricted to the left half-plane.

For such a transfer function the total phase shift throughout the range $0 < \omega < \infty$ is less than it would be if one or more of its zeros occupied positions in the right half-plane symmetrically located about the j-axis, as is evident from the fact that the transposition of a complex pair of zeros in this manner implies multiplication of $h(s)$ by the factor

$$\frac{(s + s_1)(s + \bar{s}_1)}{(s - s_1)(s - \bar{s}_1)} \tag{86}$$

where s_1 and $\bar{s}_1$ are conjugate zeros in the left half-plane as shown in Fig. 6. Although the magnitude of this factor is unity, its angle is $-(\psi - \phi)$. Hence it is clear that although such transposition of the zeros of a transfer function from the left to the right half-plane leaves its magnitude (and therefore α) unchanged, the negative of its angle which is the lag angle β has added to it the increment $\psi - \phi$ which increases from zero at $s = 0$ to 2π at $s = \infty$. Clearly the phase lag will be least for any magnitude of transfer function if its zeros are restricted to the left half-plane. We refer to such a transfer function as having *minimum-phase-shift* properties, or as being a *minimum-phase* transfer function. Only for such a one are $\alpha(\omega)$ and $\beta(\omega)$ related by the Hilbert transforms.

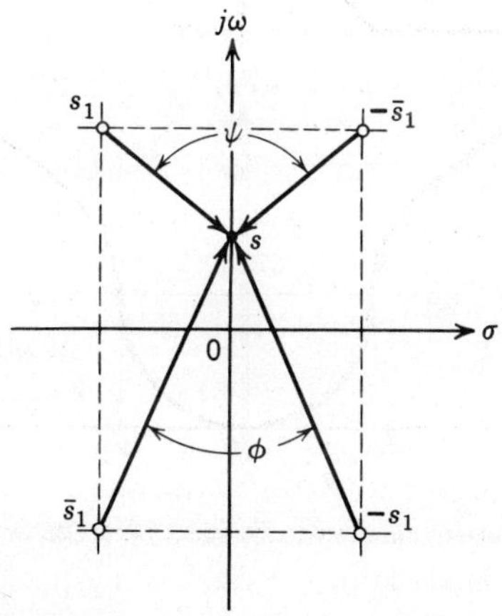

Fig. 6. Graphical interpretation of the frequency factors in expression 86.

Returning to the integration of the conjugate potential functions shown in Fig. 5, we are reminded that the asymptotic values of phase in a physical network are integer multiples of π, and so the resulting functions, interpreted as a pair of α and β functions, will be more realistic if we multiply the ramp derived from the integration of h_1 in Fig. 5 by $n\pi/\omega_1$ and, of course, multiply expression 84 by the same constant, giving

$$\alpha(\omega) = n \ln |x^2 - 1| - nx \ln \left|\frac{x - 1}{x + 1}\right| \tag{87}$$

The pair of conjugate potential functions determined in this way is shown in Fig. 7. They are useful for comparison purposes when we are interested in designing a low-pass filter with linear phase shift in its pass band. The attenuation function α has infinite slope at the break points $\omega = \pm\omega_1$ in the piecewise-linear phase β. In any actual design the phase will be rounded at these points and the slope in the α curve will there be finite. Nevertheless, the rather idealized characteristics given here are indicative of results that may be expected in such a design.

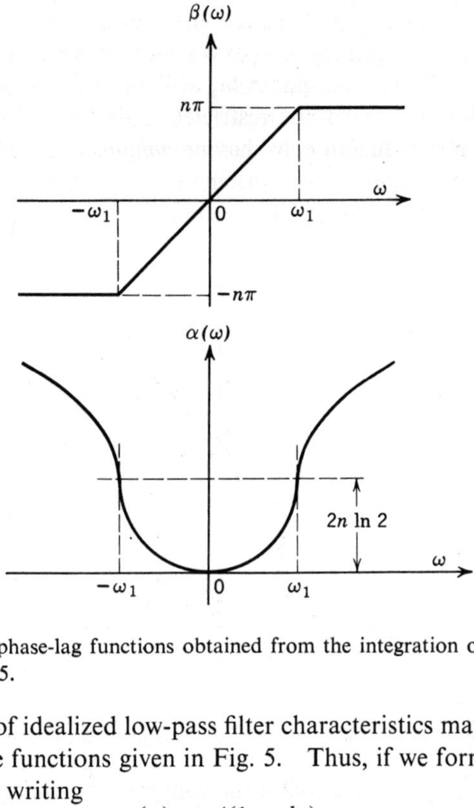

Fig. 7. Loss and phase-lag functions obtained from the integration of the conjugate pair shown in Fig. 5.

Another pair of idealized low-pass filter characteristics may be obtained directly from the functions given in Fig. 5. Thus, if we form an attenuation function by writing

$$\alpha(\omega) = A(1 - h_1) \tag{88}$$

then, since the conjugate of a constant is zero, the associated phase according to Eq. 82 is

$$\beta(\omega) = \frac{A}{\pi} \ln \left| \frac{\omega + \omega_1}{\omega - \omega_1} \right| \tag{89}$$

This pair of functions is shown in Fig. 8. The attenuation is zero within the pass band defined by $-\omega_1 < \omega < \omega_1$, and equals A nepers outside this range (the *stop* band). At the so-called *cutoff frequencies* $-\omega_1$ and ω_1, the attenuation is discontinuous. A filter having such an attenuation function is referred to as an *ideal* low-pass filter, especially if the value A is assumed to be infinite. From the associated phase characteristic, which has logarithmic infinities at the cutoff frequencies because of the assumed discontinuities in the attenuation, and has a slope at the origin proportional to A, we see that an ideal low-pass filter providing infinite attentuation in its stop band will have an associated phase that is infinite everywhere except at $\omega = 0$ and $\omega = \infty$. Such a filter does not even exist theoretically, but for a finite attenuation A its characteristics may be approximated by a physical network.

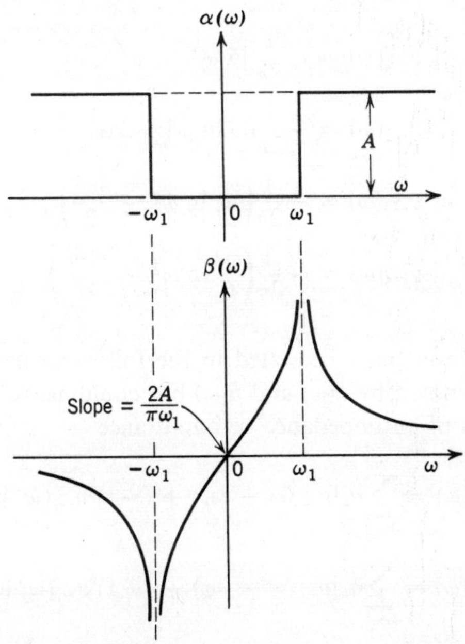

Fig. 8. A pair of loss and phase-lag functions directly obtainable from the real and imaginary parts shown in Fig. 5.

The Hilbert transforms 78 and 80 are particularly easy to evaluate when the given function has the form of a series of confluent straight-line segments; that is to say, when it is a so-called piecewise-linear function. In that case its second derivative consists of a sum of impulses. Convolution of these with the second integral of the function $-1/\pi\omega$ yields the desired result which can be written down by inspection.

Generalization of this method leads us to consider the approximation of a given real part by a set of confluent parabolic arcs instead of straight-line segments (equivalent to a piecewise-linear approximation of the first or higher derivative of the given real part) in order to achieve a smoother result and smaller tolerances. In any case, an appropriate derivative of such an approximation becomes a sum of impulses. If this sum represents the νth derivative of a given real part, then the associated imaginary part is a corresponding sum of the νth repeated integral of $-1/\pi\omega$.

It is convenient to denote the latter function by $v_0(\omega)$, and its successive integrals by $v_{-1}(\omega)$, $v_{-2}(\omega)$, and so forth. This sequence of functions that are the imaginary parts associated with real parts equal respectively to the singularity functions $u_{-\nu}(\omega)$ for $\nu = 0, 1, 2, \cdots$, are found to be

$$\begin{aligned} v_0(\omega) &= -\frac{1}{\pi\omega} \\ v_{-1}(\omega) &= -\frac{1}{2\pi}\ln \omega^2 \\ v_{-2}(\omega) &= -\frac{1}{2\pi}(\omega \ln \omega^2 - 2\omega) \\ v_{-3}(\omega) &= -\frac{1}{2\pi}\left(\frac{\omega^2}{2}\ln \omega^2 - \frac{3\omega^2}{2}\right) \\ v_{-4}(\omega) &= -\frac{1}{2\pi}\left(\frac{\omega^3}{6}\ln \omega^2 - \frac{11\omega^3}{18}\right) \end{aligned} \tag{90}$$

and so forth.

The desired result may be stated in the following terms in which the functions are denoted by $\alpha(\omega)$ and $\beta(\omega)$ but could as well be the real and imaginary parts of an impedance or admittance

$$\alpha(\omega) = \sum_{k=1}^{n} a_k[u_{-\nu}(\omega - \omega_k) + (-1)^\nu u_{-\nu}(\omega + \omega_k)] \tag{91}$$

$$\beta(\omega) = \sum_{k=1}^{n} a_k[v_{-\nu}(\omega - \omega_k) + (-1)^\nu v_{-\nu}(\omega + \omega_k)] \tag{92}$$

The radian frequencies $\pm\omega_k$ are the break points in the piecewise-linear representation of the $(\nu - 2)$th derivative of the given real part $\alpha(\omega)$; and the changes in slope at these points are denoted by the quantities a_k. The νth derivative of $\alpha(\omega)$ consists of a train of impulses located at the points $\pm\omega_k$ with values given by the coefficients a_k. The integer n equals the total number of impulses for $\omega > 0$.

Evaluation of $\beta(\omega)$ through substitution of the pertinent expression for $v_{-\nu}$ as given in Eqs. 90 is simplified if we observe that the second terms in the right-hand members of $v_{-\nu}$ for $\nu \geq 2$ contribute nothing, and may therefore be dropped. To appreciate the truth of this statement, note first, that these terms introduce corresponding ones into the sum 92 having the form

$$a_k[(\omega - \omega_k)^{\nu-1} + (-1)^\nu(\omega + \omega_k)^{\nu-1}] \tag{93}$$

If ν is even, the bracket yields terms involving only even powers of ω_k up to and including the power $\nu - 2$; and if ν is odd, only odd powers of ω_k up to $\nu - 2$ are present. Summation of the corresponding terms in 92 yields zero, for the following reason.

Without loss in generality we can assume (for the present discussion) that the given even function $\alpha(\omega)$ is finite and nonzero only for $-1 < \omega < 1$, since frequency scaling, subtraction from a constant asymptote, or similar devices can be employed if necessary to achieve this result.* Since any odd-order derivative of $\alpha(\omega)$ must then be zero both for $\omega = 0$ and $\omega = 1$, we can write

$$\int_0^1 \alpha^{(p)}(\omega)\,d\omega = 0 \quad \text{for } p = 2, 4, \cdots \tag{94}$$

where the superscript denotes a derivative of order p. Through repeated integration by parts we can obtain

$$\int_0^1 \alpha^{(p)}(\omega)\,d\omega = \frac{(-1)^q}{q!}\int_0^1 \omega^q \alpha^{(p+q)}(\omega)\,d\omega \tag{95}$$

in which p and q may have any positive integer values including zero. Combining 94 and 95 we have

$$\int_0^1 \omega^q \alpha^{(p+q)}(\omega)\,d\omega = 0 \quad \text{for } p = 2, 4, \cdots \tag{96}$$

or, letting $p + q = \nu$, $q = \nu - p$, this result reads

$$\int_0^1 \omega^{\nu-p}\alpha^{(\nu)}(\omega)\,d\omega = 0 \quad \text{for } p = 2, 4, \cdots \leq \nu \tag{97}$$

Since, for $\omega > 0$, 91 gives

$$\alpha^{(\nu)}(\omega) = \sum_{k=1}^{n} a_k u_0(\omega - \omega_k) \tag{98}$$

the last relation yields the so-called *moment conditions*†

$$\sum_{k=1}^{n} a_k \omega_k^{\nu-p} = 0 \quad \text{for } p = 2, 4, \cdots \leq \nu \tag{99}$$

according to which the summation of terms 93 are now seen to contribute nothing.

We may thus write for the phase function 92

$$\beta(\omega) = \frac{-1}{\pi}\sum_{k=1}^{n} a_k[(\omega - \omega_k)\ln|\omega - \omega_k| + (\omega + \omega_k)\ln|\omega + \omega_k|] \qquad \text{for } \nu = 2 \tag{100}$$

* Notice that α is assumed to be identically zero for $|\omega| \geq 1$.

† These matters as well as other properties of impulse-train representations are more fully discussed in Art. 2, Ch. 15.

$$\beta(\omega) = \frac{-1}{2\pi}\sum_{k=1}^{n} a_k[(\omega - \omega_k)^2 \ln|\omega - \omega_k| - (\omega + \omega_k)^2 \ln|\omega + \omega_k|] \quad \text{for } \nu = 3 \qquad (101)$$

$$\beta(\omega) = \frac{-1}{6\pi}\sum_{k=1}^{n} a_k[(\omega - \omega_k)^3 \ln|\omega - \omega_k| + (\omega + \omega_k)^3 \ln|\omega + \omega_k|] \quad \text{for } \nu = 4 \qquad (102)$$

and so forth

It is usually helpful to have an expression for the phase valid near $\omega = 0$. Evaluation of the linear terms in a Maclaurin expansion for each of the above functions leads to

$$\beta(\omega) = \frac{-2\omega}{\pi}\sum_{k=1}^{n} a_k \ln \omega_k + \cdots \quad \text{for } \nu = 2 \qquad (103)$$

$$\beta(\omega) = \frac{2\omega}{\pi}\sum_{k=1}^{n} a_k\omega_k \ln \omega_k + \cdots \quad \text{for } \nu = 3 \qquad (104)$$

$$\beta(\omega) = \frac{-2\omega}{\pi}\sum_{k=1}^{n} \frac{a_k\omega_k^2}{2} \ln \omega_k + \cdots \quad \text{for } \nu = 4 \qquad (105)$$

By means of these expressions one can readily compute the phase slope at $\omega = 0$.

An application of these results is illustrated in Fig. 9. The piecewise-linear α-function has two break points on the positive frequency scale at

$$\omega_1 = 1.0 \quad \text{and} \quad \omega_2 = 1.25 \qquad (106)$$

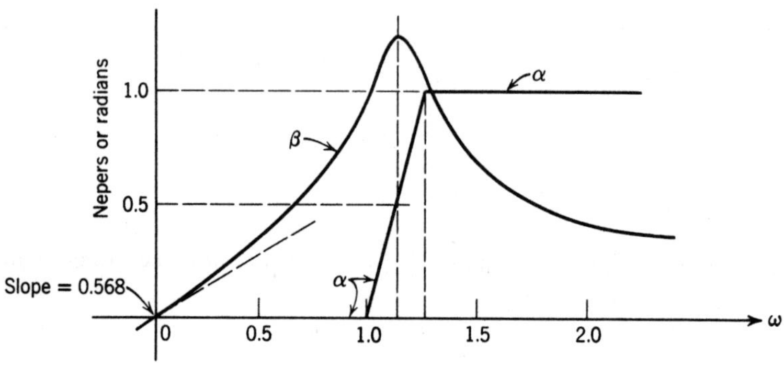

Fig. 9. Piecewise linear attenuation function of $\alpha(\omega)$ and its corresponding phase $\beta(\omega)$ calculated from Eq. 100.

where the changes in slope are respectively

$$a_1 = 4 \quad \text{and} \quad a_2 = -4 \qquad (107)$$

With these values Eq. 100 gives

$$\beta(\omega) = \frac{4}{\pi}[-(\omega - 1)\ln|\omega - 1| - (\omega + 1)\ln|\omega + 1| + (\omega - \tfrac{5}{4})\ln|\omega - \tfrac{5}{4}| + (\omega + \tfrac{5}{4})\ln|\omega + \tfrac{5}{4}|] \qquad (108)$$

and Eq. 103 yields the slope

$$\left(\frac{d\beta}{d\omega}\right)_{\omega=0} = -\frac{2}{\pi}[4\ln 1 - 4\ln\tfrac{5}{4}] = \frac{8}{\pi}\ln\tfrac{5}{4} = 0.568 \qquad (109)$$

From Eq. 108 one readily calculates that

$$\begin{aligned} &\text{for } \omega = 0.5,\ 1.0,\ 1.125,\ 1.25,\ 2.0 \\ &\beta = 0.312,\ 1.003,\ 1.230,\ 1.036,\ 0.411 \end{aligned} \qquad (110)$$

These values are plotted in Fig. 9 where the scale of ordinates is both nepers and radians. In contrast with the pair of functions in Fig. 8, we note that the phase now remains finite since the attenuation function rises at a finite rate instead of being discontinuous. The characteristics in Fig. 9 thus are more realistic.

As a second example, the phase associated with the α-function shown in Fig. 10 is computed and plotted in the same figure. Here, $\alpha(\omega)$ is assumed

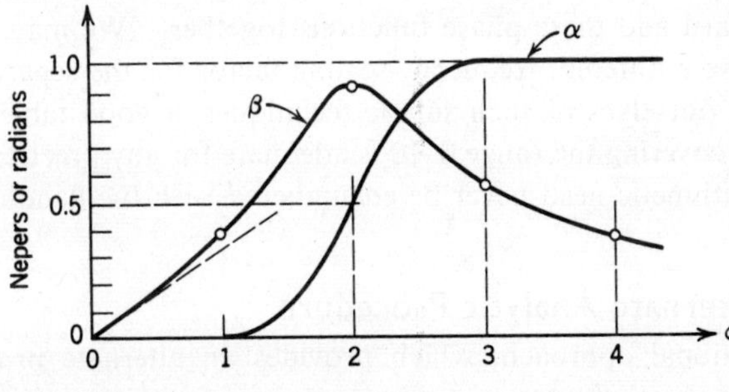

Fig. 10. Loss function $\alpha(\omega)$ whose first derivative is piecewise linear and whose third derivative is an impulse train. The associated phase is, therefore, found from Eqs. 101 and 104.

to have the form of two confluent parabolic arcs. One apex is at $\omega = 1$, the other is at $\omega = 3$, and the point of confluency is at $\omega = 2$. The amplitude of α is again normalized at 1 neper. By inspection, one has: $a_1 = 1$, $a_2 = -2$, $a_3 = 1$, and $\omega_1 = 1$, $\omega_2 = 2$, $\omega_3 = 3$. With these Eq. 101 gives

$$\beta(\omega) = \frac{-1}{2\pi}[(\omega - 1)^2\ln|\omega - 1| - (\omega + 1)^2\ln|\omega + 1| - 2(\omega - 2)^2\ln|\omega - 2| + 2(\omega + 2)^2\ln|\omega + 2| + (\omega - 3)^2\ln|\omega - 3| - (\omega + 3)^2\ln|\omega + 3|] \qquad (111)$$

and for small values of ω Eq. 104 gives

$$\beta(\omega) = \frac{1.05}{\pi}\omega + \cdots \tag{112}$$

It is possible to get a rather good idea of the phase 111 by computing its value only at the points $\omega = 1, 2$, and 3, which are particularly simple points to pick because several terms drop out. Together with the initial slope made evident by 112 one is able to plot the curve shown in Fig. 10 without difficulty. And it is significant to observe that this phase for the assumed α-function is exact.

Use of this method of phase computation is facilitated through the availability of carefully plotted curves (charts) for the functions $x \ln x$, $x^2 \ln x$, $x^3 \ln x$, $\cdots$ over the range $0 < x < 10$, for with an appropriate one of these, the numerical evaluation of an expression like 111 for ten or so points becomes a relatively rapid and effortless procedure. In this connection it is of prime importance to apply frequency scaling to the given problem for which the pertinent frequency range is not likely to lie between 0 and 10 radians/sec. If the actual frequency range covers a wider interval than can conveniently be squeezed into 0–10, we can chop it into handy chunks and compute the phase pertinent to the loss per chunk, and afterward add these phase functions together. We may also find it useful to use a different frequency-scaling factor for the separate chunks. If we avail ourselves of such simple techniques, a good table of natural logarithms covering the range 0–10 is adequate for any practical problem, and our arithmetic need never be encumbered with the handling of large numbers.

5 An Alternate Analytic Procedure

An additional approach, which provides an alternate procedure for determining the conjugate of a given potential function and is applicable to both the algebraic and graphical aspects of this problem, is obtained by introducing a transformation of the independent variable which maps conformally the right half of the s-plane upon the interior of the unit circle. This well-known transformation reads

$$w = \frac{1 - s}{1 + s} \tag{113}$$

As Fig. 11 indicates, the points $s = 0$ and $s = \infty$ correspond respectively with the points $+1$ and -1 in the w-plane, and traversal of the j-axis of the s-plane in the positive direction corresponds to traversal of the unit circle in the clockwise direction. Hence if we write

$$w = e^{-j\phi} = \frac{1 - j\omega}{1 + j\omega} \tag{114}$$

then ϕ and ω are corresponding variables. More particularly, we have

$$\phi = 2\tan^{-1}\omega \qquad \text{or} \qquad \omega = \tan\phi/2 \tag{115}$$

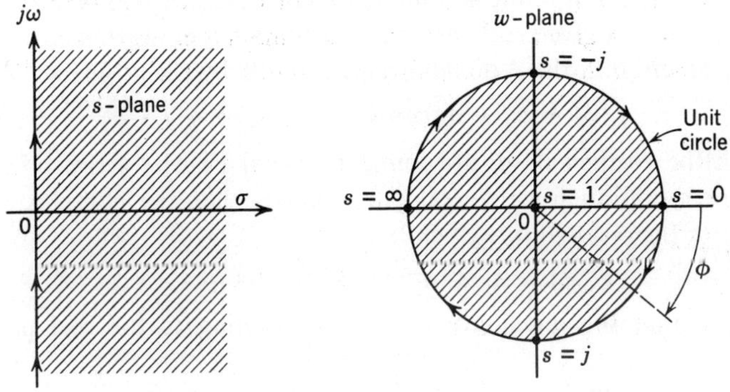

Fig. 11. Graphical interpretation of the change of variable given in Eq. 113.

from which we see that the ω-scale and the ϕ-scale correspond as indicated in Fig. 12. The origins coincide; the points $\omega = \pm 1$ correspond respectively with the points $\phi = \pm\pi/2$, and $\omega = \pm\infty$ becomes $\phi = \pm\pi$. We may regard this transformation of the independent variable from ω to ϕ as a means for changing any given function of ω into a periodic function of ϕ with the period 2π.

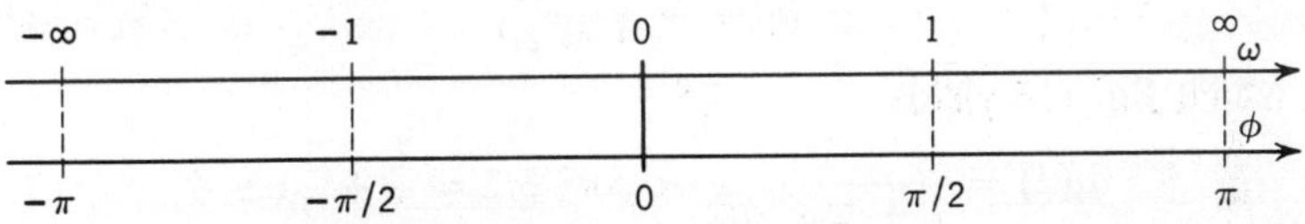

Fig. 12. Correspondence of ω and ϕ scales according to relation 115.

For example, a given plot of the real part $R(\omega)$ of an impedance may be redrawn as a function of ϕ by plotting each ordinate occurring at some ω value at the corresponding value of ϕ found from Eq. 115. As a function of ϕ, $R(\omega)$ has the same sequence of ordinate values; the abscissa values at which they occur are merely changed. We may regard this change of independent variable as a distortion of the function in the abscissa direction only; all of the ordinate values remain the same. We speak of this process as one of transforming the function $R(\omega)$ from the ω-domain to the ϕ-domain.

To any system function $h(j\omega) = h_1(\omega) + jh_2(\omega)$ there corresponds a periodic function in the ϕ-domain of the form

$$q(j\phi) = q_1(\phi) + jq_2(\phi) \tag{116}$$

with the periodic real and imaginary parts $q_1(\phi)$ and $q_2(\phi)$ corresponding respectively to $h_1(\omega)$ and $h_2(\omega)$. This picture allows the problem of finding a conjugate potential function to be stated in extremely simple terms. An even or odd function in the ω-domain remains even or odd respectively in the ϕ-domain. A given real part in the ω-domain is converted into an even periodic function in the ϕ-domain representable by the cosine series

$$q_1(\phi) = a_0 + a_1 \cos \phi + a_2 \cos 2\phi + \cdots \tag{117}$$

in which the coefficients a_k are found by usual Fourier series principles. This representation for the real part implies that the complex system function in the ϕ-domain is given by

$$q(j\phi) = a_0 + a_1 e^{-j\phi} + a_2 e^{-j2\phi} + \cdots \tag{118}$$

and hence that the imaginary part corresponding to the real part 117 reads

$$q_2(\phi) = -a_1 \sin \phi - a_2 \sin 2\phi - \cdots \tag{119}$$

In the ϕ-domain the conjugate of a given potential function may be written down without any computation whatsoever if the given function is in the form of a trigonometric polynomial involving integer multiples of the variable ϕ. In a graphically specified problem one can use ordinary Fourier methods to obtain the desired trigonometric polynomial approximation. A simple illustration of the application of this method to an algebraically given real part is provided by the example

$$h_1(\omega) = \frac{1}{1 + \omega^2} \tag{120}$$

for which Eq. 115 yields

$$q_1(\phi) = \frac{1}{1 + \tan^2 \phi/2} = \cos^2 \phi/2 = \frac{1}{2} + \frac{1}{2} \cos \phi \tag{121}$$

The corresponding imaginary part is

$$q_2(\phi) = -\frac{1}{2} \sin \phi = - \sin \frac{\phi}{2} \cos \frac{\phi}{2} \tag{122}$$

Transforming back to the ω-domain by means of Eq. 115 gives

$$h_2(\omega) = -\frac{\omega}{1 + \omega^2} \tag{123}$$

Using Eqs. 114 and 118, the system function in general is expressible as

$$h(j\omega) = a_0 + a_1\left(\frac{1 - j\omega}{1 + j\omega}\right) + a_2\left(\frac{1 - j\omega}{1 + j\omega}\right)^2 + \cdots \tag{124}$$

or in terms of the complex frequency variable as

$$h(s) = a_0 + a_1\left(\frac{1 - s}{1 + s}\right) + a_2\left(\frac{1 - s}{1 + s}\right)^2 + \cdots \tag{125}$$

When a real-part function is graphically given, this result enables one to obtain a rational approximant for the corresponding system function through use of Fourier methods.

Recalling Eq. 54 of Ch. 6 for the transfer impedance of a constant-resistance lattice network it is recognized that the function

$$\left(\frac{1-s}{1+s}\right)^n \tag{126}$$

is the transfer impedance of n identical lattices in cascade terminated in a 1-ohm resistance, each lattice having the configuration shown in Fig. 13.

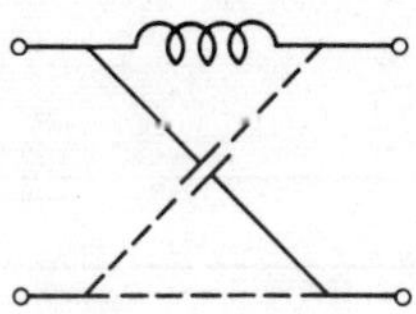

Fig. 13. Configuration of constant-resistance lattice having the transfer function $(1-s)/(1+s)$. n of these in cascade realize the function 126.

It is of interest to observe that if $h(s)$ in Eq. 125 represents a transfer impedance, a possible physical realization for it is given by the configuration shown in Fig. 14 in which the ideal transformer ratios are adjusted to the values of the coefficients $a_0, a_1, a_2, \cdots$. Another arrangement is to have a single cascade of a large number of these lattices (or some equivalent unbalanced delay network) terminated in a resistance load, and tap off voltages at successive junctions with high-resistance voltage dividers, feeding these voltages into the grids of separate tubes and combining their output voltages in series. The fundamental idea involved here (first suggested by Norbert Wiener and Y. W. Lee*) is that any desired transfer characteristic is obtainable and easily adjustable from one basic delay network alone, and that the approximation problem is solved by use of ordinary Fourier methods.

6 Complementary Impedances

As mentioned in Ch. 6 in connection with the synthesis of RC and RL networks, complementary driving-point impedances are such pairs whose sum equals a constant independent of frequency. The value of the constant is usually normalized at 1 ohm, but can have any desired value. It is clear that two impedances complement each other if the sum of their real parts equals a constant and if the sum of their imaginary parts equals zero. From what has been said in this chapter we realize that it is sufficient

* See U.S. Patent Nos. 2,024,900, Dec. 17, 1935; 2,124,599, July 26, 1938; 2,128,257, Aug. 30, 1938.

to require that the sum of their real parts shall equal a constant, since the conjugate of a constant is zero.

In any practical situation it should be obvious that the value of the constant must be finite. This fact implies that only a minimum-reactive impedance can possess a complement, since an impedance with poles on the j-axis has a real part containing impulses which are zero everywhere except at the pole locations where their values are infinite.

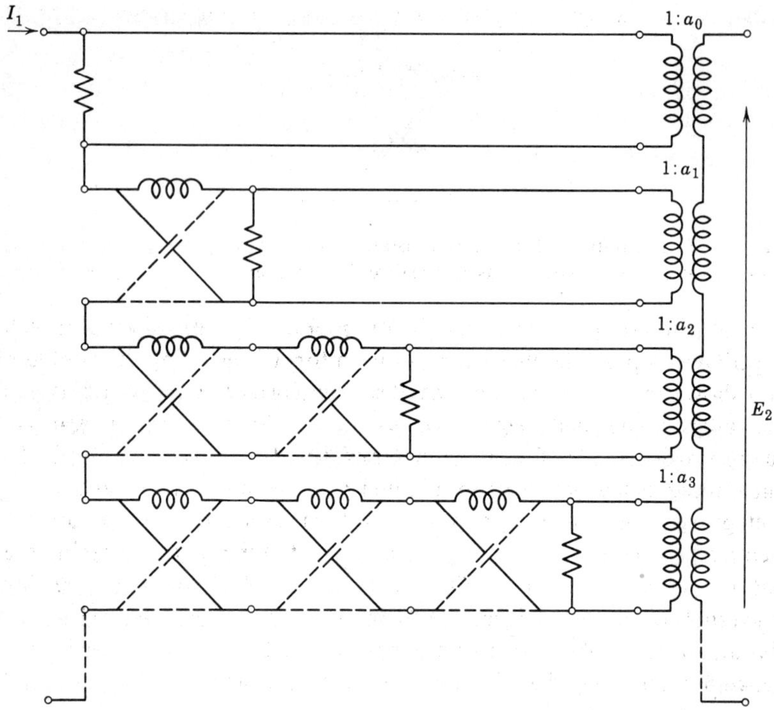

Fig. 14. Realization of the transfer function 125 according to the representation of its terms as an appropriate cascade of lattices shown in Fig. 13.

The construction of a complementary impedance may be carried out either from the real part of the given impedance or from the whole impedance. As an illustration let us consider the real part given by Eq. 15. Since its maximum value is unity we can subtract it from unity and have a remainder that is positive for all values of ω (in general one subtracts the given real part from its maximum value, whatever that is). The real part of the desired complementary impedance, therefore, is

$$1 - \frac{1}{1 + \omega^6} = \frac{\omega^6}{1 + \omega^6} \tag{127}$$

This is the real part given by Eq. 47. The corresponding impedance as expressed by Eq. 49, namely

$$Z_c(s) = \frac{s^3 + \frac{4}{3}s^2 + \frac{2}{3}s}{s^3 + 2s^2 + 2s + 1} \tag{128}$$

is the desired complement of the impedance $Z_0(s)$ in Eq. 22 having the real part 15. The same result can evidently be gotten by subtracting $Z_0(s)$ in Eq. 22 from unity.

One should not jump at the conclusion, however, that a complementary impedance can always be found by subtracting the given one from unity. First, one must make sure that the given impedance is minimum reactive, and second, one must determine the maximum value of its real part. Calling this value R, one can form $R - Z(s)$ with assurance that this difference will be a p.r. function and hence realizable as a driving-point function. One can, of course, subtract $Z(s)$ from a value larger than R, although problems in which complementary impedances are involved usually lead to the most desirable results when this contant has its smallest possible value. It is usually convenient to normalize the given impedance so that this value becomes unity.

Although one can, by essentially the same methods, also construct the complement of a transfer function, there has not been, so far at least, any practical application in which such a function becomes useful. Since the real part of a transfer function for $s = j\omega$ does not need to remain positive, all stable transfer functions possess complements, even those having j-axis poles.

7 Construction of a Transfer Impedance From Its Real Part

As has already been mentioned, the real part of a transfer impedance as a function of ω may have negative as well as positive values. Except for a finite number of impulses, it must be a bounded rational function, but it may otherwise be perfectly arbitrary. A stable rational transfer function may always be found for such a given real part by means of the same methods that are described above for the construction of driving-point functions; and the solution to such a problem is always unique if pertinent impulses are included in the real part.

When the real part is graphically or piecewise analytically specified, the Hilbert transform methods and any techniques derived therefrom are applicable in precisely the form discussed above. In other words, the construction of transfer functions from their specified real parts is accomplished by the same methods applicable to driving-point functions. The only aspect in which the problem differs from the driving-point situation is in the greater generality of the real part—the fact that it is not restricted to positive values.

It may be emphasized that when a transfer function is constructed from its real part, the question as to its possible minimum-phase or non-minimum-phase character does not arise. The process of construction uniquely determines the locations of the zeros as well as of the poles. If the zeros happen to fall in the left half-plane, the function is minimum phase; if some turn out to be in the right half-plane, then the function is nonminimum phase, as explained in Art. 3 above. Whatever the situation may be, the nature of the given real part predetermines it; the method of construction leaves no freedom of choice in this matter.

Regarding the realizability of a transfer function having a real part with negative as well as positive portions, it is interesting to consider for its synthesis the symmetrical lattice as a basic structure. According to Eqs. 44 and 48 of Ch. 6, the transfer impedance or admittance of the lattice in terms of its component impedances or admittances reads

$$z_{12} = \tfrac{1}{2}(z_b - z_a) \tag{129}$$

or

$$y_{12} = \tfrac{1}{2}(y_b - y_a) \tag{130}$$

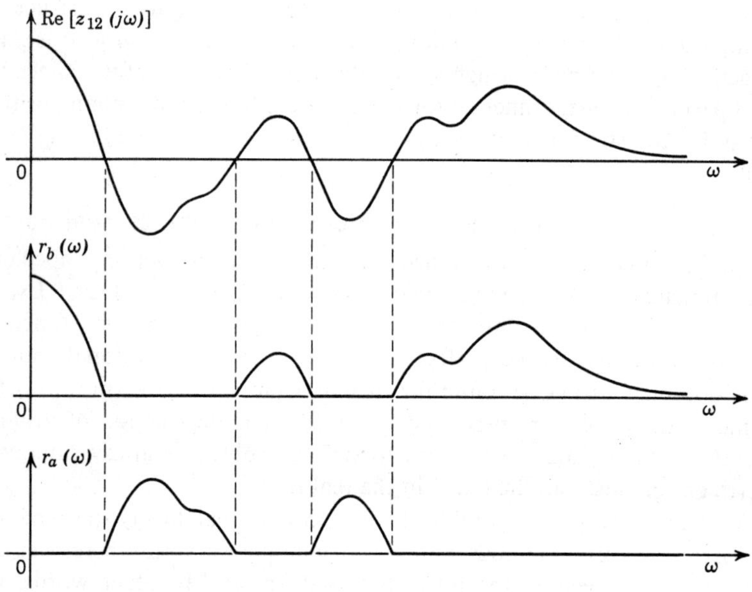

Fig. 15. Real part of an arbitrary transfer impedance is representable as the difference of two real parts r_b and r_a that are nonnegative.

As shown in Fig. 15, a given real part for $z_{12}(j\omega)$ may be represented by the difference between two real parts r_b and r_a, both of which are everywhere positive. These can be used to generate p.r. driving-point impedance functions which (after multiplication by the factor $\frac{1}{2}$) are identified respectively with the impedances z_b and z_a of a lattice structure realizing the transfer impedance with the given real part. In applying this method one need not bother to construct the transfer impedance z_{12} at all; the lattice impedances can be formulated directly from their known real parts.

Since only the difference between the lattice impedances is important, an infinite number of function pairs z_b and z_a exist, all of which yield the same transfer function. In other words, one may theoretically add to both r_b and r_a of Fig. 15 the same arbitrary function r without affecting the resulting transfer impedance or its real part. This flexibility gives one an amount of freedom that is practically useful in achieving desirable network structures and element value distributions, as will be more fully elaborated upon later.

8 Construction of an Impedance From Its Magnitude

If the impedance function is written as in Eq. 7, then its squared magnitude for $s = j\omega$ is given by

$$|Z(j\omega)|^2 = \left[\frac{P(s)\cdot P(-s)}{Q(s)\cdot Q(-s)}\right]_{s=j\omega} = \left(\frac{m_1{}^2 - n_1{}^2}{m_2{}^2 - n_2{}^2}\right)_{s=j\omega} = \frac{C(\omega^2)}{B(\omega^2)} \quad (131)$$

in which $B(\omega^2)$ is the same polynomial as in Eq. 8, and $C(\omega^2)$ is a similar one representing the squared magnitude of $P(j\omega)$ just as $B(\omega^2)$ represents the squared magnitude of $Q(j\omega)$. Construction of the impedance $Z(s)$ follows a simpler pattern than that pertinent to the formation of $Z(s)$ from its real part. The determination of $Q(s)$ from $B(\omega^2)$ is the same as in that procedure, and the determination of $P(S)$ from $C(\omega^2)$ can follow precisely the same pattern; in fact, if $Z(s)$ is a driving-point impedance, it must follow the same pattern since the zeros as well as the poles are required to lie in the left half-plane.

On the other hand, if $Z(s)$ is a transfer impedance, then some or all of the zeros of $P(s)$ may be chosen to lie in the right half-plane unless a minimum-phase transfer function is required. In the latter instance, the procedure leads to a unique result, as it does in the case of a driving-point impedance. When no minimum-phase requirement is placed upon the transfer impedance, then a number of possible solutions can be constructed according to the distribution of the zeros of $P(s)$ between the left and right half-planes.

If a driving-point impedance is involved, it is important to observe that a p.r. function does not necessarily exist for every specified magnitude function but only when the associated angle does not exceed $\pi/2$ radians in absolute value. Whether or not this condition is met depends upon the amount of variation and the rate of variation in the magnitude as a function of frequency. This point is appropriately illustrated in Figs. 8 and 9. Thus the curves in Fig. 8 show that the magnitude variation can never be abrupt, no matter how small it may be in amount, for the phase β contains logarithmic infinities for any nonzero value of the attenuation A.

When the magnitude variation is gradual, as in Figs. 9 and 10, then the maximum value of the associated angle is finite. For any amount of magnitude variation there is a certain rate of variation (best expressed in decibels per octave) for which the maximum angle does not exceed $\pi/2$; and for any given rate of magnitude variation there is a maximum amount of variation. For the rate of variation in Fig. 9 (which is 4 nepers per octave or 34.7 db per octave), the maximum angle is 1.23 radians or about 70.5° for an amount of magnitude variation corresponding to 1 neper or 8.68 db or an actual magnitude ratio of 2.718:1. For this same rate of variation we could increase the magnitude ratio up to 90/70.5 times 2.718 or a ratio of 3.47:1. For a larger magnitude ratio the rate of variation would have to be decreased, but not simply in proportion, since the relationship between α and β is more subtle than that. At any rate, one can in any specific instance determine whether a p.r. function corresponds to a stated magnitude function. When the latter is rational, one might as well form $Z(s)$ and then make a Sturm test for its real part.

9 Construction of a Driving-Point Impedance From Its Angle*

Starting from Eq. 7, and defining

$$M(s) = \tfrac{1}{2}[P(s)\cdot Q(-s) + P(-s)\cdot Q(s)] = m_1m_2 - n_1n_2 \tag{132}$$

$$N(s) = \tfrac{1}{2}[P(s)\cdot Q(-s) - P(-s)\cdot Q(s)] = m_2n_1 - m_1n_2 \tag{133}$$

the angle θ of the impedance $Z(j\omega)$ is determined from the relation

$$j \tan \theta = \frac{N(j\omega)}{M(j\omega)} \tag{134}$$

It is clear that $N(s)$ is an odd polynomial in s, and $M(s)$ is an even polynomial in s. Tan θ, therefore, is an odd rational function of ω, of the form

$$\tan \theta = \frac{\omega(G_0 + G_1\omega^2 + \cdots + G_n\omega^{2n})}{A_0 + A_1\omega^2 + \cdots + A_n\omega^{2n}} \tag{135}$$

in which the denominator polynomial is the same as the numerator polynomial in Eq. 8 for the real part.

In a practical problem in which a driving-point impedance $Z(s)$ is to be

* See "Synthesis of a Finite Two-Terminal Network Whose Phase Angle Is a Prescribed Function of Frequency," Harold Adler, S.M. Thesis, Massachusetts Institute of Technology, 1948.

found having an angle θ with a prescribed frequency variation (within the range $-\pi/2$ to $\pi/2$, of course), it is necessary first to determine in some manner the polynomials in expression 135 such that θ versus ω is acceptably close to the desired behavior. The solution to this approximation problem we shall not discuss at the moment; rather we will concern ourselves only with the question of how to construct $Z(s)$ when the polynomials in the rational function 135 are known.

From the polynomials in Eq. 135 we get $M(s)$ and $N(s)$ by setting $j\omega = s$ or $\omega^2 = -s^2$. Next we observe from Eqs. 132 and 133 that

$$M(s) + N(s) = P(s)\cdot Q(-s) \tag{136}$$

From this result we see that the zeros of the polynomial $M + N$ are those of $P(s)$ and those of $Q(-s)$ which must lie respectively in the left and in the right half-planes. The formation of the polynomials $P(s)$ and $Q(s)$ is thus clear. We solve the equation $M + N = 0$ for its roots and then assign those in the left half-plane to $P(s)$ and those in the right half-plane to $Q(-s)$. With the zeros of these polynomials known, the impedance $Z(s)$ is determined except for a constant multiplier which remains arbitrary since it has no influence upon the angle θ.

In assigning left half-plane zeros to $P(s)$ and right half-plane zeros to $Q(-s)$, we are again using the j-axis as a sorting boundary, and so the question arises: How do we assign zeros that fall upon the j-axis? Before we try to answer this question, let us understand why zeros might fall upon the j-axis in the first place, or whether the polynomial $M + N$ will ever have such zeros.

Since j-axis zeros of $M + N$ are j-axis zeros of $P(s)$ or $Q(s)$, they are j-axis zeros or poles of $Z(s)$. Now what effect do j-axis zeros or poles of $Z(s)$ have upon its angle? To answer this question we visualize the behavior of the angle of a frequency factor $(s - s_0)$ for $s_0 = j\omega_0$ as $s = j\omega$ passes over the critical frequency s_0. Clearly this angle remains constant at the value $-\pi/2$ so long as $\omega < \omega_0$ and it remains constant at the value $+\pi/2$ for $\omega > \omega_0$. At the critical frequency ω_0 the angle has a discontinuity of π radians. We cannot be sure, of course, whether the jump in θ at this point is an abrupt advance or retardation, since it is not possible to decide whether the frequency factor suddenly rotates through π radians in the clockwise or in the counterclockwise direction.

This indeterminacy in the direction of angular rotation disappears if the critical frequency s_0, instead of being on the j-axis, lies slightly to the left or to the right of it, or if, as the variable point $s = j\omega$ reaches the j-axis point s_0, it passes the latter on a vanishingly small semicircular detour that lies in the left or in the right half-plane. The point of view that is consistently taken in this regard (in dealing with analysis or synthesis problems) is the one shown in Fig. 4 where the small semicircular detours are assumed to lie in the right half-plane because of the manifestly obvious requirement that the response of a network cannot precede the excitation causing it. The angular rotation of the frequency factor $(s - s_0)$ is thus always counterclockwise as $s = j\omega$ moves in the direction of increasing ω values.

A j-axis zero in $Z(s)$ therefore introduces an abrupt positive increment of π radians in the angle θ, whereas at a pole the angle is suddenly retarded by this value. If we recall that a j-axis factor in the numerator or denominator polynomial of $Z(s)$ must be a common factor of m_1 and n_1 or of m_2 and n_2, we observe from Eqs. 132 and 133 that such factors will always be contained in both M and N, and hence will cancel out of expression 134 for $\tan\theta$. The algebraic expression for θ will therefore not yield any sudden jumps in θ, even though the impedance function $Z(s)$ produces them by virtue of its j-axis zeros or poles.

In this connection it is not correct to say that since the determination of θ requires the evaluation of an antitangent function, and the latter is multivalued, the expression for θ provides as many jumps of π in either direction as may be needed, and it is merely up to the computer to add these in as required. Such an attitude is equivalent to implying that a network provides an infinite number of values for the angle θ and we are free to choose whatever value we like, which is sheer nonsense. The network is not confused about this multivaluedness of the analytic expression for the angle; it is merely the network designer who at times becomes confused about such things. Analytically the mechanism for avoiding this kind of confusion is provided by the Riemann surface. If we remain upon the same sheet of this surface, the function θ cannot become discontinuous. A jump of π radians in the value of θ can come about only through jumping from one sheet of the Riemann surface to the next.

If $Z(s)$ is a p.r. function with a j-axis zero at some point $s_0 = j\omega_0$, then clearly this is a point where the decreasing angle θ just reaches the value $-\pi/2$ and would proceed to become less than this value were it not for the fact that the effect of the zero is suddenly to jump θ ahead to the value $+\pi/2$ so that its continued decrease does not violate the restriction $-\pi/2 \le \theta \le \pi/2$. A j-axis pole in $Z(s)$ similarly is a point where an increasing angle θ just reaches the value $+\pi/2$ and is prevented from exceeding this value by a sudden retardation of π radians. However, when θ is calculated from formula 134 or 135, the necessary jumps which prevent it from violating the condition $-\pi/2 \le \theta \le \pi/2$ are absent, and so it does violate this condition.

But this violation is only apparent, for in the construction of $Z(s)$ from a given function 135 we insert j-axis zeros and poles at the points where they are needed. The polynomial $M(s) + N(s)$ in Eq. 136 never has any j-axis zeros because these cancel in the ratio 134 or 135, and so we are not faced with the question as to which of the two polynomials $P(s)$ or $Q(-s)$ these zeros should be allotted to. We do, however, have to supply j-axis zeros to $P(s)$ or to $Q(s)$ at points where the θ-function computed from Eq. 135 threatens to violate the condition $|\theta| \leq \pi/2$. Thus, wherever this function reaches the value $\pi/2$ plus or minus an integer multiple of π radians, we insert a j-axis zero in $Q(s)$ or in $P(s)$ respectively according to the positiveness or negativeness of the slope $d\theta/d\omega$ at that point (θ being the function 135 which has no discontinuities).

A theoretically useful collateral view of this situation is provided by the requirement that residues in j-axis poles of p.r. functions must be real and positive. For a j-axis pole of $Z(s)$ at $s = s_\nu$ the residue is

$$k_\nu = [(s - s_\nu)Z(s)]_{s=s_\nu} = \left[\frac{(s^2 - s_\nu^2)Z(s)}{s + s_\nu}\right]_{s=s_\nu} \tag{137}$$

Since

$$\measuredangle(s + s_\nu) = \measuredangle j(\omega + \omega_\nu) = +\pi/2 \tag{138}$$

it is necessary that

$$\measuredangle[(s^2 - s_\nu^2)Z(s)]_{s=s_\nu} = +\pi/2 \tag{139}$$

in order that the net angle of expression 137 be zero. The function $(s^2 - s_\nu^2)Z(s) = Z'(s)$ is the partially constructed impedance before insertion of its pair of conjugate j-axis poles by assigning the frequency factor $(s^2 - s_\nu^2)$ to $Q(s)$. At the point $s = j\omega_\nu$, the angle of $Z'(s)$ has reached the value $\pi/2$. If we assign the factor $(s^2 - s_\nu^2)$ to $P(s)$, then the admittance $1/(s^2 - s_\nu^2)Z'(s)$ has a pole at $s = s_\nu$ with the residue

$$k_\nu = \left[\frac{s - s_\nu}{(s^2 - s_\nu^2)Z'(s)}\right]_{s=s_\nu} = \left[\frac{1}{(s + s_\nu)(s^2 - s_\nu^2)Z(s)}\right]_{s=s_\nu} \tag{140}$$

which is negative because its angle, according to Eqs. 138 and 139, is $-\pi$. We have no choice but to assign the factor $(s^2 - s_\nu^2)$ to $Q(s)$.

An entirely analogous argument applies if $Z(s)$ has a j-axis zero instead of a pole. We simply replace $Z(s)$ by $1/Z(s)$ in the above reasoning, the significant point being that one assigns the j-axis factor $(s^2 - s_\nu^2)$ to $Q(s)$ or to $P(s)$ respectively according to whether the resulting function $Z(s)$ or $Y(s) = 1/Z(s)$ has a positive residue at the pertinent j-axis pole. Since in one of these functions the residue turns out to be negative, the assignment of the factor $(s^2 - s_\nu^2)$ is always unique.

As an example, let us consider the function

$$\tan\theta = \frac{10\omega(16 - \omega^2)}{(1 - \omega^2)(4 - \omega^2)(9 - \omega^2)} \tag{141}$$

for which

$$M(s) = (s^2 + 1)(s^2 + 4)(s^2 + 9), \qquad N(s) = 10s(s^2 + 16) \tag{142}$$

The function θ versus ω resulting from Eq. 141 is shown plotted in Fig. 16 where the dashed portions are obtained directly from 141 and the solid curve depicts the actual behavior of the angle θ for the p.r. $Z(s)$ with its j-axis zeros and poles. We see clearly that the points $\omega = 1$ and $\omega = 3$ must be poles of $Z(s)$ and the point $\omega = 2$ must be a zero. To construct the polynomials $P(s)$ and $Q(s)$ we must determine the roots of the equation

$$M(s) + N(s) = s^6 + 14s^4 + 10s^3 + 49s^2 + 160s + 36 = 0 \tag{143}$$

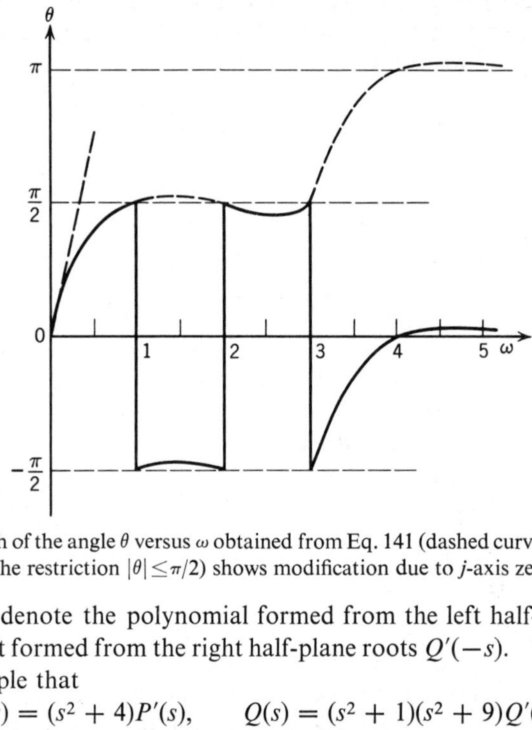

Fig. 16. Graph of the angle θ versus ω obtained from Eq. 141 (dashed curve); solid curve (which fulfills the restriction $|\theta| \leq \pi/2$) shows modification due to j-axis zeros and poles.

Suppose we denote the polynomial formed from the left half-plane roots $P'(s)$ and that formed from the right half-plane roots $Q'(-s)$. Then we see in this example that

$$P(s) = (s^2 + 4)P'(s), \qquad Q(s) = (s^2 + 1)(s^2 + 9)Q'(s) \tag{144}$$

The p.r. driving-point impedance having the angle θ is $Z(s) = P(s)/Q(s)$, and its constant multiplier is arbitrary.

10 Construction of a Transfer Impedance From Its Angle

Since a transfer impedance does not have its angle restricted to the range $-\pi/2$ to $+\pi/2$, its construction from a specified angle is simpler because the insertion of j-axis factors is not needed. Relations 132 through 136 apply as before, except that it is customary with transfer functions to define lag angles as being positive. Thus if β is the angle by which the output lags the input, then in the pertinent relations above we replace θ by $-\beta$, so

$$j \tan \beta = \left[\frac{-N(s)}{M(s)}\right]_{s=j\omega} \tag{145}$$

but Eqs. 132 and 133 remain intact, and hence, so does Eq. 136.

Assignment of the zeros of $M + N$ in Eq. 136 to $P(s)$ and to $Q(-s)$ depends upon whether the desired transfer impedance is to have minimum–phase-shift character or not. If it is, then we allot left half-plane zeros to $P(s)$ and right half-plane zeros to $Q(-s)$, just as in the construction of driving-point impedances; whereas if the transfer impedance is not to be of the minimum–phase-shift variety, then we can also allot any right half-plane zeros to $P(s)$, being mindful, however, of the requirement that the degree of $P(s)$ cannot exceed that of $Q(s)$ by more than unity.

The terms minimum phase and nonminimum phase may be somewhat misleading in this connection since we are starting with a specified phase function $\beta(\omega)$, and this function will be the phase shift of the resulting transfer function no matter how we allot zeros to $P(s)$ and to $Q(-s)$. In other words, the minimum-phase transfer function and the nonminimum-phase transfer function constructed by this procedure have identically the same phase-shift properties. Their amplitude versus frequency behavior will, however, be different depending upon the chosen disposition of the zeros of $M + N$ with regard to the polynomials $P(s)$ and $Q(-s)$.

For example, in the design of a delay network where a linear phase characteristic is desired, relation 145 shows that the zeros of $-N(s)$ and $M(s)$ may be chosen to alternate on the j-axis like the zeros and poles of the tangent function. In this case $M - N$ is a Hurwitz polynomial, and $M + N$ has zeros only in the right half-plane. If we allot all of these to $Q(-s)$, the resulting minimum-phase transfer function has all of its zeros at $s = \infty$; its numerator polynomial reduces to a constant, and its amplitude characteristic turns out to be a monotonically and rapidly decreasing function with increasing ω. On the other hand, if we allot some (up to half) of the zeros to $P(s)$, then the function is nonminimum phase, and its amplitude is a less rapidly decreasing function and may no longer be monotonic in its behavior. The choice here will depend upon the desired transient response which is markedly influenced by the character of the amplitude characteristic. These matters will be more fully discussed later on in connection with the design of delay networks.

11 Extent to Which Magnitude and Angle of a Transfer Function Are Independently Specifiable Over a Finite or Infinite Frequency Interval

In practice it is sometimes desirable to construct a transfer function having either a constant magnitude but an arbitrary phase variation or a constant phase and an arbitrary amplitude variation. Considering the expression $Z(j\omega) = P(j\omega)/Q(j\omega)$ we wish to specify on the one hand that $P(j\omega)$ should have the same magnitude as $Q(j\omega)$ for all ω, and on the other that $Q(j\omega)$ should have the same angle as $P(j\omega)$, except for an additive constant. Let us see to what extent such conditions can be met.

Representing a polynomial as a product of frequency factors, and visualizing these as vectors in the complex frequency plane, it is fairly evident (perhaps with the help of an appropriate sketch) that for any ω

$$Q(-j\omega) \text{ has the same magnitude as } Q(j\omega) \tag{146}$$

$$\measuredangle[1/P(-j\omega)] = \measuredangle[P(j\omega)] - n\pi \tag{147}$$

where n is an integer. Hence we conclude that a transfer function with constant magnitude and variable phase is possible if we can choose

$$P(s) = Q(-s) \tag{148}$$

whereas one having a constant phase and variable amplitude would require

$$Q(s) = 1/P(-s) \tag{149}$$

The first of these conditions we can easily fulfill. The second, although impossible because it requires one polynomial to be the reciprocal of another, may be approximated as closely as we please over a finite frequency interval.

Substituting Eq. 148 into Eq. 136 gives

$$M(s) + N(s) = Q^2(-s) = (m_2 - n_2)^2 \tag{150}$$

or

$$M(s) = m_2^2 + n_2^2, \qquad N(s) = -2m_2n_2 \tag{151}$$

With these expressions for M and N and the use of some familiar trigonometric identities, Eq. 145 gives

$$j \tan \frac{\beta}{2} = \left[\frac{n_2(s)}{m_2(s)}\right]_{s=j\omega} \tag{152}$$

The kind of $\beta(\omega)$ function obtainable is limited only by the restriction that the zeros of m_2 and n_2 must be simple and must separate each other on the j-axis; $Q(s) = m_2 + n_2$ must be a Hurwitz polynomial.

If we take the condition 149 literally, the expression for the impedance becomes

$$Z(s) = P(s) \cdot P(-s) \tag{153}$$

which is a polynomial instead of a quotient of polynomials. If $P(s)$ has the degree n, then the transfer function

$$Z^*(s) = \frac{P(s) \cdot P(-s)}{(1 + s/a)^{2n}} \tag{154}$$

with a sufficiently large positive real value for a, approximates $Z(s)$ in Eq. 153 as closely as we wish over a finite frequency interval $0 < \omega < \omega_0$, and is realizable.

A practical limitation in the use of this function arises from the fact that for $s \to \infty$

$$Z^*(s) \to a^{2n} \tag{155}$$

which turns out to be a rather large number unless ω_0 is moderate and the required degree with which $Z^*(s)$ needs to approximate $Z(s)$ in Eq. 153 is also moderate. The significance of this result is that since the high-frequency response is practically limited in magnitude, the low-frequency response over the band $0 < \omega < \omega_0$ by comparison with the value 155 becomes extremely small.

In other words, the constant multiplier that is needed in order that the impedance function 154 may have a realistic asymptotic magnitude for $s \to \infty$ is of the order of $1/a^{2n}$, which yields a response over the useful frequency range that is impractically small unless the value of a is not too large. In any event, one will pay for the privilege of quasi-independent amplitude control by accepting a considerable loss in gain.

Further discussion of this topic, which evidently is closely related to the process of constructing an impedance function to approximate a desired behavior in an acceptable manner, is given in Art. 10, Ch. 14.

PROBLEMS

1. Given

$$\mathrm{Re}\,[Z(j\omega)] = \frac{5\omega^8 - 40\omega^6 + 150\omega^4 - 100\omega^2 + 500}{(4\omega^4 + 16)^2}$$

by using Sturm's theorem, determine whether this is positive for all values of ω, and if so find $Z(s)$.

2. Given

$$\mathrm{Re}\,[Z(j\omega)] = \frac{\omega^2}{1 + \omega^4}$$

find $Z(s)$. Is the complementary impedance $1 - Z(s)$ physically realizable?

3. A lossless network terminated in 1 ohm has a transfer impedance whose squared magnitude is given by

$$|Z_{12}|^2 = \frac{\omega^4 + \omega^2}{\omega^6 + 1} \leq 1 \text{ for all } \omega$$

Determine the functions z_{11}, z_{22}, z_{12} for this network such that the driving-point impedance becomes minimum reactive.

4. Given

$$f(\omega^2) = \frac{1 + \omega^2}{1 + \omega^2 + \omega^4}$$

(*a*) If $f(\omega^2)$ is the real part of an impedance $Z(s)$, which may be either a driving-point or transfer function, find $Z(s)$.

(*b*) If $f(\omega^2)$ is the squared magnitude of $Z(s)$, find $Z(s)$. Is the latter p.r.?

5. The transfer impedance of a lossless two terminal-pair network terminated in a 1-ohm resistance is

$$Z_{12}(s) = \frac{1}{s^3 + s^2 + 2s + 1}$$

Find the minimum-reactive input impedance $Z_1(s)$.

6. The transfer impedance of a lossless network terminated in a 1-ohm resistance is

$$Z_{12} = \frac{1}{s^5 + s^4 + 4s^3 + 3s^2 + 3s + 1}$$

Determine the corresponding driving-point impedance having no poles on the imaginary axis.

7. Which of the following statements are correct? (There are some that are incorrect.)

(*a*) When the driving-point impedance of a lossless network terminated in 1 ohm is minimum reactive, then $(z_{11}z_{22} - z_{12}^2)$ has only simple poles.

(*b*) When the transfer impedance of a network has zeros only in the left half of the s-plane, then the network has minimum–phase-shift properties.

(*c*) When a network has minimum–phase-shift properties, then its driving-point impedance is minimum reactive.

(*d*) A rational function in ω^2 with real coefficients, having positive values for all real values of ω, can represent the squared magnitude of a transfer function only if its numerator and denominator are of like degree.

(*e*) If a real part of a transfer impedance is positive for all real frequencies, then the network has minimum–phase-shift properties.

(*f*) The converse of the previous statement is true.

(*g*) The necessary and sufficient condition that a rational function of s involving real coefficients be realizable as a transfer impedance is that its poles be restricted to the left half-plane.

(*h*) To be realizable, a driving-point impedance must have simple zeros and poles restricted to the left half-plane.

(*i*) The zeros and poles of a transfer function for an RC-network must be simple and must alternate along the negative real axis of the s-plane.

8. For which of the following networks is it possible to obtain a purely resistive input impedance through an appropriate connection of a complementary network at the driving point? Carry out the desired procedure where possible.

9. The adjacent s-plane sketch shows the pole-zero plot of a function

$$f(s) = \frac{A(-s^2)}{B(-s^2)}$$

having six poles and two zeros as shown. Let $f(s) \to 4/s^4$ for $s \to \infty$, and

$$\frac{A(\omega^2)}{B(\omega^2)} = \text{Re}\,[Z(j\omega)]$$

(*a*) Construct the impedance $Z(s)$ and synthesize it in the form of a lossless two terminal-pair network terminated in resistance.

(*b*) For this situation construct the transfer impedance $Z_{12}(s)$. Synthesize it through developing z_{22} into a ladder so as to control the zeros of z_{12} and produce the same network as before.

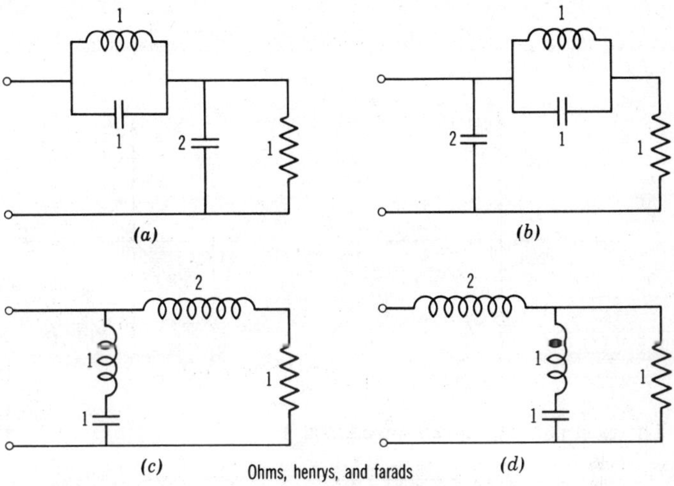

Prob. 8.

(*c*) If you wanted to connect a complementary network in series with the input so as to produce a constant-resistance input impedance, how would you go about doing this if the gain (or constant multiplier) for the resulting transfer function had to be maximized?

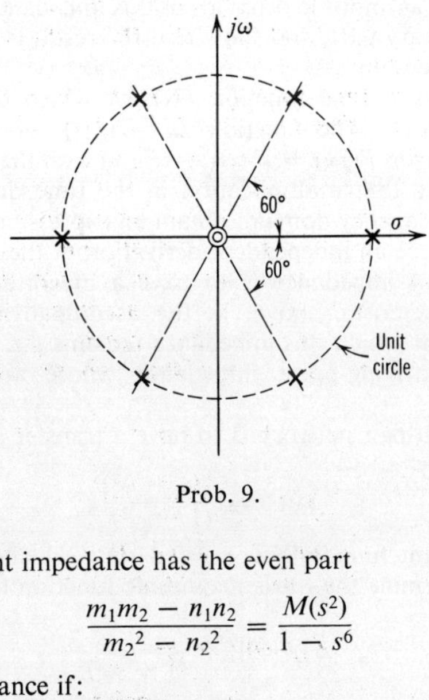

Prob. 9.

10. A driving-point impedance has the even part

$$\frac{m_1 m_2 - n_1 n_2}{m_2{}^2 - n_2{}^2} = \frac{M(s^2)}{1 - s^6}$$

Construct the impedance if:

(*a*) The zeros of this even part are all at $s = \infty$.

(*b*) The finite even-part zeros are given by the patterns shown in the *s*-plane sketches.

In each of these three cases the constant multiplier remains arbitrary.

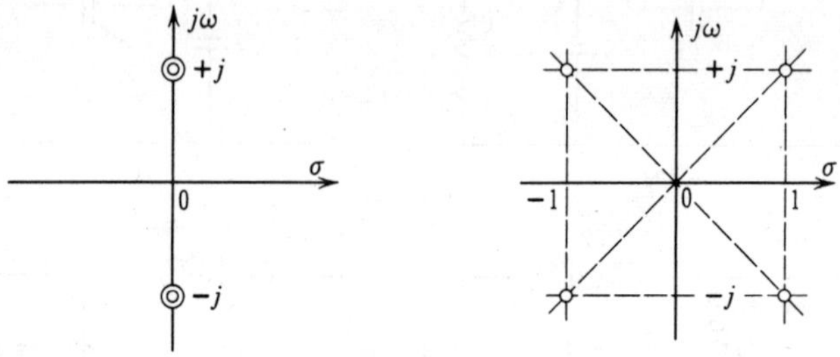

Prob. 10.

11. The even part of a given impedance is

$$f(s) = \frac{s^6 + s^4 - s^2 - 1}{(s^4 + 5s^2 + 4)^2 - (4s^3 + 4s)^2}$$

Construct this impedance function.

12. Without the aid of tables or any formal integration process evaluate the integral

$$\int_0^\infty \frac{\omega^n \, d\omega}{1 + \omega^6} \quad \text{for } n = 0, 2, 4$$

13. A driving-point impedance has a real part $R(\omega) = 1$ for $-\omega_1 < \omega < \omega_1$, and is substantially zero otherwise. Evaluate the pertinent shunt capacitance which dominates the asymptotic behavior of this impedance by means of expression 82 for its imaginary part, and show that the result is in agreement with the resistance-integral theorem.

14. Consider an even time function $f(t)$ for which the Fourier transform $F_1(\omega)$ is even and real. The function $2u_{-1}(t)f(t)$, which is zero for $t < 0$, has a complex transform $F(j\omega) = F_1(\omega) + jF_2(\omega)$ with the same real part as that of $f(t)$. Recognizing that multiplication in the time domain corresponds to convolution in the frequency domain, obtain an expression for $F(j\omega)$ in terms of $F_1(\omega)$ and thus produce an independent derivation of the Hilbert transform.

15. A driving-point impedance is to have a magnitude that approximates closely to the characteristic shown in the accompanying sketch. Find the smallest value of δ for which this impedance remains p.r.

16. Construct a driving-point impedance whose angle θ is defined by $\tan\theta = \omega^5$.

17. A two terminal-pair network is to have a transfer impedance $Z_{12}(s)$ with a lag angle defined by

$$\tan\beta = \frac{\omega(4 - \omega^2)}{(1 - \omega^2)}$$

Construct all pertinent functions $Z_{12}(s)$ and state whether they are minimum phase or not. Determine the *j*-axis magnitude function for each.

18. Given

$$|Z_{12}(j\omega)|^2 = \frac{1}{(1 + \omega^2)^4}$$

construct $Z_{12}(s)$. Is it a minimum-phase transfer function?

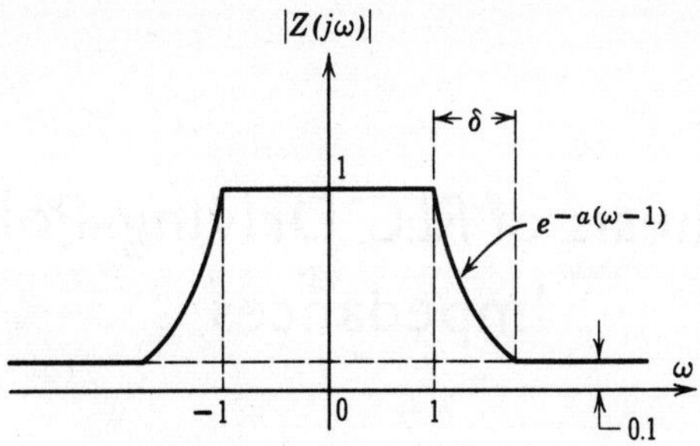

Prob. 15

19. If the function given in Prob. 18 is the j-axis real part of a driving-point impedance, construct the latter.

20. Using the result of the preceding problem, find the impedance whose j-axis real part is given by

$$\frac{1 - \omega^2 + \omega^4}{(1 + \omega^2)^4}$$

Is it p.r.?

21. Given

$$|Z_{12}(j\omega)|^2 = \frac{(4 + \omega^2)^4}{(1 + \omega^2)^4}$$

construct all transfer impedances having this squared magnitude. Sketch the pole-zero pattern for each and indicate whether or not it is minimum phase. Which one has the same phase as the transfer impedance of Prob. 18?

22. Given a minimum-phase transfer function, under what conditions is it possible to alter its j-axis magnitude variation without altering the phase?

C H A P T E R N I N E

Synthesis of *RLC* Driving-Point Impedances

1 Introductory Remarks

The problem of showing that the p.r. character of a rational function is in general sufficient to insure its realization as the driving-point impedance or admittance of a lumped, linear, finite, passive, bilateral network,* and of developing synthesis procedures that lead to physical network configurations including the determination of their element values, is one that has many facets, presenting situations that require a considerably deeper understanding of network function properties and collateral synthesis techniques than one might expect from an understanding of the two-element-kind synthesis problems alone. With few exceptions, all of the background provided by the preceding discussions is found to be essential to a clear understanding of the general *RLC* driving-point synthesis problem, which for this reason is not considered before the present basis for its presentation is established.

In order to get at once to the heart of the problem and to show why it presents difficulties of a new and profound nature, we can do no better than to study the general *RLC* situation from the standpoint of the normal co-ordinate concept which incidentally lends physical reality to our interpretive methods in terms of the natural modes of the network. Thus let us see to what extent we can carry over the methods applied to lossless networks in the discussions of Art. 7 of Ch. 3.

It is shown there that the Foster realization procedures, which are applicable in substantially unaltered form to all three of the two-element-kind networks, may be interpreted as leading to network configurations in which the normal co-ordinates become a reality. As we approach the general *RLC* synthesis problem, the question uppermost in our minds is whether essentially the same procedures in a logically extended form will again prove successful. The normal co-ordinate interpretation of our present problem shows at once that such simple tactics are inadequate; it

* We may use the abbreviation: LLFPB network.

shows basically why they are inadequate; it yields a canonic network configuration in terms of which an *RLC* driving-point impedance is always realizable; and it demonstrates the sufficiency of the positive definite character of the quadratic forms, F, T, V as realizability criteria.

In matrix notation and on the loop basis these forms are given by

$$2F = \underline{i} \times [R] \times i] \tag{1}$$

$$2T = \underline{i} \times [L] \times i] \tag{2}$$

$$2V = \underline{q} \times [S] \times q] \tag{3}$$

in which $[R]$, $[L]$, and $[S]$ are the familiar resistance, inductance, and elastance parameter matrices and $i]$ and $q]$ are column matrices for the loop currents and charges (compare with Eqs. 49, 50, and 51 of Ch. 3 for *LC* circuits). As is clear from the discussion in Art. 7 of Ch. 3, we can determine a real nonsingular transformation of the variables

$$i] = [A] \times i'], \qquad q] = [A] \times q'] \tag{4}$$

with the matrix

$$[A] = \begin{bmatrix} 1 & 1 & \cdots & 1 \\ a_{21} & a_{22} & \cdots & a_{2l} \\ \cdot & \cdot & \cdots & \cdot \\ a_{l1} & a_{l2} & \cdots & a_{ll} \end{bmatrix} \tag{5}$$

that simultaneously carries F and V over into sums of squares. In terms of the new variables, the quadratic forms 1, 2, and 3 become

$$2F = \underline{i'} \times [R'] \times i'] \tag{6}$$

$$2T = \underline{i'} \times [L'] \times i'] \tag{7}$$

$$2V = \underline{q'} \times [S'] \times q'] \tag{8}$$

in which the parameter matrices are given by

$$[R'] = [A]_t \times [R] \times [A] \tag{9}$$

$$[L'] = [A]_t \times [L] \times [A] \tag{10}$$

$$[S'] = [A]_t \times [S] \times [A] \tag{11}$$

The new resistance and elastance parameter matrices $[R']$ and $[S']$ are diagonal in form, but the inductance parameter matrix $[L']$ is not, because three quadratic forms cannot in general be simultaneously reduced to sums of squares by a real nonsingular transformation. It is not possible in the general *RLC* case to define normal co-ordinates, much less realize them in a physical network. The closest we can come to accomplishing this end is to construct a network corresponding to the matrices 9, 10, and 11. Since the first of the transformation Eqs. 4 reads

$$i_1 = i_1' + i_2' + \cdots + i_l' \tag{12}$$

we see that this network has the form shown in Fig. 1 in which R_{kk}' and

S_{kk}' are the diagonal elements of the matrices $[R']$ and $[S']$, and each inductance is mutually coupled with all other inductances, the values of the self- and mutual inductances being given by the elements of the nondiagonal matrix $[L']$.

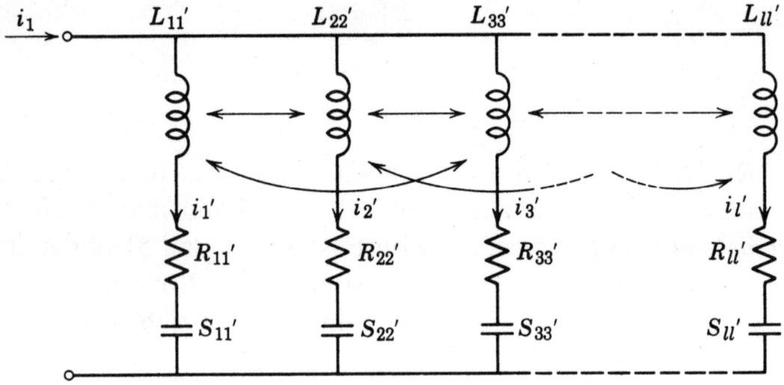

Fig. 1. Canonic realization possible for any p.r. driving-point impedance.

The positive definite character of the given quadratic forms 1, 2, and 3, together with the real nonsingular character of the matrix $[A]$, guarantees the positive realness of all the R_{kk}', S_{kk}', and L_{kk}' values as well as the physical realizability of the mutual inductive couplings since these depend only upon the condition that the instantaneous value of the stored magnetic energy be positive for all possible real values of the branch currents. Hence the circuit in Fig. 1 is a canonic form in which the network corresponding to any realizable driving-point impedance is constructible.

Because of the mutual inductive couplings involved, we see that this network realization is not a logically extended form of the Foster network of Fig. 19 in Ch. 3, nor can we expect that its method of construction from a given impedance function follows essentially the same procedure as that applicable to the two-element-kind network functions. The more profound nature of the *RLC* synthesis problem is intimately related to and directly predictable from the nonexistence of a normal co-ordinate transformation for networks having this degree of generality.

2 Limitations of the Partial-Fraction Expansion Method

Although these considerations indicate that a Foster-like procedure for the realization of general p.r. functions is not possible, it is of theoretical as well as practical interest to study this method of approach in some detail and to determine precisely the conditions under which it is applicable. In rather general terms we can see why it may fail, simply by observing that any breakdown of a given impedance into a sum of simple components like

$$Z(s) = z_1(s) + z_2(s) + \cdots + z_n(s) \tag{13}$$

cannot guarantee the p.r. character of these components merely on the basis that $Z(s)$ is p.r. If $Z(s)$ is analytic in the right half-plane then, to be sure, it follows that all of the z_ν must have this property, but the nonnegativeness of their real parts is not assured by the nonnegativeness of the real part of $Z(j\omega)$, since the latter may come about through the cancellation of positive and negative real-part values of the components.

If Eq. 13 represents the partial-fraction expansion of the function $Z(s)$, then each $z_\nu(s)$ is a combination of conjugate complex terms as given by

$$z_\nu = \frac{k_\nu}{s - s_\nu} + \frac{\bar{k}_\nu}{s - \bar{s}_\nu} = \frac{(k_\nu + \bar{k}_\nu)s - (k_\nu \bar{s}_\nu + \bar{k}_\nu s_\nu)}{(s - s_\nu)(s - \bar{s}_\nu)} \tag{14}$$

Suppose we write

$$k_\nu = \alpha + j\beta \quad \text{and} \quad s_\nu = -\sigma_\nu + j\omega_\nu \tag{15}$$

Then

$$z_\nu = \frac{a_1 s + a_0}{s^2 + b_1 s + b_0} \tag{16}$$

with

$$\begin{aligned} a_0 &= 2(\alpha\sigma_\nu - \beta\omega_\nu), & b_0 &= \sigma_\nu^2 + \omega_\nu^2 \\ a_1 &= 2\alpha, & b_1 &= 2\sigma_\nu \end{aligned} \tag{17}$$

Since s_ν is a left half-plane point, b_0 and b_1 are positive. The coefficients a_0 and a_1, however, may have either algebraic sign. In order for z_ν to be a p.r. function, its real part for $s = j\omega$ must be positive. This condition reads

$$(m_1 m_2 - n_1 n_2)_{s=j\omega} = a_0 b_0 + (a_1 b_1 - a_0)\omega^2 \geq 0 \tag{18}$$

requiring

$$a_1 b_1 \geq a_0 \geq 0 \tag{19}$$

Use of Eqs. 17 converts these conditions respectively into

$$\begin{aligned} &\alpha\sigma_\nu + \beta\omega_\nu \geq 0 \\ \text{and} \quad & \\ &\alpha\sigma_\nu - \beta\omega_\nu \geq 0 \end{aligned} \tag{20}$$

Evidently β may be positive or negative. Since σ_ν and ω_ν are positive by definition, the restriction on α is expressed by

$$\frac{\alpha}{|\beta|} \geq \frac{\omega_\nu}{\sigma_\nu} \tag{21}$$

As Fig. 2 illustrates, this condition limits the complex value of the residue k_ν to lie within the shaded region. For any given pair of complex poles s_ν and $\bar{s}_\nu$ of the function $Z(s)$, one can immediately see how the associated residue is restricted if the component impedance representing this pole-pair is to be a p.r. function.

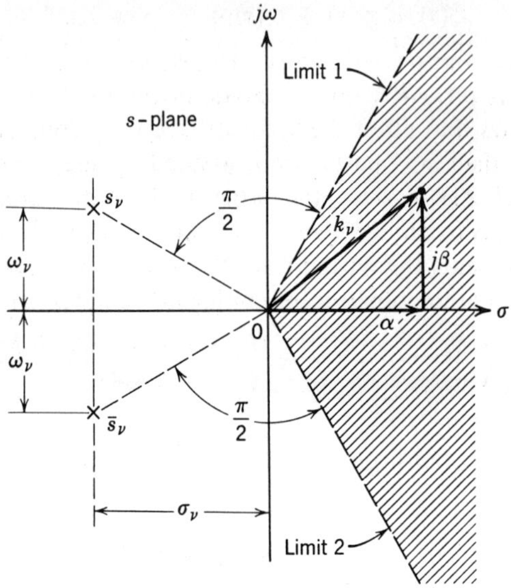

Fig. 2. Showing that condition 21 (equivalent to 18 or 19) requires the residue k_ν to lie in the shaded region.

Note that k_ν is the residue of $Z(s)$ at $s = s_\nu$. The residue at $s = \bar{s}_\nu$ is $\bar{k}_\nu$. Thus if k_ν lies upon the dashed line designated as limit 1, $a_0 = 0$, whereas if this complex value lies upon limit 2, then $a_1 b_1 = a_0$; one or the other of conditions 19 is on the verge of being violated. If the pole s_ν lies upon the j-axis, then limit 1 and limit 2 coincide with the positive real axis and the residue k_ν is required to be real and positive, as we already know from our previous discussions. The farther s_ν moves into the left half-plane, the greater amount of freedom do the permissible complex values of the residue k_ν acquire until, for s_ν equal to a negative real value, k_ν may correspond to any point in the right half-plane.

When the component impedance 16 is a p.r. function, then it is readily realized physically by observing that its reciprocal has a pole at $s = \infty$ which may be removed, yielding a parallel capacitance branch and leaving a remainder that is p.r. and *RL* realizable. Algebraically these steps are indicated by

$$\begin{array}{r|l} & \;\; s/a_1 \\ \cline{2-2} a_1 s + a_0 & s^2 + b_1 s + b_0 \\ & s^2 + \dfrac{a_0 s}{a_1} \\ \cline{2-2} & \left(b_1 - \dfrac{a_0}{a_1}\right)s + b_0 \end{array} \tag{22}$$

so that we can write

$$z_\nu = \frac{1}{(s/a_1) + (1/z_1)}, \qquad z_1 = \frac{a_1 s + a_0}{[b_1 - (a_0/a_1)]s + b_0} \tag{23}$$

where the remainder z_1 is always a realizable RL impedance because conditions 19 guarantee all coefficients positive and the lowest critical frequency is a zero since Eqs. 17 convert the condition

$$\frac{a_0}{a_1} \leq \frac{a_1 b_0}{a_1 b_1 - a_0} \tag{24}$$

into

$$\frac{\alpha\sigma_\nu - \beta\omega_\nu}{\alpha} \leq \frac{\alpha(\sigma_\nu^2 + \omega_\nu^2)}{\alpha\sigma_\nu + \beta\omega_\nu}, \qquad \text{or } \alpha^2 + \beta^2 \geq 0 \tag{25}$$

which is obviously fulfilled.

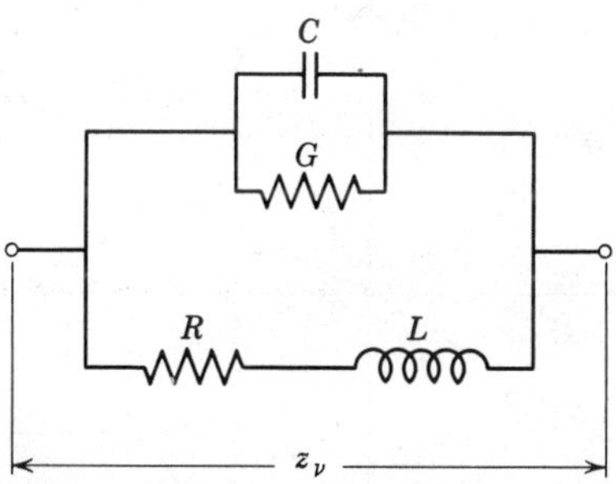

Fig. 3. Realization for the component impedance 14 or 16 when the condition expressed by 19 or 21 is fulfilled.

The resulting network may be constructed in the form shown in Fig. 3 in which

$$C = \frac{1}{a_1} \text{ farads}, \qquad L = \frac{a_1^3}{a_1^2 b_0 - a_0(a_1 b_1 - a_0)} \text{ henrys} \tag{26}$$

and

$$\frac{R}{L} = \frac{a_0}{a_1}, \qquad \frac{G}{C} = \frac{a_1 b_1 - a_0}{a_1} \tag{27}$$

We regard this circuit as consisting of a "lossy" inductance in parallel with a "lossy" capacitance, the *loss ratios* or *dissipation ratios* being given by Eqs. 27. Thus when the typical component impedance 16 in the partial-fraction expansion of the driving-point impedance $Z(s)$ is realizable, it takes the form that we would logically expect if we regard the dissipative impedance to be realizable by essentially the same methods as those applicable to reactance functions except for the appearance of lossy L's and C's in place of the lossless ones.

It is useful to observe that one can see from the location of the residue

k_ν in Fig. 2 how the loss is distributed between the inductance and the capacitance in the circuit of Fig. 3. Thus if k_ν lies upon limit 1 so that $a_0 = 0$, then according to Eqs. 27 the inductance becomes lossless, while the dissipation ratio for the capacitance becomes $G/C = b_1 = 2\sigma_\nu$. On the other hand, if k_ν lies on limit 2 so that $a_1 b_1 = a_0$, then the capacitance becomes lossless while the dissipation ratio for the inductance becomes $R/L = b_1 = 2\sigma_\nu$. Finally, if k_ν lies upon the positive real axis, Eqs. 17 show that $a_1 b_1 - a_0 = a_0$ and we get $R/L = G/C = b_1/2 = \sigma_\nu$. These results are effectively summarized by representing the loss ratios 27 geometrically as shown in Fig. 4.

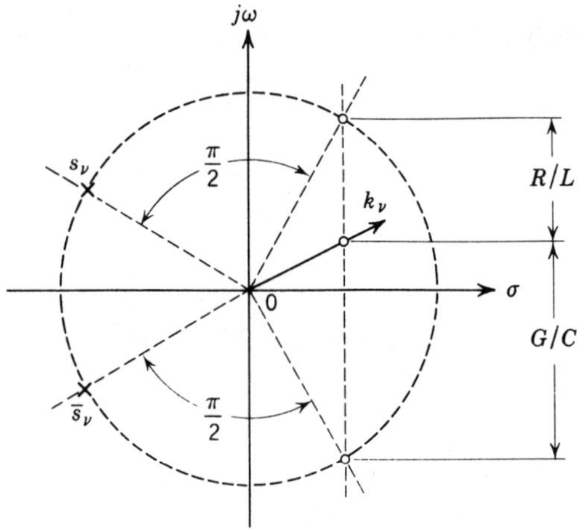

Fig. 4. Graphical determination of the loss ratios 27 from the given pole locations.

In terms of the condition illustrated in Fig. 2 it is possible to determine whether a given impedance is realizable from its partial-fraction expansion with practically no computational effort when a geometrical plot of the pertinent pole-zero pattern is available. Thus, if we visualize the residue at a given pole as evaluated from a quotient of frequency factors (vectors) pointing toward that pole, then the angle of the residue equals the sum of the angles of the zero factors minus the sum of the angles of the pole factors. In situations involving simple geometries, one does not even need a protractor to measure angles.

The pole-zero patterns in Fig. 5 represent situations of this sort. Figure 5*a* involves only poles on the real axis for which the circuit of Fig. 3 degenerates into the parallel capacitance and conductance alone, requiring positive real values for the residues. We see by inspection that the residue at $s = -2$ is positive real and residues at $s = -1$ and $s = -3$

are negative real, and so we conclude that the partial-fraction expansion method is not applicable in this straightforward manner. However, we are reminded that although negative real poles with positive residues yield an *RC* impedance, such poles with negative residues yield an *RL* impedance; and perhaps we can realize the present impedance as a series combination of an *RC* and an *RL* network. This question is readily answered by writing down the partial-fraction expansion for this impedance which we can do by inspection. We have

$$Z(s) = \frac{s(s^2 + 4s + 5)}{(s + 1)(s + 2)(s + 3)} = \frac{-1}{s + 1} + \frac{2}{s + 2} - \frac{3}{s + 3} + 1 \quad (28)$$

The important term in this expansion is the constant, for the negative terms plus this constant we are considering to be realizable as an *RL* network. The controlling factor here is the zero-frequency value of this group of terms, which, for an *RL* impedance, is the minimum value of its real part on the j-axis and hence must be positive. The constant term in Eq. 28, we observe, is not large enough to meet this condition, and so the contemplated scheme does not work in this instance. It is generally worth considering, however, for it offers a simple possible realization procedure for any impedance having real axis poles.

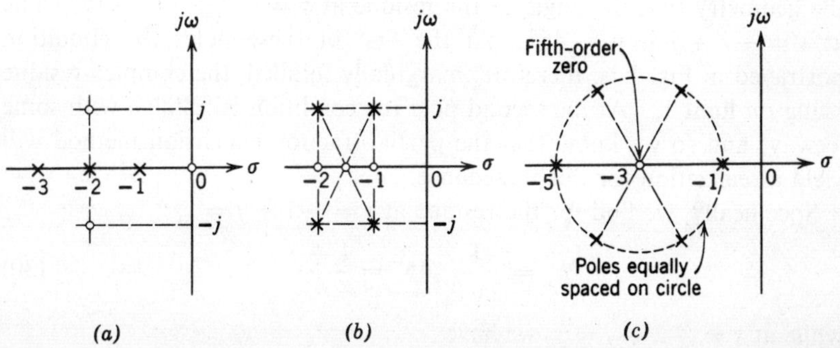

Fig. 5. Pole-zero patterns for which the realization criterion depicted in Fig. 2 may be evaluated by inspection.

This example illustrates another important point about attempting the partial-fraction expansion method of realization; namely, that the process may work on the reciprocal function although it is not applicable to the given one. Thus, if we interchange the identities of zeros and poles in the pattern of Fig. 5*a*, we see that the residue at $s = 0$ is real and positive, while at $s = -2 + j$ the angles of the pole factors produce the sum $3\pi/2 - \tan^{-1}\frac{1}{2}$ and those of the zero factors add up to $3\pi/2$, yielding a residue angle equal to $\tan^{-1}\frac{1}{2}$ which is obviously smaller than the maximum allowable angle. Specifically, we find

$$Y(s) = \frac{(s+1)(s+2)(s+3)}{s(s^2+4s+5)} = \frac{6}{5s} + \frac{4s+6}{5(s^2+4s+5)} + 1 \tag{29}$$

The corresponding network is shown in Fig. 6.

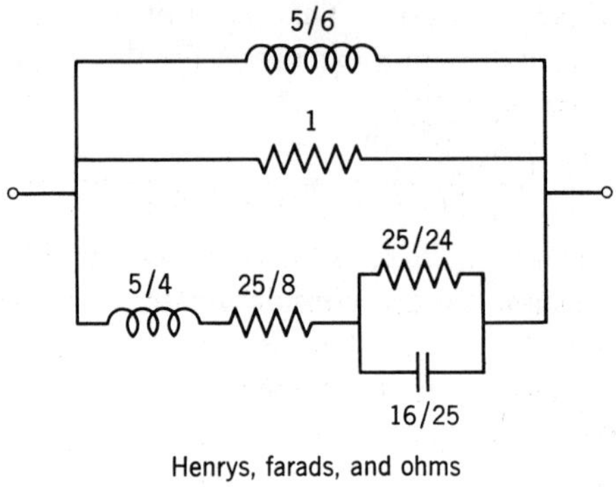

Fig. 6. Network realizing the admittance function 29 having the impedance pole-zero pattern in Fig. 5*a*.

Considering the pole-zero pattern in Fig. 5*b*, we can easily see from the geometry that the angle of the residue at $s = -1 + j$ is $+45°$ while at $s = -2 + j$ it is $-45°$. At the first of these poles the condition portrayed in Fig. 2 is, therefore, marginally fulfilled, the complex residue being on limit 1. At the second pole the condition is fulfilled with some leeway; and so we know that the partial-fraction expension method will yield a realization for this impedance.

Specifically, we find for the residue at $s = -1 + j = s_1$

$$k_1 = \frac{1}{2\sqrt{2}} \underline{/45°} = \frac{1+j}{4} \tag{30}$$

while at $s = -2 + j = s_2$ we have

$$k_2 = \frac{1}{2\sqrt{2}} \underline{/-45°} = \frac{1-j}{4} \tag{31}$$

In order to evaluate the pertinent terms in the partial-fraction expansion as in Eq. 14, we compute

$$k_1\bar{s}_1 = -j/2, \qquad k_2\bar{s}_2 = (-3+j)/4 \tag{32}$$

The impedance for the pole-zero pattern of Fig. 5*b* therefore reads

$$Z(s) = \frac{1}{2}\left(\frac{s}{s^2+2s+2} + \frac{s+3}{s^2+4s+5}\right) \tag{33}$$

and the corresponding network realization is shown in Fig. 7. We see that no dissipation is associated with the inductance in the component network corresponding to the first term in Eq. 33, as is predictable from the fact that the residue for this pole-pair lies upon limit 1 of Fig. 2, as pointed out above. Since the residue 31 for the other pole-pair has a negative angle, the representation in Fig. 4 shows the R/L ratio to be larger than G/C; and the pertinent component network in Fig. 7 bears out this result.

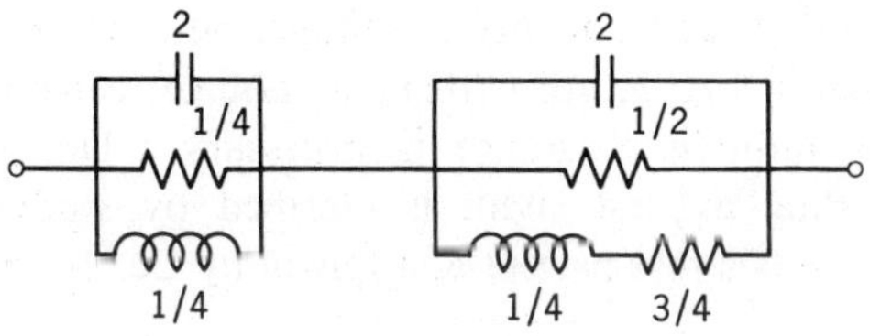

Fig. 7. Network realization for the impedance 33 having the pole-zero pattern in Fig. 5*b*.

The pole-zero pattern in Fig. 5*c* is one for which our present method is particularly suited, for it is not difficult to see that the residues in all the poles are alike and that their common value is real and positive. In the evaluation of any one of these residues, the number of pole factors equals the number of zero factors. Hence the diameter of the circular locus upon which the poles are distributed has no influence on the residue value and we may, if we like, think of the circle as having unit radius. Considering the residue at $s = -1$, we have five zero factors, all equal to $+1$; and we have five pole factors whose lengths are 1, 1, $\sqrt{3}$, $\sqrt{3}$, 2, yielding a net product equal to 6. Since their net angle evidently is zero, we see that the residue at $s = -1$ has the value $\frac{1}{6}$.

We next observe that if we consider the pole at $s = -2 + j\sqrt{3}$, all zeros and pole factors are the same as before except that their angles are advanced by 60°. The residue at this pole, and evidently at all poles, is thus readily seen to have the value $\frac{1}{6}$ also. The impedance for this pole-zero pattern, therefore, has the representation

$$Z(s) = \frac{1}{6}\left(\frac{1}{s+1} + \frac{2s+4}{s^2+4s+7} + \frac{2s+8}{s^2+8s+19} + \frac{1}{s+5}\right) \tag{34}$$

and its network realization is that shown in Fig. 8.

3 Extension of the Partial-Fraction Expansion Method of Realization Through Resistance Padding

When conditions 19 assuring the p.r. character of the component impedance z_ν are not fulfilled, its real part is negative along part of the

j-axis and assumes a minimum value at some finite nonzero frequency ω_ν. If the absolute value of this minimum real part is denoted by $R_{\min}$, then the function $z_\nu + R_{\min}$ is again a p.r. function with a j-axis real part that becomes zero at $\omega = \omega_\nu$. Through the addition of an appropriate amount of resistance to each term in the partial-fraction expansion of any p.r. impedance, one can evidently always obtain a representation for this function in the form of Eq. 13 with realizable component impedances z_ν. However, these components no longer are realizable in the configuration shown in Fig. 3, nor is their method of realization as simple as it is when no additive resistance is necessary. This result, which is somewhat puzzling at first sight, is clarified by studying the general character of the j-axis real part of z_ν as given by Eq. 16 for $s = j\omega$.

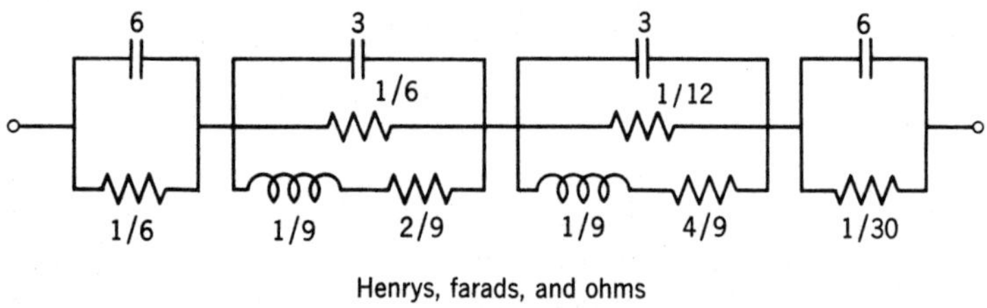

Fig. 8. Realization of the impedance 34 having the pole-zero pattern in Fig. 5c.

The behavior of this real part is most easily recognized by observing the pole-zero configuration of the even part

$$f(s) = \tfrac{1}{2}[z_\nu(s) + z_\nu(-s)] \tag{35}$$

which, for $s = j\omega$, is the pertinent real part. Using Eq. 16, we find for this even-part function

$$f(s) = \frac{(a_0 - a_1 b_1)s^2 + a_0 b_0}{(s^2 + b_1 s + b_0)(s^2 - b_1 s + b_0)} \tag{36}$$

The poles are those of z_ν and their images about the j-axis; the zeros, assuming the p.r. conditions 19 to be fulfilled, are given by a pair on the real axis at the points

$$s = \pm s_0 = \pm\sqrt{\frac{a_0 b_0}{a_1 b_1 - a_0}} \tag{37}$$

and a double-order zero at $s = \infty$, which is evident from the fact that the function 36 behaves like $1/s^2$ for large values of s. This pole-zero pattern is sketched in Fig. 9. Along the j-axis, the function $f(s)$ is seen to have the character of either of the two curves (a) or (b) shown in Fig. 10. If the poles are far from the j-axis, we evidently have simply a monotonic decrease as illustrated by curve (a), whereas a quasi-resonance effect as shown by curve (*b*) results if the poles are sufficiently close to the *j*-axis. The exact location of the finite zeros $\pm s_0$ has no critical effect upon the behavior of this real-part function. As the p.r. condition $a_1 b_1 \geq a_0$ comes closer to being violated, Eq. 37 shows that the zeros $\pm s_0$ move toward infinity, and they reach this point when $a_1 b_1 = a_0$. For this condition, the real part has a fourth-order zero at $s = \infty$, but its general character is still depicted by the curves of Fig. 10.

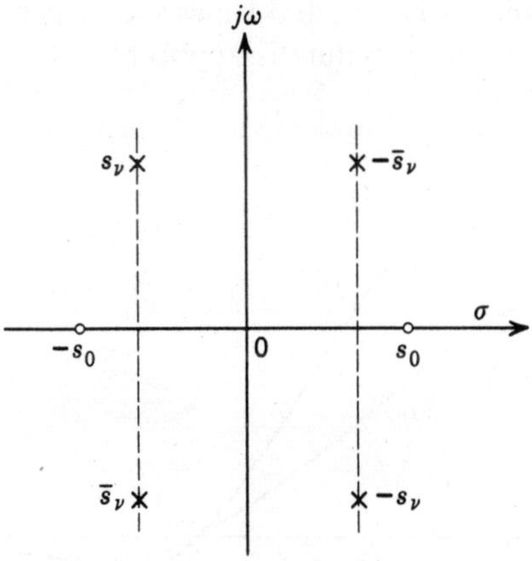

Fig. 9. Pole-zero pattern of the even-part function 35.

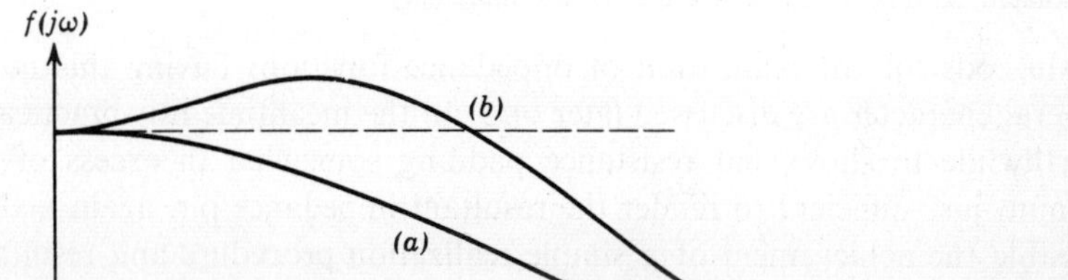

Fig. 10. *j*-axis behavior of the even-part function 35 having the pole-zero pattern in Fig. 9. (The p.r. condition $a_1 b_1 \geq a_0$ is fulfilled, so that the s_0-values, Eq. 37, are real.)

However, if the p.r. condition is violated so that $a_1 b_1 < a_0$, then Eq. 37 shows that the finite zeros lie upon the *j*-axis at the points $\pm s_0 = \pm j\omega_0$, and the behavior of the real part changes its character as illustrated by the curves in Fig. 11. Since the function still has its double-order zero at $s = \infty$ and hence must disappear like $1/\omega^2$ for large ω, it is clear that it now has a finite minimum. Although the addition of a positive resistance $R_{\min}$ to the impedance z_ν having this real-part behavior yields a resultant p.r. impedance, one does not by this simple process obtain an impedance whose real part has the character shown in Fig. 10. Rather, one obtains an impedance with a real part that approaches the nonzero value $R_{\min}$ for large values of s instead of behaving like $1/s^2$. The character of the resultant impedance z_ν is completely different than it is for the function given by Eq. 16 when it is p.r., and we cannot expect, therefore, that the same simple realization procedure is applicable.

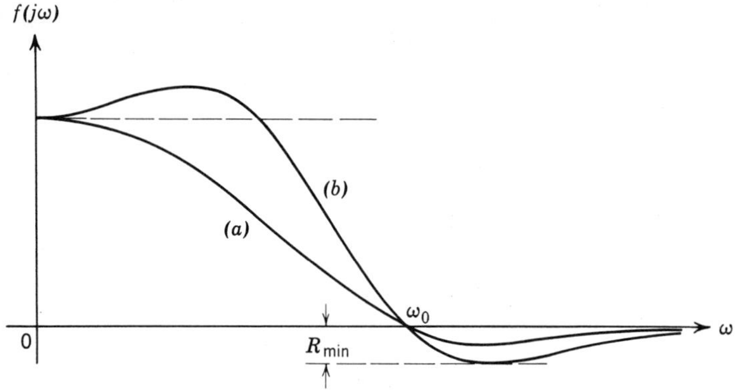

Fig. 11. j-axis behavior of the even-part function 35 when the p.r. condition $a_1b_1 \geq a_0$ is violated, so that the s_0-values, Eq. 37, are imaginary.

Methods for the realization of impedance functions having this more general character are discussed later on. In the meantime it is practically worthwhile to show that resistance padding somewhat in excess of an amount just sufficient to render the resultant impedance p.r. again makes possible the achievement of a simple realization procedure and resultant network structure. Situations arise in practice where this process can usefully be applied.

To this end, suppose we add to the component impedance z_ν as given by Eq. 16, an amount of resistance R_0, and obtain

$$z_\nu + R_0 = z^*_\nu = \frac{a_1 s + a_0}{s^2 + b_1 s + b_0} + R_0$$
$$= \frac{R_0 s^2 + (a_1 + b_1 R_0)s + (a_0 + b_0 R_0)}{s^2 + b_1 s + b_0} \tag{38}$$

We now propose to separate this expression into two additive parts, of which each has a simply recognizable realization and for which the conditions for realization can always be fulfilled through an appropriate choice of a positive R_0 value regardless of the algebraic signs of the coefficients a_1 and a_0. The detailed procedure takes one of two alternate forms, depending upon the algebraic sign of a_0. These two situations we shall distinguish as cases A and B. Specifically, for $a_0 > 0$, we separate the expression 38 as follows

$$z^*_\nu = \frac{R_0(s^2 + b_0)}{s^2 + b_1 s + b_0} + \frac{(a_1 + b_1 R_0)s + a_0}{s^2 + b_1 s + b_0} \qquad \text{(case A)} \tag{39}$$

The first term in this expression, after dividing numerator and denominator by $(s^2 + b_0)$, is seen to be realized by a resistance in parallel with a simple reactance. The second term has the form of z_ν in Eq. 16. Following the pattern of the realizability conditions 19, we have

$$(a_1 + b_1 R_0)b_1 \geq a_0 \geq 0 \tag{40}$$

and so the total realizability conditions for z^*_ν in Eq. 39 become

$$a_0 \geq 0, \qquad R_0 \geq \frac{a_0 - a_1 b_1}{b_1^2} \qquad \text{(case A)} \tag{41}$$

We see that now $a_1 b_1 \leq a_0$, which is complementary to the first of conditions 19. It should also be observed that a_1 may be positive or negative; if negative, the necessary amount of resistance padding is, of course, greater.

If a_0 is negative, one may write the expression for z^*_ν in the alternate form

$$z^*_\nu = \frac{[(a_0/b_0) + R_0](s^2 + b_0)}{s^2 + b_1 s + b_0} + \frac{[(a_1 + b_1 R_0)/b_0]s^{-1} + (-a_0/b_0^2)}{s^{-2} + (b_1/b_0)s^{-1} + (1/b_0)} \qquad \text{(case B)} \tag{42}$$

The first term is realizable as before provided

$$(a_0/b_0) + R_0 \geq 0 \tag{43}$$

and the second is the same kind of function as that considered earlier, except that s is replaced by $1/s$, making the change in the network realizing it obvious. The total realizability conditions now become

$$(-a_0) \geq 0, \qquad R_0 \geq \frac{(-a_0) - a_1 b_1}{b_1^2}, \qquad R_0 \geq \frac{(-a_0)}{b_0} \qquad \text{(case B)} \tag{44}$$

the last of these insuring the positiveness of the constant multiplier in the first term of expression 42 for z^*_ν. Again a_1 may be numerically positive or negative, and the appropriate value for R_0 is determined by the stronger of the two conditions in 44.

Reference to the relation for a_0 in Eq. 17 shows that cases A and B may be interpreted geometrically as restricting the permissible complex values of the residue k_ν to the shaded regions shown in the two parts of

Fig. 12. We thus see that these two cases complement each other and are mutually exclusive.

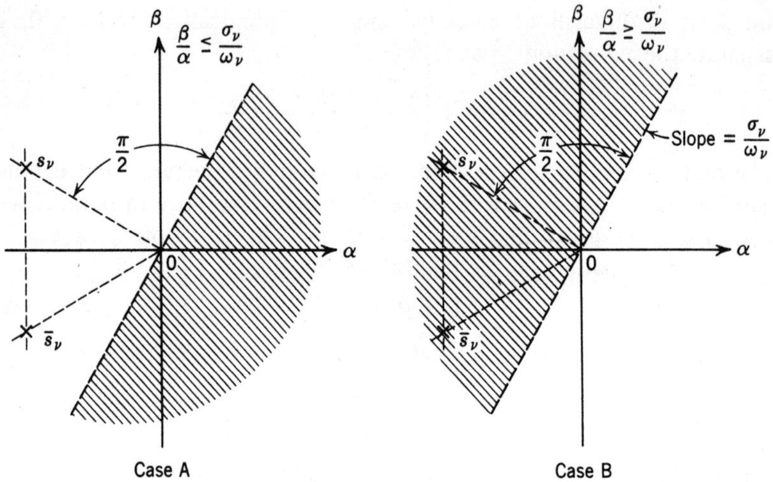

Fig. 12. Restriction of residue k_ν to shaded region is realizability condition for the padded impedance (Eq. 38) in either case A or B to which the representations 39 and 42 (respectively) refer.

It is of collateral interest to note that an impedance of the form

$$z(s) = \frac{s^2 + a_1 s + a_0}{s^2 + b_1 s + b_0} = 1 + \frac{(a_1 - b_1)s + (a_0 - b_0)}{s^2 + b_1 s + b_0} \tag{45}$$

can be interpreted as a function like z_ν in Eq. 16 with an additive resistance $R_0 = 1$. Assuming that all the coefficients in this expression are positive, the necessary and sufficient conditions that it be a p.r. function read

$$a_1 b_1 \ge (\sqrt{a_0} - \sqrt{b_0})^2 \tag{46}$$

and conditions 41 and 44 for the realizability of this function in the above simple terms are

$$a_1 b_1 \ge a_0 - b_0 \ge 0 \qquad \text{for case A} \tag{47}$$

$$a_1 b_1 \ge b_0 - a_0 \ge 0 \qquad \text{for case B} \tag{48}$$

Since the relation

$$(a_0 - b_0)^2 \ge (\sqrt{a_0} - \sqrt{b_0})^4$$

leads to

$$\frac{1}{2}\left(\sqrt{\frac{a_0}{b_0}} + \sqrt{\frac{b_0}{a_0}}\right) \ge 1$$

which is true, we see that conditions 47 and 48 are stronger than the p.r. condition 46, as we should expect since realization in the simplified manner described under cases A and B is not perfectly general and hence requires more than merely that the given impedance be a p.r. function.

4 The Brune Method of Synthesis*

The sufficiency as well as the necessity of the p.r. character of a rational function for its realizability as the driving-point impedance or admittance of a passive, lumped network was first demonstrated by O. Brune in his classic work which forms the cornerstone of network synthesis philosophy and still characterizes the basic approach to numerous other methods and procedures that have since been developed for dealing with problems in this field. Like the earlier work of Foster in connection with lossless networks, it sounded the keynote in a vitally new departure and attitude; and, in augmenting and complementing the work of Foster, it founded a mathematical discipline—the thread of rigorous thought—around which and upon which was built the coherent structure known today as the theory of network synthesis.

As we have seen in the discussions of the previous articles of this chapter, all p.r. functions do not require special methods for their realization beyond those familiar to us from the consideration of two-element-kind network synthesis. Thus, when the residues in poles of the given function satisfy certain conditions, their removal and network realization is accomplished by techniques that are essentially no different from those used in connection with reactance functions. That is to say, we can in such special situations still succeed in obtaining a network realization by use of nothing more elaborate than the Foster methods applying to impedances of lossless networks.

In this connection it is particularly important to recognize that when a p.r. function has j-axis poles we do not even need to make any inquiries as to whether the residues fulfill the special conditions called for, since the p.r. character of the given function already guarantees that such conditions are met. Thus the p.r. character of a function (as pointed out in Ch. 1) assures that any j-axis poles are simple and that the residues in these poles are real and positive. The corresponding terms in a partial-fraction expansion of the given function are surely realizable (as are those in the analogous expansion of a reactance function by the Foster method) and the remaining portion of the function involving left half-plane poles is surely p.r. and hence does not have its realizability impaired by the removal of those terms representing j-axis poles. The latter part of this

* "Synthesis of a Finite Two Terminal Network Whose Driving-Point Impedance is a Prescribed Function of Frequency," O. Brune, *J. Math. and Phys.*, Vol. 10, No. 3, Oct. 1931, pp. 191–236.

statement follows from the consideration that removal of the j-axis poles from a p.r. function cannot affect either its analyticity in the right half-plane or the values of its real part on the j-axis (since the real part contributed by such poles for $s = j\omega$ is identically zero).

j-axis poles of a p.r. function are thus always removable without question. They invariably lead to realizable circuit components; and their removal always simplifies the given function and cannot impair the realization properties of the remainder. Observe that the same statement cannot be made about left half-plane poles. As we have seen, the terms in a partial-fraction expansion representing them are not guaranteed realizable by the p.r. character of the given function, and if they are realizable, their subtraction from the function can and, except in special situations, will damage its p.r. character so that the remainder cannot be realized. In that event there is obviously no profit in removing terms even if these should be realizable separately.

In the initial approach to the synthesis of a given p.r. function, we therefore always inquire first whether any j-axis poles are present; if there are such poles, their corresponding terms as in a partial-fraction expansion of the function are subtracted. The remainder may have j-axis zeros, in which case the reciprocal remainder has removable j-axis poles. The second remainder again may have j-axis zeros so that its reciprocal can be dealt with in the same manner as just described for the given function. Just as in the synthesis of reactance functions described in Art. 4 of Ch. 3, we may thus develop the given impedance or admittance into a lossless ladder network so long as the remainder function at any stage has j-axis zeros or poles. When terminated in a network realizing the remainder function at any stage, this ladder structure evidently represents a realization for the given impedance; and it can happen that this simple process of realization continues until the remainder is a constant, in which case the total realization is given by a reactive ladder network terminated in a resistance, and no new techniques are called for beyond those provided by the Foster procedures.

In general, however, one will reach a point in the process just described where the remainder function is not a constant, and has neither j-axis zeros nor poles. This reduced function (which, according to the discussions in Art. 1 of Ch. 8, may be described as being both *minimum reactive* and *minimum susceptive*) must now be dealt with by some new and different procedure; and this is the point at which the Brune method enters our realization picture. The simpler steps by which the given function may have needed to be reduced to the minimum-reactive and minimum-susceptive condition we refer to as the "preamble" to the Brune process. It represents a simplification of the given function to a state where previously known techniques no longer suffice to produce a further reduction in the complexity of the function by the removal of realizable circuit elements in such a way as to leave a realizable remainder.

The Brune process accomplishes this further reduction; and proceeds in cycles, each of which reduces the complexity of the remainder function by a fixed amount. A discussion of the Brune method therefore involves only the demonstration that a typical cycle can always be carried through successfully.

Before entering upon this discussion, it may be helpful to illustrate the preamble process with an example for which the complete synthesis is thereby accomplished. Consider the p.r. impedance

$$Z(s) = \frac{10s^6 + 4s^5 + 200s^4 + 34s^3 + 550s^2 + 48s + 360}{s^6 + 20s^5 + 16s^4 + 200s^3 + 67s^2 + 180s + 36} \tag{49}$$

Testing the denominator polynomial for Hurwitz character according to the discussion in Art. 4 of Ch. 1 leads to a premature termination as in the continued division 48 of that chapter, and reveals $(s^2 + 9)$ to be a common factor of the even and odd parts. The impedance 49 therefore has a pair of j-axis poles at the points $s = \pm j3$. To evaluate the residue of $Z(s)$ in these poles, we make use of Eq. 88 of Ch. 1, and have

$$k_3 = \left(\frac{m_1}{n_2'}\right)_\nu = \left[\frac{s^6 + 20s^4 + 55s^2 + 36}{2(5s^4 + 30s^2 + 9)}\right]_{s^2=-9}$$

$$= \frac{-729 + 1620 - 495 + 36}{2(405 - 270 + 9)} = \frac{3}{2}$$

or alternately

$$k_3 = \left(\frac{n_1}{m_2'}\right)_\nu + \left(\frac{2s^4 + 17s^2 + 24}{3s^4 + 32s^2 + 67}\right)_{s^2=-9} = \frac{162 - 153 + 24}{243 - 288 + 67} = \frac{3}{2} \tag{50}$$

To remove these poles from $Z(s)$ we perform the following self-explanatory operations

$$Z(s) - \frac{3s}{s^2+9} = \left(\frac{10s^6 + 4s^5 + 200s^4 + 34s^3 + 550s^2 + 48s + 360}{s^4 + 20s^3 + 7s^2 + 20s + 4} - 3s\right)\frac{1}{s^2+9}$$

$$= \frac{10s^6 + s^5 + 140s^4 + 13s^3 + 490s^2 + 36s + 360}{(s^4 + 20s^3 + 7s^2 + 20s + 4)(s^2 + 9)} \tag{51}$$

$$= \frac{10s^4 + s^3 + 50s^2 + 4s + 40}{s^4 + 20s^3 + 7s^2 + 20s + 4} = Z_1(s)$$

Testing the denominator polynomial of the reciprocal remainder function $Y_1(s) = 1/Z_1(s)$ for Hurwitz character reveals in a similar manner that it contains the factor $(s^2 + 4)$ representing a pair of j-axis poles at $s = \pm j2$. Evaluation of the residue proceeds as before and yields the value $k_2 = 1$, so that we next form

$$Y_1(s) - \frac{2s}{s^2 + 4} = \left(\frac{s^4 + 20s^3 + 7s^2 + 20s + 4}{10s^2 + s + 10} - 2s\right)\frac{1}{s^2 + 4}$$

$$= \frac{s^4 + 5s^2 + 4}{(10s^2 + s + 10)(s^2 + 4)} = \frac{s^2 + 1}{10s^2 + s + 10} = Y_2(s) \qquad (52)$$

The reciprocal remainder function at this point evidently has a pair of j-axis poles at $s = \pm j$. The common residue value is found equal to $\frac{1}{2}$, and so we form

$$Z_2(s) - \frac{s}{s^2 + 1} = \frac{10s^2 + s + 10}{s^2 + 1} - \frac{s}{s^2 + 1} = 10 \qquad (53)$$

We can now summarise the development process by writing for $Z(s)$ in Eq. 49 the continued-fraction expansion

$$Z(s) = \frac{3s}{s^2 + 9} + \cfrac{1}{\cfrac{2s}{s^2 + 4} + \cfrac{1}{\cfrac{s}{s^2 + 1} + 10}} \qquad (54)$$

The resulting network is shown in Fig. 13.

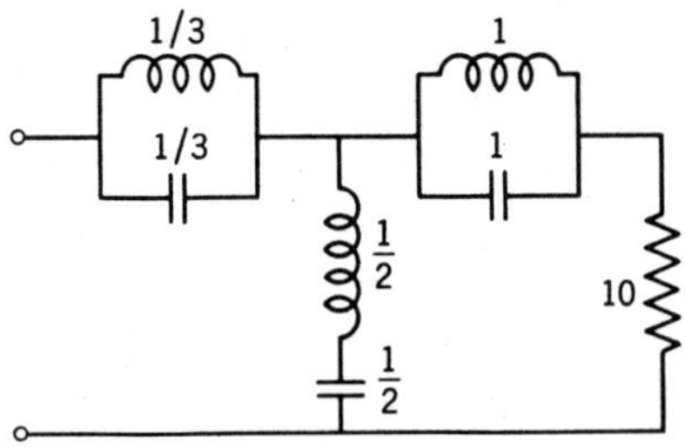

Fig. 13. Realization of the impedance 49, which is possible through the preamble to the Brune process, as summarized in the expansion 54.

For the discussion of the typical Brune cycle we will consider a p.r. function of the form

$$Z(s) = \frac{a_0 + a_1 s + a_2 s^2 + \cdots + a_n s^n}{b_0 + b_1 s + b_2 s^2 + \cdots + b_n s^n} \qquad (55)$$

in which the coefficients a_0, b_0, a_n, and b_n are nonzero, and neither numerator nor denominator polynomial has j-axis zeros. The basic thought in the initial steps of the Brune procedure is to attempt the subtraction of circuit elements in such a way as to shift a pair of left half-plane zeros onto the j-axis without damaging the p.r. character of the function. In this regard one considers first the real part of the impedance along the j-axis, which may have an appearance somewhat like the curve sketched in Fig. 14. Thus at some frequency $\omega = \omega_1$ the real part has the smallest value of several minima which this function can exhibit. If we designate this minimum by R_1, then clearly we can subtract this value from $Z(s)$ without destroying its p.r. character. The function

$$Z_1(s) = Z(s) - R_1 \tag{56}$$

then has a purely imaginary value at the frequency $s = s_1 = j\omega_1$; that is

$$Z_1(s_1) = jX \tag{57}$$

Although real, the value of X may be numerically positive or negative.

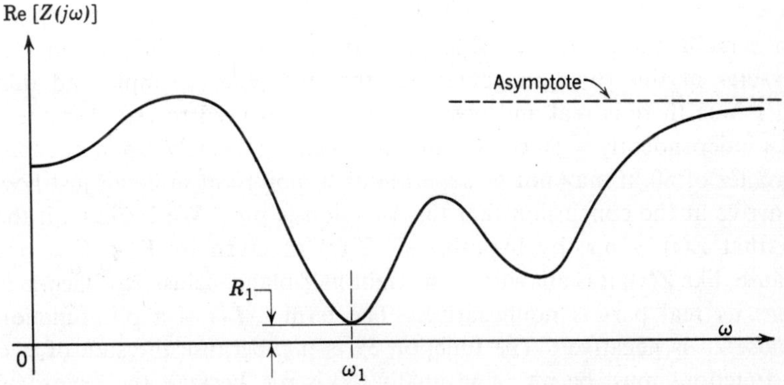

Fig. 14. Possible real-part behavior of a p.r. impedance function.

Following the basic thought expressed above, our next step is to subtract a reactance having the value jX at $s = j\omega_1$ and thus leave a remainder which is zero at this point. The simplest appropriate reactance is that of an inductance with the value

$$L_1 = X/\omega_1 \tag{58}$$

and the remainder having a j-axis zero at $s = j\omega_1$ reads

$$Z_1(s) - L_1 s \tag{59}$$

At this point we note the appearance of some difficulties. If the value of X is positive, then L_1 in Eq. 58 is positive and realizable, but the function 59 is definitely not p.r. because it has a pole at $s = \infty$ (a limiting j-axis point) with the *negative* residue $-L_1$. If X is negative so that L_1 has a negative value, then, to be sure, the function 59 has a positive residue in its pole at $s = \infty$ and hence is p.r., but now we cannot be sure that we will eventually be able to absorb the negative inductance L_1.

Although at this stage the contemplated procedure for producing a

j-axis zero does not look too encouraging, it seems too soon to abandon it altogether; and so we say to ourselves: "Suppose we assume for the moment that X is negative so that the remainder 59 is p.r. We then have a p.r. function with a j-axis zero, and we know how to proceed with its simplification. We will come back to the negative inductance L_1 later on and see if there is some way of absorbing it. What we are to do when the value of X is positive will require further thought. In the meantime we shall have the satisfaction of seeing our initial strategy through to its ultimate end."

Thus for $X < 0$ and $L_1 < 0$, 59 is a p.r. function with a j-axis zero at $s = s_1$ and

$$\frac{1}{Z_1(s) - L_1 s} \tag{60}$$

is a p.r. function with a j-axis pole at this point. The known p.r. character of this function guarantees that the pole is simple and that the residue there is real and positive; there is no need to establish these facts independently. Since we are thus leaning heavily upon the p.r. character of 60, it may not be superfluous to point out in detail just how we arrive at the conclusion that this function is p.r. We begin with the fact that $Z(s)$ is p.r. by hypothesis. $Z_1(s)$ as given by Eq. 56 is p.r. because, like $Z(s)$, it is analytic in the right half-plane inclusive of the j-axis where its real part is nonnegative. The term $-L_1 s$ is a p.r. function because L_1 is negative. The function 59 is p.r. because the sum of two p.r. functions must be p.r.; and finally, 60 is p.r. because the reciprocal of a p.r. function is p.r. also.

Removal of the j-axis poles from the admittance 60 is indicated algebraically by writing

$$\frac{1}{Z_1(s) - L_1 s} - \frac{(1/L_2)s}{s^2 + \omega_1^2} = \frac{1}{W(s)} \tag{61}$$

in which a familiar method of residue evaluation yields

$$2L_2 = \left(\frac{dZ_1}{ds}\right)_{s=s_1} - L_1 \tag{62}$$

Physically the step indicated in Eq. 61 amounts to the removal of a shunt branch in a ladder development of $Z(s)$ in which R_1 and L_1 are the initial series elements. This shunt branch consists of the inductance L_2 in series with a capacitance having the value

$$C_2 = \frac{1}{L_2 \omega_1^2} \tag{63}$$

The reciprocal remainder function $W(s)$ is surely a p.r. impedance since the removal of j-axis poles from a p.r. function as indicated in Eq. 61

must leave a p.r. remainder. This impedance $W(s)$ is seen to have a pole at $s = \infty$ since [noting that $Z_1(s)$ is regular at infinity] we get from Eq. 61 for the limit $s \rightarrow \infty$

$$\frac{1}{W(s)} \rightarrow -\frac{1}{L_1 s} - \frac{1}{L_2 s} = \frac{L_1 + L_2}{-L_1 L_2 s}$$

or

$$W(s) \rightarrow \frac{-L_1 L_2 s}{L_1 + L_2} = L_3 s \tag{64}$$

Since $W(s)$ is p.r., there can be no question about the positiveness of the inductance L_3, and removal of this pole from $W(s)$ finally leaves a remainder function

$$Z_2(s) = W(s) - L_3 s \tag{65}$$

which again is surely p.r. Like $Z(s)$, it has no j-axis zeros or poles. In fact, it has exactly the same properties as the function $Z(s)$ that we started from except (as we shall show presently) its numerator and denominator polynomials are each two degrees lower. The contemplated cycle ends with this remainder function, and since it is in all respects like the original $Z(s)$, we can presumably repeat the cycle until we finally end up with a remainder which equals a constant.

However, we have yet a number of matters to clarify before we can claim to have scored a victory. First we must see about the possibility of accepting a negative inductance L_1. In this regard it is useful to take a careful look at the network which our development cycle has produced. It is shown in Fig. 15 in which the box labeled $Z_2(s)$ represents the network realizing the remainder function.

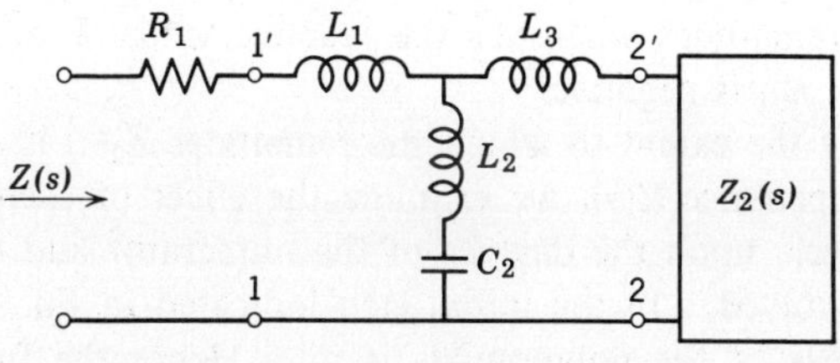

Fig. 15. Network produced by a typical cycle in the Brune procedure.

Our above argument in connection with the several steps involved in producing this portion of the ladder development of $Z(s)$ shows that all element values except that of L_1 are positive. Since this single negative inductance is associated in a T circuit with two other positive inductances, there is a good possibility that this T may be replaceable by a pair of mutually coupled positive inductances. The equivalence involved in this transformation is indicated in Fig. 16.

Its success depends upon the fulfillment of two requirements: (*a*) the algebraic sum $(L_1 + L_2)$ must be positive in spite of the numerically negative value of L_1; (*b*) the coupling coefficient for the mutually coupled coils may not exceed unity. The first of these conditions may be seen from Eq. 64 to be met since we already know that L_2 and L_3 are positive and that L_1 is negative. As for the coupling coefficient, its value is given by the expression

$$k = \frac{L_2}{\sqrt{(L_1 + L_2)(L_2 + L_3)}} = \frac{L_2}{\sqrt{L_1L_2 + L_1L_3 + L_2L_3 + L_2^2}} \tag{66}$$

Equation 64, on the other hand, shows that

$$L_1L_2 + L_1L_3 + L_2L_3 = 0 \tag{67}$$

and thus Eq. 66 yields $k = 1$. The two inductances $(L_1 + L_2)$ and $(L_2 + L_3)$ in the T equivalent shown in Fig. 16 are *closely coupled.* Absorption of the negative value of L_1 is marginally assured.

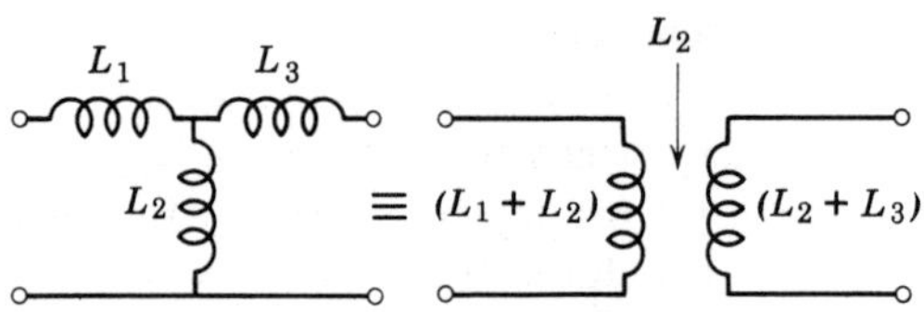

Fig. 16. Replacement of the T circuit of inductances in the network of Fig. 15 by a pair of mutually coupled inductances.

By the use of this expedient, we can conclude that the development cycle illustrated in Fig. 15 yields a physical partial development and assures a p.r. remainder whenever the reactive value X of $Z_1(s)$ at the j-axis point $s = j\omega_1$ is negative.

To determine the extent to which the remainder $Z_2(s)$ is simpler than the original impedance $Z(s)$, we examine the effect of each step in the development cycle upon the degrees of the numerator and demoninator polynomials involved. In the initial step indicated in Eq. 56 the effect upon the degrees of the polynomials is nil. Hence the function $Z_1(s)$ involves polynomials of the degree n the same as $Z(s)$ does. In the subtraction of L_1s in Eq. 59, the function becomes more complicated because a pole is inserted. The function 59 is given by an $(n + 1)$-degree polynomial divided by one of degree n. The reciprocal function 60 therefore involves the ratio of $n/(n + 1)$-degree polynomials. Removal of a pair of poles as indicated in Eq. 61 reduces the degree of each polynomial by two, and so the remainder $1/W(s)$ involves the ratio of

$(n-2)/(n-1)$- degree polynomials. $W(s)$, which has a simple pole at infinity, is given by a quotient of polynomials having the degrees $(n-1)/(n-2)$. The remainder $Z_2(s)$ after removal of the pole at $s=\infty$ therefore involves the ratio of two $(n-2)$-degree polynomials as mentioned above.

Since each polynomial in $Z_2(s)$ thus involves two less coefficients than the polynomials in the given function $Z(s)$, the characterization of $Z_2(s)$ requires four less pieces of relevant data than does the characterization of $Z(s)$. This result agrees with the fact that although the circuit development in Fig. 15 involves five elements, the constraint relation 67 among the three inductances leaves only two of their values independently specifiable. The development cycle proceeding from $Z(s)$ to $Z_2(s)$ involves the loss of four polynomial coefficients and correspondingly it results in the determination of four element values. In this sense it is a canonic development since it reduces the complexity of the given function while producing a minimum number of circuit elements.

The remaining question as to what can be done when the reactive value X in Eq. 57 is positive can be answered by switching to an admittance basis after the initial step expressed by Eq. 56 is carried out. At this point we consider the admittance $Y_1(s)=1/Z_1(s)$ which, at the j-axis point $s=j\omega_1$, has the purely susceptive value

$$Y_1(s_1)=jB=\frac{1}{jX} \tag{68}$$

Clearly, when X is positive, B is negative. A capacitance having the admittance jB at $s=j\omega_1$ is given by

$$C_1'=\frac{B}{\omega_1} \tag{69}$$

and the remainder after subtraction of this capacitive admittance from $Y_1(s)$ reads

$$Y_1(s)-C_1's \tag{70}$$

This remainder, like the one given by Eq. 59, is p.r. because C_1' is negative.

All of the remaining steps are completely dual to those just described for the impedance basis. The reciprocal of the p.r. function 70 has a pair of j-axis poles. Their removal in the form of a series branch consisting of a parallel LC combination reads

$$\frac{1}{Y_1(s)-C_1's}-\frac{(1/C_2')s}{s^2+\omega_1^2}=\frac{1}{W(s)} \tag{71}$$

The values of the pertinent inductance and capacitance are obtained from

$$2C_2' = \left(\frac{dY_1}{ds}\right)_{s=s_1} - C_1' \tag{72}$$

and

$$L_2' = \frac{1}{C_2'\omega_1^2} \tag{73}$$

The admittance $W(s)$ has a pole at $s = \infty$ as may be seen by considering Eq. 71 for the limit $s \to \infty$, giving

$$W(s) \to \frac{-C_1'C_2's}{C_1' + C_2'} = C_3's \tag{74}$$

Removal of a final shunt branch with the admittance 74 as indicated by

$$Y_2(s) = W(s) - C_3's = 1/Z_2(s) \tag{75}$$

completes the cycle and leaves the p.r. remainder $Y_2(s)$ or $Z_2(s)$. The circuit obtained in this manner is shown in Fig. 17.

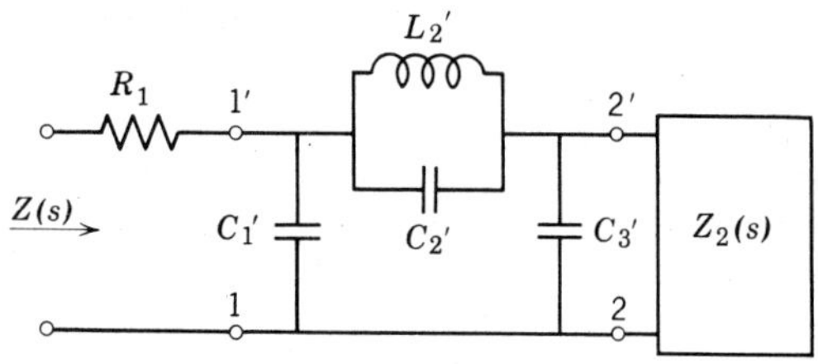

Fig. 17. Alternate form of the Brune cycle illustrated in Fig. 15 that is pertinent when the value of X in Eq. 57 is positive. All element values are positive except C_1'.

One defect remains, namely, the negative value of C_1'. This may be removed by observing that the reactive network between terminal pairs 1-1′ and 2-2′ may be replaced by the one shown in Fig. 18 in which the ideal transformer has a ratio

$$n = C_2'/(C_1' + C_2') = (C_2' + C_3')/C_2' = \sqrt{(C_2' + C_3')/(C_1' + C_2')} \tag{76}$$

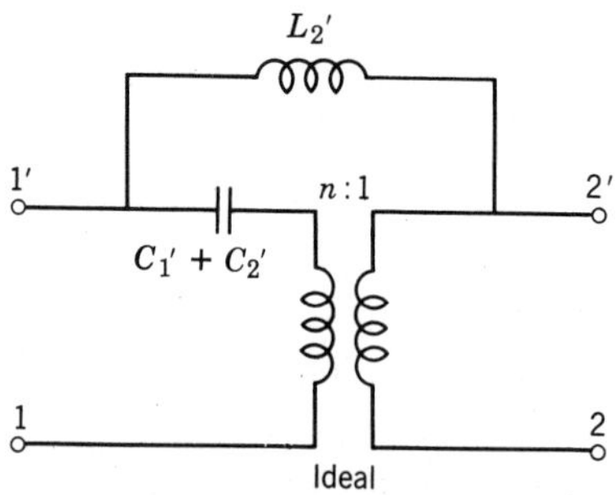

Fig. 18. Reactive two terminal-pair, equivalent to the one in Fig. 17, but containing no negative elements. The ideal transformer ratio is given by Eq. 76.

the several equivalent forms for this ratio following from the relation (Eq. 74)

$$C_1'C_2' + C_1'C_3' + C_2'C_3' = 0 \tag{77}$$

analogous to the close coupling relation 67 among the inductances in the circuit of Fig. 15. The equivalence of the reactive two terminal-pair networks in Figs. 17 and 18 is easily recognized by noting that both have the set of short-circuit driving-point and transfer admittances

$$y_{11} = (C_1' + C_2')s + \frac{1}{L_2's} \tag{78}$$

$$y_{22} = (C_2' + C_3')s + \frac{1}{L_2's} = n^2(C_1' + C_2')s + \frac{1}{L_2's} \tag{79}$$

$$-y_{12} = C_2's + \frac{1}{L_2's} = n(C_1' + C_2')s + \frac{1}{L_2's} \tag{80}$$

Since the sum $(C_1' + C_2')$ is positive because C_3' in Eq. 74 is positive, the circuit of Fig. 18 fulfills the conditions of physical realizability, although the ideal transformer is less desirable from a practical standpoint than the closely coupled inductances involved in the circuit of Fig. 15.

One can show, however, that the circuit of Fig. 17 can always be converted into an equivalent circuit of the form shown in Fig. 15 and thus the appearance of an ideal transformer can be avoided even when $X > 0$. To make this transformation we first form the determinant of the y's in Eqs. 78, 79, and 80. Noting the relation 77 we get the simple result

$$|y| = y_{11}y_{22} - y_{12}^2 = (C_1' + C_3')/L_2' \tag{81}$$

whereupon the corresponding set of open-circuit driving-point and transfer impedances is written down as follows

$$z_{11} = y_{22}/|y| = \left(\frac{C_2' + C_3'}{C_1' + C_3'}\right)L_2's + \frac{1}{(C_1' + C_3')s} \tag{82}$$

$$z_{22} = y_{11}/|y| = \left(\frac{C_1' + C_2'}{C_1' + C_3'}\right)L_2's + \frac{1}{(C_1' + C_3')s} \tag{83}$$

$$z_{12} = -y_{12}/|y| = \left(\frac{C_2'}{C_1' + C_3'}\right)L_2's + \frac{1}{(C_1' + C_3')s} \tag{84}$$

By noting the expressions that we get for these impedances in the reactive circuit of Fig. 15, we obtain without difficulty the relations

$$L_1 = \frac{L_2'C_3'}{C_1' + C_3'}, \qquad L_2 = \frac{L_2'C_2'}{C_1' + C_3'}$$
$$L_3 = \frac{L_2'C_1'}{C_1' + C_3'}, \qquad C_2 = C_1' + C_3' \tag{85}$$

From these we see that the relation 77 among the C's again yields the close coupling relation 67 among the L's; and the negative C_1' value yields a negative L_3 value, L_1 as well as L_2 now being positive. Hence the conversion from the circuit in Fig. 17 to an equivalent one like that shown in Fig. 15, in which the inductive T circuit is replaceable by the pair of closely coupled inductances shown in Fig. 16, is always possible.

It can be shown that the switch to an admittance basis followed by subsequent conversion from the circuit of Fig. 17 to that of Fig. 15 when the reactive value X is positive need actually not be done but that precisely the same end results are obtained if, at the point where the subtraction 59 is made and a non-p.r. function is obtained, one ignores the implications of this fact and methodically carries through the remaining steps on an impedance basis as though loss of the p.r. character of the function 59 did not matter. From a practical computational point of view such an attitude saves time and labor, and actual problems are ordinarily handled in this way. A simple and rigorous justification for this final result is given in the next article.

In connection with the close coupling relation among the inductances of the circuit shown in Fig. 15, it is incidentally worth mentioning that this condition is a rather obvious result of the fact that $Z(s)$ by hypothesis has no j-axis poles and hence must be regular at infinity. If we were not aware of any special relationship among the inductances of the circuit in Fig. 15, our first impression would certainly lead us to conclude that the input impedance to this circuit has a pole at $s = \infty$ because of the series inductances involved. The fact that this offhand observation is false is the result of a subtlety, the consequence of which is not obvious by inspection; namely, that the close coupling relation 67 in effect cancels the net series inductance in this circuit.

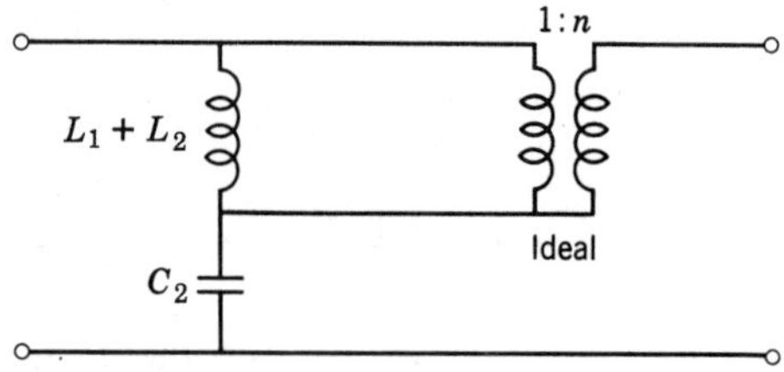

Fig. 19. Equivalent form of the reactive two terminal-pair in Fig. 15. The ideal transformer ratio is given by Eq. 86.

It is easier to see this result by inspection if we draw the reactive part of this circuit in the form shown in Fig. 19, which is comparable to (and dual to) the potentially equivalent circuit of Fig. 18, and for which the ideal transformer ratio is given by

$$n = L_2/(L_1 + L_2) = (L_2 + L_3)/L_2 = \sqrt{(L_2 + L_3)/(L_1 + L_2)} \quad (86)$$

Substantiation of the equivalence with the reactive circuit in Fig. 15 is readily accomplished by noting by inspection that the two circuits have the same open-circuit input and transfer impedances. In the equivalent circuit of Fig. 19, the absence of any series inductive effect is obvious, whereas in the circuit of Fig. 15 this fact is far from being obvious even though we are aware of the close coupling relation 67.

When a given rational p.r. function $Z(s)$, Eq. 55, is given in which the degree n is 4 or more, then several successive Brune cycles need to be carried out for its complete realization. Each of these follows the pattern outlined above. At the beginning of each cycle, the real part of the pertinent input impedance must be evaluated for $s = j\omega$ and its smallest minimum found as indicated in Fig. 14. In this regard we might be led to infer that a complete evaluation of the real part for the original impedance $Z(s)$ should suffice in the sense that the next smallest minimum together with its frequency of occurrence would be pertinent to the second cycle, and so forth. This conclusion is false, since it turns out that the minima of the real part of the remainder function in general bear no such simple relation to those of the original real part. The real-part evaluation needs to be repeated at the beginning of each cycle, the only encouraging feature in this regard being the fact that the function becomes simpler at each successive cycle.

A complete realization for a given impedance $Z(s)$ involving several cycles for its development has the appearance shown in Fig. 20. The arrangement in part (*a*) would be preferred in the actual physical construction of this impedance, whereas the circuit in part (*b*), involving the ideal transformers, is more useful in connection with analytic reasoning as to the nature and properties of the given impedance function and the circuit characteristics which these imply.

Regarding practical matters it may be pointed out that the resistances $R_1, R_2, \cdots$ can be used to absorb the resistances of coils used to realize the coupled inductances, and that the close coupling condition in all of these coil pairs is relieved if the given impedance has a pole at $s = \infty$. In this case the circuit will be provided with some series inductance at the input which may be added onto the first of the coupled inductances to relieve the close coupling situation there. The secondary inductance in this first pair of coupled coils may now be reduced in value and still leave this pair with less than unity coupling. The excess thus subtracted from the second inductance may be added onto the third so as to relieve the close coupling condition in the second pair of inductances, and so forth.

The distribution of input inductance may thus be carried out so as to yield equal coupling coefficients less than unity for all pairs of inductances. If the available series inductance is sufficient, one may thus achieve a practically reasonable total circuit for the realization of the given impedance.

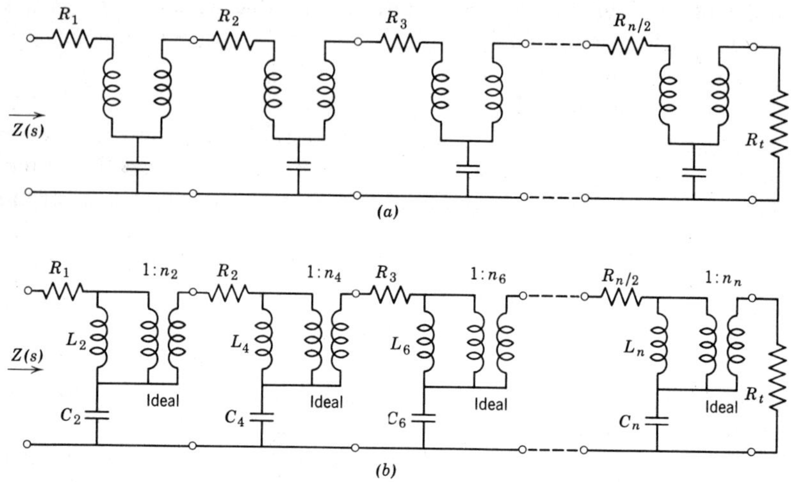

Fig. 20. Appearance of a Brune realization involving several successive cycles for its completion. Networks (*a*) and (*b*) are alternate equivalent forms.

5 An Alternate Form in Which to Express the Brune Process

The derivation and justification of the Brune process may be expressed more compactly by following the pattern used for the development of a lossless driving-point impedance in Art. 4 of Ch. 7. Figure 22 of that chapter, showing the general form of development, is changed only as indicated in Fig. 21 here, by the appearance of the resistances, R_1, R_2,

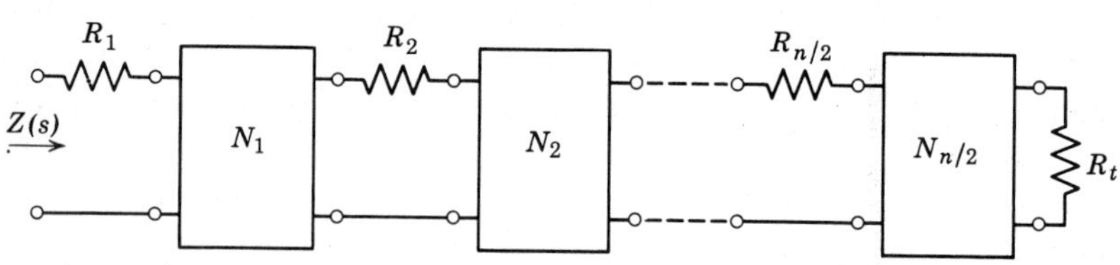

Fig. 21. Schematic form (involving lossless two terminal-pair networks) in which the Brune synthesis may be expressed.

· · · between the successive lossless two terminal-pair networks and by the resistance termination R_t. Each reactive network provides a pair of

j-axis zeros of the real part of the impedance at its respective input terminals, and these zeros collectively are zeros of the over-all transfer function* (in which we are not interested at the moment).

As indicated formally in Fig. 23 of Ch. 7, we consider a typical development cycle concerned with the extraction of one realizable lossless two terminal-pair network from the input impedance $Z_1(s)$, Eq. 56, so as to leave a p.r. remainder $Z_2(s)$.

The pertinent pair of real-part zeros (transmission zeros) we define by the factor

$$t(s) = s^2 + \omega_0^2 \tag{87}$$

and observe that the simplest set of open-circuit driving-point and transfer impedances capable of providing this zero is given by

$$z_{12} = \frac{Kt(s)}{s} = \frac{k_0}{s} + k_{12}s \tag{88}$$

$$z_{11} = \cdots = \frac{k_0}{s} + k_{11}s \tag{89}$$

$$z_{22} = \cdots = \frac{k_0}{s} + k_{22}s \tag{90}$$

in which K is a constant and Eq. 88 yields

$$k_0 = Kt(0) = K\omega_0^2 \tag{91}$$

$$k_{12} = K \tag{92}$$

The input impedance is assumed to have the specific form

$$Z_1(s) = \frac{Ap(s)}{q(s)} = \frac{A(s - s_{p1})(s - s_{p2}) \cdots (s - s_{pn})}{(s - s_1)(s - s_2) \cdots (s - s_n)} \tag{93}$$

Following the pattern in Art. 4 of Ch. 7 (specifically Eq. 67 given there), we write the relationship linking the five quantities $Z_1(s)$, $Z_2(s)$, z_{11}, z_{22}, and z_{12} in the form

$$(z_{11} - Z_1)(z_{22} + Z_2) = z_{12}^2 = \frac{K^2t^2(s)}{s^2} \tag{94}$$

from which we can immediately obtain the partial-fraction expansion

$$\frac{1}{z_{22} + Z_2} = \frac{s^2(z_{11} - Z_1)}{K^2t^2(s)} = \frac{1}{K^2}\sum_{\nu=1}^{n} \frac{-s_\nu^2 k_\nu}{t^2(s_\nu)} \frac{1}{s - s_\nu} \tag{95}$$

in which k_ν denotes the residues of $Z_1(s)$ in its poles $s_1, s_2, \cdots, s_n$. Thus the function $(z_{22} + Z_2)^{-1}$ contains only the poles of $Z_1(s)$ because the j-axis zeros defined by $t^2(s)$ must be contained in the factor $(z_{11} - Z_1)$.

We now observe that, since the j-axis poles of the z_{sk} are not contained

* The fact that real-part zeros are zeros of the over-all transfer function may easily be seen by noting that at such j-axis frequencies the input power, and hence the output power, must be zero.

in Z_1 or Z_2, the factors $(z_{11} - Z_1)$ and $(z_{22} + Z_2)$ reduce to z_{11} and z_{22} for the vicinities of these poles. This fact enables us to evaluate the residues of z_{11} and z_{22} directly from Eqs. 94 and 95. We obtain for $s \to \infty$

$$\frac{1}{k_{22}} = \frac{k_{11}}{K^2} = \frac{1}{K^2} \sum_{\nu=1}^{n} \frac{-s_\nu^2 k_\nu}{t^2(s_\nu)} \tag{96}$$

The reciprocal residue of z_{22} at $s = 0$ is given by the derivative of 95 at this point. Hence, with Eq. 91, we have

$$\frac{1}{k_0} = \frac{1}{K\omega_0^2} = \frac{1}{K^2} \sum_{\nu=1}^{n} \frac{k_\nu}{t^2(s_\nu)} \tag{97}$$

Since the terms in these sums are pairs of conjugate complex values, it follows from Eqs. 96 and 97 that the quantities K, k_0, and k_{11} are real and hence from Eq. 89 that $\text{Re}\,[z_{11}(j\omega)] \equiv 0$. Equation 95 then shows that $(z_{22} + Z_2)^{-1}$ is analytic in the right half-plane inclusive of the j-axis where its real part is positive because that of Z_1 is positive by hypothesis. Therefore, this function is p.r. and hence $(z_{22} + Z_2)$ is p.r. Its poles at $s = 0$ and $s = \infty$ belong to z_{22}; the rest is Z_2. Both of these are evidently p.r. and it follows that k_0 and k_{22} are positive. Equations 91 and 92 then yield positive values for K and k_{12}; and the residue condition $k_{11} = k_{12}^2/k_{22}$, which follows from Eqs. 92 and 96, secures a positive value for k_{11}. Thus the realizability of the lossless two terminal-pair network as well as of the remainder function are assured.

6 Darlington's Synthesis of a Driving-Point Impedance*

According to this method of realization, the p.r. driving-point impedance is shown to be representable as the input impedance to a lossless two terminal-pair network terminated in a single resistance which can always

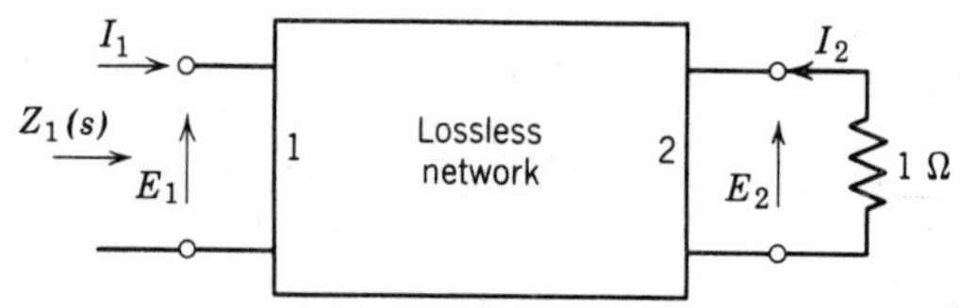

Fig. 22. Realization of a p.r. impedance by Darlington's method.

be normalized at 1 ohm. Schematically such a representation for the input impedance $Z_1(s)$ is shown in Fig. 22. The detailed discussion

* "Synthesis of Reactance 4 Poles." S. Darlington, *J. Math. and Phys.*, Vol. 18, Sept. 1939, pp. 257-353.

demonstrating the correctness of this statement and showing how such a realization may be accomplished proceeds from the familiar expression for the impedance $Z_1(s)$ given by

$$Z_1(s) = \frac{(z_{11}z_{22} - z_{12}^2) + z_{11}}{z_{22} + 1} \tag{98}$$

in which z_{11}, z_{22}, and z_{12} are the open-circuit driving-point and transfer impedances characterizing the single lossless two terminal-pair network.

If we abbreviate the determinant $z_{11}z_{22} - z_{12}^2$ of the set of z_{sk} by $|z|$, then the short-circuit driving-point and transfer admittances are given by the well-known relations

$$y_{11} = z_{22}/|z|, \qquad y_{22} = z_{11}/|z|, \qquad y_{12} = -z_{12}/|z| \tag{99}$$

Through factoring z_{11} out of 98 we can then write this expression in the form

$$Z_1(s) = z_{11}\frac{(1/y_{22}) + 1}{z_{22} + 1} \tag{100}$$

and as a quotient of polynomials we write for this impedance

$$Z_1(s) = \frac{P(s)}{Q(s)} = \frac{m_1 + n_1}{m_2 + n_2} \tag{101}$$

If we recall from the discussion in Art. 5 of Ch. 1 and Art. 1 of Ch. 2 that the ratio of any even part to any odd part of the polynomials $P(s)$ and $Q(s)$ yields a physically realizable reactance or susceptance function, a method of identifying expressions 100 and 101 becomes rather evident. Thus, if in 101 we factor m_1 out of the numerator and n_2 out of the denominator, we have

$$Z_1(s) = \frac{m_1}{n_2}\frac{(n_1/m_1) + 1}{(m_2/n_2) + 1} \qquad \text{(case A)} \tag{102}$$

whereupon a comparison with Eq. 100 suggests the set of identifications

$$z_{11} = \frac{m_1}{n_2}, \qquad z_{22} = \frac{m_2}{n_2}, \qquad y_{22} = \frac{m_1}{n_1} \qquad \text{(case A)} \tag{103}$$

Alternately one can interchange the letters m and n in expression 102. This yields for the impedance

$$Z_1(s) = \frac{n_1}{m_2}\frac{(m_1/n_1) + 1}{(n_2/m_2) + 1} \qquad \text{(case B)} \tag{104}$$

and comparison with Eq. 100 suggests the identifications

$$z_{11} = \frac{n_1}{m_2}, \qquad z_{22} = \frac{n_2}{m_2}, \qquad y_{22} = \frac{n_1}{m_1} \qquad \text{(case B)} \tag{105}$$

Although the driving-point functions 103 or 105 are guaranteed realizable simply from the fact that $Z_1(s)$ is p.r. and hence that the polynomials $P(s)$ and $Q(s)$ as well as those resulting from an interchange of their even or odd parts are Hurwitz, it does not follow that these function sets define a realizable lossless two terminal-pair network. In order to apply the pertinent results discussed in Ch. 7 to this situation, it is necessary first to determine the transfer impedance z_{12} associated with the driving-point functions given by Eqs. 103 or 105. As indicated in these two alternate sets of identifications, we find it convenient to refer to them as cases A and B. As mentioned above, the one follows from the other upon interchange of the letters m and n. Using the appropriate relation 99 we have for the z-determinant

$$z_{11}z_{22} - z_{12}^2 = z_{11}/y_{22} = n_1/n_2 \qquad \text{(case A)} \tag{106}$$

$$= m_1/m_2 \qquad \text{(case B)} \tag{107}$$

and hence the transfer impedance is found to be given by

$$z_{12} = \frac{\sqrt{m_1m_2 - n_1n_2}}{n_2} \qquad \text{for case A} \tag{108}$$

and

$$z_{12} = \frac{\sqrt{n_1n_2 - m_1m_2}}{m_2} \qquad \text{for case B} \tag{109}$$

The quantity appearing under the radical in either of these expressions we recall from the discussions in Chs. 1 and 8 to be the numerator of the even part of the impedance $Z_1(s)$ in Eq. 101. For convenience we repeat here the notation for this even polynomial as given in Eqs. 56 and 57 of Ch. 1

$$A(-s^2) = m_1m_2 - n_1n_2 = A_n(s_1^2 - s^2)(s_2^2 - s^2) \cdots (s_n^2 - s^2) \tag{110}$$

and recall further that the p.r. character of $Z_1(s)$ guarantees only that all factors $(s_\nu^2 - s^2)$ involving negative real s_ν^2 values have even multiplicity. In general this polynomial is therefore not a full square, and expression 108 or 109 will not yield a rational function for z_{12} as it should.

The cure for this difficulty is found in augmenting the given impedance 101 by means of an auxiliary Hurwitz polynomial $P_0(s) = m_0 + n_0$ as follows

$$Z_1(s) = \frac{P(s) \cdot P_0(s)}{Q(s) \cdot P_0(s)} = \frac{(m_1 + n_1)(m_0 + n_0)}{(m_2 + n_2)(m_0 + n_0)} \tag{111}$$

Trivial as this step may seem, we observe that multiplication of numerator and denominator by $(m_2 - n_2)(m_0 - n_0)$, as is usual in the formation of the even part, yields

$$\frac{(m_1 + n_1)(m_2 - n_2)}{(m_2 + n_2)(m_2 - n_2)} \frac{(m_0^2 - n_0^2)}{(m_0^2 - n_0^2)} \tag{112}$$

and since $(m_0^2 - n_0^2)$ is even, we see that the even part of $Z_1(s)$ is simply augmented by this even polynomial; that is

$$\tfrac{1}{2}[Z_1(s) + Z_1(-s)] = \frac{(m_1m_2 - n_1n_2)(m_0^2 - n_0^2)}{(m_2^2 - n_2^2)(m_0^2 - n_0^2)} \tag{113}$$

We could, of course, have started by augmenting the even part of $Z_1(s)$ in this way and then pointed out that, since the effect upon the even part is trivial, it must (by the methods discussed in Ch. 8) generate the same impedance as before, and so the effect upon it is trivial also.

In any case, the operation indicated in Eq. 113 solves our difficulty since the polynomial 110 is now multiplied by the factor $(m_0^2 - n_0^2)$ which we can choose so as to render the revised polynomial

$$\tilde{A}(-s^2) = (m_1m_2 - n_1n_2)(m_0^2 - n_0^2) \tag{114}$$

in the form of a full square. This result is evidently accomplished by letting $(m_0^2 - n_0^2)$ equal a product of those factors $(s_\nu^2 - s^2)$ in the representation 110 which are not evenly repeated, so that in the revised polynomial 114 all zero factors have even multiplicity. From the even polynomial $(m_0^2 - n_0^2)$ thus formed we construct the Hurwitz polynomial $(m_0 + n_0)$ in the same manner that we normally construct $(m_2 + n_2)$ from $(m_2^2 - n_2^2)$ in the process of finding an impedance from its real part, as discussed in Art. 1, Ch. 8. From here on, let us assume that the given impedance $Z_1(s)$ is already in this appropriately augmented form, so that when we arrive at expression 108 or 109 for z_{12} we find this function to be rational.

A few additional remarks are in order at this point regarding the augmentation process just described. In this connection we observe first of all that what distinguishes cases A and B in Eqs. 108 and 109 is evidently the question as to whether the polynomial $\sqrt{m_1m_2 - n_1n_2}$ is even or odd respectively. This conclusion follows from the observation that z_{12}, like z_{11} and z_{22}, must be given by the ratio of two polynomials that are respectively even and odd (as pointed out in Art. 1 of Ch. 2 and emphasized again in Ch. 7). In Eq. 108 for case A, the denominator polynomial is odd; therefore the numerator polynomial must be even. In Eq. 109 for case B, this situation is reversed.

A typical case B arises if in the representation 110 we find a zero s_ν^2 value. For example, if $s_1^2 = 0$, Eq. 110 reads

$$A(-s^2) = m_1m_2 - n_1n_2 = -s^2(s_2^2 - s^2) \cdots (s_n^2 - s^2)A_n \tag{115}$$

or

$$-A(-s^2) = n_1n_2 - m_1m_2 = s^2(s_2^2 - s^2) \cdots (s_n^2 - s^2)A_n \tag{116}$$

Since the tacitly assumed augmentation process has yielded all factors $(s_2^2 - s^2) \cdots (s_n^2 - s^2)$ in the form of evenly repeated sets, we see that 116 is a full square and its square root is an odd polynomial with positive coefficients as required.

In the augmentation process we can, if we wish, include a factor s^2

along with the others that are needed. If we do so, this situation reverts to a case A. There is no point in such manipulation although it is permissible. It is similarly permissible to overaugment with any evenly repeated factors $(s_\nu^2 - s^2)$ that we care to write down. Such methods merely result in unnecessarily complex expressions for the z_{sk} and hence yield more elaborate networks than are needed. In the interest of economy of network elements, one should augment with the minimum number of factors necessary to obtain a rational function for z_{12}. Thus cases A and B are automatically indicated by the character of the polynomial $(m_1m_2 - n_1n_2)$. The situation is always unique and the two cases are mutually exclusive so that when one of the relations 108 or 109 applies, the other one does not.

We must now see if the set of impedances z_{11}, z_{22}, and z_{12} defines a realizable lossless two terminal-pair network. As pointed out in Ch. 7 where the synthesis of such networks is discussed, the residue condition

$$k_{11}k_{22} - k_{12}^2 \geq 0 \tag{117}$$

as pointed out also in Eq. 29 of Ch. 2, is all that needs to be considered here in addition to the matters already disposed of. Suppose we evaluate the residues, using Eqs. 103 and 108 pertinent to case A. We have

$$k_{11} = \left(\frac{m_1}{n_2'}\right)_{s=s_\nu}, \quad k_{22} = \left(\frac{m_2}{n_2'}\right)_{s=s_\nu}, \quad k_{12} = \left[\frac{\sqrt{m_1m_2}}{n_2'}\right]_{s=s_\nu} \tag{118}$$

in which the prime on n_2 indicates differentiation with respect to s. The frequency s_ν is a zero of the polynomial n_2 and hence is a common j-axis pole of the impedances z_{sk}. Substitution of the results 118 into 117 shows that this condition is fulfilled with an equal sign. According to statement 57 of Art. 3, Ch. 7, we could have predicted this result since the impedance $Z_1(s)$, as in the Brune synthesis, is assumed to have no j-axis poles.

The conditions for the realizability of the lossless network are thus seen to be fulfilled and the method of constructing the pertinent z_{sk} is established. Hence the realization of any p.r. impedance in the form of Fig. 22 is always assured. The preamble, as discussed for the Brune process, is tacitly assumed to have been previously carried out, and the network portion which it yields can be regarded as included with the lossless network shown. Thus any p.r. impedance, whether minimum reactive and minimum susceptive or not, is always realizable in this form.

7 Some Illustrative Examples

For our first example we shall consider the simple *RC* circuit of Fig. 23 and realize its impedance by the Darlington method. There is no practical value in this exercise, of course, but it is of some theoretical interest to see an RC network represented as an LC network terminated in resistance. The impedance reads

$$Z(s) = \frac{1 + s}{1 + 2s} \tag{119}$$

We find

$$m_1m_2 - n_1n_2 = 1 - 2s^2 \tag{120}$$

and so we must choose

$$m_0^2 - n_0^2 = 1 - 2s^2 \tag{121}$$

giving for the auxiliary Hurwitz polynomial

$$P_0 = m_0 + n_0 = 1 + \sqrt{2}s \tag{122}$$

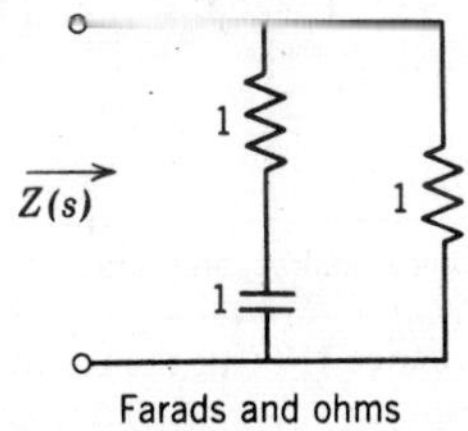

Fig. 23. Example of an RC impedance to be realized by Darlington's method.

Augmenting $Z(s)$ with this polynomial we have

$$Z(s) = \frac{(1 + s)(1 + \sqrt{2}s)}{(1 + 2s)(1 + \sqrt{2}s)} = \frac{1 + (1 + \sqrt{2})s + \sqrt{2}s^2}{1 + (2 + \sqrt{2})s + 2\sqrt{2}s^2} \tag{123}$$

As a check, we again compute $(m_1m_2 - n_1n_2)$ and find

$$m_1m_2 - n_1n_2 = (1 - 2s^2)^2 \tag{124}$$

which is a full square as required. The situation is evidently case A and so we use Eqs. 103 and 108 to get

$$z_{11} = \frac{1 + \sqrt{2}s^2}{(2 + \sqrt{2})s} = \frac{1}{(2 + \sqrt{2})s} + \frac{s}{1 + \sqrt{2}} \tag{125}$$

$$z_{22} = \frac{1 + 2\sqrt{2}s^2}{(2 + \sqrt{2})s} = \frac{1}{(2 + \sqrt{2})s} + \frac{2s}{1 + \sqrt{2}} \tag{126}$$

$$z_{12} = \frac{1 - 2s^2}{(2 + \sqrt{2})s} = \frac{1}{(2 + \sqrt{2})s} - \frac{\sqrt{2}s}{1 + \sqrt{2}} \tag{127}$$

The resulting network, obtained by the method discussed in Art 1, Ch. 7, is shown in Fig. 24. The pair of inductances are, of course, closely coupled since the residue condition is consistently fulfilled with the equal sign.

Next we shall synthesize the impedance

$$Z(s) = \frac{s^2 + s + 1}{s^2 + s + 4} = \frac{m_1 + n_1}{m_2 + n_2} \tag{128}$$

by both the Brune and Darlington methods. Starting with the Brune method, we first compute the real part for $s = j\omega$

$$\text{Re}\,[Z(j\omega)] = \left(\frac{m_1m_2 - n_1n_2}{m_2^2 - n_2^2}\right)_{s=j\omega} = \frac{\omega^4 - 4\omega^2 + 4}{\omega^4 - 7\omega^2 + 6} \tag{129}$$

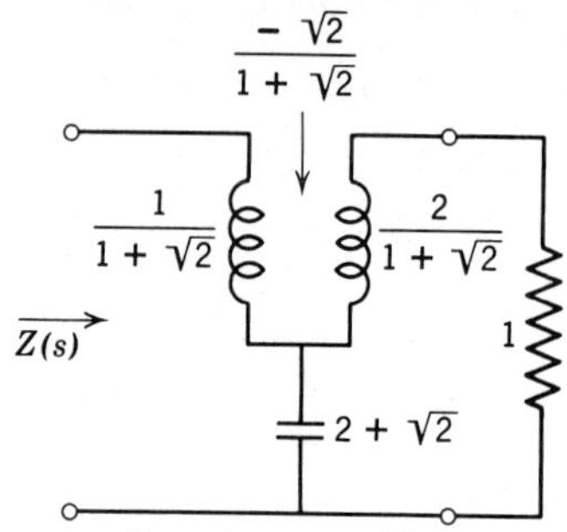

Fig. 24. Network realizing impedance 119 (for the circuit of Fig. 23) in the Darlington form.

We can see by inspection that the numerator is $(\omega^2 - 2)^2$, and hence that this real part is zero for $\omega = \pm\sqrt{2}$. Subtraction of the minimum real part as in Eq. 56 is therefore unnecessary. $Z(s)$ and $Z_1(s)$ are the same function. We next compute the reactive value of $Z_1(s)$ for $s = j\omega_1 = j\sqrt{2}$, thus

$$Z_1(j\sqrt{2}) = \frac{-2 + j\sqrt{2} + 1}{-2 + j\sqrt{2} + 4} = \frac{-1 + j\sqrt{2}}{2 + j\sqrt{2}} = \frac{j(\sqrt{2} + j)}{\sqrt{2}(\sqrt{2} + j)} = \frac{j}{\sqrt{2}} \tag{130}$$

Although X is positive, we stay with the impedance basis and compute

$$L_1 = \frac{X}{\omega_1} = \frac{1/\sqrt{2}}{\sqrt{2}} = \frac{1}{2} \tag{131}$$

Applying Eq. 59, we have

$$Z_1(s) - L_1s = \frac{s^2 + s + 1}{s^2 + s + 4} - \frac{s}{2} = \frac{-s^3 + s^2 - 2s + 2}{2(s^2 + s + 4)} = \frac{(s^2 + 2)(1 - s)}{2(s^2 + s + 4)} \tag{132}$$

The reciprocal function has the required j-axis pole. We can evaluate its residue in this pole by using Eq. 88 of Ch. 1. This gives

$$\begin{aligned} k_\nu &= \left(\frac{m_1}{n_2'}\right)_\nu = \left[\frac{2(s^2 + 4)}{-3s^2 - 2}\right]_{s^2=-2} = 1 \\ &= \left(\frac{n_1}{m_2'}\right)_\nu = \left(\frac{2s}{2s}\right)_{s^2=-2} = 1 \end{aligned} \tag{133}$$

The subtraction according to Eq. 61 reads

$$\frac{2(s^2 + s + 4)}{(s^2 + 2)(1 - s)} - \frac{2s}{s^2 + 2} = \frac{4}{1 - s} = \frac{1}{W(s)} \tag{134}$$

Finally

$$W(s) = -\frac{s}{4} + \frac{1}{4} \tag{135}$$

In summary, the pertinent element values are seen to be

$$L_1 = \tfrac{1}{2}, \qquad L_2 = \tfrac{1}{2}, \qquad C_2 = 1, \qquad L_3 = -\tfrac{1}{4} \tag{136}$$

and the remainder is a resistance equal to $\frac{1}{4}$ ohm. The circuit, after conversion of the inductive T to a pair of mutually coupled inductances, is shown in Fig. 25.

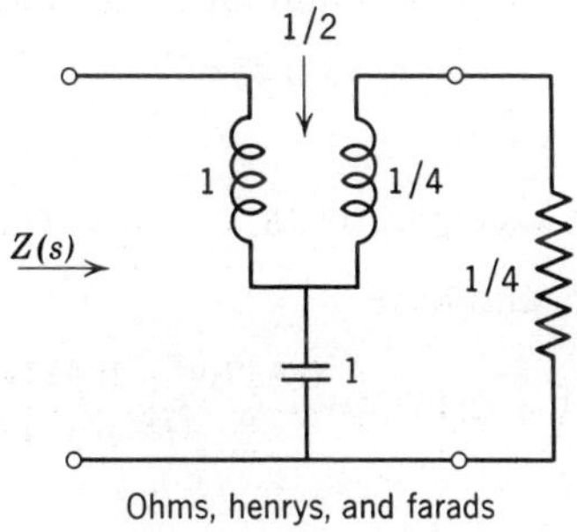

Fig. 25. Network realizing impedance 128 in the Brune form.

In solving this same problem by the Darlington method we observe from Eq. 129 that the polynomial

$$m_1 m_2 - n_1 n_2 = s^4 + 4s^2 + 4 = (s^2 + 2)^2 \tag{137}$$

is a full square and no augmentation is therefore needed. The situation is again case A, and Eqs. 103 and 108 give

$$z_{11} = \frac{s^2 + 1}{s} = \frac{1}{s} + s \tag{138}$$

$$z_{22} = \frac{s^2 + 4}{s} = \frac{4}{s} + s \tag{139}$$

$$z_{12} = \frac{s^2 + 2}{s} = \frac{2}{s} + s \tag{140}$$

To avoid an ideal transformer (according to the reasoning leading to relations 20 in Ch. 7) we multiply the impedance level on the output side of the lossless network determined by the above impedances by the factor $\frac{1}{4}$. That is to say, we multiply z_{22} by $\frac{1}{4}$, z_{12} by $\frac{1}{2}$, and terminate in $\frac{1}{4}$ ohm instead of in 1 ohm. The input impedance is obviously not affected by this process. The resultant network is immediately recognized to be the one in Fig. 25 that was just obtained by the Brune method.

Let us now consider the impedance

$$Z(s) = \frac{s^2 + s + 1}{s^2 + s + 2} \tag{141}$$

which is only slightly different from the impedance 128 used in the previous example, but for which the Brune and Darlington synthesis methods are more completely illustrated, as will presently be seen. We begin with the Brune synthesis and form the real part

$$\text{Re}\,[Z(j\omega)] = \frac{\omega^4 - 2\omega^2 + 2}{\omega^4 - 3\omega^2 + 4} \tag{142}$$

We find this real part to have a minimum value equal to

$$R_1 = 0.453 \tag{143}$$

at the frequency

$$\omega_1{}^2 = 2 - \sqrt{2} = 0.586, \qquad \omega_1 = 0.7655 \tag{144}$$

We therefore remove R_1 and have

$$Z_1(s) = Z(s) - 0.453 = \frac{0.547s^2 + 0.547s + 0.094}{s^2 + s + 2} \tag{145}$$

Next we compute

$$Z_1(j0.765) = j0.296 = j0.765L_1 \tag{146}$$

so that

$$L_1 = \frac{0.296}{0.765} = 0.387 \tag{147}$$

Subtracting this series inductance yields

$$Z_1(s) - L_1 s = \frac{-0.387s^3 + 0.160s^2 - 0.227s + 0.094}{s^2 + s + 2}$$

$$= \frac{(s^2 + 0.586)(-0.387s + 0.160)}{s^2 + s + 2} \tag{148}$$

The residue of the reciprocal function in the pole at $s = j\omega_1$ is most easily evaluated from the expression (Eq. 88, Ch. 1)

$$k_\nu = \left(\frac{m_1}{n_2'}\right)_\nu = \left(\frac{s}{0.32s}\right)_{s=j\omega_1} = \frac{1}{0.32} \tag{149}$$

Since this value equals $1/2L_2$ we have

$$L_2 = 0.16 \quad \text{and} \quad C_2 = \frac{1}{L_2\omega_1{}^2} = \frac{1}{0.586 \times 0.16} = 10.66 \tag{150}$$

The subtraction 61 yields

$$\left(\frac{s^2 + s + 2}{-0.387s + 0.160} - \frac{s}{0.16}\right)\frac{1}{s^2 + 0.586} = \frac{1}{-0.113s + 0.047} = \frac{1}{W(s)} \tag{151}$$

so

$$W(s) = -0.113s + 0.047 \tag{152}$$

The complete set of element values is given by

$$R_1 = 0.453, \quad L_1 = 0.387, \quad L_2 = 0.16, \quad C_2 = 10.66, \quad L_3 = -0.113 \tag{153}$$

and the terminating resistance equals 0.047. The resulting circuit is shown in Fig. 26.

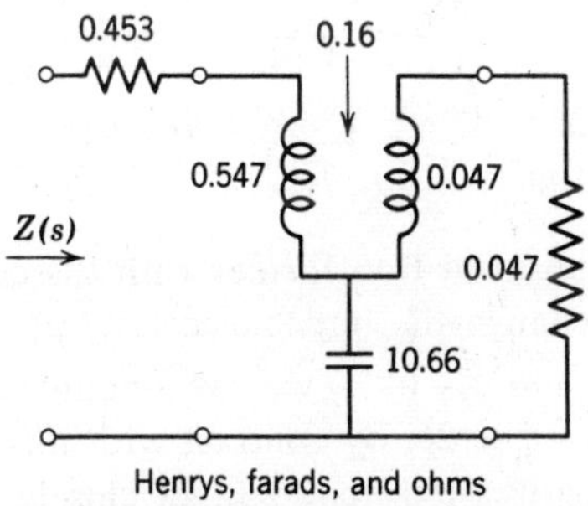

Fig. 26. Realization of impedance 141 by the Brune method.

It is interesting to note that the zero-frequency value of the impedance is given by the sum of the values 0.453 and 0.047 which equals 0.500 as Eq. 141 shows it should. At very large frequencies, the closely coupled inductances, according to the equivalent circuit in Fig. 19, act as an ideal transformer that transforms the terminal resistance of 0.047 ohm to the value 0.547 so that the net input impedance has the asymptotic value $0.453 + 0.547 = 1$ ohm, which again checks with the result obtained by inspection of Eq. 141.

In approaching the synthesis of this same impedance by Darlington's method, we first compute

$$m_1m_2 - n_1n_2 = (s^2 + 1)(s^2 + 2) - s^2 = s^4 + 2s^2 + 2 \tag{154}$$

which is not a full square, and so we must choose

$$m_0{}^2 - n_0{}^2 = s^4 + 2s^2 + 2 \tag{155}$$

If we write the auxiliary Hurwitz polynomial

$$P_0(s) = m_0 + n_0 = s^2 + 2\sigma s + \sigma^2 + \omega^2$$

then

$$s^4 + 2s^2 + 2 = P_0(s)P_0(-s) = s^4 + 2(\omega^2 - \sigma^2)s^2 + (\omega^2 + \sigma^2)^2 \tag{156}$$

from which we have

$$\omega^2 - \sigma^2 = 1, \quad \omega^2 + \sigma^2 = \sqrt{2}, \quad \omega^2 = (\sqrt{2} + 1)/2, \quad \sigma^2 = (\sqrt{2} - 1)/2 \tag{157}$$

The desired polynomial is thus easily found to be

$$P_0(s) = s^2 + 0.910s + 1.414 \tag{158}$$

Augmenting the impedance 141 with this polynomial and applying Eqs. 103 and 108, we find for the z_{sk}

$$z_{11} = \frac{s^4 + 3.324s^2 + 1.414}{1.91s(s^2 + 1.693)} = \frac{1}{2.29s} + \frac{0.416s}{s^2 + 1.693} + 0.524s \tag{159}$$

$$z_{22} = \frac{s^4 + 4.324s^2 + 2.828}{1.91s(s^2 + 1.693)} = \frac{1}{1.145s} + \frac{0.503s}{s^2 + 1.693} + 0.524s \tag{160}$$

$$z_{12} = \frac{s^4 + 2s^2 + 2}{1.91s(s^2 + 1.693)} = \frac{1}{1.617s} + \frac{0.459s}{s^2 + 1.693} + 0.524s \tag{161}$$

In order to avoid an ideal transformer with the capacitance, we again find it expedient to change the impedance level at the output; so, after multiplying z_{22} by $\frac{1}{2}$ and z_{12} by $1/\sqrt{2}$, we obtain the circuit of Fig. 27. The Brune network of Fig. 26, by contrast with this one, is considerably less elaborate and involves only one pair of closely coupled inductances instead of two.

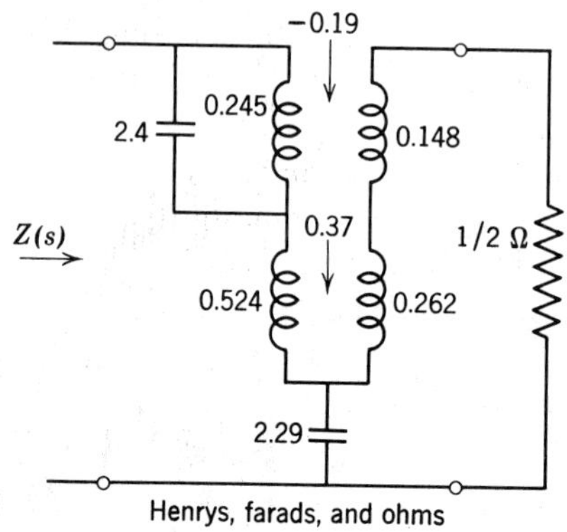

Fig. 27. Realization of impedance 141 by the Darlington method.

This example affords a good opportunity to illustrate the value of the more primitive, although less general, methods of synthesis discussed in Art. 3 above. If we consider the impedance 141 as an example of the form 45 to which conditions 47 and 48 apply, we see that the present impedance fulfills the conditions for realization as case B which allows a separation of the given function into two additive terms according to Eq. 42. Since in our present example we have

$$Z(s) = \frac{s^2 + s + 1}{s^2 + s + 2} = 1 - \frac{1}{s^2 + s + 2} \tag{162}$$

we see, in the notation of Eq. 42, that $R_0 = 1$, $a_1 = 0$, and $a_0 = -1$. Hence this equation reads

$$Z(s) = \frac{\frac{1}{2}(s^2 + 2)}{s^2 + s + 2} + \frac{(s^{-1}/2) + \frac{1}{4}}{s^{-2} + (s^{-1}/2) + \frac{1}{2}} \tag{163}$$

which yields the network shown in Fig. 28. It is quite apparent that when the simpler methods of synthesis are applicable, they are not only computationally easier but the resulting networks are practically preferable. One should use the general methods of synthesis only as a last resort, or when there is some other reason for wanting the particular structural forms which they afford.

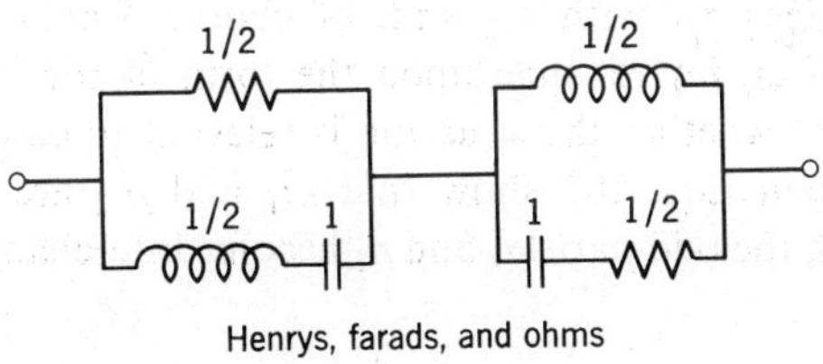

Fig. 28. Realization of impedance 141 by the more primitive method discussed in art. 3 (Eqs. 45, 47, 48).

8 Further Considerations of the Darlington Theory

Darlington's method is only incidentally of interest as a means for the synthesis of driving-point impedances. Its chief value from a practical standpoint lies in the particular form of the Darlington realization; namely, that it yields the impedance as the input to a lossless two terminal-pair network terminated in a single resistance. This fact almost immediately suggests a way of designing lossless two terminal-pair networks for prescribed transfer characteristics, and it is primarily the value of the Darlington theory in this regard that makes it practically useful and important. So far as driving-point synthesis procedures are concerned, we shall later discuss methods whereby more practical configurations avoiding closely coupled inductances, or avoiding inductive coupling altogether, can always be obtained (although at the expense of additional circuit elements). For the sake of providing an adequate background for these methods as well as for a subsequent discussion of transfer function synthesis, it is useful at this time to present a fuller picture of the principles which are part of the so-called *Darlington theory*.

In this regard it is interesting to compare the situation in Fig. 22 with that in Fig. 29 and inquire as to how the two impedances $Z_1(s)$ and $Z_2(s)$ are related when the lossless network is the same in both cases. This question is equivalent to asking how the input impedance changes if the lossless network is reversed so that its ends 1 and 2 become interchanged.

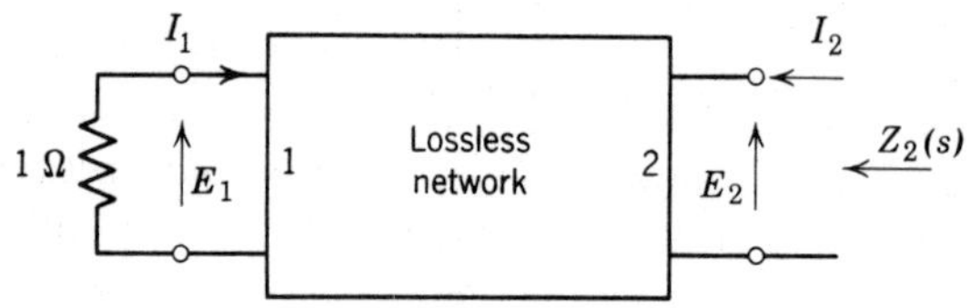

Fig. 29. Situation to be compared with that shown in Fig. 22 for evaluation of the manner in which $Z_1(s)$ and $Z_2(s)$ are related.

The answer is easily found by noting that such reversal of the lossless network interchanges z_{11} with z_{22} and, of course, leaves z_{12} the same. The effect of such an interchange upon the form of the impedance 101 differs according to whether the situation is relevant to case A or case B. If it is case A, then Eqs. 103 show that m_1 and m_2 are interchanged, whereas for case B, the odd parts n_1 and n_2 become interchanged. Thus if, as in Eq. 101

$$Z_1(s) = \frac{m_1 + n_1}{m_2 + n_2} \tag{164}$$

then for case A

$$Z_2(s) = \frac{m_2 + n_1}{m_1 + n_2} \tag{165}$$

and for case B

$$Z_2(s) = \frac{m_1 + n_2}{m_2 + n_1} \tag{166}$$

These results are interesting and useful in several ways. For example, if, in an appropriate situation, we are developing $Z_1(s)$ into a lossless ladder (or other) network, and we are interested in how this development ends (we might, for example, be interested in the value of the last reactive element), then we can form $Z_2(s)$ without any computational effort at all and begin the development from the opposite end.* We shall have use for this technique later on.

The following special situations are likewise readily formulated.

* Z_1 must, of course, be in the properly augmented form.

Suppose, under case A, we have $m_1 \equiv m_2$: Eqs. 164 and 165 at once show that then $Z_1(s) \equiv Z_2(s)$, which implies that turning the lossless network end for end has no effect upon driving-point or transfer impedances. The network is electrically symmetrical. On the other hand, if $n_1 \equiv n_2$ and the situation is case A, Eqs. 164 and 165 show that $Z_1(s)$ and $Z_2(s)$ are reciprocal, which is the same as saying that reciprocating the network (forming its dual) and turning it end for end are equivalent. In contrast to the electrically symmetrical network, such a one is referred to as being *antimetrical* or *anti*symmetrical. We can summarize these results as follows:

For case A: If $m_1 \equiv m_2$, then $Z_2(s) \equiv Z_1(s)$.
The network is symmetrical.
If $n_1 \equiv n_2$, then $Z_2(s) = 1/Z_1(s)$.
The network is antimetrical. (167)

When the situation is case B, we have the same conclusions but with an interchange of the even and odd parts m and n. Thus

For case B: If $m_1 \equiv m_2$, then $Z_2(s) \equiv 1/Z_1(s)$.
The network is antimetrical.
If $n_1 \equiv n_2$, then $Z_2(s) \equiv Z_1(s)$.
The network is symmetrical. (168)

Since symmetrical and antimetrical structures play an important part in the practical design of lossless coupling networks, it is useful to be able to predict or control the occurrence of these special results. The properties expressed in statements 167 and 168 provide a simple means for accomplishing this end.

Additional useful results are obtained by observing the manner in which the driving-point and transfer impedances for the resistance-loaded lossless network of Fig. 22 are related. While $Z_1(s) = E_1/I_1$, the loaded transfer impedance is defined as $Z_{12}(s) = E_2/I_1$. By use of Thévenin's theorem, it is not difficult to see that

$$Z_{12}(s) = \frac{z_{12}}{1 + z_{22}} \tag{169}$$

Thus, if we assume the current I_1 applied, the open-circuit Thévenin voltage produced by it is $I_1 z_{12}$. The current $-I_2$ is given by this voltage divided by $1 + z_{22}$, and the load voltage E_2 is numerically equal to $-I_2$.

Use of Eqs. 103 and 108 pertinent to case A now shows that

$$Z_{12}(s) \times Z_{12}(-s) = \frac{m_1 m_2 - n_1 n_2}{m_2^2 - n_2^2} = \tfrac{1}{2}[Z_1(s) + Z_1(-s)] \tag{170}$$

and it is obvious that the same result also is obtained for case B. Since z_{12} and z_{22} have the same denominator polynomial, Eq. 169 shows that the finite zeros of Z_{12} are those of z_{12}, and these must form a pattern in the s-plane that is completely symmetrical with respect to both the real and imaginary axes because the numerator polynomial of z_{12} is either even or odd. $Z_{12}(s)$ and $Z_{12}(-s)$ therefore have the same zeros, and these are the double zeros of $(m_1m_2 - n_1n_2)$—double because this polynomial, as has been seen above, must be a full square. These are also the double-order zeros of the even part of the input impedance which is given by the right-hand side of Eq. 170. The poles of $Z_1(s)$ and those of $Z_{12}(s)$ are, of course, the same, since they are the natural frequencies of the total network in Fig. 22.

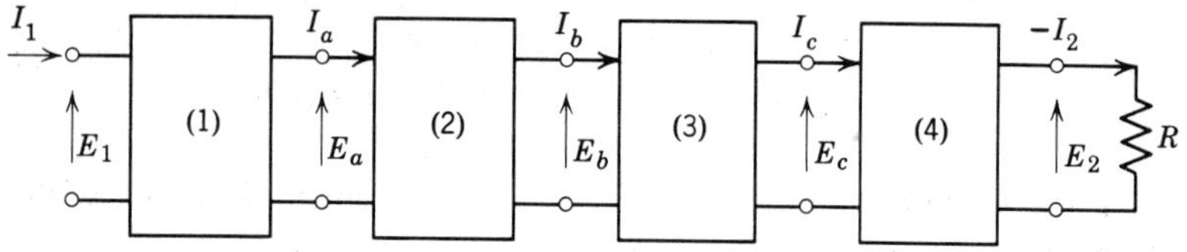

Fig. 30. Cascade of lossless coupling networks (terminated in resistance) to which relation 172 applies.

Suppose we now consider a cascade of lossless coupling networks terminated in a resistance, as shown in Fig. 30. For this arrangement we wish to write the over-all voltage or current transfer ratio expressed in terms of the impedances or admittances of the component coupling networks. The quantities

$$\begin{aligned} Z_1 &= E_1/I_1, \qquad Z_a = E_a/I_a, \qquad Z_b = E_b/I_b \\ Z_c &= E_c/I_c, \qquad R = E_2/(-I_2) \end{aligned} \tag{171}$$

are the various impedances looking to the right at the respective terminal pairs. If superscripts on z_{sk} are used to denote the component network to which they refer, we obtain

$$\frac{-I_2}{I_1} = \frac{z_{12}^{(1)}}{z_{22}^{(1)} + Z_a} \times \frac{z_{12}^{(2)}}{z_{22}^{(2)} + Z_b} \times \frac{z_{12}^{(3)}}{z_{22}^{(3)} + Z_c} \times \frac{z_{12}^{(4)}}{z_{22}^{(4)} + R} = \frac{Z_{12}}{R} \tag{172}$$

where Z_{12} is the over-all transfer impedance E_2/I_1.

We can easily recognize the correctness of this expression by means of Thévenin's theorem. Thus $I_1 z_{12}^{(1)}$ is the open-circuit voltage at the terminal pair a, and $z_{22}^{(1)} + Z_a$ is the impedance which this voltage must overcome to produce the current I_a. Hence the first factor in Eq. 172 is I_a/I_1. The second factor similarly is I_b/I_a, and so forth.

What the total expression shows is that the zeros of Z_{12} (the transmission zeros of the total network) are collectively the zeros of $z_{12}^{(1)}$, $z_{12}^{(2)}$, and so forth. Each network inserts some transmission zeros, and these taken together are the transmission zeros of the total network.

There may be some thought that expression 172 can have some additional zeros where a denominator factor like $z_{22}^{(1)} + Z_a$ or $z_{22}^{(2)} + Z_b$ has poles. This is an erroneous thought, as the following reasoning shows. The poles of $z_{22}^{(1)} + Z_a$ are those of $z_{22}^{(1)}$ and those of Z_a. The poles of $z_{22}^{(1)}$ are also poles of $z_{12}^{(1)}$, so the first factor in Eq. 172 does not become zero at these frequencies. The poles of Z_a are the zeros of $z_{22}^{(2)} + Z_b$ since these are natural frequencies of the network of Fig. 30 with the first box removed. The impedances Z_a and $z_{22}^{(2)} + Z_b$ are respectively soldering-iron–type and pliers-type impedances of this same network portion. Next, the poles of $z_{22}^{(2)}$ are canceled by the same poles in $z_{12}^{(2)}$, and the poles of Z_b are zeros of $z_{22}^{(3)} + Z_c$. The poles of $z_{22}^{(3)}$ are canceled by those of $z_{12}^{(3)}$, and the poles of Z_c are the zeros of $z_{22}^{(4)} + R$. Finally, the poles of $z_{22}^{(4)}$ are canceled by those of $z_{12}^{(4)}$. So in the end, the only zeros of Z_{12} are, as stated above, the various zeros produced by the separate transfer impedances $z_{12}^{(1)}$, $z_{12}^{(2)}$, and so forth. One can, of course, create degenerate situations in which some trivial exceptions to this conclusion are possible, but these are of no importance in the present argument since, in an intelligent approach to our problem, such matters are always controllable.

If now we are mindful of the conclusion drawn in connection with Eq. 170—namely, that the zeros of the transfer impedance of a network are the even-part zeros of its input impedance—then in the network of Fig. 30 we see that the even-part zeros of $Z_1(s) = E_1/I_1$ are collectively the transmission zeros of boxes 1, 2, 3, and 4. The even-part zeros of $Z_a(s) = E_a/I_a$ are collectively the transmission zeros of boxes 2, 3, and 4; the even-part zeros of $Z_b(s) = E_b/I_b$ are the transmission zeros of boxes 3 and 4; and the even-part zeros of $Z_c(s) = E_c/I_c$, finally, are simply the zeros of $z_{12}^{(4)}$—the transmission zeros of the last box.

These conclusions are useful if (as will be discussed in the following article) we are interested in making a cascade development of a driving-point impedance. In the first lossless network certain even-part zeros of this impedance are extracted and placed in evidence as the transmission zeros of this network. The remainder function (Z_a in Fig. 30) has even-part zeros which are those of the even part of Z_1 not produced by the first network. In other words, even-part zeros stay with us; those that are not extracted by the first component network in the development appear as even-part zeros in the remainder function. The various cycles in the development amount to successive steps in which certain ones of the total even-part zeros of the given input function are removed. The process terminates when there are no more even-part zeros left, in which case the remainder impedance is reduced to a constant—the terminating resistance.

Observe that everything that we are saying here pertains to situations in which the only resistance present is the terminating resistance. It is not, for example, pertinent to the Brune development except in the very special case in which all even-part zeros lie upon the j-axis. In this case the real-part plot shown in Fig. 14 exhibits all minima equal to zero. The cascade development shown schematically in Fig. 21 then involves no series resistances $R_1, R_2, \cdots$, and the total realization is equivalent to that obtained by the Darlington method. To such a situation one needs to study the real part only for the *given* impedance to determine its j-axis zeros, since these do not shift as we proceed from one cycle to the next. If not all but several of the j-axis real-part minima are zero, these conclusions still hold for those that are zero.

It is useful to obtain also the complete dual to the expression in Eq. 172. This reads

$$\frac{E_2}{E_1} = \frac{-y_{12}^{(1)}}{y_{22}^{(1)} + Y_a} \times \frac{-y_{12}^{(2)}}{y_{22}^{(2)} + Y_b} \times \frac{-y_{12}^{(3)}}{y_{22}^{(3)} + Y_c} \times \frac{-y_{12}^{(4)}}{y_{22}^{(4)} + G} = \frac{Y_{12}}{G} \tag{173}$$

where $G = R^{-1}$, the y_{sk} are the familiar short-circuit admittances, and the $Y_a, Y_b, \cdots$ are the reciprocals of the $Z_a, Z_b, \cdots$ appearing in Eq. 172.

A mixed form is also possible and useful in certain situations. For example, we have

$$\frac{E_2}{I_1} = \frac{z_{12}^{(1)}Z_a}{z_{22}^{(1)} + Z_a} \times \frac{-y_{12}^{(2)}Y_b}{y_{22}^{(2)} + Y_b} \times \frac{z_{12}^{(3)}Z_c}{z_{22}^{(3)} + Z_c} \times \frac{-y_{12}^{(4)}}{y_{22}^{(4)} + G} = Z_{12} \tag{174}$$

9 Cascade Synthesis Applied to the Darlington Theory

Here we shall apply the method of cascade synthesis, as discussed for lossless network functions in Art. 4 of Ch. 7 and for the Brune development in Art. 5 above, to the realization of a p.r. driving-point impedance according to the Darlington theory.

As mentioned before, the zeros of a lossless transfer impedance form a pattern in the s-plane that has complete symmetry with respect to both the real and imaginary axes. Specifically, these zeros may come in pairs on the real axis, pairs on the j-axis, or in the form of sets of quadruplets in the complex plane. We shall discuss the typical development cycle with respect to a quadruplet of transmission zeros since this is the most general situation.

The even-part zeros of the driving-point impedance to be produced by the cycle under discussion are defined by the factors

$$\begin{aligned} t(s) &= (s - s_0)(s - \bar{s}_0) = t_+ \\ t(-s) &= (s + s_0)(s + \bar{s}_0) = t_- \end{aligned} \tag{175}$$

with

$$s_0 = -\sigma_0 + j\omega_0, \qquad \bar{s}_0 = -\sigma_0 - j\omega_0, \qquad \sigma_0 > 0, \qquad \omega_0 > 0$$

Thus $t(s)$ defines the left half-plane pair and $t(-s)$ the right half-plane pair in this quadruplet set of zeros.

The impedances of the component lossless network are assumed in the same form as in Art. 4 of Ch. 7, namely

$$z_{12} = \frac{Kt(s)t(-s)}{s(s^2 - s_a^2)} = K_{12}s + \frac{k_0}{s} - \frac{2k_{12}s}{s^2 - s_a^2} \tag{176}$$

$$z_{11} = \quad \cdots \quad = K_{11}s + \frac{k_0}{s} + \frac{2k_{11}s}{s^2 - s_a^2} \tag{177}$$

$$z_{22} = \quad \cdots \quad = K_{22}s + \frac{k_0}{s} + \frac{2k_{22}s}{s^2 - s_a^2} \tag{178}$$

with the important detailed difference that the residues at $s = \infty$ are here not assumed to be equal. K is a constant multiplier. From Eq. 176 we have

$$K_{12} = K \tag{179}$$

$$k_0 = \frac{Kt^2(0)}{-s_a^2} \tag{180}$$

$$k_{12} = \frac{Kt(s_a)t(-s_a)}{-2s_a^2} \tag{181}$$

The given input impedance is written in the same form as for the analogous Brune development cycle (Eq. 93) except that we must consider its augmentation with the polynomial $t(s)$ since we are assuming that the complex zeros of $m_1m_2 - n_1n_2$ are not initially of double order. Observe, however, that in the present procedure we do not augment with all the complex even-part zero factors so as to make a full square out of $m_1m_2 - n_1n_2$. We augment only with the factor that is pertinent to the particular cycle that we are dealing with. Our present method is piecemeal in this regard, in contrast to the procedure discussed in Art. 6 where all even-part zeros are produced in a single step. Thus we write the given driving-point impedance

$$Z_1(s) = \frac{Ap(s)}{q(s)} = \frac{A(s - s_{p1}) \cdots (s - s_{pn})}{(s - s_1) \cdots (s - s_n)} = \frac{AP(s)}{Q(s)} \tag{182}$$

where

$$P(s) = p(s)t(s) \qquad \text{and} \qquad Q(s) = q(s)t(s) \tag{183}$$

are the appropriately augmented polynomials.

The relationship between $Z_1(s)$, the z_{sk}, and the remainder function $Z_2(s)$, as in Eq. 94 for the Brune development or in Eq. 67 of Ch. 7 for the corresponding lossless network synthesis, is written in the form

$$(z_{11} - Z_1)(z_{22} + Z_2) = z_{12}^2 = \frac{K^2t_+^2t_-^2}{s^2(s^2 - s_a^2)^2} \tag{184}$$

The separate factors are given by the expressions

$$(z_{11} - Z_1) = \frac{K^2 t_+^2 t_-^2 h}{s(s^2 - s_a^2)Q} = \frac{K^2 t_+ t_-^2 h}{s(s^2 - s_a^2)q} \tag{185}$$

$$(z_{22} + Z_2) = \frac{Q}{s^2(s^2 - s_a^2)h} = \frac{t_+ q}{s(s^2 - s_a^2)h} \tag{186}$$

in which $h(s)$ is seen to be a polynomial of degree $n - 2$ since each of these factors has a simple pole at $s = \infty$. The factor t_-^2, which represents right half-plane zeros, must be allotted to $(z_{11} - Z_1)$. The disposition of the factor t_+^2 is determined so as to yield polynomials in the remainder Z_2 that are appropriately two degrees lower than those in Z_1. Since the factor t_+ cancels in Eq. 185, we see that the augmentation process indicated in Eqs. 183 is actually not required at all in the present development method.

The polynomial $h(s)$ is most conveniently found by solving Eq. 185 for the rational function h/q and writing its partial-fraction expansion, thus

$$\frac{h(s)}{q(s)} = \frac{s(s^2 - s_a^2)(z_{11} - Z_1)}{K^2 t_+ t_-^2} = \frac{1}{K^2} \sum_{\nu=1}^{n} \frac{-s_\nu(s_\nu^2 - s_a^2)k_\nu}{t(s_\nu)t^2(-s_\nu)} \times \frac{1}{s - s_\nu} \tag{187}$$

Here the quantities k_ν are the residues of $Z_1(s)$ in its poles $s_1, s_2, \cdots, s_n$. Using the abbreviation

$$m_\nu = \frac{-s_\nu k_\nu}{t(s_\nu)t^2(-s_\nu)} \tag{188}$$

we have

$$\frac{h(s)}{q(s)} = \frac{1}{K^2} \sum_{\nu=1}^{n} m_\nu \frac{s_\nu^2 - s_a^2}{s - s_\nu} \tag{189}$$

For $s \to \infty$, this rational function must behave like $1/s^2$, since $h(s)$ and $q(s)$ have the degrees $n - 2$ and n respectively. In order that the behavior of the function 189 for large s may more easily be seen, it is effective to write

$$\frac{1}{s - s_\nu} = \frac{1}{s}\left(\frac{1}{1 - s_\nu/s}\right) = \frac{1}{s}\left(1 + \frac{s_\nu}{s} + \frac{s_\nu^2}{s^2} + \cdots\right) \tag{190}$$

from which it is clear that

$$\sum_{\nu=1}^{n} m_\nu(s_\nu^2 - s_a^2) = 0 \tag{191}$$

and that the constant multiplier in the polynomial

$$h(s) = H(s - s_{h1})(s - s_{h2}) \cdots (s - s_{hn-2}) \tag{192}$$

is given by

$$H = \frac{1}{K^2} \sum_{\nu=1}^{n} m_\nu s_\nu (s_\nu^2 - s_a^2) \tag{193}$$

The nonzero pole of the z_{sk} is determined from Eq. 191 which yields

$$s_a^2 = \frac{\sum m_\nu s_\nu^2}{\sum m_\nu} \tag{194}$$

This value must evidently be real since the terms in the sums occur in conjugate pairs. Equation 187 therefore shows that z_{11} is real for real values of s. The residues K_{11}, k_0, and k_{11} in Eq. 177 must be real, and we may conclude that

$$\text{Re}\,[z_{11}(j\omega)] \equiv 0 \tag{195}$$

From Eqs. 186 and 187, the function

$$\frac{1}{z_{22} + Z_2} = \frac{s(s^2 - s_a^2)h}{t_+q} = \frac{s^2(s^2 - s_a^2)^2(z_{11} - Z_1)}{K^2t_+^2t_-^2} \tag{196}$$

is now recognized to be analytic in the right half-plane inclusive of the j-axis where its real part is positive. Hence the function 196 is p.r. and, therefore, so is $(z_{22} + Z_2)$.

The poles of $(z_{22} + Z_2)$ at $s = \pm s_a$ must lie on the j-axis since they would otherwise have to lie in the right and left half-planes respectively. Hence the quantity $-s_a^2$ must be positive.

The j-axis poles of $(z_{22} + Z_2)$, which are guaranteed to be simple and have positive real residues, belong to z_{22}; the rest is the remainder function $Z_2(s)$ which is thus shown to be p.r. Since the residues K_{22}, k_0, and k_{22} are real and positive, Eq. 180 shows K to be positive real also, and Eqs. 179 and 181 yield positive values for K_{12} and k_{12}. The residue conditions

$$K_{11}K_{22} - K_{12}^2 = 0 \tag{197}$$

and

$$k_{11}k_{22} - k_{12}^2 = 0 \tag{198}$$

which must be fulfilled because $Z_1(s)$ has none of the j-axis poles, finally secure positive real values for the residues K_{11} and k_{11} also. The realizability conditions for the lossless network as well as for the remainder function are thus assured.

An expression for the constant multiplier K may be found by first obtaining from Eqs. 180 and 186 the result

$$k_0 = \frac{Kt^2(0)}{-s_a^2} = \frac{t(0)q(0)}{-s_a^2h(0)} \tag{199}$$

Noting that $t(0) = |s_0|^2$, Eq. 189 then gives

$$K = |s_0|^2 \sum_{\nu=1}^{n} \frac{m_\nu}{-s_\nu}(s_\nu^2 - s_a^2) \tag{200}$$

The computational procedure for carrying out the present development is to obtain the m_ν values from Eq. 188, s_a^2 from 194, and K from 200.

Then form the rational function 189, and the expression for $(z_{22} + Z_2)$ in Eq. 186, from the partial-fraction expansion of which z_{22} and $Z_2(s)$ are separately known. With the help of Eqs. 179 and 181 as well as the residue conditions 197 and 198, the impedances z_{11} and z_{12} are then found, and one is ready to construct the lossless network which the development cycle produces.

This network is shown in Fig. 31. Figure 31*a* gives the network in its conventional form according to the synthesis method described in Art. 1 of Ch. 7; Fig. 31*b* shows a slightly revised arrangement that makes use of the equivalence which Fig. 32*a* is intended to demonstrate with the help of the relationship illustrated by Fig. 32*b*. The arrangement in Fig. 31*b* is Darlington's so-called Type-D section.

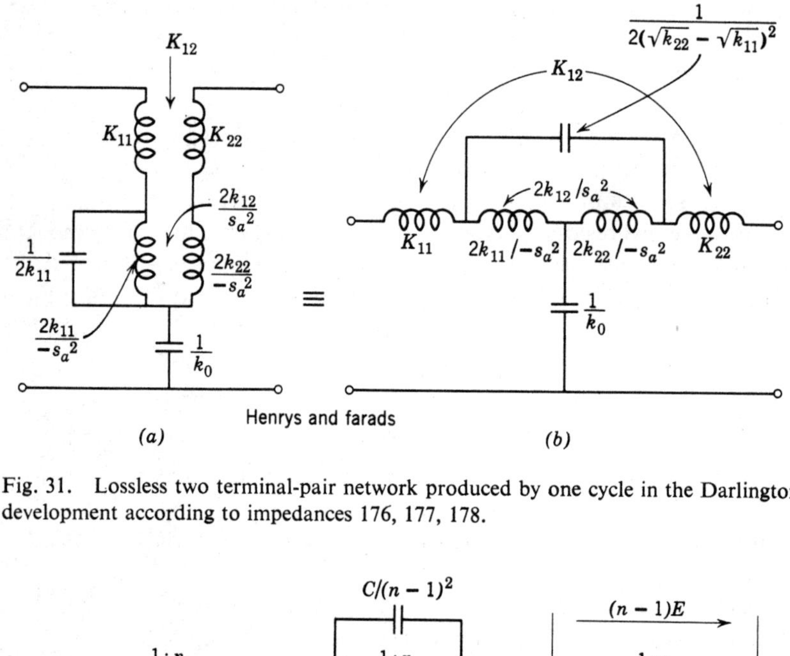

Fig. 31. Lossless two terminal-pair network produced by one cycle in the Darlington development according to impedances 176, 177, 178.

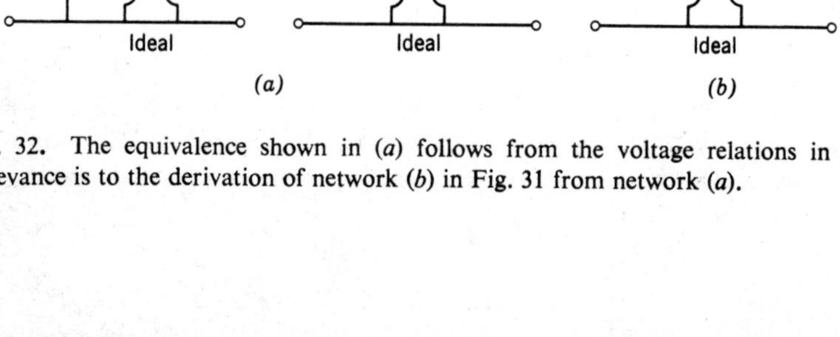

Fig. 32. The equivalence shown in (*a*) follows from the voltage relations in (*b*). Relevance is to the derivation of network (*b*) in Fig. 31 from network (*a*).

Two pairs of closely coupled inductances need not always be involved in this lossless network. If we subtract the terms in Eqs. 176, 177, and 178 for the pole at $s = \infty$, we can apply to the remaining terms in these expressions for the z_{sk} the realization method discussed in Art. 6, Ch. 6, and used in the synthesis of lossless networks described in Art. 4 of Ch. 7. According to Eqs. 104, 105, and 106 given there, we get for the pertinent impedance level factor

$$a = \frac{k_{22} + k_{12}}{k_{11} + k_{12}} = \frac{k_{22}}{k_{12}} = \frac{k_{12}}{k_{11}} \tag{201}$$

and for the lattice impedances

$$z_a = \frac{2(k_{11} + k_{12})s}{s^2 - s_a^2}, \qquad z_b = \frac{k_0}{s}\left(1 + \frac{k_{11}}{k_{12}}\right) \tag{202}$$

Figure 33 shows the lattice and its bridged-T equivalent which is realizable if $C_a \geq C_b$, or if

$$2k_{12} \leq k_0 \tag{203}$$

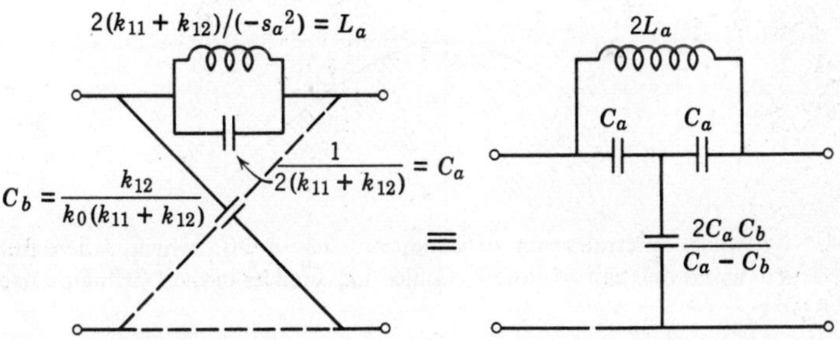

Fig. 33. Realization of the second and third terms in Eqs. 176, 177, 178 so as to avoid the closely coupled inductances in the bottom half of Fig. 31*a* when condition 203 is fulfilled. (Right-hand half of this network must be multiplied by the level factor 201.)

After this bridged-T is bisected, its right-hand half multiplied by the impedance level factor a in Eq. 201, and the halves rejoined, the pair of closely coupled coils K_{11} and K_{22} in Fig. 31*b* must be added in order to form the desired lossless network. Thus, when condition 203 is fulfilled, only one pair of closely coupled inductances is needed to realize the Darlington Type-D section. Use of Eqs. 180 and 181 converts condition 203 into

$$t(s_a)t(-s_a) \leq t^2(0) = |s_0|^4 \tag{204}$$

Using the notation

$$s_a = j\omega_a \tag{205}$$

Eqs. 175 enable us to write condition 203 as

$$[\sigma_0^2 + (\omega_a - \omega_0)^2]\,[\sigma_0^2 + (\omega_a + \omega_0)^2] \leq (\sigma_0^2 + \omega_0^2)^4 \tag{206}$$

which leads to

$$\frac{\omega_a^2}{2} \leq \omega_0^2 - \sigma_0^2 \tag{207}$$

The geometrical interpretation of this condition is represented in the s-plane sketch of Fig. 34 which shows that the even-part zeros defined by the factors $t(s)$ and $t(-s)$ must lie in the shaded regions above and below the hyperbolas if one pair of closely coupled inductances is to be avoided.

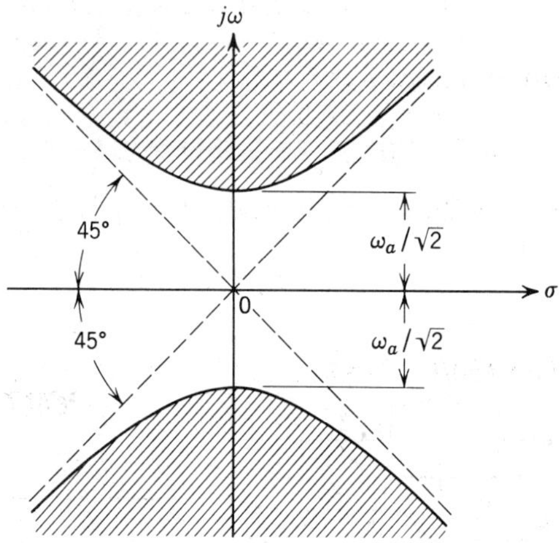

Fig. 34. Graphical interpretation of condition 203 or 207 which, when fulfilled, enables one to avoid one pair of closely coupled inductances in the Darlington network of Fig. 31.

When even-part zeros lie upon the real axis, one can use precisely the same procedure as just described for quadruplet zeros in the complex plane if the real zeros are taken four at a time. The quadratic factor $t(s)$ in Eqs. 175 can just as well represent a pair of negative real zeros. If the total pairs of real zeros is an odd number, then one such pair is left until the end of the cascade development. The remainder function which it implies has one negative real pole and hence presents no problem. Any j-axis zeros of the even part are taken care of by the method in Art. 5 pertinent to the Brune synthesis. Hence all possible kinds of even-part zeros are provided for in the cascade synthesis method.

10 A Quasi Darlington Development That Avoids Mutual Inductive Coupling

Under certain restricted conditions it is possible to achieve a cascade development for a p.r. impedance that involves no mutual inductive coupling. To obtain this result we modify the process discussed in the preceding article to conform more nearly with the method presented in Art. 4 of Ch. 7 for the synthesis of lossless two terminal-pair networks. To obtain the bridged-T and twin-T networks shown there in Figs. 27 and 28, it is necessary to choose the z_{sk} in Eqs. 176, 177, and 178 to have equal residues

$$K = K_{12} = K_{11} = K_{22} \tag{208}$$

in the pole at $s = \infty$, as is done in Eqs. 60, 61, and 62 of Ch. 7.

In order to meet this condition it is necessary to assume that the given input impedance has poles at $s = 0$ and $s = \infty$. Instead of Eq. 182, we must have

$$Z_1(s) = \frac{Ap(s)}{q(s)} + \frac{\rho_0}{s} + \rho_\infty s \tag{209}$$

The condition 208 can then be met if

$$(sz_{11})_{s=0} \geq k_0, \qquad [s(z_{22} + Z_2)]_{s=0} > k_0 \tag{210}$$

$$\left(\frac{z_{11}}{s}\right)_{s=\infty} \geq K, \qquad \left(\frac{z_{22} + Z_2}{s}\right)_{s=\infty} > K \tag{211}$$

Through use of Eqs. 185 and 186 these are converted into

$$0 < K - |s_0|^2 \frac{K^2h(0)}{q(0)} \leq \frac{-s_a^2}{|s_0|^4} \rho_0 \tag{212}$$

and

$$0 < K - HK^2 \leq \rho_\infty \tag{213}$$

Substitution from Eqs. 189 and 193 yields these conditions in the form

$$0 < K - |s_0|^2 \sum_{\nu=1}^{n} \frac{m_\nu}{-s_\nu}(s_\nu^2 - s_a^2) \leq \frac{-s_a^2}{|s_0|^4} \rho_0 \tag{214}$$

and

$$0 < K - \sum_{\nu=1}^{n} m_\nu s_\nu (s_\nu^2 - s_a^2) \leq \rho_\infty \tag{215}$$

With slight modification, these are the same as conditions 86 and 87 of Ch. 7 applicable to lossless driving-point and transfer impedance synthesis. Following the same argument as given there, we find that the present development cycle will be successful if we can fulfill the condition

$$-\left|\frac{s_a}{s_0}\right|^2 \rho_0 < |s_0|^3(\hat{s}_a^2 - s_a^2) \sum_{\nu=1}^{n} \hat{m}_\nu < |s_0|^2 \rho_\infty \tag{216}$$

in which

$$\hat{m}_\nu = m_\nu w_\nu \tag{217}$$

and

$$w_\nu = \frac{|s_0|}{-s_\nu} + \frac{-s_\nu}{|s_0|} \tag{218}$$

$$\hat{s}_a^{\,2} = \frac{\sum \hat{m}_\nu s_\nu^{\,2}}{\sum \hat{m}_\nu} \tag{219}$$

Depending upon the circumstances, it may be possible to meet condition 216 without requiring $Z_1(s)$ to have a pole at both $s = 0$ and $s = \infty$. Since realization of the desired twin-T network involving no inductive coupling requires the further condition expressed by Eq. 114 in Ch. 7, the restriction expressed by Eq. 119 and illustrated there in Fig. 29 applies here also. Hence the success of the development contemplated here necessitates, in addition to possible poles at $s = 0$ and $s = \infty$, that the given input impedance function have its even-part zeros restricted to lie not more than 45° from the j-axis of the s-plane. When such conditions cannot be met, it is still possible to realize any p.r. driving-point impedance without mutual inductive coupling by applying one of the several procedures discussed in the following chapter.

PROBLEMS

1. For each of the following impedance functions sketch, in the s-plane, the corresponding constellation of zeros and poles and tell which are realizable as driving-point functions through their partial-fraction expansions (similar to the Foster procedure used with reactances). Find networks for those that can be realized in this manner.

$$(a)\ \frac{s+1}{s^2+2s+2} \qquad (b)\ \frac{s^2+2s+2}{(s+1)(s^2+2s+5)} \qquad (c)\ \frac{s+3}{s^2+2s+2}$$

$$(d)\ \frac{(s+2)^2}{(s+1)(s^2+2s+2)} \qquad (e)\ \frac{(s+2)^2}{(s+3)(s^2+2s+2)}$$

2. Which of the following driving-point impedances are realizable through their partial-fraction expansions in the manner that reactance functions are realizable by the Foster procedure?

$$(a)\ Z(s) = \frac{(s+1)(s+2)}{s^2+2s+2} \qquad (b)\ Z(s) = \frac{(s+2)(s+1+\sqrt{3})[s+1+(1/\sqrt{3})]}{(s+1)(s^2+2s+2)}$$

$$(c)\ Z(s) = \frac{[2(s+1)]^4+1}{[(s+1)^4+1](s+1)} \qquad (d)\ Z(s) = \frac{[2(s+1)]^4-1}{[(s+1)^4+1](s+1)}$$

Find networks for those that are realizable in this manner.

3. Realize the following *RLC* impedance functions

$$(a)\ Z(s) = \frac{s^3+s^2+2s+1}{s^2+s+1} \qquad (b)\ Z(s) = \frac{s^4+s^3+3s^2+2s+1}{s^3+s^2+2s+1}$$

4. If realizable, find a network having the following driving-point impedance

$$Z(s) = \frac{2s^4+5s^3+7s^2+5s+2}{2s^3+3s^2+2s+1}$$

5. If realizable, find networks for the following driving-point impedance functions

$$(a)\ Z(s) = \frac{s^2}{(s+1)^3} \qquad (b)\ Z(s) = \frac{s^3 + 2s^2 + 2s}{(s+1)^2}$$

6. A driving-point impedance has three simple poles and one third-order zero as shown in the accompanying s-plane sketch. $Z(0) = 4$. Construct the impedance and if p.r., find its realization in the simplest way possible.

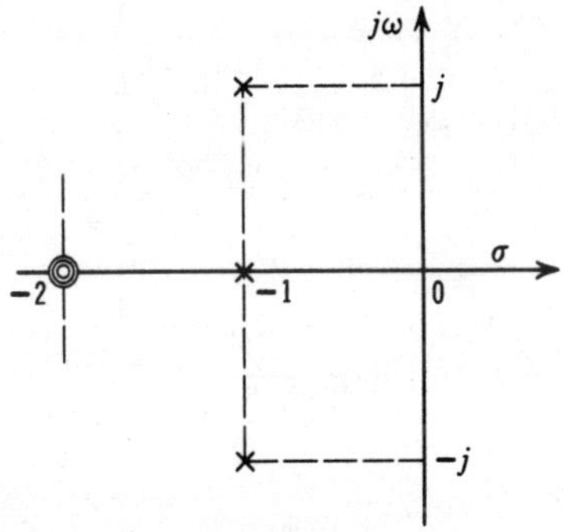

Prob. 6.

7. An RLC driving-point impedance $Z(s)$ has three poles and a double-order zero as shown in the adjacent figure. Find the smallest and the largest values of the dimension a for which $Z(s)$ is realizable from its partial-fraction expansion. Determine corresponding networks.

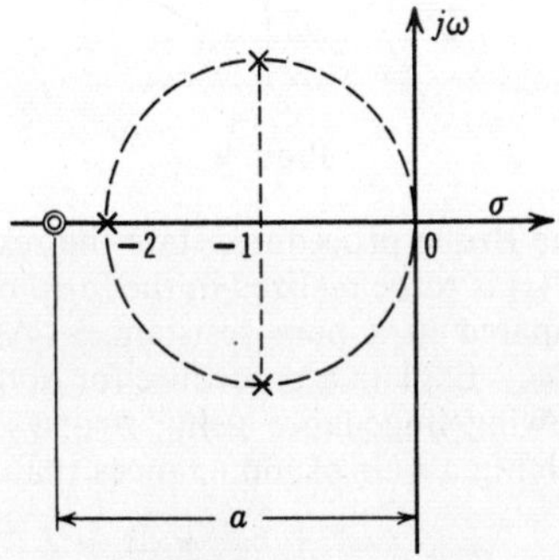

Prob. 7.

8. A given driving-point impedance function $Z(s)$ has a double-order zero at $s = -1$ and three poles as shown in the s-plane sketch. $Z(0) = \frac{2}{9}$. Construct $Z(s)$ and find the most practical network you can for its realization.

9. The pole-zero plot of the even part of a driving-point impedance $Z(s)$ is shown in the adjacent figure. Construct this impedance and approach its

synthesis by the Brune method. Carry through the synthesis in the simplest manner possible so as to obtain a realization without mutual inductive coupling.

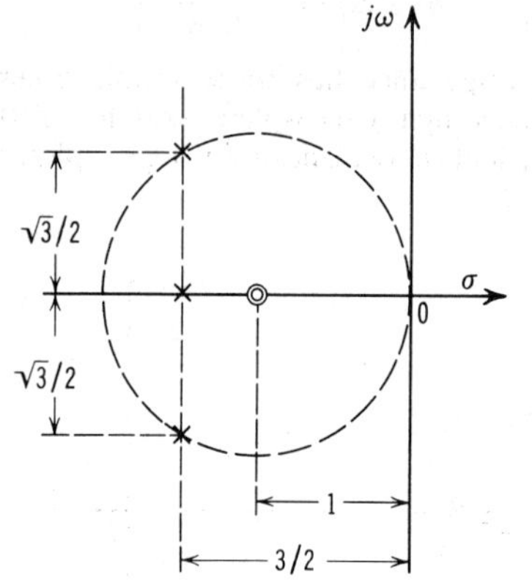

Prob. 8.

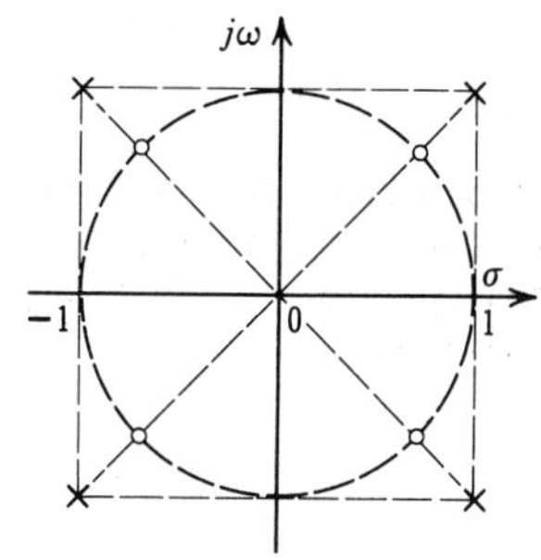

Prob. 9.

10. In the light of the Brune procedure, state the restrictions on the analytic form of Re $[Z(j\omega)]$ if $Z(s)$ is to be realized in the form of an unbalanced entirely reactive network terminated in a pure resistance. Assume that $Z(s)$ has no poles or zeros for $s = j\omega$. Explain a procedure for obtaining this realization.

11. Realize the following impedance using Brune's procedure and put the result into a form involving a pair of inductances that are not closely coupled.

$$Z(s) = \frac{s^3 + 6s^2 + 3s + 2}{s^2 + s + 1}$$

12. For a lossless two terminal-pair network terminated in 1 ohm it is known that

$$\text{Re}\,[Z_1(j\omega)] = |Z_{12}(j\omega)|^2 = \frac{\omega^2}{\omega^4 - \omega^2 + 1}$$

If the minimum reactive input impedance $Z_1(s)$ were to be synthesized by Brune's method, could you tell from the form of its real part what the structure of the network would turn out to be? Find $Z_1(s)$ and check your conclusions through synthesizing the pertinent network.

13. A given driving-point impedance has the form

$$Z(s) = \frac{a_0 + a_1 s + a_2 s^2 + a_3 s^3 + a_4 s^4}{b_0 + b_1 s + b_2 s^2 + b_3 s^3 + b_4 s^4}$$

Re $[Z(j\omega)]$ has two minima, both equal to zero; one of these occurs at $\omega = 1$, the other at $\omega = 2$. At $\omega = 0$ this real part equals unity and its asymptotic value for $\omega \to \infty$ is unity also. If it is known that

$$Z(j1) = j(1 - \sqrt{2}), \qquad Z(j2) = j2(10 - 7\sqrt{2}), \qquad (dZ/ds)_{s=j} = 1 + \sqrt{2}$$

synthesize a network according to the Brune procedure.

14. Given the following data and those illustrated in the accompanying figure

$$Z(j1) = j1, \qquad Z(j2) = j8, \qquad \left(\frac{dZ}{j\,d\omega}\right)_{\omega=1} = 2$$

If the driving-point impedance to which these data pertain is realized in the Brune manner, sketch the complete network and show that the number of given data exactly equal the number of unknown element values. Determine all element values.

In the following four examples test for p.r. character, and treat the realizable cases by (*a*) the Brune method, (*b*) the partial-fraction expansion method using an appropriate amount of resistance padding to avoid mutual inductive coupling.

In the nonrealizable cases add (*a*) just enough resistance padding to make them p.r. and then realize by Brune's method, and (*b*) enough resistance padding to make them realizable from a partial-fraction expansion by networks involving no mutual inductive coupling.

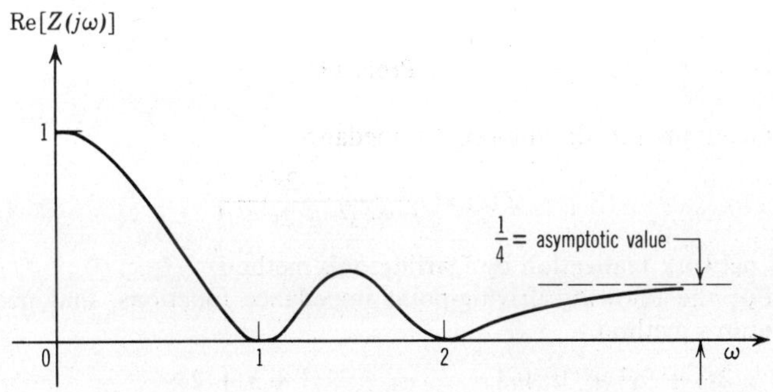

Prob. 14.

15. $Z(s) = \dfrac{s^2 + s + 1}{s^2 + s + 4}$ 16. $Z(s) = \dfrac{s^2 + s + 1}{s^2 + s + 9}$

17. $Z(s) = \dfrac{s^2 + s + 1}{s^2 + s + 2}$ 18. $Z(s) = \dfrac{5s^2 + s + 1}{s^2 + 4s + 2}$

19. The adjacent s-plane sketch shows the pole-zero constellation of a

function

$$f(s) = \frac{A(-s^2)}{B(-s^2)}$$

having four poles and two double-order j-axis zeros as shown. If $f(s) \to 4$ for $s \to \infty$, and

$$\frac{A(\omega^2)}{B(\omega^2)} = \text{Re}\,[Z(j\omega)]$$

construct the impedance $Z(s)$.

Now realize this impedance by Brune's method and by Darlington's method. Compare the results. For the lossless network terminated in 1 ohm, write down the transfer impedance $Z_{12}(s) = E_2/I_1$ as a quotient of polynomials without further computation.

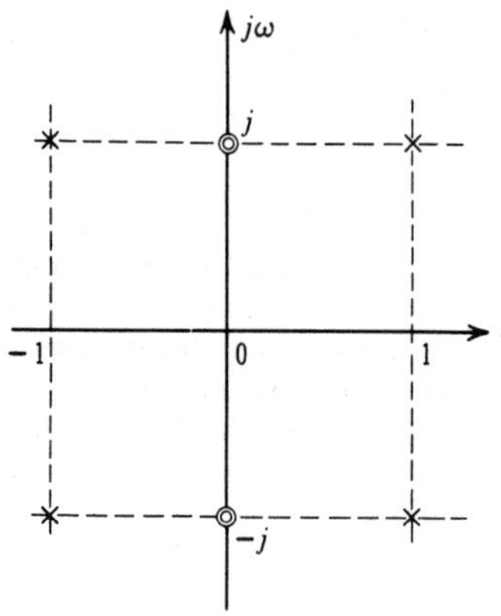

Prob. 19.

20. Given the p.r. driving-point impedance

$$Z(s) = \frac{s^3 + 2s}{s^3 + s^2 + 2s + 1}$$

find its network realization by Darlington's method.

21. For the folowing driving-point impedance functions, find networks by Darlington's method

(a) $\dfrac{2s^3 + 9s^2 + 3s + 1}{4s^4 + 18s^3 + 8s^2 + 3s + 1}$ (b) $\dfrac{s^2 + s + 2}{2s^2 + 3s + 1}$

22. The real part of a driving-point impedance is given by

(a) $\dfrac{(1 + \omega^2)^2}{1 + \omega^6}$ (b) $\dfrac{(1 - \omega^2)^2}{1 + \omega^6}$

In each case find $Z(s)$ and synthesize by Darlington's method.

23. For the impedance functions in Probs. 15 and 17, find network realizations by Darlington's method.

24. Realize by Darlington's method

$$(a)\ Z(s) = \frac{s^3 + 6s^2 + 3s + 2}{s^2 + s + 1} \qquad (b)\ Z(s) = \frac{s^4 + s^3 + 3s^2 + 2s + 1}{s^3 + s^2 + 2s + 1}$$

In the latter realization make use of a symmetrical lattice and compare the result with that found in Prob. 3.

25. The parameter values for the network shown are in ohms and henrys. Realize the same $Z(s)$ by means of a nondissipative two terminal-pair network terminated in 1 ohm.

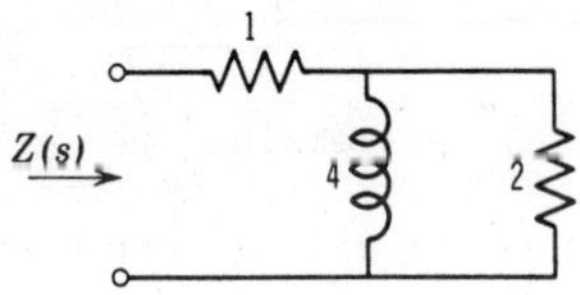

Prob. 25.

26. Given the transfer impedance

$$Z_{12} = \frac{As}{(s+1)(s+2)}$$

find a solution in the form of a nondissipative two terminal-pair network terminated in a 1-ohm resistance such that the driving-point impedance is minimum reactive, and choose a value for the constant A which yields the simplest form for this network.

27. The transfer impedance of a lossless network terminated in a 1-ohm resistance is

$$Z_{12} = \frac{1}{s^5 + s^4 + 4s^3 + 3s^2 + 3s + 1}$$

Determine the corresponding minimum-reactive driving-point impedance. Find a practical network possessing the above transfer impedance, first utilizing this result, and then without using the driving-point impedance.

28. (*a*) If the function $\omega^2/(1 + 2\omega^2 + \omega^4)$ is the real part of $Z(j\omega)$, find $Z(s)$ and its network realization.

(*b*) If the same function is the squared absolute value of $Z(j\omega)$, find $Z(s)$ and its network realization.

29. From the following real parts find the corresponding driving-point impedances and synthesize by Darlington's method

$$(a)\ \text{Re}\,[Z(j\omega)] = \frac{\omega^2}{1 + \omega^4} \qquad (b)\ \text{Re}\,[Z(j\omega)] = \frac{\omega^6}{1 + \omega^6}$$

30. Given

$$\text{Re}\,[Z_1(j\omega)] = \frac{1}{(1 + \omega^2)^2}$$

(*a*) Realize $Z_1(s)$ in the form of a lossless network terminated in 1 ohm.

(*b*) Determine a network for the complementary impedance $Z_2 = 1 - Z_1$.

31. In the accompanying sketch a network N is to be determined which will make the resultant impedance of the two networks in series equal a pure

resistance R, with a minimum value of R. Element values are in henrys, farads, and ohms.

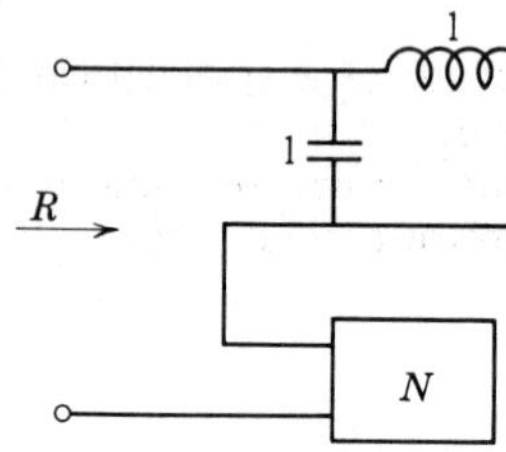

Prob. 31.

32. For networks (a), (b), and (c) find the complementary ones which, placed in series or in parallel (as the situation requires), yield a resultant constant-resistance network. The element values are in ohms, henrys, and farads.

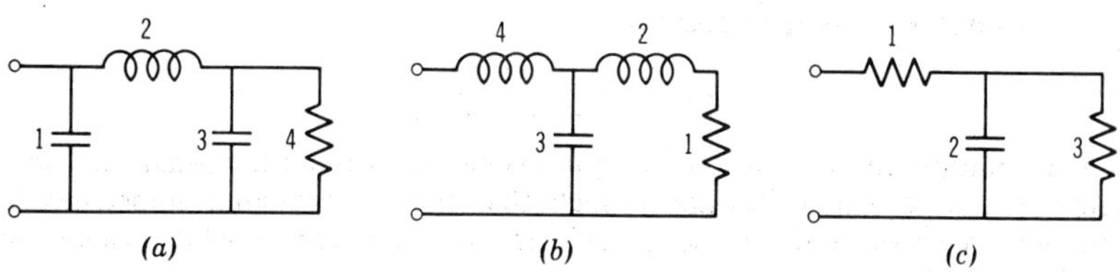

Prob. 32.

33. Given the impedance

$$Z(s) = \frac{(s - s_1)(s - s_3)}{(s - s_2)(s - s_4)}$$

for which the pair of conjugate left half-plane poles s_2 and s_4 are assumed to be fixed, define the boundaries of a region within which the zeros s_1 and s_3 must lie if $Z(s)$ is to be realizable from its partial-fraction expansion.

34. Consider Prob. 33 in which the denominator contains an additional factor $(s - s_6)$ which may be chosen at will.

C H A P T E R T E N

Transformerless Driving-Point Impedance Synthesis

1 The Bott and Duffin Procedure*

This method does not follow a logical argument or pattern of straightforward reasoning as do those discussed in the preceding chapter. In its presentation it is difficult, therefore, to make the succession of steps appear natural and meaningful except that they do ultimately lead to the desired result. One can, however, point to a basic motivation for the several initial steps in the process by calling attention to the fact that what is desired through proper manipulation of the given minimum-reactive and minimum-susceptive impedance is the construction of a related p.r. function that has j-axis zeros or poles and involves polynomials of no higher degree than those in the given impedance. Through removal of such j-axis poles, this function then can surely be simplified and leave a remainder that can be dealt with in the same manner so as to effect a continued simplification.

Essentially, this is the motivation for the initial steps in the Brune process. Since the given impedance has no j-axis zeros or poles, we try to manipulate it somehow until it does have such zeros or poles. Unfortunately, in the Brune process the necessary manipulative steps create a situation that cannot be resolved without resorting to the use of closely coupled inductances. In the procedure to be discussed here, a more elaborate series of manipulative steps are employed that avoid mutual inductive coupling altogether, but lead to the appearance of two remainder functions instead of one, thus requiring ultimately a much larger total number of elements in the realization. Apparently this price must be paid in order to avoid mutual inductance, although some of the alternate methods discussed in the subsequent articles of this chapter accomplish, for some functions of higher degree, the same objective with a cost in element numbers considerably less than is demanded by the Bott and Duffin method.

---

* "Impedance Synthesis without Use of Transformers," R. Bott and R. J. Duffin, *J. Appl. Phys.*, Vol. 20, No. 8, Aug. 1949, p. 816.

Like the Brune and Darlington synthesis procedures, the Bott and Duffin method is classic in the sense that it represents the first successful demonstration of the fact that mutual inductive coupling is not an essential accompaniment of driving-point impedance synthesis, and that it always can be avoided.

Starting from a p.r. function

$$Z(s) = \frac{a_0 + a_1 s + a_2 s^2 + \cdots + a_n s^n}{b_0 + b_1 s + b_2 s^2 + \cdots + b_n s^n} \tag{1}$$

having no j-axis zeros or poles, we first remove the smallest value R_1 of its real part on the j-axis, as in the Brune process, and leave a p.r. remainder

$$Z_1(s) = Z(s) - R_1 \tag{2}$$

that has a purely reactive value

$$Z_1(s_1) = jX \tag{3}$$

at the frequency $s_1 = j\omega_1$. We now consider the function

$$R(s) = \frac{1 - w(s)}{1 + w(s)} \tag{4}$$

with

$$w(s) = \frac{1 - [Z_1(s)/Z_1(k)]}{1 + [Z_1(s)/Z_1(k)]} \times \frac{k + s}{k - s} \tag{5}$$

in which the quantity k is fixed by the relation

$$Z_1(s_1)/Z_1(k) = s_1/k, \qquad \text{or} \qquad Z_1(k) = (X/\omega_1)k \tag{6}$$

Success of the Bott and Duffin method rests upon $R(s)$ having the following properties: (*a*) it is p.r.; (*b*) it involves polynomials that have the same degree as those in $Z_1(s)$; (*c*) it has j-axis zeros at the points $s = \pm j\omega_1$. We shall first show that these statements are true, and then base a realization procedure for $Z_1(s)$ upon them.

Concerning proof of the p.r. character of $R(s)$, we make use of the following property of p.r. functions discussed on p. 417, Ch. VI of *MCA*. Regarding the linear fractional transformation

$$z_a = \frac{1 - z_b}{1 + z_b} \tag{7}$$

z_a is there proved to be a p.r. function if it can be shown that z_b is analytic in the right-half s-plane, and if $|z_b(j\omega)| \le 1$ for all real values of ω. The p.r. character of $R(s)$ in Eq. 4 is therefore established if we can show that $w(s)$ is analytic in the right half-plane, and on the j-axis has a magnitude that does not exceed unity.

Accepting for the moment the fact that a positive real k value can be found to satisfy Eq. 6, we observe that $w(s)$ in Eq. 5 is analytic in the right half-plane because $Z_1(k)$ has a positive real value and hence $1 + Z_1(s)/Z_1(k)$ can have only left half-plane zeros, while the factor $(k - s)$ is contained in $1 - Z_1(s)/Z_1(k)$. So far as the magnitude of $w(j\omega)$ is concerned, we note in Eq. 5 that the magnitude of $(k + j\omega)/(k - j\omega)$ is unity and that the magnitude of the rest of this expression cannot exceed unity on the j-axis because of the p.r. character of $Z_1(s)/Z_1(k)$ and the fact that this linear fractional function maps the right half of the $[Z_1(s)/Z_1(k)]$ plane upon the interior of the unit circle in the function plane (the function being this linear fractional expression).

With regard to Eq. 6 we recall the property of p.r. functions (Eq. 741, p. 420 in *MCA*) which states that the angle of $Z_1(s)$ cannot equal the angle of s anywhere in the right half-plane except where this equality holds identically as it evidently does on the positive real axis. Hence for $X > 0$, Eq. 6 can be satisfied by a positive real k value. As in the Brune procedure, we must switch to an admittance basis when X has the wrong algebraic sign, except that here the impedance basis is appropriate to $X > 0$ instead of to $X < 0$. Like the function 59 (for $L_1 > 0$) in the previous chapter, we note that $Z_1(k) - (X/\omega)k$ can have only one positive real zero because the restriction of poles and zeros to the left half-plane causes the variation of $Z_1(\sigma)$ with increasing σ to be such that $|dZ_1/d\sigma|$ is continuously decreasing.

Item (*b*) above is easily seen to be true because $w(s)$ has the same degree polynomials as those in $Z_1(s)$ since, as already mentioned, the factor $(k - s)$ in Eq. 5 cancels. Item (*c*) follows because the determination of k in Eq. 6 yields $w(s) = 1$ for $s = s_1$.

To proceed with the realization process, we must express $Z_1(s)$ in terms of $R(s)$. This step is facilitated by use of the fact that the transformation 7 is its own inverse so that $z_b = (1 - z_a)/(1 + z_a)$. Applying this result to 4, substituting into 5, and multiplying both sides by $(k - s)/(k + s)$ gives

$$\frac{1 - [Z_1(s)/Z_1(k)]}{1 + [Z_1(s)/Z_1(k)]} = \frac{1 - R}{1 + R} \times \frac{1 - s/k}{1 + s/k} = \frac{1 - \dfrac{R + s/k}{1 + sR/k}}{1 + \dfrac{R + s/k}{1 + sR/k}} \tag{8}$$

whereupon it is evident that

$$\frac{Z_1(s)}{Z_1(k)} = \frac{R + s/k}{1 + sR/k} = \frac{R}{1 + sR/k} + \frac{s/k}{1 + sR/k} = \frac{1}{(1/R) + (s/k)} + \frac{1}{(k/s) + 1/(1/R)} \tag{9}$$

Since $R(s)$ has j-axis zeros at $s = \pm j\omega_1$, the reciprocal function has poles there. Hence we can write

$$\frac{1}{R(s)} = \frac{as}{s^2 + \omega_1^2} + r(s) \tag{10}$$

in which $r(s)$ is a p.r. remainder function involving polynomials that are respectively two degrees lower than those in $Z_1(s)$. It is not difficult to see that Eq. 9 now yields the network configuration in Fig. 1 in which Z_2' and Z_2'' are remainder impedance functions given by

$$Z_2' = 1/r(s) \qquad \text{and} \qquad Z_2'' = r(s) \tag{11}$$

To obtain $Z_1(s)$, the impedance level of this circuit must be multiplied by the factor $Z_1(k)$, and to obtain the impedance $Z(s)$ we must then insert the resistance R_1 in series.

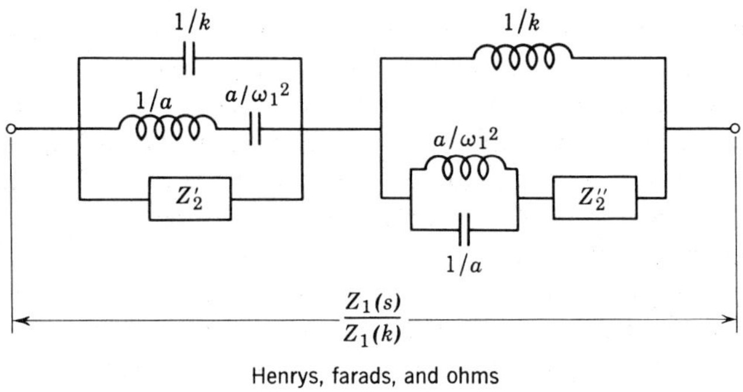

Fig. 1. Network obtained from one cycle of the Bott and Duffin procedure according to the result expressed by Eq. 9.

The appearance of two remainder functions shows that, after a second similar development cycle is carried out on each of these, we will have four remainders; and after the application of another cycle we will have eight, and so forth. The number of remainder functions grows exponentially with the number of successive cycles needed to reduce the given function to constant remainders. Since each cycle reduces the degree n in Eq. 1 by two, the number of successive cycles needed for complete reduction is $n/2$ (assuming for simplicity that n is even). The last of these involves $2^{n/2}$ components like the two in Fig. 1, each of which requires three reactive elements. The total number of reactive elements becomes

$$3(2^{n/2} + 2^{(n/2)-1} + \cdots + 2) \tag{12}$$

For $n = 6$ we get a total of 42 reactive elements plus 8 resistances; for $n = 8$ these numbers are respectively 90 and 16; for $n = 10$ they are 186 and 32, and so forth. The price that must be paid in order to avoid mutual inductive coupling by this method becomes rather high.

In this connection F. M. Reza* and others have recently pointed out that one reactive element in the Bott and Duffin circuit of Fig. 1 is superfluous, as may be seen by first noting that the latter can be redrawn as shown in Fig. 2 since it is a balanced Wheatstone bridge and hence the connection of points a'-b' is immaterial. For the same reason we may connect any impedance we wish across these points; and if we make this an inductance L (as indicated in a dashed fashion in Fig. 2) which we resonate with the capacitance in a-a' at the frequency $s = j\omega_1$, then we can avail ourselves of the equivalence shown in Fig. 3 and, by combining identically resonant parallel LC circuits, obtain the alternate bridge arrangement illustrated in Fig. 4.

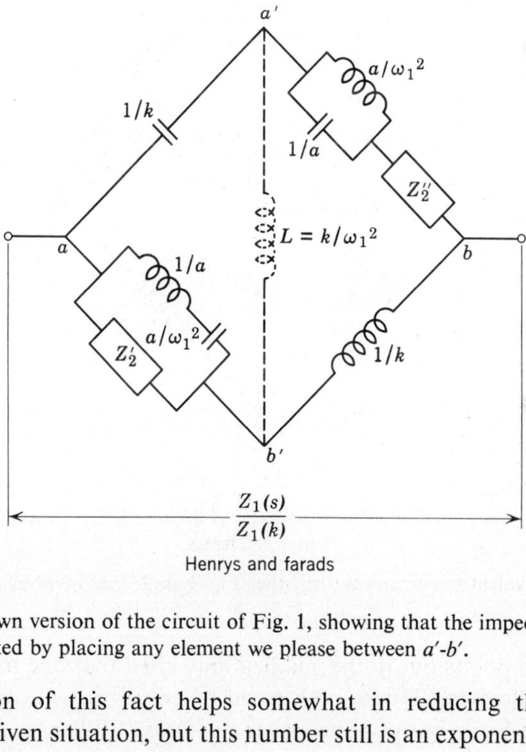

Fig. 2. Redrawn version of the circuit of Fig. 1, showing that the impedance between a-b is not affected by placing any element we please between a'-b'.

Recognition of this fact helps somewhat in reducing the elements needed in a given situation, but this number still is an exponential function of the degree n in the impedance 1, and hence the method remains fundamentally uneconomical.

---

* *Quart. Prog. Rept.*, Mass. Inst. Technol., Research Lab., Electronics, July 15, 1953, pp. 82–87; also *J. Appl. Phys.*, Vol. 25, No. 6, June 1954, pp. 807–808.

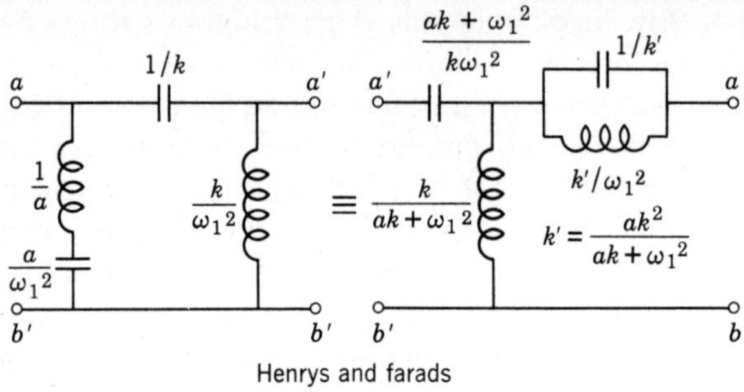

Fig. 3. Equivalence which verifies conversion of the circuit of Fig. 2 to the form shown in Fig. 4.

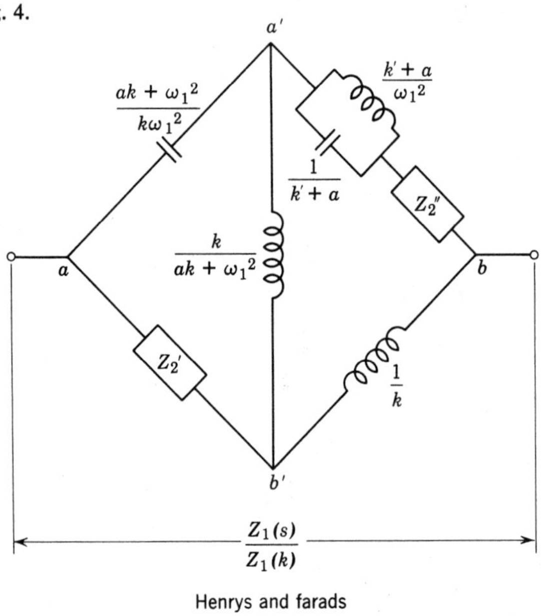

Fig. 4. Equivalent circuit to that in either Fig. 1 or 2, that involves one less reactive element.

Reza also points out in the publications cited that one may alternately obtain the Bott and Duffin realization by converting the Brune network to this form by means of appropriate relations. When several successive cycles are involved, this process may circumvent some tedious computation. As shown later on in this chapter, there are alternative ways of achieving results equivalent to those which the Bott and Duffin method yields.

It is incidentally of interest to mention that Eq. 6 in the variable k

always possesses the j-axis roots $k = \pm s_1$. When the impedance $Z(s)$ involves polynomials of the degree n as in Eq. 1, it turns out that Eq. 6 has the degree $n + 1$, but since we always know two of the roots we need only be concerned with the determination of the single positive real root contained in an $(n - 1)$-degree equation. For $n = 2$, the well-known relation between the equation coefficients and the roots, together with the fact that two of these roots are given by $k = \pm j\omega_1$, results in k being equal to the negative of the coefficient of the square term in the cubic Eq. 6.

As an illustrative example suppose we consider the impedance

$$Z(s) = \frac{s^2 + s + 1}{s^2 + s + 4} \tag{13}$$

used in Art. 7 of the previous chapter (Eq. 128) to illustrate the Brune and Darlington methods and for which these yield the same circuit shown there in Fig. 25. We may initially make use of the results given there, showing that the real part of $Z(j\omega)$ is zero at $\omega_1 = \pm\sqrt{2}$, and that $Z_1(s)$ computed at this frequency (Eq. 130) yields the value $j/\sqrt{2}$. $X = 1/\sqrt{2}$ is positive, and so the impedance basis is relevant in the Bott and Duffin method. We have

$$X/\omega_1 = \tfrac{1}{2} \tag{14}$$

so that Eq. 6 becomes

$$2(k^2 + k + 1) = k(k^2 + k + 4)$$

or

$$k^3 - k^2 + 2k - 2 = 0 \tag{15}$$

Making use of the statement in the above paragraph about the roots of this equation, we immediately know that $k = 1$ is the desired root in this example. With it we compute

$$Z_1(k) = \tfrac{1}{2} \tag{16}$$

and from Eq. 5 find

$$w(s) = \frac{-s^2 - s + 2}{3s^2 + 3s + 6} \times \frac{1 + s}{1 - s} = \frac{s^2 + 3s + 2}{3s^2 + 3s + 6} \tag{17}$$

Eq. 4 then yields

$$R(s) = \frac{s^2 + 2}{2s^2 + 3s + 4} \tag{18}$$

which has the j-axis zeros at $s = \pm j\sqrt{2}$ as required. According to Eq. 9 we can now write

$$2Z_1(s) = \frac{1}{\dfrac{2s^2 + 3s + 4}{s^2 + 2} + s} + \frac{1}{\dfrac{1}{s} + \dfrac{1}{\dfrac{2s^2 + 3s + 4}{s^2 + 2}}} \tag{19}$$

which yields the realization for $Z_1(s) = Z(s)$ shown in Fig. 5.

As a second example, consider

$$Z(s) = \frac{s^2 + s + 8}{s^2 + 2s + 2} \tag{20}$$

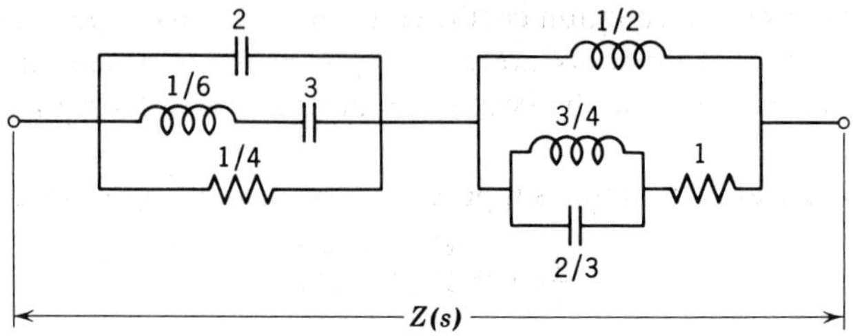

Fig. 5. Realization of the impedance 13 by the Bott and Duffin method. (Compare with the realization of this same impedance in Fig. 25 of Ch. 9 by the Brune or Darlington method.)

for which the real part is readily seen to be zero at $s_1 = \pm j2$. Again $R_1 = 0$, and $Z_1(s) = Z(s)$. Here we find

$$Z_1(s_1) = \frac{4 + j2}{-2 + j4} = -j \tag{21}$$

so that $X = -1$, and we must switch to an admittance basis for which $Y_1(s) = 1/Z_1(s)$ and we have

$$Y_1(s_1) = jB = j \tag{22}$$

which yields a positive $B = 1$, and $B/\omega_1 = \frac{1}{2}$, so that Eq. 6 becomes

$$Y_1(k) = \frac{k^2 + 2k + 2}{k^2 + k + 8} = \frac{k}{2}$$

or

$$k^3 - k^2 + 4k - 4 = 0 \tag{23}$$

Again we have $k = 1$ as the desired positive real root.

Next we compute

$$Y_1(k) = Y_1(1) = \frac{1 + 2 + 2}{1 + 1 + 8} = \frac{1}{2} \tag{24}$$

from which we form, according to Eq. 5

$$w(s) = \frac{4 - 3s - s^2}{12 + 5s + 3s^2} \times \frac{1 + s}{1 - s} = \frac{s^2 + 5s + 4}{3s^2 + 5s + 12} \tag{25}$$

and obtain from Eq. 4

$$R(s) = \frac{s^2 + 4}{2s^2 + 5s + 8} \tag{26}$$

Finally the representation in Eq. 9 yields

$$\frac{Y_1(s)}{Y_1(k)} = 2Y_1(s) = \frac{1}{s + \dfrac{2s^2 + 5s + 8}{s^2 + 4}} + \frac{1}{\dfrac{1}{s} + \dfrac{s^2 + 4}{2s^2 + 5s + 8}} \tag{27}$$

from which the network in Fig. 6 is obtained for $Z(s)$ in Eq. 20.

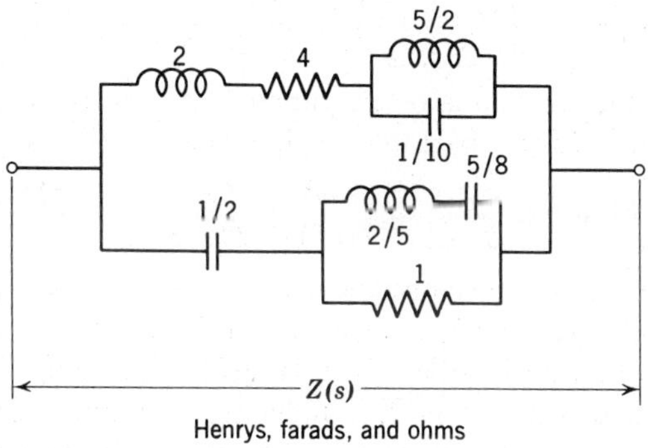

Fig. 6. Realization of the impedance 20 by the Bott and Duffin method.

2 Miyata's Method for Transformerless Synthesis*

From the form of the result given in Eq. 9, it is clear that the Bott and Duffin theory essentially provides a method for separating the given impedance into additive components which individually satisfy certain special conditions that permit their realization in simple terms. However, that method (as has already been pointed out) fails completely to provide an insight into the mechanism whereby such a separation of the given impedance is made possible.

The Miyata method, although subject to a fundamental restriction, and therefore not as completely general as the procedure of Bott and Duffin, approaches the problem from the point of view of separation into additive components and hence succeeds in shedding much more light upon the mechanism of transformerless synthesis. It bases this method of separation upon the real part or even part of the given impedance, since the methods discussed in Art. 1 of Ch. 8 for the generation of an impedance from its even part, together with principles of the Darlington theory pointed out in Art. 8 of the preceding chapter, enable one to state clearly the sufficient conditions for transformerless synthesis in terms of properties of the real part. The following discussion will show how this method operates.

* F. Miyata, "A New System of Two-Terminal Synthesis," *J. Inst. Elec. Engrs. (Japan)*, Vol. 35, May 1952, pp. 211–218.

Suppose we consider the impedance in Eq. 1, having no j-axis zeros or poles, written in the familiar notation

$$Z(s) = \frac{P(s)}{Q(s)} = \frac{m_1 + n_1}{m_2 + n_2} \tag{28}$$

and its real part for $s = j\omega$ evaluated

$$\text{Re}\,[Z(j\omega)] = \left(\frac{m_1 m_2 - n_1 n_2}{m_2{}^2 - n_2{}^2}\right)_{s=j\omega} = \frac{A_0 + A_1\omega^2 + \cdots A_n\omega^{2n}}{B_0 + B_1\omega^2 + \cdots B_n\omega^{2n}} \tag{29}$$

Abbreviating the denominator polynomial in this expression to $B(\omega^2)$, we can separate this real part into components as follows

$$\text{Re}\,[Z(j\omega)] = \frac{A_0}{B(\omega^2)} + \frac{A_1\omega^2}{B(\omega^2)} + \frac{A_2\omega^4}{B(\omega^2)} + \cdots + \frac{A_n\omega^{2n}}{B(\omega^2)} \tag{30}$$

Each of the terms on the right-hand side of this equation, regarded as the real part of a separate impedance, may (if we assume the coefficients $A_0 \cdots A_n$ to be positive) be shown to generate an impedance that can easily be realized without mutual inductive coupling. The series connection of such separate realizations evidently yields the desired synthesis of $Z(s)$.

In order to gain a clear picture of how this result comes about, we begin by asking the following question. If we regard each term in expression 30 as the real part of a component impedance (assuming again that all coefficients A_k are positive), what can we say about the special properties (if any) of the component impedances generated by these particular real parts?

An answer to this question is provided by observing by inspection of the terms in Eq. 30 the nature and distribution of the zeros of these real parts. Thus it is evident that the first term has n double-order zeros at $s = \infty$; the second term has one double zero at $s = 0$ and $n - 1$ of them at $s = \infty$; the third term has two double-order zeros at $s = 0$ and $n - 2$ at $s = \infty$; finally the last term has all of its n double-order zeros at $s = 0$. Thus these terms, regarded as real parts of impedances, have their double-order zeros located exclusively at $s = 0$ and at $s = \infty$, the total number being n in each case but the distribution of them between the points $s = 0$ and $s = \infty$ varying all the way from a situation for which all zeros are located at $s = \infty$ to one for which they are all at $s = 0$.

When the real part of an impedance is zero at some j-axis point, it does not necessarily follow, of course, that the whole impedance is zero there. However, when the j-axis point is at $s = 0$ or at $s = \infty$, then the whole impedance *is* zero there because (having no j-axis poles) the imaginary part of this impedance is zero at $s = 0$ and at $s = \infty$ anyway. That is to say, a minimum-reactive impedance must be purely real at $s = 0$ and at $s = \infty$; hence if its real part is zero at these points, the whole impedance must likewise be zero there. We see, therefore, that the component impedances generated by the terms in Eq. 30 will have zeros at $s = 0$ and/or at $s = \infty$; but how can we tell whether these impedances are completely developable in simple terms? Here is where the Darlington theory discussed in Art. 8 of the preceding chapter yields the complete answer.

When we make a cascade development of an impedance in the form illustrated schematically in Fig. 30, Ch. 9, the Darlington theory shows that each component lossless network provides one set of zeros out of the total complement of zeros which the even part (real part on the j-axis) of that impedance contains. Thus the network generated by the first development cycle provides one zero or pair of quadruplet of zeros (according to their nature) and leaves a remainder whose even part contains the remaining even-part zeros. The next network in the chain provides another set of zeros out of these remaining ones and leaves a second remainder with the leftover even-part zeros, and so forth.

Applying this reasoning to the real part given by the first term in Eq. 30, for example, we can predict that the corresponding impedance will have a zero at $s = \infty$ and its reciprocal (the corresponding admittance) has a pole at this point. Removal of this pole as a shunt capacitance amounts in this degenerate situation to a complete development cycle. The pertinent shunt capacitance represents the first lossless two terminal-pair network in the schematic of Fig. 30, Ch. 9. The admittance remainder after this pole at $s = \infty$ is removed has an even part that still has $n - 1$ double-order zeros at $s = \infty$, because only one has so far been subtracted or "used up" out of the original complement of n double-order zeros at $s = \infty$. Being minimum susceptive, this admittance remainder must therefore have a zero at $s = \infty$, and its reciprocal, the impedance remainder, must have a pole there. Removal of this pole in the form of a series inductance constitutes carrying out the second development cycle, the pertinent series inductance branch alone being the component lossless two terminal-pair network generated by this second cycle.

The remainder function now has an even part with $n - 2$ double-order zeros at $s = \infty$, and being a minimum-reactive impedance this remainder must be zero at $s = \infty$. Its reciprocal (an admittance) must have a pole at this point, and its removal in the form of a shunt capacitance, as in the first degenerate development cycle, constitutes completion of the third cycle. The process continues in this same manner, yielding alternate series inductive and shunt capacitive branches, the total number of these being n, the number of double-order zeros possessed by the original real part. After the last of these zeros has been accounted for, the remainder is reduced to a constant because its even part has no more zeros. The complete development consists of a simple unbalanced ladder network with series L's and shunt C's, terminated in a resistance, the total number of L's plus C's being n. We observe in this development process that we alternatively shift from an impedance to an admittance basis and back again as we proceed from one cycle to the next. The representation in Eq. 174 of Ch. 9 is therefore relevant to this process.

It should be observed carefully that although the real part given by the first term in Eq. 30 has n double-order zeros at $s = \infty$, the impedance which it generates can have only one simple zero there, since j-axis zeros in a p.r. function must be simple. The reciprocal of this impedance has only a simple pole at $s = \infty$; but note! It has the property that when this pole is removed, the next remainder has a simple zero at $s = \infty$ again; and if we reciprocate this remainder and remove its pole at $s = \infty$, the next remainder comes up with another simple zero at $s = \infty$, and so forth. Obviously the original impedance generated by the real part which is the first term in Eq. 30 is more than just an impedance with a simple zero at $s = \infty$. It is a very special brand of such an impedance in order that all the properties mentioned here should be possessed by it; and it is truly remarkable that we should be able to tell all this just from the nature of the real part. Darlington's theory is a pretty remarkable and useful theory!

The second term in Eq. 30 has one double zero at $s = 0$ and the rest at $s = \infty$. The impedance which this real part generates has simple zeros at $s = 0$ and at $s = \infty$. Its reciprocal has simple poles there; and if we remove them as shunt inductance and capacitance branches, the remainder has a simple zero at $s = \infty$ again. From here on the development process follows the same pattern as with the impedance generated by the first term in Eq. 30. We again obtain a simple unbalanced ladder with single inductances and capacitances in its branches, terminated in a resistance, the total number of reactive elements being n. All series branches are inductances and all but one of the shunt branches are capacitances.

In the development of this second component impedance, we do not have to remove the pole at $s = 0$ in the first step. We can manipulate exactly as with the first component impedance, taking out only shunt capacitances and series inductances, and leaving the critical frequency at $s = 0$ alone. It will continue to be present in each remainder function, and we can remove it at any stage of the total development that we choose. If we remove it as a shunt branch, it becomes an inductance as in the process described in the previous paragraph; if we remove it as a series branch, it is represented by a capacitance. We see that many alternative developments are possible for the impedance generated by the second term in Eq. 30. All involve the same total number of elements, but the element values will be different, and for this reason one development may be preferable to another.

The third term in Eq. 30, like the second, again generates an impedance that has simple zeros at $s = 0$ and at $s = \infty$; but if we remove a shunt inductance and a shunt capacitance as in the development of the second component impedance, we find for the third one that the remainder function (an admittance) again has zeros at both $s = 0$ and $s = \infty$, so that, from its reciprocal, we can extract a series inductance and a series capacitance. The next remainder (an impedance) now has only a zero at $s = \infty$, since the two even-part zeros at $s = 0$ have at this stage been produced (as well as two at $s = \infty$) so that only $n - 4$ even-part zeros (of double order) remain. Our ladder network now has one shunt inductance and one series capacitance, plus remaining branches that are series inductances or shunt capacitances as in the ladder network realizing the first component impedance.

As with the second impedance, we can vary the development procedure in many ways, by removing the two branches accounting for the transmission zeros at $s = 0$, at various stages in the total development. Thus these may both become series capacitances or they may both take the form of shunt inductances instead of being respectively a series capacitance and a shunt inductance as in the process described in the previous paragraph. In any case, the total number of reactive elements again is n, and the realization has the form of a simple unbalanced ladder terminated in a resistance.

The last term in Eq. 30, finally, generates an impedance whose development yields a ladder network with series capacitances and shunt inductances exclusively, because this development must produce even-part zeros solely at the point $s = 0$.

Regarding the total number of elements required by this form of realization, it is significant to point out that we can always eliminate one of the $n + 1$ terms* in expression 30 by subtracting from $Z(s)$ or from its real part 29 an appropriate constant R_0. The result of such a subtraction is evidently that we have, in place of the coefficients $A_0 \cdots A_n$, the quantities $(A_0 - R_0B_0)$, $(A_1 - R_0B_1)$, and so forth. The value of R_0 is chosen just large enough to make the smallest of these revised coefficient values equal to zero.

If this tactic is utilized, we see that the total number of reactive elements

---

* Pointed out by S. N. Hunt in his S.M. thesis entitled "An Evaluation of Miyata's Driving-Point Impedance Synthesis Procedure," Massachusetts Institute of Technology, June 1955.

in the realization of an impedance by this method equals n^2, where n equals the degree of the polynomials in $Z(s)$. As compared with the total number of reactive elements required by the Bott and Duffin method which (if the saving of one element in the normal cycle, as pointed out above, is taken advantage of) is $5(2^{(n/2)-1} + 2^{(n/2)-2} + \cdots + 2^0)$, the Miyata method as discussed so far shows no advantage until the degree n is 8 or better, but for large values of n the Bott and Duffin method really loses ground fast in this respect. For example, for $n = 20$ the Bott and Duffin method requires 5115 reactive elements as compared with 400 by the Miyata procedure discussed above. In this connection we should not, of course, lose sight of the fact that the Miyata method is restricted by the requirement that all coefficients in the polynomial $(m_1m_2 - n_1n_2)_{s=j\omega}$ must be positive. Let us consider this condition in some detail.

A sufficient condition that the coefficients $A_0 \cdots A_n$ be positive is readily obtained if we are reminded of the fact that a Hurwitz polynomial has positive coefficients. Thus if

$$A(\omega^2) = A_0 + A_1\omega^2 + \cdots + A_n\omega^{2n} \tag{31}$$

has no zeros in the right-half ω^2-plane, then the $A_0 \cdots A_n$ are surely positive. Or we can say that if

$$A(-s^2) = m_1m_2 - n_1n_2 \tag{32}$$

has no zeros in the left-half s^2-plane, then the $A_0 \cdots A_n$ are positive. Since the left-half of the s^2-plane maps upon a region of the s-plane containing all points that are not more than $\pm 45°$ from the j-axis, we can say that the coefficients in the polynomial 31 are surely positive so long as there are no even-part zeros of the given impedance $Z(s)$ closer than 45° to the j-axis.

Although the coefficients in 31 may still be positive when some even-part zeros lie closer to the j-axis than this sufficient condition permits, the chances of the desired condition being fulfilled when even-part zeros are relatively close to the j-axis become rather remote, and they vanish completely as soon as a single pair of even-part zeros lie upon the j-axis.

In a crude way we can say that a polynomial 32 that has some zeros above the 45° lines in the s-plane may still yield all positive coefficients in 31 if there are a sufficient number of zeros below the 45° lines to produce a compensating effect. Hence in a situation in which the polynomial 31 has some negative coefficients [these can never be A_0 or A_n, for the real part of $Z(s)$, which is p.r. by hypothesis, must be positive for $\omega = 0$ and $\omega = \infty$], we can augment the real part 29 with appropriately chosen factors and thus bring about the desired compensation.*

* For a discussion of this item, see the appendix to this chapter.

This method of augmentation fails if the even part of $Z(s)$ has any j-axis zeros, since the number of necessary compensating factors tends toward infinity as one or more pairs of even-part zeros approach the j-axis. In any case, such a compensation process increases the number of terms in the real-part expression 30 as well as the degree of the polynomial $B(\omega^2)$ appearing in each term, and hence the number of necessary elements increases rapidly. It is necessary to devise a better way to salvage the Miyata method in such situations. Possibilities in this regard are discussed presently.

In the meantime a few examples may be helpful to illustrate the principles just described. Suppose we choose the impedance

$$Z(s) = \frac{s^2 + \frac{1}{2}s + 1}{s^2 + 2s + 1} \tag{33}$$

for which we find

$$m_1 m_2 - n_1 n_2 = s^4 + s^2 + 1 \tag{34}$$

so that we get

$$A(\omega^2) = 1 - \omega^2 + \omega^4 \tag{35}$$

Since the polynomial 34 represents a quadruplet of zeros on the unit circle, 30° from the j-axis, we could have predicted that a negative coefficient would be encountered in $A(\omega^2)$. We will try the augmentation process by multiplying 35 by the factor $(a + \omega^2)$, thus

$$(1 - \omega^2 + \omega^4)(a + \omega^2) = a + (1 - a)\omega^2 + (a - 1)\omega^4 + \omega^6 \tag{36}$$

A single factor just suffices to accomplish the desired end, the obvious choice for the constant a being unity. Evidently the impedance 33 is augmented by the factor $(s + 1)$, and Eq. 30 becomes in this instance

$$\text{Re}\,[Z(j\omega)] = \frac{1}{B(\omega^2)} + \frac{\omega^6}{B(\omega^2)} \tag{37}$$

in which

$$B(\omega^2) = [(1 + s)^3(1 - s)^3]_{s=j\omega} \tag{38}$$

since the denominator in the given impedance 33 is $(1 + s)^2$ and we have just augmented with the factor $(1 + s)$.

Corresponding to the breakdown of the real part given in Eq. 37, we regard the impedance 33 to be represented as

$$Z(s) = Z_0(s) + Z_6(s) \tag{39}$$

These component impedances are now generated from the respective real parts in 37 by any of the methods given in Art. 1, Ch. 8. Using the Gewertz method on the first term, we find

$$Z_0(s) = \frac{\frac{3}{8}s^2 + \frac{9}{8}s + 1}{s^3 + 3s^2 + 3s + 1} \tag{40}$$

which, in the Miyata method of impedance construction (also discussed in Art. 1, Ch. 8), we can also regard as the auxiliary impedance

$$Z'(s) = \frac{P'(s)}{Q(s)} = \frac{m_1' + n_1'}{m_2 + n_2} \tag{41}$$

since the condition $m_1'm_2 - n_1'n_2 = 1$ characterizing this impedance is evidently fulfilled by the form of the first term in Eq. 37. Hence the impedance $Z_6(s)$, corresponding to the second term in Eq. 37, is obtained through multiplying the impedance 40 by $-s^6$ and dividing the denominator polynomial into the numerator until the degree of the remainder no longer exceeds that of the denominator. This process is indicated as follows

$$\begin{array}{r l}
 & -\tfrac{3}{8}s^5 + \tfrac{1}{8}s^3 - \tfrac{3}{8}s \\
s^3 + 3s^2 + 3s + 1 \,| & -\tfrac{3}{8}s^8 - \tfrac{9}{8}s^7 - s^6 \\
 & -\tfrac{3}{8}s^8 - \tfrac{9}{8}s^7 - \tfrac{9}{8}s^6 - \tfrac{3}{8}s^5 \\ \hline
 & \tfrac{1}{8}s^6 + \tfrac{3}{8}s^5 \\
 & \tfrac{1}{8}s^6 + \tfrac{3}{8}s^5 + \tfrac{3}{8}s^4 + \tfrac{1}{8}s^3 \\ \hline
 & -\tfrac{3}{8}s^4 - \tfrac{1}{8}s^3 \\
 & -\tfrac{3}{8}s^4 - \tfrac{9}{8}s^3 - \tfrac{9}{8}s^2 - \tfrac{3}{8}s \\ \hline
 & s^3 + \tfrac{9}{8}s^2 + \tfrac{3}{8}s
\end{array} \tag{42}$$

Hence we have

$$Z_6(s) = \frac{s^3 + \frac{9}{8}s^2 + \frac{3}{8}s}{s^3 + 3s^2 + 3s + 1} \tag{43}$$

The Miyata method of impedance construction is evidently well suited to the present method of impedance synthesis since the same auxiliary impedance, which is that generated by the first term in Eq. 30 for $A_0 = 1$, readily yields the impedances generated by the remaining terms in this decomposition of the real part.

It is interesting at this point to form from Eqs. 40 and 43

$$Z(s) = Z_0(s) + Z_6(s) = \frac{s^3 + \frac{3}{2}s^2 + \frac{3}{2}s + 1}{s^3 + 3s^2 + 3s + 1} = \frac{(s^2 + \frac{1}{2}s + 1)(s + 1)}{(s^2 + 2s + 1)(s + 1)} \tag{44}$$

which checks with Eq. 33 as it should. We will now see that the components $Z_0(s)$ and $Z_6(s)$ into which the given impedance has been decomposed are developable into the simple ladder networks described above.

In this connection we observe that Z_0 in Eq. 40 and Z_6 in Eq. 43 follow, one from the other, upon replacing s by $1/s$. As soon as we have a network for one of these impedances, that for the other is obtained simply by replacing an inductance by a capacitance having the reciprocal value, and vice versa. The appropriate ladder development of $Z_0(s)$ is indicated in the calculations

$$\begin{array}{l} \qquad\qquad\qquad\quad 8s/3 \\ \frac{3}{8}s^2 + \frac{9}{8}s + 1 \,\big|\, \overline{s^3 + 3s^2 + 3s + 1} \\ \qquad\qquad\qquad\quad \underline{s^3 + 3s^2 + \frac{8}{3}s \qquad\quad} \big| 9s/8 \\ \qquad\qquad\qquad\qquad\qquad\quad \frac{1}{3}s + 1 \,\big|\, \frac{3}{8}s^2 + \frac{9}{8}s + 1 \\ \qquad\qquad\qquad\qquad\qquad\qquad\qquad\quad \underline{\frac{3}{8}s^2 + \frac{9}{8}s \qquad\quad} \big| \\ \qquad\qquad\qquad\qquad\qquad\qquad\qquad\qquad\qquad 1 \,\big|\, \frac{1}{3}s + 1 \end{array} \tag{45}$$

from which the networks in Fig. 7 are obtained. Their series connection yields the desired realization for $Z(s)$ in Eq. 33.

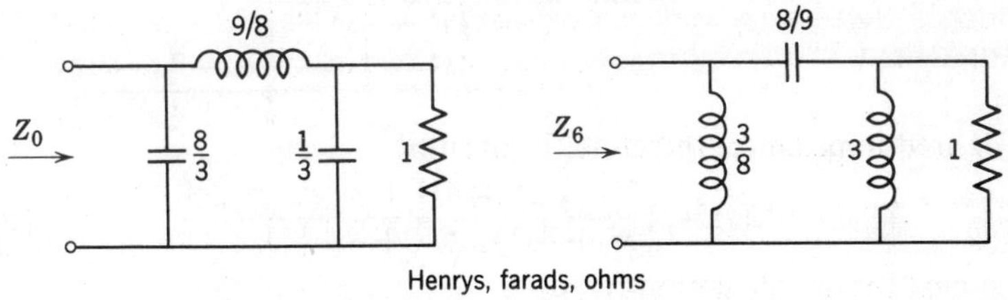

Fig. 7. Networks (obtained by Miyata method) whose series connection realizes the impedance 33.

Note in the calculations 45 that two terms cancel at each subtraction, which fact is necessary in order that the remainder function at each step shall again have a zero at $s = \infty$. As pointed out above, this property of the impedance function is revealed by the form of its real part. The impedance itself does not place the same information in evidence. Manipulation of the real part, therefore, is the key which solves the problem of breaking an impedance into components with predictable special properties.

Let us now consider the function

$$Z(s) = \frac{s^4 + \frac{9}{4}s^3 + 3s^2 + \frac{9}{4}s + 1}{s^4 + 4s^3 + 6s^2 + 4s + 1} \tag{46}$$

which is constructed to have a fourth-order pole at $s = -1$ and a double quadruplet of even-part zeros lying upon the intersections of the unit circle and the 45° lines, as is evident from the fact that

$$m_1 m_2 - n_1 n_2 = (s^4 + 1)^2 \tag{47}$$

Here we have

$$A(\omega^2) = 1 + 2\omega^4 + \omega^8 \tag{48}$$

The coefficients in this polynomial are all positive, as can be predicted in terms of the previous discussion.

Again we begin by generating the auxiliary impedance 41 which corresponds to a real part equal to $1/B(\omega^2)$ and represents the first of the three component impedances into which the net impedance 46 will be decomposed according to the general Eq. 30 and the particular form of the polynomial 48 pertinent to this example. We find by the methods of Art. 1, Ch. 8,

$$Z'(s) = Z_0(s) = \frac{\frac{5}{16}s^3 + \frac{5}{4}s^2 + \frac{29}{16}s + 1}{s^4 + 4s^3 + 6s^2 + 4s + 1} \tag{49}$$

The second impedance, which we shall denote by $Z_4(s)$, is found by multiplying 49 by $2s^4$ and then dividing, as follows

$$\begin{array}{r|l}
 & \frac{5}{8}s^3 - \frac{1}{8}s \\
\hline
s^4 + 4s^3 + 6s^2 + 4s + 1 & \frac{5}{8}s^7 + \frac{5}{2}s^6 + \frac{29}{8}s^5 + 2s^4 \\
 & \frac{5}{8}s^7 + \frac{5}{2}s^6 + \frac{30}{8}s^5 + \frac{5}{2}s^4 + \frac{5}{8}s^3 \\
\hline
 & \qquad - \frac{1}{8}s^5 - \frac{1}{2}s^4 - \frac{5}{8}s^3 \\
 & \qquad - \frac{1}{8}s^5 - \frac{1}{2}s^4 - \frac{6}{8}s^3 - \frac{1}{2}s^2 - \frac{1}{8}s \\
\hline
 & \qquad\qquad\qquad \frac{1}{8}s^3 + \frac{1}{2}s^2 + \frac{1}{8}s
\end{array} \tag{50}$$

The desired impedance, therefore, is given by

$$Z_4(s) = \frac{1}{8}\,\frac{s^3 + 4s^2 + s}{s^4 + 4s^3 + 6s^2 + 4s + 1} \tag{51}$$

The next set of calculations reads

$$\begin{array}{r|l}
 & \frac{1}{16}s^3 - \frac{5}{16}s \\
\hline
s^4 + 4s^3 + 6s^2 + 4s + 1 & \frac{1}{16}s^7 + \frac{1}{4}s^6 + \frac{1}{16}s^5 \\
 & \frac{1}{16}s^7 + \frac{1}{4}s^6 + \frac{6}{16}s^5 + \frac{1}{4}s^4 + \frac{1}{16}s^3 \\
\hline
 & \qquad - \frac{5}{16}s^5 - \frac{1}{4}s^4 - \frac{1}{16}s^3 \\
 & \qquad - \frac{5}{16}s^5 - \frac{5}{4}s^4 - \frac{30}{16}s^3 - \frac{5}{4}s^2 - \frac{5}{16}s \\
\hline
 & \qquad\qquad\qquad s^4 + \frac{29}{16}s^3 + \frac{5}{4}s^2 + \frac{5}{16}s
\end{array} \tag{52}$$

yielding the third component impedance

$$Z_8(s) = \frac{s^4 + \frac{29}{16}s^3 + \frac{5}{4}s^2 + \frac{5}{16}s}{s^4 + 4s^3 + 6s^2 + 4s + 1} \tag{53}$$

Addition of the components 49, 51, and 53 is easily seen to verify the decomposition

$$Z(s) = Z_0(s) + Z_4(s) + Z_8(s) = \frac{s^4 + \frac{9}{4}s^3 + 3s^2 + \frac{9}{4}s + 1}{s^4 + 4s^3 + 6s^2 + 4s + 1} \tag{54}$$

The development of $Z_0(s)$, in which we consistently remove series inductances and shunt capacitances, leads to the calculations

$$\begin{array}{r|l}
 & 16s/5 \\
\hline
\frac{5}{16}s^3 + \frac{5}{4}s^2 + \frac{29}{16}s + 1 & s^4 + 4s^3 + 6s^2 + 4s + 1 \\
 & s^4 + 4s^3 + \frac{29}{5}s^2 + \frac{16}{5}s \\
\hline
\end{array}$$

$$\begin{array}{r|l}
 & 25s/16 \\
\hline
\frac{1}{5}s^2 + \frac{4}{5}s + 1 & \frac{5}{16}s^3 + \frac{5}{4}s^2 + \frac{29}{16}s + 1 \\
 & \frac{5}{16}s^3 + \frac{5}{4}s^2 + \frac{25}{16}s \\
\hline
 & \qquad \frac{1}{4}s + 1
\end{array} \tag{55}$$

$$\begin{array}{r|l}
 & 4s/5 \\
\hline
\frac{1}{4}s + 1 & \frac{1}{5}s^2 + \frac{4}{5}s + 1 \\
 & \frac{1}{5}s^2 + \frac{4}{5}s \\
\hline
 & 1 \,|\, \frac{1}{4}s + 1
\end{array}$$

The way in which two terms disappear consistently at every step in this division process is almost uncanny. That this very special result should take place is certainly not evident from inspection of expression 49 for $Z_0(s)$ as it is from inspection of its real part.

In the development of $Z_4(s)$ in Eq. 51 the process must change midway, from pole removals at $s = \infty$ to pole removals at $s = 0$. Those at $s = \infty$ lead to the calculations

$$
\begin{array}{r|l}
 & s \\
\cline{2-2}
s^3 + 4s^2 + s & s^4 + 4s^3 + 6s^2 + 4s + 1 \\
 & s^4 + 4s^3 + s^2 \qquad\qquad | s/5 \\
\cline{2-2}
 & \qquad 5s^2 + 4s + 1 \,|\, s^3 + 4s^2 + s \\
 & \qquad\qquad\qquad\qquad s^3 + \tfrac{4}{5}s^2 + \tfrac{1}{5}s \\
\cline{2-2}
 & \qquad\qquad\qquad\qquad\quad \tfrac{16}{5}s^2 + \tfrac{4}{5}s
\end{array}
\tag{56}
$$

while the continued divisions corresponding to pole removals at $s = 0$ take the form

$$
\begin{array}{r|l}
 & 5/4s \\
\cline{2-2}
\tfrac{4}{5}s + \tfrac{16}{5}s^2 & 1 + 4s + 5s^2 \\
 & 1 + 4s \qquad\quad | 4/25s \\
\cline{2-2}
 & \qquad\quad 5s^2 \,|\, \tfrac{4}{5}s + \tfrac{16}{5}s^2 \\
 & \qquad\qquad\quad \tfrac{4}{5}s \qquad\quad | \tfrac{25}{16} \\
\cline{2-2}
 & \qquad\qquad\qquad \tfrac{16}{5}s^2 \,|\, 5s^2
\end{array}
\tag{57}
$$

where we have tentatively ignored the level factor $\frac{1}{8}$ in expression 51.

In the development of $Z_8(s)$, Eq. 53, we remove poles consistently at $s = 0$. This process follows the pattern of calculations 55 except that the polynomials are written in the opposite orientation. At this point we observe by inspection of Eqs. 49 and 53 that $Z_8(s) = Z_0(1/s)$, which, as in the previous example, is a result of the symmetrical character of the assumed denominator polynomial. A separate set of calculations for the development of $Z_8(s)$ is therefore not necessary. We simply obtain the network for Z_8 from that of Z_0 by replacing each inductance by a capacitance having the reciprocal value, and vice versa.

In this manner calculations 55 and 57 (remembering the level factor $\frac{1}{8}$ needed in the latter) lead to the final network in Fig. 8 for the realization of the impedance 46. Because of the special pole-zero pattern assumed for the even part of this impedance, its realization by the Miyata method results in a greater economy of elements than would ordinarily be the case for an impedance involving fourth-degree polynomials. This fact is in evidence in expression 48 for $A(\omega^2)$ where two terms are missing (there would ordinarily be five terms in this expression, of which we could eliminate one by the method discussed above).

Thus in this example the realization turns out to involve fewer elements than would be required by the Bott and Duffin method. A realization of this same impedance by that method would require 15 reactive elements and 4 resistances whereas the one in Fig. 8 involves 12 reactive elements and 3 resistances, besides being computationally easier to obtain.

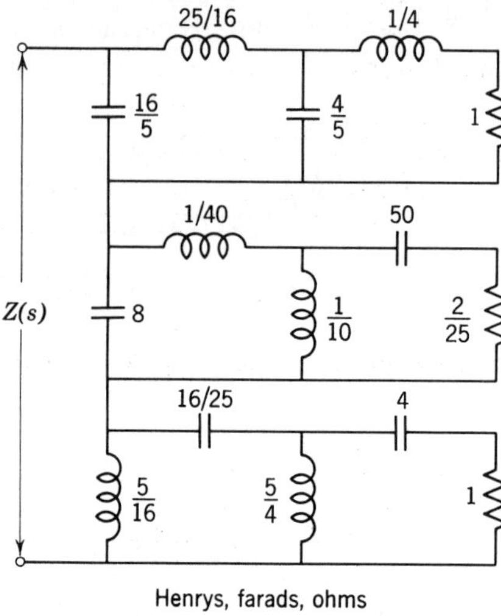

Fig. 8. Realization of the impedance 46 by the Miyata method.

3 Variations in the Miyata Procedure

As pointed out by Miyata, the detailed procedure for carrying out his basic method of decomposition may be varied in a large number of ways by separating the numerator polynomial $A(\omega^2)$ in Eq. 29 into parts consisting of groups of terms rather than separate terms. We may, for example, divide this numerator polynomial essentially into two equal parts and write instead of Eq. 30

$$\text{Re}\,[Z(j\omega)] = \frac{A_0 + A_1\omega^2 + \cdots + A_{n-k}\omega^{2(n-k)}}{B(\omega^2)} + \frac{A_{n-k+1}\omega^{2(n-k+1)} + \cdots + A_n\omega^{2n}}{B(\omega^2)} \tag{58}$$

in which k can have any integer value from 1 to n but is for the present assumed to have the value $n/2$ or $(n + 1)/2$ according to whether n is even or odd respectively. In writing the real part in this form, we are tentatively decomposing $Z(s)$ into just two components. Let us see what the forms of the real parts of these separate components, as given by the two terms in Eq. 58, can tell us about the characters of the components themselves with regard to their realization.

The first term in Eq. 58 for $\omega^2 = -s^2$ is an even part that has k double-order zeros at $s = \infty$, and $(n - k)$ pairs of zeros in the finite s-plane (in general none of these occur at $s = 0$). The second term for $\omega^2 = -s^2$ is an even part that has $(n - k + 1)$ double-order zeros at $s = 0$, and $(k - 1)$ pairs of zeros in the finite s-plane. The impedance generated by the first term in Eq. 58 is therefore developable into a simple unbalanced ladder with series L's and shunt C's (their total number being k) terminated in a remainder whose even-part zeros are the same finite nonzero ones that are contained in the even part of this impedance. The impedance generated by the second term in Eq. 58 is seen to be developable into an unbalanced ladder with series C's and shunt L's [their total number being $(n - k + 1)$] and a terminal impedance whose even-part zeros are those in the finite s-plane that are contained in the even part of this second impedance.

To these terminal impedances which appear as remainder functions, we can apply a similar treatment, and continue in this manner until the ultimate remainder functions are resistances. S. N. Hunt (*loc. cit.*), who refers to this particular decomposition procedure as the "half-n" synthesis method, finds that it yields a smaller total number of elements than any other division of the numerator polynomial $A(\omega^2)$ into parts. When it is applicable, it requires fewer elements than the Bott and Duffin method for any degree n equal to or greater than 2. Table 1 (for which the data are taken from the thesis referred to above) illustrates a comparison of the two Miyata methods discussed here with the Bott and Duffin method in this regard. N_{LC} and N_R denote the numbers of reactive elements and resistances respectively. In the Bott and Duffin method it is assumed that we avail ourselves of the one-element saving per typical cycle that was pointed out above.

The geometric character of the Bott and Duffin method, compared with the arithmetic character of the Miyata method as regards element numbers, becomes strongly apparent in this tabular presentation. Even though we need to resort to special devices (like augmentation or the techniques discussed in the following articles) in order to salvage the Miyata method when all coefficients in the polynomial $A(\omega^2)$ are not initially positive, it would still seem advantageous to explore such tactics in conjunction with the Miyata procedure in situations where more elaborate impedance functions are involved.

In the half-n synthesis, or similar scheme, one may note that if the coefficients $A_0 \cdots A_n$ are initially positive, then no negative coefficients can be encountered in the $A(\omega^2)$ polynomials associated with the first pair of remainder functions because their even-part zeros are the same sets in the finite s-plane as are determined by the numerator polynomials (for $\omega^2 = -s^2$) in the separate terms in 58, and hence the coefficients A_k in the real-part expressions for these remainder functions must be the same. A similar argument shows that no negative coefficients can be encountered in a subsequent decomposition of the first pair of remainder functions, and so forth.

Table I

n	Single-Term Synthesis		Half-N Synthesis		Modified Bott and Duffin Synthesis	
	N_{LC}	N_R	N_{LC}	N_R	N_{LC}	N_R
1	1	2	1	2	1	2
2	4	3	4	3	5	3
3	9	4	6	4	7	5
4	16	5	10	5	15	7
5	25	6	14	6	19	11
6	36	7	17	7	35	15
8	64	9	25	9	75	31
10	100	11	35	11	155	63
12	144	13	44	13	315	127
14	196	15	52	15	635	255
16	256	17	62	17	1,275	511
18	324	19	74	19	2,555	1,023
20	400	21	86	21	5,115	2,047
30	900	31	139	31	163,835	65,535

Initially it may be necessary to do some augmentation in order to arrive at a polynomial $A(\omega^2)$ with positive coefficients. Thereafter we can separate the real part in any manner we wish, and follow by subsequent decompositions of the real parts of remainder functions in any manner, with complete assurance that no negative coefficients will appear in any real-part expressions.

As an example of the half-n procedure, suppose we reconsider the impedance 46 and, in view of $A(\omega^2)$ given by Eq. 48, separate the real part as follows

$$\text{Re}\,[Z(j\omega)] = \frac{1}{B(\omega^2)} + \frac{\omega^4(2+\omega^4)}{B(\omega^2)} \tag{59}$$

What this separation amounts to is a decomposition of $Z(s)$ into two additive components, of which the first is $Z_0(s)$ in Eq. 49 as before, and the second is given by the sum of $Z_4(s)$ in Eq. 51 and $Z_8(s)$ in Eq. 53. Let us call this impedance $Z_a(s)$ and have

$$Z_a(s) = Z_4(s) + Z_8(s) = \frac{s^4 + \frac{31}{16}s^3 + \frac{7}{4}s^2 + \frac{7}{16}s}{s^4 + 4s^3 + 6s^2 + 4s + 1} \tag{60}$$

According to the real part given by the second term in Eq. 59, this impedance should be developable by removing poles at $s = 0$ for two successive steps. The calculations read

$$\begin{array}{r|l} & 16/7s \\ \hline \frac{7}{16}s + \frac{7}{4}s^2 + \frac{31}{16}s^3 + s^4 & 1 + 4s + 6s^2 + 4s^3 + s^4 \\ & 1 + 4s + \frac{31}{7}s^2 + \frac{16}{7}s^3 \qquad |49/176s \\ \hline & \frac{11}{7}s^2 + \frac{12}{7}s^3 + s^4 \,|\, \frac{7}{16}s + \frac{7}{4}s^2 + \frac{31}{16}s^3 + s^4 \\ & \qquad \frac{7}{16}s + \frac{21}{44}s^2 + \frac{49}{176}s^3 \\ \hline & \qquad \frac{14}{11}s^2 + \frac{73}{44}s^3 + s^4 \end{array} \tag{61}$$

and leave the impedance remainder

$$Z_b = \frac{s^2 + \frac{73}{44}s + \frac{14}{11}}{s^2 + \frac{12}{7}s + \frac{11}{7}} \tag{62}$$

with the real part

$$\text{Re}\,[Z_b(j\omega)] = \frac{2 + \omega^4}{B(\omega^2)} = \frac{2}{B(\omega^2)} + \frac{\omega^4}{B(\omega^2)} \tag{63}$$

in which $B(\omega^2)$ represents the squared magnitude of the denominator of Eq. 62 for $s = j\omega$. The impedances generated by the separate terms in the right-hand side of Eq. 63 are readily found to be

$$Z_c(s) = \frac{\frac{49}{66}s + \frac{14}{11}}{s^2 + \frac{12}{7}s + \frac{11}{7}} \tag{64}$$

and

$$Z_d(s) = \frac{s^2 + \frac{11}{12}s}{s^2 + \frac{12}{7}s + \frac{11}{7}} \tag{65}$$

Their sum is the desired decomposition of $Z_b(s)$. The respective ladder developments of Z_c and Z_d are given by the calculations

$$\begin{array}{r|l} & \frac{66}{49}s \\ \hline \frac{49}{66}s + \frac{14}{11} & s^2 + \frac{12}{7}s + \frac{11}{7} \\ & s^2 + \frac{12}{7}s \qquad |\frac{343}{726}s + \frac{98}{121} \\ \hline & \qquad \frac{11}{7} \,|\, \frac{49}{66}s + \frac{14}{11} \end{array} \tag{66}$$

and

$$\begin{array}{r|l} & 12/7s \\ \hline \frac{11}{12}s + s^2 & \frac{11}{7} + \frac{12}{7}s + s^2 \\ & \frac{11}{7} + \frac{12}{7}s \qquad |11/12s + 1 \\ \hline & \qquad s^2 \,|\, \frac{11}{12}s + s^2 \end{array} \tag{67}$$

The two-branch ladder network corresponding to calculations 61, terminated in the series combination of the circuits given by 66 and 67, represents the impedance Z_a in Eq. 60. In series with Z_0, given by the top network in Fig. 8, this result, shown in Fig. 9, is an alternative realization for the impedance 46. It involves 10 reactive elements as compared with 12 in the realization given in Fig. 8. Even in this simple example, the half-n method shows an advantage.

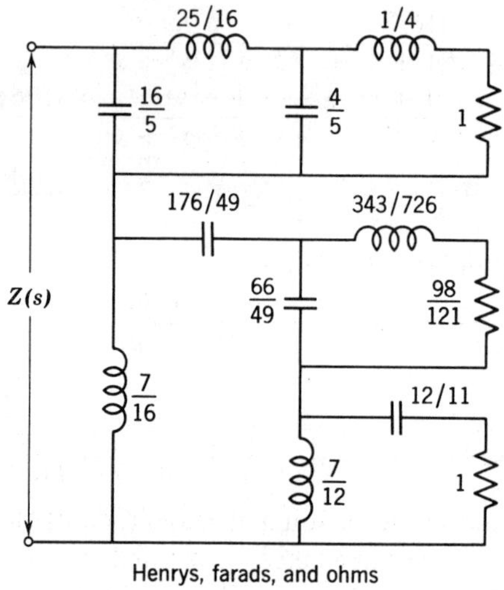

Fig. 9. Realization of the impedance 46, alternate to that shown in Fig. 8, represents a variation in the Miyata method whereby a further reduction in the number of elements may be achieved.

We can derive an alternate half-n synthesis for this same impedance by regrouping the terms in Eq. 48 so as to consider the sum $Z_0 + Z_4$ as a component to be dealt with in the manner that Z_b is treated in the example just completed. This variation is left as an exercise for the reader.

It remains to show how we can salvage the Miyata method when even-part zeros lie near or upon the j-axis so that augmentation becomes either very uneconomical or completely impossible as a salvaging means.

4 Extensions of the Miyata Procedure*

The occurrence of negative coefficients in the polynomial $A(\omega^2)$ indicates that it contains quadratic factors in the variable ω^2 of the form

$$\omega^4 - 2\omega_\nu^2\omega^2 + a^4 = (\omega^2 - \omega_\nu^2)^2 + a^4 - \omega_\nu^4 \tag{68}$$

* Presented at the Symposium on Modern Network Synthesis, Apr. 13–15, 1955, and published in the *Transactions*, Polytechnic Institute of Brooklyn, 1956.

in which the value of $(a^4 - \omega_\nu^4) > 0$ is relatively close to zero. The positive character of $A(\omega^2)$ for real ω, of course, does not allow this quantity to become negative, but it may become zero. When it does, then the even part of $Z(s)$ has a pair of double-order zeros at the j-axis points $s = \pm j\omega_\nu$.

So long as $(a^4 - \omega_\nu^4)$ has a nonzero value, one may overcome the negative coefficients in $A(\omega^2)$ through the process of augmentation, as already mentioned, but if this value is very small then the number of factors required by the augmentation method becomes extremely large, and one may find some other procedure for achieving the same end to be less uneconomical.

Suppose we group the troublesome factors 68 into one factor $A^*(\omega^2)$, and write the polynomial $A(\omega^2)$ in the form

$$A(\omega^2) = (A_0 + A_1\omega^2 + \cdots + A_k\omega^{2k})A^*(\omega^2) \tag{69}$$

in which $k = n - 2r$ if there are r factors 68 contained in $A^*(\omega^2)$. We now apply the "single-term" or half-n synthesis procedure as described above except that each component real part contains the factor $A^*(\omega^2)$. Instead of ultimately having our simple ladder expansions terminated in resistances, we now find that the common termination is given by some positive constant times an impedance with the real part $A^*(\omega^2)/B^*(\omega^2)$ in which the detailed character of $B^*(\omega^2)$ is unimportant.

If we multiply out the r factors 68 contained in $A^*(\omega^2)$, we see that this function equals a sum of terms of which each consists of a positive constant times a product of j-axis factors $(\omega^2 - \omega_\nu^2)^2$. That is to say, the common remainder function breaks down into a sum of components whose real parts have j-axis zeros exclusively. Our problem, therefore, is solved as soon as we devise a method for dealing with this special type of impedance function.

Suppose $Z_1(s) = (m_1 + n_1)/(m_2 + n_2)$ is such an impedance. Then the polynomial $(m_1m_2 - n_1n_2)$ is composed exclusively of factors $(s_\nu^2 - s^2)^2$ in which the s_ν^2 values are negative real. If this impedance is augmented with a Hurwitz polynomial $P_0 = m_0 + n_0$ as indicated by writing

$$Z_1(s) = \frac{(m_1 + n_1)(m_0 + n_0)}{(m_2 + n_2)(m_0 + n_0)} = \frac{(m_1m_0 + n_1n_0) + (m_1n_0 + n_1m_0)}{(m_2m_0 + n_2n_0) + (m_2n_0 + n_2m_0)}$$
$$= \frac{\hat{m}_1 + \hat{n}_1}{\hat{m}_2 + \hat{n}_2} \tag{70}$$

and if we specify that the polynomial P_0 be determined such that

$$\hat{m}_2 = m_2m_0 + n_2n_0 = 0, \qquad \text{or} \quad \frac{m_2}{n_2} = -\frac{n_0}{m_0} \quad \text{for all } s^2 = s_\nu^2 \tag{71}$$

then the fact that the s_ν^2 are zeros of $(m_1m_2 - n_1n_2)$ evidently yields simultaneously the result

$$\hat{n}_1 = m_1 n_0 + n_1 m_0 = 0, \qquad \text{or} \quad \frac{n_1}{m_1} = -\frac{n_0}{m_0} \quad \text{for all } s^2 = s_\nu^2 \tag{72}$$

We have yet to see whether a polynomial P_0 can be found to fulfill the condition demanded by 71, but if it can, then the Darlington theory discussed in Art. 8, Ch. 9, suggests an interesting possibility for the synthesis of $Z_1(s)$. Namely, it is shown there that if we regard $Z_1(s)$ synthesized in the Darlington fashion as the input impedance to a lossless two terminal-pair network terminated in a 1-ohm resistance, and if we turn this network end for end so as to obtain the related driving-point impedance $Z_2(s)$, then (assuming case A) we get $Z_2(s)$ from $Z_1(s)$ simply by interchanging the even parts of the pertinent polynomials. That is to say, from 70 we get

$$Z_2(s) = \frac{\hat{m}_2 + \hat{n}_1}{\hat{m}_1 + \hat{n}_2} \tag{73}$$

which the results 71 and 72 show has j-axis zeros at all of the points $s = \pm s_\nu$. If this scheme works it looks as if we can kill all of our feathered fowl with a single whack. A number of details need to be ironed out, however, before we can claim to have achieved such a sweeping result.

Let us study the even part of the impedance 70. If we make use of the abbreviation

$$M = m_1 m_2 - n_1 n_2 \tag{74}$$

(as we have done previously, Ch. 8), then the even part of $Z_1(s)$ can be written

$$Z_{1E} = \tfrac{1}{2}[Z_1(s) + Z_1(-s)] = \frac{M(m_0^2 - n_0^2)}{\hat{m}_2^2 - \hat{n}_2^2} \tag{75}$$

where Z_{1E} is a convenient notation for the even part of Z_1 that we shall also use in connection with other impedance functions from now on. For $s = j\omega$, the polynomial $(m_0^2 - n_0^2)$ has the form

$$(m_0^2 - n_0^2)_{s=j\omega} = \alpha_0 + \alpha_1\omega^2 + \cdots + \alpha_k\omega^{2k} \tag{76}$$

in which k is the degree of $P_0 = m_0 + n_0$. The real part of $Z_1(j\omega)$ therefore has the representation

$$\text{Re}\,[Z_1(j\omega)] = \frac{\alpha_0 M}{\hat{m}_2^2 - \hat{n}_2^2} + \frac{\alpha_1 M\omega^2}{\hat{m}_2^2 - \hat{n}_2^2} + \cdots + \frac{\alpha_k M\omega^{2k}}{\hat{m}_2^2 - \hat{n}_2^2} \tag{77}$$

in which M and $(\hat{m}_2^2 - \hat{n}_2^2)$ are tacitly understood to be evaluated for $s = j\omega$.

If we now regard the terms in this expression as real parts of component impedances, we see that all of these real parts have the double-order

j-axis zeros of the factor M and that the latter is contained in $\hat{m}_2^2$. Therefore, all of these real parts will, first of all, generate impedances for which the odd part of the numerator polynomial and the even part of the denominator polynomial contain the factor $\sqrt{M}$; that is to say, they contain the simple factors $(s_\nu^2 - s^2)$ for all of the j-axis zeros of M. This much follows from the determination of the polynomial P_0 as expressed in conditions 71 and 72 which must be fulfilled simultaneously if they are fulfilled at all.

The first term in the decomposition 77 has additionally k double-order zeros at $s = \infty$; the second term has one such zero at $s = 0$ and $k - 1$ remaining ones at $s = \infty$, and so forth. Finally the last term has k additional double-order zeros at $s = 0$.

We may conclude that each component impedance generated by a real-part term in 77 is developable into a simple unbalanced ladder having a total of k single inductance or capacitance branches as is each component impedance in the Miyata single-term synthesis method. Each development yields the same terminal function (except perhaps for different constant multipliers) involving polynomials of degree n, the same as those in $Z_1(s)$. The even part of the denominator polynomial in this common remainder contains all n pairs of j-axis zeros of that function's even part (the zeros of $\sqrt{M}$). Such a function must reduce to the form

$$\frac{m}{m+n} = \frac{1}{1+(n/m)} \tag{78}$$

Since the ladder developments corresponding to successive terms in Eq. 77 terminate in remainder functions that alternately are impedances and admittances, the function 78, which consistently represents the remainder whether it be an impedance or an admittance, is realized alternately and respectively as a resistance in parallel or in series with a reactance.

Each component impedance corresponding to a term in Eq. 77, if regarded as synthesized in the Darlington manner, can equally well be realized by developing the lossless network from its output end. In this connection it should be observed that the terms in 77 lead alternately to case A and case B situations as defined in Art. 6, Ch. 9. If, for a case A component impedance $Z_1^{(\nu)}(s)$, we interchange the odd parts of the polynomials, then we have the reciprocal of the impedance $Z_2^{(\nu)}(s)$ pertinent to a reversed orientation of the lossless network relative to its input and output terminal pairs. The j-axis poles of this reciprocal impedance are precisely the poles of the terminal admittance equal to the reciprocal of the function 78. In a case B situation, an interchange of the odd parts in $Z_1^{(\nu)}(s)$ yields the impedance $Z_2^{(\nu)}(s)$ whose j-axis poles are those of the terminal impedance which now equals the reciprocal of the function 78.

If we approach the synthesis of the lossless network in this manner, these finite nonzero j-axis poles are removed at the beginning of the development. If, on the other hand, we begin at the input end with the ladder development of $Z_1^{(\nu)}(s)$, then these same j-axis poles are removed at the end of the development as part of the terminal function given by the reciprocal of expression 78.

Essential to the success of this process is the condition that the coefficients $\alpha_0, \alpha_1, \cdots, \alpha_k$ in Eqs. 76 and 77 be positive. Thus the requirements upon the auxiliary polynomial P_0 are twofold. First, it must fulfill condition 71 or 72 (if it fulfills one of these, it automatically fulfills the other); second, its squared magnitude for $s = j\omega$, Eq. 76, can have no negative coefficients. Before entering upon a general discussion of how such a polynomial can be found, it will be helpful to consider a simple example for which these conditions can be met without any elaborate analysis.

Suppose we consider for this purpose the impedance 13 used to illustrate the Bott and Duffin method in Art. 1. For it we have

$$m_1m_2 - n_1n_2 = (s^2 + 2)^2 \tag{79}$$

We will augment this impedance with the simple polynomial $P_0 = (s + a)$ and see if the desired result can be accomplished. The augmented impedance reads

$$Z(s) = \frac{(s^2 + s + 1)(s + a)}{(s^2 + s + 4)(s + a)} = \frac{[(1 + a)s^2 + a] + [s^3 + (1 + a)s]}{[(1 + a)s^2 + 4a] + [s^3 + (4 + a)s]} \tag{80}$$

Condition 71 requires in this example that

$$(1 + a)s^2 + 4a = 0 \quad \text{for } s^2 = -2 \tag{81}$$

which yields

$$a = 1 \tag{82}$$

Therefore

$$m_0 + n_0 = s + 1 \tag{83}$$

and we find

$$\hat{m}_2 + \hat{n}_2 = s^3 + 2s^2 + 5s + 4 \tag{84}$$

The augmented even part of $Z(s)$ and its decomposition into components reads

$$\begin{aligned} Z_E &= \frac{(s^2 + 2)^2(1 - s^2)}{4(s^2 + 2)^2 - s^2(s^2 + 5)^2} \\ &= \frac{(s^2 + 2)^2}{4(s^2 + 2)^2 - s^2(s^2 + 5)^2} + \frac{-s^2(s^2 + 2)^2}{4(s^2 + 2)^2 - s^2(s^2 + 5)^2} \end{aligned} \tag{85}$$

Denoting by Z_a and Z_b the impedances generated respectively by these two terms, we have by inspection

$$Z_a(s) = \frac{\frac{1}{2}(s^2+2)}{2(s^2+2)+s(s^2+5)} = \frac{\frac{1}{2}(s^2+2)}{s^3+2s^2+5s+4} \tag{86}$$

The impedance Z_b is found through multiplying Z_a by $-s^2$ and dividing, as follows

$$\begin{array}{r|l} & -\frac{1}{2}s \\ \cline{2-2} s^3+2s^2+5s+4 & -\frac{1}{2}s^4 + 0s^3 - s^2 \\ & -\frac{1}{2}s^4 - s^3 - \frac{5}{2}s^2 - 2s \\ \cline{2-2} & \qquad\quad s^3 + \frac{3}{2}s^2 + 2s \end{array} \tag{87}$$

Hence we have

$$Z_b(s) = \frac{s^3+\frac{3}{2}s^2+2s}{s^3+2s^2+5s+4} \tag{88}$$

The respective realizations for these two component impedances are immediately obtained from the developments

$$Z_a(s) = \cfrac{1}{2s + \cfrac{1}{\cfrac{\frac{1}{2}(s^2+2)}{2(s^2+2)+3s}}} \tag{89}$$

and

$$Z_b(s) = \cfrac{1}{\cfrac{2}{s} + \cfrac{2(s^2+2)}{2(s^2+2)+3s}} \tag{90}$$

The ladder networks in this simple example consist of single branches (the degree k of P_0 being unity); and the common terminal function, within a constant multiplier, is given by the expression

$$\frac{2(s^2+2)}{2(s^2+2)+3s} \tag{91}$$

which has the form 78, as it should.

It is interesting to note that the sum of Eqs. 89 and 90 yields identically the same expression for the net impedance as is given by Eq. 19 obtained by the Bott and Duffin method. The final network, represented by a series connection of those corresponding to 89 and 90 is therefore the same as that shown in Fig. 5. Our present method and that of Bott and Duffin yield identical results when the impedance involves second-degree polynomials.

In this connection it may be mentioned that if the determination of the constant a in Eq. 81 yields a negative value, one can always get a positive one by switching to an admittance basis. It is not difficult to show that a positive value is always obtainable on one basis or the other.

In terms of the impedances Z_a and Z_b given by Eqs. 86 and 88, we can also check further details about the Darlington theory pointed out above. Thus $Z_a(s)$, being derived from a case A real part, should, apart from a constant multiplier, become equal to the reciprocal of the input impedance with the lossless network turned end for end if we interchange the odd parts of its polynomials. Denoting this latter impedance by Z_{a2}, we get from Eq. 86

$$\frac{1}{Z_{a2}} = \frac{\frac{1}{2}(s^2 + 2) + s(s^2 + 5)}{2(s^2 + 2)} \tag{92}$$

which has the j-axis poles that the preceding general discussion requires it to have. Similarly, an interchange of the odd parts of $Z_b(s)$, derived from a case B real part, should yield an impedance $Z_{b2}(s)$ pertinent to a reversed orientation of the lossless network. We thus find from Eq. 88

$$Z_{b2} = \frac{s^3 + \frac{3}{2}s^2 + 5s}{s^3 + 2s^2 + 2s + 4} = \frac{s^3 + \frac{3}{2}s^2 + 5s}{(s^2 + 2)(s + 2)} \tag{93}$$

and again note that the required j-axis poles are present.

When the given impedance involves polynomials of higher degree, then it is in general no longer possible to satisfy condition 71 with a polynomial P_0 given by the linear factor $(s + a)$. It is therefore necessary that we understand the demands of this condition more fully in order to devise a more adequate method for the construction of the auxiliary polynomial P_0.

If we regard the ratios m_2/n_2 and n_0/m_0 appearing in condition 71 as reactance functions and visualize these as plotted versus the radian frequency ω in the usual manner, then this condition asks us to find a reactance n_0/m_0, the negative of which will intersect the curve for m_2/n_2 at the frequencies ω_ν where the real part of our given impedance has zeros. For the impedance 13 in the example just considered, this graphical interpretation of condition 71 is illustrated in Fig. 10.

It is clear that a linear factor $(s + a)$ for $m_0 + n_0$ will always suffice when $m_2 + n_2$ is of second degree. It is also clear from this graphical interpretation that if ω_ν were larger than the zero of m_2, then the required straight line $-n_0/m_0$ would need to have a positive slope, which is not permissible because $m_0 + n_0$ must be a Hurwitz polynomial. However, since $n_1/m_1 = -m_2/n_2$ at $\omega = \omega_\nu$, we see that a switch to an admittance basis, which interchanges $m_2 + n_2$ with $m_1 + n_1$, solves the difficulty.

When $m_2 + n_2$ is of higher degree so that the reactance m_2/n_2 has several alternating zeros and poles, and there are a number of frequencies ω_ν at which the negative of the unknown reactance n_0/m_0 is required to intersect the curve for m_2/n_2, this method of graphical interpretation, although still helpful in clarifying the nature of the problem involved, becomes less effective as a computational means for constructing the desired polynomial $m_0 + n_0$.

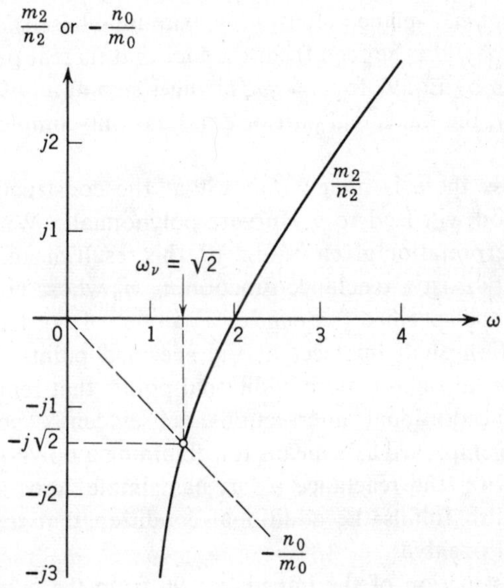

Fig. 10. Graphical interpretation of the condition 71 for the impedance 13.

Using the abbreviation in Eq. 74, the double-order j-axis zeros of the even part of the given impedance $Z(s)$ are indicated by writing

$$M = (s^2 + \omega_1^2)^2(s^2 + \omega_2^2)^2 \cdots (s^2 + \omega_r^2)^2 \tag{94}$$

in which the even integer $2r$ equals the degree n of the polynomials in $Z(s)$. Condition 71 may then be written

$$\hat{m}_2 = (m_2m_0 + n_2n_0) = \sqrt{M} = (s^2 + \omega_1^2) \cdots (s^2 + \omega_r^2) \tag{95}$$

We now observe that the demand made by this equation is equivalent to that encountered in the construction of an impedance

$$Z^*(s) = \frac{m_0 - n_0}{m_2 + n_2} \tag{96}$$

when its even part is specified to be

$$\frac{m_2m_0 + n_2n_0}{m_2^2 - n_2^2} = \frac{\sqrt{M}}{m_2^2 - n_2^2} \tag{97}$$

We have thus succeeded in reducing the problem of constructing the polynomial $P_0 = m_0 + n_0$ to that of constructing an impedance from its specified even part—a problem with which we are well acquainted and for the solution of which we have several alternative procedures at our disposal.

Observe that Eq. 96 is not a driving-point impedance. Its zeros are all in the right half-plane. It is a "maximum–phase-shift" transfer impedance, as may also be seen from the fact that its real part on the j-axis, which is given by Eq. 97 for $s = j\omega$, changes sign at all of the zeros of M. That is to say, the j-axis real part of $Z^*(s)$ has only simple zeros restricted to that axis.

Nevertheless there is no guarantee that the construction of $m_0 + n_0$ by this method will lead to a Hurwitz polynomial. With regard to the graphical interpretation given in Fig. 10, this result means that there does not necessarily exist a reactance function n_0/m_0 whose negative intersects m_2/n_2 at just the specified frequencies ω_ν alone. In order that a negative (n_0/m_0) function shall intersect at the specified points, it may have to intersect m_2/n_2 at one or more additional points that remain arbitrary.

Since such additional intersections are evidently permissible, their choice can be employed as a means for obtaining a polynomial P_0 that not only is Hurwitz (the reactance n_0/m_0 has simple zeros and poles alternating) but also fulfills the additional condition that the coefficients in Eq. 76 are all positive.

In the construction of the impedance 96 from the even part 97, these additional arbitrary intersections take the form of surplus factors $(s^2 + \omega_k^2)$ injected into the expression for $\sqrt{M}$. The Miyata method for the determination of $m_0 - n_0$ in the numerator of 96 is ideally suited to the present situation. Thus we tentatively find $Z^*(s)$ by using only the specified factors in $\sqrt{M}$. If this process yields an acceptable P_0, we are finished. If it does not, then we multiply the tentative $Z^*(s)$ by a factor $(s^2 + \omega_k^2)$ with the value of ω_k^2 unspecified, and carry through the familiar division process. In the resulting remainder we can, by trial, readily see if a solution is possible. If not, we would need to inject additional surplus factors† and enlarge our trial-and-error procedure. Few practical situations will arise (we hope) in which this process reaches such an awkward stage. We will now illustrate the method with several examples.

Consider the impedance function

$$Z(s) = \frac{s^4 + \frac{3}{2}s^3 + \frac{13}{2}s^2 + 6s + 6}{s^4 + s^3 + 11s^2 + 5s + 24} \tag{98}$$

which was constructed assuming poles at $s = -1 \pm j$ and $s = -1 \pm j2$, and even-part zeros given by

$$M = m_1m_2 - n_1n_2 = (s^2 + 2)^2(s^2 + 6)^2 \tag{99}$$

† When the insertion of surplus factors renders 97 an improper fraction, $Z^*(s)$ in Eq. 96 is still formed by the same process, differing only in that the quotient contains even terms. These are part of $Z^*(s)$ which then is also an improper fraction.

Preliminary to constructing the impedance 96, we determine the Miyata auxiliary impedance $Z'(s)$ corresponding to an M function equal to unity. This process, which is straightforward and familiar, yields

$$Z'(s) = \frac{s^3 + s^2 + 7s + 1}{s^4 + s^3 + 11s^2 + 5s + 24} \tag{100}$$

Construction of the polynomial $m_0 - n_0$ in Eq. 96 requires that we multiply $Z'(s)$ by $\sqrt{M} = (s^2 + 2)(s^2 + 6) = s^4 + 8s^2 + 12$ and divide, as follows

$$\begin{array}{r|l} & s^3 + 4s \\ \hline s^4 + s^3 + 11s^2 + 5s + 24 & s^7 + s^6 + 15s^5 + 9s^4 + 68s^3 + 20s^2 + 84s + 12 \\ & s^7 + s^6 + 11s^5 + 5s^4 + 24s^3 \\ \hline & \qquad 4s^5 + 4s^4 + 44s^3 + 20s^2 + 84s + 12 \\ & \qquad 4s^5 + 4s^4 + 44s^3 + 20s^2 + 96s \\ \hline & \qquad\qquad -12s + 12 \end{array} \tag{101}$$

A constant multiplier is trivial, and so we can conclude that $m_0 - n_0 = 1 - s$, or that $m_0 + n_0 = s + 1$ is an appropriate auxiliary polynomial. Figure 11 illustrates the situation graphically. As in the previous simpler example, a linear function n_0/m_0 suffices to fulfill the required condition.

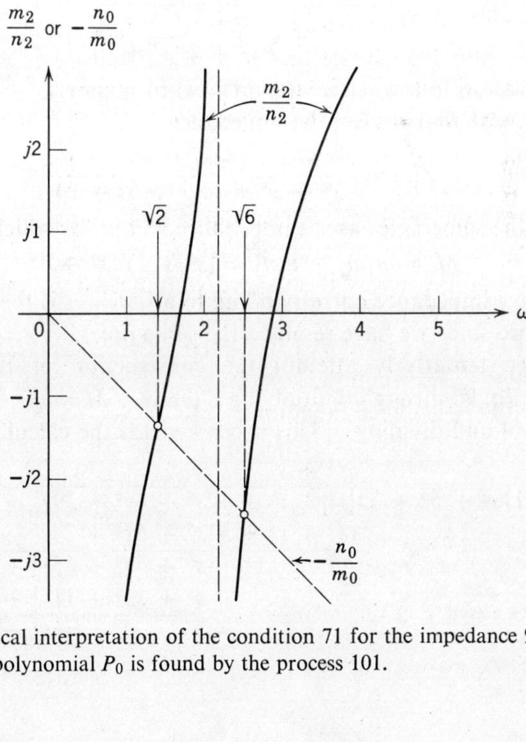

Fig. 11. Graphical interpretation of the condition 71 for the impedance 98 for which the appropriate polynomial P_0 is found by the process 101.

The even part of the augmented impedance is

$$Z_E = \frac{(s^2 + 2)^2(s^2 + 6)^2(1 - s^2)}{[(s^4 + 11s^2 + 24)^2 - s^2(s^2 + 5)^2](1 - s^2)} = Z_{aE} + Z_{bE} \tag{102}$$

which is decomposed into

$$Z_{aE} = \frac{(s^2 + 2)^2(s^2 + 6)^2}{[(s^4 + 11s^2 + 24)^2 - s^2(s^2 + 5)^2](1 - s^2)} \tag{103}$$

and

$$Z_{bE} = \frac{-s^2(s^2 + 2)^2(s^2 + 6)^2}{[(s^4 + 11s^2 + 24)^2 - s^2(s^2 + 5)^2](1 - s^2)} \tag{104}$$

The impedances constructed from these even parts are

$$\begin{aligned} Z_a(s) &= \frac{\frac{1}{2}s^4 + 4s^2 + 6}{s^5 + 2s^4 + 12s^3 + 16s^2 + 29s + 24} \\ &= \cfrac{1}{2s + 4 + \cfrac{2s(4s^2 + 17)}{(s^2 + 2)(s^2 + 6)}} \end{aligned} \tag{105}$$

and

$$\begin{aligned} Z_b(s) &= \frac{s^5 + 2s^4 + 8s^3 + \frac{17}{2}s^2 + 12s}{s^5 + 2s^4 + 12s^3 + 16s^2 + 29s + 24} \\ &= \cfrac{1}{\cfrac{2}{s} + \cfrac{1}{1 + \cfrac{s(4s^2 + 17)}{2(s^2 + 2)(s^2 + 6)}}} \end{aligned} \tag{106}$$

Their realization follows in a straightforward manner.

Suppose next we consider the impedance

$$Z(s) = \frac{s^4 + \frac{20}{3}s^3 + \frac{17}{3}s^2 + \frac{62}{3}s + \frac{2}{3}}{s^4 + s^3 + 11s^2 + 5s + 24} \tag{107}$$

which has the same poles as the impedance 98 but for which

$$M = m_1m_2 - n_1n_2 = (s^2 + 1)^2(s^2 + 4)^2 \tag{108}$$

The auxiliary impedance corresponding to an $M = 1$ is the same function 100 as before since we have retained the same poles.

Again we tentatively attempt the construction of the polynomial $m_0 - n_0$ in Eq. 96 through multiplying $Z'(s)$ by $\sqrt{M} = (s^2 + 1)(s^2 + 4) = s^4 + 5s^2 + 4$ and dividing. This process yields the calculations

$$\begin{array}{r|l} & s^3 + s \\ \hline s^4 + s^3 + 11s^2 + 5s + 24 & s^7 + s^6 + 12s^5 + 6s^4 + 39s^3 + 9s^2 + 28s + 4 \\ & s^7 + s^6 + 11s^5 + 5s^4 + 24s^3 \\ \hline & \qquad s^5 + s^4 + 15s^3 + 9s^2 + 28s + 4 \\ & \qquad s^5 + s^4 + 11s^3 + 5s^2 + 24s \\ \hline & \qquad\qquad 4s^3 + 4s^2 + 4s + 4 \end{array} \tag{109}$$

The remainder obtained here is obviously not an acceptable $m_0 - n_0$ since the corresponding polynomial $m_0 + n_0$ is not Hurwitz. We therefore multiply by the surplus factor $(s^2 + a^2)$ and divide again. Since

$$(s^3 + s^2 + s + 1)(s^2 + a^2) = s^5 + s^4 + (a^2 + 1)s^3 + (a^2 + 1)s^2 + a^2 s + a^2 \tag{110}$$

this division reads

$$\begin{array}{r|l} & s \\ \hline s^4+s^3+11s^2+5s+24 & s^5 + s^4 + (a^2 + 1)s^3 + (a^2 + 1)s^2 + a^2 s + a^2 \\ & s^5 + s^4 + \quad 11s^3 + \quad 5s^2 + 24s \\ \cline{2-2} & \quad (a^2 - 10)s^3 + (a^2 - 4)s^2 + (a^2 - 24)s + a^2 \end{array} \tag{111}$$

At this point we not only try to choose a value for a^2 such that the remainder for $s \to -s$ becomes a Hurwitz polynomial and that $(m_0{}^2 - n_0{}^2)$ for $s = j\omega$ has all positive coefficients; we try at the same time to get all this as cheaply as possible. In this example we find (by happy accident) that if we choose $a^2 = 10$, the polynomial becomes

$$P_0 = m_0 + n_0 = 3s^2 + 7s + 5 \tag{112}$$

which is Hurwitz, and

$$m_0{}^2 - n_0{}^2 = 9s^4 - 19s^2 + 25 \tag{113}$$

has positive coefficients for $s = j\omega$. This choice for the value of a^2 incidentally reduces P_0 to a quadratic polynomial and simplifies the remaining calculations as well as the resulting network.

We now compute

$$\hat{m}_2 + \hat{n}_2 = (s^4 + s^3 + 11s^2 + 5s + 24)(3s^2 + 7s + 5) \tag{114}$$

and form the even part of the augmented impedance

$$Z_E = \frac{(s^2 + 1)^2(s^2 + 4)^2(25 - 19s^2 + 9s^4)}{\hat{m}_2{}^2 - \hat{n}_2{}^2} = Z_{aE} + Z_{bE} + Z_{cE} \tag{115}$$

with the components

$$Z_{aE} = \frac{25(s^2 + 1)^2(s^2 + 4)^2}{\hat{m}_2{}^2 - \hat{n}_2{}^2} \tag{116}$$

$$Z_{bE} = \frac{-19s^2(s^2 + 1)^2(s^2 + 4)^2}{\hat{m}_2{}^2 - \hat{n}_2{}^2} \tag{117}$$

$$Z_{cE} = \frac{9s^4(s^2 + 1)^2(s^2 + 4)^2}{\hat{m}_2{}^2 - \hat{n}_2{}^2} \tag{118}$$

Construction of the corresponding impedances Z_a, Z_b, Z_c, and their realization, then follows a familiar pattern.

Graphical interpretation of the determination of the polynomial $m_0 + n_0$ in this example is illustrated in Fig. 12 which it is interesting to compare with Fig. 11. The reactance m_2/n_2 is the same for both. However, the altered location of the desired intersections with $-n_0/m_0$ makes it quite apparent that a linear function for $-n_0/m_0$ could not be expected to suffice. The additional intersection at $\omega = \sqrt{10}$ due to the addition of the surplus factor is also obviously a result of the more general character of the $-n_0/m_0$ function needed to produce the desired intersections at $\omega = 1$ and $\omega = 2$.

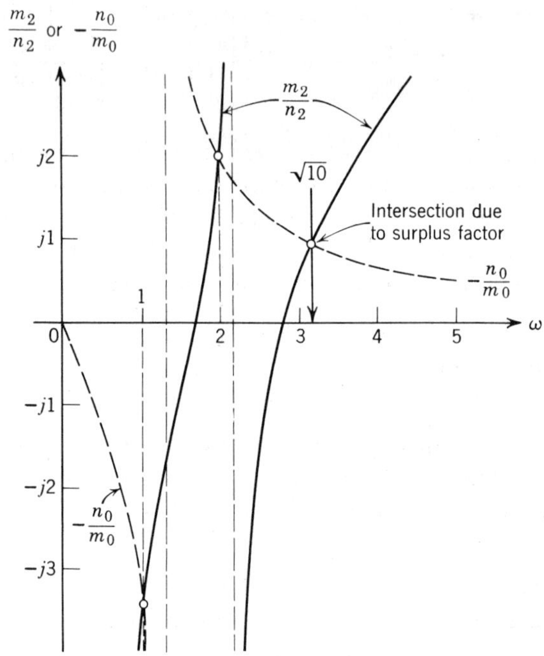

Fig. 12. Graphical interpretation of the condition 71 for the impedance 107 having the same poles as impedance 98 (to which Fig. 11 is pertinent) but altered even-part zeros ($\omega = 1$, $\omega = 2$).

Sketches like the illustrations in Figs. 11 and 12 are very helpful in giving one a quick appreciation of the approximate degree of complexity needed in the function $-\, n_0/m_0$ for any given situation. This information may then be used as a guide in monitoring the cut-and-try process involved in the long-division method of determining the polynomial $m_0 - n_0$.

We can also see from this graphical presentation that the present method evidently fails when $\omega_\nu{}^2$ values in Eq. 94 for M are coincident, that is, when the real part of the given impedance has j-axis zeros of multiplicity four, or six, and so forth. In order to deal with this situation we consider the following extension of the Miyata method to the special case in which the even-part zeros of a given impedance are all located at the same finite nonzero j-axis point.

In this case it is effective to normalize our frequency scale in such a way that the single multiple-order even-part zero falls at the point $\omega = 1$. The given impedance $Z(s) = P(s)/Q(s)$, with polynomials of degree n, then has an even part of the form

$$Z_E(s) = \frac{(s^2 + 1)^n}{Q(s)Q(-s)} \tag{119}$$

Provided the poles of $Z(s)$ are properly located, this function is readily synthesized by the Miyata method through the introduction of a transformation of the frequency variable that reads

$$\lambda = s + \frac{1}{s} \tag{120}$$

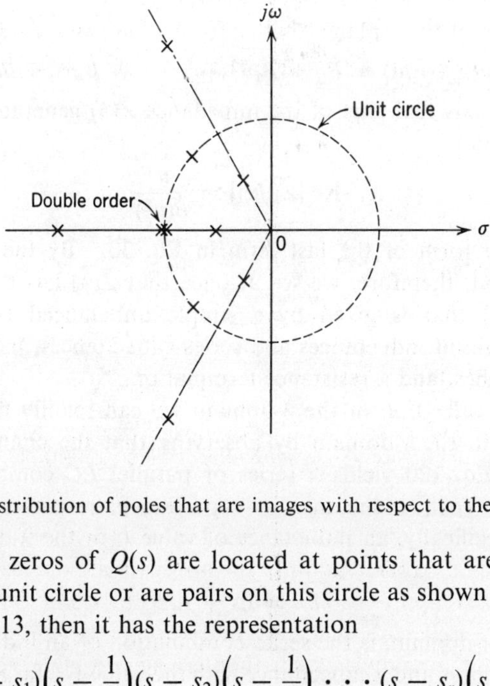

Fig. 13. Distribution of poles that are images with respect to the unit circle.

Thus, if the zeros of $Q(s)$ are located at points that are images with respect to the unit circle or are pairs on this circle as shown in the s-plane sketch of Fig. 13, then it has the representation

$$Q(s) = (s - s_1)\left(s - \frac{1}{s_1}\right)(s - s_2)\left(s - \frac{1}{s_2}\right) \cdots (s - s_r)\left(s - \frac{1}{s_r}\right) \tag{121}$$

with $2r = n$. If we recognize that

$$(s - s_k)\left(s - \frac{1}{s_k}\right) = s\left[\left(s + \frac{1}{s}\right) - \left(s_k + \frac{1}{s_k}\right)\right] \tag{122}$$

then the transformation 120 together with

$$\lambda_k = s_k + \frac{1}{s_k} \tag{123}$$

enables us to write

$$Q(s) = s^r(\lambda - \lambda_1)(\lambda - \lambda_2) \cdots (\lambda - \lambda_r) = s^r\tilde{Q}(\lambda) \tag{124}$$

and

$$Q(-s) = s^r(\lambda + \lambda_1)(\lambda + \lambda_2) \cdots (\lambda + \lambda_r) = (-1)^r s^r \tilde{Q}(-\lambda) \tag{125}$$

Since we have

$$(s^2 + 1)^n = s^n\left(s + \frac{1}{s}\right)^n = s^n\lambda^n \tag{126}$$

the even part 119, expressed in terms of the new frequency variable λ, reads

$$\tilde{Z}_E(\lambda) = \frac{(-1)^r\lambda^n}{\tilde{Q}(\lambda)\tilde{Q}(-\lambda)} \tag{127}$$

On the j-axis of the λ-plane (that is, for $\lambda = ju$) we may write

$$\tilde{Q}(ju)\tilde{Q}(-ju) = \tilde{B}_0 + \tilde{B}_1u^2 + \cdots + \tilde{B}_ru^n = \tilde{B}(u^2) \tag{128}$$

and, for the j-axis real part of the impedance $\tilde{Z}(\lambda)$ generated by the even part 127, we find

$$\text{Re}\,[\tilde{Z}(ju)] = \frac{u^n}{\tilde{B}(u^2)} \tag{129}$$

which has the form of the last term in Eq. 30. By the theory of the Miyata method, therefore, we see at once that $\tilde{Z}(\lambda)$ has a realization (in the λ-domain) that is given by a simple unbalanced ladder network consisting of shunt inductances and series capacitances, having a total of $r = n/2$ branches, and a resistance termination.

From this realization in the λ-domain we can readily find the corresponding one in the s-domain by observing that the change of variable expressed in Eq. 120 yields a series or parallel LC combination in the s-domain for each inductance or capacitance, respectively, in the λ-domain. Specifically, an inductance of value L in the λ-domain has the impedance

$$L\lambda = Ls + L/s \tag{130}$$

which, in the s-domain, is the series combination of an inductance having a value of L henrys and a capacitance with the value $1/L$ farads. Similarly, a capacitance C in the λ-domain has the admittance

$$C\lambda = Cs + C/s \tag{131}$$

which becomes, in the s-domain, the parallel combination of a capacitance of value C farads and an inductance having a value of $1/C$ henrys.

When the poles of the given impedance do not have the s-plane distribution shown in Fig. 13, then we can evidently correct this defect by augmenting the impedance with frequency factors $(s - s_k)$ for which the points s_k are images of those poles in the given impedance that do not already possess such images. The aggregate of all such factors we now define as the appropriate auxiliary polynomial

$$P_0 = m_0 + n_0 = (s - s_a)(s - s_b) \cdots (s - s_i) \tag{132}$$

and, as in the augmentation procedures discussed previously, require that $(m_0^2 - n_0^2)$ for $s = j\omega$ (Eq. 76) be a polynomial in ω^2 with positive coefficients. If the latter condition is at first not fulfilled, we can further augment with pairs of real-axis pole factors until it is.

If we denote the number of factors in P_0 by k (always an even integer), then we can let $Q(s)$ and $\tilde{Q}(\lambda)$ in Eqs. 124 and 125 represent the correspondingly augmented polynomials, where now the integer $r = (n + k)/2$. In place of Eq. 119 we have for the even part of our given impedance

$$Z_E(s) = \frac{(s^2 + 1)^n(m_0^2 - n_0^2)}{Q(s)Q(-s)} \tag{133}$$

and in place of Eq. 127 we find

$$\tilde{Z}_E(\lambda) = \frac{(-1)^{n/2}\lambda^n}{\tilde{Q}(\lambda)\tilde{Q}(-\lambda)} \times \frac{m_0^2 - n_0^2}{(-1)^{k/2}s^k} \tag{134}$$

If we write

$$\left[\frac{m_0^2 - n_0^2}{(-1)^{k/2}s^k}\right]_{s=j\omega} = \frac{\beta_{-k}}{\omega^k} + \cdots + \beta_0 + \cdots + \beta_k\omega^k \tag{135}$$

then the j-axis real part of our given impedance assumes the form

$$\text{Re}\,[\tilde{Z}(ju)] = \text{Re}\,[Z(j\omega)] = \frac{u^n}{\tilde{B}(u^2)}\left(\frac{\beta_{-k}}{\omega^k} + \cdots + \beta_0 + \cdots + \beta_k\omega^k\right) \tag{136}$$

The polynomial $\tilde{B}(u^2)$ has the same form as in Eq. 128, but its degree is now $n + k$ instead of n. The factor in front of the large parentheses therefore varies like ω^k for small ω, and like $1/\omega^k$ for large ω. This factor times the first term in the large parentheses is a component real part having k double-order zeros at $\omega = \infty$; the next term has one such zero at $\omega = 0$ and $k - 1$ of them at $\omega = \infty$; and so forth; the last term has k double-order zeros at $\omega = 0$. In addition, each component real part has n coincident double-order zeros at $\omega = 1$, as does the real part in Eq. 129.

The component impedances generated by the terms in Eq. 136 therefore are realized in the s-domain by simple unbalanced ladder networks involving a total of k branches consisting of single inductances or capacitances, all terminated in the same basic impedance (except for possible differences in constant multipliers) whose realization follows precisely the pattern applied to the impedance generated by the real part 129.

In the process of generating the impedances from the real-part terms in Eq. 136, it is useful to note that we need generate only the basic impedance for the term with β_0, and impedances for the terms with $\beta_1 \cdots \beta_k$. The ones corresponding to terms with β_{-i} are then obtained from the respective ones for terms with β_i, through simply replacing s by $1/s$ and multiplying by β_{-i}/β_i for $i = 1, 2, \cdots, k$. The correctness of this conclusion follows from the obvious fact that the even parts of these impedances are so related.

As a simple example, consider the impedance

$$Z(s) = \frac{s^4 + \frac{2}{3}s^3 + 2s^2 + \frac{2}{3}s + 1}{s^4 + 3s^3 + 4s^2 + 3s + 1} \tag{137}$$

Its even-part zeros are given by

$$M = m_1m_2 - n_1n_2 = (s^2 + 1)^4 \tag{138}$$

and its pole positions are recognized from the following factorization of the denominator polynomial

$$s^4 + 3s^3 + 4s^2 + 3s + 1 = (s^2 + s + 1)(s + 1)^2 \tag{139}$$

They obviously fulfill the necessary requirement since they are in the form of pairs on the unit circle. Incidentally, one may recognize without factoring the denominator polynomial that its zero locations are images about the unit circle, by observing that the coefficient values exhibit symmetry with regard to their location about the central term. It is easy to show that this condition is sufficient to insure the desired zero distribution in the s-plane.

The even part of this impedance evidently reads

$$Z_E(s) = \frac{(s^2 + 1)^4}{(s^2 + s + 1)(s^2 - s + 1)(s^2 + 2s + 1)(s^2 - 2s + 1)} \tag{140}$$

In the λ-domain it becomes

$$\tilde{Z}_E(\lambda) = \frac{\lambda^4}{(\lambda^2 - 1)(\lambda^2 - 4)} \tag{141}$$

and generates the impedance

$$\tilde{Z}(\lambda) = \frac{\lambda^2 + \frac{2}{3}\lambda}{\lambda^2 + 3\lambda + 2} \tag{142}$$

The following continued-fraction development of this impedance

$$\tilde{Z}(\lambda) = \cfrac{1}{\cfrac{3}{\lambda} + \cfrac{1}{\cfrac{2}{3\lambda} + 1}} \tag{143}$$

immediately leads to the network of Fig. 14*a* as a realization in the λ-domain and to the network in Fig. 14*b* as the corresponding one in the s-domain.

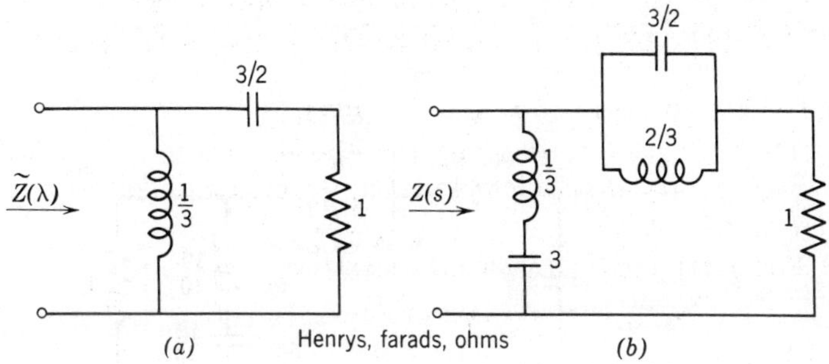

Fig. 14. λ- and s-domain realizations of the impedance 137.

Recognizing that the given impedance 137 fulfills the special conditions for realizability in these terms, we could alternately synthesize it directly from its expansion in the s-domain, which reads

$$Z(s) = \cfrac{1}{\cfrac{3s}{s^2+1} + \cfrac{1}{\cfrac{\frac{2}{3}s}{s^2+1} + 1}} \tag{144}$$

and leads directly to the realization in Fig. 14*b*.

A more interesting example is provided by the function

$$Z(s) = \frac{s^4 + \frac{115}{126}s^3 + \frac{208}{126}s^2 + \frac{29}{63}s + \frac{1}{2}}{s^4 + 4s^3 + 6s^2 + 5s + 2} \tag{145}$$

which has the same even-part zeros as in the previous example but for which the denominator polynomial

$$s^4 + 4s^3 + 6s^2 + 5s + 2 = (s^2 + s + 1)(s + 1)(s + 2) \tag{146}$$

does not have the desired distribution of zeros. It is evidently necessary to augment with the auxiliary polynomial

$$P_0 = (s + 1)(s + \tfrac{1}{2}) \tag{147}$$

The resulting j-axis real part, according to the representation in Eq. 136, becomes

$$\text{Re}\,[Z(j\omega)] = \frac{u^4}{(u^2+1)(u^2+4)(u^2+\frac{25}{4})}\left(\frac{1}{4\omega^2} + \frac{5}{4} + \omega^2\right) \tag{148}$$

The basic real part is

$$\frac{u^4}{(u^2+1)(u^2+4)(u^2+\frac{25}{4})} = \frac{\omega^{10} - 4\omega^8 + 6\omega^6 - 4\omega^4 + \omega^2}{[Q(s)Q(-s)]_{s=j\omega}} \tag{149}$$

in which $Q(s)$ is the augmented denominator polynomial given by the product of the factors in Eqs. 146 and 147. This real part generates the basic impedance

$$Z_b(s) = \frac{2}{189} \times \frac{38s^5 + 20s^4 + 76s^3 + 20s^2 + 38s}{2s^6 + 11s^5 + 25s^4 + 32s^3 + 25s^2 + 11s + 2} \tag{150}$$

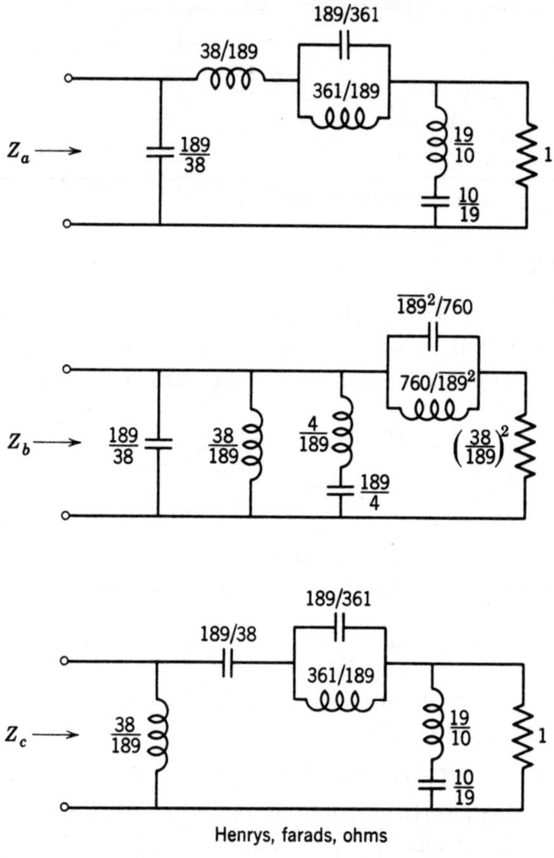

Fig. 15. Realizations for the impedances 150, 151, 152 in terms of which the impedance 145 is represented according to Eq. 153.

Multiplying by $-s^2$ and dividing in the Miyata fashion yields the impedance for the term with ω^2 in Eq. 148. This one is found to be

$$Z_c(s) = \frac{2}{189} \times \frac{189s^6 + 399s^5 + 588s^4 + 437s^3 + 209s^2 + 38s}{2s^6 + 11s^5 + 25s^4 + 32s^3 + 25s^2 + 11s + 2} \tag{151}$$

Replacing s by $1/s$ we obtain (except for the multiplier $\frac{1}{4}$) the impedance appropriate to the first term in Eq. 148, which reads

$$Z_a(s) = \frac{2}{189} \times \frac{38s^5 + 209s^4 + 437s^3 + 588s^2 + 399s + 189}{2s^6 + 11s^5 + 25s^4 + 32s^3 + 25s^2 + 11s + 2} \tag{152}$$

The given impedance in Eq. 145 thus has the decomposition

$$Z(s) = \tfrac{1}{4}Z_a(s) + \tfrac{5}{4}Z_b(s) + Z_c(s) \tag{153}$$

The separate impedances permit the developments

$$\frac{189}{38} \times Z_a(s) = \cfrac{1}{s + \cfrac{1}{s + f(s)}} \tag{154}$$

$$\frac{189}{38} \times Z_b(s) = \cfrac{1}{s + \cfrac{1}{s} + f(s)} \tag{155}$$

$$\frac{189}{38} \times Z_c(s) = \cfrac{1}{\cfrac{1}{s} + \cfrac{1}{\cfrac{1}{s} + f\left(\cfrac{1}{s}\right)}} \tag{156}$$

in which the common function $f(s)$ is given by

$$\begin{aligned} f(s) = f\left(\frac{1}{s}\right) &= \frac{189s^5 + 361s^4 + 568s^3 + 361s^2 + 189s}{38s^5 + 20s^4 + 76s^3 + 20s^2 + 38s} \\ &= \frac{189\lambda^2 + 361\lambda + 190}{38\lambda^2 + 20\lambda} = \frac{19}{2\lambda} + \cfrac{1}{\cfrac{20}{189\lambda} + \cfrac{38}{189}} \end{aligned} \tag{157}$$

Networks for the impedances Z_a, Z_b, and Z_c according to these developments are given in Fig. 15. From these, the desired realization for $Z(s)$ in Eq. 153 can readily be formed.

5 An Alternate Method of Obtaining the Bott and Duffin Result*

In the light of the background provided by these special procedures it is possible to devise a way of obtaining the Bott and Duffin result that is more logical and direct as well as computationally less tedious. We begin our considerations with the impedance $Z_1(s)$ that has a zero real part at a pair of j-axis points which we again can normalize at $\omega = \pm 1$. The even-part zeros for this impedance may therefore be represented by

$$M = m_1m_2 - n_1n_2 = (s^2 + 1)^2p(s^2) \tag{158}$$

in which the even polynomial $p(s^2)$ contains the remaining zeros other than the pair in question.

---

* Presented at the Symposium on Modern Network Synthesis, Apr. 13–15, 1955, and published in the *Transactions*, Polytechnic Institute of Brooklyn, 1956.

We again augment the impedance $Z_1(s)$ with an auxiliary polynomial $m_0 + n_0$, obtaining the even part

$$Z_{1E} = \frac{(s^2 + 1)^2(m_0^2 - n_0^2)p(s^2)}{(m_2m_0 + n_2n_0)^2 - (m_2n_0 + n_2m_0)^2} \tag{159}$$

in which we regard $(m_0^2 - n_0^2)$ for $s = j\omega$ as given by the polynomial 76 with positive coefficients, and thus we obtain a decomposition of the real part into additive components. Hence we consider the basic even part

$$Z_{BE} = \frac{(s^2 + 1)^2p(s^2)}{(m_2m_0 + n_2n_0)^2 - (m_2n_0 + n_2m_0)^2} = \frac{(s^2 + 1)^2p(s^2)}{(m_2^2 - n_2^2)(m_0^2 - n_0^2)} \tag{160}$$

from which we wish to generate a corresponding basic impedance having a j-axis zero, as is assumed by writing it in the form

$$Z_B(s) = \frac{(s^2 + 1)(m_b + n_b)}{(m_2 + n_2)(m_0 + n_0)} \tag{161}$$

Our problem is to determine $(m_0 + n_0)$ such that the even part 160 generates the impedance 161. In the normal process of generating an impedance from its even part, we first determine the denominator polynomial of that impedance from the denominator of its even part. This step is simple since it amounts merely to discarding the right half-plane zeros of the even-part denominator. The essential problem, then, is to determine the numerator polynomial of the desired impedance, given its denominator polynomial and its even-part zeros. In the present situation, however, we cannot construct the denominator polynomial independently because the factor $(m_0 + n_0)$ is unknown. Instead of essentially having only the unknown numerator, we have an unknown polynomial in both numerator and denominator since we must generate both $(m_b + n_b)$ and $(m_0 + n_0)$ in the impedance 161. We can, however, convert this problem into the normal situation in the following way.

To generate the impedance 161 from the even-part 160 we multiply numerator and denominator in 161 by $(m_2 - n_2)(m_0 - n_0)$ and obtain the equation

$$\text{Ev}\,[(s^2 + 1)(m_b + n_b)(m_2 - n_2)(m_0 - n_0)] = (s^2 + 1)^2p(s^2) \tag{162}$$

where the symbol Ev is an abbreviation for "even-part of." The Gewertz method, for example, amounts to equating coefficients of like powers of s^2 on the two sides of this equation after the common even factor $(s^2 + 1)$ has been canceled. Compare this situation with that of having to generate the impedance

$$\frac{(m_b + n_b)(m_0 - n_0)}{m_2 + n_2} \tag{163}$$

from the specified even part

$$\frac{(s^2 + 1)p(s^2)}{m_2{}^2 - n_2{}^2} \tag{164}$$

for which the determining equation reads

$$\text{Ev}\,[(m_b + n_b)(m_0 - n_0)(m_2 - n_2)] = (s^2 + 1)p(s^2) \tag{165}$$

It is clear that this alternate problem may be used to determine both the auxiliary polynomial $(m_0 + n_0)$ and the factor $(m_b + n_b)$ in the numerator of the basic impedance 161. Thus with $(m_2 + n_2)$ known, Eq. 165 enables us to find, in the normal Gewertz fashion, a polynomial which is the product $(m_b + n_b)(m_0 - n_0)$. Since both $(m_b + n_b)$ and $(m_0 + n_0)$ must be Hurwitz, we can easily separate the polynomial thus found into its two factors by assigning to $(m_0 - n_0)$ the right half-plane zeros.

Regarding the degrees of polynomials, we observe that if $(m_1 + n_1)$ and $(m_2 + n_2)$ in the given impedance are of degree n, and we denote the degrees of $(m_b + n_b)$ and $(m_0 + n_0)$ by r and k respectively, then

$$r + k \leq n, \qquad \text{or } k \leq n - r \tag{166}$$

because the numerator degree of the impedance 163 cannot exceed n. The basic impedance 161, according to the formation of its even-part 160, must have a simple zero at $s = \infty$, so that

$$r + 2 = n + k - 1 \tag{167}$$

Substituting the value of r gotten from this equation into 166 gives $k \leq \frac{3}{2}$; and since k must be an integer we conclude that

$$k = 1 \qquad \text{and} \qquad r = n - 2 \tag{168}$$

Hence we see that the desired auxiliary polynomial is always of first degree, which implies that Eq. 165 always leads to a polynomial $(m_b + n_b)(m_0 - n_0)$ having no more than one right half-plane zero.

This result may alternatively be seen by noting that $(s^2 + 1)p(s^2)$ for $s = j\omega$ changes sign only once for $0 < \omega < \infty$, and hence that the real part of the impedance 163 changes sign only once within this infinite interval also. If more than one right half-plane zero were involved, the net change in the angle of this impedance would exceed $3\pi/2$ radians and the real part would change sign at least twice within the range $0 < \omega < \infty$.

However, the mere fact that the real part of an impedance changes sign somewhere along the j-axis does not necessarily imply that the impedance has a right half-plane zero. Its real part, as we know, may not remain positive on the j-axis in spite of the fact that all zeros and poles are in the left half-plane. Hence, construction of an impedance from the even part 164 need not lead to a numerator polynomial having a right half-plane zero, in which case the desired polynomial $(m_0 + n_0)$ does not appear to be obtained from the contemplated procedure. We shall now show that achievement of the desired result may require switching from an impedance to an admittance basis, and that if one basis fails, the other is sure to succeed, and vice versa.

Since we know that the desired polynomial $(m_0 + n_0)$ is of first degree, we can represent it by the factor $(a + s)$ with $a > 0$. The numerator polynomial in Eq. 163, which we will abbreviate by

$$(m_a + n_a) = (m_b + n_b)(a - s) \tag{169}$$

is of odd degree, and its leading coefficient is negative. Noting that the even part of the given impedance $Z_1(s)$ is the even part 164 multiplied by $(s^2 + 1)$, the Miyata method of impedance construction enables us to write

$$\frac{(m_a + n_a)(s^2 + 1)}{m_2 + n_2} = -bs + \frac{m_1 + n_1}{m_2 + n_2} \tag{170}$$

in which the coefficient b is positive. We thus obtain

$$(m_a + n_a) = \frac{(m_1 - bsn_2) + (n_1 - bsm_2)}{s^2 + 1} \tag{171}$$

and hence the requirement

$$\frac{m_1}{n_2} = \frac{n_1}{m_2} = bs \quad \text{for } s = j \tag{172}$$

The first equality sign in this relation is automatically fulfilled since $(s^2 + 1)$ is contained in $(m_1m_2 - n_1n_2)$. Fulfillment of the second equality requires that

$$\left(\frac{m_1}{n_2}\right)_{s=j} = \left(\frac{n_1}{m_2}\right)_{s=j} = Z_1(j) = jX \tag{173}$$

shall yield $X > 0$.

When X is negative, we must switch to an admittance basis, which is equivalent to an interchange of the indexes 1 and 2 in the above analysis. We then have in place of 172

$$\frac{m_2}{n_1} = \frac{n_2}{m_1} = bs \quad \text{for } s = j \tag{174}$$

Since (n_2/m_1) and (m_1/n_2) have opposite signs at $s = j$, we see that the impedance and admittance bases are mutually exclusive, as in the Bott and Duffin method.

As an illustrative example of this method of synthesis, we will consider the impedance

$$Z(s) = \frac{m_1 + n_1}{m_2 + n_2} = \frac{s^4 + 1.5s^3 + 4s^2 + 2.4s + 1.6}{s^4 + 4s^3 + 11s^2 + 14s + 10} \tag{175}$$

for which

$$M = m_1m_2 - n_1n_2 = (s^2 + 4)^2(s^4 + s^2 + 1) \tag{176}$$

Following the pattern in forms 163 and 164, we set for ourselves the problem of generating an impedance having the even part

$$\frac{(s^2+4)(s^4+s^2+1)}{(s^4+11s^2+10)^2-s^2(4s^2+14)^2} \tag{177}$$

This impedance, of course, has the denominator in Eq. 175, and by writing and solving the normal Gewertz equations or by the Miyata method, we obtain for the numerator polynomial

$$(m_b+n_b)(m_0-n_0)=(4+s+2s^2-2s^3)/10 \tag{178}$$

The one positive real zero of this polynomial is found by the usual successive approximation method to have the value 1.8525. Hence we obtain

$$m_0+n_0=s+1.8525 \tag{179}$$

$$m_b+n_b=(s^2+0.8525s+1.0793)/5 \tag{180}$$

The denominator of the basic impedance 161 is now found to be

$$\begin{aligned}(m_2+n_2)(m_0+n_0)&=(s^4+4s^3+11s^2+14s+10)(s+1.8525)\\ &=s^5+5.8525s^4+18.41s^3+34.378s^2+35.935s+18.525\end{aligned} \tag{181}$$

so that we have

$$Z_B(s)=\frac{(s^2+4)(s^2+0.8525s+1.0793)/5}{s^5+5.8525s^4+18.41s^3+34.378s^2+35.935s+18.525} \tag{182}$$

Since $(m_0^2-n_0^2)$ in the even part 159 consists of two terms, the given impedance is correspondingly decomposed into two additive components as expressed by

$$Z(s)=Z_a(s)+Z_b(s) \tag{183}$$

In view of

$$m_0^2-n_0^2=(1.8525)^2-s^2 \tag{184}$$

the result 182 yields

$$\begin{aligned}Z_a(s)&=\frac{0.6863(s^2+4)(s^2+0.8525s+1.0793)}{s^5+5.8525s^4+18.41s^3+34.378s^2+35.935s+18.525}\\ &=\cfrac{0.6863}{s+\cfrac{5(s^4+2.666s^3+6.1935s^2+6.3236s+3.705)}{(s^2+4)(s^2+0.8525s+1.0793)}}\end{aligned} \tag{185}$$

The other component impedance is best found by Miyata's method through multiplying $Z_B(s)$ in Eq. 182 by $-s^2$ and carrying out one step of the long-division process. The remainder is the numerator polynomial of $Z_b(s)$ which reads

$$\begin{aligned}Z_b(s)&=\frac{s^5+2.666s^4+6.1935s^3+6.3236s^2+3.705s}{s^5+5.8525s^4+18.41s^3+34.378s^2+35.935s+18.525}\\ &=\cfrac{1}{\cfrac{5}{s}+\cfrac{(s^2+4)(s^2+0.8525s+1.0793)}{s^4+2.666s^3+6.1935s^2+6.3236s+3.705}}\end{aligned} \tag{186}$$

As should be expected, the component impedance Z_a has a zero at $s = \infty$ and the component Z_b has a zero at $s = 0$. After the corresponding poles in the reciprocal functions are removed, as is indicated in Eqs. 185 and 186, there is left a remainder having the j-axis pole or zero, respectively, as placed in evidence by the factor $(s^2 + 4)$. To remove a corresponding pair of j-axis poles from this remainder, we observe that

$$\left[\frac{s^4 + 2.666s^3 + 6.1935s^2 + 6.3236s + 3.705}{(s^2 + 4)(s^2 + 0.8525s + 1.0793)}\right]_{s^2 \to -4} \to \frac{1.486s}{s^2 + 4} \tag{187}$$

In the further development of Z_a and Z_b respectively, this term represents an admittance (a shunt branch consisting of a series LC combination) or an impedance (a series branch consisting of a parallel LC combination). The second-degree remainder after subtraction of this term from the bracketed expression in 187 is found to be

$$r(s) = \frac{s^2 + 1.18s + 0.926}{s^2 + 0.8525s + 1.0793} = \frac{m_1 + n_1}{m_2 + n_2} \tag{188}$$

It is an admittance in the development of $Z_a(s)$ and an impedance in the development of $Z_b(s)$. Its even-part zeros are found by computing the polynomial

$$m_1 m_2 - n_1 n_2 = s^4 + s^2 + 1 \tag{189}$$

As we should expect, this factor represents the remaining zeros of those assumed in Eq. 176 for the even part of the given impedance.

As in the Bott and Duffin procedure, we get two remainder functions. Each of these can be treated by the same method after the minimum value of its j-axis real part is subtracted. Since the remainder in this example is of second degree, we can use the method of the preceding article, applicable to situations for which all even-part zeros are on the j-axis because the even part has only one pair of double-order j-axis zeros. To handle a second-degree rational function one does not need either the Bott and Duffin procedure or the one discussed in this article in order to get the form of realization that these methods normally yield. A second-degree function is too simple to require any elaborate process of synthesis.

The particular second-degree function given by Eq. 188, incidentally, is seen to meet one of the conditions 47 or 48 of Art. 3, Ch. 9, and hence may be synthesized by even more elementary methods than any of those discussed here. In the development of $Z_a(s)$ in Eq. 185, in which this remainder is an admittance, we find that the corresponding impedance fulfills the case A condition 47 in Ch. 9, and hence may immediately be decomposed as follows

$$\frac{s^2 + 0.8525s + 1.0793}{s^2 + 1.18s + 0.926} = \frac{s^2 + 0.926}{s^2 + 1.18s + 0.926} + \frac{0.8525s + 0.1533}{s^2 + 1.18s + 0.926}$$

$$= \frac{1}{1 + \dfrac{1.18s}{s^2 + 0.926}} + \frac{1}{1.173s + 1.173 + \dfrac{1}{1.143s + 0.2055}} \tag{190}$$

Its realization either as an impedance or as an admittance is now obvious by inspection.

6 Use of Duality in Connection With the Miyata and Related Methods

In applying the methods discussed in Arts. 2, 3, and 4, one may achieve a variety of additional variants by use of the duality principle. First of all, it should be clear that all that is said about impedances and their decomposition in terms of their real parts applies equally well to admittances. If we choose an admittance rather than an impedance basis, the final network is in the form of ladder components connected in parallel instead of in series. In the half-n synthesis where the first two ladders terminate in remainder functions which are then separately dealt with in a similar fashion, we can treat these remainders as impedances or as admittances, independently of which basis was used in the first step. Since we can switch from one basis to the other at the beginning of each new step, many variants are obviously possible.

When switching from an impedance to an admittance basis or vice versa, it is significant to observe that the pertinent polynomial $(m_1m_2 - n_1n_2) = A(-s^2)$, being symmetrical in the subscripts 1 and 2, is the same for either function. The positiveness of the coefficients $A_0, A_1, \cdots$ in the polynomial $A(\omega^2)$ therefore is satisfied for the admittance function if it is satisfied for the corresponding impedance. Hence there can be no question as to the possibility of switching from one basis to another at any stage of the total development process.

APPENDIX

On the Augmentation of Polynomials to Secure All Positive Coefficients

For nonnegative coefficients a_k, we want to know whether the polynomial

$$a_0 + a_1z + a_2z^2 + \cdots + a_nz^n \tag{1}$$

can have zeros in the right-half z-plane and, if so, how close these may be to the positive real axis. To investigate this question, we assume such a zero to be on the unit circle and write

$$z = e^{j\theta} \tag{2}$$

since any zero may thus be accommodated by appropriate magnitude scaling. We then inquire whether the equation

$$a_0 + a_1 e^{j\theta} + a_2 e^{j2\theta} + \cdots + a_n e^{jn\theta} = 0 \tag{3}$$

can be satisfied for $0 \le |\theta| < \pi/2$.

Because the a_k are positive, it evidently cannot be satisfied for $\theta = 0$; and if it is to be satisfied for the smallest possible θ value, then we must have

$$a_1 = a_2 = \cdots = a_{n-1} = 0 \tag{4}$$

because the terms a_0 and $a_n e^{jn\theta}$ can then cancel for $n\theta = \pi$ by choosing $a_0 = a_n$, whereas cancelation of any intermediate terms would require a larger θ value than this one. (For example, if the terms were to become any other symmetrically oriented set of vectors, then $n\theta$ would have to be 2π or very nearly this value.) Hence we see that the nth-degree polynomial having a zero closest to the positive real axis is simply $(1 + z^n)$, having zeros uniformly spaced on the unit circle. Specifically, we have

$$(z^n + 1) = \left(z^2 - 2z\cos\frac{n}{\pi} + 1\right)\left(z^2 - 2z\cos\frac{3\pi}{n} + 1\right)\cdots \left[z^2 - 2z\cos\frac{(n-1)\pi}{n} + 1\right] \quad \text{for } n \text{ even} \tag{5}$$

$$= \left(z^2 - 2z\cos\frac{\pi}{n} + 1\right)\cdots \left[z^2 - 2z\cos\frac{(n-2)\pi}{n} + 1\right](z + 1) \quad \text{for } n \text{ odd} \tag{6}$$

If a given polynomial fails to have all positive coefficients because it contains one factor $(z^2 - 2z\cos\theta + 1)$ with $0 < \theta < \pi/2$, and if θ happens to equal π/n with n equal to an integer, then the appropriate augmentation factors are simply those in Eq. 5 or 6 except the first, and (so far as this one factor is concerned) the resulting polynomial after augmentation is $(z^n + 1)$. The same factors are still appropriate if $(\pi/n) < \theta < (\pi/n - 1)$, but the resulting nth-degree polynomial then contains all terms.

It is possible for any value of θ to write down at once an augmenting polynomial which will yield a resultant having but three nonzero positive terms. Poincaré* gives the following expression for this augmenting polynomial

$$z^{n-1} + \frac{\sin 2\theta}{\sin\theta} z^{n-2} + \cdots + \frac{\sin(n-1)\theta}{\sin\theta} z + \frac{\sin n\theta}{\sin\theta} \tag{7}$$

in which n is the *largest* integer for which $\sin n\theta \ge 0$. Multiplication

* J. H. Poincaré, *Comptes Rendus*, Vol. 97, 1883, p. 1418.

by the single factor $(z^2 - 2z\cos\theta + 1)$ yields

$$z^{n+1} - \frac{\sin(n+1)\theta}{\sin\theta}z + \frac{\sin n\theta}{\sin\theta} \tag{8}$$

P. M. Lewis devised the following alternate scheme for accomplishing a similar result. This one is based upon the set of identities

$$\begin{aligned} (z^2 - 2z\cos\theta + 1)(z^2 + 2z\cos\theta + 1) &\equiv z^4 - 2z^2\cos 2\theta + 1 \\ (z^4 - 2z^2\cos 2\theta + 1)(z^4 + 2z^2\cos 2\theta + 1) &\equiv z^8 - 2z^4\cos 4\theta + 1 \\ \cdots\cdots\cdots\cdots\cdots\cdots\cdots\cdots\cdots\cdots\cdots\cdots &\cdots\cdots\cdots\cdots\cdots\cdots \end{aligned} \tag{9}$$

The augmenting polynomial is chosen to be

$$(z^2 + 2z\cos\theta + 1)(z^4 + 2z^2\cos 2\theta + 1)\cdots(z^n + 2z^{n/2}\cos n\theta/2 + 1) \tag{10}$$

and the resultant becomes

$$z^{2n} - 2z^n\cos n\theta + 1 \tag{11}$$

in which n is the *smallest* even integer for which $\cos n\theta \leq 0$.

A rather simple-minded approach to this problem is to say that if the given polynomial is $A(z)$, the augmenting polynomial $B(z)$, and the resultant $C(z)$, then we can find $B(z)$ by dividing $A(z)$ into $C(z)$ by long division, and the desired character of $C(z)$ can be more or less specified at the outset. Thus the leading coefficient in $C(z)$ can be chosen equal to unity and as many of the following ones as possible are chosen to be zero. The scheme is best illustrated by a numerical example. Suppose we have the factor $z^2 - \frac{3}{2}z + 1$. The pertinent division process is indicated as follows

$$\begin{array}{r|l} & 1 + \frac{3}{2} + \frac{5}{4} + \frac{3}{8} \\ \hline 1 - \frac{3}{2} + 1 & 1 + 0 + 0 + 0 + c_4 + c_5 \\ & 1 - \frac{3}{2} + 1 \\ \cline{2-2} & \quad \frac{3}{2} - 1 \\ & \quad \frac{3}{2} - \frac{9}{4} + \frac{3}{2} \\ \cline{2-2} & \qquad \frac{5}{4} - \frac{3}{2} \\ & \qquad \frac{5}{4} - \frac{15}{8} + \frac{5}{4} \\ \cline{2-2} & \qquad\quad \frac{3}{8} - \frac{5}{4} \\ & \qquad\quad \frac{3}{8} - \frac{9}{16} + \frac{3}{8} \\ \cline{2-2} & \qquad\qquad - \frac{11}{16} - \frac{3}{8} \end{array} \tag{12}$$

By choosing the coefficients c_4 and c_5 in $C(z)$ equal to $\frac{11}{16}$ and $\frac{3}{8}$ respectively, the division terminates, and the appropriate augmenting polynomial is seen to be

$$z^3 + \tfrac{3}{2}z^2 + \tfrac{5}{4}z + \tfrac{3}{8} \tag{13}$$

The resulting polynomial has three nonzero terms, and reads

$$z^5 + \tfrac{11}{16}z + \tfrac{3}{8} \tag{14}$$

The division process in 12 terminates as soon as both remainder terms are negative. The second of these is negative at each step, but the first does not turn negative until the fourth step is reached. Comparison with the Poincaré method in 7 and 8 shows that results 13 and 14 are exactly the same.

While the Poincaré and Lewis methods are restricted to the augmentation of a single quadratic factor, the division process can be applied in more general cases and sometimes yields rather simple results with little effort. Take, for example, the situation to which the following division is pertinent

$$\begin{array}{r|l} & 1 + 2 + 1 \\ \hline 1 - 2 + 3 - 4 + 5 & 1 + 0 + 0 + 0 + 0 + c_5 + c_6 \\ & 1 - 2 + 3 - 4 + 5 \\ \cline{2-2} & \quad 2 - 3 + 4 - 5 \\ & \quad 2 - 4 + 6 - 8 + 10 \\ \cline{2-2} & \qquad 1 - 2 + 3 - 10 \\ & \qquad 1 - 2 + 3 - 4 + 5 \\ \cline{2-2} & \qquad\qquad\qquad - 6 - 5 \end{array} \tag{15}$$

The given polynomial is

$$z^4 - 2z^3 + 3z^2 - 4z + 5 \tag{16}$$

The augmenting polynomial is found to be

$$z^2 + 2z + 1 \tag{17}$$

and the resulting polynomial reads

$$z^6 + 6z + 5 \tag{18}$$

Again, let us consider the given polynomial

$$(z^2 - z + 1)(z^2 - \tfrac{3}{2}z + 1) = z^4 - \tfrac{5}{2}z^3 + \tfrac{7}{2}z^2 - \tfrac{5}{2}z + 1 \tag{19}$$

We can, of course, take the factors separately. For the first we readily find the proper augmenting polynomial to be $(z + 1)$; for the second the polynomial 13 is appropriate. The resulting polynomial by this method is of eighth degree and has six nonzero terms. Considering the polynomial 19 as a whole, however, leads to the division process

$$\begin{array}{r|l} & 1 + \tfrac{5}{2} + \tfrac{11}{4} + \tfrac{5}{8} \\ \hline 1 - \tfrac{5}{2} + \tfrac{7}{2} - \tfrac{5}{2} + 1 & 1 + 0 + 0 + 0 + c_4 + c_5 + c_6 + c_7 \\ & 1 - \tfrac{5}{2} + \tfrac{7}{2} - \tfrac{5}{2} + 1 \\ \cline{2-2} & \quad \tfrac{5}{2} - \tfrac{7}{2} + \tfrac{5}{2} - 1 \\ & \quad \tfrac{5}{2} - \tfrac{25}{4} + \tfrac{35}{4} - \tfrac{25}{4} + \tfrac{5}{2} \\ \cline{2-2} & \qquad \tfrac{11}{4} - \tfrac{25}{4} + \tfrac{21}{4} - \tfrac{5}{2} \\ & \qquad \tfrac{11}{4} - \tfrac{55}{8} + \tfrac{77}{8} - \tfrac{55}{8} + \tfrac{11}{4} \\ \cline{2-2} & \qquad\quad \tfrac{5}{8} - \tfrac{35}{8} + \tfrac{35}{8} - \tfrac{11}{4} \\ & \qquad\quad \tfrac{5}{8} - \tfrac{25}{16} + \tfrac{35}{16} - \tfrac{25}{16} + \tfrac{5}{8} \\ \cline{2-2} & \qquad\qquad - \tfrac{10}{16} + \tfrac{35}{16} - \tfrac{19}{16} - \tfrac{5}{8} \end{array} \tag{20}$$

requiring a negative value for c_5, which is unacceptable.

This difficulty is ascribable to having arbitrarily set too many of the coefficients in $C(z)$ equal to zero. If we leave the value of c_3 undetermined and denote the last quotient in 20 (the term $\frac{5}{8}$) by b_3, then the last step in the division yields the lines

$$\begin{array}{l} \frac{5}{8} - \frac{35}{8} + \frac{35}{8} - \frac{11}{4} \\ (1 - \frac{5}{2} + \frac{7}{2} - \frac{5}{2} + 1)b_3 \end{array} \tag{21}$$

Termination of the process implies that

$$\begin{aligned} c_3 &= b_3 - \tfrac{5}{8} \geq 0 \\ c_4 &= -\tfrac{5}{2}b_3 + \tfrac{35}{8} \geq 0 \\ c_5 &= \tfrac{7}{2}b_3 - \tfrac{35}{8} \geq 0 \\ c_6 &= -\tfrac{5}{2}b_3 + \tfrac{11}{4} \geq 0 \\ c_7 &= b_3 \geq 0 \end{aligned} \tag{22}$$

It is not difficult to see that simultaneous fulfillment of all inequalities with a single b_3 value is not possible, and so we conclude that the augmenting polynomial must be at least of fourth degree (as it turns out to be when the factors in 19 are considered separately).

An additional step in the division process revises 21 to

$$\begin{array}{l} \frac{5}{8} - \frac{35}{8} + \frac{35}{8} - \frac{11}{4} \\ (1 - \frac{5}{2} + \frac{7}{2} - \frac{5}{2} + 1)b_3 \\ \qquad (1 - \frac{5}{2} + \frac{7}{2} - \frac{5}{2} + 1)b_4 \end{array} \tag{23}$$

and relations 22 are replaced by

$$\begin{aligned} c_3 &= b_3 - \tfrac{5}{8} \geq 0 \\ c_4 &= b_4 - \tfrac{5}{2}b_3 + \tfrac{35}{8} \geq 0 \\ c_5 &= -\tfrac{5}{2}b_4 + \tfrac{7}{2}b_3 - \tfrac{35}{8} \geq 0 \\ c_6 &= \tfrac{7}{2}b_4 - \tfrac{5}{2}b_3 + \tfrac{11}{4} \geq 0 \\ c_7 &= -\tfrac{5}{2}b_4 + b_3 \geq 0 \\ c_8 &= b_4 \geq 0 \end{aligned} \tag{24}$$

Here numerous solutions are possible, among them the one resulting from the separate consideration of the factors in 19, which yields $b_3 = \frac{13}{8}$, $b_4 = \frac{3}{8}$, and hence the augmenting polynomial

$$z^4 + \tfrac{5}{2}z^3 + \tfrac{11}{4}z^2 + \tfrac{13}{8}z + \tfrac{3}{8} \tag{25}$$

The resultant in this case has the coefficients

$$1 + 0 + 0 + 1 + \tfrac{11}{16} + \tfrac{3}{8} + 0 + \tfrac{11}{16} + \tfrac{3}{8} \tag{26}$$

which may be seen to satisfy 24.

Two solutions are readily found for which an additional coefficient in the resultant polynomial becomes zero. Thus we see that 24 can be satisfied by assuming $c_5 = c_6 = 0$, yielding augmenting and resultant polynomials with the coefficients

$$1 + \tfrac{5}{2} + \tfrac{11}{4} + \tfrac{45}{32} + \tfrac{7}{32} \tag{27}$$

$$1 + 0 + 0 + \tfrac{25}{32} + \tfrac{69}{64} + 0 + 0 + \tfrac{55}{64} + \tfrac{7}{32} \tag{28}$$

Again, for $c_5 = c_7 = 0$, solution of 24 leads to polynomials with the coefficients

$$1 + \tfrac{5}{2} + \tfrac{11}{4} + \tfrac{7}{4} + \tfrac{7}{10} \tag{29}$$

$$1 + 0 + 0 + \tfrac{9}{8} + \tfrac{7}{10} + 0 + \tfrac{33}{40} + 0 + \tfrac{7}{10} \tag{30}$$

An equivalent solution can be obtained with less effort if we begin by tentatively multiplying the given polynomial 19 by the factor $(z + 1)$ giving

$$z^5 - \tfrac{3}{2}z^4 + z^3 + z^2 - \tfrac{3}{2}z + 1 \tag{31}$$

This augmentation, in any case, should do some good; and so it does, as we now see from the following division

$$\begin{array}{r|lllllllll}
 & 1 & +\tfrac{3}{2} & +\tfrac{5}{4} & +b_3 & & & & & \\
\hline
1 - \tfrac{3}{2} + 1 + 1 - \tfrac{3}{2} + 1 & 1 & +0 & +0 & +c_3 & +c_4 & +c_5 & +c_6 & +c_7 & +c_8 \\
 & 1 & -\tfrac{3}{2} & +1 & +1 & -\tfrac{3}{2} & +1 & & & \\
\cline{2-7}
 & & \tfrac{3}{2} & -1 & -1 & +\tfrac{3}{2} & -1 & & & \\
 & & \tfrac{3}{2} & -\tfrac{9}{4} & +\tfrac{3}{2} & +\tfrac{3}{2} & -\tfrac{9}{4} & +\tfrac{3}{2} & & \\
\cline{3-8}
 & & & \tfrac{5}{4} & -\tfrac{5}{2} & +0 & +\tfrac{5}{4} & -\tfrac{3}{2} & & \\
 & & & \tfrac{5}{4} & -\tfrac{15}{8} & +\tfrac{5}{4} & +\tfrac{5}{4} & -\tfrac{15}{8} & +\tfrac{5}{4} & \\
\cline{4-9}
 & & & & -\tfrac{5}{8} & -\tfrac{5}{4} & +0 & +\tfrac{3}{8} & -\tfrac{5}{4} & \\
 & & & & (1 & -\tfrac{3}{2} & +1 & +1 & -\tfrac{3}{2} & +1)b_3
\end{array} \tag{32}$$

Termination of the division process at this point requires that

$$\begin{aligned}
c_3 &= b_3 + \tfrac{5}{8} \geq 0 \\
c_4 &= -\tfrac{3}{2}b_3 + \tfrac{5}{4} \geq 0 \\
c_5 &= b_3 \geq 0 \\
c_6 &= b_3 - \tfrac{3}{8} \geq 0 \\
c_7 &= -\tfrac{3}{2}b_3 + \tfrac{5}{4} \geq 0 \\
c_8 &= b_3 \geq 0
\end{aligned} \tag{33}$$

It is not difficult to see that these conditions can be fulfilled with a single b_3 value. In fact, if we choose $b_3 = \frac{5}{6}$, then $c_4 = c_7 = 0$, and the corresponding augmenting polynomial reads

$$z^3 + \tfrac{3}{2}z^2 + \tfrac{5}{4}z + \tfrac{5}{6} \tag{34}$$

The resultant polynomial, like 28 and 30, is of eighth degree and contains five nonzero terms.

These augmentation procedures are therefore preferable to that leading to the resultant 26 which is obtained by considering the quadratic factors in 19 separately. In situations where the given polynomial is not in factored form, the latter method, moreover, involves considerably more computational work.

If, in this example, we choose to satisfy the conditions 33 with $b_3 = \frac{3}{8}$, the augmenting polynomial is the same as 13. We could have recognized this result at the outset since the divisor in 32 is the one in 12 twice repeated.

Such a decomposition scheme (which may sometimes be used advantageously) leads to a resultant polynomial of the same degree (eighth) but with six nonzero terms instead of five.

A feature of this method that can be additionally useful is the fact that the augmenting polynomial produced by it always has nonnegative coefficients. It may even be Hurwitz, as are the polynomials 13, 17, 25, 27, 29, and 34. In the Miyata method, this added constraint is not needed, but in transfer function synthesis where augmentation may be called for (see Art. 4, Ch. 11, and Art. 3, Ch. 13) the augmenting polynomial is additionally required to be Hurwitz. In fact, in the pertinent *RC* problem the augmenting polynomial is required to have only simple negative real zeros. Determination of an appropriate polynomial in such situations is, of course, more difficult, and no established mathematical methods are at hand although ideas brought out by the above discussion will be found helpful.

An alternate interpretation may be collaterally useful here. Suppose we regard the coefficients in the pertinent polynomials as the uniformly spaced samples of corresponding time functions. Specifically, let the coefficients $a_1, a_2, a_3 \cdots$ of $A(z)$ be the successive values of a uniform train of time-domain impulses; and let the time function thus represented in sampled form be regarded as the input to some network. The time functions whose impulse-train representations are similarly determined by the coefficients of $B(z)$ and $C(z)$ are regarded respectively as the impulse response and the output of this network. Such interpretation is evidently possible since the algebraic relationship yielding the coefficients of $C(z)$ in terms of those of $A(z)$ and $B(z)$ is precisely what we would get for the samples of the output time function by expressing it as a convolution of the input and impulse response functions.

The input time function has negative as well as positive samples. The first and last of these as well as the algebraic sum of all samples must be positive and nonzero [since the given polynomial $A(z)$ can have no positive real zeros].* The output, which is a superposition of regularly spaced and constant-multiplied input replicas (the constant multipliers being the samples of the impulse response), may contain no negative samples. The problem is to determine the samples of an impulse response meeting this requirement, while incidentally yielding as large a number of zero samples in the output as possible.

* We may easily see the truth of this statement if we regard $A(z)$ to be in the form of a product of factors as in Eqs. 5 and 6. Since augmentation can be successful only so long as the angle θ in any factor $(z^2 - 2z\cos\theta + 1)$ is nonzero, we see that for each factor the coefficient sum (the value of the factor for $z = 1$) is positive. Hence the coefficient sum for the resultant polynomial, which is $A(1)$, must likewise be positive.

Since convolution is a mutual process, we can also regard the output as a superposition of regularly spaced and constant-multiplied replicas of a desired impulse response, the constant multipliers being the samples (some of them negative) of the given input.

When the negative samples in the input are few and not too large, we may readily visualize how an appropriate impulse response is constructible. Similarly, when a few large negative samples are embedded in a large number of positive ones, a decomposition of the input into a sum of components of which each involves some portion of the negative samples can be seen to facilitate the formulation of an acceptable result. In other words, the pictorial representation lent by this alternate way of interpreting the pertinent algebraic equations involved in this problem gives to our analysis a clarity which not only results in a better understanding of the demands made but also stimulates the formulation of possible solutions.

PROBLEMS

1. Form the j-axis real parts of the impedances in Prob. 3 of Ch. 9 and show by the Miyata theory that the realizations found earlier could have been predicted immediately.

2. Realize the impedance of Prob. 11, Ch. 9, by Miyata's method, using augmentation of the real part if necessary.

3. Synthesize the impedances of Probs. 15 and 17 of Ch. 9 by the Bott and Duffin method.

4. After adding just enough resistance to make the impedances of Probs. 16 and 18 of Ch. 9 p.r., synthesize these by the Bott and Duffin method.

5. Synthesize the impedances of Prob. 3 by Miyata's method, using augmentation if necessary.

6. Synthesize by Miyata's method the impedances whose j-axis real parts are given in Prob. 22, Ch. 9.

7. Synthesize by Miyata's method the impedances whose j-axis real parts are given in Prob. 29, Ch. 9.

8. Synthesize the following impedance by an extended form of Miyata's method

$$Z(s) = \frac{5s^4 + 2.674s^3 + 50.7s^2 + 8.94s + 80.67}{5s^4 + 20s^3 + 86s^2 + 80s + 96}$$

9. The even part of an impedance is given by

$$Z_E = \frac{s^4(s^2 + 1)^2}{(s^4 + 3s^2 + 1)^2 - 4s^2(s^2 + 1)^2}$$

Construct the impedance and synthesize it by an extended form of Miyata's method.

10. The even part of an impedance is

$$Z_E = \frac{(s^2 + 1)^2}{(1 + s)^3(1 - s)^3}$$

Construct the impedance and synthesize it by an extended form of Miyata's method.

CHAPTER ELEVEN

Conventional Methods of Transfer Function Synthesis

1 Synthesis of the Transfer Impedance of a Resistance-Terminated Lossless Network

In the discussions of this chapter, our attention will be devoted primarily to the consideration of transfer functions that are either impedances or admittances. Our attitude in this regard is that the transfer impedance is the most practical form of transfer function, since a current input (from a high impedance source like a pentode) and a voltage output are the commonly occurring quantities of physical interest. The transfer admittance is less immediately practical for the reason that a pure voltage source (one with zero internal impedance) is more difficult to approximate and hence less appropriate to most physical situations, whereas an output current is usually not the quantity of physical interest. For the same reason, a transfer voltage or current ratio is likewise of less practical interest.

From a theoretical point of view, however, one cannot very well omit the discussion of transfer admittances, since this attitude would simultaneously rule out of consideration the duality principle and hence obscure more than half of the total situation that we wish to present. Special emphasis upon the properties and synthesis methods appropriate to transfer voltage or current ratios, on the other hand, appears not to be needful either from a general theoretical point of view or from a practical standpoint inasmuch as the special properties of such functions and any unique characteristics pertinent to their synthesis become rather obvious once an adequate background relative to transfer impedances and admittances has been established.

As mentioned in connection with the discussion of driving-point impedance synthesis in Ch. 9, the Darlington method, and the insight which it provides with regard to the interrelation between driving-point and transfer impedances associated with resistance-terminated lossless coupling networks, is primarily of practical importance because of its relevance to transfer function synthesis. Our initial attention is therefore directed toward exploiting the Darlington theory in this regard.

Specifically, we shall consider Eq. 170 (for case A) in Art. 8, Ch. 9, as our starting point. With reference to the situation in Fig. 1, in which the driving-point impedance is written in the familiar form

$$Z_1(s) = \frac{p(s)}{q(s)} = \frac{m_1 + n_1}{m_2 + n_2} \tag{1}$$

and the transfer impedance is defined as $Z_{12}(s) = E_2/I_1$, this equation reads

$$Z_{12}(s)Z_{12}(-s) = \frac{m_1 m_2 - n_1 n_2}{m_2{}^2 - n_2{}^2} = \tfrac{1}{2}[Z_1(s) + Z_1(-s)] \tag{2}$$

For $s = j\omega$ it yields the particular result

$$|Z_{12}(j\omega)|^2 = \text{Re}\,[Z_1(j\omega)] \tag{3}$$

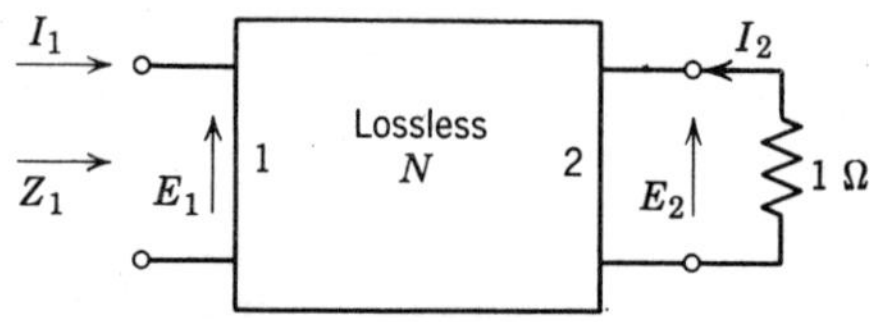

Fig. 1. Resistance-terminated lossless network whose transfer impedance synthesis is obtainable by Darlington's theory, or more directly, by Cauer's.

which immediately suggests that the Darlington method of driving-point synthesis provides the means for synthesizing a resistance-terminated lossless coupling network when the magnitude of the transfer impedance at real frequencies ($s = j\omega$) is prescribed. Before elaborating upon this observation, it is helpful to note that the result expressed by Eq. 3 may also be obtained from the physical consideration that the average power input to the lossless network of Fig. 1 must (in the sinusoidal steady state) equal the average power delivered to the 1-ohm load impedance. This fact is expressed by the equation

$$|I_1|^2\,\text{Re}\,[Z_1(j\omega)] = |E_2|^2/1\text{ ohm} \tag{4}$$

from which

$$\left|\frac{E_2}{I_1}\right|^2 = |Z_{12}(j\omega)|^2 = \text{Re}\,[Z_1(j\omega)] \tag{5}$$

as given by Eq. 3.

Using the notation for the representation of the real part of a driving-point impedance that is familiar from the detailed discussions in the preceding chapters, and assuming as usual that the polynomials $p(s)$ and $q(s)$ have the degree n, we observe that

$$|Z_{12}(j\omega)|^2 = \frac{A_0 + A_1\omega^2 + \cdots + A_n\omega^{2n}}{B_0 + B_1\omega^2 + \cdots + B_n\omega^{2n}} = \frac{A(\omega^2)}{B(\omega^2)} \tag{6}$$

This detailed form for the squared magnitude of a transfer impedance is, of course, quite generally appropriate, since, as discussed in Ch. 1, a transfer impedance is a rational function, the same as a driving-point impedance, and hence can be written in the form

$$Z_{12}(s) = \frac{g(s)}{q(s)} = \frac{m + n}{m_2 + n_2} \tag{7}$$

in which the denominator polynomial $q(s)$ is, incidentally, the same as that in the driving-point impedance pertinent to the same network because the poles of these functions are the open-circuit natural frequencies. The squared magnitude for $s = j\omega$ is thus given by

$$\begin{aligned} |Z_{12}(j\omega)|^2 &= \left[\frac{g(s)g(-s)}{q(s)q(-s)}\right]_{s=j\omega} = \left(\frac{m^2 - n^2}{m_2{}^2 - n_2{}^2}\right)_{s=j\omega} \\ &= \frac{A_0 + A_1\omega^2 + \cdots + A_n\omega^{2n}}{B_0 + B_1\omega^2 + \cdots + B_n\omega^{2n}} = \frac{A(\omega^2)}{B(\omega^2)} \end{aligned} \tag{8}$$

Identification of this expression with the real part of the driving-point impedance is pertinent only to the situation depicted in Fig. 1. The Darlington theory is useful only for the design of *lossless* coupling networks.

The only restrictions on expression 6 relevant to the lossless network are that it be real and bounded for all real ω, including the point at $\omega = \infty$. In the more general case to which Eqs. 7 and 8 are pertinent, the boundedness requirement may, theoretically at least, be dropped, although, as pointed out in Ch. 1, its omission is of little practical importance, and so we may for all practical purposes confine our discussions (whether directed specifically to the situation in Fig. 1 or not) to the consideration of bounded transfer functions alone.

These restrictions (as elaborated upon in Ch. 8) are considerably more lenient than those pertaining to the magnitude of a driving-point impedance. In contrast to the latter, the present expression 6 or 8 may have any missing coefficients in the numerator. A common situation of this sort is that in which the numerator polynomial reduces to a constant.

A method of constructing the network N in Fig. 1, when a particular function is specified for the squared magnitude of $Z_{12}(s)$ in Eq. 6, is to regard this function as the real part of a driving-point impedance, construct that impedance by any of the methods discussed in Ch. 8, and then synthesize it using Darlington's procedure. Although this method provides the desired solution, it is somewhat roundabout, and imposes a needless restriction upon the synthesis of the lossless two terminal-pair network that leads us in many cases to a less advantageous form of solution than is actually necessary in view of the demands of the problem.

Thus, when we use any of the algebraic procedures for the construction of the driving-point impedance from its real part, we automatically arrive at a minimum-reactive impedance; and the resulting lossless network is characterized by a set of z_{sk} for which the residue condition is fulfilled with an equal sign at all poles (see Art. 6, Ch. 9). Actually, the transfer impedance $Z_{12}(s)$ depends only upon the real part of the input impedance and is the same whether the latter is minimum reactive or not. Stated in another way, $Z_{12}(s)$, as Eq. 169 of Ch. 9 shows, and as we can see by inspection of the physical situation involved, is independent of the open-circuit input impedance z_{11} of the lossless network. Thus we recall that the expression for $Z_{12}(s)$ in terms of the z_{sk} reads

$$Z_{12}(s) = \frac{z_{12}}{1 + z_{22}} \tag{9}$$

Its independence of z_{11} is physically obvious in view of the fact that the input is tacitly a current source whose value is independent of the input impedance.

In following the Darlington method as described above for the synthesis of a given $|Z_{12}(j\omega)|^2$, we specify not only z_{12} and z_{22} but also z_{11} and automatically demand that the residue condition be fulfilled without any leeway. If we recall the discussion in Art. 1, Ch. 7, in regard to synthesis of the lossless network under these circumstances, we are reminded that a solution in general is not possible without closely coupled inductances, which are impractical. On the other hand, when only z_{12} and z_{22} are specified, the procedures in Arts. 2, 3, and 4 of Ch. 7 show that more practical structures like ladder networks and other unbalanced structures, mostly without mutual inductive coupling, are obtainable. For this reason, and also because of its more direct nature, an approach to the design of a lossless coupling network based upon the representation 9 for the desired transfer impedance proves to be worthwhile. An additional advantage derives from the fact that this alternate approach* yields practically useful information regarding restrictions which the lossless character of the coupling network places upon the phase shift that accompanies the specified magnitude of the transfer impedance.

Before we can make use of the representation given in Eq. 9, it is necessary to discuss the process of forming the function $Z_{12}(s)$ from its specified squared magnitude for $s = j\omega$ as given by Eq. 6. In general terms the construction of an impedance from its squared magnitude is discussed in Art. 8 of Ch. 8, and the detailed procedure given there is relevant to our present problem. However, we do not have the freedom of disposing of the zeros as we please when constructing the numerator

* First suggested by W. Cauer in his article entitled "Ausgangsseitig leerlaufende Filter," *Elek. Nachr.-Tech.*, Vol. 16, No. 6, 1939, pp. 161–163.

polynomial from its squared magnitude because these zeros are also zeros of z_{12} (except those due to poles of z_{22} not also contained in z_{12}, which can occur in degenerate situations), and zeros of z_{12} for a lossless network form a pattern in the s-plane that is symmetrical about both real and imaginary axes (since the polynomials in the lossless z_{sk} must be odd or even).

It follows, therefore, that if a transfer impedance $Z_{12}(s)$ is to be realized by a resistance-terminated lossless network, its zeros must be equally distributed between the right and left half-planes. Such a network clearly cannot be used to realize a minimum-phase transfer function for which the zeros are restricted to the left half-plane (Art. 4, Ch. 8) except in the special case for which all zeros lie upon the j-axis. In this limiting situation we are permitted to interpret the zeros as being in the left half-plane since the effect of incidental dissipation in the practical realization of such an impedance displaces these j-axis zeros slightly into the left half-plane (see Art. 6, Ch. 12).

In order to see more specifically what these limitations imply with regard to the process of constructing the lossless network from a specified magnitude of $Z_{12}(s)$, we observe first of all that the condition requiring the z_{sk} in Eq. 9 to be ratios of respectively odd and even polynomials results in representation 7 being forced to have either of the following two special forms

$$Z_{12}(s) = \frac{m}{m_2 + n_2} \qquad \text{(case A)} \tag{10}$$

or

$$Z_{12}(s) = \frac{n}{m_2 + n_2} \qquad \text{(case B)} \tag{11}$$

which correspond to cases A and B in the Darlington procedure because these also are pertinent to the numerator polynomial in z_{12} being even or odd respectively.

In order that the quotient of polynomials $g(s)/q(s)$ in Eq. 7 may represent the transfer impedance of a resistance-terminated lossless network, the numerator polynomial $g(s)$ must be either even or odd. The denominator polynomial is restricted only by the condition that it be Hurwitz, since this is necessary and sufficient to insure the stability of the network which the given transfer function is intended to characterize.

Comparison of the special forms 10 and 11 with Eq. 9 suggests the identifications

$$z_{12} = m/n_2, \qquad z_{22} = m_2/n_2 \qquad \text{for case A} \tag{12}$$

and

$$z_{12} = n/m_2, \qquad z_{22} = n_2/m_2 \qquad \text{for case B} \tag{13}$$

As in the Darlington synthesis, these situations are mutually exclusive.

We are now in a position to see how this particular type of transfer impedance may be formulated directly from its specified squared magnitude as given by Eq. 6 or 8. In view of Eqs. 10 and 11 we see that

$$B(-s^2) = m_2^2 - n_2^2 \tag{14}$$

for either case A or case B, but

$$A(-s^2) = m^2 \qquad \text{for case A} \tag{15}$$

and

$$-A(-s^2) = n^2 \qquad \text{for case B} \tag{16}$$

The polynomial $A(-s^2)$ must evidently be a full square, which is not a surprising result since this same condition is encountered in the Darlington synthesis and our present method yields equivalent results except for the residue condition amongst the z_{sk} not necessarily being fulfilled with an equal sign. If $A(-s^2)$ is not a full square to begin with, we must augment the quotient $A(-s^2)/B(-s^2)$ by multiplying numerator and denominator by those factors $(s_\nu^2 - s^2)$ in $A(-s^2)$ that are not of even multiplicity, just as is required in the Darlington synthesis.

Assuming that the polynomials $A(-s^2)$ and $B(-s^2)$ in Eqs. 14, 15, and 16 are the resultant ones after any needful augmentation has been done, the construction of the polynomials $(m_2 + n_2)$ and either m or n for cases A or B respectively is straightforward. Thus we have

$$m = \sqrt{A(-s^2)}, \qquad n = \sqrt{-A(-s^2)} \tag{17}$$

for the determination of the even or odd polynomial as the case may be; and the Hurwitz polynomial $(m_2 + n_2)$, as usual, is constructed from the left half-plane zeros of $B(-s^2)$. The functions z_{22} and z_{12} are then obtained from Eq. 12 or 13, and a lossless network is constructed from these impedances according to any of the methods discussed in Arts. 2, 3, and 4 of Ch. 7.

In this construction of the transfer function $Z_{12}(s)$ from its squared magnitude, and in the network synthesis that follows, it should be observed that no control is exercised over the constant multiplier. We ignore this multiplier completely for the reason that none of the relevant synthesis procedures include the ability to control such a constant multiplier. In a transfer impedance this multiplier is meaningful only if expressed with reference to a 1-ohm termination, since the output voltage per ampere of excitation can theoretically be adjusted at will through subsequent revision of the over-all impedance level of the network inclusive of its load resistance. What is practically more important than the constant multiplier is the configuration of the network and its element value distribution. One ordinarily chooses the synthesis procedure that yields an acceptable result on these terms and accepts whatever multiplier one gets. There are too many variables involved in this situation to make the consideration of an optimization problem practically feasible.

As a simple example to illustrate these methods we consider the squared magnitude

$$|Z_{12}(j\omega)|^2 = \frac{1}{1+\omega^6} \tag{18}$$

To use the Darlington method, we regard this rational function as the real part of an impedance. Its construction begins by noting that the zeros of

$$B(-s^2) = m_2{}^2 - n_2{}^2 = 1 - s^6 = 0 \tag{19}$$

are the sixth roots of unity, which are given by

$$s_\nu = e^{j2\pi\nu/6} \qquad \text{for } \nu = 1, 2, \cdots, 6 \tag{20}$$

The left half-plane zeros are those for $\nu = 2, 3, 4$, or

$$s_{2,4} = -\frac{1}{2} \pm j\frac{\sqrt{3}}{2}, \qquad s_3 = -1 \tag{21}$$

Hence we get for the denominator of the input impedance

$$(m_2 + n_2) = (s^2 + s + 1)(s + 1) = s^3 + 2s^2 + 2s + 1 \tag{22}$$

This impedance itself is evidently the same as the auxiliary impedance $Z'(s)$ in the Miyata method, since we have $M = 1$. By the Gewertz procedure we find this impedance to be

$$Z_1(s) = \frac{\frac{2}{3}s^2 + \frac{4}{3}s + 1}{s^3 + 2s^2 + 2s + 1} \tag{23}$$

In terms of what we know about driving-point impedance synthesis by the Miyata method, we can predict from the real part of Eq. 18 that $Z_1(s)$ in Eq. 23 is developable into a simple lossless ladder terminated in resistance. Hence this ladder network is the desired lossless coupling network. We obtain it immediately from the development of $Z_1(s)$ based upon the calculations

$$\begin{array}{r|l}
 & \tfrac{3}{2}s \\
\tfrac{2}{3}s^2 + \tfrac{4}{3}s + 1 & s^3 + 2s^2 + 2s + 1 \\
 & s^3 + 2s^2 + \tfrac{3}{2}s \quad \big|\ \tfrac{4}{3}s \\
 & \quad \tfrac{1}{2}s + 1 \,\big|\, \tfrac{2}{3}s^2 + \tfrac{4}{3}s + 1 \\
 & \qquad\qquad \tfrac{2}{3}s^2 + \tfrac{4}{3}s \quad \big| \\
 & \qquad\qquad\qquad 1 \,\big|\, \tfrac{1}{2}s + 1
\end{array} \tag{24}$$

Figure 2 shows the resulting network. Because of the special form of the assumed function 18, it turns out to have a practically advantageous structure involving no mutual inductive coupling.

Let us now solve this same problem by the direct method based upon Eq. 9. The first step here is to see whether any augmentation is necessary.

Noting that $A(-s^2) = 1$ in this example, we observe that none is needed. The situation is case A, and Eqs. 15 and 17 yield $m = 1$. Formation of the denominator polynomial $m_2 + n_2$ from $B(-s^2)$ follows the same pattern as for the Darlington approach, and so we have

$$Z_{12}(s) = \frac{1}{s^3 + 2s^2 + 2s + 1} \tag{25}$$

Use of Eqs. 12 then gives

$$z_{12} = \frac{1}{s^3 + 2s}, \qquad z_{22} = \frac{2s^2 + 1}{s^3 + 2s} \tag{26}$$

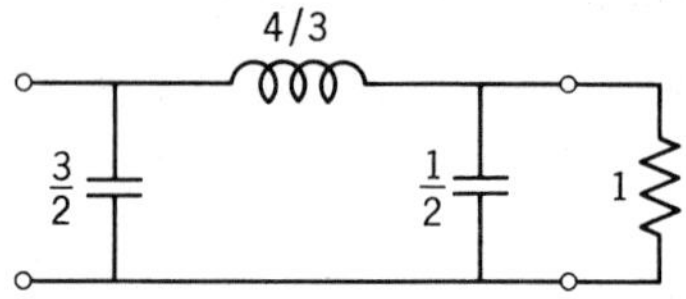

Fig. 2. Realization of the transfer impedance 18 by either Darlington's or Cauer's method.

Since all of the zeros of z_{12} lie at $s = \infty$, the method of ladder development discussed in Art. 3, Ch. 7, shows that the desired network is obtained through the continued-fraction development of z_{22} indicated in

$$z_{22} = \cfrac{1}{\cfrac{s}{2} + \cfrac{1}{\cfrac{4s}{3} + \cfrac{1}{\cfrac{3s}{2}}}} \tag{27}$$

which yields the familiar Cauer form consisting of an unbalanced ladder with shunt capacitances and series inductances. The lossless network is recognized to be that shown in Fig. 2.

2 Relevance of These Results to the Construction of an Impedance From Its Real Part

This direct method for the synthesis of $Z_{12}(s)$ is of collateral interest in connection with the problem of constructing an impedance from its real part or even part. The three algebraic procedures discussed in Art. 1, Ch. 8, for accomplishing this end are: (*a*) the partial-fraction expansion method of Bode; (*b*) the Gewertz method requiring the solution of a set of simultaneous equations; (*c*) the Miyata method involving a continued-division process. Of these, the Miyata method is computationally the least tedious, except for the fact that the auxiliary impedance $Z'(s)$, corresponding to an $M = m_1m_2 - n_1n_2 = 1$, must first be found by either method (*a*) or (*b*). Here the Gewertz method is usually preferable since it yields the result in polynomial form directly, which we need in order to proceed with the Miyata method. When a real part of higher degree is involved, the solution of the pertinent set of simultaneous equations becomes rather tedious, even though we use a systematic elimination process. Here is where the results of the previous article suggest a computationally more compact approach.

The specification $M = m_1m_2 - n_1n_2 = A(-s^2) = 1$, which is made in the construction of the Miyata auxiliary impedance $Z'(s)$, leads in Eqs. 17 always to the case A result: $m = 1$, and hence consistently yields a $Z_{12}(s)$ of the form

$$Z_{12}(s) = \frac{1}{m_2 + n_2} \tag{28}$$

like Eq. 25 in the above example. Equations 12 are applicable, and give

$$z_{12} = \frac{1}{n_2}, \qquad z_{22} = \frac{m_2}{n_2} \tag{29}$$

The zeros of z_{12} are all at $s = \infty$, and so the appropriate development of z_{22} is always the Cauer continued-fraction expansion (obtained from a continued-division process as described in Art. 4, Ch. 3) leading to a ladder with series L's and shunt C's.

According to Eqs. 103 (Art. 6, Ch. 9) we see that the impedance z_{22} looking into the output terminals of this reactive ladder is converted into the impedance

$$\frac{1}{y_{22}} = \frac{n_1}{m_1} \tag{30}$$

when its input terminal pair is short-circuited. Applying such a short circuit to the physical network is equivalent, in the continued-fraction development of z_{22}, to dropping the term corresponding to the impedance of the last shunt branch (which is the first shunt branch on the "number 1" end of this network). We observe, incidentally, that this last branch in the development of z_{22} is always *shunt* (never a series branch) because excitation is a current, and a series branch at the input end of the lossless coupling network in Fig. 1 has no influence upon the function $Z_{12}(s)$.

In the example of the preceding article, Eq. 27 immediately yields

$$y_{22} = \frac{m_1}{n_1} = \frac{s}{2} + \frac{1}{\frac{4s}{3}} = \frac{2s^2 + 3}{4s} \tag{31}$$

from which we can conclude that

$$(m_1 + n_1) = 2s^2 + 4s + 3 \tag{32}$$

within a constant multiplier. With the denominator $(m_2 + n_2)$ already known, as in Eq. 22, we have for the Miyata auxiliary impedance in this instance

$$Z'(s) = \frac{2s^2 + 4s + 3}{s^3 + 2s^2 + 2s + 1} \tag{33}$$

Apart from a constant multiplier, this result checks the input impedance $Z_1(s)$ given by Eq. 23. The proper value of the constant multiplier can always be seen by inspection by noting that the zero-frequency value of an impedance must equal the zero-frequency value of its specified real part.

This method of constructing the polynomial $(m_1 + n_1)$, corresponding to a given denominator polynomial $(m_2 + n_2)$ and an assumed $M = 1$, requires less computational tedium than is involved in the solution of the pertinent simultaneous equation set. Together with the Miyata method of constructing an impedance appropriate to any other even polynomial $M = m_1m_2 - n_1n_2$ once the auxiliary impedance has been found, this scheme offers perhaps the most direct and convenient way of constructing an impedance from its specified *j*-axis real part.

3 Practical Uses of the Resistance-Terminated Lossless Coupling Network

An identical procedure to that described in Art. 1 is applicable on an admittance basis to the design of a resistance-terminated lossless coupling network for a prescribed transfer admittance defined with reference to Fig. 1 by the function

$$Y_{12} = \frac{y_{12}}{1 + y_{22}} = \frac{I_2}{E_1} \tag{34}$$

in which the y_{sk} are the familiar short-circuit driving-point and transfer admittances. Since the voltage across the 1-ohm load resistance is numerically the negative of the current I_2, we have

$$\frac{E_2}{E_1} = \frac{-y_{12}}{1 + y_{22}} \tag{35}$$

from which we see that this same synthesis procedure is also available as a means for designing a coupling network for a prescribed voltage ratio.

The impedance basis is appropriate for the design of an output network to follow a pentode amplifier or wherever we have a high-impedance source (essentially a current source) and a resistance load. The admittance basis, on the other hand, is suitable for the design of an output network where the source is a low-impedance voltage source like that provided by the cathode follower.

It is significant to note in this connection that the impedance basis is also suitable for the design of a network to couple a voltage source with finite internal resistance to a high-impedance load (essentially an open circuit). This situation arises, for example, at the output end of a transmission line where we wish to couple into a vacuum-tube amplifier. The Thévenin equivalent of the transmission line is represented by the voltage source and series internal resistance, whereas the grid circuit of the vacuum tube presents essentially an open circuit or infinite load impedance.

Figure 3 illustrates how the design in Fig. 1 on an impedance basis becomes appropriate to this situation. Figure 3*a* shows the total network of Fig. 1 in reversed orientation. By the reciprocity theorem, the ratio E_2/I_1 there is the same as in Fig. 1 for a given design. In Fig. 3*b* we have converted the current source I_1 of Fig. 3*a* into an equivalent voltage source E_1, which now appears in series with the 1-ohm resistance. Numerically, E_1 and I_1 are equal; and so we see that the voltage ratio E_2/E_1 in the arrangement of Fig. 3*b* is given by the transfer impedance $Z_{12}(s)$ considered in Art. 1.

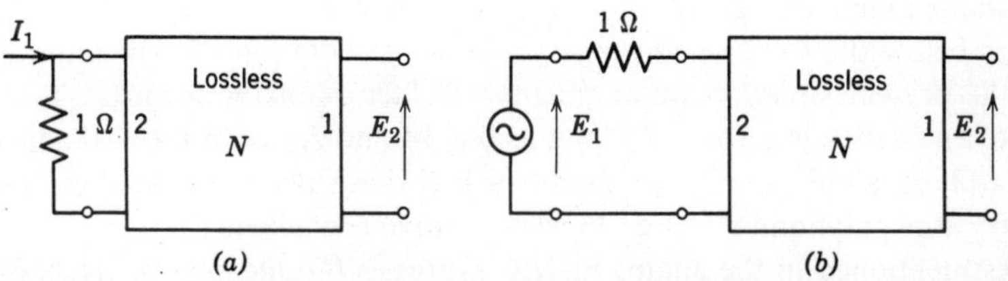

Fig. 3. Rearrangement of the circuit in Fig. 1 to accommodate a situation involving a lossy voltage source and essentially infinite load impedance.

4 An Alternate Way of Synthesizing the Lossless Network

In situations where the admittance basis, yielding the transfer admittance 34, is appropriate, an additional method of synthesis for the lossless network is possible. For the sake of its collateral interest, a brief discussion of this method is in order at this point.

Following the pattern in Eqs. 12 and 13, we write for the short-circuit transfer admittance of the lossless network (assuming case A)

$$y_{12} = \frac{m}{n_2} = \frac{\alpha_0 + \alpha_1 s^2 + \alpha_2 s^4 + \cdots + \alpha_k s^{2k}}{n_2} \tag{36}$$

Our strategy in this method is to decompose the y_{12} function into additive components, each of which has zeros only on the j-axis so that its separate realization is obtained by the ladder development of y_{22} discussed in Art. 3, Ch. 7. Realization of the whole function Y_{12} is then accomplished through an appropriate parallel combination of such ladder networks.

If the coefficients α_k in Eq. 36 are all positive, a possible decomposition of this sort is evidently obtained by writing

$$y_{12} = \frac{\alpha_0}{n_2} + \frac{\alpha_1 s^2}{n_2} + \frac{\alpha_2 s^4}{n_2} + \cdots + \frac{\alpha_k s^{2k}}{n_2} \tag{37}$$

which parallels the method originally used in *RC* network synthesis to obtain an unbalanced form of realization.* Each component transfer admittance, given by a term in this expression, has all of its zeros either at zero or infinite frequency, so that the appropriate development of y_{22} consists of a ladder network with single inductance or capacitance branches, as in the Miyata synthesis procedure (Art. 2, Ch. 10) in which the decomposition of the real part of an *RLC* driving-point impedance is based upon this same scheme.

Positiveness of the coefficients α_k in 36 is assured if m is a Hurwitz polynomial in the variable s^2, which requires the zeros of the prescribed Y_{12} function to be not more than 45° from the j-axis of the s-plane. As mentioned in the discussion of the Miyata procedure, we can produce a resultant numerator polynomial with positive coefficients even if Y_{12} has zeros below the 45° lines through proper augmentation of expression 36. In the present situation we augment with factors of the form $(s^2 + \omega_i{}^2)$ representing j-axis zeros. So long as Y_{12} has no zeros on the real axis of the s-plane, a finite number of factors will always suffice to yield a resultant numerator polynomial in Eq. 36 with positive coefficients.

As mentioned in the analogous *RC* synthesis (Guillemin, *loc. cit.*), one can evidently reduce the number of parallel ladder networks in this form of realization by forming components with numerators consisting of two consecutive terms instead of single terms. This decomposition has the form

$$y_{12} = \frac{(\alpha_0 + \alpha_1 s^2)}{n_2} + \frac{s^4(\alpha_2 + \alpha_3 s^2)}{n_2} + \cdots + \frac{s^{2(k-1)}(\alpha_{k-1} + \alpha_k s^2)}{n_2} \tag{38}$$

in which the last numerator has only one term if k is an even integer. Since the parenthesis expression in each numerator represents a pair of j-axis zeros, an unbalanced ladder realization without mutual inductive coupling is achievable for each term in this decomposition by the method of ladder development of y_{22} discussed in Art. 3, Ch. 7. One of the branches in this ladder now consists of a parallel *LC* combination (if series) or a series *LC* combination (if shunt). The remaining branches are single *L*'s or *C*'s as for the decomposition 37.

In order further to economize in the number of elements required for the total realization, one should seek to find a decomposition of the poly-

* "Synthesis of *RC* Networks," E. A. Guillemin, *J. Math. and Phys.*, Vol. 28, No. 1, Apr. 1949. (This method is discussed in Ch. 13.)

nomial m in Eq. 36 into the fewest number of components of which each has exclusively j-axis zeros (or negative real zeros if regarded as a polynomial in s^2). Although it is always possible to decompose any even polynomial into the *difference* of two having only j-axis zeros,* its decomposition into the sum of two such component polynomials cannot in general be done. Certain special situations, on the other hand, are easily dealt with. For example, if a plot of $m(s^2)$ versus real values of s^2 has an appearance like the curve shown in Fig. 4, then it is easy to see that a straight line having positive slope and a negative real zero itself can be found such that all of its intersections with $m(s^2)$ occur at negative real values of s^2. Subtraction of the straight line from the given curve therefore leaves a polynomial $m(s^2) - (s^2 + a^2)$ having all negative real s^2 zeros.† In less favorable situations it may be necessary to choose a parabola or higher degree curve (rather than a straight line) in order to obtain the desired intersections with $m(s^2)$ at negative real s^2 values. The polynomial representing this higher degree curve must, of course, have all negative real s^2 zeros also, besides having a positive leading coefficient.

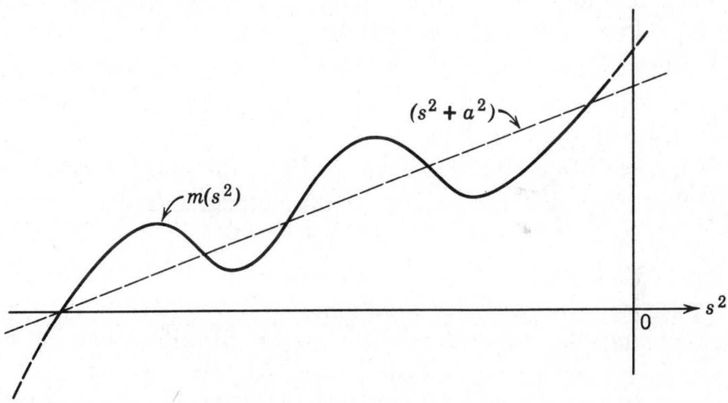

Fig. 4. Graphical interpretation of the process of decomposing $m(s^2)$ into a sum of polynomials each of which has only negative real s^2 zeros.

Whether the decomposition of y_{12} is carried out in the single-term or double-term fashion shown respectively in Eqs. 37 and 38, or in some more elaborate manner, the primary purpose is to obtain a representation

* With the left half-plane zeros of $m(s^2)$ we construct a Hurwitz polynomial $m_0 - n_0$, whereupon $m = m_0{}^2 - n_0{}^2$ is the desired decomposition.

† This scheme was first suggested in connection with the *RC* synthesis problem pertinent to a servo design method discussed by J. G. Truxal in *Tech. Rept.* 162, Mass. Inst. Technol., Research Lab. Electronics, Aug. 25, 1950. See also his book *Automatic Feedback Control System Synthesis*, McGraw-Hill Book Co., New York, 1955, p. 310.

for y_{12} of the form

$$y_{12} = y_{12}^{(1)} + y_{12}^{(2)} + \cdots + y_{12}^{(r)} \tag{39}$$

in which each $y_{12}^{(i)}$ has only j-axis zeros so that a development of y_{22} according to the methods discussed in Art. 3, Ch. 7, yields a simple unbalanced ladder with a short-circuit transfer admittance equal to $y_{12}^{(i)}$, apart from a constant multiplier.

Thus each component ladder has the short-circuit driving-point admittance y_{22} and a transfer admittance which we can write $K_i y_{12}^{(i)}$, where the resulting multiplier is denoted by K_i. If we multiply the admittance level of each of these ladder networks by K/K_i where

$$\frac{1}{K} = \frac{1}{K_1} + \frac{1}{K_2} + \cdots + \frac{1}{K_r} \tag{40}$$

then their parallel combination (parallel connection of all their input terminal pairs as well as of all their output terminal pairs) yields a resultant lossless two terminal-pair network having the driving-point admittance y_{22} and a transfer admittance equal to y_{12} in Eq. 39 multiplied by K, which then becomes the resultant multiplier for the desired admittance Y_{12} in Eq. 34 or for the voltage ratio 35 as well.

This method of realization evidently enables one to obtain an unbalanced lossless network without mutual inductive coupling even when some of the specified transmission zeros (zeros of Y_{12}) are more than 45° from the j-axis, whereas the method of cascade synthesis discussed in Art. 4, Ch. 7, requires that some inductances be mutually coupled unless all transmission zeros are at least 45° from the real axis. Although mutual inductive coupling is not desirable from a practical standpoint, the structure discussed here, consisting of a parallel combination of component networks, is practically much less desirable. The reason for this situation derives from the fact that the transmission zeros for the total network come about through a cancelation of the transfer admittances for the components, so that a small misalignment of one component can cause *all* transmission zeros to be in error. Hence the possibility of being able to align such a network properly is very remote, whereas the cascade structure, which permits the separate alignment of each pair or set of transmission zeros, is much more likely to be successfully constructible. The method of cascade synthesis, moreover, is appropriate equally well to the realization of a transfer impedance or admittance, whereas the parallel ladder scheme is pertinent only to the admittance basis, requiring an impedanceless voltage generator at the input.

5 The Double-Loaded Lossless Coupling Network

There are situations where it is desirable to be able to design a lossless coupling network that is resistance-loaded at both ends, as shown in Fig. 5 and for which the specified transfer function is the ratio E_2/I_{10}. If the current source I_{10} with its associated 1-ohm resistance is converted into a voltage source E_{10} (numerically equal to I_{10}) in series with this resistance, as is done on the input side of the network shown in Fig. 3, then the present situation again becomes appropriate to the synthesis of a voltage ratio, but now the source may have a finite nonzero internal resistance.

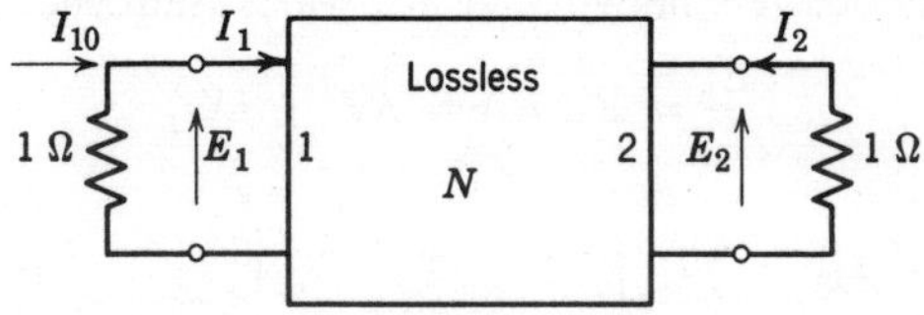

Fig. 5. The double-loaded lossless coupling network.

Making use of the fact that

$$I_1/I_{10} = 1/(1 + Z_1) \tag{41}$$

in which $Z_1 = E_1/I_1$ is the input impedance to the lossless network, we see that the desired transfer impedance is given by

$$\frac{E_2}{I_{10}} = \frac{E_2}{I_1} \times \frac{I_1}{I_{10}} = \frac{Z_{12}}{1 + Z_1} \tag{42}$$

Noting from Eqs. 2 and 9 (or from Eqs. 103, 108, and 169 of Ch. 9) that

$$Z_{12}(s) = \frac{\sqrt{m_1 m_2 - n_1 n_2}}{m_2 + n_2} \tag{43}$$

for case A (and the same expression with the letters m and n interchanged for case B), and making use also of Eq. 1, we get for the ratio 42

$$\frac{E_2}{I_{10}} = \frac{\sqrt{m_1 m_2 - n_1 n_2}}{(m_1 + m_2) + (n_1 + n_2)} \tag{44}$$

and for its squared j-axis magnitude

$$\left|\frac{E_2}{I_{10}}\right|^2_{s=j\omega} = \left[\frac{m_1 m_2 - n_1 n_2}{(m_1 + m_2)^2 - (n_1 + n_2)^2}\right]_{s=j\omega} \tag{45}$$

If we observe that

$$\left|\frac{1 - Z_1}{1 + Z_1}\right|^2_{s=j\omega} = \left[\frac{(m_2 - m_1)^2 - (n_2 - n_1)^2}{(m_2 + m_1)^2 - (n_2 + n_1)^2}\right]_{s=j\omega} \tag{46}$$

from which

$$1 - \left|\frac{1 - Z_1}{1 + Z_1}\right|^2_{s=j\omega} = \left[\frac{4(m_1 m_2 - n_1 n_2)}{(m_2 + m_1)^2 - (n_2 + n_1)^2}\right]_{s=j\omega} \tag{47}$$

we see that Eq. 45 can also be written

$$\left|\frac{2E_2}{I_{10}}\right|^2_{s=j\omega} = 1 - \left|\frac{1 - Z_1}{1 + Z_1}\right|^2_{s=j\omega} \tag{48}$$

Before continuing with the discussion of how the lossless network is designed in this case, it is useful to understand the physical significance of this result. In this connection it is additionally helpful to consider the unnormalized situation regarding associated load resistances as shown in Fig. 6 where these are denoted by R_1 and R_2. If we write for the input impedance to the lossless network at any j-axis frequency

$$\frac{E_1}{I_1} = Z_1(j\omega) = R_{11} + jX_{11} \tag{49}$$

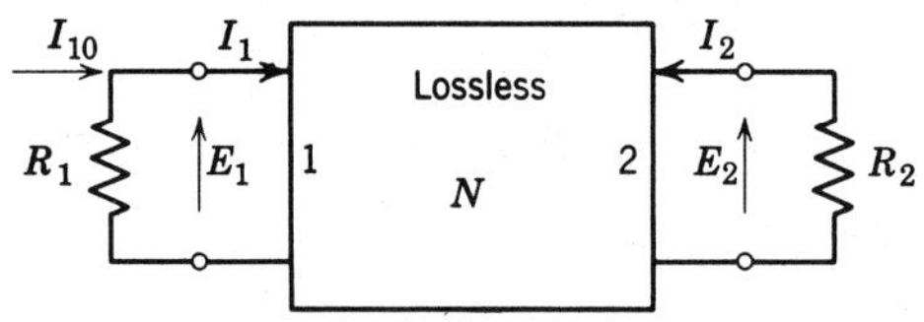

Fig. 6. The situation in Fig. 5 in which the associated source and load resistances are not normalized.

then the equality of the average power entering and leaving this network is expressed by

$$|I_1|^2 R_{11} = |E_2|^2/R_2 \tag{50}$$

Since Eq. 41 in this case reads

$$I_1/I_{10} = R_1/(R_1 + Z_1) \tag{51}$$

giving

$$|I_1|^2 = |I_{10}|^2 R_1^2/|R_1 + Z_1|^2 \tag{52}$$

Eq. 50 yields

$$\frac{|E_2|^2}{R_1 R_2 |I_{10}|^2} = \frac{R_{11} R_1}{|R_1 + Z_1|^2} = \frac{1}{4}\left(1 - \left|\frac{R_1 - Z_1}{R_1 + Z_1}\right|^2\right) \tag{53}$$

The final form here is more easily seen to follow if we recognize that

$$\begin{aligned} |R_1 - Z_1|^2 &= (R_1 - R_{11})^2 + X_{11}^2 \\ |R_1 + Z_1|^2 &= (R_1 + R_{11})^2 + X_{11}^2 \end{aligned} \tag{54}$$

If the current source I_{10} paralleled by R_1 is converted into an equivalent voltage source $E_{10} = R_1 I_{10}$, then we can write

$$\frac{4|E_2|^2}{R_1 R_2 |I_{10}|^2} = \frac{|E_2|^2/R_2}{|E_{10}|^2/4R_1} = |t(j\omega)|^2 \tag{55}$$

This quantity is evidently the ratio of power delivered to the load R_2 to the maximum power deliverable by the source. Physically its value can at most equal unity. We shall regard it as the squared absolute value of a *transmission coefficient* $t(s)$ for $s = j\omega$.

In connection with this physical interpretation it is usual to regard the quantity

$$\rho(s) = \frac{R_1 - Z_1(s)}{R_1 + Z_1(s)} \tag{56}$$

as a *reflection coefficient* appropriate to the input end of the network in Fig. 6, and write in place of Eq. 53

$$|t(j\omega)|^2 = 1 - |\rho(j\omega)|^2 \tag{57}$$

which expresses the physically readily understandable fact that whatever average power is undeliverable to the load must be returned to the source.

Note from Eq. 55 that

$$t(s) = 2\sqrt{\frac{R_1}{R_2}}\left(\frac{E_2}{E_{10}}\right) \tag{58}$$

Hence $t(s) = 1$ implies

$$\frac{E_2}{E_{10}} = \frac{1}{2}\sqrt{\frac{R_2}{R_1}} \tag{59}$$

which, it will be recalled, is the maximum voltage ratio obtainable. Under these circumstances the lossless network is equivalent to an ideal transformer having the optimum ratio $\sqrt{R_2/R_1}$.

In passing it is useful to note that all of the above analytic relationships pertaining to the unnormalized network in Fig. 6 are different from the corresponding ones derived earlier for the normalized situation only in that Z_1 is replaced by Z_1/R_1 and (E_2/I_{10}) by $(E_2/I_{10})/\sqrt{R_1R_2}$ or (E_2/E_{10}) by $(E_2/E_{10})\sqrt{R_1/R_2}$ since $I_{10} = E_{10}/R_1$. Similarly, a relation involving $Z_2 = E_2/I_2$ will be changed by having this driving-point impedance replaced by Z_2/R_2. The analytic relationships characterizing an unnormalized situation are thus seen to differ in that the driving-point impedances are divided by their respective associated terminal resistances, and transfer impedances by the geometric mean of the two terminal resistances. For example, in an unnormalized situation both Z_{12} and z_{12} in Eq. 9 would be divided by $\sqrt{R_1R_2}$ and z_{22} by R_2.

These matters are particularly relevant to the design of a double-loaded lossless network since the resistance loading on the two sides need not be the same. When only one terminal resistance is associated with the network, there is evidently no loss in generality in assuming its value to be unity. The present considerations show that when two resistance terminations are considered, we can nevertheless normalize both at 1 ohm as in the situation shown in Fig. 5 without restricting the generality of the pertinent analytic relationships, if we properly interpret the normalized quantities with respect to the values of the terminal resistances that we ultimately wish to associate with the lossless network.

In the normalized situation, the transmission and reflection coefficients are defined with the help of Eqs. 44 and 48 as

$$t(s) = \frac{2E_2}{I_{10}} = \frac{2\sqrt{m_1m_2 - n_1n_2}}{(m_2 + m_1) + (n_2 + n_1)} \tag{60}$$

and

$$\rho(s) = \frac{1 - Z_1}{1 + Z_1} = \frac{(m_2 - m_1) + (n_2 - n_1)}{(m_2 + m_1) + (n_2 + n_1)} \tag{61}$$

The synthesis begins with the specified squared magnitude of $t(j\omega)$ having the form

$$|t(j\omega)|^2 = \frac{A(\omega^2)}{D(\omega^2)} = \frac{A_0 + A_1\omega^2 + \cdots + A_n\omega^{2n}}{D_0 + D_1\omega^2 + \cdots + D_n\omega^{2n}} \tag{62}$$

where the pertinent input impedance $Z_1(s)$ is, as usual, assumed to be given by the ratio of two nth-degree polynomials.

It is incidentally worth noting that these results are pertinent to cases A and B alike, since an interchange of the letters m and n in the expressions for $t(s)$ and $\rho(s)$ leaves them essentially unchanged. Turning the network end for end, or interchanging the identities of the input and output ends, likewise has no effect upon $t(s)$, as we should expect according to the reciprocity theorem. Thus, following the principles discussed in Art. 8, Ch. 9, we note that an interchange of m_1 and m_2 or of n_1 and n_2 has no effect upon $t(s)$. The reflection coefficient $\rho(s)$, on the other hand, is affected by such an interchange; indeed, there is every reason to expect that the reflection coefficient $(1 - Z_2)/(1 + Z_2)$ pertinent to the opposite end of the network should be different. But note that the squared absolute value for $s = j\omega$, as expressed by Eq. 46, is the same for both reflection coefficients. This fact, of course, must follow since $t(s)$ is insensitive to reversal of the network orientation, and so is the average power relationship expressed by Eq. 57.

The necessary and sufficient conditions to be imposed upon the specified squared magnitude 62 in order that realization of the desired network be assured are readily obtained by observing from Eq. 61 that

$$Z_1(s) = \frac{1 - \rho(s)}{1 + \rho(s)} \tag{63}$$

and recalling from the discussion in Art. 5, Ch. 1, that the p.r. character of $Z_1(s)$ requires $\rho(s)$ to be analytic in the right half-plane and have a j-axis magnitude that does not exceed unity. The latter condition in a sense is automatically assured by Eq. 57 since the squared absolute value of $t(j\omega)$ cannot become negative. To make sure of this fact, however, and also that the squared absolute value of $\rho(j\omega)$ likewise does not become negative it is necessary and sufficient to require in Eq. 62 that

$$0 \leq \frac{A(\omega^2)}{D(\omega^2)} \leq 1 \tag{64}$$

From Eqs. 46, 57 and 62 we now have

$$|\rho(j\omega)|^2 = \frac{A(\omega^2) - D(\omega^2)}{D(\omega^2)} = \left[\frac{(m_2 - m_1)^2 - (n_2 - n_1)^2}{(m_2 + m_1)^2 - (n_2 + n_1)^2}\right]_{s=j\omega} \tag{65}$$

The required analyticity of $\rho(s)$ in the right half-plane is controlled by the manner in which we construct the polynomial $(m_2 + m_1) + (n_2 + n_1)$ from $D(-s^2) = (m_2 + m_1)^2 - (n_2 + n_1)^2$ to make sure that it has Hurwitz character. This process, which at the same time assures the stability of the network since it also determines the right half-plane analyticity of $t(s)$ in Eq. 60, is a familiar one, and imposes no further conditions upon the specified rational function 62. The simple condition 64 is both necessary and sufficient to assure success in the present design procedure.

Before we can continue with the synthesis, however, we must make sure that expression 62 fulfills the collateral condition required by the lossless character of the coupling network, namely, that the polynomial $A(-s^2) = m_1m_2 - n_1n_2$ be a full square. As usual, we can remedy this situation, if necessary, by augmentation; and we will now proceed on the assumption that this item has been taken care of.

Having formed the denominator polynomial $(m_2 + m_1) + (n_2 + n_1)$ in $\rho(s)$, we may in the same manner form the numerator polynomial $(m_2 - m_1) + (n_2 - n_1)$ from

$$D(-s^2) - A(-s^2) = (m_2 - m_1)^2 - (n_2 - n_1)^2 \tag{66}$$

Here, however, the process is not unique, for we have no Hurwitz criterion to meet. The numerator polynomial may have zeros in the right as well as in the left half-plane; and we can form a variety of alternative possible and admissible functions $\rho(s)$ according to the choice we make regarding their zero distributions. This choice, incidentally, has nothing to do with the minimum- or nonminimum-phase character of the resulting transfer function $t(s)$. Its zeros must in any case be equally distributed between the right and left half-planes because of the required even or odd character of this function's numerator polynomial.

It may be mentioned in this connection, however, that if the design is for an interstage coupling network and we are interested in maximizing the gain-bandwidth product for a given amount of parasitic shunt capacitance, then an optimum result is achieved if the zeros as well as the poles of $\rho(s)$ are restricted to the left half-plane.* Discussion of these matters, which

* In this connection see H. W. Bode's *Network Analysis and Feedback Amplifier Design*, D. Van Nostrand Co., New York, 1945, pp. 360–368.

are not directly relevant to the implementation of our present synthesis method, is not considered at this time.

Having formed the function $\rho(s)$, we can readily complete the synthesis either by computing $Z_1(s)$ from Eq. 63 and determining the lossless network following the Darlington method, or more directly, by noting from Eq. 61 that the even and odd parts of numerator and denominator polynomials in $\rho(s)$ enable us to calculate m_1, m_2, n_1, and n_2 at once. Thus the sum of the even parts yields m_2, and their difference yields m_1. The odd parts similarly determine n_2 and n_1. The z_{sk} of the lossless network are then obtained from the familiar Darlington equations (Art. 6, Ch. 9, case A)

$$z_{11} = \frac{m_1}{n_2}, \qquad z_{22} = \frac{m_2}{n_2}, \qquad z_{12} = \frac{\sqrt{A(-s^2)}}{n_2} \tag{67}$$

and the pertinent network is then determined by the methods given in Ch. 7.

It should be noted that this design, calling for resistance loading at both ends of the lossless network, fixes all three of the z_{sk} and, therefore, is more rigid than the design involving a single-loaded lossless network, which leaves the input impedance z_{11} free. Synthesis of the lossless network when all three z_{sk} are specified, it will be recalled, does not permit the freedom that enables one in most cases to obtain unbalanced networks without mutual inductive coupling. On the contrary, the general situation always involves mutual inductive coupling and may even call for some closely coupled inductances. Only in special cases does the design for resistance loading at both ends of the lossless network lead to practically desirable structures.

An important special case of this sort is that for which the numerator polynomial in Eq. 61 reduces to a constant. The zeros of the transfer impedance z_{12}, and hence those of the even part of $Z_1(s)$, then all lie at $s = \infty$, and the discussion of Miyata's method of driving-point impedance synthesis (Art. 2, Ch. 10) shows that $Z_1(s)$ is developable into an unbalanced ladder with series inductances and shunt capacitances, terminated in a resistance.

A similar possibility obviously exists if the numerator polynomial in Eq. 62 reduces to any one of its single terms. The even part of $Z_1(s)$ then has all or some of its zeros at $s = 0$ and the rest at $s = \infty$; and the lossless ladder network also contains some series capacitance branches or shunt inductance branches, or it contains only branches of this kind.

As an example, let us choose for $|t(j\omega)|^2$ the function 18 used in Art. 1 to illustrate the design of a single-loaded lossless network. We have

$$|t(j\omega)|^2 = \frac{1}{1 + \omega^6} \tag{68}$$

and by Eq. 57

$$|\rho(j\omega)|^2 = \frac{\omega^6}{1 + \omega^6} \tag{69}$$

The left half-plane roots of the equation

$$D(-s^2) = 1 - s^6 = 0 \tag{70}$$

are the values 21 as determined in Art. 1 for Eq. 19. The corresponding Hurwitz polynomial 22 now is identified as

$$(m_2 + m_1) + (n_2 + n_1) = s^3 + 2s^2 + 2s + 1 \tag{71}$$

For the determination of the numerator polynomial in $\rho(s)$ we have from Eq. 69

$$(m_2 - m_1)^2 - (n_2 - n_1)^2 = -s^6 \tag{72}$$

Hence

$$(m_2 - m_1) + (n_2 - n_1) = s^3 \tag{73}$$

from which we see that

$$m_2 = m_1 \tag{74}$$

Since this is a case A situation, we expect the resulting lossless network to be symmetrical. Equations 71 and 73 now readily yield the polynomials m_1, m_2, n_1, and n_2; or we can form

$$\rho(s) = \frac{s^3}{s^3 + 2s^2 + 2s + 1} \tag{75}$$

and from Eq. 63 we get

$$Z_1(s) = \frac{2s^2 + 2s + 1}{2s^3 + 2s^2 + 2s + 1} \tag{76}$$

Knowing that this impedance is developable as a simple ladder, we perform the calculations

$$\begin{array}{r|l}
 & s \\
2s^2 + 2s + 1 & 2s^3 + 2s^2 + 2s + 1 \\
 & 2s^3 + 2s^2 + s \quad |\, 2s \\
 & \quad s + 1 \,|\, 2s^2 + 2s + 1 \\
 & \qquad 2s^2 + 2s \quad | \\
 & \qquad\qquad 1 \,|\, s + 1
\end{array} \tag{77}$$

and recognize the network to be that shown in Fig. 7. As predicted, the lossless network is symmetrical.

As a second example, we assume

$$|t(j\omega)|^2 = \frac{1}{1 + \omega^8} \tag{78}$$

and have

$$|\rho(j\omega)|^2 = \frac{\omega^8}{1 + \omega^8} \tag{79}$$

Formation of the denominator polynomial requires solution of the equation

$$D(-s^2) = 1 + s^8 = 0 \tag{80}$$

leading to the eighth roots of -1 which are

$$s_\nu = e^{j\nu\pi/8} \quad \text{for } \nu = 1, 3, 5, \cdots, 15 \tag{81}$$

The left half-plane roots are those for $\nu = 5, 7, 9, 11$, yielding the polynomial

$$\begin{aligned}(m_2 + m_1) + (n_2 + n_1) &= (s^2 + 0.7654s + 1)(s^2 + 1.8478s + 1) \\ &= s^4 + 2.613s^3 + 3.414s^2 + 2.613s + 1\end{aligned} \tag{82}$$

For the determination of the numerator polynomial in $\rho(s)$ we now have

$$(m_2 - m_1)^2 - (n_2 - n_1)^2 = s^8 \tag{83}$$

from which we recognize at once that

$$(m_2 - m_1) = s^4, \qquad \text{and } (n_2 - n_1) = 0 \tag{84}$$

The last result indicates that the network this time will be antimetrical. We have from Eqs. 82 and 84

$$\rho(s) = \frac{s^4}{s^4 + 2.613s^3 + 3.414s^2 + 2.613s + 1} \tag{85}$$

and hence

$$Z_1(s) = \frac{2.613s^3 + 3.414s^2 + 2.613s + 1}{2s^4 + 2.613s^3 + 3.414s^2 + 2.613s + 1} \tag{86}$$

The calculations corresponding to the ladder development of this impedance read

$$\begin{array}{r|l} & 0.766s \\ 2.613s^3 + 3.414s^2 + 2.613s + 1 & 2s^4 + 2.613s^3 + 3.414s^2 + 2.613s + 1 \\ & 2s^4 + 2.613s^3 + 2.000s^2 + 0.766s \\ \hline & \qquad\qquad 1.414s^2 + 1.847s + 1 \end{array}$$

$$\begin{array}{r|l} & 1.847s \\ 1.414s^2 + 1.847s + 1 & 2.613s^3 + 3.414s^2 + 2.613s + 1 \\ & 2.613s^3 + 3.414s^2 + 1.847s \\ \hline & \qquad\qquad 0.766s + 1 \end{array}$$

$$\begin{array}{r|l} & 1.847s \\ 0.766s + 1 & 1.414s^2 + 1.847s^2 + 1 \\ & 1.414s^2 + 1.847s \qquad |0.766s \\ \hline & \qquad\qquad 1|0.766s + 1 \\ & \qquad\qquad\;\; 0.766s \\ \hline & \qquad\qquad\qquad 1 \end{array} \tag{87}$$

They lead to the network shown in Fig. 8. We observe that if we bisect it at *a-a*, then one half is the reciprocal of the other in reversed orientation. Hence the network is antimetrical as predicted.

These examples suggest the thought that it must be possible to simplify the calculations in the design of symmetrical and antimetrical situations by determining the impedance looking into either half as at the cut *a-a* in Fig. 8. For the symmetrical network the impedance at this point of bisection is the same in either direction; for the antimetric network the impedance looking in one direction at this point is the reciprocal of that looking in the opposite direction. Since the impedance at the point of bisection in any case is half as complicated as the impedance $Z_1(s)$ for the whole network, the computations needed in the realization of the network will be simplified materially if we can devise a method for deriving the expression for the impedance of the bisected network directly from the specified $|t(j\omega)|$. The following discussions show how this can be done.

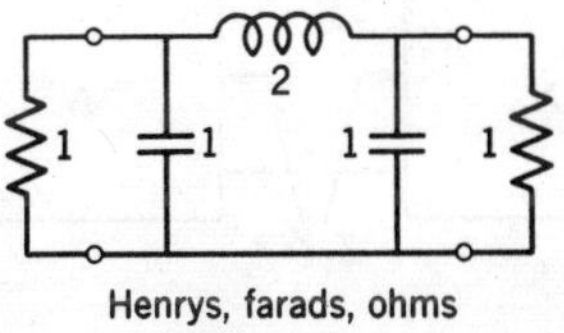

Fig. 7. Realization of the transfer impedance 18 as a double-loaded lossless network.

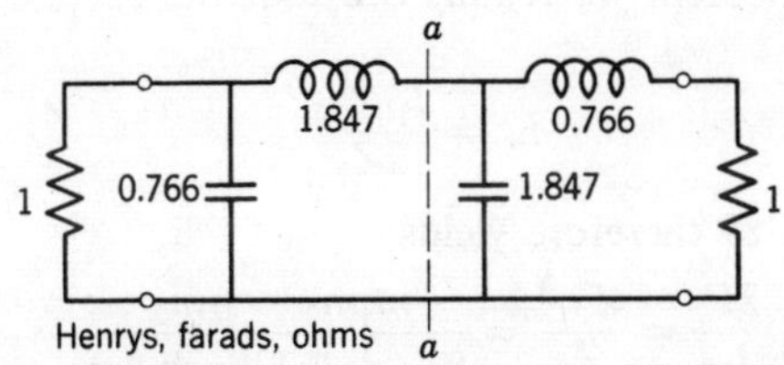

Fig. 8. Realization of the transfer function 78 as a double-loaded lossless network.

6 Methods That Simplify the Design of Symmetrical and Antimetrical Networks*

The symmetrical network is advantageously represented by the back-to-back arrangement of two identical networks, as shown in Fig. 9. In the following analysis the polynomials m_1, m_2, n_1, and n_2 refer to either one of these identical networks for which $Z_1(s) = (m_1 + n_1)/(m_2 + n_2)$ is the impedance at terminal pair 1 with the 1-ohm load at terminal pair 2. The impedance $Z_2(s)$, shown in Fig. 9, according to Eq. 165 of Ch. 9 (assuming

---

* Based upon results given in the Massachusetts Institute of Technology S.M. theses by H. Adler (1948) and E. H. Bradley (1950) as well as suggestions by D. F. Tuttle.

case A), is written

$$Z_2(s) = \frac{P(s)}{Q(s)} = \frac{m_2 + n_1}{m_1 + n_2} \tag{88}$$

Correspondingly, the transfer impedance for one of the networks in reversed orientation is given by Eq. 43 with m_1 and m_2 interchanged. For this reversed orientation we will denote the transfer impedance as $Z_{21}(s)$, indicating that it pertains to a transfer of energy in the $2 \to 1$ direction as contrasted to the notation $Z_{12}(s)$ pertinent to transfer in the $1 \to 2$ direction. With reference to Fig. 9 we thus write

$$Z_{21}(s) = \frac{E_2}{I_m} = \frac{\sqrt{m_1 m_2 - n_1 n_2}}{m_1 + n_2} \tag{89}$$

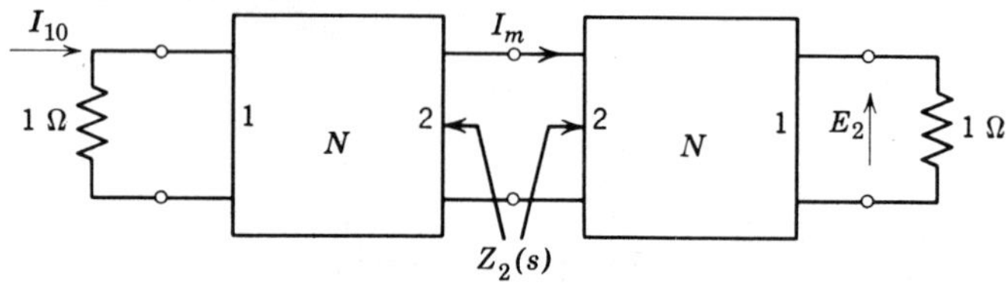

Fig. 9. Symmetrical network represented as a back-to-back arrangement of identical halves.

By Thévenin's theorem we readily see that the current I_m in Fig. 9 is given by

$$I_m = \frac{I_{10} Z_{21}}{2Z_2} \tag{90}$$

Substitution of Eq. 89 therefore yields

$$\frac{2E_2}{I_{10}} = \frac{Z_{21}{}^2}{Z_2} = \frac{m_1 m_2 - n_1 n_2}{(m_2 + n_1)(m_1 + n_2)} \tag{91}$$

If we abbreviate $P(s)$ and $P(-s)$ by P_+ and P_- respectively, and use a similar notation for the polynomial $Q(s)$, we can write Eq. 91 in the form

$$\frac{2E_2}{I_{10}} = \frac{(P_+Q_- + P_-Q_+)}{2P_+Q_+} \tag{92}$$

The squared magnitude for $s = j\omega$ becomes

$$\left|\frac{2E_2}{I_{10}}\right|^2 = \frac{(P_+Q_- + P_-Q_+)^2}{4P_+Q_+P_-Q_-} = \frac{(P_+Q_- + P_-Q_+)^2}{(P_+Q_- + P_-Q_+)^2 - (P_+Q_- - P_-Q_+)^2} \tag{93}$$

where evaluation of these expressions for $s = j\omega$ is tacitly understood. If we use the abbreviations

$$\tfrac{1}{2}(P_+Q_- + P_-Q_+) = M = m_1 m_2 - n_1 n_2 \tag{94}$$

$$\tfrac{1}{2}(P_+Q_- - P_-Q_+) = N = m_1 n_1 - m_2 n_2 \tag{95}$$

we can write

$$\left|\frac{2E_2}{I_{10}}\right|^2 = \frac{M^2}{M^2 - N^2} = \frac{1}{1 - (N^2/M^2)} \tag{96}$$

where, again, evaluation for $s = j\omega$ is understood.

Since M and N are respectively even and odd polynomials in s, we see that $-N^2/M^2$ for $s = j\omega$ is the square of an odd rational function of ω, thus

$$\left(\frac{-N^2}{M^2}\right)_{s=j\omega} = \left(\frac{\alpha_1\omega + \alpha_3\omega^3 + \cdots + \alpha_{2k+1}\omega^{2k+1}}{\beta_0 + \beta_2\omega^2 + \cdots + \beta_{2k}\omega^{2k}}\right)^2 \tag{97}$$

We may conclude that any given squared transfer function $|t(j\omega)|^2$ is realizable in the form of a symmetrical network if it is expressible as the reciprocal of the quantity: unity plus the square of an odd rational function like 97. In this function, the coefficients α_i and β_i may be either positive or negative, and any coefficients may be zero. Thus if each polynomial reduces to a single term, we have a function like 68 for $|t(j\omega)|^2$.

From this odd rational function in ω we get M and N by inspection and then note from Eqs. 94 and 95 that

$$M + N = P_+Q_- \tag{98}$$

$$M - N = P_-Q_+ \tag{99}$$

The method of finding the polynomials $P(s)$ and $Q(s)$ and hence the impedance $Z_2(s)$ in Eq. 88 is made evident by these equations. We simply assign the left half-plane zeros of $M + N$ to $P(s)$ and the right half-plane zeros to $Q(-s)$.

For the funtion 68, for example, we see by inspection that $M = 1$ and $N = s^3$. Hence the polynomial 98 has zeros determined by the equation

$$1 + s^3 = 0 \tag{100}$$

having the roots

$$s_\nu = e^{j\nu\pi/3} \quad \text{for } \nu = 1, 3, 5 \tag{101}$$

The right half-plane zeros are those for $\nu = 1$ and 5; the left half-plane zero corresponds to $\nu = 3$. The polynomials P and Q are evidently

$$P(s) = s + 1, \qquad Q(s) = s^2 + s + 1 \tag{102}$$

and the impedance 88 at the point of bisection becomes

$$Z_2(s) = \frac{s+1}{s^2+s+1} = \cfrac{1}{s + \cfrac{1}{s+1}} \tag{103}$$

yielding the network shown in Fig. 10 which we recognize as half of the network in Fig. 7.

It is incidentally interesting to observe that all of the above relations apply alike to cases A and B although we started the derivations with Eq. 88 assuming the situation to be case A. Thus we note that all of the relations following Eq. 88 remain valid if we interchange the letters m and n. We also observe that interchanging the polynomials P and Q in Eqs. 94 and 95 merely reverses the algebraic sign of N and hence interchanges Eqs. 98 and 99. Hence we see that $Z_2(s)$ may be regarded as Q/P as well as P/Q. In other words, we can reciprocate the networks in Figs. 7 and 10 if we like. This is, of course, a familiar result which is generally true whether the network is symmetrical or not (see the discussion of this item in Art. 6, Ch. 1, specifically Eqs. 112 through 116).

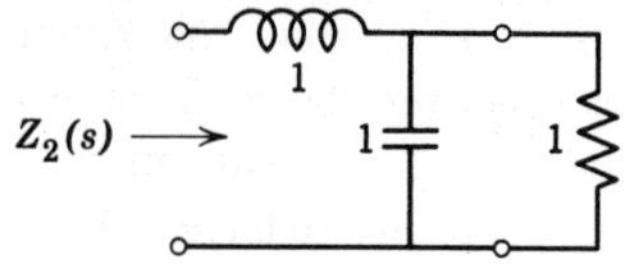

Fig. 10. Half of the symmetrical network in Fig. 7 as determined from the impedance $Z_2(s)$ of Fig. 9 given by Eq. 103.

Now let us consider the antimetrical situation shown in Fig. 11. Here we again have two lossless networks in a back-to-back arrangement, but one of these now is the dual or reciprocal of the other, so that the impedance at the point of bisection is $Z_2(s)$ looking to the left, and $1/Z_2(s)$ looking to the right.

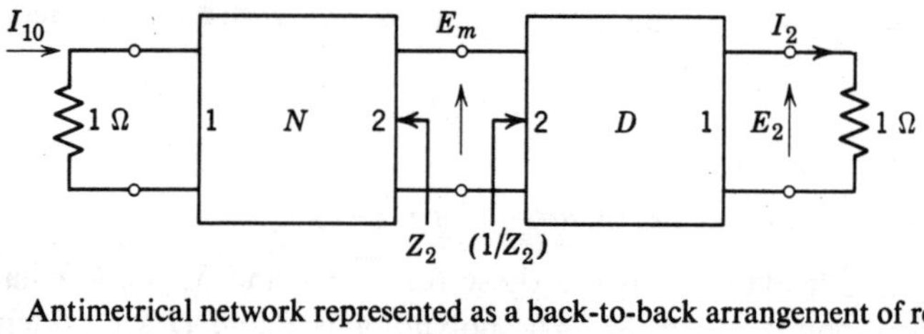

Fig. 11. Antimetrical network represented as a back-to-back arrangement of reciprocal halves.

Equation 88 applies as before, but in place of Eq. 89 we now have

$$Z_{21}(s) = \frac{I_2}{E_m} = \frac{E_2}{E_m} = \frac{\sqrt{m_1 m_2 - n_1 n_2}}{m_1 + n_2} \tag{104}$$

because the dual network D has a transfer admittance (ratio of output current to input voltage) that equals the transfer impedance (ratio of output voltage to input current) of the network N. By Thévenin's theorem we see that the voltage E_m is given by the expression

$$\frac{E_m}{I_{10}} = \frac{Z_{21}(1/Z_2)}{Z_2 + (1/Z_2)} = \frac{Z_{21}}{Z_2^2 + 1} \tag{105}$$

so that substitution from Eq. 104 yields

$$\frac{2E_2}{I_{10}} = \frac{2Z_{21}^2}{Z_2^2 + 1} = \frac{2(m_1 m_2 - n_1 n_2)}{(m_2 + n_1)^2 + (m_1 + n_2)^2} = \frac{P_+ Q_- + P_- Q_+}{P_+^2 + Q_+^2} \tag{106}$$

The squared magnitude of this function for $s = j\omega$ reads

$$\begin{aligned} \left|\frac{2E_2}{I_{10}}\right|^2 &= \frac{(P_+ Q_- + P_- Q_+)^2}{(P_+^2 + Q_+^2)(P_-^2 + Q_-^2)} \\ &= \frac{(P_+ Q_- + P_- Q_+)^2}{(P_+ Q_- + P_- Q_+)^2 + (P_+ P_- - Q_+ Q_-)^2} \end{aligned} \tag{107}$$

where the evaluation is understood to be for $s = j\omega$. We now use the same abbreviation 94 as before, but in place of 95 we have

$$\tfrac{1}{2}(P_+ P_- - Q_+ Q_-) = W = \tfrac{1}{2}[(m_2^2 - n_1^2) - (m_1^2 - n_2^2)] \tag{108}$$

which, like M, is an even polynomial in s. The squared magnitude in Eq. 107 may then be written

$$\left|\frac{2E_2}{I_{10}}\right|^2 = \frac{M^2}{M^2 + W^2} = \frac{1}{1 + (W^2/M^2)} \tag{109}$$

where evaluation for $s = j\omega$ is again understood.

From this result we see that a given squared transfer function is realizable in the form of an antimetrical network if it is expressible as the reciprocal of: unity plus the square of an even rational function (like 97 divided by ω^2). In terms of W and M, which can be formed by inspection of this even rational function, we now find in place of relations 98 and 99 the following

$$2(W + jM) = (P_+ + jQ_+)(P_- + jQ_-) \tag{110}$$

$$2(W - jM) = (P_+ - jQ_+)(P_- - jQ_-) \tag{111}$$

The method of finding the polynomials $P(s)$ and $Q(s)$ now involves two steps. First we form the polynomial $(P_+ + jQ_+)$ from the left half-plane zeros of $(W + jM)$; the real and imaginary parts of the polynomial thus found are respectively $P(s)$ and $Q(s)$.

The novel feature in this situation is the appearance of a polynomial with complex coefficients like $P(s) + jQ(s)$. Ordinarily such polynomials do not occur in network theory because all impedance or admittance functions have the property of being real for real values of the frequency variable s. In this connection it will be recalled that the adjective "real" in the expression "positive real function" connotes that the function is real for real values of s. The adjective "positive" indicates that the real part of the function is positive in the right-half of the s-plane. A "positive function" has the latter property but is not necessarily real for real values of s. Thus if $Z_2 = P(s)/Q(s)$ is p.r. (as the impedance 88 must be), the function

$$Z = Z_2 + j = \frac{P_+ + jQ_+}{Q_+} \tag{112}$$

although obviously still a positive function, is not p.r.

The necessary and sufficient condition that a function be "positive" is that its zeros and poles be in the left half-plane and its real part be positive on the j-axis. The correctness of this statement may be established by precisely the same argument as that used in Ch. 1 in connection with p.r. functions. Since the p.r. character of Z_2 in Eq. 112 requires Z to be a positive function, the zeros of the complex polynomial $P_+ + jQ_+$ must lie in the left half-plane. Hence in Eq. 110 the j-axis may be used as a "sorting boundary" to separate the polynomial $2(W + jM)$ into its two components $(P_+ + jQ_+)$ and $(P_- + jQ_-)$. The component $(P_+ + jQ_+)$ can then further be separated into the parts P_+ and Q_+ by virtue of its complex character. Thus the impedance $Z_2(s)$ at the point of bisection can again be found.

As in the symmetrical case, interchange of $P(s)$ and $Q(s)$ is possible since it merely reverses the algebraic sign of W which has no effect upon the function 109. So far as the resulting network in Fig. 9 is concerned, it has the effect of turning it end for end.

As an example of this method, suppose we reconsider the function 78 for which $W^2/M^2 = s^8$ so that we have $W = s^4$ and $M = 1$. Zeros of the polynomial 110 are roots of the equation

$$W + jM = s^4 + j = 0 \tag{113}$$

which are

$$s_\nu = e^{-j\nu\pi/8} \quad \text{for } \nu = 1, 5, 9, 13 \tag{114}$$

The left half-plane zeros are those for $\nu = 5$ and 9. These yield the polynomial

$$\begin{aligned} P(s) + jQ(s) &= s^2 + s\sqrt{2}e^{j\pi/8} + e^{j\pi/4} \\ &= (s^2 + 1.306s + 0.707) + j(0.5414 + 0.707) \end{aligned} \tag{115}$$

The desired impedance is

$$Z_2(s) = \frac{s^2 + 1.306s + 0.707}{0.5414s + 0.707} = 1.847s + \frac{1}{0.766s + 1} \tag{116}$$

which yields half of the network in Fig. 8. Even in this simple example the reduction in computational work is significant.

Returning to the symmetrical network in Fig. 9, and the derivation of the pertinent transfer impedance 91, suppose the right-hand network N, inclusive of the terminal resistance, has its impedance level multiplied by a factor r^2. The transfer impedance $Z_{21}(s)$ in Eq. 89 becomes multiplied by r^2, and Eq. 90 for the current I_m is changed to

$$I_m = \frac{r^2 I_{10} Z_{21}}{(1 + r^2) Z_2} \tag{117}$$

so that we get in place of Eq. 91

$$\frac{2E_2}{I_{10}} = \frac{2r^2}{1 + r^2} \times \frac{Z_{21}^2}{Z_2} \tag{118}$$

Except for the constant factor $2r^2/(1 + r^2)$, this result is the same as before the impedance level change in the right-hand half of the network in Fig. 9 was made. That is to say, the resulting network now is capable of accommodating any ratio of the finite source and load resistances, while yielding the same function for the desired over-all transfer ratio $2E_2/I_{10}$. The usefulness of this result is obvious in view of the fact that the normal procedure in practice is to use a transformer at the input or output terminals to accomplish the same result.

In this connection it is essential to note, however, that one does not avoid the added cost of a transformer without a compensating loss—in this instance, a loss in gain. Thus, the resultant over-all constant multiplier is less by this method of accommodating a given ratio of source-to-load resistance than with the use of an ideal coupling transformer. An ideal transformer with the impedance ratio r^2 has a voltage ratio r, so that its insertion at the output terminals multiplies the resultant load voltage, and hence the transfer function $2E_2/I_{10}$, by r. The ratio of this transfer function to the one given by Eq. 118 has the value

$$\frac{1 + r^2}{2r} = \frac{1}{2}\left(r + \frac{1}{r}\right) \tag{119}$$

This familiar function has a value larger than unity for all values of r except $r = 1$ for which its value equals unity. Hence, use of an ideal transformer yields a higher over-all gain; but in many cases this increase in gain is not sufficient to justify the cost of a transformer, quite apart from the fact that the actual transformer would not be ideal and hence would distort the desired transfer function to some extent.

A similar artifice cannot be applied to the antimetric network. Hence this simple device for accommodating arbitrary impedance levels at the source and load is available only when the desired transfer function satisfies the above-mentioned condition permitting its realization as a symmetrical network.

7 An Alternate Way of Accommodating Both Source and Load Resistances

The design procedure discussed in the previous two articles is (as mentioned there) a means for accommodating a situation in which the source is a voltage with finite nonzero internal resistance. It is also pointed out that this design procedure requires the lossless network to meet a specification of all three of its characterizing impedances, and hence materially limits the achievement of practically desirable structures. In this article, and in the following ones, we will discuss alternative ways in which the required source accommodation may be realized.

Returning our attention to the single-loaded lossless coupling network discussed in Art. 1, the problem of accommodating a voltage source with finite nonzero internal resistance appears to be related to that of controlling the input impedance $Z_1(s)$. Thus, if this impedance were a constant, it would evidently not matter in the least whether the source were a current or a voltage, and, if a voltage, whether a series source resistance were involved or not. The input current would be proportional to the voltage in any case, and hence the same transfer function, apart from a constant multiplier, would necessarily result.

The input impedance cannot, except in a trivial case, be a constant, since this result requires that its real part be a constant, and Eq. 3 shows this real part to be equal to the squared magnitude of the desired transfer impedance. However, we may consider the series addition of an impedance complementary to $Z_1(s)$ in order to secure a resultant input impedance having the desired constant character. Indeed, this process, which is spoken of as "complementing the input impedance" of the single-loaded lossless coupling network, turns out to be a useful way of solving the stated problem, although a solution by this means is not always possible in the contemplated manner since it requires that the conditions for the existence of a complementary impedance be fulfilled.

According to the discussion of this topic in Art. 6, Ch. 8, it is necessary and sufficient if the given input impedance $Z_1(s)$ to the resistance-terminated lossless network is minimum reactive, that is, if it has no j-axis poles. Now the methods of synthesis, from specified z_{22} and z_{12} functions, that we like to use for the realization of the lossless network for the most part do not guarantee an associated z_{11} function fulfilling the residue condition with an equal sign at all poles (see the relevant discussion in Art. 3, Ch. 7), and hence the resultant input impedance $Z_1(s)$ in such a situation is usually not minimum reactive. Where the choice of a particular transfer function permits synthesis of the lossless network on these terms without yielding an input impedance $Z_1(s)$ having j-axis poles, the method of synthesizing a double-loaded lossless network usually leads to an equally desirable structure, so that little if anything is gained by adopting this alternate approach which requires a more elaborate total network in the bargain.

The significant feature about this situation, however, is the fact that the existence of a complementary impedance merely requires the minimum-reactive character of $Z_1(s)$; it does not also require that the corresponding admittance $Y_1(s) = 1/Z_1(s)$ be minimum susceptive. Thus, if on an admittance basis the design of a single-loaded lossless coupling network yields an input admittance $Y_1(s)$ with j-axis poles, it is unlikely that $Y_1(s)$ will have j-axis zeros also. In this situation $Z_1(s)$ usually has no j-axis poles, and a complementary impedance can be constructed. Similarly, although $Z_1(s)$ on an impedance basis is likely to have j-axis poles, it is not likely also to have j-axis zeros. In this case $Y_1(s)$ can usually be complemented by placing an appropriate network in parallel with the input terminals to the lossless network.

The complementary impedance or admittance can be synthesized by any of the variety of methods discussed in Chs. 9 and 10 for the realization of driving-point impedances. One of the methods of transformerless synthesis described in Ch. 10 will probably be desirable in most cases. However, if the Darlington method is used, we observe, in view of Eq. 3, that the complementary network not only has an input impedance that complements the input impedance of the given network but the squared magnitude of its transfer impedance complements that of the given network as well. Thus, if we regard the complementary network as a coupling network in its own right, then its transfer function and that of the given coupling network are complementary in the sense that the sum of their squared magnitudes equals a constant (unity in any normalized situation). If one network is a low-pass filter, for example, the other is a high-pass filter that passes where the low-pass blocks, and vice versa. The combination of the two networks, because of the constant character of the resultant input impedance, is in this situation referred to as a *constant-resistance filter group*.

As an example, let us consider complementing the network of Fig. 2 having the squared transfer impedance 18, which simultaneously is the real part of its input impedance. The real part of the input impedance to the complementary network, or the squared magnitude of its transfer impedance if that network is realized as a lossless two terminal-pair loaded with a 1-ohm resistance, is given by

$$\operatorname{Re}\,[Z_1^c(j\omega)] = |Z_{12}^c(j\omega)|^2 = 1 - \frac{1}{1+\omega^6} = \frac{\omega^6}{1+\omega^6} \tag{120}$$

where the superscript c indicates that the pertinent quantity characterizes the complementary network. The corresponding input impedance $Z_1^c(s)$ can most easily be obtained by subtracting the impedance $Z_1(s)$, Eq. 23, pertinent to this example, from unity. Thus we have at once

$$Z_1^c(s) = \frac{\frac{1}{3}s^3 + \frac{4}{3}s^2 + \frac{2}{3}s}{s^3 + 2s^2 + 2s + 1} \tag{121}$$

which is seen to be the function $Z_1(s)$ with s replaced by $1/s$, as may have been predicted from the form of the real part 120 since it follows from the function 18 through replacing ω by $1/\omega$.

The network, which is that in Fig. 2 with L's replaced by C's having reciprocal values, and vice versa, may alternately be gotten by considering Eq. 120 as a specification of its transfer function. The poles of the corresponding function $Z_{12}{}^c(s)$ are evidently the same as those of the function $Z_{12}(s)$ in Eq. 25; and the numerator polynomial is seen by inspection to be s^3. We have

$$Z_{12}{}^c(s) = \frac{s^3}{s^3 + 2s^2 + 2s + 1} \tag{122}$$

Since this is a case B situation, Eqs. 13 are applicable. They give

$$z_{12} = \frac{s^3}{2s^2 + 1}, \qquad z_{22} = \frac{s^3 + 2s}{2s^2 + 1} \tag{123}$$

The zeros of z_{12} all occur at $s = 0$. Hence the Cauer development of z_{22} yielding a ladder with shunt inductances and series capacitances is appropriate. The continued-fraction expansion of z_{22} for this situation reads

$$z_{22} = \cfrac{1}{\cfrac{1}{2s} + \cfrac{1}{\cfrac{4}{3s} + \cfrac{1}{\cfrac{3}{2s}}}} \tag{124}$$

The corresponding network is shown in Fig. 12a together with that of Fig. 2 which it complements. The complete dual of this situation is

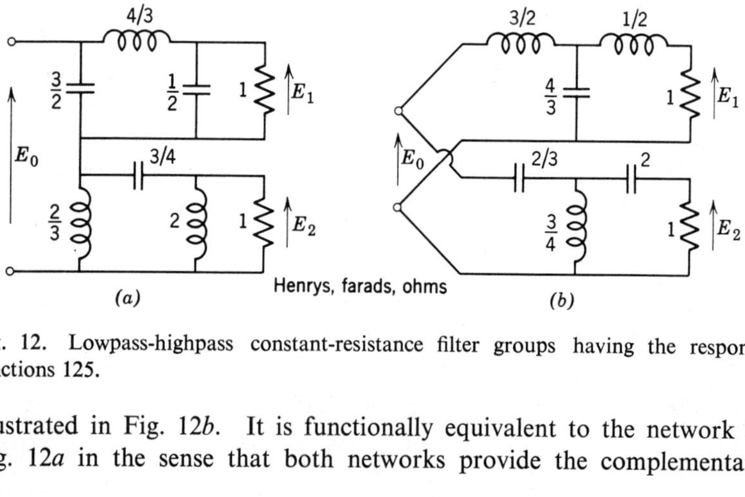

Fig. 12. Lowpass-highpass constant-resistance filter groups having the response functions 125.

illustrated in Fig. 12b. It is functionally equivalent to the network in Fig. 12a in the sense that both networks provide the complementary response functions

$$\left|\frac{E_1}{E_0}\right|^2 = \frac{1}{1 + \omega^6}, \qquad \left|\frac{E_2}{E_0}\right|^2 = \frac{\omega^6}{1 + \omega^6} \tag{125}$$

and present a pure resistance of 1 ohm at their input terminals. They represent simple examples of lowpass-highpass constant-resistance filter groups.

8 Use of the Constant-Resistance Lattice

The lossless coupling network is evidently appropriate in situations where only the magnitude of the transfer function is specified and its phase variation with frequency is unimportant. The phase-shift properties of the lossless coupling network are always midway between minimum and maximum, for we are forced to accept a transfer function with a completely symmetrical zero pattern whether we like it or not. When the zero distribution as well as the magnitude must be controlled, the lossless coupling network is in general not suitable.

A minimum–phase-shift network, for which the transfer function is permitted to have no zeros in the right half-plane, can be realized as a lossless network only in some rather special cases. The network in Fig. 2 for the function given by Eq. 25, for example, may be regarded as being minimum phase since the zeros of this function all lie at infinity. A transfer function having zeros exclusively on the j-axis is realizable as a lossless network, and is in a limiting sense regarded as belonging to the minimum-phase category since the effect of incidental dissipation (as is shown in Art. 6, Ch. 12) is to shift the pole-zero pattern slightly to the left, so that the physical network which results from such a design has all of its transmission zeros in the left half-plane. The lossless networks in Figs. 7, 8, and 12 thus are all regarded as having minimum–phase-shift character.*

These special situations, however, are borderline cases rather than full-fledged examples of minimum–phase-shift networks. The latter must in general be lossy in character, and their synthesis must be accomplished by other means, which proceed from a specification of the complex transfer function rather than from its magnitude.

Foremost as a basis for the realization of transfer functions specfied in this manner is the constant-resistance lattice network discussed in Art. 3, Ch. 6, and shown there in Fig. 8. For convenience, this figure is given

* It is pertinent in this connection to note that a ladder network without mutual inductive coupling must of necessity have a minimum-phase transfer impedance, because its transmission zeros are physically seen to be zeros of shunt impedances or poles of series impedances, which must certainly be restricted to the left half-plane since the branch impedances are p.r. functions.

here as Fig. 13, and the relevant characterizing relations (Eqs. 53, 54, and 55 of Ch. 6) are repeated below

$$z_a z_b = 1 \tag{126}$$

$$\frac{E_2}{E_1} = \frac{E_2}{I_1} = Z_{12} = \frac{1 - z_a}{1 + z_a} \tag{127}$$

$$z_a = \frac{1 - Z_{12}}{1 + Z_{12}} \tag{128}$$

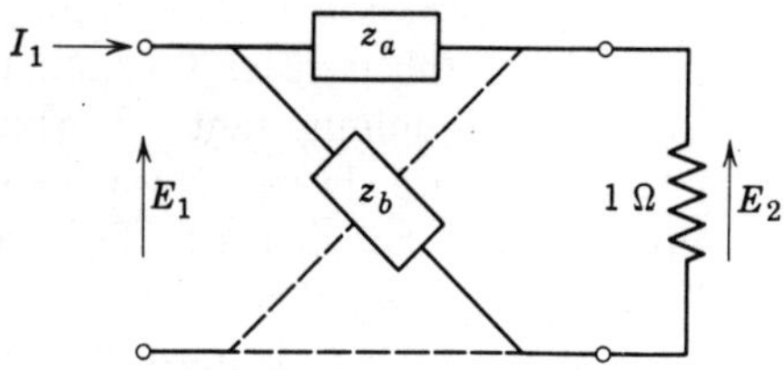

Fig. 13. The symmetrical lattice, which becomes a constant-resistance network if $z_a z_b = 1$ and is then characterized by Eqs. 127 and 128.

Because of the reciprocal relationship between the lattice impedances expressed by Eq. 126, it follows that $E_1/I_1 = 1$; the input impedance is a pure resistance equal to unity. This fact, which accounts for the adjective "constant-resistance" in the designation of this particular lattice network, results in the important property that the over-all transfer impedance of a cascade of such lattices equals simply the product of their individual transfer impedances.

Application of the converse of this property leads us to see that a given complicated transfer impedance may be decomposed in any way into multiplicative factors whose individual lattice realizations, when connected in cascade, represent a synthesis of the given impedance. Although it is analytically just as easy to synthesize the complicated transfer impedance as a single lattice with correspondingly elaborate networks for its impedances z_a and z_b, it is practically much easier to construct and align the separate simpler lattices. The facility with which a single complicated lattice may thus be decomposed into a cascade of simple ones is, from a practical standpoint, one of the outstanding advantages of the constant-resistance lattice.

Before going into a detailed discussion of the synthesis of transfer functions as constant-resistance lattices, it is helpful to have a closer look at the reasons for its special properties from a somewhat different point of view. To this end, suppose we consider the familiar equivalence shown in Fig. 14 which is discussed in Art. 5, Ch. 6, and illustrated there in Fig. 19. The left-hand lattice in Fig. 14 differs from the one in Fig. 13 only in that a 1-ohm resistance is bridged across the input terminals. The resultant input impedance is, on this account, equal to $\frac{1}{2}$ ohm, and the over-all transfer impedance likewise is one-half of its former value as given in Eq. 127.

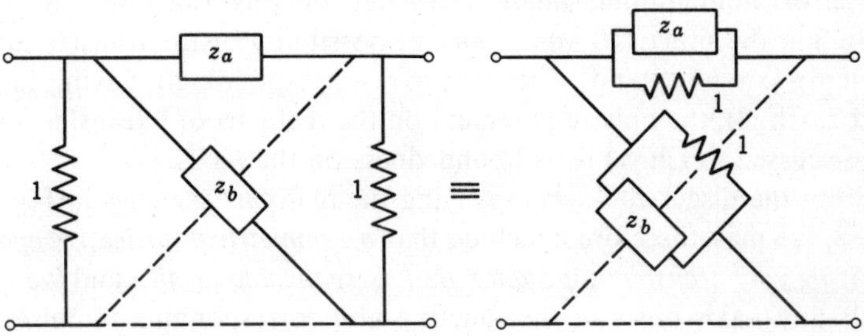

Fig. 14. Constant-resistance lattice with 1-ohm terminations and its equivalent in the form of an unterminated lattice.

If we denote the resultant lattice impedances in the right-hand structure of Fig. 14 by z^*_a and z^*_b, their relation to z_a and z_b is expressed by

$$z^*_a = \frac{z_a}{1 + z_a} \quad \text{and} \quad z^*_b = \frac{z_b}{1 + z_b} \tag{129}$$

If we analogously denote the open-circuit driving-point and transfer impedances of this resultant lattice by z^*_{11} and z^*_{12}, the fundamental Eqs. 44 of Ch. 6 enable us to write

$$z^*_{11} = \tfrac{1}{2}(z^*_b + z^*_a) \tag{130}$$

$$z^*_{12} = \tfrac{1}{2}(z^*_b - z^*_a) \tag{131}$$

The first of these relations is illustrated in Fig. 15. The circuit shown there we recognize as the simple constant-resistance arrangement discussed in Ch. 6 of *ICT* (see Fig. 21, p. 307). The important point is that the

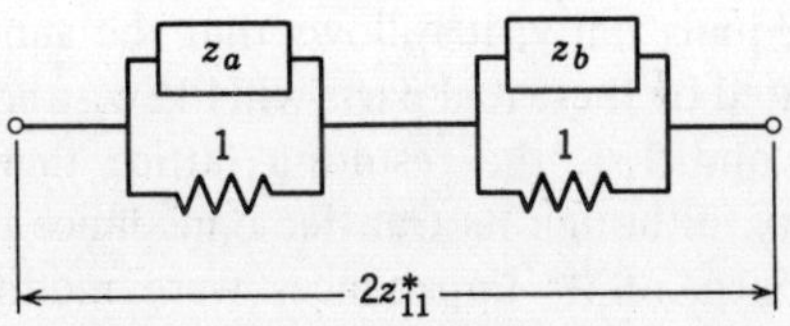

Fig. 15. Interpretation of the input impedance to the lattice of Fig. 14.

impedances z^*_a and z^*_b complement each other. Since complementary impedances are of necessity minimum reactive, and since minimum-reactive impedances are uniquely constructible by ordinary algebraic means from their respective real parts, we can study the situation expressed by Eqs. 130 and 131 by considering only the real-part behavior of the various impedances involved here.

As pointed out in Art. 7 of Ch. 8 where the construction of a transfer impedance from its specified real part is discussed, the characterization of this impedance through its real part has the important advantage that there can be no implied or expressed restriction regarding the minimum-phase or nonminimum-phase character of this function. Moreover, excluding the practically unimportant possibility of the transfer function having j-axis poles and of its real part correspondingly having impulses (Art 2, Ch. 8), the only requirement on the real part of a transfer function (as discussed in Ch. 1) is its boundedness on the j-axis.

From the discussion accompanying the real-part sketches in Fig. 15 of Ch. 8, we may therefore conclude that *a symmetrical lattice is capable of realizing any transfer impedance that is realizable at all*; and we can include any j-axis poles of the function and corresponding impulses in its real part if we wish.

The question as to how a constant-resistance lattice fits into this picture is now easily answered in terms of relations 130 and 131 pertinent to the double-loaded lattice of Fig. 14. If the lattice which we construct according to the discussion accompanying Fig. 15 of Ch. 8 is to have a constant input impedance, then the real parts r_a and r_b of its impedances z_a and z_b must add up to a constant value. It is an easy matter to accomplish this result since (as pointed out in the pertinent discussion) one may add the same arbitrary function $r(\omega)$ to each real part $r_a(\omega)$ and $r_b(\omega)$ without affecting the real part of the transfer impedance which depends only upon the difference $(r_b - r_a)$. Hence (assuming that we have normalized r_a and r_b so that their maximum values may equal but do not exceed unity) we can compute the function

$$r(\omega) = \tfrac{1}{2}(1 - r_a - r_b) \tag{132}$$

and then form the real parts

$$r^*_a = r_a + r \qquad \text{and} \qquad r^*_b = r_b + r \tag{133}$$

Since their sum equals unity, it follows that the sum of the impedances z^*_a and z^*_b generated by these real parts will likewise equal unity. According to Eqs. 130 and 131, the resulting lattice then has the constant-resistance property, although its transfer impedance is the same as it was before the real parts of its impedances were modified as indicated in relations 133. Figure 16 summarizes and illustrates these results.

The important conclusion follows that a constant-resistance lattice is no less general than any other lattice; any realizable transfer impedance can be realized in the form of a constant-resistance lattice.†

† This statement ignores the unimportant possibility of a transfer impedance having j-axis poles. Although such a one is realizable by an ordinary lattice, it is not realizable by a constant-resistance lattice, which requires the j-axis boundedness of the transfer impedance.

Another interesting attitude may be advanced as a result of the interpretation illustrated graphically in Fig. 16. Visualize first a lattice having the impedances generated by the real parts r_a and r_b shown in this figure.

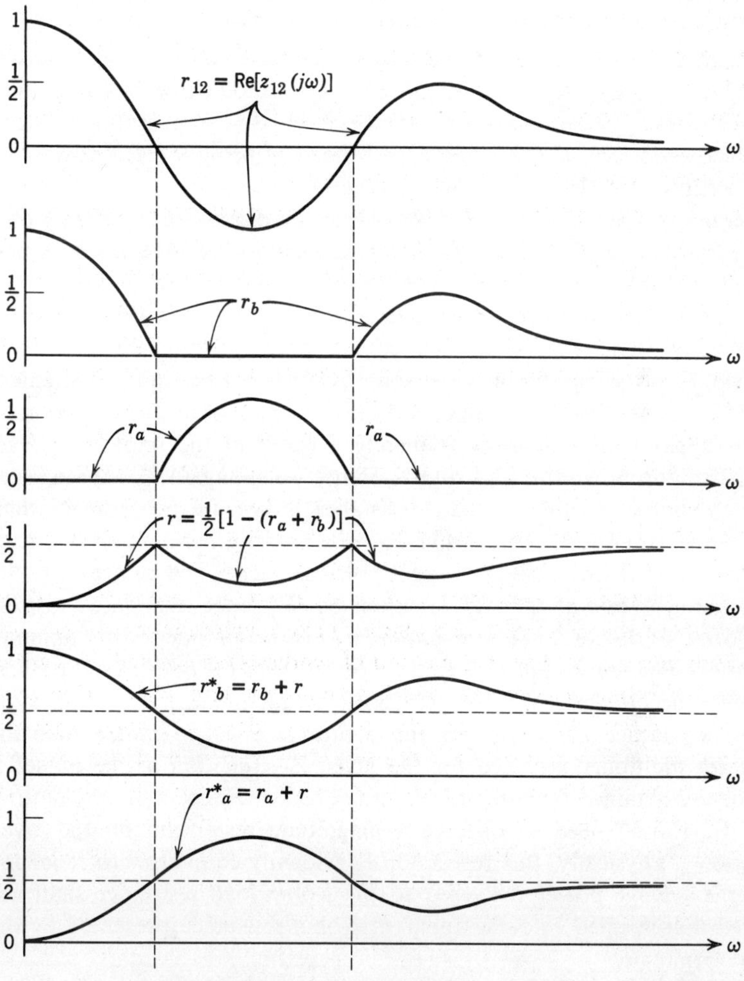

Fig. 16. Showing how the real parts of a constant-resistance lattice impedances z^*_a and z^*_b are constructible from the given real part of an arbitrary transfer impedance z_{12}, thus proving that (apart from a constant multiplier and j-axis poles) any transfer impedance is realizable as a constant-resistance lattice.

This, of course, is not a constant-resistance lattice, since these real parts do not add up to give a constant value. The open-circuit input impedance $z_{11} = z_{22}$ of this lattice is not equal to a constant. However, the complementary impedance to $z_{11} = z_{22}$ is the impedance z, generated by the real part $r = \frac{1}{2}(1 - r_a - r_b)$. If a network realizing z is connected in series at the input and output ends of this lattice, the resultant structure has constant-resistance lattice properties. Hence we may say that a constant-resistance lattice is equivalent to an ordinary lattice whose input impedance is complemented at both terminal pairs. The transfer impedance is obviously not affected by the series addition of the compensating networks. In the constant-resistance lattice, the compensating network is added onto each lattice impedance instead of being located externally. It is a network with built-in compensation.

Returning now to Eqs. 127 and 128, let us discuss the synthesis of a given transfer function Z_{12}. As mentioned previously (Art. 1, Ch. 1, and Art. 3, Ch. 6), the necessary and sufficient condition to insure that z_a is p.r. requires Z_{12} to be analytic in the right half-plane and to have a j-axis magnitude that does not exceed unity. The right half-plane analyticity of the transfer impedance is obviously not a unique requirement of the constant-resistance lattice; and the restriction of its j-axis magnitude to an upper bound of unity is merely a result of the impedance level normalization to a value of 1 ohm. Hence we see again that the realization conditions for the constant-resistance lattice are no stronger than realization conditions for transfer impedances in general (leaving out of consideration the possibility of j-axis poles, as mentioned before).

In the synthesis of transfer functions as resistance-terminated lossless networks, we do not, to be sure, impose a j-axis magnitude restriction upon the given impedance; but that method of synthesis (as pointed out earlier) exercises no control over the resulting constant multiplier. Constant-resistance lattice synthesis, on the contrary, does guarantee that the constant multiplier assumed for the given Z_{12} function at the outset is ultimately obtained from the resulting network. Hence it is only natural that we should need to observe a magnitude restriction on the given function. Physically, this restriction is evidently equivalent to requiring that the average power delivered to the 1-ohm load resistance shall not exceed that injected into the apparent 1-ohm resistance presented by the input terminals.

The j-axis magnitude restriction on the given transfer impedance evidently presents no difficulty, since one can correct this situation, if necessary, by the choice of an appropriate constant multiplier. Realization of a given transfer impedance as a single constant-resistance lattice is thus immediately obtained from Eq. 128 plus the already established methods of synthesis for driving-point impedances.

In this connection it is significant to observe that we need never require mutual inductive coupling, much less closely coupled inductances; and we do not necessarily have to use the Bott and Duffin procedure in order to accomplish this result. We can, for example, use the synthesis method described in Art. 3, Ch. 9, which requires resistance padding of the given impedance. Thus if we subtract a resistance $R_0 < 1$ from the 1-ohm load in Fig. 13, and add an equal series resistance R_0 at the input terminals, the effect, according to the well-known property illustrated in Fig. 18, Ch. 6, is to replace z_a and z_b respectively by $z_a + R_0$ and $z_b + R_0$. The net effect upon the voltage ratio E_2/E_1 is evidently to multiply its value by the factor $(1 - R_0)/(1 + R_0)$. The Miyata method of synthesis (Arts. 2 and 3, Ch. 10), which requires all positive coefficients in the numerator polynomial of the j-axis real part, is also a possibility if sufficient resistance padding can be made available. In this process we are evidently trading gain in the transfer function for simplicity and ease of synthesis. When the cost is too high, or the method fails because of the restriction $R_0 < 1$, or both, then we must either use the Bott and Duffin synthesis or look for other means of simplification.

In this connection we can try the decomposition into a cascade of simpler lattices mentioned earlier. If the stated transfer impedance is written in the factored form

$$Z_{12} = \frac{H(s - s_1)(s - s_3) \cdots (s - s_{2n-1})}{(s - s_2)(s - s_4) \cdots (s - s_{2n})} \tag{134}$$

we can regard it as the product of functions

$$Z_{12} = Z_{12}^{(1)} \times Z_{12}^{(2)} \times \cdots Z_{12}^{(k)} \tag{135}$$

in which each component $Z_{12}^{(i)}$ is composed of a combination of numerator and denominator factors selected from expression 134. In this selection we must observe the obvious requirement that factors representing conjugate complex zeros cannot be separated, and the numerator polynomial in any $Z_{12}^{(i)}$ cannot be of higher degree than the denominator polynomial (otherwise the boundedness condition would obviously be violated). It may additionally be necessary to supply each $Z_{12}^{(i)}$ with a suitable constant multiplier to insure that its j-axis magnitude does not exceed unity, for this condition does not necessarily hold for each component impedance in expression 135 even though it is met by the whole function Z_{12}.

Except for these simple rules of procedure, we can separate the given impedance 134 into multiplicative factors in any way we wish. Each factor determines its own constant-resistance lattice, and the cascade of all of these yields a realization for the stated Z_{12}. This process provides many possibilities, of which some will lead to simpler and practically more acceptable structures for the lattice impedances than others. Also, the resultant multiplier will vary according to how the factorization indicated in Eq. 135 is carried out, since the needed multipliers for the separate

$Z_{12}^{(i)}$ functions will vary according to the particular choice made for their frequency factors.

The number of possible distinct results obtainable from the decomposition of a given transfer impedance may be increased by augmenting expression 134 with single or conjugate pairs of surplus factors of the form $(s - s_k)$. This process is especially useful if we are interested in converting the resultant lattices to equivalent unbalanced forms according to the various techniques discussed in Art. 5, Ch. 6. Since such conversions require that the lattice impedances fulfill certain conditions regarding structure and element values, a considerable degree of flexibility in the initial synthesis is essential. Although the realization of a contemplated transfer function in unbalanced form is always possible, the conditions which such realization imposes may not be practically acceptable. The following discussion is concerned with some of the detailed aspects of this problem.

We shall consider several simple Z_{12} functions and determine the conditions under which their corresponding lattices are convertible to unbalanced forms. We begin with the function

$$Z_{12} = \frac{k_1}{s + a} \tag{136}$$

for which

$$z_a = \frac{s + (a - k_1)}{s + (a + k_1)} \tag{137}$$

So long as

$$k_1 \leq a \tag{138}$$

this is a realizable RL impedance. The corresponding networks for z_a and z_b are shown in Fig. 17. According to the methods discussed in Art. 5,

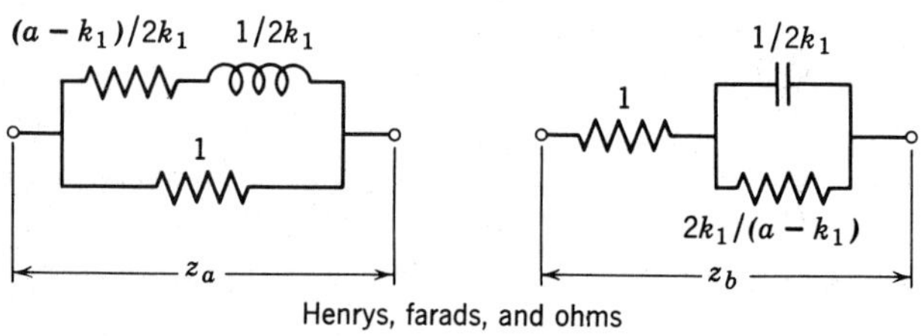

Fig. 17. Networks realizing the component impedances in a constant-resistance lattice having the transfer impedance 136.

Ch. 6, this lattice is immediately transformable to the unbalanced form shown in Fig. 18. Condition 138 on the constant multiplier of Z_{12} merely expresses the requirement that its maximum absolute j-axis value, which occurs at $s = 0$, shall not exceed unity.

It is significant to point out that the lattice in this example is representative of a special class of constant-resistance lattices having the structure shown in Fig. 19*a*. The component impedances z_1 and z_2 fulfill the reciprocal condition

$$z_1 z_2 = 1 \tag{139}$$

and the conversion of this lattice to the equivalent bridged-T form shown in Fig. 19*b* is straightforward.

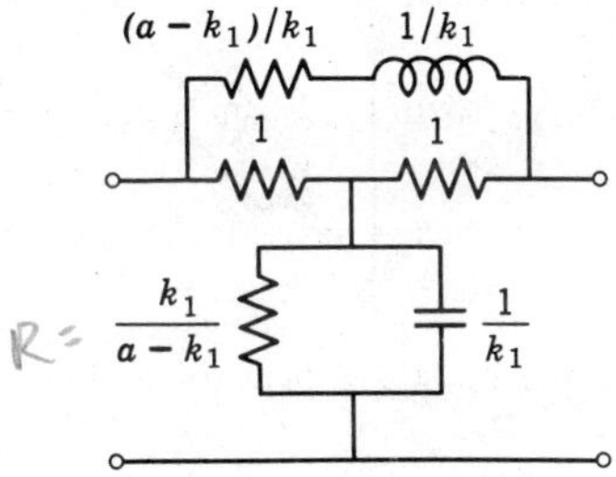

Henrys, farads, and ohms

Fig. 18. Unbalanced equivalent form for a constant-resistance lattice having the transfer impedance 136.

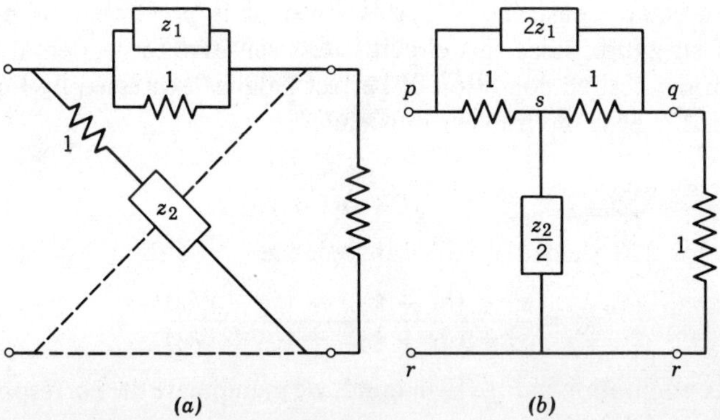

Fig. 19. Restricted class of constant-resistance lattices that always possess simple bridged-T equivalents.

An interesting feature of this bridged-T circuit becomes apparent if we redraw it in the Wheatstone bridge form shown in Fig. 20. Because of relation 139, this bridge circuit is balanced. Hence an excitation across the points *p*-*r* produces no current in the resistance *q*-*s*. This resistance in the bridge circuit of Fig. 20 or in the bridged-T circuit if Fig. 19*b* is superfluous and could be dispensed with. If we remove it in the circuit of

Fig. 20, the latter is clearly seen to reduce to the simple constant-resistance circuit discussed in Ch. 6 of *ICT* and shown there in Fig. 21*a* on p. 307.

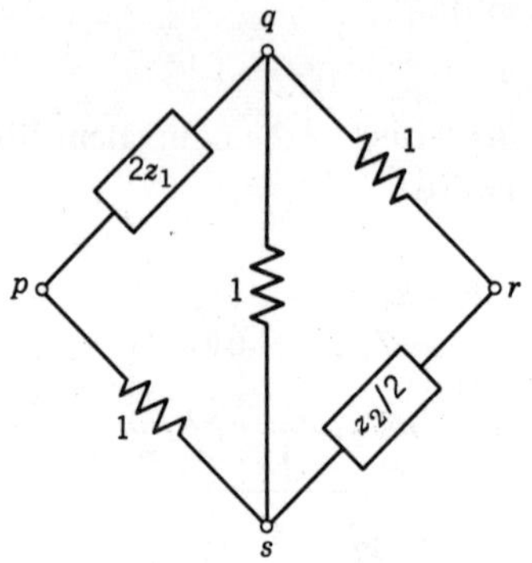

Fig. 20. Redrawn version of the bridged-T network of Fig. 19 reveals it to be a Wheatstone bridge.

Whenever the impedances z_a and z_b have the special forms shown in the lattice of Fig. 19*a*, and condition 139 is fulfilled, we see that this lattice is actually equivalent to the much simpler constant-resistance circuit with which we are familiar from our elementary circuit theory studies. Although we could construct it in this form, it is preferable to use the bridged-T structure, since this circuit is less sensitive to incidental deviations from the desired condition of perfect balance expressed by Eq. 139.

We consider next the transfer impedance

$$Z_{12} = \frac{k_2(s + a)}{s^2 + b_1 s + b_0} \tag{140}$$

for which Eq. 128 yields the lattice impedance

$$z_a = \frac{s^2 + (b_1 - k_2)s + (b_0 - ak_2)}{s^2 + (b_1 + k_2)s + (b_0 + ak_2)} \tag{141}$$

With the configuration in Fig. 19 in mind, we manipulate the corresponding admittance as follows

$$y_a = 1 + \frac{2k_2(s + a)}{s^2 + (b_1 - k_2)s + (b_0 - ak_2)} \tag{142}$$

Taking $a > 0$ for granted, the p.r. character of the second term in this expression requires first of all that

$$k_2 \leq b_1 \qquad \text{and} \qquad k_2 \leq b_0/a \tag{143}$$

and, according to conditions 19 of Ch. 9, that

$$b_1 - k_2 \geq a, \qquad \text{or } k_2 \leq b_1 - a \tag{144}$$

These conditions limit the constant multiplier of Z_{12} in Eq. 140 to a smaller value than necessary to insure the p.r. character of z_a. We are paying for a desired form of realization by accepting less gain. The resulting circuit for z_a is shown in Fig. 21. Construction of the pertinent bridged-T network according to the principles illustrated in Fig. 19 follows without difficulty.

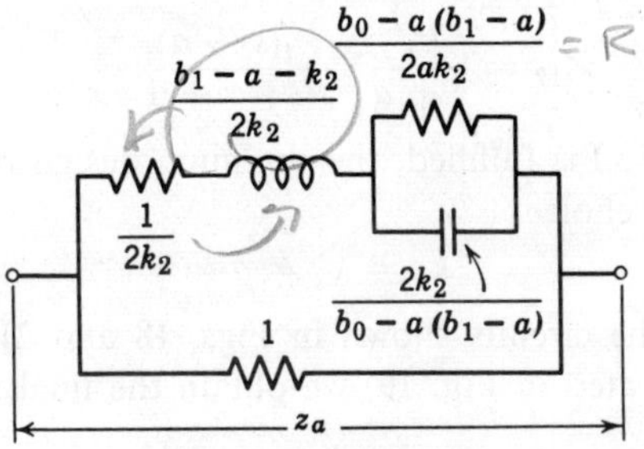

Fig. 21. Constant-resistance lattice impedance yielding the transfer function 140 and permitting transformation to the bridged-T form of Fig. 19.

Since a product of the two transfer impedances 136 and 140 yields the function

$$Z_{12} = \frac{k}{s^2 + b_1 s + b_0} \tag{145}$$

we may conclude that a transfer impedance of the form

$$Z_{12} = \frac{k}{(s - s_1)(s - s_2) \cdots (s - s_n)} \tag{146}$$

can always be realized as an unbalanced constant-resistance network provided we are willing to accept a sufficiently small value for the multiplier k to make such realization possible.

In the interest of making the multiplier k in the function 145 as large as possible, suppose we tentatively ignore conditions 143 and consider only 144 plus condition 138 for k_1. We then have

$$k = k_1 k_2 = ab_1 - a^2 \tag{147}$$

which yields the maximum value

$$k_{max} = (b_1/2)^2 \tag{148}$$

and we have

$$k_1 = k_2 = a = b_1/2 \tag{149}$$

If we now consider conditions 143, we see that the first is always fulfilled, and the second one is if

$$b_0 \geq (b_1/2)^2 \tag{150}$$

which is likely to be satisfied in most cases

Let us illustrate with the function 25 having the squared magnitude 18 and the realization shown in Fig. 12, which also provides a pure resistance at its input terminals. The transfer impedance in this example is written in the form

$$Z_{12} = \frac{k_1}{s+a} \times \frac{k_2(s+a)}{s^2+s+1} \times \frac{1}{s+1} \tag{151}$$

Since condition 150 is fulfilled, the optimum design results, according to Eqs. 149, for the choice

$$k_1 = k_2 = a = \tfrac{1}{2} \tag{152}$$

Making use of the circuits shown in Figs. 18 and 21, together with the conversion illustrated in Fig. 19, we obtain the final network realization given in Fig. 22.

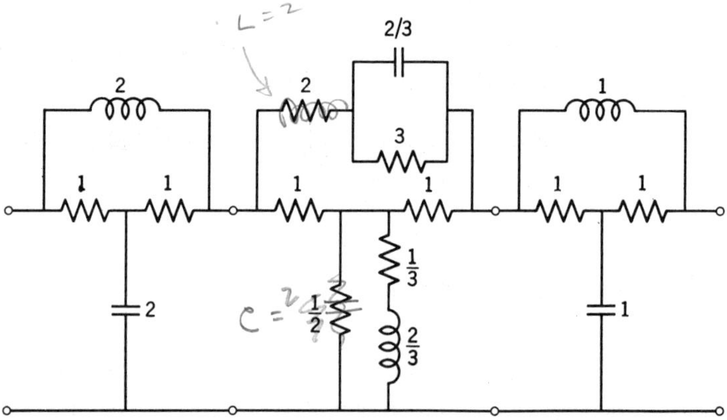

Fig. 22. Realization of the transfer impedance 25 (with the squared magnitude 18) as a constant-resistance network in contrast with the circuit of Fig. 12 which realizes the same transfer function but with the resistive input property in the forward direction only.

Comparison with the networks of Fig. 12 reveals that the same number of reactive elements are involved. The number of resistances is appreciably larger in the realization of Fig. 22, and the over-all multiplier $k = \frac{1}{4}$, whereas in the circuit of Fig. 12 it has the value unity.

Before deciding in favor of the circuit in Fig. 12, one should consider two things: (*a*) resistances are relatively inexpensive as compared with reactive elements; (*b*) the circuit of Fig. 22 has the constant-resistance property in both directions, whereas the circuit of Fig. 12 has it only in the forward direction. This fact may, under certain circumstances, lead one to choose the design in Fig. 22 in spite of its lower gain factor.

The next transfer impedance to be considered is the ratio of quadratic factors given by

$$Z_{12} = k \times \frac{s^2 + 2as + \omega_a^2}{s^2 + 2bs + \omega_b^2} \tag{153}$$

yielding the lattice impedance

$$z_a = \frac{(1 - k)s^2 + 2(b - ak)s + \omega_b^2 - k\omega_a^2}{(1 + k)s^2 + 2(b + ak)s + \omega_b^2 + k\omega_a^2} \tag{154}$$

Suppose we try setting $k = 1$. The impedance z_a then assumes the form of z_ν in Eq. 16, Ch. 9; and the condition that it be p.r. is expressed by

$$2(b^2 - a^2) \geq \omega_b^2 - \omega_a^2 \geq 0 \tag{155}$$

The circuit which realizes z_a is given in Fig. 23, and its parameters are determined from

$$C = 1/(b - a) \tag{156}$$

$$G = \frac{2(b^2 - a^2) - (\omega_b^2 - \omega_a^2)}{2(b - a)^2} \tag{157}$$

$$\frac{1}{R} - G + \frac{\omega_b^2 + \omega_a^2}{\omega_b^2 - \omega_a^2} \tag{158}$$

and

$$R/L = (\omega_b^2 - \omega_a^2)/2(b - a) \tag{159}$$

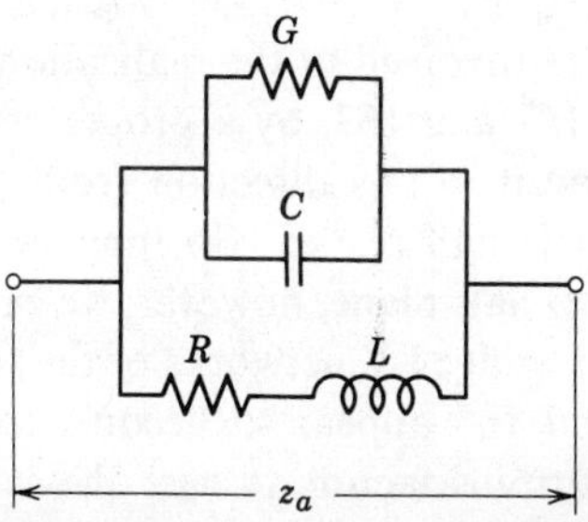

Fig. 23. Circuit realizing z_a in Eq. 154 for $k = 1$. Its parameter values are given in Eqs. 156 through 159.

We now observe that the transformation illustrated in Fig. 19 again becomes available if $G \geq 1$, which leads to the condition

$$4a(b - a) \geq \omega_b^2 - \omega_a^2 \tag{160}$$

Since, in any event, b must be larger than a, we see that this last condition is stronger than that required by the first inequality in 155. Hence if the second inequality in 155 plus condition 160 are fulfilled, realization of the lattice is possible and leads to an unbalanced bridged-T equivalent.

The second inequality in 155 requires that the pole in the given transfer impedance Z_{12} be farther from the origin than the zero; and condition 160 adds that the pole must also be farther from the j-axis by a certain amount than the zero. If these requirements are not acceptable, then we must return to Eq. 154 and consider values of k less than unity.

Switching to an admittance basis, we obtain

$$y_a = 1 + \frac{2k}{1-k} \times \frac{s^2 + 2as + \omega_a^2}{s^2 + [2(b-ak)/(1-k)]s + [(\omega_b^2 - k\omega_a^2)/(1-k)]} \tag{161}$$

In order for the second term in this expression to be realizable in simple terms (according to the discussion in Art. 3, Ch. 9, leading to conditions 47 and 48), the following conditions are found to apply

$$\omega_b^2 - k\omega_a^2 \geq 0 \tag{162}$$

$$4a(b - ak) \geq |\omega_b^2 - \omega_a^2| \tag{163}$$

It is evident from the form of expression 161 that the lattice transformation shown in Fig. 19 again applies. The present conditions 162 and 163, compared with 155 and 160 obtained for $k = 1$, show some additional lattitude in the permissible pole-zero pattern of Z_{12}, although the price of obtaining this result is twofold: (*a*) we sacrifice gain because k is less than unity; and (*b*) we need a more elaborate network because realization of the second term in 161 requires twice as many reactive elements and twice as many resistances as are involved in the realization of z_a in Fig. 23.

Although conditions 162 and 163, by a proper choice of k, permit the pole of Z_{12} to be displaced in any direction from the zero, the latter is restricted to lie in the left half-plane. So long as the zeros of a given Z_{12} function are in the left half-plane, however, we can always synthesize it by means of a cascade of bridged-T networks of the sort considered above, for if conditions 162 and 163 appear to become too restrictive, we can always inject a pair of surplus factors to ease the situation.

In order to accommodate right half-plane zeros in a given Z_{12} function and still obtain an unbalanced equivalent to the constant-resistance lattice, we return to the functions 153 and 154, and set $k = 1$, $\omega_b = \omega_a = \omega_0$. We then get

$$\begin{aligned} z_a &= \frac{(b-a)s}{s^2 + (b+a)s + \omega_0^2} \\ &= \frac{1}{[s/(b-a)] + [(b+a)/(b-a)] + [\omega_0^2/(b-a)s]} \end{aligned} \tag{164}$$

The corresponding networks for the lattice impedances are those in Fig. 24.

Conversion of this lattice to an unbalanced form follows the pattern used in Art. 5, Ch. 6, to convert the *RC* lattice given there in Fig. 23 to the twin-T circuit of Fig. 26. The resultant unbalanced network in the present instance becomes a "triple-T." Denoting the parameters in z_a and z_b by R_a, L_a, C_a, and R_b, L_b, C_b respectively, the impedance z_b is represented by a parallel combination of three proportional series *RLC* branches as shown in Fig. 25 in which α and β are suitably chosen positive

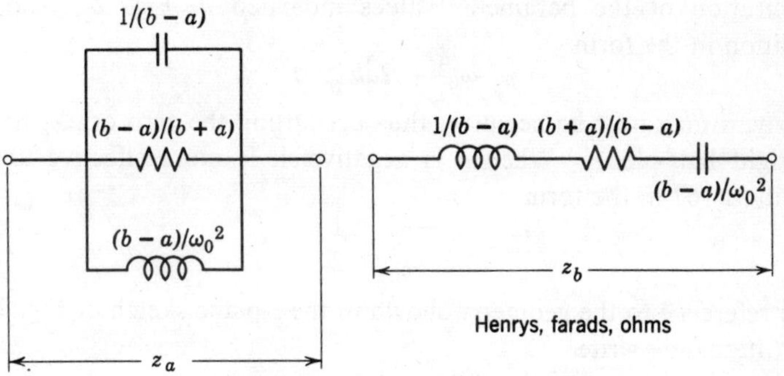

Fig. 24. Networks for the realization of impedance 164 and its reciprocal. The pertinent lattice yields the transfer impedance 153 for $k = 1$ and $\omega_a = \omega_b = \omega_0$.

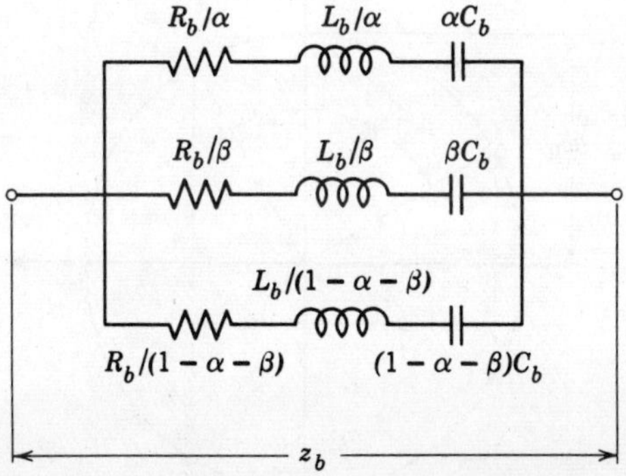

Fig. 25. Representation of z_b in Fig. 24 preparatory to transformation of the pertinent lattice to an equivalent trip-T circuit.

numerics less than unity. The lattice with z_a of Fig. 24 and z_b of Fig. 25 is now regarded as a parallel combination of three lattices, corresponding to each of the three elements in the z_a circuit respectively associated with one of the three parallel components of z_b. The condition that each lattice be convertible to a transformerless T circuit reads respectively

$$R_a \leq R_b/\alpha, \qquad L_a \leq L_b/\beta, \qquad C_a \geq (1 - \alpha - \beta)C_b$$

or

$$\alpha \leq R_b/R_a, \quad \beta \leq L_b/L_a, \quad (1 - \alpha - \beta) \leq C_a/C_b \tag{165}$$

Simultaneous fulfillment of these conditions is possible if

$$\frac{R_b}{R_a} + \frac{L_b}{L_a} + \frac{C_a}{C_b} \geq 1 \tag{166}$$

which is a logical extension of the condition expressed by Eq. 69 in Ch. 7. Substitution of the parameter values indicated in Fig. 24 yields this condition in the form

$$\omega_0^2 + 2ab \geq 0 \tag{167}$$

Clearly, a may now be negative, thus permitting the zero of Z_{12} to be in the right half-plane. When a is negative, it is more effective to write condition 167 in the form

$$\frac{-ab}{\omega_0^2} \leq \frac{1}{2} \tag{168}$$

With reference to the geometry shown in the s-plane sketch of Fig. 26, we may alternately write

$$\sin \theta_1 \sin \theta_2 \leq \tfrac{1}{2} \tag{169}$$

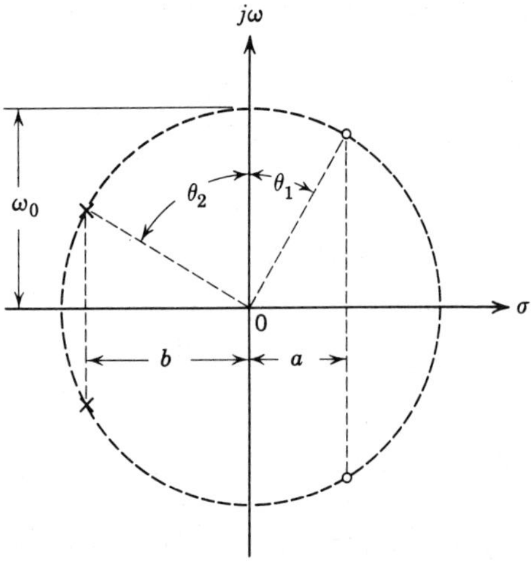

Fig. 26. Graphical interpretation of the conditions 170 which are required if the transfer function having this pole-zero pattern is to be realizable as a triple-T circuit without mutual inductive coupling.

Since b must be larger than the absolute value of a (whether positive or negative) in order that the resistances in z_a and z_b of Fig. 24 may remain positive, we must couple condition 169 with the requirement

$$\theta_1 \leq \theta_2, \quad \text{or } \sin \theta_1 \leq \sin \theta_2 \tag{170}$$

It thus becomes clear that the angle θ_1 must be less than 45° and can equal this value only in a limiting sense.

We can, of course, obtain zeros closer to the real axis than this limiting value permits by allowing one of the T circuits in the triple-T to involve mutual inductive coupling. The condition $L_a \leq L_b/\beta$ in 165 then drops out. The corresponding lattice conversion is similar to that shown in Fig. 21 of Ch. 6 if we ignore the capacitance C_1 shown there. The required mutual inductive coupling is negative and the coupling coefficient has the value

$$k = \frac{\beta - (L_b/L_a)}{\beta + (L_b/L_a)} = \frac{\beta - \omega_0^2/(b-a)^2}{\beta + \omega_0^2/(b-a)^2} \tag{171}$$

The other two conditions in 165 require

$$1 - \beta - (C_a/C_b) \leq \alpha \leq (R_b/R_a), \quad \text{or } \beta \geq 1 - (R_b/R_a) - C_a/C_b) \tag{172}$$

which yields

$$\beta \geq \frac{-4ab - \omega_0^2}{(b-a)^2} \tag{173}$$

The smallest value for the coupling coefficient in 171 occurs for the smallest β value in 173; thus we find

$$k_{\text{min}} = 1 - \omega_0^2/(-2ab) \tag{174}$$

This has its largest value when both pole and zero of Z_{12} fall upon the real axis, for then $-a = b = \omega_0$ and $k = \frac{1}{2}$. Even for this limiting pole-zero pattern the required coupling does not become close.*

In summary we can say that the constant-resistance lattice offers a realization for any desired transfer impedance having left half-plane poles, and zeros anywhere in the s-plane, including the point at infinity. A reduction of the lattice to an equivalent unbalanced form is always possible provided we can accept the loss in gain which this conversion implies. Mutual inductive coupling in this unbalanced equivalent structure can be avoided if all right half-plane zeros are less than 45° from the j-axis. Where mutual inductive coupling is required, the coupling coefficient need never exceed 50 per cent.†

* The absolute value of a may, of course, exceed that of ω_0, corresponding to a pair of zeros at discrete points on the positive real axis with the geometric mean value ω_0. The coupling becomes close only in the limiting situation for which one of these zeros lies at $s = 0$ and the other at $s = \infty$.

† This statement is based upon the practical attitude that the extreme condition to be met is that for which the right half-plane zeros are required to be 90° from the j-axis. In the design of delay networks, which commonly require right half-plane zeros (see Art. 9, Ch. 14), this attitude is certainly justified.

PROBLEMS

1. The function

$$Z_1(s) = \frac{s^3 + s^2 + 2s + 1}{s^4 + s^3 + 3s^2 + 2s + 1}$$

is the driving-point impedance of a lossless two terminal-pair network terminated in 1 ohm. Determine the corresponding transfer impedance $Z_{12}(s) = E_2/I_1$.

2. Given the minimum-phase transfer impedance

$$Z_{12}(s) = \frac{s + 1}{s^2 + s + 1}$$

find a transfer impedance $Z^*_{12}(s)$ having the same magnitude as $Z_{12}(s)$ for $s = j\omega$, and realizable in terms of a lossless network terminated in 1 ohm, with a minimum-reactive input impedance $Z_1(s)$. Find the network and $Z_1(s)$.

3. The open-circuit transfer impedance of a symmetrical lossless two terminal-pair is $z_{12} = 1/(s^3 + 3s)$. When terminated in 1 ohm, the input impedance is to be minimum reactive. Find this input impedance and synthesize the network.

4. Determine lossless ladder networks whose transfer impedances when terminated in 1 ohm are respectively:

$$Z_{12} = \frac{1}{(s + 1)^4} \qquad \text{and} \qquad Z_{12} = \frac{s^4}{(s + 1)^4}$$

5. The desired transfer characteristic of a lossless network terminated in a 1-ohm resistor is specified to be

$$|Z_{12}| = \left|\frac{E_2}{I_1}\right| = \frac{1}{\sqrt{1 + \omega^8}}$$

(*a*) Using the relation Re $[Z(j\omega)] = |Z_{12}|^2$, find $Z(s)$ and synthesize the lossless network by Darlington's method.

(*b*) Form $Z_{12}(s)$ directly and from this function synthesize the lossless network. Compare results with part (*a*).

6. Show that the impedance complementary to that found in part (*a*) of Prob. 5 is physically realizable, and that the associated Z_{12} function is complementary to the one given in Prob. 5 in the sense that these two transfer functions differ by the change of variable $\omega \to 1/\omega$ and hence represent characteristics that are respectively low-pass and high-pass. Synthesize the network and, by combining it with that found in Prob. 5, form a low-pass, high-pass, constant-resistance filter group.

7. A lossless network is to be driven by a voltage source having an internal resistance of 2 ohms, as shown in the accompanying sketch. Synthesize N so that the transfer voltage ratio has the following squared magnitude for $s = j\omega$

$$\left|\frac{E_2}{E_1}\right|^2 = \frac{(1 + \omega^2)^2}{4 + \omega^4}$$

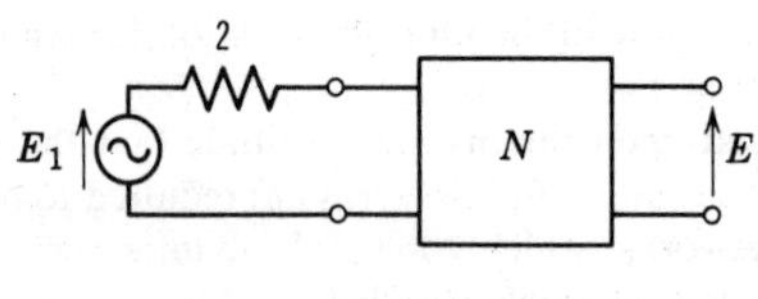

Prob. 7.

8. A desired transfer impedance is to have the squared magnitude

$$|Z_{12}(j\omega)|^2 = \frac{1}{(1 + \omega^2)^4}$$

Find its realization in the form of a lossless unbalanced ladder network terminated in 1 ohm such that the input impedance possesses a complement. Determine the analytic expression for this input impedance and for its complement. Do not synthesize the complementary network but state whether it is likely to be realizable in the form of a lossless unbalanced ladder without mutual coupling, terminated in 1 ohm.

9. A lossless two terminal-pair network has the transfer impedance

$$z_{12} = \frac{3}{s(s^2 + 2)(s^2 + 4)}$$

When terminated in a 1-ohm resistance, the transfer impedance $Z_{12}(s)$ has the properties

$$Z_{12}(0) = 1, \qquad Z_{12}(j\sqrt{2}) = -3, \qquad Z_{12}(j2) = 1$$

Find $Z_{12}(s)$ as a quotient of polynomials and synthesize the network in a simple and practical form.

10. $Z(s) = [P(s)/(s + 1)^3]$ is the *minimum-reactive* driving-point impedance at one terminal pair of a lossless *symmetrical* two terminal-pair network terminated in a 1-ohm resistance. The problem is to determine the lossless two terminal-pair network without specific knowledge of the numerator polynomial $P(s)$.

(*a*) State whether the problem has a unique solution. If not, is the number of possible solutions finite (apart from those differing merely by a constant multiplier), and if so how many distinct solutions are there?

(*b*) Synthesize a possible solution and determine the corresponding zeros of $Z(s)$.

11. The transfer admittance Y_{12} of a network has the pole-zero pattern shown in the sketch on page 496.

(*a*) Is the pertinent network minimum phase shift?

(*b*) Could the network be a lossless two terminal-pair terminated in resistance (1 ohm)?

(*c*) Find such a lossless two terminal-pair having the same $|Y_{12}|$ for $s = j\omega$ if $Y_{12}(0) = \frac{1}{2}$.

(*d*) Show in an s-plane sketch the pole-zero pattern for the Y_{12} of this network. Is it minimum phase shift?

12. A network to be realized for practical reasons in the form shown in the figure on page 496 is to have the transfer impedance

$$Z_{12} = \frac{As}{s^4 + s^3 + 12s^2 + 7s + 30} = \frac{z_{12}}{1 + z_{22}}$$

The load must be 1 ohm and the shunt capacitances must be equal. Synthesize a lossless ladder network with a z_{22} corresponding to the above Z_{12} so as to control the zeros of z_{12} as required. Convert the ladder network into the practical equivalent form shown in the figure, and evaluate A, M, L_1, L_2, and C.

13. $z_{12} = 1/(s^3 + 3s)$ is the open-circuit transfer impedance of a *symmetrical* lossless two terminal-pair network which, when terminated in a 1-ohm resistance, has an input impedance $Z_1(s)$ that is minimum reactive. Find $Z_1(s)$ and synthesize the network.

14. With reference to Prob. 3, Ch. 8, determine the analytic expressions for impedances z_a and z_b of a constant-resistance lattice which

(*a*) has the same transfer impedance as the lossless network of that problem.

(*b*) has the same $|Z_{12}|$ but is a minimum–phase-shift network.

15. The transfer impedance Z_{12} of a lossless network terminated in a 1-ohm load has poles at $s = -1$, $s = -1 - j1$, and $s = -1 + j1$. It has a double zero at $s = 0$ and a simple one at $s = \infty$. A constant multiplier is arbitrary. Find that network having a minimum-reactive input impedance, and put it into the simplest form.

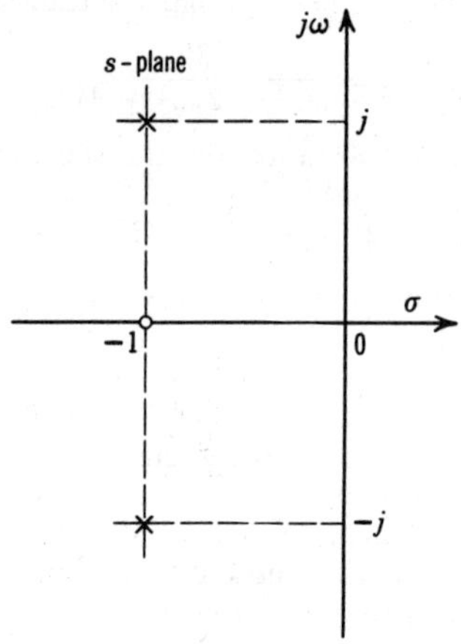

Prob. 11.

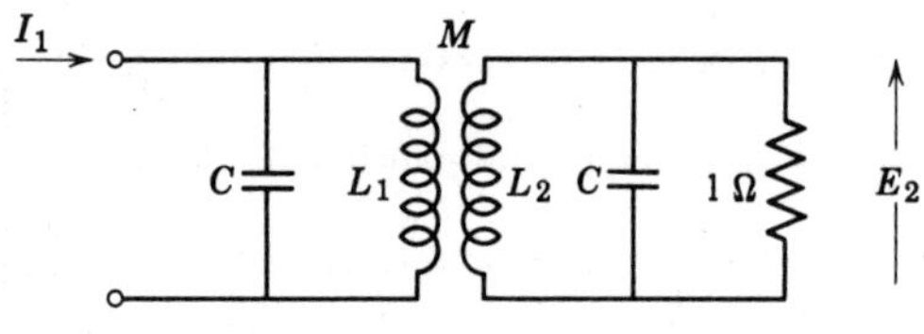

Prob. 12.

16. Given the network shown here:

(*a*) Find another one having the same transfer impedance $Z_{12}(s) = E_2/I_1$ but with a minimum-reactive input impedance $Z_1(s)$.

(*b*) Determine a network for the difference $Z(s) - Z_1(s)$, and, by placing this in series with the input to the network found in part (*a*), obtain an exact equivalent to the given network.

17. The transfer impedance of a lossless two terminal-pair network terminated in 1 ohm is

$$Z_{12}(s) = \frac{g(s)}{4s^3 + 3s^2 + 4s + 2}$$

in which $g(s)$ is an even polynomial. The input impedance to this network is to be minimum reactive and the lossless network should be symmetrical and unbalanced with three distinct terminals. Find the network and the input impedance function.

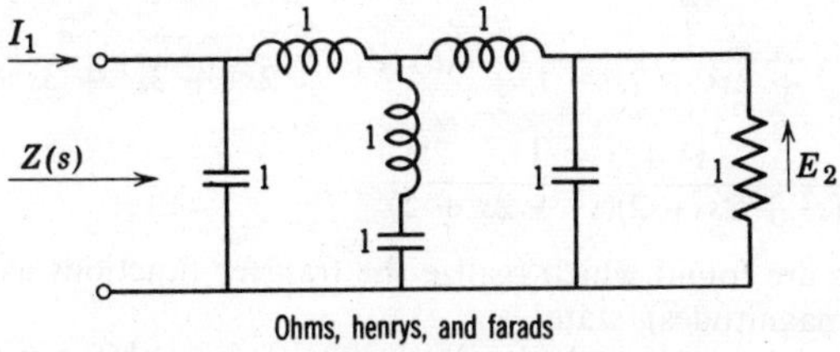

Prob. 16.

18. It is desired to form a low-pass, high-pass, constant-resistance filter group, the low-pass $|Z_{12}(j\omega)|^2$ being that of the network shown. Determine an over-all network meeting this specification and find its input impedance.

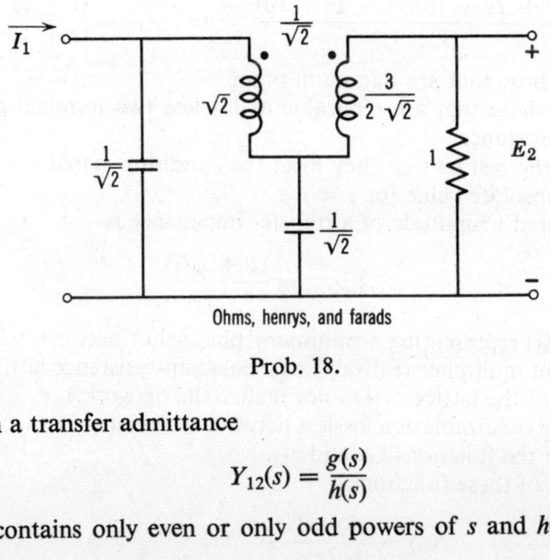

Prob. 18.

19. Given a transfer admittance

$$Y_{12}(s) = \frac{g(s)}{h(s)}$$

where $g(s)$ contains only even or only odd powers of s and $h(s)$ is a Hurwitz polynomial:

(*a*) Show that we can always represent this transfer function as

$$Y_{12}(s) = Y_{12}^{(1)} - Y_{12}^{(2)}$$

in such a way that $Y_{12}^{(1)}$ and $Y_{12}^{(2)}$ are realizable by simple, lossless, unbalanced ladder networks involving no mutual couplings. How can these networks be combined to obtain a realization for $Y_{12}(s)$? Illustrate with the function

$$Y_{12}(s) = \frac{(s^2 - 1)^2}{(s + 1)^4} = \left(\frac{s - 1}{s + 1}\right)^2$$

(*b*) Show that this same Y_{12} function may alternately be realized as a cascade of two identical constant-resistance lattices and convert this network into an equivalent unbalanced form.

20. The following transfer impedances are to be realized in *magnitude* by means of *nondissipative* networks terminated in a 1-ohm resistance. In which cases will the resulting network have minimum–phase-shift properties?

$$(a)\ Z_{12} = \frac{1}{s^3 + 2s^2 + 2s + 1} \qquad (b)\ Z_{12} = \frac{s^2 + 2}{2s^3 + 3s^2 + 3s + 2}$$

$$(c)\ Z_{12} = \frac{s^2 + s + 1}{(s^2 + 2s + 2)(s^2 + 2s + 5)}$$

If networks are found which realize the transfer functions as they stand (not merely their magnitudes), state:

(i) Which of these networks have minimum–phase-shift properties.

(ii) Which of the functions may be realized by nondissipative networks terminated in a 1-ohm resistance.

21. From the following transfer impedance functions

$$(a)\ Z_{12} = \frac{s^2 + 2s + 10}{(s + 1)^4} \qquad (b)\ Z_{12} = \frac{s^2 - 2s + 2}{(s + 1)^4}$$

$$(c)\ Z_{12} = \frac{(s^2 + 2s + 10)(s^2 - 2s + 10)}{(s + 1)^6} \qquad (d)\ Z_{12} = \frac{(s + 2)^4}{(s + 1)^6}$$

(i) Pick out those that are minimum phase.

(ii) Pick out those that are realizable as lossless two terminal-pair networks terminated in resistance.

(iii) Modify the rest so that they meet the condition stated under (ii) but do not change in absolute value for $s = j\omega$.

22. The squared magnitude of a transfer impedance is

$$|Z_{12}|^2_{s=j\omega} = \frac{(1 + \omega^2)^2}{4 + \omega^4}$$

(*a*) Find $Z_{12}(s)$ representing a minimum–phase-shift network with the largest possible constant multiplier realizable as a constant-resistance lattice, and determine z_a and z_b of the lattice. (Do not realize the network.)

(*b*) Find $Z_{12}(s)$ realizable as a lossless network terminated in a 1-ohm resistance and write down the functions z_{22} and z_{12}.

23. For each of these functions

$$(a)\ \frac{s + 1}{(s + 2)(s + 3)} \qquad (b)\ \frac{s^2 - 3}{s^2 + 2s + 2}$$

$$(c)\ \frac{s^2 - s}{s^2 + 2s + 2} \qquad (d)\ \frac{(s^2 + 2)^2}{2s^4 + 2s^3 + 5s^2 + s + 2}$$

answer the following questions:

(i) Is the function realizable as the transfer impedance $Z_{12}(s)$ of a lossless two terminal-pair network terminated in resistance? If not, why not?

(ii) If the answer to question (i) is negative, and if there exists a different function $Z^*_{12}(s)$ which has the same magnitude as the given one for $s = j\omega$ and which *is* realizable as the transfer impedance of a lossless network terminated in resistance, find that function $Z^*_{12}(s)$.

24. The squared magnitude of a given transfer impedance is

$$|Z_{12}(j\omega)|^2 = \frac{1 + 4\omega^2}{(1 + \omega^2)^2}$$

Find:

(*a*) A minimum-phase transfer impedance having this magnitude.

(*b*) A $Z_{12}(s)$ realizable as a lossless network terminated in a 1-ohm load.

Synthesize both networks.

25. A lossless two terminal-pair network terminated in 1-ohm resistances at both ends has the transfer function

$$t = \frac{2E_2}{I_{10}} = \frac{1}{(1 + s)^2}$$

Find the network.

26. A lossless coupling network with 1-ohm resistances at both ends is to have a transfer function $t(s)$ with the squared magnitude

$$|t(j\omega)|^2 = \frac{\omega^4 - \omega^2 + 1}{\omega^4 + 2\omega^2 + 1}$$

(*a*) Is a realization in the form of a simple unbalanced ladder without mutual coupling possible? Give reason.

(*b*) Is a realization in the form of a symmetrical network possible? Why or why not?

(*c*) Is the transfer function $t(s)$ minimum phase? Explain.

(*d*) Synthesize any network to realize the specified $|t(j\omega)|^2$.

27. Given $f(s) = s^2/(s + 1)^2$:

(*a*) If this is the transfer function $t = 2E_2/I_{10}$ of a lossless two terminal-pair network terminated in a resistance at both ends, find the network.

(*b*) If this is the transfer impedance $Z_{12} = E_2/I_1$ of a lossless two terminal-pair network terminated in 1 ohm, find the network.

28. With reference to the adjacent sketch, the lossless coupling network N is to be designed so that

$$\left|\frac{E_L}{E_s}\right| = \frac{K}{\sqrt{1 + (\omega/\omega_c)^6}}$$

with the largest possible values of K and ω_c.

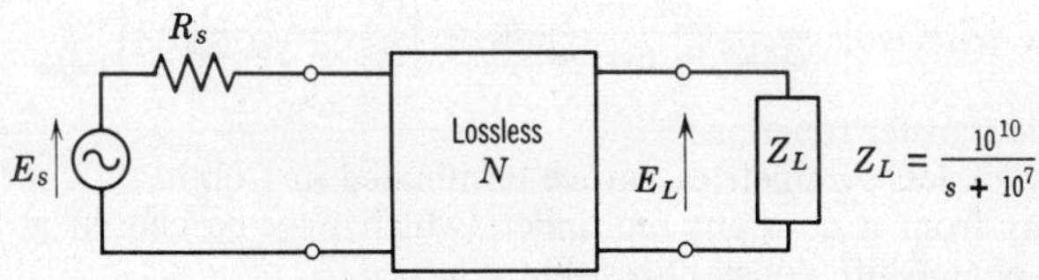

Prob. 28.

29. Synthesize a lossless network terminated in 1 ohm at both ends realizing the transfer function:

$$\left|\frac{2E_2}{E_1}\right|^2 = \frac{(2 - \omega^2)^2}{4(1 + \omega^6)}$$

30. For the network shown, $|E_2/E_{10}|^2_{s=j\omega}$ is to be $K(\omega^8)/(1 + \omega^8)$. Choose an appropriate K and realize the network as a ladder without transformers.

31. A lossless network N with 1-ohm resistances at both ends has the transfer function

$$\left|\frac{2E_2}{I_{10}}\right|^2 = \frac{1}{1 + (\omega^2 - \frac{1}{2})^2}$$

Find the network.

32. For a situation like that in Prob. 30 with both resistances equal to 1 ohm

$$\left|\frac{2E_2}{E_{10}}\right|^2_{s=j\omega} = \frac{4(\omega^2 - 1)^2}{(9\omega^4 - 11\omega^2 + 4)(\omega^2 + 1)} = \frac{4\omega^4 - 8\omega^2 + 4}{9\omega^6 - 2\omega^4 - 7\omega^2 + 4}$$

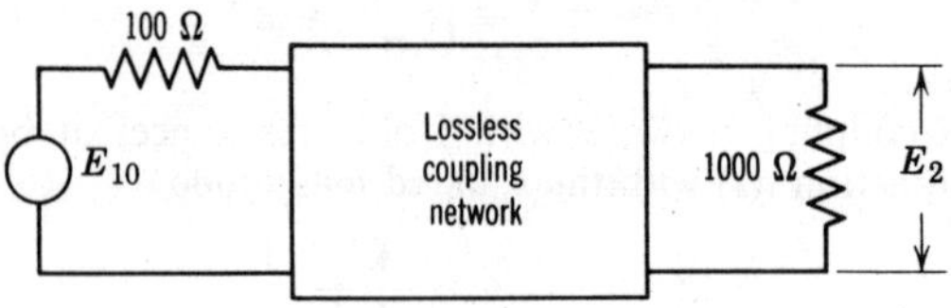

Prob. 30.

(*a*) Determine the appropriate open-circuit impedances z_{11}, z_{12}, and z_{22} for the lossless network.

(*b*) Synthesize the network.

33. Given

$$\left|\frac{2E_2}{E_1}\right|^2 = \left[\frac{1}{1 - s^2(1 + s^2)^2}\right]_{s=j\omega} = \frac{1}{1 + \omega^2(1 - \omega^2)^2}$$

and the information that the s-roots of the denominator are

$$s = \pm 0.6830, \qquad s = \pm 0.3415 \pm j0.6702\sqrt{3}$$

(*a*) Find a nondissipative network with 1-ohm resistances at both ends realizing this transfer function.

(*b*) If the same function is the $|Z_{12}|^2$ of a lossless network terminated in a 1-ohm load, find the network.

34. The squared magnitude of a transfer impedance is

$$|Z_{12}(j\omega)|^2 = \frac{\omega^4 + 4}{\omega^4 - 6\omega^2 + 25} = \left[\frac{(s^2 + 2)^2 - 4s^2}{(s^2 + 5)^2 - 4s^2}\right]_{s=j\omega}$$

Realize this transfer function:

(*a*) As a lossless symmetrical lattice terminated in 1 ohm.

(*b*) Apart from a constant multiplier (which may be chosen at will), as a minimum–phase-shift, constant-resistance lattice.

35. $Z_{12} = (s - 1)^2/(s^2 + s + 1)$

(*a*) Write Z_{12} as the product of two factors, one of which is minimum phase and another which has a constant magnitude for $s = j\omega$.

(*b*) Realize the transfer impedance by two constant-resistance lattices in cascade terminated in 1 ohm.

36. Given the transfer impedance

$$Z_{12}(s) = \frac{s^4 + 4}{(s + 1)^4}$$

(*a*) If kZ_{12} is to be realized through a (1-ohm) constant-resistance lattice, find the largest real k value and the impedance functions z_a and z_b of the corresponding lattice.

(*b*) If the lattice thus found is not minimum phase, find the z_a and z_b of a minimum-phase lattice that has the same magnitude of transfer impedance.

37. (*a*) Find a constant-resistance lattice which when terminated in 1 ohm has a transfer impedance equal to the impedance of 1 ohm, 1 henry, and 1 farad in parallel. Transform this lattice to an equivalent unbalanced network.

(*b*) Alternatively achieve a constant-resistance network in another way.

38. A given transfer impedance $Z_{12}(s)$ is realized by a lossless symmetrical lattice terminated in 1 ohm. The impedances of the lattice have poles at the frequencies: $s = 0$, $s = \pm j2$, $s = \pm j4$, and the residues of z_{22} in these poles are all equal to unity. If

$$Z_{12}(0) = 1, \qquad Z_{12}(\pm j2) = -1, \qquad Z_{12}(\pm j4) = 1$$

find the lattice network.

39. Carry through the double-loaded lossless network design discussed in Art. 5 for a Butterworth function

$$|t(j\omega)|^2 = \frac{K^2}{1 + \omega^{2n}} \quad \text{with } K < 1$$

Normalizing the input end by assuming $R_1 = 1$, show that now the appropriate load resistance $R_2 = (1 - \sqrt{1 - K^2})/(1 + \sqrt{1 - K^2})$, and that the shunt capacitance adjacent to the source is given by: $C_1 = 2/a[1 - (1 - K^2)^{1/2n}]$ with $a = 1/\sin(\pi/2n)$, the sum of the distances between the j-axis and the poles.

40. By making use of the Darlington theory discussed in Art. 8, Ch. 9, show that for odd-order Butterworth functions the shunt capacitance adjacent to the load in the preceding problem has the value

$$C_2 = 2/aR_2[1 + (1 - K^2)^{1/2n}]$$

41. In the synthesis of a lossless network terminated in a 1-ohm resistance from given z_{12} and z_{22} functions, if z_{12} has all zeros on the j-axis, show that the lossless network is realizable in the form of a simple inverted L provided one is willing to accept a sufficiently small constant multiplier.

CHAPTER TWELVE

Other Methods of Realizing Transfer Functions

Since practical situations requiring the synthesis of networks for prescribed transfer characteristics may impose a wide variety of collateral conditions in addition to those inherent in the realization methods themselves, it is essential that the latter be as flexible and as varied in character as possible. For this reason it is well to be amply familiar with the advantageous use of basic principles and techniques, and, above all, to have a clear appreciation of the mechanism that enables one to control the transmission properties of networks. The following discussion is devoted to an elaboration of both of these ideas.

1 Additional Methods Based Upon the Lattice Structure

Requiring a lattice to have the constant-resistance property and possibly also to possess an unbalanced equivalent form may, for certain specified transfer characteristics, restrict the obtainable gain of this network to such an extent as to make its use impractical. In a situation of this sort, the obvious expedient is to relinquish the constant-resistance requirement but retain the lattice as a point of departure, both for its ability to realize any achievable transfer impedance and for the simplicity and directness with which a given design is accomplished, especially when there is no objection to the use of a balanced structure. If an unbalanced network is essential, then a study of the possible conversion of the lattice design to such an equivalent form must again be considered, but now some artifices become available that are ruled out if the lattice impedances are required to be mutually reciprocal.

Starting from the familiar relation

$$z_{12} = \tfrac{1}{2}(z_b - z_a) \tag{1}$$

one can make use of either of two procedures to construct the lattice impedances z_b and z_a from a specified transfer function z_{12}. Both of these methods make use of the obvious fact that the function z_{12} remains unchanged if the same quantity is added to both z_b and z_a.

Suppose we consider the partial-fraction expansion

$$2z_{12}(s) = \frac{k_1}{s - s_1} + \frac{\bar{k}_1}{s - \bar{s}_1} + \frac{k_2}{s - s_2} + \frac{\bar{k}_2}{s - \bar{s}_2} + \cdots \tag{2}$$

in which the bar indicates a conjugate complex value. Recalling the discussion in Art. 1, Ch. 9, a pair of conjugate terms in this expansion represents a p.r. function provided the real part of the pertinent residue is positive and sufficiently large. Repeating Eqs. 15 and 21 of Ch. 9, if we write

$$k_\nu = \alpha_\nu + j\beta_\nu \qquad \text{and} \qquad s_\nu = -\sigma_\nu + j\omega_\nu \tag{3}$$

the condition in question reads

$$\frac{\alpha_\nu}{|\beta_\nu|} \geq \frac{\omega_\nu}{\sigma_\nu} \tag{4}$$

If all the residues in expansion 2 fulfill this condition, then z_{12} is p.r. We choose $z_b = 2z_{12}$, $z_a = 0$, and the lattice degenerates into a "ladder" with one shunt branch equal to the impedance $z_b/2 = z_{12}$. This degenerate situation is hardly of any practical importance.

If a residue in expansion 2 has a real part that is not large enough to satisfy condition 4, then we can add to this real part an appropriate positive value δ_ν until the sum

$$\alpha_\nu' = \alpha_\nu + \delta_\nu \tag{5}$$

does fulfill this condition (with a margin if we wish), and compensate for this addition by subtracting from $2z_{12}$ the conjugate terms

$$\frac{\delta_\nu}{s - s_\nu} + \frac{\delta_\nu}{s - \bar{s}_\nu} \tag{6}$$

which we assign to the impedance z_a, the rest of the expansion (including the terms with the modified real part 5) being identified with z_b as before. This process is evidently applicable whether the given real part α_ν has a negative or positive value, albeit we can simplify things if a negative α_ν has a large enough absolute value so that the condition

$$\frac{-\alpha_\nu}{|\beta_\nu|} \geq \frac{\omega_\nu}{\sigma_\nu} \tag{7}$$

is fulfilled, for then we assign the pertinent pair of conjugate terms in expansion 2 to z_a outright, and have no modification of this real part to do.

The impedances z_a and z_b thus constructed from a given z_{12} function are not only p.r. but, according to the discussion in Art. 1, Ch. 9, are immediately realizable without mutual inductive coupling and without further computational effort.

Since the construction of z_a and z_b in this manner is evidently not unique, the number of distinct lattices realizing a given transfer impedance is unlimited. Thus we can pad the real parts of residues, as indicated in Eq. 5, by any additional amount beyond the minimum required, and we can pad real parts that do not need padding if we want to. We can evidently also pad the imaginary parts of residues, and thus increase still further the variety of detailed results obtainable. Such practice may conceivably be used to affect element value distributions in the total realization or for other reasons dictated by practical requirements of the design.

Let us consider a few examples of this method of synthesis. Suppose we begin with the transfer function 25, Ch. 11, having the squared magnitude 18. Its partial-fraction expansion, after combining a pair of conjugate complex terms, reads

$$2z_{12} = \frac{2}{(s+1)(s^2+s+1)} = \frac{2}{s+1} - \frac{2s}{s^2+s+1} \tag{8}$$

We are condensing the procedure here somewhat by not bothering to study residue values since we can see immediately that a solution is had by assuming

$$z_b = \frac{2}{s+1} \quad \text{and} \quad z_a = \frac{2s}{s^2+s+1} \tag{9}$$

The resultant lattice and its ladder development (according to methods discussed in Art. 5, Ch. 6) are shown in Fig. 1. Except for an impedance

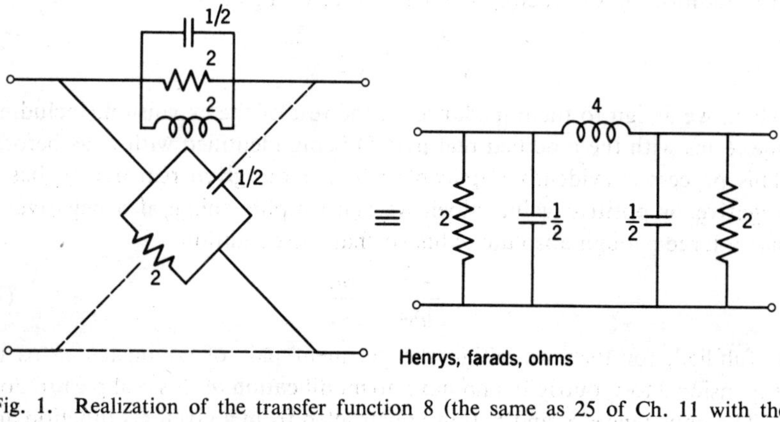

Fig. 1. Realization of the transfer function 8 (the same as 25 of Ch. 11 with the squared magnitude 18) by a direct lattice approach. Except for an impedance level factor, the result is the same as in Fig. 7, Ch. 11.

level factor of 2, this is the same result as that obtained by the method discussed in Art. 5, Ch. 11, and illustrated there for this same transfer impedance in Fig. 7.

Next let us consider the transfer function

$$2z_{12} = \frac{1}{(s^2 + 1.8478s + 1)(s^2 + 0.7645s + 1)} \tag{10}$$

which, by the method of synthesis discussed in Art. 5, Ch. 11, leads to the antimetric network shown there in Fig. 8. Its squared magnitude is the function 78, and the computations yielding the pertinent denominator polynomial are given in Ch. 11 by Eqs. 80, 81, and 82.

The partial-fraction expansion of 10, after combining conjugate terms, reads

$$2z_{12} = \frac{0.92387s + 1.7071}{s^2 + 1.8478s + 1} - \frac{0.92387s + 0.7071}{s^2 + 0.7645s + 1} \tag{11}$$

from which we immediately have

$$z_b = \frac{0.9239s + 1.707}{s^2 + 1.848s + 1} = \cfrac{1}{1.082s + \cfrac{1}{0.9239s + 1.707}} \tag{12}$$

and

$$z_a = \frac{0.9239s + 0.707}{s^2 + 0.7654s + 1} = \cfrac{1}{1.082s + \cfrac{1}{0.9239s + 0.707}} \tag{13}$$

yielding the unbalanced ladder development shown in Fig. 2. The somewhat unusual feature about this result is the fact that the resistive loading appears in the center instead of at the terminations. For this reason it is practically less desirable than the network of Fig. 8 in Ch. 11 but nevertheless theoretically interesting. Although symmetrical, as it must be since it is derived from a lattice, its transfer function is the same as that for the antimetric network obtained in Ch. 11.

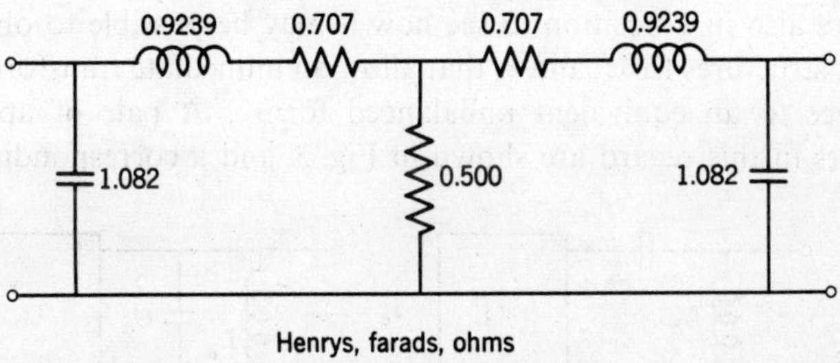

Fig. 2. Ladder development of lattice realization for the transfer impedance 10 having the squared magnitude 78 in Ch. 11. It is there realized by the antimetric network shown in Fig. 8.

Although the resulting lattices for these two examples are readily developable into ladder structures, this method of synthesis (based upon a partial-fraction expansion of the given transfer impedance) does not offer good possibilities for an unbalanced realization since the impedances z_a and z_b in general cannot be expected to have the rather special properties required to make the ladder development procedure work successfully. The following alternate method, on the other hand, guarantees the existance of an unbalanced equivalent structure for a large class of transfer functions whose characteristics may be defined in a straightforwardly identifiable manner.

This method proceeds from the j-axis real part of a given transfer function, having the familiar form

$$r_{12}(\omega) = \frac{A_0 + A_1\omega^2 + \cdots + A_n\omega^{2n}}{B_0 + B_1\omega^2 + \cdots + B_n\omega^{2n}} = \frac{A(\omega^2)}{B(\omega^2)} \tag{14}$$

If $r_a(\omega)$ and $r_b(\omega)$ denote the real parts of $z_a(j\omega)$ and $z_b(j\omega)$ respectively, then, except for a factor of 2 which we shall not bother to write from now on, we have

$$r_{12}(\omega) = r_b(\omega) - r_a(\omega) \tag{15}$$

Construction of the lattice impedances z_a and z_b is accomplished as soon as we have separated the rational function 14 into the difference of two parts which separately are positive for all real ω values. Such a separation is easy, and can be done in an infinite number of ways. Moreover, it can just as easily be done to yield positive coefficients in the numerators of both $r_a(\omega)$ and $r_b(\omega)$, so that the Miyata method of transformerless synthesis (Arts. 2 and 3, Ch. 10) is applicable. A simple way of accomplishing this result is to let $r_a(\omega)$ and $r_b(\omega)$ have the same denominator $B(\omega^2)$, and assign the positive terms in the numerator of 14 to $r_b(\omega)$, the negative ones to $r_a(\omega)$. A lattice realization without mutual inductive coupling is thus at once obtainable.

We are also in a position to see how it may be possible to obtain such network structures for z_a and z_b that allow an immediate transformation of the lattice to an equivalent unbalanced form. A pair of appropriate structures in this regard are shown in Fig. 3, and a corresponding reduc-

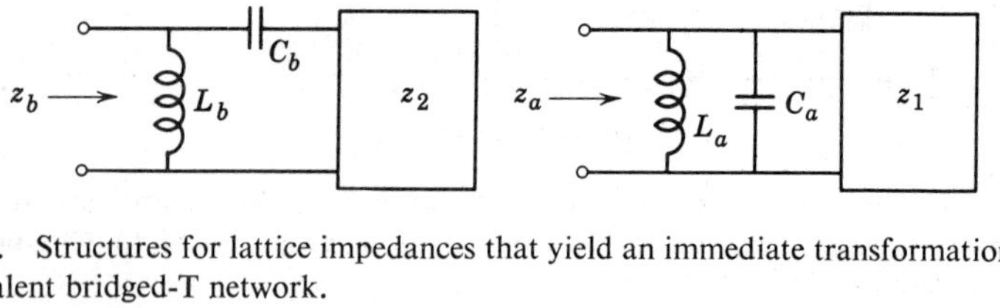

Fig. 3. Structures for lattice impedances that yield an immediate transformation to an equivalent bridged-T network.

tion of the lattice to an equivalent bridged-T, according to the methods given in Art. 5, Ch. 6, is seen without difficulty to be that in Fig. 4.

Positiveness of all the element values in this bridged-T requires that

$$L_b \geq L_a \tag{16}$$

and

$$C_a \geq C_b \tag{17}$$

The shunt inductances in the circuits shown in Fig. 3 actually are not essential to the process of transforming the lattice to an equivalent bridged-T structure. In fact, their presence adds to the complexity of the contemplated procedure since we must not only carry out the separation of $r_{12}(\omega)$ into parts $r_a(\omega)$ and $r_b(\omega)$ in a way that yields the shunt inductances as well as the shunt and series capacitances C_a and C_b, but we have also the additional condition 16 to deal with. However, without the shunt inductance in the circuit for z_b, this impedance (and hence also z_{12}) is required to have a pole at $s = 0$, which is not only undesirable but actually impossible since z_b must be generated from its real part $r_b(\omega)$.

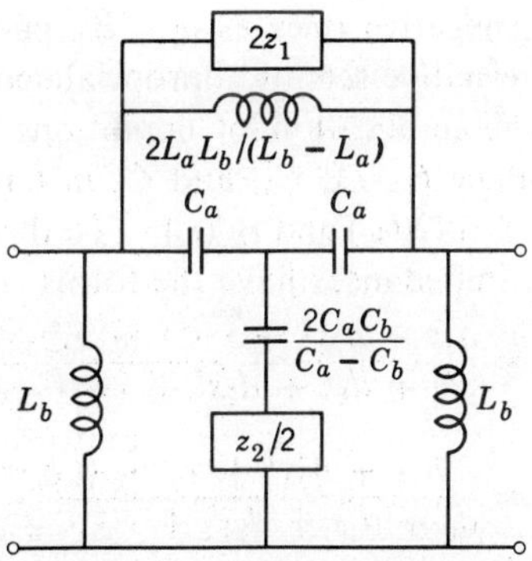

Fig. 4. Bridged-T equivalent of a lattice having the impedances shown in Fig. 3.

The problem now is to find that separation of expression 14 into r_a and r_b functions that yields structures like those shown in Fig. 3, and to interpret conditions 16 and 17 in the form of straightforwardly expressible properties which the given r_{12} function is required to possess. A solution to the first part of this problem is suggested by the reasoning accompanying the Miyata method of synthesis discussed in Arts. 2 and 3 of Ch. 10. Thus, if we write

$$r_a(\omega) = \frac{\alpha_1\omega^2 + \alpha_2\omega^4 + \cdots + \alpha_{n-1}\omega^{2(n-1)}}{B(\omega^2)} \tag{18}$$

$$r_b(\omega) = \frac{\beta_0 + \beta_1\omega^2 + \beta_2\omega^4 + \cdots + \beta_n\omega^{2n}}{B(\omega^2)} \tag{19}$$

and separate this real part into the additive parts

$$r_b'(\omega) = \frac{\beta_0 + \beta_1\omega^2}{B(\omega^2)} \quad \text{and} \quad r_b''(\omega) = \frac{\beta_2\omega^4 + \cdots + \beta_n\omega^{2n}}{B(\omega^2)} \tag{20}$$

then it is clear that the impedances $z_a(s)$ and $z_b''(s)$ generated by the real parts r_a and r_b'' respectively have realizations of the form shown in Fig. 3. The impedance $z_b'(s)$, generated by $r_b'(\omega)$, is a series component in the lattice crossarm, and may be dealt with as shown in Fig. 16, Ch. 6, for the component denoted there by z_{b1}.

A tacit assumption in this specification of the real parts r_a and r_b is that

$$\beta_0 = A_0 \geq 0 \quad \text{and} \quad \beta_n = A_n \geq 0 \tag{21}$$

which for most practical purposes amounts to restricting the coefficients A_0 and A_n in Eq. 14 to have the same algebraic sign, since the sign of the entire function is usually of no importance. The remaining coefficients $\beta_1 \cdots \beta_{n-1}$ as well as $\alpha_1 \cdots \alpha_{n-1}$ are likewise all positive, which is a condition that obviously can always be met. The normal procedure is to denote the positive coefficients in the numerator of Eq. 14 as β_i and the absolute values of the negative ones as α_i. By padding these, one can obtain innumerable alternative sets of appropriate coefficients.

In order to study the implication of conditions 16 and 17 we must express the element values L_a, L_b, C_a, and C_b in terms of the quantities characterizing the real parts $r_a(\omega)$ and $r_b''(\omega)$. To this end we first observe that the corresponding impedances have the forms

$$z_a(s) = \frac{a_1 s + a_2 s^2 + \cdots + a_{n-1}s^{n-1}}{d_0 + d_1 s + d_2 s^2 + \cdots + s^n} \tag{22}$$

and

$$z_b''(s) = \frac{b_1 s + b_2 s^2 + \cdots + b_n s^n}{d_0 + d_1 s + d_2 s^2 + \cdots + s^n} \tag{23}$$

in which the common denominator polynomial

$$D(s) = d_0 + d_1 s + d_2 s^2 + \cdots + s^n \tag{24}$$

is related to and determined by the denominator in Eq. 14 by the familiar expression

$$B(\omega^2) = \Big[D(s) \cdot D(-s)\Big]_{s=j\omega} \tag{25}$$

By the method of ladder development applied to the impedances 22 and 23, we find that the elements of interest in the circuits of Fig. 3 are given by

$$L_a = a_1/d_0 \tag{26}$$

$$C_a = 1/a_{n-1} \tag{27}$$

$$L_b = b_1/d_0 \tag{28}$$

and

$$C_b = \frac{d_2 b_1 - d_0 b_3}{b_1^2} \tag{29}$$

The coefficients d_i appearing in these expressions are obtainable directly from the denominator in Eq. 14 by relation 25. The a_i and b_i can be determined in the following manner. Let the basic impedance functions

$$Z_k(s) = \frac{a_{k0} + a_{k1}s + a_{k2}s^2 + \cdots + a_{kn}s^n}{d_0 + d_1 s + d_2 s^2 + \cdots + s^n} \tag{30}$$

for $k = 0, 1, 2, \cdots, n$ be those generated by the real parts $\omega^{2k}/B(\omega^2)$. The impedance $Z_0(s)$ in this family is most easily found for any given denominator polynomial $D(s)$ by the method described in Art. 2, Ch. 11, and the remaining impedances are then constructed with a minimum of computational labor by the Miyata procedure discussed in Art. 1, Ch. 8.

The coefficients a_s and b_s in the impedances 22 and 23 may then be computed from

$$a_s = \sum_{k=1}^{n-1} \alpha_k a_{ks} \tag{31}$$

and

$$b_s = \sum_{k=2}^{n} \beta_k a_{ks} \tag{32}$$

as is clear from the fact that the impedance corresponding to a sum of real parts equals the sum of the impedances corresponding to these real parts considered separately.

Before combining these results with relations 26 through 29 and conditions 16 and 17, it is useful to observe certain interrelationships between the numerator coefficients in the impedance 30. From the Miyata method of forming the impedance Z_{k+1} from Z_k, we have

$$-s^2 Z_k = -a_{k(n-1)}s + Z_{k+1} \quad \text{for } k = 0, 1, \cdots, n-1 \tag{33}$$

More specifically, this yields the result

$$\begin{aligned} -s^2(a_{k(n-1)}s^{n-1} + a_{k1}s) = -a_{k(n-1)}s(s^n + \cdots + d_1 s + d_0) \\ + (a_{(k+1)n}s^n + \cdots + a_{(k+1)1}s) \end{aligned} \tag{34}$$

which can be written

$$\begin{aligned} a_{k(n-1)}(d_{n-1}s^n + \cdots + d_0 s) - (a_{k(n-2)}s^n + \cdots + a_{k1}s) \\ = (a_{(k+1)n}s^n + \cdots + a_{(k+1)1}s) \end{aligned} \tag{35}$$

Equating coefficients of like powers of s leads to the relation

$$a_{k(n-1)}d_{n-i} - a_{k(n-i-1)} = a_{(k+1)(n-i+1)} \tag{36}$$

for $k = 0, 1, \cdots, n-1$, and $i = 1, 2, \cdots, n$, where a coefficient a_{kt} for a negative integer value of t is tacitly understood to be zero, and $a_{(k+1)n}$ is zero for all k except $k = n - 1$ since only the impedance $Z_n(s)$ has a numerator of degree n.

For $i = n$, this relation gives

$$a_{k(n-1)} = a_{(k+1)1}/d_0 \tag{37}$$

and for $i = n - 2$ we get

$$a_{k(n-1)}d_2 - a_{k1} = a_{(k+1)3} \tag{38}$$

Making use of Eq. 37, and replacing k by $k - 1$, we have

$$d_2 a_{k1} - d_0 a_{k3} = d_0 a_{(k-1)1} \tag{39}$$

These results, together with Eqs. 31 and 32, enable us to write

$$a_1 = \sum_{k=1}^{n-1} \alpha_k a_{k1} \tag{40}$$

$$b_1 = \sum_{k=2}^{n} \beta_k a_{k1} \tag{41}$$

$$a_{n-1} = \sum_{k=1}^{n-1} \alpha_k a_{k(n-1)} = \frac{1}{d_0} \sum_{k=1}^{n-1} \alpha_k a_{(k+1)1} \tag{42}$$

and

$$d_2 b_1 - d_0 b_3 = d_0 \sum_{k=2}^{n} \beta_k a_{(k-1)1} \tag{43}$$

Their substitution into Eqs. 26 through 29 finally yields conditions 16 and 17 respectively in the form

$$\sum_{k=2}^{n} \beta_k a_{k1} \geq \sum_{k=1}^{n-1} \alpha_k a_{k1} \tag{44}$$

and

$$\left(\sum_{k=2}^{n} \beta_k a_{k1}\right)^2 \geq \sum_{k=2}^{n} \alpha_{k-1} a_{k1} \sum_{k=1}^{n-1} \beta_{k+1} a_{k1} \tag{45}$$

For a given transfer function or its real part, the coefficients α_k and β_k are readily available, and the coefficients a_{k1} for $k = 1 \cdots n$ can be computed from the known denominator polynomial 24 with a moderate amount of effort. If both conditions 44 and 45 are fulfilled, an unbalanced lattice equivalent involving no mutual inductive coupling is possible. If only 45 is fulfilled, an unbalanced equivalent is still possible, but the Π arrangement of inductances in the bridged-T circuit of Fig. 4 must be replaced by a pair of mutually coupled inductances having the common value $(L_b + L_a)/2$ and the mutual inductance $(L_b - L_a)/2$. The coupling coefficient

$$k = \left|\frac{L_b - L_a}{L_b + L_a}\right| \tag{46}$$

always remains less than unity except in a degenerate limiting situation in which $L_b = 0$ (the left-hand sum in 44 becomes zero). However, no unbalanced equivalent structure, according to this process, is possible unless condition 45 (corresponding to 17) is fulfilled.

When this condition is not fulfilled we can vary the procedure by considering the impedance z_b'' to have the form of z_b in Fig. 3 with the branches L_b and C_b interchanged. With this revision of its crossarm impedance, the lattice (except for the crossarm component z_b') is then convertible to the bridged-T circuit shown in Fig. 5. The conditions insuring positiveness of all element values are still given by the relations 16 and 17.

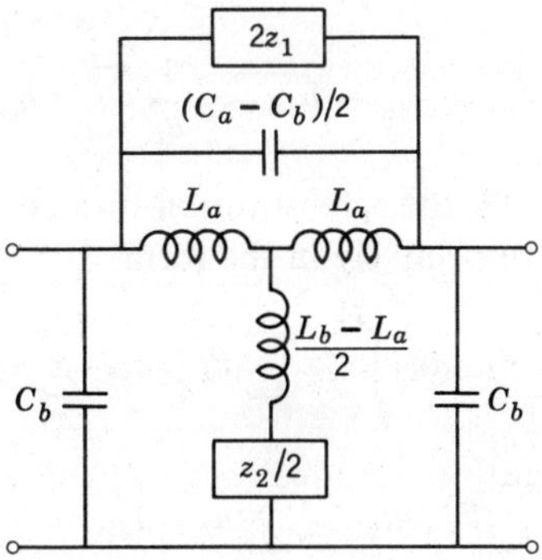

Fig. 5. Bridged-T equivalent when the branches L_b and C_b in the lattice impedances of Fig. 3 are interchanged.

In order to achieve this result, we separate the real part $r_b(\omega)$ given by Eq. 19 into the additive components

$$r_b'(\omega) = \frac{\beta_{n-1}\omega^{2(n-1)} + \beta_n\omega^{2n}}{B(\omega^2)} \quad \text{and} \quad r_b''(\omega) = \frac{\beta_0 + \cdots + \beta_{n-2}\omega^{2(n-2)}}{B(\omega^2)} \tag{47}$$

in place of those shown in Eqs. 20. The real part $r_a(\omega)$ in Eq. 18 and the impedance z_a in Eq. 22, which it generates, retain the same forms. Equation 23 for $z_b''(s)$, however, is replaced by

$$z_b'' = \frac{b_0 + b_1 s + b_2 s^2 + \cdots + b_{n-1}s^{n-1}}{d_0 + d_1 s + d_2 s^2 + \cdots + s^n} \tag{48}$$

The impedance $z_b'(s)$ generated by the component $r_b'(\omega)$ is likewise changed, but its detailed character does not interest us in the present discussion since its disposition is straightforward and has nothing to do with the lattice transformation problem.

The parameters L_a and C_a are given by Eqs. 26 and 27 as before, but the expressions for L_b and C_b are changed to read

$$C_b = 1/b_{n-1} \tag{49}$$

and

$$L_b = \frac{b_{n-1}^2}{d_{n-2}b_{n-1} - b_{n-3}} \tag{50}$$

Use of relation 36 for $i = 2$, together with 37 in which k is replaced by $k + 1$, gives

$$d_{n-2}a_{k(n-1)} - a_{k(n-3)} = a_{(k+2)1}/d_0 \tag{51}$$

This result, plus relations 31 and 32 as well as 37, enables us to write

$$b_{n-1} = \sum_{k=0}^{n-2} \beta_k a_{k(n-1)} = \frac{1}{d_0} \sum_{k=0}^{n-2} \beta_k a_{(k+1)1} \tag{52}$$

and

$$d_{n-2}b_{n-1} - b_{n-3} = \frac{1}{d_0} \sum_{k=0}^{n-2} \beta_k a_{(k+2)1} \tag{53}$$

Together with 40 and 42, their substitution into 26, 27, 49, and 50 yields conditions 16 and 17 respectively in the form

$$\left(\sum_{k=1}^{n-1} \beta_{k-1}a_{k1}\right)^2 \geq \sum_{k=2}^{n} \beta_{k-2}a_{k1} \sum_{k=1}^{n-1} \alpha_k a_{k1} \tag{54}$$

$$\sum_{k=1}^{n-1} \beta_{k-1}a_{k1} \geq \sum_{k=2}^{n} \alpha_{k-1}a_{k1} \tag{55}$$

Here condition 55 is the essential one, since 54, like condition 44 in the alternate procedure, is necessary only to avoid mutual inductive coupling. If 54 is not fulfilled, the T-arrangement of inductances in the bridged-T circuit of Fig. 5 can, in a familiar manner, be replaced by a pair of mutually coupled inductances with the coupling coefficient 46. Condition 55, on the other hand, must be fulfilled in order for the present procedure to yield a realizable unbalanced network.

This question may further be clarified through use of the resistance integral theorem discussed in Art. 3, Ch. 8, and expressed there by the integral relation 69. Use of this relation enables us to write

$$\frac{1}{C_a} = \frac{2}{\pi} \int_0^\infty r_a(\omega)\, d\omega \tag{56}$$

and

$$\frac{1}{C_b} = \frac{2}{\pi} \int_0^\infty r_b''(\omega)\, d\omega \tag{57}$$

whereupon condition 17 (equivalent to 55) becomes

$$\int_0^\infty r_a(\omega)\, d\omega \leq \int_0^\infty r_b''(\omega)\, d\omega \tag{58}$$

This form for condition 17 or 55, incidentally, shows that the division of

$r_b(\omega)$ in Eq. 19 into the portions $r_b'(\omega)$ and $r_b''(\omega)$ given in Eqs. 47 is most advantageous, since including terms with $\beta_{n-2}, \beta_{n-3}, \cdots$ in $r_b'(\omega)$ rather than limiting this function to the essential minimum of the last two terms in $r_b(\omega)$ robs the integrand on the right of 58 of some of its much-needed area.

Most interesting about condition 58, however, is the fact that it shows the existence of an unbalanced realization (according to this procedure) to depend upon the relative extent to which the real part of a given transfer impedance becomes negative throught the frequency interval $0 < \omega < \infty$. A graphical sketch of the function

$$r_{12}(\omega) - r_b'(\omega) = r_b''(\omega) - r_a(\omega) = \frac{A_0 + \cdots + A_{n-2}\omega^{2(n-2)} + \alpha_{n-1}\omega^{2(n-1)}}{B_0 + B_1\omega^2 + \cdots + B_n\omega^{2n}} \tag{59}$$

may give one a quick idea, in a given situation, as to the probable fulfullment of condition 55. Namely, this condition is fulfilled if the net area enclosed by the function 59 is nonnegative.

If the given transfer impedance $z_{12}(s)$ behaves like K/s for large s, then the resistance-integral theorem (specifically, Eq. 68, Ch. 8) shows that the net area enclosed by its real part equals $\pi K/2$. The coefficient A_n in Eq. 14 is then zero, and A_{n-1} can be positive or negative independent of the sign of K. If $A_{n-1} < 0$, then we may set $r_b' = 0$ in 47; and if $K > 0$, we conclude that the net area under the function 59 is positive. On the other hand, if $z_{12}(s) \to 1/s^2$ for $s \to \infty$, then $A_n = A_{n-1} = 0$. We may again set $r_b' = 0$; and since K (the coefficient of the $1/s$ term in the asymptotic behavior of z_{12}) is now zero, the net area under the function 59 is zero. In both of these situations condition 17 or 55 is fulfilled, and the pertinent lattice is sure to possess an unbalanced equivalent (perhaps with one pair of mutually coupled inductances involving less than close coupling).

Since these conclusions are independent of the pole-zero configuration of $z_{12}(s)$, they apply alike to nonminimum-phase and to minimum-phase transfer functions. So-called "all-pass" functions, having the form $z_{12}(s) = P(-s)/P(s)$, in which $P(s)$ is of even degree, may be made to conform to the first set of conditions described in the above paragraph through multiplication by the factor $1/(s + a)$. For a suitably chosen value of a, the all-pass character of $z_{12}(s)$ is practically undisturbed, while its unbalanced realization is assured. If $P(s)$ is of odd degree, a similar result is obtained by inserting a *pair* of appropriately chosen left half-plane pole factors. Through expedients such as these, one can often assure beforehand that a contemplated form of realization will be possible.

When the variant of this procedure to which conditions 44 and 45 apply is possible (that is, if 45 is fulfilled), one can express the possibility of avoiding mutual inductive coupling in an analogous fashion. The values of the inductances L_a and L_b in Eqs. 26 and 28 are determined by the behavior of the impedances $z_a(s)$ and $z_b''(s)$ in Eqs. 22 and 23 at $s \to 0$. We can therefore obtain the desired results by applying the resistance-integral theorem to the functions $z_a(1/s)$ and $z_b''(1/s)$. Observe that the real part of a function $f(1/j\omega)$ is the real part of $f(j\omega)$ with ω replaced by $1/\omega$. Use of Eq. 69 of Ch. 8 thus enables us to write

$$L_a = \frac{2}{\pi}\int_0^\infty r_a(1/\omega)\, d\omega \tag{60}$$

and

$$L_b = \frac{2}{\pi}\int_0^\infty r_b''(1/\omega)\, d\omega \tag{61}$$

If we now replace ω by $1/\omega$, noting that $d(1/\omega) = -(1/\omega^2)\, d\omega$, and also noting the effect upon the limits of integration, we see that condition 44 is expressible as

$$\int_0^\infty \frac{r_a(\omega)}{\omega^2}\, d\omega \le \int_0^\infty \frac{r_b''(\omega)}{\omega^2}\, d\omega \tag{62}$$

From this result we see, incidentally, that keeping the number of terms in $r_b'(\omega)$ to a minimum, as is done in specifications 20, is advantageous since it maximizes the area enclosed by $r_b''(\omega)/\omega^2$. The weighting factor $1/\omega^2$ shows, moreover, that negative loops in the real part of $z_{12}(j\omega) - z_b'(j\omega)$ occurring at higher frequencies are less apt to violate condition 44 than are such negative regions near the origin ($\omega = 0$).

As an illustrative example of this synthesis method, suppose we consider a transfer impedance with the real part

$$r_{12}(\omega) = \frac{A_0 - A_1\omega^2 + A_2\omega^4 + A_3\omega^6}{1 + \omega^6} \tag{63}$$

in which the $A_0 \cdots A_3$ are positive. Here we choose

$$r_a(\omega) = \frac{A_1\omega^2}{1 + \omega^6} \tag{64}$$

and for the first variant of our procedure to which the separation of $r_b(\omega)$ given by 20 applies

$$r_b'(\omega) = \frac{A_0}{1 + \omega^6}, \qquad r_b''(\omega) = \frac{A_2\omega^4 + A_3\omega^6}{1 + \omega^6} \tag{65}$$

For the common denominator polynomial we have

$$D(s) = 1 + 2s + 2s^2 + s^3 \tag{66}$$

which generates the family of basic impedances

$$Z_0(s) = \frac{1 + \frac{4}{3}s + \frac{2}{3}s^2}{D(s)} \tag{67}$$

$$Z_1(s) = \frac{\frac{2}{3}s + \frac{1}{3}s^2}{D(s)} \tag{68}$$

$$Z_2(s) = \frac{\frac{1}{3}s + \frac{2}{3}s^2}{D(s)} \tag{69}$$

$$Z_3(s) = \frac{\frac{2}{3}s + \frac{4}{3}s^2 + s^3}{D(s)} \tag{70}$$

For the evaluation of conditions 44 and 45 we note first that

$$a_{01} = \tfrac{4}{3}, \qquad a_{11} = \tfrac{2}{3}, \qquad a_{21} = \tfrac{1}{3}, \qquad a_{31} = \tfrac{2}{3} \tag{71}$$

$$\alpha_1 = A_1, \qquad \alpha_2 = 0 \tag{72}$$

and

$$\beta_0 = A_0, \qquad \beta_1 = 0, \qquad \beta_2 = A_2, \qquad \beta_3 = A_3 \tag{73}$$

Hence we find

$$\sum_1^{n-1} \alpha_k a_{k1} = 2A_1/3, \qquad \sum_2^{n} \alpha_{k-1} a_{k1} = A_1/3 \tag{74}$$

$$\sum_2^{n} \beta_k a_{k1} = \tfrac{1}{3}A_2 + \tfrac{2}{3}A_3, \qquad \sum_1^{n-1} \beta_{k+1} a_{k1} = \tfrac{2}{3}A_2 + \tfrac{1}{3}A_3 \tag{75}$$

Conditions 44 and 45 read respectively

$$A_2 + 2A_3 \geq 2A_1 \qquad \text{and} \qquad (A_2 + 2A_3)^2 \geq A_1(2A_2 + A_3) \tag{76}$$

In the second variant for which we separate $r_b(\omega)$ according to Eqs. 47, we have in place of 65

$$r_b'(\omega) = \frac{A_2\omega^4 + A_3\omega^6}{1 + \omega^6}, \qquad r_b''(\omega) = \frac{A_0}{1 + \omega^6} \tag{77}$$

All of the relations 67 through 74 apply unaltered, but for the evaluation of conditions 54 and 55 we compute in place of the sums 75

$$\sum_1^{n-1} \beta_{k-1} a_{k1} = 2A_0/3, \qquad \sum_2^{n} \beta_{k-2} a_{k1} = A_0/3 \tag{78}$$

Conditions 54 and 55 are thus found to yield the same result, namely

$$A_1 \leq 2A_0 \tag{79}$$

According to conditions 76 pertinent to the first variant of our procedure, the value of A_0 is unrestricted, while in the second variant to which the condition 79 applies, A_2 and A_3 are unrestricted. If we choose $A_3 = 0$, condition 79 is unaffected while 76 pertinent to the first variant reduces to the single condition

$$A_1 \leq A_2/2 \tag{80}$$

This situation now applies to a transfer impedance $z_{12}(s)$ that behaves like $1/s$ for large s.

Under these circumstances (for $A_3 = 0$), suppose the given function 63 involves nonzero values for A_1 and A_2 but calls for $A_0 = 0$. The second variant is then not possible. For the first, let condition 80 be satisfied by the choice of $A_1 = 1$ and $A_2 = 2$. We then have

$$r_a = \frac{\omega^2}{1 + \omega^6}, \qquad r_b' \equiv 0, \qquad r_b'' = \frac{2\omega^4}{1 + \omega^6} \tag{81}$$

From the known basic impedances 68 and 69 we have immediately

$$z_a = \frac{\frac{2}{3}s + \frac{1}{3}s^2}{1 + 2s + 2s^2 + s^3}, \qquad z_b'' = z_b = \frac{\frac{2}{3}s + \frac{4}{3}s^2}{1 + 2s + 2s^2 + s^3} \tag{82}$$

The circuits for these lattice impedances and the resultant unbalanced network are shown in Fig. 6. The pertinent transfer impedance is given by

$$2z_{12}(s) = z_b(s) - z_a(s) = \frac{s^2}{1 + 2s + 2s^2 + s^3} \tag{83}$$

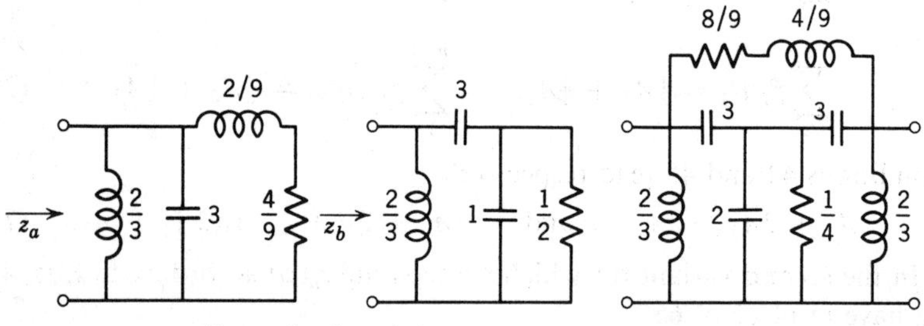

Fig. 6. Lattice impedances and resultant unbalanced equivalent network for the realization of the transfer impedance 83.

Again for $A_3 = 0$, suppose the given real-part function 63 involves nonzero values for A_0 and A_1 but calls for $A_2 = 0$. Now we must use the second variant since 80 cannot be fulfilled. For the choice of $A_0 = 0$, $A_1 = 2$, we get

$$r_a = \frac{2\omega^2}{1 + \omega^6}, \qquad r_b' = 0, \qquad r_b'' = \frac{1}{1 + \omega^6} \tag{84}$$

The lattice impedances are

$$z_a = \frac{\frac{4}{3}s + \frac{2}{3}s^2}{1 + 2s + 2s^2 + s^3}, \qquad z_b'' = z_b = \frac{1 + \frac{4}{3}s + \frac{2}{3}s^2}{1 + 2s + 2s^2 + s^3} \tag{85}$$

Figure 7 illustrates the resultant networks, and the pertinent transfer impedance is determined from

$$2z_{12}(s) = z_b(s) - z_a(s) = \frac{1}{1 + 2s + 2s^2 + s^3} \tag{86}$$

which is the same (except for a factor of 2) as the function 25 in Ch. 11, realized by the network given there in Fig. 2.

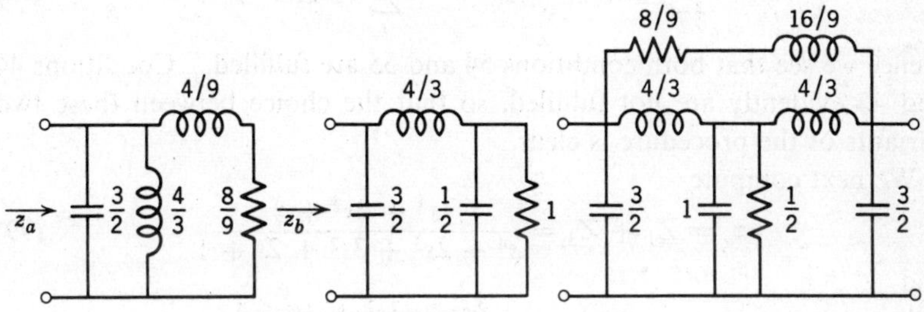

Henrys, farads, and ohms

Fig. 7. Lattice impedances and resultant unbalanced equivalent network for the realization of the transfer impedance 86.

A somewhat more interesting example is given by the function

$$2z_{12}(s) = \frac{s^2 + \frac{2}{3}s + 1}{s^4 + 2s^3 + 3s^2 + 2s + 1} \tag{87}$$

with the real part

$$2r_{12}(\omega) = \frac{1 - \omega^2 + \omega^4 - \omega^6}{(1 - \omega^2 + \omega^4)^2} \tag{88}$$

From the denominator polynomial in 87 we compute the family of basic impedances

$$Z_0(s) = \frac{s^3 + 2s^2 + \frac{5}{2}s + 1}{s^4 + 2s^3 + 3s^2 + 2s + 1} \tag{89}$$

$$Z_1(s) = \frac{\frac{1}{2}s^3 + s^2 + s}{s^4 + 2s^3 + 3s^2 + 2s + 1} \tag{90}$$

$$Z_2(s) = \frac{\frac{1}{2}s^3 + s^2 + \frac{1}{2}s}{s^4 + 2s^3 + 3s^2 + 2s + 1} \tag{91}$$

$$Z_3(s) = \frac{s^3 + s^2 + \frac{1}{2}s}{s^4 + 2s^3 + 3s^2 + 2s + 1} \tag{92}$$

$$Z_4(s) = \frac{s^4 + \frac{5}{2}s^3 + 2s^2 + s}{s^4 + 2s^3 + 3s^2 + 2s + 1} \tag{93}$$

We have

$$a_{01} = \tfrac{5}{2}, \quad a_{11} = 1, \quad a_{21} = \tfrac{1}{2}, \quad a_{31} = \tfrac{1}{2}, \quad a_{41} = 1 \tag{94}$$

$$\beta_0 = 1, \quad \beta_1 = 0, \quad \beta_2 = 1, \quad \beta_3 = 0, \quad \beta_4 = 0 \tag{95}$$

$$\alpha_1 = 1, \quad \alpha_2 = 0, \quad \alpha_3 = 1 \tag{96}$$

from which we find

$$\sum_{1}^{3} \beta_{k-1} a_{k1} = \tfrac{3}{2}, \qquad \sum_{2}^{4} \alpha_{k-1} a_{k1} = \tfrac{3}{2} \tag{97}$$

$$\sum_{2}^{4} \beta_{k-2} a_{k1} = \tfrac{3}{2}, \qquad \sum_{1}^{3} \alpha_{k} a_{k1} = \tfrac{3}{2} \tag{98}$$

Hence we see that both conditions 54 and 55 are fulfilled. Conditions 44 and 45 evidently are not fulfilled, so that the choice between these two variants of the procedure is clear.

We next compute

$$z_a = Z_1 + Z_3 = \frac{\tfrac{3}{2}s^3 + 2s^2 + \tfrac{3}{2}s}{s^4 + 2s^3 + 3s^2 + 2s + 1} \tag{99}$$

and

$$z_b = Z_0 + Z_2 = \frac{\tfrac{3}{2}s^3 + 3s^2 + 3s + 1}{s^4 + 2s^3 + 3s^2 + 2s + 1} \tag{100}$$

and find the developments

$$z_a = \cfrac{1}{\cfrac{2s}{3} + \cfrac{2}{3s} + \cfrac{1}{z_1}} \tag{101}$$

$$z_b = \cfrac{1}{\cfrac{2s}{3} + \cfrac{1}{\cfrac{3s}{2} + z_2}} \tag{102}$$

in which

$$z_2 = \frac{1}{\tfrac{4}{9}z_1} = \frac{s^2 + \tfrac{3}{2}s + 1}{s^2 + \tfrac{4}{3}s + 1} \tag{103}$$

This remainder function has a real-part numerator equal to $1 + \omega^4$ as we should expect from the theory of the Miyata method. The networks for

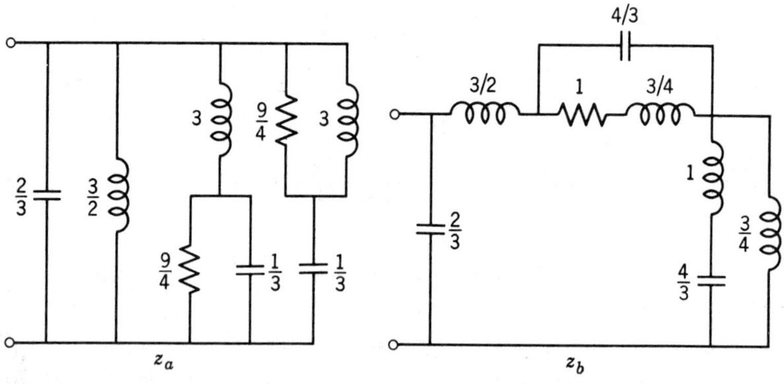

Fig. 8. Networks realizing the lattice impedances 101 and 102 pertinent to the transfer impedance 87.

z_a and z_b are shown in Fig. 8 and the resultant unbalanced nework realizing this transfer impedance (Eq. 87) is given in Fig. 9.

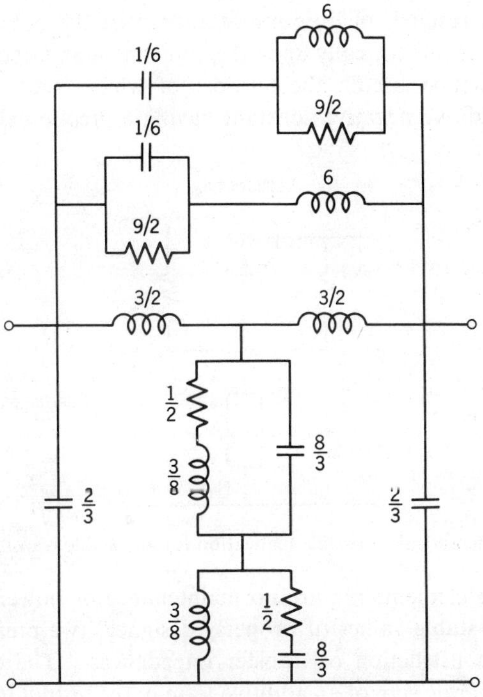

Fig. 9. Unbalanced equivalent of the lattice realizing the transfer impedance 87.

2 Construction From Pole-Zero Requirements

A given transfer impedance is placed in one of three categories according to its relative zero and pole locations. To begin with, we may separate all realizable transfer functions into two classes according to whether or not they are p.r.; then we subclassify those that are not p.r. into two groups in which one allows right half-plane zeros and the other does not. If a transfer function has right half-plane zeros, then it obviously cannot be p.r. If it has no right half-plane zeros, then it may or may not be p.r.

If a given transfer function happens to be p.r., then its realization does not require a two terminal-pair network, for it can more simply be realized as a driving-point impedance. When it is not p.r., it can evidently be regarded as a p.r. function in series with a negative resistance, since its failure to be p.r. merely means that its real part has some negative values for $s = j\omega$. Theoretically, at least, we can say that any realizable transfer function, whether minimum phase or not, can be realized with the help of one vacuum tube by the simple circuit shown in Fig. 10, in which the resistances R_1 and R_2 are appropriately chosen and Z is a p.r. impedance. Practically this result is of little importance since the obtainable transfer function depends too critically upon the adjustment and operating stability of the vacuum-tube circuit, the purpose of which is to inject into this function an additive negative constant having a precise value.

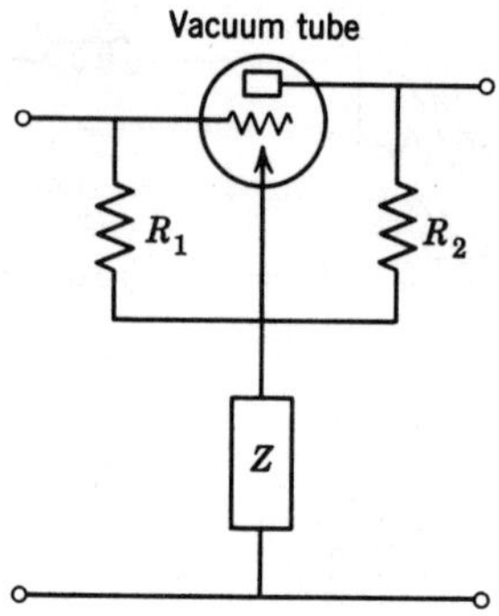

Fig. 10. A theoretically possible realization for any stable transfer impedance.

Since passive elements require no maintenance or power supplies, and have relatively stable values (if properly designed), we prefer to use only these for the construction of transfer impedances. The easiest way to change the algebraic sign of an additive term in the output voltage of some circuit is to incorporate into it a pair of crossed lead wires, as the lattice structure does by its crossarm impedances. Another means for accomplishing the same end is to incorporate a transformer with a negative voltage ratio. It thus becomes logically clear why the lattice structure plays such a prominent part in transfer function synthesis, and why its transformation to an equivalent unbalanced form so frequently involves negative mutual inductive coupling.

When the desired transfer impedance has no right half-plane zeros—that is to say, if it is a minimum-phase function—then its realization is always possible in the form of a simple unbalanced ladder network. When it is a nonminimum-phase function, its realization in this simple form is no longer possible; but neither will an unbalanced realization necessarily involve mutual inductive coupling. It may take the form of an unbalanced ladder in cascade with a bridged-T or twin-T or triple-T circuit of some sort. These matters are now discussed in some detail.

Our approach here will follow a very logical pattern by establishing procedures of network construction that enable us to exercise a direct and independent control over the production of zeros and poles in the resultant transfer impedance. Although the network obtained in this way may not always be as economical in the use of elements as one obtainable by some more specialized approach of limited applicability, the procedure to be discussed here has the advantage of being generally applicable and of being very fundamental in character, in the sense that its principle of operation is closely related to the response mechanism of circuits. Its value in shedding light upon the general problem of transfer function synthesis and thus being collaterally useful in clarifying and suggesting alternative methods is therefore of equal importance to its value as a possible means for the solution of a given realization problem.

Our thinking in connection with our present problem is influenced in no small measure by our knowledge of and experience with the methods of realizing transfer functions of lossless networks as discussed in Ch. 7, and especially by the principles involved in the ladder development procedure discussed there in Art. 3. Given a desired transfer impedance, our approach there is to choose a driving-point impedance having the same poles, and develop it into a ladder network in such a way that the series impedances or shunt admittances of certain branches have poles at the desired transmission zeros. As a means for maneuvering the poles of the developing function so that their frequencies have the desired values, we make use of the fact that only total pole removals produce transmission zeros. Partial pole removals are used as zero-shifting steps, these zeros becoming poles of the developing function after its subsequent inversion.

Characteristic of this process is the fact that we need give no heed to the production of the proper poles of the desired transfer function. Since these are the open-circuit natural frequencies of the resultant network, they are simultaneously poles of the driving-point impedance and hence are automatically produced because the developing function is this impedance.

Attempts to apply an analogous procedure to the synthesis of transfer impedances of lossy networks have, so far at least, met with very limited success,* although some useful results of specialized applicability are achieved. Two major difficulties arise that are absent in the corresponding lossless situation. There, when a pole is partially or totally removed, both this term and the remainder function are always automatically realizable. As also pointed out in Art. 1 of Ch. 9, neither the pole thus removed nor the remainder function in an analogous *RLC* situation have

* In this connection see "New Synthesis Procedures for Realizing Transfer Functions of RLC and RC Networks," Louis Weinberg, *Tech. Rept.* 210, Mass. Inst. Technol., Research Lab. Electronics, Sept. 14, 1951.

any such a priori assurance of being p.r. Secondly, it is not clear how the zero-shifting process in the complex plane is to be controlled, since the zeros are not automatically constrained to move along a fixed path as provided in the *LC* case by the *j*-axis. There is no assurance that the required zero-shifting can always be done under the requirement that both the removed pole and the remainder function be p.r.

Since the production of desired transmission zeros in the development of a driving-point function is thus seen to be extremely difficult, if not impossible, in a general sense, we try an approach that begins by choosing a function for which a realization yielding the desired transmission zeros is readily obtainable, even though a particular pole distribution, not necessarily the desired one, may tentatively need to be accepted. We then see if it is possible to shift the poles without affecting the zeros.

In other words, our approach "turns the tables," so to speak, with regard to poles and zeros. Instead of letting the poles take care of themselves and concentrating upon a method of producing the desired transmission zeros, we begin either with a network structure that places these zeros in evidence or with a function that has the desired zeros and is realizable in an obvious manner, and then devise ways in which we can subsequently shift the poles of this function to the desired positions without affecting its zero pattern.

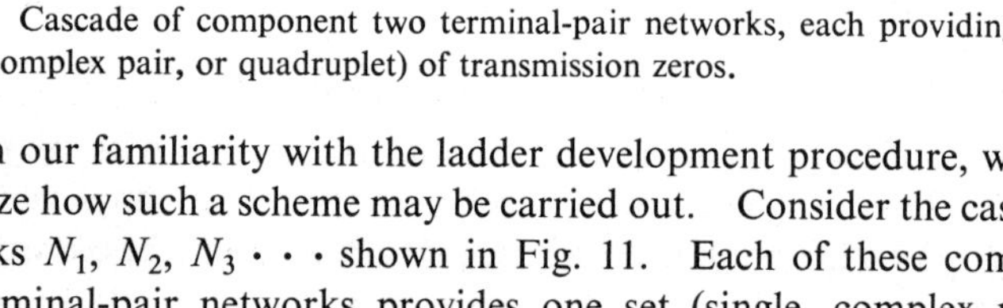

Fig. 11. Cascade of component two terminal-pair networks, each providing one set (single, complex pair, or quadruplet) of transmission zeros.

From our familiarity with the ladder development procedure, we easily recognize how such a scheme may be carried out. Consider the cascade of networks N_1, N_2, N_3 $\cdots$ shown in Fig. 11. Each of these component two terminal-pair networks provides one set (single, complex pair, or quadruplet) of transmission zeros. Collectively these, plus the zeros of the terminal impedance Z_1, represent all of the zeros of the over-all transfer impedance. If these zeros are all in the left half-plane, then the networks N_1, N_2, $\cdots$ may be replaced by single series or shunt branches that are respectively parallel or series resonant at the desired transmission zeros. Two terminal-pair networks are called for only if such zeros are required to lie in the right half-plane. If only the desired transmission zeros of the over-all transfer function are specified, and no restriction is placed upon the pole locations (except that they be in the left half-plane, of course), then a structure like that in Fig. 11 can be determined without difficulty, and the networks N_1, N_2, $\cdots$ can all be unbalanced and at most will involve a pair of not-closely-coupled inductances.

The poles of the resultant transfer impedance can now be shifted, without influencing either the zero locations or the total number of poles, by proceeding in one of a variety of ways. Thus if we make a "pliers-type" entry into any one of the meshes separating the networks Z_1, N_1, N_2 $\cdots$ and insert an impedance whose poles are none other than those of the respective pliers-type impedance created by that entry, then, by the theory of the ladder development process (Art. 3, Ch. 7), the over-all transmission zeros are not affected, and the poles of the transfer impedance are changed in position but not in number. Figure 12 illustrates such

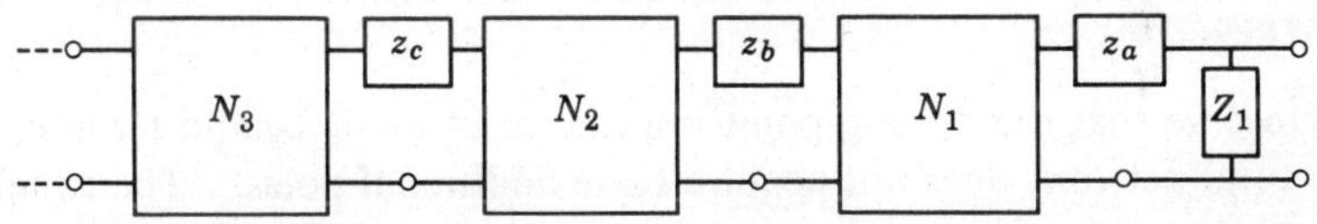

Fig. 12. Insertion of impedances z_a, z_b, . . . into the network of Fig. 11 may be done so as to shift the poles of the associated transfer impedance but leave its zeros unchanged.

insertion of impedances z_a, z_b, z_c, $\cdots$ into the meshes separating the boxes in Fig. 11. Thus z_a contains some or all of the poles of the pliers-type impedance created by cutting into the right-hand mesh. Similarly, z_b contains only poles contained in the pliers-type impedance created by cutting into the central mesh; and if z_a and z_b are now both present, the poles of the pertinent pliers-type impedance for the central mesh are those resulting in the presence of z_a, which may or may not be the same as the poles of this impedance in the absence of z_a. This point (which is of minor importance) will be elaborated upon presently.

Similarly, as shown in Fig. 13, we may add admittances in parallel with the networks of Fig. 11 without affecting any of the transmission zeros if the poles of each admittance are none other than the poles of the soldering-iron type admittance created by attaching leads to the terminal pair in question.

The poles of an impedance created by cutting into a mesh between boxes in the circuit of Fig. 11 fall into two groups: those of the impedance looking into the right-hand portion of the cascade, and those of the impedance looking into the left-hand portion, since the mesh impedance equals the sum of these two impedances. If we insert an impedance like

z_b in Fig. 12 into the central mesh of Fig. 11, and if z_b contains only poles of the impedance E_2/I_2 looking to the right at this junction, then the driving-point impedance at the right-hand terminals of the total network, after insertion of z_b, contains these poles in addition to the poles that this driving-point impedance had to begin with. In other words, the input impedance at the right-hand end of the total network acquires some additional poles because of the insertion of z_b (namely, the poles of z_b), which are some or all of the natural frequencies of the open-circuited portion to the right of the central mesh.

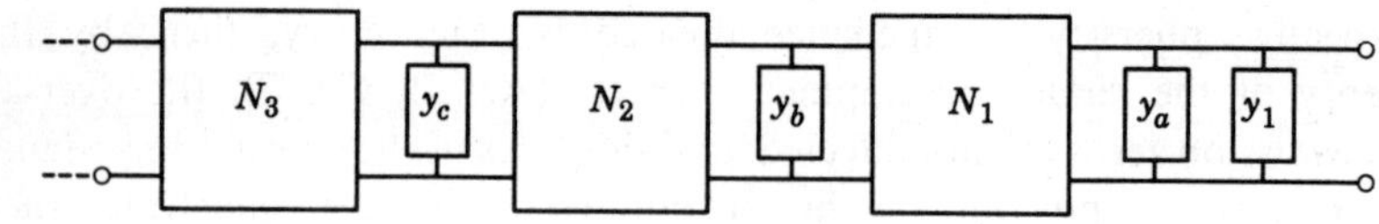

Fig. 13. Alternate method (dual to that illustrated in Fig. 12) for accomplishing the same end.

Observe that the driving-point impedance at the left-hand terminals of the total network does not acquire these additional poles. The situation here is exactly analogous to that discussed in Art. 3, Ch. 7, where, in the ladder development of z_{11}, it is pointed out that partial pole removals (done for the purpose of zero-shifting) cause the impedance z_{22} of the resulting network to contain these poles in addition to the poles of z_{11} and z_{12}; but z_{11} obviously does not acquire such additional poles because this is the developing function, and neither does z_{12} acquire them because z_{12} cannot contain poles that are not also contained in *both* z_{11} and z_{22}. The same condition applies to lossy networks since for physical reasons the transfer impedance evidently cannot contain natural frequencies that are not excitable at either terminal pair because it must obey the demands of the reciprocity theorem.

If, with reference to Fig. 11, we define the impedances

$$Z_1 = E_1/I_1, \qquad Z_2 = E_2/I_2, \qquad Z_3 = E_3/I_3, \cdots \tag{104}$$

and insert an impedance z_a as in Fig. 12 having no poles other than those of Z_1, then the impedances $Z_2, Z_3, \cdots$ do not acquire these or any other additional poles, and the transmission zeros are not affected. However, the pole positions of the impedances $Z_2, Z_3, \cdots$ as well as those of the over-all transfer impedance are shifted, these latter poles being the zeros of the mesh impedances like the one in the right-hand mesh that has z_a added to it. This addition of z_a to the right-hand mesh impedance shifts its zeros and hence shifts the poles of the over-all transfer impedance.

By next inserting z_b having only poles of Z_2 (for z_a present, because the insertion of z_a shifts these poles although it does not augment their number), we can produce a further shift in the poles of the transfer impedance without affecting either the number of these poles or the positions of the transmission zeros. Similarly, with reference to Fig. 13, we can accomplish the same result by the insertion of shunt admittances y_a, y_b, $\cdots$, the argument being completely dual to the one given for the insertion of impedances.

These are procedures whereby we can attempt to produce a desired pole pattern, having fixed the prescribed transmission zeros to start with. By choosing for the poles of z_a, z_b, $\cdots$ only those of the impedances Z_1, Z_2, $\cdots$, successively and respectively, we avoid having the successively encountered mesh impedances become increasingly complicated. For each inserted impedance we choose zeros that: (*a*) leave that impedance p.r., and (*b*) yield zeros for the resultant mesh impedance that are located as near to the desired transfer impedance poles as is consistent with requirement (*a*).

When many zeros and poles are involved, this process becomes computationally somewhat tedious unless appropriate computing aids are available. Simple problems, on the other hand, are most effectively dealt with in this manner. As an example, suppose we want a transfer function with a pair of *j*-axis zeros (which we may normalize at the points $\omega = \pm 1$) and a pair of left half-plane poles at prescribed points. The network is to terminate in a 1-ohm load resistance.

We begin tentatively with the circuit shown in Fig. 14*a*. This evidently provides the desired pair of transmission zeros and the 1-ohm terminal resistance. The impedance level factor k for the series LC branch remains arbitrary. By varying k, the poles of the impedance of this simple circuit can, in a familiar manner, be made to lie anywhere upon the dotted semicircle in the s-plane sketch of Fig. 15, or at points on the real axis whose geometric mean value equals -1, but no other locations in the left half-plane are possible pole positions.

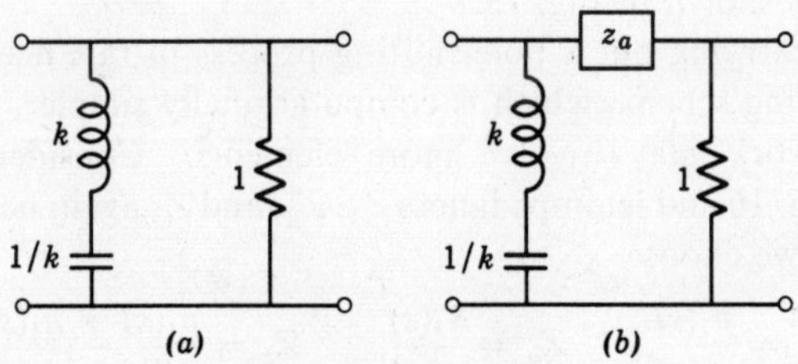

Fig. 14. Tentative circuit (*a*) having desired *j*-axis transmission zeros; modified circuit (*b*) having transfer impedance with same zeros as in (*a*) but with altered poles.

We therefore insert the impedance z_a as shown in Fig. 14*b*. Since the mesh impedance has poles at $s = 0$ and at $s = \infty$, z_a may be an inductance or a capacitance or their series combination. The zeros of the mesh impedance with z_a equal to an inductive reactance evidently lie within the dotted semicircle of Fig. 15, whereas for z_a equal to a capacitive reactance they lie outside the circle. It is a simple matter to choose the values of k and the additional L or C so as to make these zeros (which are the transfer function poles) fall at any desired points. The transmission zeros meanwhile remain at their originally chosen j-axis positions.

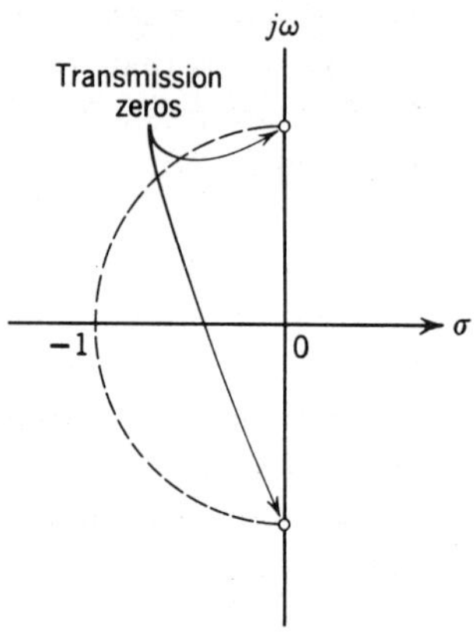

Fig. 15. Pertinent to the discussion of possible pole position of the transfer impedance function for the circuit of Fig. 14*a*.

If we want left half-plane zeros, we simply introduce some resistance in the series-resonant shunt branch; and if we want more than one pair of such zeros we consider a tentative network with several series-resonant shunt branches. Alternately, we can use a completely dual circuit consisting of parallel-resonant series branches, into which we insert shunt admittances as shown in Fig. 13.

Instead of carrying out a pole-shifting process in this manner, we can use the following scheme which is computationally simpler, although the resulting network may involve more elements. Consider the ladder network of Fig. 16 and let impedances Z_1, Z_2, and Z_3 again be defined as in Eqs. 104. If we choose

$$Z_1 = \frac{p_1(s)}{q_1(s)}, \qquad Z_a = \frac{h_1(s)}{q_1(s)}, \qquad Z_b = \frac{p_1(s) + h_1(s)}{q_2(s) - q_1(s)} \tag{105}$$

then we find

$$\frac{I_1}{I_2} = \frac{Z_b}{Z_1 + Z_a + Z_b} = \frac{q_1(s)}{q_2(s)} \tag{106}$$

and

$$\frac{E_1}{I_2} = \frac{E_1}{I_1} \times \frac{I_1}{I_2} = \frac{p_1(s)}{q_1(s)} \times \frac{q_1(s)}{q_2(s)} = \frac{p_1(s)}{q_2(s)} \tag{107}$$

Thus, if we regard $Z_1(s)$ as a tentative form of transfer impedance, then the branches Z_a and Z_b leave its zeros intact but shift its poles from the zeros of $q_1(s)$ to those of $q_2(s)$.

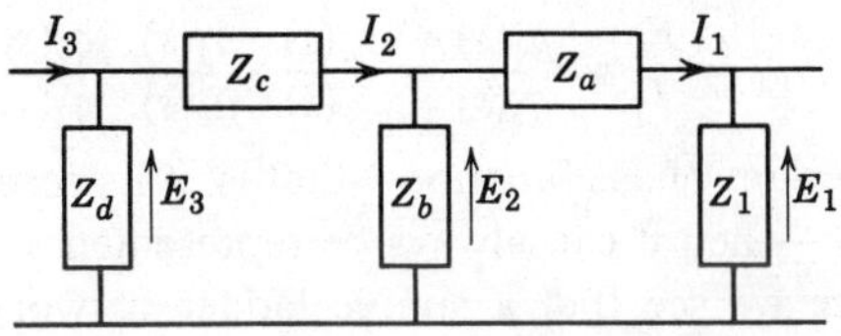

Fig. 16. Ladder network pertinent to the pole-shifting process given by Eqs. 105 through 111.

The polynomial $h_1(s)$, which plays the role of a mediator and drops out of the end result, must be chosen so that both impedances Z_a and Z_b become p.r. for the desired $q_2(s)$ determining the new pole locations. Essentially it must be possible to find a polynomial $h_1(s)$ such that both h_1/q_1 and $h_1/(q_2 - q_1)$ are p.r., for one can always give h_1 a sufficiently large constant multiplier to make $(p_1 + h_1)/(q_2 - q_1)$ p.r. if $h_1/(q_2 - q_1)$ is p.r. to start with. The polynomial $h_1(s)$ evidently exists if the new pole locations determined by $q_2(s)$ are not too far from the original ones determined by $q_1(s)$. One can always secure a Hurwitz polynomial for $(q_2 - q_1)$ by weighting q_2 with a sufficiently large multiplier, although the use of this expedient should be minimized since it tends to lower the over-all gain.

The important feature about this process is the fact that it may be applied repeatedly in the same simple fashion. We have

$$Z_2 = \frac{Z_b(Z_1 + Z_a)}{Z_1 + Z_a + Z_b} = \frac{p_1(s) + h_1(s)}{q_2(s)} = \frac{p_2(s)}{q_2(s)} \tag{108}$$

where p_2 is an abbreviation for $p_1 + h_1$. We then choose

$$Z_c = \frac{h_2(s)}{q_2(s)}, \qquad Z_d = \frac{p_2(s) + h_2(s)}{q_3(s) - q_2(s)} \tag{109}$$

and have

$$\frac{I_2}{I_3} = \frac{Z_d}{Z_2 + Z_c + Z_d} = \frac{q_2(s)}{q_3(s)} \tag{110}$$

The next input impedance is

$$Z_3 = \frac{Z_d(Z_2 + Z_c)}{Z_2 + Z_c + Z_d} = \frac{p_2(s) + h_2(s)}{q_3(s)} = \frac{p_3(s)}{q_3(s)} \tag{111}$$

and so forth.

Observe that Z_a has the poles of Z_1, and Z_b has the zeros of $Z_1 + Z_a$, as the preceding discussion shows is necessary and sufficient to keep the transmission zeros intact and to assure that the resultant input impedance does not increase in complexity as the process continues. Similarly, Z_c has the poles of Z_2 and Z_d has the zeros of $Z_2 + Z_c$.

The over-all transfer impedance for the structure in Fig. 16 appears broken down into a product of factors, in the form

$$Z_{13} = \frac{E_1}{I_3} = \frac{p_1(s)}{q_3(s)} = \frac{p_1(s)}{q_1(s)} \times \frac{q_1(s)}{q_2(s)} \times \frac{q_2(s)}{q_3(s)} \tag{112}$$

If $Z_{13}(s)$ is a minimum-phase function—that is, if its zeros are restricted to the left half-plane—then it can always be represented as a product of p.r. functions. Hence we see that a simple ladder network like the one in Fig. 16 can realize any minimum-phase transfer function, apart from a constant multiplier. The converse of this statement is obvious; namely, that a simple ladder network must yield a minimum-phase transfer impedance because its branch impedances (whose zeros and poles determine the transmission zeros) are p.r. Now we see that any given minimum-phase transfer impedance is always so realizable.

When the given transfer impedance is not minimum phase, we can begin its synthesis by constructing a two terminal-pair network having the desired transmission zeros and any convenient set of poles that expedite the realization of this network in unbalanced form. We should seek in this step to pick the poles in such a way that the two terminal-pair has a desired realization, while the chosen poles are as near to the prescribed poles as is consistent with this objective. The synthesis procedure described in the preceding article can be used to carry out this step. The input impedance to this two terminal-pair network we then identify with the impedance Z_1 in Fig. 16, and apply the process just described to shift the poles into their prescribed positions.

A similar procedure is necessary for the realization of a minimum-phase function if it has multiple-order zeros at $s = \infty$, or anywhere else on the j-axis. The following example serves to illustrate the thought expressed here. Suppose we want a transfer impedance having a double-order zero at $s = \infty$ and poles at $s = -\frac{1}{2} \pm j(\sqrt{3}/2)$. In order to show how arbitrary we can be to start with, we choose as a tentative network the *RC* circuit shown in Fig. 17, which clearly has two transmission zeros at $s = \infty$ because of the two shunt capacitance branches. A possible input impedance to this network is

$$Z_1(s) = \frac{s+2}{(s+1)(s+3)} = \frac{s+2}{s^2+4s+3} \tag{113}$$

where we have deliberately not even tried to choose poles close to the desired ones. Of course, since the tentative network is RC, the poles of its impedance must lie on the real axis and hence could not be very close to the desired poles anyway.

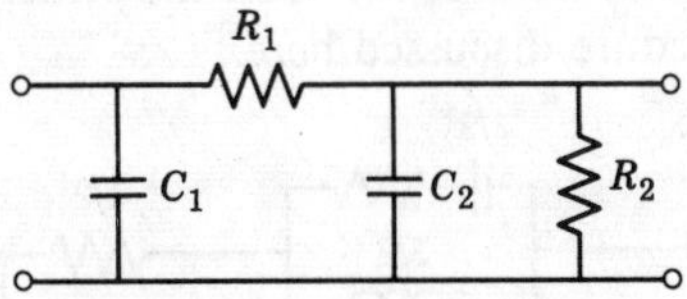

Fig. 17. Tentative circuit having a transfer impedance with zeros at $s = \infty$.

We now shift the poles by identifying Z_1 in Eq. 113 with Z_1 in Fig. 16. Assuming that the desired shift is possible in one step, we write

$$q_2(s) = K(s^2 + s + 1) \tag{114}$$

in which K is a multiplier. Writing the auxiliary polynomial as

$$h_1(s) = a_2s^2 + a_1s + a_0 \tag{115}$$

Eqs. 105 yield

$$Z_a = \frac{a_2s^2 + a_1s + a_0}{s^2 + 4s + 3} \tag{116}$$

and

$$Z_b = \frac{a_2s^2 + (1 + a_1)s + 2 + a_0}{(K-1)s^2 + (K-4)s + K - 3} \tag{117}$$

The constant K must be at least 4. If we choose its value equal to 4, then the discussion in Art. 3, Ch. 9, shows that Z_b has a simple realization for $a_2/(2 + a_0) = 3$. Letting $a_0 = 0$ is evidently possible and yields a simple Z_a. Since we now have $a_2 = 6$, the realization of Z_a requires $a_1 \geq \frac{9}{2}$. Choosing the equal sign here, we have

$$Z_a = \frac{6s^2 + \frac{9}{2}s}{s^2 + 4s + 3} = \cfrac{1}{\cfrac{2}{3s} + \cfrac{1}{\cfrac{9}{2s} + 6}} \tag{118}$$

and

$$Z_b = \frac{6s^2 + \frac{5}{2}s + 2}{3s^2 + 1} = 2 + \frac{\frac{5}{2}s}{3s^2 + 1} \tag{119}$$

The total network is shown in Fig. 18, and the over-all transfer impedance turns out to be

$$Z_{12} = \frac{\frac{1}{8}}{s^2 + s + 1} \tag{120}$$

If we raise the over-all impedance level by a factor of 6 so as to normalize the load resistance at 1 ohm, the constant multiplier is $\frac{3}{4}$, which is quite acceptable. The total network is, of course, rather uneconomical regarding the number of required circuit elements (one can obtain the same result with a total of three elements by the method of Art. 1, Ch. 11), but the example serves nevertheless to illustrate the principles and the flexibility of the synthesis procedure discussed here.

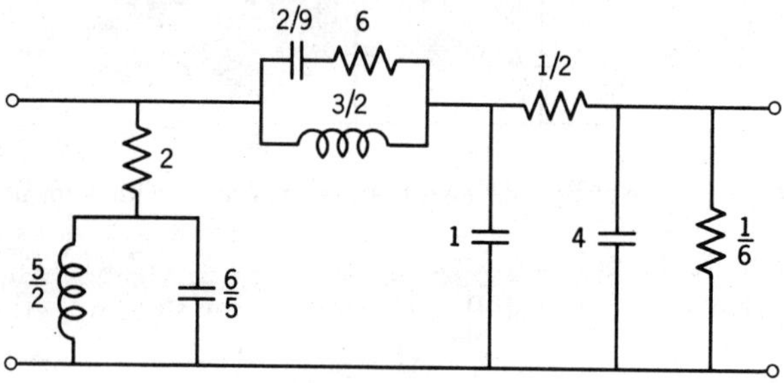

Fig. 18. The circuit of Fig. 17 with the poles of its transfer impedance shifted by the scheme of Fig. 16 to yield the result 120.

As a second example, consider the desired transfer impedance

$$Z_{12}(s) = \left(\frac{s + \frac{1}{2}}{s + 2}\right)^3 \tag{121}$$

The p.r. impedance

$$Z_1(s) = \left(\frac{s + \frac{1}{2}}{s + \frac{3}{2}}\right)^3 \tag{122}$$

has the same zeros, and poles as close to those of $Z_{12}(s)$ as the p.r. requirement will allow. We identify it with the impedance Z_1 in Fig. 16. Therefore

$$p_1(s) = s^3 + \tfrac{3}{2}s^2 + \tfrac{3}{4}s + \tfrac{1}{8} \tag{123}$$

and

$$q_1(s) = s^3 + \tfrac{9}{2}s^2 + \tfrac{27}{4}s + \tfrac{27}{8} \tag{124}$$

Again assuming that one step in the pole-shifting procedure will suffice, we have

$$q_2(s) = K(s + 2)^3 = K(s^3 + 6s^2 + 12s + 8) \tag{125}$$

and hence

$$q_2 - q_1 = (K - 1)s^3 + (6K - \tfrac{9}{2})s^2 + (12K - \tfrac{27}{4})s + (8K - \tfrac{27}{8}) \tag{126}$$

For $K = 1$

$$q_2 - q_1 = \tfrac{3}{2}s^2 + \tfrac{21}{4}s + \tfrac{37}{8} \tag{127}$$

which has zeros very close to $s = -\frac{3}{2}$ where the zeros of $q_1(s)$ lie. We therefore see that we might try setting $h_1 = Rq_1$ since $Z_a = h_1/q_1$ is then simply a constant, and $h_1/(q_2 - q_1)$ is surely p.r. because its zeros and poles are close together. By choosing a sufficiently large value for R we can then make Z_b p.r. also, and achieve a very simple end result. The scheme works for $R = 2$, and we get

$$Z_a = 2, \qquad Z_b = 2s + \cfrac{10}{3s + \cfrac{51s + 74}{8s + 11}} \tag{128}$$

Since realization of the p.r. impedance $Z_1(s)$ in Eq. 122 is a straightforward matter, and the circuits for Z_a and Z_b in 128 are almost obvious by inspection, we shall not bother to carry through the rest of the synthesis.

The choice $h_1 = Rq_1$ which yields $Z_a = R$ can always be made if the quotient q_2/q_1 is p.r. with a nonzero j-axis real part. For, assigning to q_2 a positive real multiplier k, we obtain from Eq. 105 $Z_b = (p_1 + Rq_1)/(kq_2 - q_1)$, which then can be rendered p.r. for sufficiently large values of k and R. The condition that q_2/q_1 be p.r. may not be necessary in some cases, but it is seen to be sufficient.*

3 Some Closely Related Procedures

With reference to Fig. 16 and the impedances defined by Eqs. 104, suppose we introduce the additional admittances

$$Y_\alpha = I_1/E_2, \qquad Y_\beta = I_2/E_3 \tag{129}$$

We then have

$$Z_1 \times Y_\alpha \times Z_2 \times Y_\beta \times Z_3 = \frac{E_1}{I_1} \times \frac{I_1}{E_2} \times \frac{E_2}{I_2} \times \frac{I_2}{E_3} \times \frac{E_3}{I_3} = \frac{E_1}{I_3} = Z_{13} \tag{130}$$

and by inspection

$$\begin{aligned} Z_a &= \frac{1}{Y_\alpha} - Z_1 \\ \frac{1}{Z_b} &= \frac{1}{Z_2} - Y_\alpha \\ Z_c &= \frac{1}{Y_\beta} - Z_2 \\ \frac{1}{Z_d} &= \frac{1}{Z_3} - Y_\beta \end{aligned} \tag{131}$$

In Eq. 130 the desired transfer impedance is expressed as a product of p.r. functions, introducing, if necessary, some surplus factors into the numerator and denominator polynomials of Z_{13} in order to make such a representation possible.† The impedance functions Z_a, Z_b, Z_c, Z_d, according to Eqs. 131, can then be rendered p.r. by incorporating appropriate constant multipliers in the p.r. factors Z_1, Y_α, $Z_2 \cdots$ of Z_{13}.

* In judging the adequacy of tentative choices for the multipliers k and R, the property discussed in connection with Fig. 9, Ch. 1, will here be found to save computational labor.

† In this connection see Prob. 31 of Ch. 1.

The method can obviously be applied to the synthesis of a voltage transfer ratio as well; for, by omitting the last factor in Eq. 130 we get*

$$\frac{E_1}{E_3} = Z_1 \times Y_\alpha \times Z_2 \times Y_\beta \tag{132}$$

Applying this method to the realization of the transfer impedance 121, we write

$$\frac{Z_1}{A_1} = \frac{Y_\alpha}{A_\alpha} = \frac{Z_2}{A_2} = \frac{s+\frac{1}{2}}{s+2} \tag{133}$$

and have by Eqs. 131

$$Z_a = \frac{1}{A_\alpha} \times \frac{s+2}{s+\frac{1}{2}} - A_1 \times \frac{s+\frac{1}{2}}{s+2} \tag{134}$$

$$\frac{1}{Z_b} = \frac{1}{A_2} \times \frac{s+2}{s+\frac{1}{2}} - A_\alpha \times \frac{s+\frac{1}{2}}{s+2} \tag{135}$$

The constant multipliers A_α, A_1, A_2 may here be chosen equal to unity, and we get

$$Z_a = \frac{1}{Z_b} = \frac{3(s+\frac{5}{4})}{(s+2)(s+\frac{1}{2})} \tag{136}$$

Together with Z_1 given by Eq. 133, this result yields the network shown in Fig. 19.

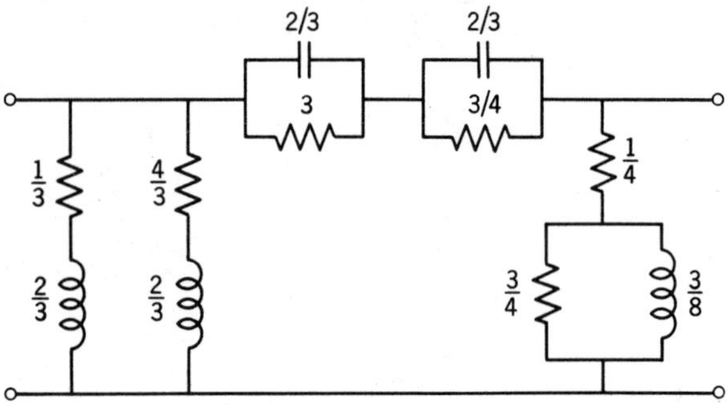

Fig. 19. Realization of the transfer impedance 121 by the method to which Eqs. 129, 130, 131 and Fig. 16 are pertinent.

A rather simple-minded method of realizing minimum-phase transfer functions (pointed out on p. 323 of *ICT*) bears a close relation to the one just given. If Z_{13} in Eq. 130 has no j-axis poles, then the functions Z_1 and Z_2 possess complements. The functions Y_α and Y_β in Eqs. 131 may

* P. M. Lewis pointed out that the method discussed by R. H. Pantell in "Minimum-Phase Transfer Function Synthesis," *IRE Trans.*, Vol. CT-2, No. 2, June 1955, arrives at this same result, although through a more elaborate matrix factorization process.

therefore be chosen equal to constants, so that Z_a complements Z_1 and Z_c complements Z_2.

Specifically, if the largest j-axis real parts of Z_1 and Z_2 are R_1 and R_2 respectively, then we choose $Y_\alpha = 1/R_1$ and $Y_\beta = 1/R_2$. Eqs. 131 then read

$$\begin{aligned} Z_a &= R_1 - Z_1 \\ \frac{1}{Z_b} &= \frac{1}{Z_2} - \frac{1}{R_1} \\ Z_c &= R_2 - Z_2 \\ \frac{1}{Z_d} &= \frac{1}{Z_3} - \frac{1}{R_2} \end{aligned} \tag{137}$$

Through the proper choice of multipliers in Z_2 and Z_3, the p.r. character of Z_b and Z_d can evidently be assured.

For the same example considered above (for which the desired transfer impedance is given by Eq. 121) we write

$$\frac{Z_1}{A_1} = \frac{Z_2}{A_2} = \frac{Z_3}{A_3} = \frac{s + \frac{1}{2}}{s + 2} \tag{138}$$

which is an RL impedance. Hence its maximum j-axis real part occurs at $s = \infty$. It is not difficult to see that the constant multipliers may all be chosen equal to unity, and Eqs. 137 yield

$$Z_a = Z_c = 1 - \frac{s + \frac{1}{2}}{s + 2} = \frac{\frac{3}{2}}{s + 2} \tag{139}$$

$$\frac{1}{Z_a} = \frac{1}{Z_d} = \frac{s + 2}{s + \frac{1}{2}} - 1 = \frac{\frac{3}{2}}{s + \frac{1}{2}} \tag{140}$$

The resulting network is shown in Fig. 20. It contains the same number of elements as the one in Fig. 19.

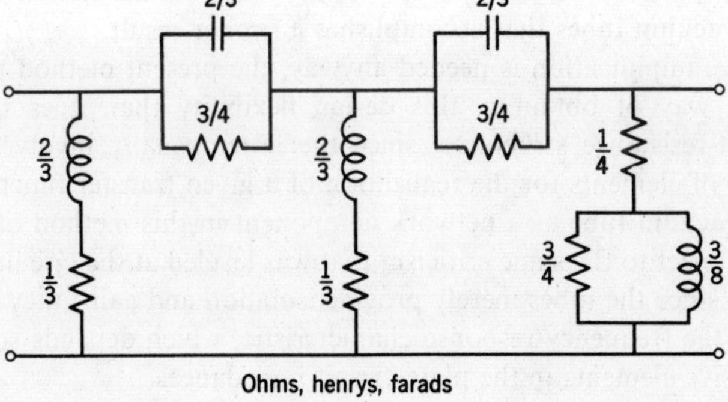

Fig. 20. Alternate realization of the transfer impedance 121 in which each series branch complements the impedance following it. (See *ICT*, p. 323, or Eq. 137 relative to Fig. 16.)

Either of these procedures may, of course, be regarded as a special form of the more general one discussed in the previous article. We will let the reader write down the pertinent relations for the polynomial $h_1(s)$ in terms of $p_1(s)$ and $q_1(s)$ for a typical cycle as an exercise.

4 Realization as a Cascade of Pentode Amplifier Stages

The representation of a minimum-phase transfer function as a product of p.r. functions (Eq. 112 or 130) immediately suggests the procedure for designing a multistage amplifier to have a response characteristic of this sort. Since a pentode vacuum tube is essentially a device whose output current is proportional to the input voltage, it follows that the voltage across an impedance in the output (plate) circuit is proportional to the product of the input voltage and that impedance. In other words, such a device has a voltage transfer ratio (or output-voltage to input-current ratio if we connect a resistance across the input terminals) which, apart from a constant multiplier, is simply equal to the impedance in its plate circuit. Hence, a cascade of such stages yields an over-all transfer impedance or voltage ratio that is a constant times a product of the pertinent plate impedances.

The method of decomposing a given minimum-phase transfer impedance into a product of p.r. functions in general involves the use of surplus factors as in the synthesis of transfer functions by means of the constant-resistance lattice (Art. 8, Ch. 11). In fact, the present procedure is like the synthesis of a cascade of constant-resistance lattices in that each stage contributes its share to the over-all transfer function independently of its association with the others. In the lattice procedure it is the constant-resistance property of each network in the cascade that provides this independence in the behavior of each component; in the cascade of pentode amplifier stages, it is the isolation or "buffer action" provided by the vacuum tubes that accomplishes a similar result.

Where amplification is needed anyway, the present method provides a cheaper way of obtaining this design flexibility than does the use of constant-resistance structures, since the latter usually involves a larger number of elements for the realization of a given transfer function. Use of the vacuum tube as a network component in this method of synthesis is not subject to the same criticism that was leveled at the circuit shown in Fig. 10, since the tubes merely provide isolation and gain; they in no way control the frequency response characteristic, which depends solely upon the passive elements in the plate-circuit impedances.

When surplus factors are used in the process of reducing a given transfer impedance to a product of p.r. functions, we can interpret the action of stages involving such factors as pole-shifting stages or zero-shifting-stages. For example, suppose the prescribed transfer impedance is

$$Z_{12}(s) = \frac{(s+1)^2}{s^2 + 10s + 50} \tag{141}$$

If we rewrite this function as

$$Z_{12}(s) = \frac{(s+3)^2}{s^2 + 10s + 50} \times \frac{(s+1)^2}{(s+3)^2} \tag{142}$$

we find that each factor is not only p.r. but is realizable in the simple terms to which conditions 47 and 48 in Art. 3, Ch. 9, pertain. The first factor in the product 142 may be regarded as shifting the poles from the points $s = -5 \pm j5$ to $s = -3$; or the second factor may be regarded as shifting the zeros from the point $s = -1$ to $s = -3$.

Suppose we illustrate the method with a synthesis of the transfer impedance 25 in Ch. 11 having the magnitude 18 and the network realization given there in Fig. 2. We again need to introduce a surplus factor in order to produce the desired reduction. A possible form for the transfer impedance reads

$$Z_{12}(s) = \frac{s+1}{s^2 + s + 1} \times \frac{1}{s+1} \times \frac{1}{s+1} \tag{143}$$

and its realization (apart from a constant multiplier) as a three-stage pentode amplifier is shown in Fig. 21.

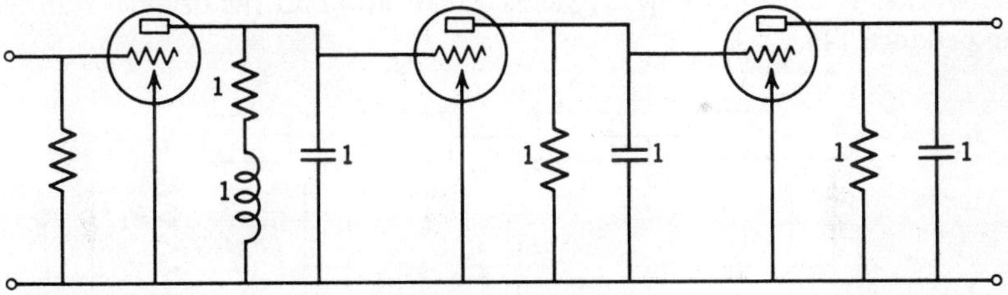

Fig. 21. Realization of the transfer impedance 25 in Ch. 11 (with the squared magnitude given there by Eq. 18 and the network of Fig. 2) by a cascade of pentode amplifier stages according to the representation 143.

5 Another Method of Realizing Transfer Impedances

Suppose a given minimum-phase function is written

$$Z_{12}(s) = \frac{m_1 + n_1}{m_2 + n_2} = \frac{m_1}{m_2 + n_2} + \frac{n_1}{m_2 + n_2} \tag{144}$$

Since $m_1 + n_1$ as well as $m_2 + n_2$ is a Hurwitz polynomial, the even and odd parts m_1 and n_1 have simple zeros restricted to the j-axis. The terms in Eq. 144 may therefore be realized individually by means of resistance-terminated lossless ladder networks using the method discussed in Art. 1, Ch. 11, in conjunction with the ladder development process described in Art. 3, Ch. 7. Specifically, these terms may be identified respectively with the forms 10 and 11 given in Art. 1, Ch. 11. We may thus write

$$Z_{12}^{(1)}(s) = \frac{m_1}{m_2 + n_2} = \frac{z_{12}^{(1)}}{1 + z_{22}^{(1)}}, \qquad z_{12}^{(1)} = \frac{m_1}{n_2}, \qquad z_{22}^{(1)} = \frac{m_2}{n_2} \tag{145}$$

and

$$Z_{12}^{(2)}(s) = \frac{n_1}{m_2 + n_2} = \frac{z_{12}^{(2)}}{1 + z_{22}^{(2)}}, \qquad z_{12}^{(2)} = \frac{n_1}{m_2}, \qquad z_{22}^{(2)} = \frac{n_2}{m_2} \tag{146}$$

Since the zeros of each z_{12} function are restricted to the j-axis, the simple method of developing z_{22} into a lossless ladder, with series-resonant shunt branches or parallel-resonant series branches producing zeros of transmission, is possible for the separate realization of each of these two component transfer impedances.

An unbalanced lossless ladder network, terminated in its appropriate load resistance, is thus readily found for each of the component Z_{12} functions given in 145 and 146. The terminal resistances for these networks are not necessarily alike since their impedance levels are the means for independently controlling the multipliers called for by the expressions for $Z_{12}^{(1)}$ and $Z_{12}^{(2)}$. A back-to-back arrangement of these two ladder networks, as shown in Fig. 22, yields a realization for the original transfer impedance 144.

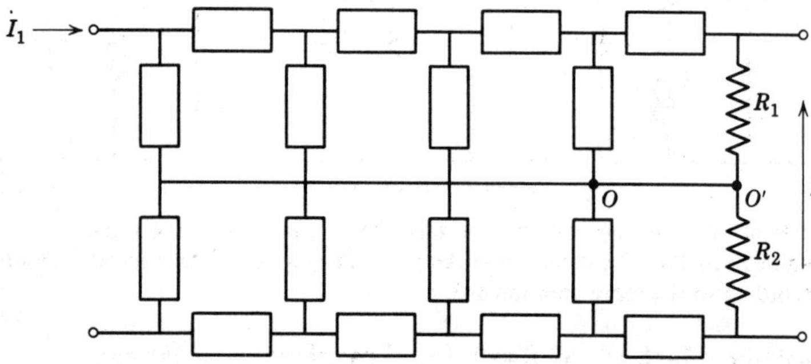

Fig. 22. A possible realization for the minimum-phase transfer impedance 144 according to the representation of its components as given by Eqs. 145 and 146.

Observe that the z_{22} functions in 145 and 146 are reciprocal. If they were alike, one could omit the central wire between the points O–O' in the circuit of Fig. 22 and terminate the back-to-back ladder arrangement in a single resistance. However, since the z_{22} of the two ladders are not alike, this simplification is not possible. One can, of course, expect that termination in a single resistance is not possible since it was pointed out in Art. 1, Ch. 11, that a minimum-phase transfer function cannot be realized in a single resistance-terminated lossless network.

When the given transfer impedance 144 is not minimum phase, one can still represent it as a sum of the components 145 and 146. Although not realizable as ladder networks, these are still realizable as unbalanced lossless two terminal-pair networks (possibly involving mutual inductive coupling as shown in Art. 4, Ch. 7) terminated in resistance loads. Their back-to-back arrangement, similar to that shown for ladder networks in Fig. 22, yields a pertinent realization.

6 Precorrection for Incidental Dissipation

In the practical application of synthesis methods, it is essential to be able in some measure to cope with the perturbative effect of so-called parasitic elements. These unavoidable circuit elements, which are actually but not intentionally present in the physical embodiment of a network realization, usually produce undesirable if not intolerable modifications in the resulting response characteristics.

There are in general two ways in which one attempts to deal with this problem. One is to choose, out of an array of equivalent networks of varying topological configurations and element-value distributions, that one in which these effects become least prominent in the corresponding physical circuit. The other method anticipates the distorting effect of the parasitic elements and alters accordingly the called-for response function in a compensating manner.

The effect of parasitic elements in a passive circuit is simply to change slightly the expected pole-zero pattern of the desired response function. If these shifts in the pole and zero positions can be predetermined, an obvious corrective measure is to preshift the poles and zeros of the desired response function by like amounts in opposite directions. This process, which is referred to as "precorrecting" the given response function, is particularly useful as a means for combating the effect of incidental losses associated with the inductance and capacitance elements. The extent to which it can be used, and the detailed method of carrying out this compensating scheme, are discussed in the following paragraphs.

The approximate effect of incidental dissipation in an inductance is to associate with it a series resistance; in a capacitance element it effectively produces a parallel conductance. If the pertinent network schematic happens to associate resistances in series with all the inductances, and conductances in parallel with all the capacitance elements, and if these resistances and conductances have values at least as large as those called for by the incidental dissipation anticipated in the physical embodiment of this network schematic, then we obviously do not need to carry out any compensation scheme. Unless we deliberately plan the synthesis procedure so as to bring about this fortuitous situation, however, it cannot be counted upon as a method of solving our difficulty; and if we do carry out the synthesis with proper forethought plus a clever trick or two to force the hand of fate, so to speak, then, in effect, we *are* carrying out a precorrection scheme; and all such schemes, as we shall see, are equivalent in the sense that none can do any better than the character of the desired pole-zero pattern will allow. That is to say, when one such method fails to provide a sufficient amount of compensation for the anticipated incidental loss, then all other methods will likewise fail* (assuming, of course, that they are all adjusted to provide optimum compensation).

We shall see that the criterion that sets an upper bound upon the amount of incidental loss that can be absorbed by any precorrection scheme is the nearness of the poles of the pertinent response function to the j-axis. The farther the poles are from the j-axis, the larger the amount of incidental loss may be that we can compensate for by appropriate maneuvering of our total synthesis procedure. Wherever this maximum amount of incidental loss absorption is insufficient, we must find an acceptable response function with poles farther from the j-axis. This process, which involves the solution to an approximation problem, is discussed in Ch. 14, where we show that an acceptable response function with poles farther from the j-axis has a larger number of poles and inherently a lower gain factor. Thus it turns out that theoretically any amount of incidental loss can be absorbed if we are willing to pay the price in added network complexity and loss in gain.

The basis for carrying out a precorrection scheme is readily recognized if, in a given impedance function $Z(s)$, we consider the effect upon the associated network of replacing the variable s by $s + \delta$ in which δ is a positive real quantity. The result produced by this change of variable is more easily seen if we regard the impedance as a function of the network resistances R_k, its inductive impedances $L_k s$, and capacitive admittances $C_k s$, which collectively determine the values of the given impedance for different values of s. We thus become aware of the fact that the variable s is always associated multiplicatively with an L_k or a C_k. The following

---

* A few very special exceptions to this statement do occur. In one of these, inductances are involved only in series with the source and load resistances, and other losses are negligible. Replacing the pure inductances by lossy inductances in this situation merely lowers somewhat the anticipated gain. Such devices are too restricted in their applicability to exert an influence upon any statement regarding more generally achievable results.

changes, therefore, are evident

$$\begin{aligned} s &\to s + \delta \\ L_k s &\to L_k s + \delta L_k \\ C_k s &\to C_k s + \delta C_k \end{aligned} \tag{147}$$

Physically these results imply that each inductance has associated with it a proportionate amount of series resistance, and each capacitance is paralleled by a proportionate amount of conductance. The ratios R/L and G/C, which are referred to as the *dissipation ratios*, have the common value δ.

The corresponding effect upon the critical frequencies of the impedance $Z(s)$ is recognized by observing that any particular point $s = s_k$ is changed to a point where $s + \delta = s_k$, or $s = s_k - \delta$. Hence all zeros and poles of the function $Z(s)$ are displaced horizontally to the left by the amount δ. The entire pole-zero pattern in the s-plane is displaced bodily to the left by the value of δ.

Hence we have the simple result that adding proportionate amounts of series resistance to all the L's and like amounts of parallel conductance to all the C's in a given network (leaving the values of all existing R's as well as the network geometry unchanged) has the effect of displacing the pole-zero patterns of all associated driving-point and transfer impedance or admittance functions to the left by an amount equal to the pertinent proportionality constant. The converse of this statement is, of course, also true.

The method of precorrecting for an anticipated amount of incidental dissipation is equally clear. If we wish to provide for series resistances in all the L's and parallel conductances in all the C's with a specified dissipation ratio δ, then the synthesis is tentatively carried out for a pole-zero pattern that is displaced to the right by an amount δ. Or we may say that the synthesis of a given impedance $Z(s)$ is tentatively done for the function $Z(s - \delta)$ instead. After the proportionate amounts of resistance and conductance are associated with the L's and C's of the network thus found, the pertinent impedance function reverts from $Z(s - \delta)$ to $Z(s)$, which we wanted in the first place.

The necessary and sufficient condition for being able to carry out this corrective procedure is evidently that the poles of the function $Z(s - \delta)$ must still be in the left half-plane; and if a driving-point impedance is involved, $Z(s - \delta)$ must be p.r. If $Z(s)$ is a p.r. function whose real part has zeros on the j-axis, then $Z(s - \delta)$ for any $\delta > 0$ is clearly no longer p.r. even though its zeros and poles may still be in the left half-plane. Similarly, a p.r. function $Z(s)$ having j-axis poles cannot remain p.r. under the transformation $s \to s - \delta$ with $\delta > 0$, even though the j-axis real part of $Z(s)$ is positive by an appreciable margin, for the real part associated with the j-axis poles, although identically zero in the given function (neglecting impulses), becomes nonzero and negative for the displacement $s \to s - \delta$, which moves these poles into the right half-plane.

In applying this method to a transfer function, it is the pole or pole-pair nearest the j-axis that sets the limit upon the largest tolerable value of δ for the entire realization unless we use a method of synthesis that involves the breakdown of this function into multiplicative factors which are separately identified with component networks in a cascade arrangement. In a situation of this sort, the pole pattern of each factor independently sets its upper limit upon δ, and the maximum amount of incidental dissipation tolerable in each network is then correspondingly different. Hence the nearest pole in the pattern of the given function demands low-loss elements only in its corresponding network and not in the entire realization.

A similar scheme can be applied to the realization of a driving-point function only if it is decomposable into additive components that are separately p.r., minimum reactive, and have j-axis real parts that are positive with a margin to spare. When compensation for incidental losses under these circumstances is inadequate, the only available procedure is to reconsider the approximation problem yielding the given impedance function, so as to obtain an acceptable substitute function that provides larger margins in the necessary conditions.

Although this statement is true in a general sense, it is possible in many practical situations to squeeze a little more dissipation absorption ability out of a given impedance function (a maximum of a factor of 2 in δ) by recognizing that the necessary dissipation ratio G/C for the capacitances is less than the required ratio R/L for the inductances; and that the following considerations allow us, in an acceptably approximate fashion, to trade one for the other since the effective value of δ turns out essentially to depend only upon the sum of $R/2L$ and $G/2C$. For a given maximum allowable δ value, the ratio R/L may thus become twice as large if the anticipated incidental loss in the capacitance elements is essentially zero.

In order to obtain this result, we begin by recognizing that the driving-point or transfer impedance of any RLC network may be regarded as a function of R's and quantities having the form $(Ls + 1/Cs)$ or $(Cs + 1/Ls)$, the first of these occurring wherever there is an inductance alone or in series with a capacitance, and the second, wherever there is a capacitance alone or in parallel with an inductance. The functional dependence of an impedance upon the frequency variable s and its sensitivity to changes in this variable clearly are not concerned with the resistance branches in the network but only with the branches whose impedances or admittances are given by the forms $(Ls + 1/Cs)$ or $(Cs + 1/Ls)$ respectively. Since these are duals, it will suffice to consider only one of them.

Suppose the L's are replaced by series combinations of R and L with a uniform dissipation ratio $\alpha = R/L$, and the capacitances C by parallel combinations of C and G with a uniform dissipation ratio $\beta = G/C$. The quantity $(Ls + 1/Cs)$ correspondingly becomes replaced by $[L(s + \alpha) + 1/C(s + \beta)]$. If $\alpha \neq \beta$, the effect of adding incidental dissipation apparently cannot be interpreted as a simple frequency shift of the form $s \rightarrow s + \delta$. Nevertheless, we may inquire whether such an interpretation is still possible on an approximate basis, and if so, what the approximate value of the committed error is.

The actual quantity in question may be written

$$L(s + \alpha) + \frac{1}{C(s + \beta)} = \frac{1}{C(s + \beta)}[LC(s + \alpha)(s + \beta) + 1] \tag{148}$$

while a corresponding quantity involving a uniform shift reads

$$L(s + \delta) + \frac{1}{C(s + \delta)} = \frac{1}{C(s + \delta)}[LC(s + \delta)^2 + 1] \tag{149}$$

The question is: For what value of δ are these two expressions nearly alike in value, and how can we estimate the error? We will approach the answer to this question by considering first whether a value of δ can be found for which

$$(s + \delta)^2 = (s + \alpha)(s + \beta) \tag{150}$$

or

$$s^2 + 2\delta s + \delta^2 = s^2 + (\alpha + \beta)s + \alpha\beta \tag{151}$$

On an exact basis the answer is obviously "no," for, equating coefficients of like powers of s in 151 shows that the value of δ must simultaneously equal the arithmetic mean of α and β and their geometric mean, which is possible only if $\alpha = \beta$. However, let us consider the square root of 150, and make use of the binomial expansions

$$\sqrt{s + \alpha} = \sqrt{s}\left(1 + \frac{\alpha}{2s} - \frac{\alpha^2}{8s^2} + \cdots\right) \tag{152}$$

and

$$\sqrt{s + \beta} = \sqrt{s}\left(1 + \frac{\beta}{2s} - \frac{\beta^2}{8s^2} + \cdots\right) \tag{153}$$

to obtain

$$s + \delta = s\left(1 + \frac{\alpha}{2s} - \frac{\alpha^2}{8s^2} + \cdots\right)\left(1 + \frac{\beta}{2s} - \frac{\beta^2}{8s^2} + \cdots\right) \tag{154}$$

Multiplying out and collecting terms with like powers of s, we find

$$\delta = \frac{\alpha + \beta}{2} - \frac{(\alpha - \beta)^2}{8s} + \cdots \tag{155}$$

or

$$\delta \approx \frac{\alpha + \beta}{2}\left(1 - \frac{\alpha - \beta}{4s} \times \frac{\alpha - \beta}{\alpha + \beta}\right) \tag{156}$$

Since the maximum absolute value of the factor $(\alpha - \beta)/(\alpha + \beta)$ is unity, we may conclude that 150 is approximately fulfilled for the value

$$\delta = \frac{\alpha + \beta}{2} \tag{157}$$

with an error not in excess of $|(\alpha - \beta)/4s|$. So far as the first factor in each of the expressions 148 and 149 is concerned, their ratio for the value of δ given in Eq. 157 is

$$\frac{s + \beta}{s + \delta} = \frac{s + \delta - (\alpha - \beta)/2}{s + \delta} \approx 1 - \frac{\alpha - \beta}{2s} \tag{158}$$

Hence we conclude that the quantity 148 is replaceable by 149 with the value of δ given by Eq. 157 in all situations where an error of the approximate magnitude

$$\epsilon = \left|\frac{\alpha - \beta}{2s}\right| = \frac{1}{2}\left|\frac{R}{L\omega} - \frac{G}{C\omega}\right| \tag{159}$$

is tolerable.

If we assume the capacitances to be essentially lossless, and recall that $L\omega/R$ is the Q factor for the inductances, this error is expressible as

$$\epsilon = 1/2Q \tag{160}$$

In many practical situations (particularly those involving band-pass characteristics) this error is entirely tolerable, and we may apply the simple change of variable $s \to s + \delta$ to consider the effect upon an impedance of adding uniform dissipation to the inductances alone. The pertinent value of δ in this case equals $R/2L$.

These results are frequently useful also in analysis, especially where we wish to compare values of an impedance $Z(j\omega)$ before and after a specified amount of uniform dissipation is added to the inductances and/or capacitances. From the stated dissipation ratios α and β we compute δ from Eq. 157. If $Z(j\omega)$ refers to the impedance before addition of the uniform dissipation, then $Z(j\omega + \delta)$ yields values of the corresponding impedance after the additional uniform dissipation is included. The given network may, of course, contain resistances in any random manner in addition to the inductance and capacitance elements. The present method enables one to compute the *change* in impedance values resulting from incremental series resistances and shunt conductances added to the L's and C's, some of which may already have resistances and conductances associated with them.

Usefulness of this result for analysis problems stems from the fact that one does not need to derive a revised impedance function in order to be able to compute impedance values of the altered network resulting from the addition of incremental resistances and conductances. One uses the *same* impedance function that characterizes the network before addition of the incremental dissipation, but instead of computing it along the j-axis to find its values for various sinusoidal steady-state frequencies, one evaluates it for points along a line parallel to, and δ units to the right of, the j-axis. The original impedance evaluated along this line in the s-plane equals (approximately) the revised impedance evaluated along the j-axis. One is thus able to compute values of the revised impedance without having an expression for this impedance.

In a number of practical situations, we deal with networks in which the total dissipative character is incidental. Wave guides and transmission-line sections used at ultrahigh frequencies, as well as many filter networks, resonant cavities, or tank circuits, and numerous similar devices fall into this category. Here one achieves a simplification in the pertinent analysis or synthesis problem by ignoring dissipative elements entirely as a tentative procedure, since one is then dealing with the simpler two-element-kind situation presented by LC networks and reactance functions. After carrying through the pertinent problem on this tentative basis, one determines the perturbations caused by the addition of an appropriate amount of uniform incremental dissipation by the methods discussed here. The relevant effects of the incremental dissipation are in this way much more effectively introduced into the analysis or design considerations than through their initial inclusion in all of the pertinent performance relations.

In dealing with problems of this sort we can avail ourselves of a further computational advantage through use of a Taylor expansion of the pertinent nondissipative impedance function about j-axis points, for the computation of its values δ units to the right of this axis. If $z_0(s)$ denotes the lossless impedance, or reactance function, and $z(s)$ the corresponding uniformly dissipative one, then such a Taylor series reads

$$z(j\omega) = z_0(j\omega) + \frac{dz_0}{dj\omega}\cdot\delta + \frac{1}{2}\frac{d^2z_0}{dj\omega^2}\cdot\delta^2 + \cdots \tag{161}$$

Observing that

$$z_0(j\omega) = jx_0(\omega), \qquad \frac{dz_0}{dj\omega} = \frac{dx_0}{d\omega}, \qquad \frac{d^2z_0}{dj\omega^2} = -j\frac{d^2x_0}{d\omega^2} \tag{162}$$

Eq. 161 becomes

$$z(j\omega) = jx_0(\omega) + \frac{dx_0}{d\omega}\cdot\delta - \frac{j}{2}\frac{d^2x_0}{d\omega^2}\cdot\delta^2 + \cdots \tag{163}$$

Writing

$$z(j\omega) = r(\omega) + jx(\omega) \tag{164}$$

one thus finds, to a first approximation, that

$$r(\omega) = \frac{dx_0}{d\omega} \cdot \delta \tag{165}$$

and

$$x(\omega) = x_0(\omega) \tag{166}$$

These approximate relations for the determination of the uniformly dissipative impedance from the corresponding nondissipative one are usually sufficiently close, so that additional terms in the Taylor expansion need not be considered.

According to the principle of duality it follows at once that the same situation applies also to admittances. Thus if $y_0(j\omega) = jb_0(\omega)$ is a nondissipative admittance (susceptance function), and $g(\omega) + jb(\omega)$ is the corresponding uniformly dissipative one, then to a first approximation

$$g(\omega) = \frac{db_0}{d\omega} \cdot \delta \tag{167}$$

and

$$b(\omega) = b_0(\omega) \tag{168}$$

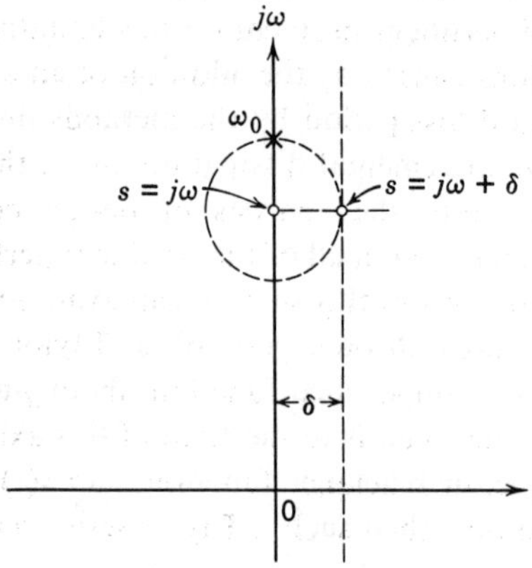

Fig. 23. Convergence circle for the Taylor expansion 161 illustrating the closest approach to a j-axis pole of $z_0(s)$ for which the approximate relations 165 and 166 are valid.

Since the Taylor expansion, Eq. 161, converges uniformly only within a circle (about the pertinent j-axis point in the s-plans) extending to the nearest singularity of $z_0(s)$, these results are not applicable to reactance functions for values of $s = j\omega$ which lie closer than δ radians per second to any poles of $z_0(s)$. Figure 23 illustrates a limiting value of $s = j\omega$ according to this restriction, assuming that the reactance function $z_0(s)$ has a pole at the point $s = j\omega_0$.

To overcome this restriction we can apply the method to the admittance $y_0 = 1/z_0$, which has a zero where z_0 has a pole. After the corresponding dissipative function $y(s)$ is evaluated for the vicinity of the pole of z_0, it is inverted to obtain the pertinent values of the dissipative $z(s)$ for this vicinity.

To illustrate the method we shall consider the nondissipative circuit of Fig. 24. Let

$$\omega_1 = 1/\sqrt{L_1 C_1} \quad \text{and} \quad \omega_3 = 1/\sqrt{L_3 C_3} \tag{169}$$

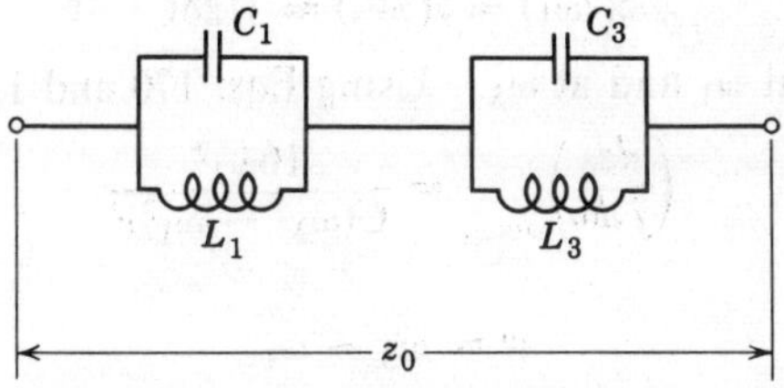

Fig. 24. Circuit for which use of the approximate relations 165 through 168 is illustrated.

denote the antiresonant frequencies of the two parallel LC networks. The reactance function z_0 is given by

$$z_0 = j\omega\left(\frac{1/C_1}{\omega_1^2 - \omega^2} + \frac{1/C_3}{\omega_3^2 - \omega^2}\right) \tag{170}$$

The frequencies ω_1 and ω_3 are poles of z_0. At the intermediate frequency

$$\omega_2^2 = \frac{1}{C_1 + C_3}\left(\frac{1}{L_1} + \frac{1}{L_3}\right) \tag{171}$$

the reactance is zero. If we assume for simplicity that $C_1 = C_3$, then

$$\omega_2^2 = \frac{\omega_1^2 + \omega_3^2}{2} \tag{172}$$

In a practical problem the circuit of Fig. 24 may be used to provide an impedance whose magnitude is zero at the frequency ω_2, and infinite at a pair of neighboring frequencies ω_1 and ω_3. The closer the spacing between these frequencies, the more rapidly the reactance z_0 varies with frequency in this vicinity. Because of the unavoidable presence of some dissipation, however, the actual dissipative impedance z will have finite values at the frequencies ω_1 and ω_3, and it will have a nonzero value at the intermediate frequency ω_2. Unless the dissipation is small enough, one may find that the actual impedance does not exhibit the desired critical behavior. In fact, it is possible for the magnitude of the dissipative impedance actually to be larger at ω_2 than at ω_1 and ω_3 unless the dissipation ratios are sufficiently small. Let the problem be to determine the necessary values of the dissipation ratios such that, for a given spacing of the critical frequencies ω_1 and ω_3, the ratio of the impedance magnitudes at ω_1 and ω_2, or at ω_3 and ω_2, shall equal some prescribed value.

To determine the magnitude of the dissipative impedance at $\omega = \omega_1$ and $\omega = \omega_3$, we compute

$$(dy_0/j\,d\omega)_{\omega=\omega_1} = 2C_1, \qquad (dy_0/j\,d\omega)_{\omega=\omega_3} = 2C_3 \tag{173}$$

in which $y_0 = 1/z_0$. Since it is assumed that $C_1 = C_3 = C$, the relations 167 and 168 give

$$z(j\omega_1) = z(j\omega_3) = 1/2\delta C \tag{174}$$

because $y_0 = 0$ at ω_1 and at ω_3. Using Eqs. 170 and 172 we find

$$\left(\frac{dz_0}{j\,d\omega}\right)_{\omega=\omega_2} = \frac{16\omega_2^2}{C(\omega_3^2 - \omega_1^2)^2} \tag{175}$$

Letting

$$w = \omega_3 - \omega_1 \tag{176}$$

and assuming that

$$(w/\omega_2) << 1 \tag{177}$$

Eq. 175 yields very nearly

$$(dz_0/j\,d\omega)_{\omega=\omega_2} = 4/Cw^2 \tag{178}$$

Since $z_0(j\omega_2) = jx_0(\omega_2) = 0$, formulas 165 and 166 give

$$z(j\omega_2) = 4\delta/Cw^2 \tag{179}$$

This result, together with that expressed by Eq. 174, finally yields

$$\frac{z(j\omega_{1,3})}{z(j\omega_2)} = \frac{1}{2}\left(\frac{w}{2\delta}\right)^2 \tag{180}$$

In terms of the Q factor of the resonant circuits defined by

$$Q^{-1} = \left(\frac{R}{L\omega_2} + \frac{G}{C\omega_2}\right) = \frac{2\delta}{\omega_2} \tag{181}$$

the result 180 may be written

$$\frac{z(j\omega_{1,3})}{z(j\omega_2)} = \frac{1}{2}\left(\frac{w}{\omega_2}\right)^2 Q^2 \tag{182}$$

Suppose we wish to have this ratio be at least as large as 100 for a problem in which $w/\omega_2 = 1$ per cent. Equation 182 then shows that the Q requirement reads

$$Q \geq 1414 \tag{183}$$

It is interesting to observe, incidentally, that the restriction to which Fig. 23 pertains yields in this example the limitation

$$\frac{w}{2} > \delta \qquad \text{or} \qquad \frac{w}{2\delta} > 1 \tag{184}$$

This condition implies that the present method of analysis fails if the impedance ratio 180 is not larger than $\frac{1}{2}$. Any practical application of this example will be well within that requirement.

PROBLEMS

1. The transfer admittance

$$Y_{12} = \frac{\lambda^4 + 4}{(\lambda + 1)^4}$$

is to be realized in the form of a lossless network terminated in a 1-ohm resistance. Show that the desired result may be found in the form shown in the figure, in which N_1 and N_2 are ladder networks whose z_{12} functions have zeros at real frequencies only, and N_2 is provided with the indicated ideal transformer at its output. Carry through the synthesis of the complete network such that the stated Y_{12} function is realized except for a constant multiplier. The method used should be applicable to Y_{12} functions of arbitrary complexity so long as the numerator polynomial is odd or even, the resultant network always consisting of two components N_1 and N_2 as shown in the sketch. Develop two different basic procedures for accomplishing the desired end.

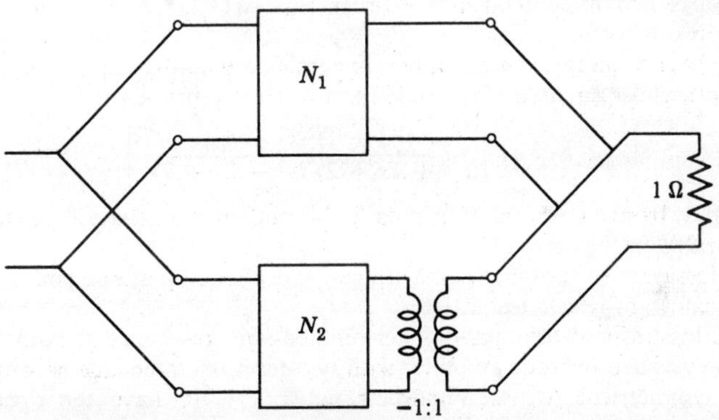

Prob. 1.

2. For the circuit shown on the next page

$$\frac{E_2}{E_1} = \frac{A}{s^2 + 2s + 2} \quad \text{and} \quad \frac{E_2}{I_1} = \frac{B}{s^2 + 2s + 2}$$

in which A and B are constants.

(*a*) Find a symmetrical lattice network which realizes the desired transfer function with maximum values of A and B.

(*b*) Using a synthesis procedure which is fundamentally different from that of part of (*a*), find another passive coupling network which meets the given specifications. If the use of transformers is forbidden, what is the maximum value of B obtainable?

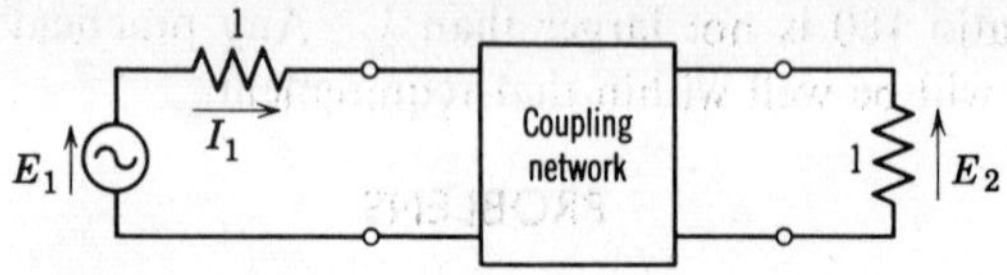

Prob. 2.

3. The transfer function of a constant-resistance lattice, between 1-ohm terminations, is to have the squared magnitude

$$|t|^2 = \frac{1}{1 + \omega^4}$$

(*a*) Find the lattice. Is it minimum phase?

(*b*) If possible, realize this transfer function in terms of a lossless symmetrical lattice operating between 1-ohm resistances.

(*c*) Synthesize this transfer function in the form of a simple cascade of pentode amplifiers. The constant multiplier is of no importance, but loss should be associated with each inductance and capacitance in such a way that $R/L = G/C =$ constant.

4. Generalizing the result of the previous problem, show that a Butterworth or Tschebyscheff filter of arbitrary order (see Art. 2, Ch. 14) can always be realized in this form.

5. (*a*) In the s-plane, sketch all possible pole-zero configurations of a transfer impedance whose squared magnitude for $s = j\omega$ is expressed by

$$|Z_{12}(j\omega)|^2 = \left[\frac{(s^2 + s + 1)^2(s^2 - s + 1)^2}{(s^2 + 2s + 2)^2(s^2 - 2s + 2)^2}\right]_{s=j\omega}$$

(*b*) Apart from a constant multiplier, point out which of these $Z_{12}(s)$ functions are realizable in the form of:

(i) A lossless two terminal-pair with resistance loading at one end.
(ii) A constant-resistance lattice.
(iii) A lossless two terminal-pair terminated with resistance at both ends.
(iv) A cascaded pentode amplifier with two-terminal interstage networks.

6. A symmetrical two terminal-pair network is to have the open-circuit transfer impedance

$$z_{12} = \frac{s}{(s^2 + 1)(s^2 + 4)(s^2 + 9)}$$

(*a*) Find a network.

(*b*) Now modify this network so that the pole distribution becomes that shown in the accompanying sketch and the numerator z_{12} becomes $s + 1$.

(*c*) Is this a minimum–phase-shift network?

7. A low-pass filter is to be designed with a transfer impedance having the pole distribution shown in the adjacent sketch (all zeros are at infinity). In order to avoid lossless elements and the disturbing effects of parasitic loss in the actual network, it is decided first to design a tentative lossless structure with a z_{12} having poles on the j-axis at $s = 0$, $s = \pm j$, $s = \pm 2j$, and then to associate resistances with the elements in such a way that the desired pole positions are achieved.

(*a*) Construct the $z_{12}(s)$ for the appropriate lossless network choosing a constant multiplier such that $z_{12} \to 1/s^5$ for $s \to \infty$.

(*b*) Synthesize this z_{12} into a symmetrical network, preferably an unbalanced ladder.

(*c*) Insert the appropriate resistances and draw the final network indicating all element values in ohms, henrys, and farads.

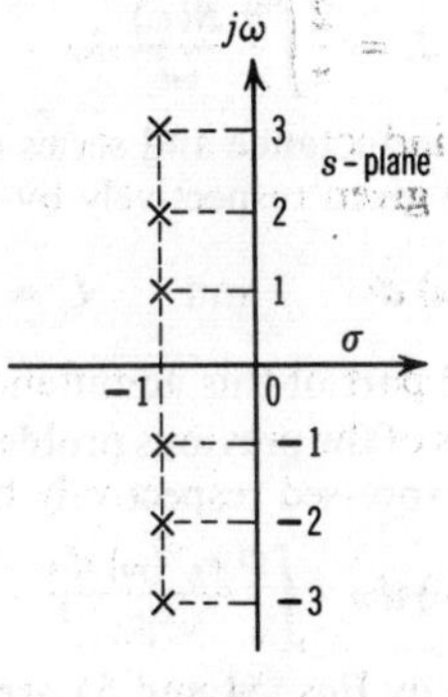

Prob. 6.

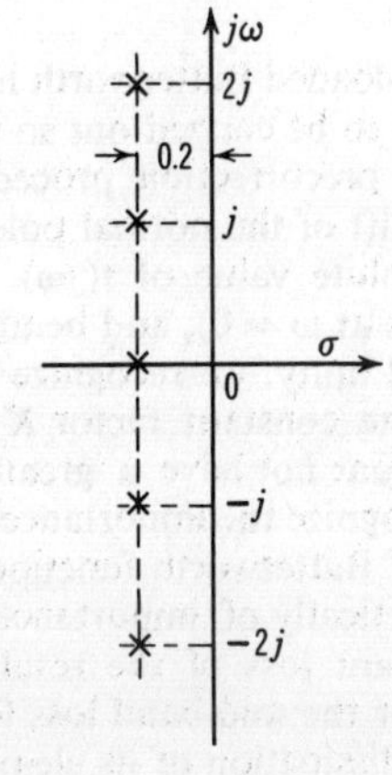

Prob. 7.

8. Given the transfer impedance

$$Z_{12}(s) = \frac{(s+1)^2}{s^2 + 2s + 2}$$

determine an appropriate network in the form of a cascade of pentode amplifier tubes having driving-point impedances in their plate circuits. The resulting constant multiplier of Z_{12} is of no consequence, but an attempt should be made to obtain the simplest practical form of solution.

9. Design a cascade of pentode amplifier stages such that the voltage transfer ratio (apart from a constant multiplier) is given by

$$|E_2/E_1| = 1/\sqrt{1 + \omega^8}$$

10. In terms of the resistance-integral theorem (Art. 3, Ch. 8) and a change of the independent variable amounting to replacing s by $1/s$, show:

(*a*) That the inherent shunt inductance associated with a driving-point impedance is given by

$$L = \frac{2}{\pi}\int_0^\infty \frac{R(\omega)}{\omega^2}\,d\omega$$

(*b*) That the inherent series inductance and series capacitance associated with a driving-point admittance are given respectively by

$$\frac{1}{L} = \frac{2}{\pi}\int_0^\infty G(\omega)\,d\omega \qquad \text{and} \qquad C = \frac{2}{\pi}\int_0^\infty \frac{G(\omega)}{\omega^2}\,d\omega$$

in which $G(\omega)$ is the j-axis real part of this admittance.

11. Making use of the results of the previous problem, show that the conditions stated by Eqs. 44 and 45 are expressed respectively by relation 62 together with

$$\int_0^\infty r_a(\omega)\,d\omega \times \int_0^\infty \frac{r_b''(\omega)\,d\omega}{\omega^2|z_b''|^2} \leq \frac{\pi^2}{4}$$

and that the conditions stated by Eqs. 54 and 55 are expressed respectively by

$$\int_0^\infty \frac{r_a(\omega)}{\omega^2}\,d\omega \times \int_0^\infty \frac{r_b''(\omega)\,d\omega}{|z_b''|^2} \leq \frac{\pi^2}{4}$$

together with relation 58.

12. The design of a double-loaded Butterworth filter according to the method discussed in Art. 5, Ch. 11, is to be carried out so as to allow for the effect of incidental dissipation by the precorrection procedure given in Art. 6 of this chapter. Observing that a shift of the normal pole pattern closer to the j-axis results in the maximum absolute value of $t(j\omega)$ becoming larger than unity (the value that it normally has at $\omega = 0$), and being mindful of the fact that the value of $t(j\omega)$ may not exceed unity, we recognize the necessity of multiplying this function by an appropriate constant factor $K < 1$. In order that the resulting filter may on this account not have a greater loss in its pass band than is absolutely necessary, we recognize the importance of determining just the right K-values for various orders of Butterworth functions.

In this connection it is practically of importance to have a simple relation that will give the added constant loss of the resulting Butterworth filter (the loss at $\omega = 0$ for a low-pass or the mid-band loss for a band-pass filter) that is attributable to the incidental dissipation of its elements. To this end show first that a value for K that is exact for Butterworth functions of order $n = 1$ and 2, and a good approximation for higher orders, reads

$$K = 1 - (\delta/a)$$

with δ equal to the pole shift given by Eq. 157 and a equal to the horizontal displacement of the nearest pole from the j-axis. Thus $a = \omega_c \sin(\pi/2n)$ for a low-pass filter with cutoff at $\omega = \omega_c$ and $a = (w/2)\sin(\pi/2n)$ for a band-pass filter with a bandwidth w. For the latter show that the above relation yields

$$K = 1 - \frac{Q_f/Q_c}{\sin(\pi/2n)}$$

and hence that the mid-band loss for either filter because of its dissipative elements must be at least

$$\alpha_0 = 2 \ln (1/K) \quad \text{nepers}$$

$Q_f = \omega_0/w$ is the " Q " of the band-pass filter with mid-band frequency ω_0 and

$$Q_c = \left(\frac{R}{L\omega_0} + \frac{G}{C\omega_0}\right)^{-1}$$

is the so-called " unloaded Q " of its resonant circuits. The pertinent low-pass prototype filter (before precorrection) is obtained by the method of Prob. 39, Ch. 11.

CHAPTER THIRTEEN

RC Transfer Function Synthesis

1 Origin of the Problem

There are situations in the practical use of circuits where it is desirable or necessary to avoid inductances. In a problem involving a very low frequency range, the conventional design procedures yield circuits with excessively large inductance values, and their physical realization even for a moderately small loss ratio leads to unwieldy results regarding physical size and weight. Spurious inductive couplings add to the undesirability of such physical realizations, with the result that one is led to inquire whether the inductive element can be dispensed with, even at the cost of increased circuit complexity and a loss in over-all gain.

A strong interest in this question did not develop until sometime during the last war, when it was principally motivated by the space and weight requirements placed upon air-borne equipment. At that time it was generally held that *RC* circuits by themselves could not be expected to provide sharp frequency discrimination characteristics, since the normal resonance properties of *RLC* circuits are absent. The latter are in evidence only when poles of the pertinent response function are relatively close to the j-axis; and the poles of *RC* circuits are required to lie along the negative real axis of the s-plane.

Since the zeros of *RC* transfer functions may lie anywhere in the complex plane,* it was suggested that one might capitalize upon the well-known property of feedback circuits to invert the transfer function of the feedback link (usually called the β-circuit), and thus obtain poles in the over-all transfer function that could be placed anywhere in the left-half s-plane even though the circuit contained no inductances. A widely used "wave analyzer" provides a sharp resonance effect in this way.

Although this principle was well understood at that time, and its exploitation for the realization of *RC* circuits, to provide more general

* See Arts. 2 and 3, Ch. 2.

kinds of transfer characteristics, was rather obvious,* it was considered impractical because of the required vacuum tubes and power supplies, whose parameters critically affect the resultant frequency response characteristics and render such circuits less rugged than the familiar passive ones.

Appreciation of the fact that *RC* circuits could (contrary to popular belief at that time) provide arbitrarily sharp frequency discrimination characteristics was originally brought about through the following approach to the construction of an appropriate low-pass transfer impedance. Suppose we start with a polynomial $Q(s)$ having only simple, negative real zeros and form its squared j-axis magnitude in the usual way, so as to get

$$B(\omega^2) = Q(j\omega)Q(-j\omega) = B_0 + B_1\omega^2 + B_2\omega^4 + \cdots + B_n\omega^{2n} \tag{1}$$

If we now write for the squared magnitude of a transfer impedance

$$|Z_{12}(j\omega)|^2 = \frac{B_0 + B_1\omega^2 + B_2\omega^4 + \cdots + B_k\omega^{2k}}{B_0 + B_1\omega^2 + B_2\omega^4 + \cdots + B_n\omega^{2n}} \tag{2}$$

with the index k roughly equal to $n/2$, then it is obvious that this function becomes very similar to a Butterworth characteristic.† Specifically, if we choose the zeros of $Q(s)$ in pairs, with the geometric mean value $s = -1$, it turns out that $|Z_{12}(j\omega)|$ (for a proper choice of constant multiplier) has the value 0.707 at $\omega = 1$ for any integer n. It is maximally flat at $\omega = 0$, and approximates the ideal low-pass rectangle arbitrarily closely with increasing n, much in the same manner that the Butterworth function does in the design of an *RLC* low-pass filter.

From Eq. 2 one can construct $Z_{12}(s)$ in the normal manner (Art. 8, Ch. 8). Its realization as an *RC* network is assured by the initial choice of simple negative real poles, except for the fact that we will have to accept whatever constant multiplier results. Since the impedance level of the network is adjustable, and we are willing to provide a flat gain, this latter item is regarded as being of secondary importance. Theoretically, at least, this simple procedure convinces us that *RC* networks need not be regarded as inferior to *RLC* circuits in their ability to provide a desired critical behavior. The detailed discussions in Arts. 6 and 10 of Ch. 14 demonstrate more adequately the truth of this contention. In the following articles of this chapter we shall not be further concerned with the construction of an appropriate transfer impedance or admittance function, but only with the methods for its realization.

---

* This form of solution has recently been rediscovered by D. B. Armstrong and F. M. Reza, "Synthesis of Transfer Functions by Active *RC* Networks," *IRE PGCT Trans.*, Vol. CT-1, No. 2, June 1954, pp. 8–17.

† See Art. 2, Ch. 14.

2 Lattice Methods of *RC* Synthesis*

This realization procedure follows the same general pattern as that discussed for *RLC* transfer function synthesis in Art. 1, Ch. 12. The *RC* situation is somewhat simpler because the residues of the pertinent transfer impedance are real.† Since the open-circuit transfer impedance of a symmetrical lattice with impedances z_a and z_b is

$$z_{12} = \tfrac{1}{2}(z_b - z_a) \tag{3}$$

and the residues of *RC* driving-point impedances are both real and positive, the determination of z_a and z_b from a partial-fraction expansion of the prescribed $z_{12}(s)$ amounts simply to assigning the positive terms to z_b and the negative ones to $-z_a$, the constant multiplier being of no importance anyway. Foster-like networks for the realization of z_a and z_b are thus obtained at once by the methods of Art. 1, Ch. 4, and the lattice design is thus completed.

Since the lattice itself is practically less desirable than an unbalanced equivalent structure, this method of synthesis, although extremely simple and straightforward, is usually not attractive unless one can subsequently unbalance the lattice. If we consider the equivalence relations illustrated

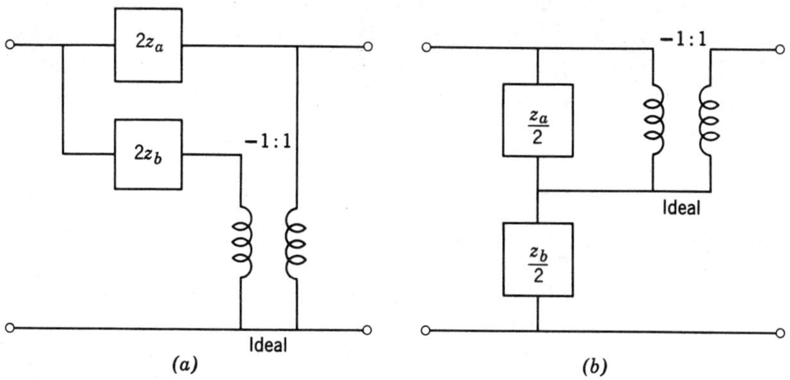

Fig. 1. Unbalanced lattice equivalents obtainable from Fig. 16, Ch. 6, by letting either $z_{a2} = \infty$ and $z_{b1} = 0$ or $z_{a1} = \infty$ and $z_{b2} = 0$.

in Fig. 16, Ch. 6, for the special conditions $z_{a2} = \infty$, $z_{b1} = 0$, or $z_{a1} = \infty$, $z_{b2} = 0$, the well-known lattice equivalents shown respectively in Fig. 1*a* and *b* are obtained. Since the use of an actual transformer to approximate the ideal transformer appearing here is inconsistent with the objective of avoiding inductances, these equivalent circuits are not particularly helpful

* This realization method was first presented by the writer in *Report* 43, Mass. Inst. Technol., Radiation Lab., Oct. 16, 1944.

† See Arts. 2 and 3, Ch. 2.

unless we devise a more acceptable substitute for the ideal transformer. An obvious thought here is to use a vacuum tube to obtain the $-1:1$ voltage ratio called for in the circuit of Fig. 1*b*. Although *RC* synthesis methods based upon this principle have been suggested,* they are practically not very useful because the frequency response of the circuit is too critically dependent upon the tube behavior which does not remain sufficiently constant.

There are, of course, situations in which the conversion methods discussed in Art. 5, Ch. 6, are successful in yielding an unbalanced equivalent, but these are far from being straightforward; and if at first one does not succeed, it is not clear whether continued efforts along the same lines will be rewarding.

As we shall see from the following discussion, an unbalanced form of realization for a transfer *admittance* function is always possible so long as no zeros on the positive real axis are called for. Similar resulting networks could presumably be obtained also through a lattice approach, although an alternate procedure discussed in a succeeding article has the advantage of being more direct.

3 Realization by Parallel Ladder Networks†

With reference to the situation shown in Fig. 2, the pertinent transfer admittance function is given by

$$Y_{12}(s) = \frac{I_2}{E_1} = \frac{-E_2}{E_1} = \frac{y_{12}}{1 + y_{22}} \tag{4}$$

in which y_{22} and y_{12} are the familiar short-circuit driving-point and transfer admittances of the two terminal-pair network N.

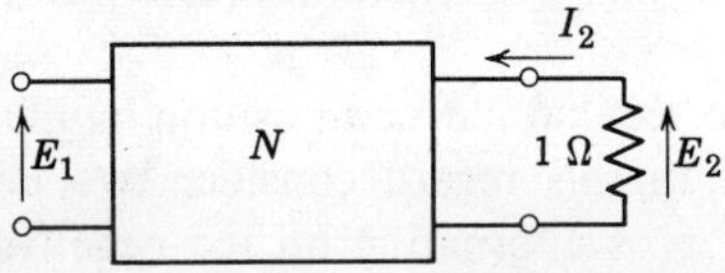

Fig. 2. Situation to which relation 4 is relevant.

As a rational function, the specified transfer admittance has the form

$$Y_{12}(s) = \frac{p(s)}{q(s)} = \frac{a_0 + a_1 s + a_2 s^2 + \cdots + a_n s^n}{b_0 + b_1 s + b_2 s^2 + \cdots + b_n s^n} \tag{5}$$

* See "Transfer-Function Synthesis with Computer Amplifiers and Passive Networks," M. V. Mathews and W. W. Seifert, *Proc. Western Joint Computer Conference*, Mar. 1955.

† "Synthesis of *RC* Networks," E. A. Guillemin, *J. Math. and Phys.*, Vol. 28, No. 1, Apr. 1949, pp. 22–42. The procedure for *RLC* transfer function synthesis given in Art. 4, Ch. 11, is based upon this method.

The denominator polynomial $q(s)$ has simple zeros restricted to the negative real axis. The numerator polynomial, as we shall see, is required to have positive coefficients. If its zeros are restricted to the left half-plane (that is to say, if Y_{12} is a minimum-phase transfer function), then this condition is automatically fulfilled. If $p(s)$ has some right half-plane zeros, its coefficients may still be positive; and so long as $p(s)$ does not have zeros on the positive real axis, one can obtain a resultant polynomial $p(s)$ with positive coefficients by augmenting $Y_{12}(s)$ with factors $(s + a)$ in a manner similar to that discussed in connection with the Miyata synthesis method in Art. 2 of Ch. 10 (see appendix to Ch. 10).

The procedure at this point is somewhat different from (although similar to) that used in the analogous RLC situation (Arts. 1 and 4, Ch. 11) where one makes use of the fact that the even and odd parts of the denominator polynomial automatically have simple zeros alternating on the j-axis. Here we must analogously split $q(s)$ into two additive parts having simple zeros alternating upon the negative real axis. If such a decomposition is indicated by

$$q(s) = q_1(s) + q_2(s) \tag{6}$$

then Eq. 5 may be written in the form

$$Y_{12}(s) = \frac{p(s)}{q_1(s) + q_2(s)} = \frac{p(s)/q_2(s)}{1 + q_1(s)/q_2(s)} \tag{7}$$

whereupon a comparison with Eq. 4 suggests the identifications

$$y_{12} = p(s)/q_2(s) \qquad \text{and} \qquad y_{22} = q_1(s)/q_2(s) \tag{8}$$

In order that these admittances be RC realizable, it is necessary and sufficient that $q_1(s)$ and $q_2(s)$ have simple zeros alternating upon the negative real axis and that the critical frequency of y_{22} nearest the origin $(s = 0)$ be a zero.

It is not difficult to see that a decomposition 6 fulfilling these conditions is always possible. In this regard consider two polynomials $q_1(s)$ and $q_2(s)$ having simple zeros alternating on the negative real axis. If these polynomials have the factored forms

$$q_1(s) = (s + \alpha_1)(s + \alpha_2) \cdot \cdot \cdot (s + \alpha_n) \tag{9}$$

$$q_2(s) = (s + \beta_1)(s + \beta_2) \cdot \cdot \cdot (s + \beta_n) \tag{10}$$

then α_k and β_k have positive real values fulfilling the conditions

$$0 < \alpha_1 < \beta_1 < \alpha_2 < \beta_2 < \cdot \cdot \cdot < \alpha_n < \beta_n < \infty \tag{11}$$

From a graphical sketch of these polynomials for negative real values of s, it becomes clear that zeros of the sum $q(s)$ are located between pairs of zeros belonging respectively to $q_1(s)$ and $q_2(s)$. That is to say, if we write

$$q(s) = (s + \sigma_1)(s + \sigma_2) \cdot \cdot \cdot (s + \sigma_n) \tag{12}$$

with positive real σ_k values numbered as indicated in the sequence

$$0 < \sigma_1 < \sigma_2 < \sigma_3 < \cdots < \sigma_n < \infty \tag{13}$$

then we find that

$$\alpha_k < \sigma_k < \beta_k \qquad \text{for } k = 1, 2, \cdots, n \tag{14}$$

It follows that if we form a polynomial $q_1(s)$ by choosing its zeros so as to fulfill the conditions

$$0 < \alpha_1 < \sigma_1 < \alpha_2 < \sigma_2 < \cdots < \alpha_n < \sigma_n \tag{15}$$

then the difference $q(s) - q_1(s)$ will automatically yield a polynomial $q_2(s)$ with zeros β_k fulfilling conditions 11 and 14 provided only that

$$q(s) > q_1(s) \qquad \text{for both } s = 0 \text{ and } s \to \infty \tag{16}$$

which can always be accomplished through multiplying one of these polynomials by an appropriate constant factor if necessary.

After the functions y_{12} and y_{22} are thus constructed according to relations 8, realization of the transfer admittance 4 (within a constant multiplier) can usually take a variety of detailed forms depending upon the zeros of $Y_{12}(s)$ as determined by the polynomial $p(s)$. If these zeros all lie upon the negative real axis (including the point at infinity), then the situation is exactly parallel to that encountered in the synthesis of a resistance-loaded lossless two terminal-pair network for which the zeros of transmission lie upon the j-axis or at $s = \infty$. Following the pattern described for the latter situation in Art. 3, Ch. 7, one can expand y_{22} into a ladder network so as to produce the required transmission zeros.

The process is simplest if these zeros fall at $s = 0$ or $s = \infty$, for then they are produced respectively by series or shunt capacitance branches. The development of the y_{22} function then takes one of the two canonic Cauer forms (Art. 2, Ch. 4) or an appropriate combination of these according to whether all zeros are at one point or distributed between the two points $s = 0$ and $s = \infty$. When some zeros are located at finite nonzero points on the negative real axis, then series branches consisting of parallel RC combinations or shunt branches consisting of series RC combinations are needed, and in this event it will in general be necessary to "zero-shift" with partial pole removals in order to produce the transmission zeros at the proper locations.

When the zeros of $p(s)$ are not all located upon the negative real axis, then one must decompose this polynomial into additive parts, each of which does fulfill this condition. Let us indicate this decomposition by

$$p(s) = p_1(s) + p_2(s) + \cdots + p_k(s) \tag{17}$$

The transfer admittance y_{12} in 8 is correspondingly written

$$y_{12}(s) = y_{12}^{(1)} + y_{12}^{(2)} + \cdots + y_{12}^{(k)} \tag{18}$$

where

$$y_{12}^{(\nu)} = p_\nu(s)/q_2(s) \tag{19}$$

The strategy now is to expand y_{22} into ladder networks that realize the transfer admittances $y_{12}^{(\nu)}$ for $\nu = 1, 2, \cdots, k$, and, after multiplying these by suitable admittance level factors, combine them in parallel in the same manner that this is done in Art. 4, Ch. 11, for the LC ladders considered there. Suppose each ladder development of y_{22} in 8, carried out so as to produce the zeros of the pertinent transfer admittance $y_{12}^{(\nu)}$, yields this function with a multiplier equal to K_ν times the desired value. If we let

$$\frac{1}{K} = \frac{1}{K_1} + \frac{1}{K_2} + \cdots + \frac{1}{K_k} \tag{20}$$

and multiply the admittance level of each respective ladder network by the factor K/K_ν, then their parallel combination evidently has a resultant short-circuit driving-point admittance equal to y_{22} and a transfer admittance equal to Ky_{12}. When terminated in a load resistance of 1 ohm, this parallel ladder arrangement is a realization for the desired $Y_{12}(s)$ function in Eq. 4, except for the multiplier K.

The remaining question to be resolved is concerned with the manner in which the decomposition 17 can be carried out. The simplest procedure here is to regard each term in the polynomial $p(s)$ as a component $p_\nu(s)$. The positiveness of the coefficients a_ν is seen to be essential to the success of the method. An alternate scheme, which yields half as many ladder networks and is always possible, is to consider the terms in $p(s)$ two at a time. The components $p_\nu(s)$ in 17 then become

$$\begin{aligned} p_1(s) &= a_0 + a_1 s \\ p_2(s) &= (a_2 + a_3 s)s^2 \\ p_3(s) &= (a_4 + a_5 s)s^4 \\ &\cdots\cdots\cdots\cdots \end{aligned} \tag{21}$$

Each of these has one negative real zero, and all but $p_1(s)$ have additional zeros at $s = 0$. The appropriate ladder developments of y_{22} can always be accomplished.

In certain cases it may be possible to decompose $p(s)$ into fewer additive components and hence obtain a realization involving correspondingly fewer ladder networks. It is easy to construct such situations. For example, if we add the two polynomials

$$p_1(s) = (s + 1)(s + 2)(s + 3)(s + 4) \tag{22}$$

and

$$p_2(s) = s + 5 \tag{23}$$

the result is a polynomial having no real zeros, yet it is decomposable into the sum of two polynomials having only negative real zeros. Nothing of a general nature can be said about this problem, however, except that it is definitely not possible always to decompose into just two additive components, even if we do not permit $p(s)$ to have right half-plane zeros. Thus the polynomial

$$p(s) = (s^2 + 1)(s^2 + 2)(s^2 + 3) \cdots \tag{24}$$

having only j-axis zeros defies such a decomposition; and augmentation is of no help here either.

It may be of interest to mention that decomposition of any polynomial into the *difference* of two polynomials having only negative real zeros is easily seen to be possible. Thus if $p(s)$ is the given polynomial and $p_1(s)$ is one of like degree having only negative real zeros, then

$$p_2(s) = p(s) + Kp_1(s) \tag{25}$$

for the choice of a sufficiently large value of K, evidently is also a polynomial having only negative real zeros, since its zeros approach closer and closer to those of $p_1(s)$ as the value of K is increased.* On the basis of this fact, it is obvious that a synthesis is always possible in terms of two parallel ladders, if one of these is provided with a $-1:1$ ratio ideal transformer at one of its terminal pairs; and it is possible also to use a vacuum tube instead of the ideal transformer in order to provide the necessary minus sign. These schemes, however, have little, if any, practical value for reasons already mentioned.

As an example of the parallel ladder method of synthesis, let us consider the transfer admittance

$$Y_{12}(s) = \frac{s^2 + 1}{2(s + 1)(s + 2)} = \frac{p(s)}{q(s)} \tag{26}$$

which provides j-axis zeros at the points $\omega = \pm 1$. Decomposition of the denominator polynomial is accomplished by assuming

$$q_1(s) = (s + \tfrac{1}{2})(s + \tfrac{3}{2}) = s^2 + 2s + \tfrac{3}{4} \tag{27}$$

Subtraction from $q(s)$ yields

$$q_2(s) = s^2 + 4s + \tfrac{13}{4} \tag{28}$$

According to Eqs. 8, we then have

$$y_{12} = \frac{s^2 + 1}{s^2 + 4s + \frac{13}{4}}, \qquad y_{22} = \frac{s^2 + 2s + \frac{3}{4}}{s^2 + 4s + \frac{13}{4}} \tag{29}$$

and the decomposition of y_{12} as indicated in Eq. 18 is in this case given by

$$y_{12}^{(1)} = s^2/q_2, \qquad y_{12}^{(2)} = 1/q_2 \tag{30}$$

---

* Another scheme is to let $s = \lambda^2$ and, in the familiar manner, determine a Hurwitz polynomial $p_0(\lambda) = m_0 + n_0$ such that $p(\lambda^2) = m_0^2 - n_0^2$. As functions of s, the polynomials m_0 and n_0 then represent the desired decomposition of $p(s)$.

The first of these calls for a ladder development of y_{22} involving series capacitances and shunt resistances. The corresponding continued-fraction expansion reads

$$y_{22} = 0.2308 + \cfrac{1}{(1/0.3312s) + \cfrac{1}{0.6414 + \cfrac{1}{(1/0.076s) + \cfrac{1}{0.1277}}}} \tag{31}$$

and leads to the ladder network shown in Fig. 3. By considering the asymptotic behavior for large s, we see that the transfer admittance equals K_1 times $y_{12}^{(1)}$ in 30, with $K_1 = 0.1277$.

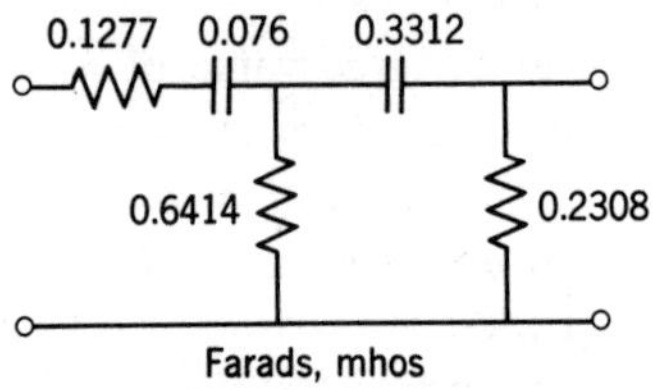

Fig. 3. Ladder network pertinent to the continued-fraction expansion 31 of the function y_{22} in Eq. 29.

Since the zeros of $y_{12}^{(2)}$ are all at $s = \infty$, the pertinent ladder network is one having series resistances and shunt capacitances. The appropriate continued-fraction expansion of y_{22} takes the form

$$y_{22} = \cfrac{1}{1 + \cfrac{1}{s/2 + \cfrac{1}{\frac{8}{3} + \cfrac{1}{3s/2 + \cfrac{1}{\frac{2}{3}}}}}} \tag{32}$$

and the corresponding ladder network is given in Fig. 4. Considering

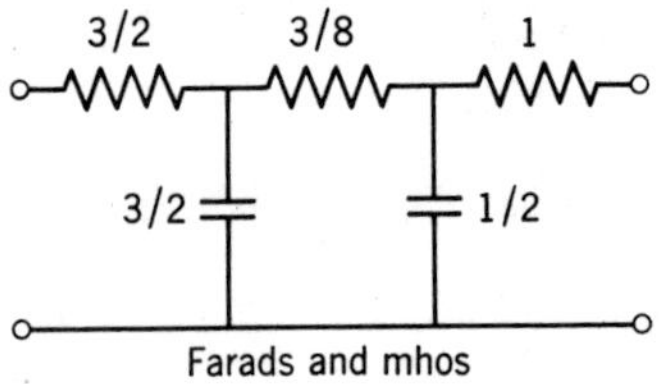

Fig. 4. Ladder network pertinent to the continued-fraction expansion 32 of the function y_{22} in Eq. 29.

the behavior for small s shows that the transfer admittance equals $K_2 y_{12}^{(2)}$ with $K_2 = \frac{3}{4}$.

Use of Eq. 20 yields $K = 0.1092$. Hence the admittance level multiplier for the network of Fig. 3 becomes $K/K_1 = 0.855$ and that for the ladder of Fig. 4 is $K/K_2 = 0.1455$. The resulting ladders, after parallel combination and termination in a 1-ohm load resistance, are shown in Fig. 5. The voltage ratio E_2/E_1 equals the assumed function 26 multiplied by the value of K.

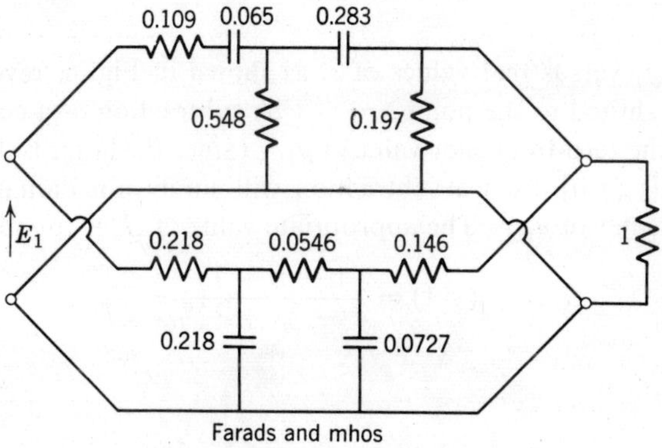

Fig. 5. Realization of the transfer admittance 26 (equal to the voltage ratio E_2/E_1) apart from a constant multiplier $K = 0.1092$.

As a second example we shall consider a transfer admittance having a third-order zero at $s = -1$. Instead of assuming some denominator polynomial and subsequently separating it into the parts q_1 and q_2, our problem will be equally illustrative of the present synthesis method if we simplify this process by assuming

$$q_1(s) = (s+1)(s+3)(s+5) = s^3 + 9s^2 + 23s + 15 \tag{33}$$

$$q_2(s) = (s+2)(s+4)(s+6) = s^3 + 12s^2 + 44s + 48 \tag{34}$$

and forming

$$q(s) = q_1(s) + q_2(s) = 2s^3 + 21s^2 + 67s + 63 \tag{35}$$

The desired transfer function therefore reads

$$Y_{12} = \frac{(s+1)^3}{2s^3 + 21s^2 + 67s + 63} \tag{36}$$

and Eqs. 8 yield

$$y_{12} = \frac{(s+1)^3}{s^3 + 12s^2 + 44s + 48}, \qquad y_{22} = \frac{s^3 + 9s^2 + 23s + 15}{s^3 + 12s^2 + 44s + 48} \tag{37}$$

Since $q_1(s)$ has a zero at $s = -1$, $1/y_{22}$ has a pole there. Removal of this pole produces one of the desired transmission zeros. To this end we compute the residue of $1/y_{22}$ at $s = -1$, thus

$$k_1 = \left(\frac{s+1}{y_{22}}\right)_{s=-1} = \frac{15}{8} \tag{38}$$

Subtraction of this pole leaves the remainder

$$\frac{1}{y_{22}} - \frac{\frac{15}{8}}{s+1} = \frac{1}{s+1}\left(\frac{s^3+12s^2+44s+48}{s^2+8s+15} - \frac{15}{8}\right) = \frac{s^2+\frac{73}{8}s+\frac{159}{8}}{s^2+8s+15}$$

$$= \frac{1}{y_1} \tag{39}$$

A sketch of y_1 versus real values of s, as shown in Fig. 6, reveals that a zero may be shifted to the point $s = -1$ by subtraction of a conductance G_1 less than the zero-frequency value of y_1. (Since the latter is the smallest value of Re $[y_1(j\omega)]$, such a subtraction will surely not damage the p.r. and RC character of y_1.) The appropriate value of G_1 is found from

$$G_1 = y_1(-1) = \frac{1-8+15}{1-\frac{73}{8}+\frac{159}{8}} = \frac{32}{47} \tag{40}$$

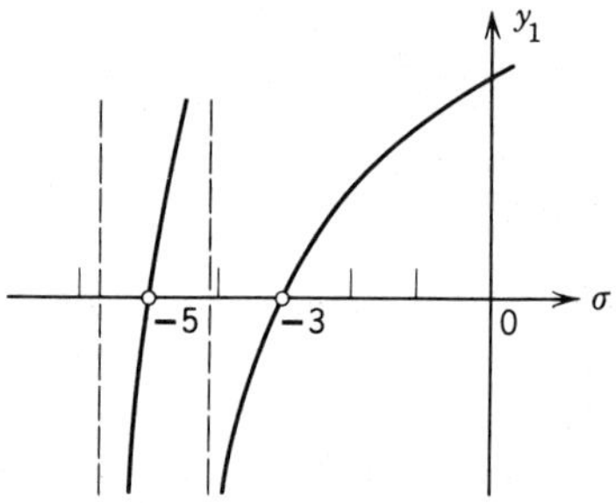

Fig. 6. Sketch of the admittance given in Eq. 39 for negative real values of s.

Subtraction of this value from $y_1(s)$ in 39 yields

$$y_1(s) - \frac{32}{47} = \frac{15(s+1)(s+\frac{69}{15})}{47(s^2+\frac{73}{8}s+\frac{159}{8})} = y_2 = \frac{1}{z_2} \tag{41}$$

The residue of z_2 at $s = -1$ is found to be

$$k_2 = \left(\frac{s+1}{y_2}\right)_{s=-1} = 10.23 \tag{42}$$

and subtraction of this pole from z_2 leaves

$$z_2 - \frac{10.23}{s+1} = \frac{47}{15} \times \frac{s+4.86}{s+4.6} = z_3 = \frac{1}{y_3} \tag{43}$$

Again we can shift a zero of y_3 to the point $s = -1$ by the subtraction of a conductance. Its value is

$$G_3 = y_3(-1) = 0.2973 \tag{44}$$

and subtraction from y_3 yields

$$y_3(s) - 0.2973 = \frac{s + 1}{46.35(s + 4.86)} = y_4 = \frac{1}{z_4} \tag{45}$$

Finally we have

$$z_4 = \frac{179}{s + 1} + 46.35 \tag{46}$$

These calculations lead to the ladder network shown in Fig. 7. The ratio E_2/E_1 equals the function 36 multiplied by the factor 0.082.

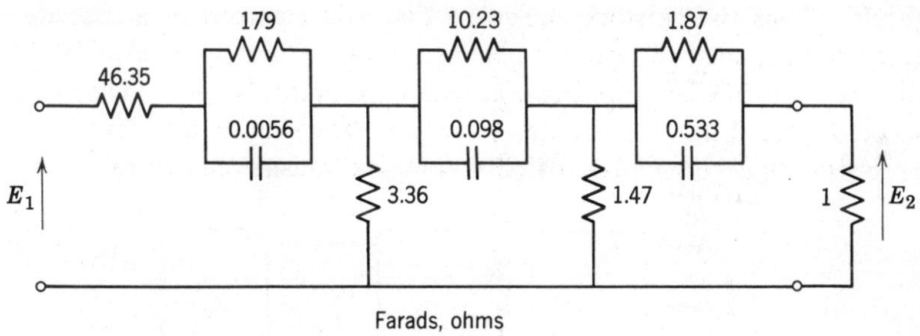

Fig. 7. Ladder network obtained for the realization of the transfer admittance 36 by the calculations given in Eqs. 38 through 46.

4 Cascade Synthesis of *RC* Networks

Except in simple situations leading to networks like those in the examples just given, a major practical disadvantage of the *RC* synthesis methods discussed so far is due to the difficulty of their proper alignment. This difficulty is especially in evidence when the network is intended to realize a frequency characteristic exhibiting critical behavior, like that of a filter near a cutoff point. Since the poles of an *RC* response function cannot contribute to such critical behavior, the burden to provide a sharp rate of discrimination falls upon the zeros. Unless these are accurately located, the resulting behavior may deviate markedly from its expected pattern.

In a lattice or parallel ladder form of realization, the zeros of transmission result from a cancelation of the values of complex impedance or admittance functions, and all zeros are in general affected by the adjustment of a single element value. Unless all zeros happen, by some stroke of good fortune, to fall initially at exactly the right values, their alignment through subsequent readjustment of separate element values is practically hopeless except in very simple situations where only one or two zeros are involved.

As in the case of *RLC* transfer function synthesis, this difficulty is overcome by devising a method of realization in the form of a cascade of component two terminal-pair networks of which each provides one transmission zero or pair of conjugate complex zeros. As we shall show, such a method of cascade synthesis, following the pattern discussed for *LC* networks in Art. 4, Ch. 7, and for the Brune and Darlington synthesis problems in Arts. 5 and 9 of Ch. 9, is possible in suitably modified form for the synthesis of *RC* transfer functions so long as their zeros do not fall in the right half-plane.*

So far as left half-plane zeros on the real axis are concerned, the method of ladder development described in the previous article yields the desired result. Thus the network shown in Fig. 7 is regarded as a cascade of three component two terminal-pair networks (separately adjustable) each of which provides one transmission zero at the point $s = -1$. The method to be discussed now is in a sense a logical extension of this development procedure to accommodate pairs of complex transmission zeros.

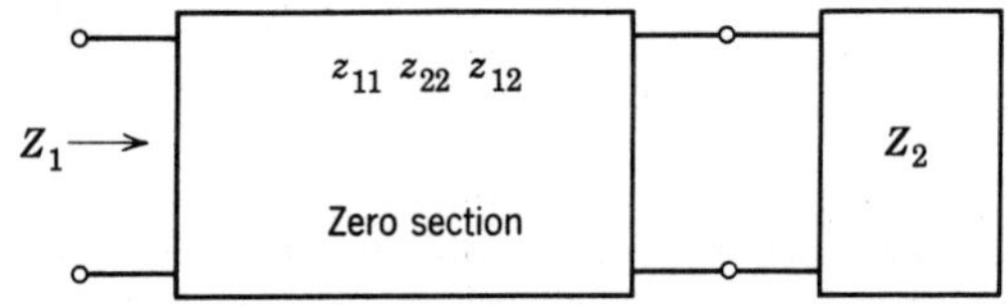

Fig. 8. Typical cycle in the cascade synthesis of an *RC* transfer impedance.

As in the analogous *LC* case (Art. 4, Ch. 7), a typical cycle in this development is illustrated in Fig. 8 in which Z_1 is the given driving-point impedance to be developed (like $1/y_{22}$ in the previous article), z_{11}, z_{22}, and z_{12} are the open-circuit driving-point and transfer impedances of the component two terminal-pair network (also referred to as a zero section, since it places in evidence a typical pair of complex zeros of the given over-all-transfer impedance), and Z_2 is the remainder function, having the same character as Z_1. The complete realization is obtained by carrying out a second cycle in terms of Z_2, and so forth, until only an open circuit or a constant is left as a remainder function.

In the present problem we assume that the pair of driving-point and transfer impedances Z_1 and Z_{12}, having identical negative real poles, are prescribed. Realization of Z_{12} is accomplished by developing the driving-point function Z_1 into a cascade of component networks which individually

* This result was first demonstrated by B. J. Dasher, *Tech. Rept. 215*, Mass. Inst. Technol., Research Lab. Electronics, Nov. 20, 1951, or *IRE PGCT Trans.*, Vol. CT-1, Dec. 1952, pp. 20–34. The procedure described here is a modified form of Dasher's method.

realize pairs of zeros of Z_{12}. The development is completed when all of the zeros of Z_{12} are realized, at which point the remainder function is reduced to a constant. The transfer impedance Z_{12} plays no part in the actual development process except to prescribe the transmission zeros which are to be produced. The order in which these are considered during the development of Z_1 is immaterial except as it may influence the simplicity of the resultant structure in some relatively small degree.

There may be situations in which only Z_{12} is specified. In this instance one must construct an appropriate driving-point impedance Z_1, which is a rational function having the same denominator as Z_{12} and simple negative real zeros alternating with the poles, the critical frequency nearest the point $s = 0$ being a pole (see Art. 3, Ch. 2). A driving-point impedance fulfilling these simple specifications is easily formed.

There is an important fundamental difference between the *RC* development problem and the analogous *RLC* driving-point impedance development procedure according to the Darlington theory as discussed in Art. 9 of Ch. 9. As pointed out there, the transmission zeros of the resultant transfer impedance are collectively the even-part zeros of the given *RLC* driving-point impedance. Each successive cycle realizes one set of these even-part zeros (or transmission zeros) and the even part of the remainder function at any stage contains those of the original even-part zeros not yet produced or placed in evidence by the developed portion of the total network.

In the analogous *RC* problem the picture is completely different in this regard. The even-part zeros of the driving-point impedance are totally unrelated to the transmission zeros of the resulting network or any component thereof. Whereas in the *RLC* development we have no choice but to make the transmission zeros equal to the even-part zeros of the given input impedance, the *RC* problem allows us to choose the transmission zeros with complete freedom. In fact, the process completely ignores the even-part zeros of the input impedance, the reason for all this being attributable to the lossy character of the present component networks as contrasted with the lossless character of these networks in the corresponding Darlington development.

The driving-point impedance is now assumed in the form

$$Z_1(s) = \frac{Ap(s)}{q(s)} + \frac{\rho_0}{s} + \rho_\infty \tag{47}$$

in which A, ρ_0, and ρ_∞ are constants, and

$$p(s) = (s - s_{p1})(s - s_{p2}) \cdots (s - s_{pn-1}) \tag{48}$$

$$q(s) = (s - s_1)(s - s_2) \cdots (s - s_n) \tag{49}$$

with

$$s_n < s_{pn-1} < s_{n-1} < \cdots < s_1 < s_{p1} < 0 \tag{50}$$

This form for Z_1 appears at first sight to be very special, but actually it is hardly more so than if we had written it in the simple form of the right-hand side of Eq. 5. Thus the term ρ_∞ merely represents the minimum value of the j-axis real part of $Z_1(s)$ which we can always remove, leaving the numerator polynomial one degree lower than the denominator. Presence of the term ρ_0/s places in evidence a pole at $s = 0$, a condition that can always be brought about (if necessary) through the initial removal of a shunt-resistive branch. In fact, such a branch might be desirable to transform the implied current source at the input into an equivalent lossy voltage source.

The desired complex pair of transmission zeros is given by the quadratic factor

$$t(s) = (s - s_0)(s - \bar{s}_0) = (s + \sigma_0)^2 + \omega_0{}^2 \tag{51}$$

with

$$s_0 = -\sigma_0 + j\omega_0, \qquad \bar{s}_0 = -\sigma_0 - j\omega_0 \tag{52}$$

The simplest form for the transfer impedance of the zero section reads

$$z_{12} = \frac{Kt(s)}{s(s - s_a)} = K + \frac{k_0}{s} - \frac{k_{12}}{s - s_a} \tag{53}$$

since it must have at least two poles, and letting one of these coincide with the point $s = 0$ leaves only one unknown pole at $s = s_a$ to be determined. K is assumed to be a positive real constant, and Eq. 53 yields

$$k_0 = Kt(0)/(-s_a) \tag{54}$$

$$k_{12} = Kt(s_a)/(-s_a) \tag{55}$$

For reasons that depend upon the desired physical structure of the zero section, we assume that its driving-point impedances have the form

$$z_{11} = K + \frac{k_0}{s} + \frac{k_{11}}{s - s_a} \tag{56}$$

$$z_{22} = K + \frac{k_0}{s} + \frac{k_{22}}{s - s_a} \tag{57}$$

with the tacit assumption that

$$k_{11}k_{22} - k_{12}{}^2 = 0 \tag{58}$$

which must automatically be fulfilled since $Z_1(s)$ does not contain the pole at $s = s_a$. (This conclusion may be seen from Eq. 55, Ch. 7, which applies to the situation in Fig. 8 above if we replace R by Z_2. Thus, if Z_1 does not contain the pole at s_a, then the z-determinant must have a simple pole at this point, and hence the residue condition must there be fulfilled with the equal sign, as pointed out in the analogous LC situation leading to statements 56 and 57 in Ch. 7.)

The analytic relationship between the five quantities z_{11}, z_{12}, z_{22}, Z_1, and Z_2 is, as in the analogous LC situation, written in the form

$$(z_{11} - Z_1)(z_{22} + Z_2) = z_{12}^2 = \frac{K^2 t^2(s)}{s^2(s - s_a)^2} \tag{59}$$

From this expression we can deduce that the separate factors on the left-hand side are given by the forms

$$(z_{11} - Z_1) = \frac{K^2 t^2 h}{s(s - s_a)q} \tag{60}$$

and

$$(z_{22} + Z_2) = \frac{q}{s(s - s_a)h} \tag{61}$$

Thus it is easy to see that the factor t^2 cannot be allotted to $(z_{22} + Z_2)$ because this is an RC driving-point impedance, and as such it cannot have complex zeros. The denominator in 60 is easily justified by inspection of the forms of z_{11} and Z_1, and h in the numerator of this factor is a polynomial of degree $n - 2$, since the numerator must have the same degree as the denominator (z_{11} and Z_1 behave like nonzero constants for $s \to \infty$). The factor 61, then, is chosen to fulfill the demand of Eq. 59.

The major problem in our present procedure is the determination of $h(s)$ which we shall write

$$h(s) = H(s - s_{h1})(s - s_{h2}) \cdots (s - s_{hn-2}) \tag{62}$$

Once this polynomial is known and the common nonzero pole s_a for the z_{ik} is determined, these impedances characterizing the zero section as well as the remainder function $Z_2(s)$ will be extractable from the pertinent factors in Eqs. 60 and 61.

The desired determination of $h(s)$ is most easily accomplished through a partial-fraction expansion of the rational function h/q, which can be obtained directly from Eq. 60. We thus obtain

$$\frac{h(s)}{q(s)} = \frac{s(s - s_a)(z_{11} - Z_1)}{K^2 t^2(s)} = \frac{1}{K^2} \sum_{\nu=1}^{n} m_\nu \left(\frac{s_\nu - s_a}{s - s_\nu}\right) \tag{63}$$

with the abbreviation

$$m_\nu = \frac{-s_\nu k_\nu}{t^2(s_\nu)} \tag{64}$$

in which k_ν, the residue of $Z_1(s)$ in any of its nonzero poles, is given by

$$k_\nu = A\left[\frac{(s - s_\nu)p(s)}{q(s)}\right]_{s=s_\nu} \qquad \text{for } \nu = 1, 2, \cdots, n \tag{65}$$

Since the degree of $h(s)$ is $n - 2$ while that of $q(s)$ is n, the rational function 63 should behave like $1/s^2$ for large s. To place in evidence the asymptotic behavior of 63, we write

$$\frac{1}{s - s_\nu} = \frac{1}{s}\left(\frac{1}{1 - s_\nu/s}\right) = \frac{1}{s}\left(1 + \frac{s_\nu}{s} + \frac{s_\nu^2}{s^2} + \cdots\right) \tag{66}$$

and then observe that the required asymptotic behavior demands

$$\sum_{\nu=1}^{n} m_\nu(s_\nu - s_a) = 0 \tag{67}$$

and that the constant multiplier in the representation 62 for $h(s)$ is given by

$$H = \frac{1}{K^2}\sum_{\nu=1}^{n} m_\nu s_\nu(s_\nu - s_a) \tag{68}$$

Equation 67 yields the nonzero pole in the impedances z_{ik}, thus

$$s_a = \frac{\sum m_\nu s_\nu}{\sum m_\nu} \tag{69}$$

which is seen to have a required negative real value since the quantities s_ν are negative real numbers while the quantities m_ν determined by Eq. 64 are positive. If we interpret the quantities m_ν as a set of masses located at the points s_ν on the negative real axis, then the point s_a becomes their center of gravity.

In order to see that the multiplier H as given by Eq. 68 is positive, we regard the quantities $m_\nu(s_\nu - s_a)$ as a set of forces which are negative to the left of s_a and positive to the right of this point, their sum being zero. Multiplication of these forces by $s_\nu < 0$, followed by addition, must lead to a positive value for H.

We now formulate the conditions necessary and sufficient to obtain driving-point impedances having the forms assumed in Eqs. 56 and 57. In this connection we observe that it is sufficient if at $s = 0$

$$(sz_{11})_{s=0} \geq k_1 \qquad \text{and} \qquad [s(z_{22} + Z_2)]_{s=0} > k_0 \tag{70}$$

while at $s = \infty$

$$(z_{11})_{s=\infty} \geq K \qquad \text{and} \qquad (z_{22} + Z_2)_{s=\infty} > K \tag{71}$$

because any excess in the value of each expression over the required minimum may in the case of z_{11} be removed as a series branch preceding the zero section, while in z_{22} it can be regarded as an allotment to the remainder function Z_2. In fact, the remainder should have such an allotment in order that it may have the same character as Z_1; hence the equality sign is not included in the conditions pertaining to the function $(z_{22} + Z_2)$.

Use of Eqs. 47, 54, 60, and 61 enable us to write conditions 70 and 71 in the form

$$\frac{K^2t^2(0)h(0)}{-s_aq(0)} + \rho_0 \geq \frac{Kt(0)}{-s_a}, \qquad \frac{q(0)}{-s_ah(0)} > \frac{Kt(0)}{-s_a} \tag{72}$$

and

$$K^2H + \rho_\infty \geq K, \qquad \frac{1}{H} > K \tag{73}$$

Noting that $t(0) = |s_0|^2$, and making use of Eqs. 63 and 68, these may be written respectively

$$0 < K - |s_0|^2 \sum_{\nu=1}^{n} \frac{m_\nu}{-s_\nu}(s_\nu - s_a) \leq \frac{-s_a}{|s_0|^2}\rho_0 \tag{74}$$

and

$$0 < K - \sum_{\nu=1}^{n} m_\nu s_\nu(s_\nu - s_a) \leq \rho_\infty \tag{75}$$

They have the algebraic form

$$\begin{aligned} 0 &< K - a \leq b \\ 0 &< K - c \leq d \end{aligned} \tag{76}$$

In order that a single K value may satisfy them, it is found necessary and sufficient that

$$-b < (a - c) < d \tag{77}$$

From Eqs. 74 and 75 (noting that $s_\nu < 0$) we get

$$(a - c) = |s_0| \sum_{\nu=1}^{n} \left(\left|\frac{s_0}{s_\nu}\right| + \left|\frac{s_\nu}{s_0}\right| \right) m_\nu(s_\nu - s_a) \tag{78}$$

If we define a weighting function

$$w_\nu = \left|\frac{s_0}{s_\nu}\right| + \left|\frac{s_\nu}{s_0}\right| \tag{79}$$

and the weighted masses

$$\hat{m}_\nu = w_\nu m_\nu \tag{80}$$

with the center of gravity

$$\hat{s}_a = \frac{\sum \hat{m}_\nu s_\nu}{\sum \hat{m}_\nu} \tag{81}$$

then Eq. 78 may be written

$$(a - c) = |s_0|(\hat{s}_a - s_a)\sum \hat{m}_\nu \tag{82}$$

and condition 77 becomes

$$-\left|\frac{s_a}{s_0}\right|\rho_0 < |s_0|^2(\hat{s}_a - s_a)\sum \hat{m}_\nu < |s_0|\rho_\infty \tag{83}$$

Since the left-hand quantity in this condition is negative and the right-hand one positive, we see that the center of gravity $\hat{s}_a$ of the weighted masses may lie on either side of the center of gravity of the unweighted masses.

If we regard s_ν in the weighting function 79 as a continuous variable for the moment, we see that this function has the character shown in Fig. 9. In terms of this picture, careful consideration reveals that if $|s_0|$ is small compared with $|s_a|$, then $(\hat{s}_a - s_a)$ is negative; if $|s_0|$ is large compared with $|s_a|$, then $(\hat{s}_a - s_a)$ is positive. In absolute value, $(\hat{s}_a - s_a)$ cannot, however, exceed the value of $|s_n - s_a|$ for small $|s_0|$ or of $|s_a - s_1|$ for large values of $|s_0|$; and for values of $|s_0|$ that are about the same as $|s_a|$, the value of $(\hat{s}_a - s_a)$ must be rather small since the weighting function shifts the center of gravity of the masses by only a small amount.

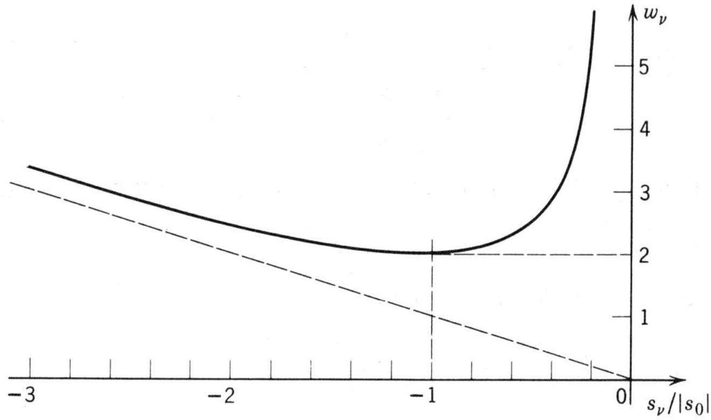

Fig. 9. The weighting function 79 considering s_ν to be a continuous variable.

In considering the probability of being able to fulfill condition 83, it is significant to observe the dependence of m_ν upon $|s_0|$ as expressed by Eqs. 51 and 64. For large values of $|s_0|$, $t^2(s_\nu) \approx |s_0|^4$ and hence m_ν varies inversely as $|s_0|^4$. Since w_ν, according to Eq. 79, varies as $|s_0|$ for large values of $|s_0|$, the weighted masses $\hat{m}_\nu$ are seen to be inversely proportional to $|s_0|^3$. The central quantity in condition 83, therefore, varies inversely as $|s_0|$ and hence becomes small but remains positive so that the right-hand inequality governs. Since the right-hand quantity is proportional to $|s_0|$, we see that the condition will easily be met.

For small values of $|s_0|$, on the other hand, Eq. 51 shows that $t^2(s_\nu) \approx |s_\nu|^4$. w_ν now varies inversely as $|s_0|$, and so does $\hat{m}_\nu$ since m_ν is substantially independent of the value of $|s_0|$. The central quantity in condition 83 becomes proportional to $|s_0|$ and again is small, but now it is negative so that the left-hand inequality governs. Since the left-hand quantity varies inversely as $|s_0|$, we see that the condition will easily be met for this case also.

Finally, if the value of $|s_0|$ is about the same as $|s_a|$ so that $(\hat{s}_a - s_a)$ is small compared with $|s_0|$, the condition is again likely to be fulfillable.

In this connection it is also significant to observe that the values of both ρ_0 and ρ_∞ can at any stage in the synthesis development be increased by the partial removal of a pole of the admittance $Y_1(s) = 1/Z_1(s)$. Thus, if for the moment we indicate a partial-fraction expansion of $Z_1(s)$ by the expression

$$Z_1(s) = \frac{k_1}{s + \sigma_1} + \frac{k_3}{s + \sigma_3} + \cdots + \frac{\rho_0}{s} + \rho_\infty \tag{84}$$

then the expansion of the corresponding admittance has the form

$$Y_1(s) = -\frac{k_2}{s + \sigma_2} - \frac{k_4}{s + \sigma_4} - \cdots + \frac{1}{\rho_\infty} \tag{85}$$

in which the k_ν values are positive. Since we know that $Y_1(0) = 0$, because $Z_1(s)$ has a pole at $s = 0$, we have

$$\frac{1}{\rho_\infty} = \frac{k_2}{\sigma_2} + \frac{k_4}{\sigma_4} + \cdots \tag{86}$$

The partial-fraction expansion of $Y_1(s)/s$ (in which the s-multiplied terms represent admittances of series RC branches) reads

$$\frac{Y_1(s)}{s} = \frac{k_2/\sigma_2}{s + \sigma_2} + \frac{k_4/\sigma_4}{s + \sigma_4} + \cdots \tag{87}$$

It contains no constant term since it must be zero for $s = \infty$, and it contains no term representing a pole at $s = 0$ because $sZ_1(s)$, by Eq. 84, becomes equal to ρ_0 for $s = 0$. Hence Eq. 87 yields

$$\frac{1}{\rho_0} = \frac{k_2}{\sigma_2^2} + \frac{k_4}{\sigma_4^2} + \cdots \tag{88}$$

From Eqs. 86 and 88 we see that reduction of a k value yields smaller values for both $1/\rho_0$ and $1/\rho_\infty$, and hence is a means for expanding the permissible range in condition 83.

Returning to conditions 74 and 75, we observe that fulfilling 74 with the equal sign (equivalent to making the residue of z_{11} at $s = 0$ equal to k_0) avoids the necessity of removing a series-capacitance branch on the input side of the zero section. In terms of the form 76 of conditions 74 and 75, this possibility is found to exist if

$$-b < (a - c) < d - b \tag{89}$$

leading to the following extended form of 83

$$-\left|\frac{s_a}{s_0}\right|\rho_0 < |s_0|^2(\hat{s}_a - s_a)\sum \hat{m}_\nu < |s_0|\rho_\infty - \left|\frac{s_a}{s_0}\right|\rho_0 \tag{90}$$

For small values of $|s_0|$ for which $(\hat{s}_a - s_a)$ is negative, this condition is fulfillable whenever 83 is; for large values of $|s_0|$ for which the value of $|s_a/s_0|\rho_0$ is small, it is only moderately more restrictive than 83. For intermediate values of $|s_0|$, fulfillment of 90 may require the partial removal of a pole of $Y_1(s)$ as just described, but on the whole the possibility of avoiding a series capacitance without resorting to this expedient is rather good. In any event the possibility of trading the removal of a shunt branch for the removal of a series capacitance may be useful in certain situations.

Finally it is necessary to show that the remainder function $Z_2(s)$ is p.r. and RC. This will be so if $(z_{22} + Z_2)$ is p.r. and RC because conditions 74 and 75 ensure that the residue of $(z_{22} + Z_2)$ in the pole at $s = 0$ and its value for $s = \infty$ are larger than the corresponding quantities for z_{22} alone. According to Eqs. 60 and 61 we can form the function

$$\frac{1}{z_{22} + Z_2} = \frac{s^2(s - s_a)^2(z_{11} - Z_1)}{K^2t^2(s)} \tag{91}$$

which has only the nonzero poles of $Z_1(s)$ with residues given by the expression

$$\frac{-s_\nu^{\,2}(s_\nu - s_a)^2k_\nu}{K^2t^2(s_\nu)} \tag{92}$$

Since these values are negative, the function 91 is p.r. and $(z_{22} + Z_2)$ must be p.r. and RC realizable. As pointed out above, conditions 74 and 75 assure the separate p.r. and RC character of z_{22} and Z_2. The residues k_0 and k_{22} of z_{22} as well as the value of K are therefore positive; k_{12} in Eq. 55 is positive, and the residue condition 58 secures a positive value for k_{11}.

The realizability conditions for the zero section and for the remainder function are thus established; and condition 83, which as we have seen is always fulfillable, assures the success of the development cycle in its contemplated form.

The computational procedure for determining the pertinent impedance functions is to expand $(z_{22} + Z_2)$ into partial fractions after the polynomial $h(s)$ is determined and its zeros as well as the value of s_a in Eq. 69 are found. The term for the pole at $s = s_a$ plus $(k_0/s) + K$ (with k_0 determined from Eq. 54) belong to z_{22}; the rest is $Z_2(s)$ with all of its residues computed, ready for carrying out the next cycle.* z_{12} is completely known from Eq. 53, and z_{11} as given by Eq. 56 is determined as soon as the residue k_{11} is computed from Eq. 58.

* An appropriate value for K is chosen to satisfy conditions 74 and 75, which can always be done if 83 is fulfilled.

We come now to the realization of the zero section from its z_{ik}. Although this component two terminal-pair network is not symmetrical, it is derivable from a symmetrical lattice according to the method discussed in Art. 6, Ch. 6. For convenience, Eqs. 73, 74, and 78 as given there are repeated below

$$z_a = \frac{z_{22} - z_{12}}{a} \tag{93}$$

$$z_b = \frac{z_{22}}{a} + z_{12} \tag{94}$$

$$a = \frac{z_{22} - z_{12}}{z_{11} - z_{12}} \tag{95}$$

Substitution from Eqs. 53, 56, and 57 gives for the impedance level factor

$$a = \frac{k_{22} + k_{12}}{k_{11} + k_{12}} = \frac{k_{22}}{k_{12}} = \frac{k_{12}}{k_{11}} \tag{96}$$

where use is also made of condition 58. The reason for choosing the z_{ik} in the forms given by Eqs. 53, 56, and 57 is now clear: these are the most general forms that the z_{ik} can have if the value of the impedance level factor is to be positive real. Since the k_{ik} in Eq. 96 have been shown to be positive real, such a value for the factor a is assured.

Substitution of Eqs. 53, 56, and 57 into Eqs. 93 and 94 yields for the lattice impedances the expressions

$$z_a = \frac{k_{11} + k_{12}}{s - s_a} \tag{97}$$

$$z_b = \left(K + \frac{k_0}{s}\right) \times \frac{k_{11} + k_{12}}{k_{12}} \tag{98}$$

The corresponding lattice is that shown in Fig. 23, Ch. 6, and the element values are

$$R_1 = \frac{k_{11} + k_{12}}{-s_a}, \qquad C_1 = \frac{1}{k_{11} + k_{12}} \tag{99}$$

$$R_2 = \frac{K(k_{11} + k_{12})}{k_{12}}, \qquad C_2 = \frac{k_{12}}{k_0(k_{11} + k_{12})} \tag{100}$$

The process of converting this lattice to an unbalanced form is given in Art. 5, Ch. 6. As pointed out there, condition 69 leading to the twin-T circuit of Fig. 26 is most lenient. Substitution for the element values given here by Eqs. 99 and 100 converts this condition into the form

$$\frac{k_0 - s_a K}{k_{12}} \geq 1 \tag{101}$$

and through use of Eqs. 54 and 55, noting that $t(0) = |s_0|^2$, we get

$$s_a^2 + |s_0|^2 \geq t(s_a) \tag{102}$$

Interpretation of this result is aided by the s-plane sketch in Fig. 10, from which it is seen that the desired transmission zero is restricted to lie in the left half-plane or upon the j-axis.

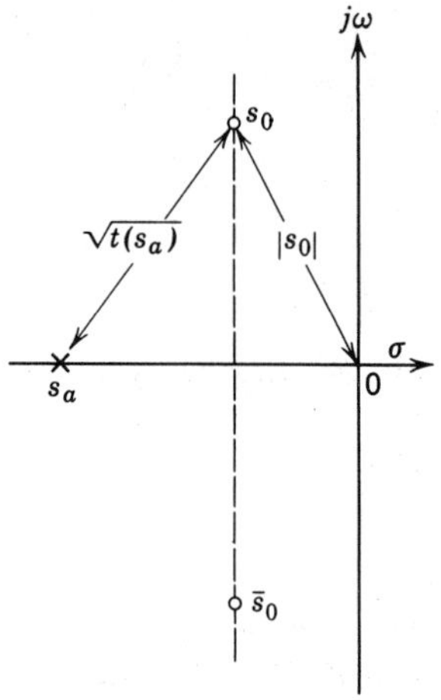

Fig. 10. Relative to the graphical interpretation of condition 102 which assures realization of the zero section in Fig. 8 as a twin-T circuit.

In terms of the parameter values given by Eqs. 99 and 100 it is collaterally of interest to see that the conditions for the realizability of the bridged-T networks shown in Fig. 24a and b, Ch. 6, are respectively expressed by

$$t(s_a) \leq s_a^2 \qquad \text{and} \qquad t(s_a) \leq |s_0|^2 \tag{103}$$

The s-plane significance of each of these conditions is evident from Fig. 10.

The desired zero section is finally formed through bisecting the pertinent unbalanced structure (twin-T or bridged-T), multiplying the impedance level of the right-hand half by the factor a, Eq. 96, and rejoining the halves.

It remains to point out that a completely analogous procedure may be carried out on an admittance basis for which the given driving-point and transfer functions (having identical negative real poles) are given by $Y_1(s)$ and $Y_{12}(s)$. All of the foregoing relations, starting with Eq. 47, remain the same if we consistently replace Z_1 by Y_1/s, z_{11} and z_{22} by y_{11}/s and y_{22}/s, and z_{12} by $-y_{12}/s$. The remainder function Z_2 is correspondingly replaced by Y_2/s, in which $Y_2(s)$ is the actual remainder as an admittance.

Use of the admittance basis may be appropriate in either of two distinctly different situations: (1) We may be given the over-all short-circuit driving-point and transfer admittances $Y_{11}(s)$ and $Y_{12}(s)$ and be asked to find a corresponding two terminal-pair network, just as on the previously assumed impedance basis our problem was to associate a network with the over-all open-circuit driving-point and transfer impedances $Z_{11}(s)$ and $Z_{12}(s)$. (2) We may be given the impedances $Z_{11}(s)$ and $Z_{12}(s)$ and be asked to determine the pertinent two terminal-pair network, but $Z_{11}(s)$ has no pole at $s = 0$ and its asymptotic value for $s \to \infty$ is zero, so that the terms with ρ_0 and ρ_∞ in Eq. 47 are absent.

Although in the latter event we can create a pole at $s = 0$ through removal of an appropriate shunt resistance, it may be simpler to reciprocate $Z_{11}(s)$ and proceed on an admittance basis as though this reciprocated $Z_{11}(s)$ were a short-circuit driving-point admittance which, of course, it is not. Nevertheless, if we develop this Y_{11} function into a cascade of component two terminal-pair networks characterized by admittances y_{ik}, choosing the same transmission zeros as those specified for the over-all transfer impedance $Z_{12}(s)$, then we will obtain a resultant development that is just as pertinent to the given $Z_{11}(s)$ and $Z_{12}(s)$ as we would if these had been developable on an impedance basis.

The truth of this statement follows from the fact that the development of $Y_{11} = 1/Z_{11}$ on an admittance basis must just as surely yield the desired Z_{11} function; and the z_{12} function for any component network has the same zeros as the y_{12} function for that network because $z_{12} = -y_{12}/(y_{11}y_{22} - y_{12}^2)$ in which the y_{ik} fulfill the residue condition with an equal sign at all finite poles. Hence the development of $Y_{11} = 1/Z_{11}$ on an admittance basis provides: (1) the correct input impedance, and (2) the desired over-all transmission zeros, which is neither more nor less than is provided by a development of $Z_{11}(s)$ on an impedance basis.

These ideas tell us that either condition 74 or 75 can at some stage in the total development be fulfilled on the left by an equality if we wish. By doing so, a remainder function $Z_2(s)$ is obtained in which either the pole at the origin or the constant term ρ_∞ is missing. We may then not be able to proceed further on an impedance basis in the normal manner, but we can reciprocate the remainder and proceed on an admittance basis. This ability to switch from one basis to another, perhaps repeatedly during an entire development, injects a degree of flexibility that may prove useful under certain circumstances.

Synthesis of the zero section on an admittance basis is entirely analogous to that on the impedance basis, making use of the relations expressed by Eqs. 80 through 86 in Ch. 6. Thus we have for the lattice admittances

$$y_a = y_{11} - ay_{12} \tag{104}$$

$$y_b = y_{11} + y_{12} \tag{105}$$

and for the impedance level factor

$$a = \frac{y_{11} + y_{12}}{y_{22} + y_{12}} \tag{106}$$

The latter yields the result expressed by Eq. 96; for the lattice admittances we get

$$y_a = (k_0 + Ks)\left(1 + \frac{k_{11}}{k_{12}}\right) \tag{107}$$

$$y_b = \frac{(k_{11} + k_{12})s}{s - s_a} \tag{108}$$

These again lead to the lattice of Fig. 23, Ch. 6, but instead of the element values 99 and 100 we have

$$R_1 = \frac{k_{12}}{k_0(k_{11} + k_{12})}, \qquad C_1 = \frac{K(k_{11} + k_{12})}{k_{12}} \tag{109}$$

$$R_2 = \frac{1}{k_{11} + k_{12}}, \qquad C_2 = \frac{k_{11} + k_{12}}{-s_a} \tag{110}$$

Transformation to an equivalent unbalanced form, and conditions under which it is possible, are the same as before.

5 A Useful Artifice in Transfer Function Synthesis

When the prescribed transfer function is relatively complicated so that a fair number of successive cycles are needed for its complete cascade realization, one can save a considerable amount of computational labor by making use of a so-called partitioning theorem* whereby the synthesis is effectively reduced to two problems involving functions of half the complexity. The basic idea involved in this method is contained in the transfer function representation given in Art. 8, Ch. 9, by Eq. 172 for the cascade of component networks shown there in Fig. 30. Application of this result to the pair of networks in Fig. 11 (having no external input or output resistances) gives

$$Z_{12} = \frac{E_2}{I_1} = \frac{z_{12}^{(1)} \times z_{12}^{(2)}}{z_{22}^{(1)} + z_{11}^{(2)}} \tag{111}$$

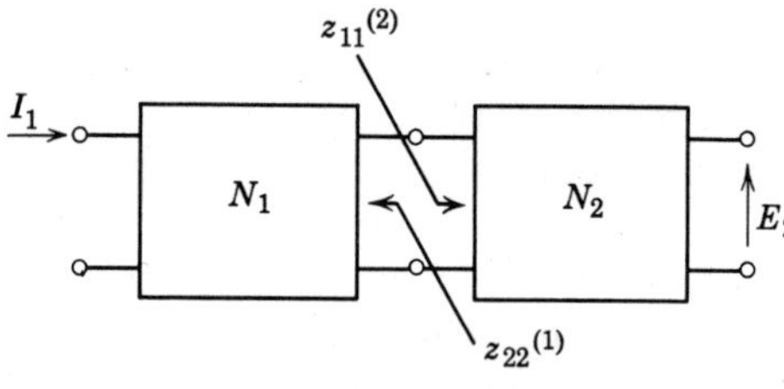

Fig. 11. Illustrating a partitioning scheme that simplifies the synthesis of more elaborate transfer functions.

* H. Adler, *Master's thesis*, Electrical Engineering Department, Massachusetts Institute of Technology, 1948.

If we write $Z_{12}(s) = p(s)/q(s)$, with

$$p(s) = (s - s_{p1})(s - s_{p2}) \cdots (s - s_{p2n}) \tag{112}$$

$$q(s) = (s - s_1)(s - s_2) \cdots (s - s_{2n}) \tag{113}$$

then Dasher suggests* the following identifications

$$z_{12}^{(1)} = \frac{(s - s_{p1})(s - s_{p3}) \cdots (s - s_{p2n-1})}{(s - s_1)(s - s_3) \cdots (s - s_{2n-1})} \tag{114}$$

$$z_{12}^{(2)} = \frac{(s - s_{p2})(s - s_{p4}) \cdots (s - s_{p2n})}{(s - s_1)(s - s_3) \cdots (s - s_{2n-1})} \tag{115}$$

$$z_{22}^{(1)} = z_{11}^{(2)} = \frac{(s - s_2)(s - s_4) \cdots (s - s_{2n})}{2(s - s_1)(s - s_3) \cdots (s - s_{2n-1})} \tag{116}$$

where

$$s_{2n} < s_{2n-1} < \cdots < s_2 < s_1 < 0 \tag{117}$$

These are obtained by dividing numerator and denominator in $Z_{12}(s)$ by $(s - s_1)^2(s - s_3)^2 \cdots (s - s_{2n-1})^2$ and then comparing with Eq. 111. If $q(s)$ is of odd degree, the impedances 116 will be zero for $s = \infty$; and if $p(s)$ is of lower degree than $q(s)$, then a variety of possible transfer impedance pairs 114 and 115 may be constructed having different relative numerator and denominator degrees. The zeros of $Z_{12}(s)$ can be distributed between $z_{12}^{(1)}$ and $z_{12}^{(2)}$ in any way, so long as conjugate pairs are not separated.

The same procedure applies also to an admittance basis with the variant that the common denominator is chosen to be $(s - s_2)(s - s_4) \cdots (s - s_{2n})$ so that the lowest critical frequency in the driving-point functions $y_{11}^{(2)}$ and $y_{22}^{(1)}$ becomes a zero, as required for *RC* admittance functions.

The foregoing cascade development is applied separately to the functions $z_{22}^{(1)}$ and $z_{11}^{(2)}$ so as to produce the transmission zeros called for by $z_{12}^{(1)}$ and $z_{12}^{(2)}$ respectively. A back-to-back connection of the resulting networks, as shown in Fig. 11, then yields the desired realization (within a constant multiplier) for the stated Z_{12} function.

Since no control is exercised over the input impedance of the resultant network, this method of synthesis is appropriate only when $Z_{12}(s)$ alone is specified.

6 Some Examples of Cascade Synthesis

Suppose we are given the functions

$$Z_{12}(s) = \frac{(s + 1)^2 + 4}{(s + 1)(s + 3)} = \frac{s^2 + 2s + 5}{s^2 + 4s + 3} \tag{118}$$

---

* B. J. Dasher, *Tech. Rept.*, *215*, Mass. Inst. Technol., Research Lab. Electronics, Nov. 20, 1951, p. 40.

and

$$Z_{11}(s) = \frac{(s+2)(s+4)}{(s+1)(s+3)} = \frac{s^2+6s+8}{s^2+4s+3} \tag{119}$$

The complex pair of transmission zeros are

$$s_0 = -1 + j2, \qquad \bar{s}_0 = -1 - j2 \tag{120}$$

and the corresponding quadratic factor reads

$$t(s) = (s+1)^2 + 4 = s^2 + 2s + 5 \tag{121}$$

We begin by expanding $Z_{11}(s)$ in partial fractions

$$Z_{11}(s) = \frac{\frac{3}{2}}{s+1} + \frac{\frac{1}{2}}{s+3} + 1 \tag{122}$$

According to the form 47 for the input impedance, we see that $\rho_0 = 0$ and $\rho_\infty = 1$. We could remedy the absence of a ρ_0 term by switching to an admittance basis, but we shall first see whether the development cycle will succeed in spite of this defect. Preparatory to computing the m_ν from Eq. 64, we find

$$t(-1) = 4, \qquad t(-3) = 8 \tag{123}$$

and then obtain

$$m_1 = \frac{\frac{3}{2}}{16} = \frac{3}{32}, \qquad m_2 = \frac{3 \times \frac{1}{2}}{64} = \frac{3}{128} \tag{124}$$

Next we find

$$\sum m_\nu = \frac{15}{128}, \qquad \sum m_\nu s_\nu = \frac{-21}{128} \tag{125}$$

whence Eq. 69 yields

$$s_a = -\tfrac{7}{5} \tag{126}$$

We can now compute

$$\sum m_\nu s_\nu (s_\nu - s_a) = \frac{3}{40} \quad \text{and} \quad \sum \frac{m_\nu}{-s_\nu}(s_\nu - s_a) = \frac{1}{40} \tag{127}$$

and with $t(0) = |s_0|^2 = 5$, obtain for conditions 74 and 75 respectively

$$0 < K - \tfrac{1}{8} \leq 0 \qquad \text{and} \qquad 0 < K - \tfrac{3}{40} \leq 1 \tag{128}$$

The first of these allows no choice other than $K = \frac{1}{8}$, and the second is fulfilled with this value for K. Hence the contemplated cycle will succeed.

Equation 63 now yields

$$\frac{h(s)}{q(s)} = \frac{64 \times 3}{80}\left(\frac{1}{s+1} - \frac{1}{s+3}\right) = \frac{24}{5(s+1)(s+3)} \tag{129}$$

so that we get by Eq. 61

$$(z_{22} + Z_2) = \frac{5(s+1)(s+3)}{24s(s+\frac{7}{5})} = \frac{\frac{25}{56}}{s} + \frac{\frac{2}{21}}{s+\frac{7}{5}} + \frac{5}{24} \tag{130}$$

From Eqs. 54 and 55 we find

$$k_0 = \tfrac{25}{56} \qquad \text{and} \qquad k_{12} = \tfrac{13}{35} \tag{131}$$

Noting in Eq. 130 that $k_{22} = \frac{2}{21}$, the residue condition 58 gives

$$k_{11} = \tfrac{507}{350} \tag{132}$$

Since condition 74 (equivalent to 70) is fulfilled with an equal sign at both ends, we know that the residue of z_{11} at $s = 0$ is equal to the required value k_0 with nothing to spare, but condition 75 (equivalent to 71) is fulfilled with an inequality at both ends. One of these inequalities accounts for the constant term in Eq. 130 being larger than the minimum required $K = \frac{1}{8}$; the other implies that $z_{11}(\infty)$ is also larger than this minimum value. In this regard we find from Eqs. 60 and 68 (making use of the value of the first sum in 127)

$$z_{11}(\infty) = K^2H + 1 = \tfrac{3}{40} + 1 \tag{133}$$

which is larger than the required minimum value K by the amount

$$\tfrac{43}{40} - \tfrac{1}{8} = \tfrac{38}{40} = \tfrac{19}{20} \tag{134}$$

A series resistance of this value precedes the zero section. If we are interested only in realizing the transfer impedance 118, this series branch may be omitted.

We thus obtain for the zero section, the impedances

$$z_{11} = \frac{1}{8} + \frac{\frac{25}{56}}{s} + \frac{\frac{507}{350}}{s + \frac{7}{5}} \tag{135}$$

$$z_{22} = \frac{1}{8} + \frac{\frac{25}{56}}{s} + \frac{\frac{2}{21}}{s + \frac{7}{5}} \tag{136}$$

$$z_{12} = \frac{1}{8} + \frac{\frac{25}{56}}{s} - \frac{\frac{13}{35}}{s + \frac{7}{5}} \tag{137}$$

and for the remainder

$$Z_2(s) = \tfrac{1}{12} \tag{138}$$

With the values of k_{11}, k_{12}, K, k_0, and s_a thus found, Eqs. 99 and 100 yield for the lattice parameters

$$\begin{aligned} R_1 &= \tfrac{13}{10}, & C_1 &= \tfrac{50}{91} \\ R_2 &= \tfrac{49}{80}, & C_2 &= \tfrac{16}{35} \end{aligned} \tag{139}$$

and for the impedance level factor 95, the value

$$a = \tfrac{10}{39} \tag{140}$$

The resulting twin-T is now easily constructible according to the methods discussed in Art. 6, Ch. 6. Its terminal resistance (which is not affected by the impedance level change of the right-hand half of the bisected twin-T) has the value 138.

As a second example we shall consider a slight modification of the preceding one. We will retain the same complex pair of transmission zeros, but increase the degree of the denominator polynomial from 2 to 3, so that the transfer function has one zero at $s = \infty$. Specifically we assume

$$Z_{12}(s) = \frac{s^2 + 2s + 5}{2s^3 + 11s^2 + 15s + 3} \tag{141}$$

and

$$Z_{11}(s) = \frac{s^3 + 7s^2 + 12s + 3}{2s^3 + 11s^2 + 15s + 3} \tag{142}$$

Again the driving-point impedance has no pole at $s = 0$. This time we will cure this defect by removing an appropriate shunt resistance. Its value equals $Z_{11}(0) = 1$ ohm. The remainder function is found to be

$$Z_1(s) = \frac{1}{(1/Z_{11}) - 1} = \frac{s^3 + 7s^2 + 12s + 3}{s^3 + 4s^2 + 3s} \tag{143}$$

The cascade development is begun by expanding this input impedance into partial fractions. We obtain

$$Z_1(s) = \frac{1}{s} + \frac{\frac{3}{2}}{s + 1} + \frac{\frac{1}{2}}{s + 3} + 1 \tag{144}$$

which is the same as the expansion 122 for the preceding problem except for the additive term $1/s$. Assuming the same transmission zeros as defined by 121, relations 123 through 127 are again applicable, and only the first of conditions 128 (corresponding to 74) changes because ρ_0 now equals unity instead of being zero. These conditions now read

$$0 < K - \tfrac{1}{8} \leq \tfrac{7}{25} \quad \text{and} \quad 0 < K - \tfrac{3}{40} \leq 1 \tag{145}$$

In order to avoid a series capacitance at the input, we should now choose

$$K - \tfrac{1}{8} = \tfrac{7}{25} \quad \text{or} \quad K = \tfrac{81}{200} = 0.405 \tag{146}$$

which the second condition 145 permits.

Since

$$z_{11}(\infty) = K^2 H + 1 = 1.0123 \tag{147}$$

we will have a series resistance at the input with the value

$$z_{11}(\infty) - K = 1.0123 - 0.405 = 0.6073 \tag{148}$$

The expression for $(z_{22} + Z_2)$, which is proportional to K^2, changes from the form given in Eq. 130 only by the square of the factor $0.405/\frac{1}{8}$. We thus find for the present problem that

$$(z_{22} + Z_2) = \frac{2.187(s + 1)(s + 3)}{s(s + \frac{7}{5})} = 2.187 + \frac{4.686}{s} + \frac{0.9998}{s + \frac{7}{5}} \tag{149}$$

Equations 54 and 55 yield

$$k_0 = 1.446 \quad \text{and} \quad k_{12} = 1.203 \tag{150}$$

and with $k_{12} = 0.9998$, the residue condition 58 gives

$$k_{11} = 1.447 \tag{151}$$

The impedances characterizing the zero section are now evident, and the remainder is

$$Z_2(s) = 1.782 + \frac{3.24}{s} \tag{152}$$

Since the transfer impedance 141 calls for one zero at $s = \infty$, we regard the first term in this remainder function as a series resistance and the second as a final shunt capacitance.

We will now consider an example requiring two successive cycles. Let the given impedances be

$$Z_{12}(s) = \frac{[(s+1)^2 + 4][(s+1)^2 + 9]}{(s+1)(s+3)(s+5)(s+7)} \tag{153}$$

$$Z_{11}(s) = \frac{(s+2)(s+4)(s+6)(s+8)}{(s+1)(s+3)(s+5)(s+7)} \tag{154}$$

In the first cycle we will produce the transmission zeros defined by

$$t(s) = (s+1)^2 + 4 \tag{155}$$

Proceeding on an impedance basis, we determine the partial-fraction expansion

$$Z_{11}(s) = \frac{\frac{115}{48}}{s+1} + \frac{\frac{15}{16}}{s+3} + \frac{\frac{9}{16}}{s+5} + \frac{\frac{15}{48}}{s+7} + 1 \tag{156}$$

and compute, according to Eq. 64

$$\begin{aligned} m_1 &= 0.1497395, \quad & m_2 &= 0.0439453 \\ m_3 &= 0.00703125, \quad & m_4 &= 0.00136718 \end{aligned} \tag{157}$$

From these we find

$$\sum m_\nu = 0.2020832, \qquad \sum m_\nu s_\nu = 0.3263019 \tag{158}$$

and thus have

$$s_a = -\frac{0.3263019}{0.2020832} = -1.61469 \tag{159}$$

Preparatory to evaluating conditions 74 and 75 we next compute

$$\sum \frac{m_\nu}{-s_\nu}(s_\nu - s_a) = 0.0659385, \quad \sum m_\nu s_\nu (s_\nu - s_a) = 0.2611436 \tag{160}$$

whereupon these conditions read respectively

$$0 < K - 0.0659 \leq 0 \quad \text{and} \quad 0 < K - 0.261 \leq 1 \tag{161}$$

The first of these is rigid (because the term with ρ_0 is absent in 156) and requires $K = 0.0659$, which makes fulfillment of the second condition impossible. We might have anticipated this situation at the outset in view of the relatively small value of $|s_0|$, yielding $(\hat{s}_a - s_a)$ in 83 negative according to the previous discussion of this item. A cure for the difficulty thus presented lies in switching to an admittance basis.

Accordingly, we make the partial-fraction expansion

$$\frac{Y_{11}(s)}{s} = \frac{(s+1)(s+3)(s+5)(s+7)}{s(s+2)(s+4)(s+6)(s+8)}$$
$$= \frac{\frac{35}{128}}{s} + \frac{\frac{5}{32}}{s+2} + \frac{\frac{9}{64}}{s+4} + \frac{\frac{5}{32}}{s+6} + \frac{\frac{35}{128}}{s+8} \tag{162}$$

We next compute

$$t(-2) = 5, \quad t(-4) = 13, \quad t(-6) = 29, \quad t(-8) = 53 \tag{163}$$

and from Eq. 64

$$m_1 = \frac{5}{16 \times 25}, \quad m_2 = \frac{9}{16 \times 169}, \quad m_3 = \frac{15}{16 \times 841}, \quad m_4 = \frac{35}{16 \times 2809} \tag{164}$$

From these we find

$$\sum m_\nu = \frac{1}{16}(0.2 + 0.0532544 + 0.0178359 + 0.0124599) = \frac{0.2835502}{16} \tag{165}$$

$$\sum m_\nu s_\nu = \frac{-1}{16}(0.4 + 0.2130176 + 0.1070154 + 0.0996795) = \frac{-0.8197125}{16} \tag{166}$$

and

$$s_a = -\frac{0.8197125}{0.2835502} = -2.89089 \tag{167}$$

from which we get

$$\sum \frac{m_\nu}{-s_\nu}(s_\nu - s_a) = \frac{0.057123}{16} \quad \text{and} \quad \sum m_\nu s_\nu(s_\nu - s_a) = \frac{0.721899}{16} \tag{168}$$

Conditions 74 and 75 are thus found to be

$$0 < K - 0.017851 \leq 0.15809 \quad \text{and} \quad 0 < K - 0.045118 \leq 0 \tag{169}$$

The second of these requires

$$K = 0.045118 \tag{170}$$

and the first is fulfilled by this choice.

Use of Eq. 63 now yields straightforwardly

$$\frac{h(s)}{q(s)} = \frac{22.16(s+4.47)(s+6.80)}{(s+2)(s+4)(s+6)(s+8)} \tag{171}$$

and hence Eq. 61, modified to suit the admittance basis, gives

$$\frac{y_{22} + Y_2}{s} = \frac{0.04512(s + 2)(s + 4)(s + 6)(s + 8)}{(s + 2.891)(s + 4.47)(s + 6.80)}$$
$$= \frac{0.197}{s} + \frac{0.03965}{s + 2.891} + \frac{0.0172}{s + 4.47} + \frac{0.0094}{s + 6.80} + 0.04512 \quad (172)$$

Equations 54, 55, and 58 now produce the values

$$k_0 = 0.04512 \times 5/2.891 = 0.078 \quad (173)$$
$$k_{12} = 0.04512 \times 7.576/2.891 = 0.118 \quad (174)$$
$$k_{11} = 0.118 \times 0.118/0.03965 = 0.3526 \quad (175)$$

Fulfillment of condition 75 by the K value 170 implies that $(y_{11}/s)_{s=\infty} = K$, but $y_{11}(0)$ will have a value larger than k_0 since the first condition 169 (equivalent to 74) is fulfilled with a margin. We find that

$$y_{11}(0) = \frac{K^2t^2(0)h(0)}{-s_a q(0)} + Y_{11}(0) = 0.0308 + 0.2733 = 0.3041 \quad (176)$$

so that a shunt conductance of the value

$$0.3041 - 0.078 = 0.2261 \text{ mho} \quad (177)$$

appears at the input terminals of the zero section, which is characterized by

$$\frac{y_{11}}{s} = 0.0451 + \frac{0.078}{s} + \frac{0.3526}{s + 2.89} \quad (178)$$
$$\frac{y_{22}}{s} = 0.0451 + \frac{0.078}{s} + \frac{0.0396}{s + 2.89} \quad (179)$$
$$\frac{-y_{12}}{s} = 0.0451 + \frac{0.078}{s} - \frac{0.118}{s + 2.89} \quad (180)$$

and the remainder function, by Eq. 172, reads

$$\frac{Y_2(s)}{s} = \frac{0.119}{s} + \frac{0.0172}{s + 4.47} + \frac{0.0094}{s + 6.80} \quad (181)$$

We now consider the following cycle for the transmission zeros defined by the factor

$$t(s) = (s + 1)^2 + 9 \quad (182)$$

Although $|s_0|$ here is somewhat larger, we will stay with the admittance basis tentatively since the residues of $Y_2(s)/s$ are already known. We have

$$t(-4.47) = 21.041, \qquad t(-6.80) = 42.64 \quad (183)$$

and Eq. 64 yields

$$m_1 = 0.0001736, \qquad m_2 = 0.00003513 \quad (184)$$

whence

$$\sum m_\nu = 0.0002087 \quad \text{and} \quad \sum m_\nu s_\nu = -0.0010149$$

so that we find

$$s_a = -4.8624 \tag{185}$$

Next we compute

$$\sum \frac{m_\nu}{-s_\nu}(s_\nu - s_a) = 0.00000522 \qquad \text{and} \qquad \sum m_\nu s_\nu (s_\nu - s_a) = 0.0001586 \tag{186}$$

and have for conditions 74 and 75 respectively

$$0 < K - 0.0000522 \leq 0.0579 \qquad \text{and} \qquad 0 < K - 0.0001586 \leq 0 \tag{187}$$

Again the second condition fixes the value of K

$$K = 0.0001586 \tag{188}$$

and the first condition remains satisfied.

Equation 63 now produces the result

$$\frac{h(s)}{q(s)} = \frac{6300}{(s + 4.47)(s + 6.80)} \tag{189}$$

so that from Eq. 61 we have

$$\frac{y_{22} + Y_2}{s} = \frac{0.0001586(s + 4.47)(s + 6.80)}{s(s + 4.86)}$$

$$= \frac{0.000992}{s} + \frac{0.0000248}{s + 4.86} + 0.0001586 \tag{190}$$

Equations 54, 55, and 58 give

$$k_0 = 0.0001586 \times 10/4.862 = 0.000326 \tag{191}$$

$$k_{12} = 0.0001586 \times 23.92/4.862 = 0.000781 \tag{192}$$

$$k_{11} = 0.000781 \times 0.000781/0.0000248 = 0.0246 \tag{193}$$

and

$$y_{11}(0) = \frac{100}{4.862} \sum \frac{m_\nu}{-s_\nu}(s_\nu - s_a) + \rho_0 = \frac{0.000522}{4.862} + 0.119 = 0.119107 \tag{194}$$

shows that a shunt conductance branch of

$$y_{11}(0) - k_0 = 0.118781 \text{ mho} \tag{195}$$

is required at the input.

The admittances of the zero section are now evident, and the remainder is a conductance

$$Y_3 = 0.000992 - 0.000326 = 0.000666 \text{ mho} \tag{196}$$

It may be of interest to see how one would approach this latter problem through use of the partitioning theorem discussed in Art. 5 above. In this case, we would not expect to obtain a realization that has the particular impedance 154, but this fact is usually of no practical importance.

After dividing numerator and denominator in expression 153 for $Z_{12}(s)$ by the square of the factors $(s+1)(s+5)$, the identifications 114, 115, and 116 appropriate to this example become

$$z_{12}^{(1)} = \frac{(s+1)^2+4}{(s+1)(s+5)}, \qquad z_{12}^{(2)} = \frac{(s+1)^2+9}{(s+1)(s+5)},$$

$$z_{22}^{(1)} = z_{11}^{(2)} = \frac{(s+3)(s+7)}{2(s+1)(s+5)} \tag{197}$$

Realization of each network in the back-to-back arrangement shown in Fig. 11 requires no further elaboration.

PROBLEMS

1. (*a*) Derive the restrictions on $p(s)$ and $q(s)$ under which the short-circuit transfer admittance

$$y_{12}(s) = \frac{p(s)}{q(s)}$$

potentially represents y_{12} of an *RC* network.

(*b*) Develop the complete procedure for synthesis of a two terminal-pair *RC* network given the partial-fraction expansions of y_{11}, y_{12}, and y_{22} analogous to the method given in Art. 1, Ch. 7. Point out the conditions under which no ideal transformers will be required, and the form of the resulting network. How is the generality of y_{12} restricted in this case?

2. The open-circuit transfer impedance of an *RC* two terminal-pair network is

$$z_{12} = \frac{1}{s(s+2)(s+4)}$$

Find the simplest z_{11} function having integer values for its zeros and from this determine the resultant network in the form of a ladder, giving all element values.

3. The open-circuit transfer impedance of an *RC* network is specified to be

$$z_{12} = \frac{s}{(s+1)(s+2)}$$

Associate with this function the z_{11} of a symmetrical network such that the residue condition is fulfilled with the equal sign at each pole. Determine a corresponding lattice network and develop this into an equivalent ladder.

4. (*a*) Find a two terminal-pair network with the *RL* impedances

$$z_{11} = \frac{s}{s+1} + \frac{2s}{s+2} + 3s$$

$$z_{12} = \frac{s}{s+1} - \frac{s}{s+2} + 6s$$

$$z_{22} = \frac{s}{s+1} + \frac{s}{2(s+2)} + 15s$$

(*b*) Using the same z_{12} as in part (*a*), set $z_{11} = z_{22}$ to be K times the z_{11} of part (*a*), selecting the smallest K to make the network realizable as a symmetrical lattice.

5. Synthesize an RC lattice having the transfer impedance

$$z_{12} = \frac{(s-1)(s-2)(s-3)}{(s+1)(s+2)(s+3)}$$

6. Synthesize an RC lattice having the transfer impedance

$$z_{12} = \frac{s^2+1}{(s+1)(s+3)}$$

7. Synthesize an RC network in the form of paralleled unbalanced ladder networks terminated in resistances at both ends, having the transfer impedance of Prob. 6.

8. Use the method of cascade synthesis to realize the transfer impedance of Prob. 6, associating the driving-point function

$$z_{11} = \frac{(s+2)(s+4)}{(s+1)(s+3)}$$

9. By the parallel ladder method, realize the transfer impedance

$$z_{12} = \frac{(s^2+s+1)^2}{(s+1)(s+3)(s+5)(s+7)}$$

10. With the transfer impedance of Prob. 9, associate the driving-point impedance

$$z_{11} = \frac{(s+2)(s+4)(s+6)(s+8)}{(s+1)(s+3)(s+5)(s+7)}$$

and realize by the method of cascade synthesis, making use of the partitioning scheme discussed in Art. 5.

11. By the method of cascade synthesis, realize the functions

$$z_{11} = \frac{(s+2)(s+4)(s+6)(s+8)}{(s+1)(s+3)(s+5)(s+7)}, \qquad z_{12} = \frac{s^2+s+1}{(s+1)(s+3)(s+5)(s+7)}$$

12. Reconsider Prob. 11 with z_{12} replaced by the function

$$z_{12} = \frac{(s^2+1)^2}{(s+1)(s+3)(s+5)(s+7)}$$

13. Reconsider Prob. 11 with z_{12} replaced by the function

$$z_{12} = \frac{(s+2)^3}{(s+1)(s+3)(s+5)(s+7)}$$

14. Realize the functions of Prob. 13 in the parallel ladder method.

15. Realize the functions of Prob. 13 in a symmetrical lattice network and see if you can convert the result into an unbalanced equivalent form.

CHAPTER FOURTEEN

The Approximation Problem

1 Introductory Remarks

In this chapter we present various available methods whereby one may determine an appropriate rational function for the representation of a driving-point or transfer impedance (more specifically, its j-axis magnitude or real part or angle) when the desired character of this function is specified by a plotted curve or by a set of discrete values or by any other analytic or graphical means. Practical problems arise in which one wishes specifically to control the magnitude or real part or angle of a network function singly or jointly. The extent to which these characteristics of a driving-point or transfer impedance can simultaneously be controlled is discussed in Ch. 8, and we shall make frequent use of the results given there to supply the theoretical background needed in the present considerations.

Solution to an appropriate approximation problem is the first step in any synthesis problem. It yields the rational function representation for the desired network response characteristic. The second step involves the realization of this rational function. In our discussions so far, we have been concerned primarily with realization techniques. We shall now give some detailed attention to the manner in which the so-called given impedance function originates.

Distinctly different about this part of the synthesis problem is the fact that it involves inexact processes. In contrast with the synthesis procedures which *exactly* realize a given network function, the approximation methods, as their name implies, are inexact in character.

The given data in a network design problem are usually idealized in some graphical manner, or stated only in terms of a set of upper and lower bounds. They are seldom given analytically in the form of a conventional impedance or admittance, and, if so stated, they are not likely to have the form of a rational function fulfilling the necessary conditions for physical realizability. The approximation problem is concerned with the construction of a rational function fulfilling, on the one hand, the appropriate realizability conditions, and on the other, the demands of the given data regarding performance characteristics of the desired network. It is here that all questions regarding allowable tolerances and the acceptability of possible solutions are settled. Once an appropriate rational function has thus been established, there is no need to make further concessions or to revise tolerances except as may be needful because of the effects of unavoidable parasitic elements whose presence may noticeably interfere with the predicted response.

An experienced designer will not consider the approximation and the realization problems separately, but will make judicious use of the flexible character of the approximation problem to arrive at a rational function that not only fulfills the necessary realizability conditions but also incorporates certain special properties which make possible a preferred form of realization, lending itself easily to construction and alignment while minimizing parasitic effects.

The desirability of thus absorbing into the approximation problem some of the anticipated difficulties of the subsequent realization procedure shows that these two parts of the total synthesis problem are, or should be, rather intimately interwoven if we wish to achieve the maximum in performance from linear, passive, lumped networks. In the following discussion we shall not attempt to go into such detailed interrelationships between the approximation and realization problems, but shall present only some of the presently known, and more generally useful, techniques for obtaining rational approximants appropriate to certain commonly occurring response specifications. The background thereby provided is in any case essential to a more intimate correlation of the several steps which collectively form a solution to the total synthesis problem.

2 The Use of Butterworth and Tschebyscheff Functions

A commonly occurring problem is that of finding a rational function, specifically a quotient of polynomials in ω^2 as given by Eq. 6 in Ch. 11, that approximates the so-called normalized and idealized low-pass characteristic shown in Fig. 1. The normalization pertains to the fact that we choose the value unity for both the impedance level and the radian-frequency cutoff, and the idealization pertains to the choice of a perfectly rectangular shape for this low-pass characteristic which is constant throughout the pass region $-1 < \omega < 1$ (the characteristic must be an even function) and identically zero elsewhere. By means of magnitude and frequency scaling (see *ICT*, p. 309) we can subsequently accommodate any desired impedance level and cutoff; and by the specification of appropriate tolerances, we shall indicate the degree to which this idealized form of the response is to be approximated.

A simple, yet effective, function that affords any desired degree of approximation to the idealized curve in Fig. 1 is the so-called *Butterworth function of order n* given by

$$|Z_{12}(j\omega)|^2 = \frac{1}{1 + \omega^{2n}} \tag{1}$$

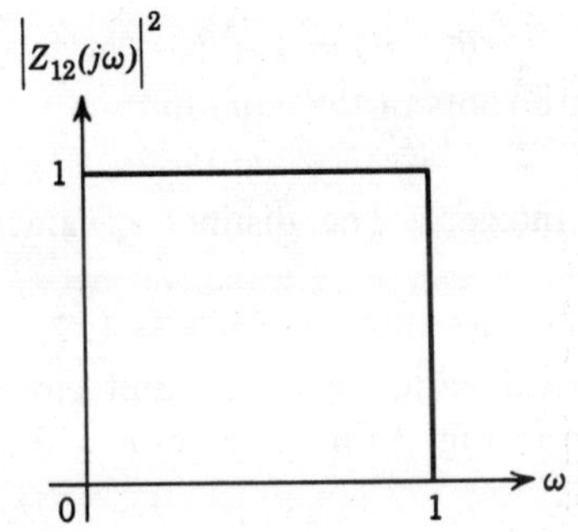

Fig. 1. Idealized and normalized low-pass filter characteristic.

The actual magnitude of the transfer impedance obtained by this rational approximant to its squared magnitude is shown for several orders n in the curves of Fig. 2. We see that the curves for all orders pass through the 0.707 point at $\omega = 1$. Like the simple resonance curve (which, incidentally, is a first-order Butterworth function) all the functions in this family pass through their half-power points at the theoretical cutoff frequency.

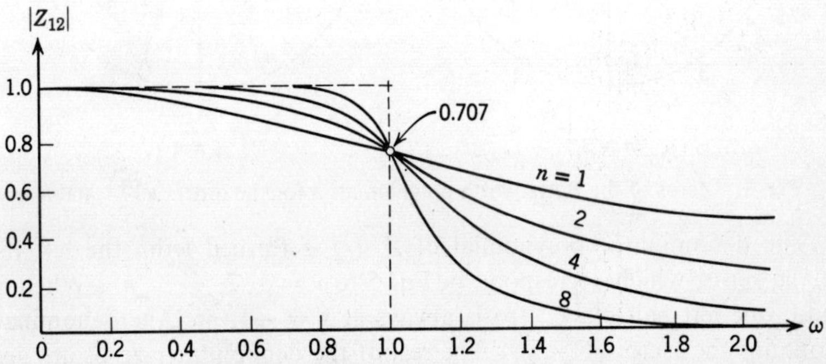

Fig. 2. Butterworth functions of varying order (Eq. 1) are a simple and effective means for approximating the idealized characteristic of Fig. 1 by a rational function.

The character of the approximation is monotonic throughout. For the vicinity of $\omega = 0$ we have

$$1 - |Z_{12}(j\omega)| = \tfrac{1}{2}\omega^{2n} - \tfrac{3}{8}\omega^{4n} + \tfrac{5}{16}\omega^{6n} - + \cdots \tag{2}$$

which shows that the first $2n - 1$ derivatives are zero at the origin. This fact results in an extremely flat or nearly constant behavior for low frequencies, and for this reason the Butterworth characteristic is referred to as being "flat-flat" or "maximally flat."

The complex transfer impedance $Z_{12}(s)$ corresponding to the squared j-axis magnitude 1 is readily generated. Following the notation in Eq. 6, Ch. 11, we have

$$B(-s^2) = 1 + (-1)^n s^{2n} \tag{3}$$

the zeros of which are roots of the equation

$$s^{2n} = e^{j\pi(n-1)+j2\pi\nu} \tag{4}$$

for ν equal to an integer. The distinct s values obtained from this equation are

$$s_\nu = e^{j\pi(2\nu+n-1)/2n} \qquad \text{for } \nu = 1, 2, \cdots, 2n \tag{5}$$

They form symmetrical patterns on the unit circle about the origin in the s-plane as shown in Fig. 3 for the cases $n = 3$ and $n = 4$.

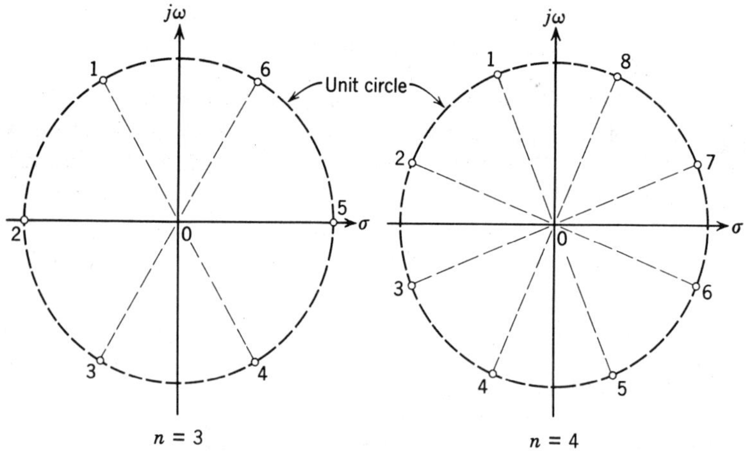

Fig. 3. Zeros of the Butterworth polynomials 3 for the orders $n = 3$ and $n = 4$.

The denominator polynomial of $Z_{12}(s)$ is formed from the left half-plane zeros, which correspond in Eq. 5 to $\nu = 1, 2, \cdots, n$. When n is odd, one left half-plane zero is always at $s = -1$, and the denominator contains the factor $(1 + s)$. The rest of the denominator zeros are given by quadratic factors, representing conjugate complex pairs. Since all s_ν values have unit magnitude, the constant term in the denominator polynomial equals unity. The complex transfer impedance appropriate to the squared magnitude 1 of order n, therefore, has the form

$$Z_{12}(s) = \frac{1}{1 + a_1 s + a_2 s^2 + a_3 s^3 + \cdots + a_n s^n} \tag{6}$$

Values of the coefficients in this denominator polynomial and its factored form are given for several values of the order n in the following two tables.*

n	a_1	a_2	a_3	a_4	a_5	a_6	a_7	a_8
1	1.000							
2	1.414	1.000						
3	2.000	2.000	1.000					
4	2.613	3.414	2.613	1.000				
5	3.236	5.236	5.236	3.236	1.000			
6	3.864	7.464	9.141	7.464	3.864	1.000		
7	4.494	10.103	14.606	14.606	10.103	4.494	1.000	
8	5.126	13.138	21.848	25.691	21.848	13.138	5.126	1.000

(7)

n	Denominator Polynomial
1	$(1 + s)$
2	$(1 + 1.414s + s^2)$
3	$(1 + s)(1 + s + s^2)$
4	$(1 + 0.7653s + s^2)(1 + 1.8477s + s^2)$
5	$(1 + s)(1 + 0.6180s + s^2)(1 + 1.6180s + s^2)$
6	$(1 + 0.5176s + s^2)(1 + 1.4142s + s^2)(1 + 1.9318s + s^2)$
7	$(1 + s)(1 + 0.4449s + s^2)(1 + 1.2465s + s^2)(1 + 1.8022s + s^2)$
8	$(1 + 0.3896s + s^2)(1 + 1.1110s + s^2)(1 + 1.6630s + s^2)(1 + 1.9622s + s^2)$

(8)

Additional polynomials for higher orders may be computed in a straightforward manner. We observe that the coefficient values have a symmetrical distribution like that of binomial coefficients, which is always true when the zeros of a polynomial lie upon the unit circle (not necessarily uniformly spaced) or are located at points that are images with respect to the unit circle.

From the standpoint of obtaining the best approximation to the ideal rectangle of Fig. 1 for a polynomial of given degree, the Butterworth function does not do as well as may be done because it concentrates all of the approximating ability of the polynomial near $\omega = 0$ instead of distributing it uniformly over the range $0 < \omega < 1$. A better result in this regard may be obtained if we look for a rational function that approximates the constant value unity throughout this range in an oscillatory rather than a monotonic manner.

---

* Taken from the paper by R. F. Baum entitled "A Contribution to the Approximation Problem," *Proc. I.R.E.*, Vol. 36, No. 7, July 1948.

In searching for this kind of a result, let us first look more closely at the Butterworth function in Eq. 1 and observe that the term ω^{2n} in the denominator, if plotted separately, may be regarded as approximating zero in the interval $0 < \omega < 1$, and infinity beyond this range. Indeed, if we write in place of Eq. 1

$$|Z_{12}(j\omega)|^2 = \frac{1}{1 + F^2(\omega)} \tag{9}$$

and ask that the function $F^2(\omega)$ be substantially zero for $0 < \omega < 1$, and substantially infinite (large compared with unity) for $\omega > 1$, then the function 9 will have (very nearly) the form of the rectangle in Fig. 1. The term ω^{2n} comes closer and closer to fulfilling this requirement with increasing values of the integer n, but it does so in a monotonic manner and with major emphasis upon the points $\omega = 0$ and $\omega = \infty$. What we need is a polynomial in ω that approximates zero in a uniform oscillatory manner for $0 < \omega < 1$ (or rather for $-1 < \omega < 1$, since the characteristic we are approximating is an even function), and assumes large values rapidly outside of this interval. The so-called Tschebyscheff polynomial has precisely this desired behavior.

This polynomial (as a function of any real variable x) is defined by

$$T_n(x) = \frac{1}{2^{n-1}} \times \cos(n \cos^{-1} x) \tag{10}$$

which looks more like a transcendental function than a polynomial, but it is readily recognized to be a finite polynomial in x if we let

$$\cos^{-1} x = \phi \qquad \text{or} \qquad x = \cos\phi \tag{11}$$

and recall that $\cos n\phi$ [which is $T_n(x)$ except for the factor $1/2^{n-1}$] is expressible as a sum of terms involving powers of $\cos\phi$ up to $\cos^n\phi$ (odd powers only if n is odd, and even powers only, if n is even). For example

$$\begin{aligned} \cos 2\phi &= -1 + 2\cos^2\phi \\ \cos 3\phi &= -3\cos\phi + 4\cos^3\phi \\ \cos 4\phi &= 1 - 8\cos^2\phi + 8\cos^4\phi \\ &\cdots\cdots\cdots\cdots \end{aligned} \tag{12}$$

Substituting x for $\cos\phi$, these clearly are polynomials in x.

It is simultaneously clear that the polynomial oscillates with a uniform amplitude for the range $-1 < x < 1$ where ϕ, according to 11, is real, since it is effectively $\cos n\phi$ with an amplitude $1/2^{n-1}$. The oscillations, although uniform in amplitude, are not uniform in the length of successive cycles when plotted on a linear x-scale since the associated ϕ-scale is nonlinear. For our purpose, this circumstance is, of course, of no consequence.

Since we prefer, in the following analysis, to deal with a polynomial having a unit amplitude of oscillation, we shall use, in place of the standard form of Tschebyscheff polynomial, the following modified form

$$V_n(x) = \cos(n \cos^{-1} x) \tag{13}$$

differing only by the factor $1/2^{n-1}$.

For the purpose of generating polynomials of any order as well as for the computation of their values for given values of the argument x, an effective representation for this polynomial is found by noting from Eq. 11 that

$$\sqrt{1-x^2} = \sin\phi \qquad \text{and} \qquad x + j\sqrt{1-x^2} = e^{j\phi} \tag{14}$$

whence

$$e^{\pm jn\phi} = (x \pm j\sqrt{1-x^2})^n \tag{15}$$

The desired polynomial 13 is therefore given by the expression

$$V_n(x) = \cos n\phi = \frac{(x + j\sqrt{1-x^2})^n + (x + j\sqrt{1-x^2})^{-n}}{2} \tag{16}$$

which we can also write

$$V_n(x) = \cos n\phi = \frac{(x + j\sqrt{1-x^2})^n + (x - j\sqrt{1-x^2})^n}{2} \tag{17}$$

as is clear from 15.

Instead of 11 and 14, we also can write

$$x = \cosh\psi, \qquad \sqrt{x^2-1} = \sinh\psi \tag{18}$$

and have instead of 15

$$e^{\pm\psi} = x \pm \sqrt{x^2-1} \tag{19}$$

whereupon the Tschebyscheff polynomial is expressed by

$$V_n(x) = \cosh n\psi = \frac{(x + \sqrt{x^2-1})^n + (x \pm \sqrt{x^2-1})^{\mp n}}{2} \tag{20}$$

which is obviously equivalent to 16 or 17, but better suited for the calculation of $V_n(x)$ for $x > 1$.

Expressions 17 and 20 are convenient also for the construction of polynomials of any order, since it is clear that $V_n(x)$ is given either by the real part of $(x + j\sqrt{1-x^2})^n$ or by the rational part of $(x + \sqrt{x^2-1})^n$, which equals the sum of the first, third, fifth, · · ·, terms in the binomial expansion of either of these two expressions. We readily find in this manner

$$\begin{aligned}
V_1(x) &= x \\
V_2(x) &= 2x^2 - 1 \\
V_3(x) &= 4x^3 - 3x \\
V_4(x) &= 8x^4 - 8x^2 + 1 \\
V_5(x) &= 16x^5 - 20x^3 + 5x \\
V_6(x) &= 32x^6 - 48x^4 + 18x^2 - 1 \\
&\cdots\cdots\cdots
\end{aligned} \tag{21}$$

A recursion formula exists which may be used to compute additional polynomials from those given above. This reads

$$V_{n+1} - 2xV_n + V_{n-1} = 0 \tag{22}$$

Polynomials of odd order are seen to be odd, and those of even order are even.

Except for an additive constant and an appropriate multiplier, the square of a Tschebyscheff polynomial is evidently equal to one of double order. Specifically, it is not difficult to see that

$$V_n^2(x) = (V_{2n} + 1)/2 \tag{23}$$

For example, the square of $V_3(x)$ in 21 yields $16x^6 - 24x^4 + 9x^2$, which equals $(V_6 + 1)/2$.

In order to compute and plot the values of a given polynomial, we consider separately the range $-1 < x < 1$ and that corresponding to $|x| > 1$. In the range $-1 < x < 1$ we need only to find those x values at which the polynomial equals ± 1 or 0. The ϕ values at which $\cos n\phi$ equals ± 1 or 0 are $\phi = \nu\pi/2n$ for ν equal to integer values. Use of Eq. 11 then yields the corresponding x values as

$$x = \cos(\nu\pi/2n) \qquad \text{for } \nu = 0, 1, 2, \cdots, 2n \tag{24}$$

In the range $|x| > 1$, Eq. 20 affords the best means for calculating values of $V_n(x)$. Here it will usually suffice to calculate one value for x near unity and additional values for large values of x. The x value near unity is conveniently chosen as $x = \frac{5}{4}$, since 20 yields

$$V_n(\tfrac{5}{4}) = (2^n + 2^{-n})/2 \approx 2^{n-1} \tag{25}$$

We also have from 20

$$V_n(x) \approx 2^{n-1}x^n \qquad \text{for } x \gg 1 \tag{26}$$

which is simply the leading term in the polynomial.

Figures 4 and 5 show plots of the polynomials $V_3(x)$ and $V_4(x)$ computed in this manner. The higher order polynomials have this same general character, with more oscillations within the interval $-1 < x < 1$, and a more rapid rate of increase in their absolute values beyond this range. The frequency of oscillation is relatively low near $x = 0$, and increases as x approaches the points ± 1, as is clear also from the x values 24 at which the function $V_n(x)$ passes through the values ± 1 and 0.

Returning now to expression 9, we observe that the function $F^2(\omega)$ evidently remains positive. The simplest way of meeting this condition is to square the polynomial $V_n(x)$, which is simpler than (but entirely equivalent to) adding appropriate constants to polynomials of even order. The variable x we may identify with ω; and if we multiply $V_n(x)$ by a quantity $\epsilon < 1$, the tolerance with which the resultant $F^2(\omega)$ approximates zero for $-1 < \omega < 1$ is easily controllable. Hence we write

$$F^2(\omega) = \epsilon^2 V_n^2(\omega) \tag{27}$$

and, for the squared magnitude of our transfer impedance, Eq. 9, have

$$|Z_{12}(j\omega)|^2 = \frac{1}{1 + \epsilon^2 V_n^2(\omega)} \tag{28}$$

A plot of the magnitude of $Z_{12}(j\omega)$ for $\epsilon^2 = \frac{1}{5}$ and $n = 4$ obtainable in

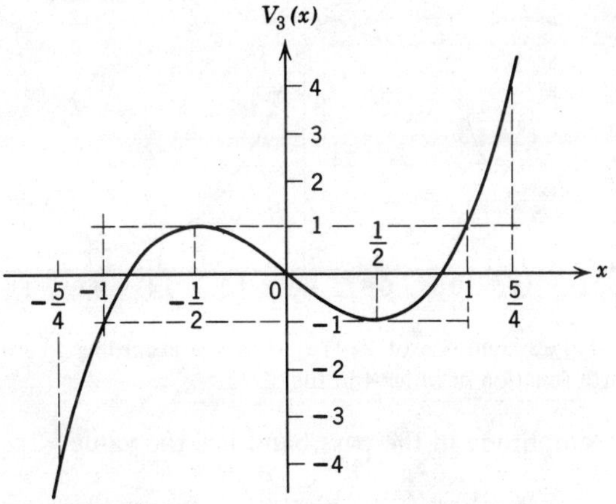

Fig. 4. Tschebyscheff polynomial of order 3 according to Eqs. 11, 13, and 20, together with the approximate relations 25 and 26.

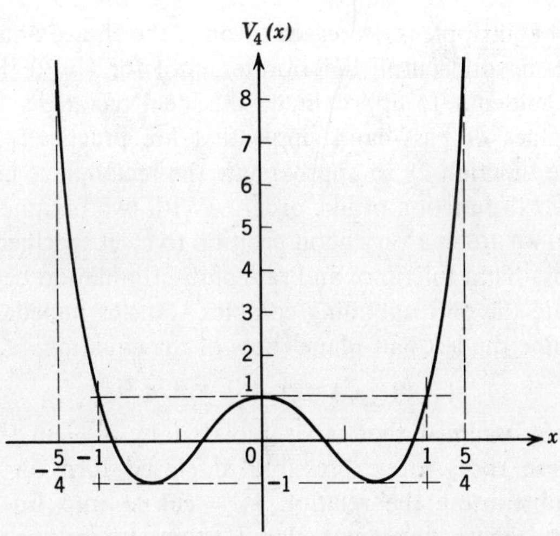

Fig. 5. Tschebyscheff polynomial of order 4 determined as stated in Fig. 4.

this way is shown in Fig. 6. We observe that it drops off considerably faster beyond the cutoff than the fourth-order Butterworth function in Fig. 2 which leads to a transfer impedance of the same complexity. In fact, it is almost as good in this respect as the Butterworth function of order eight.

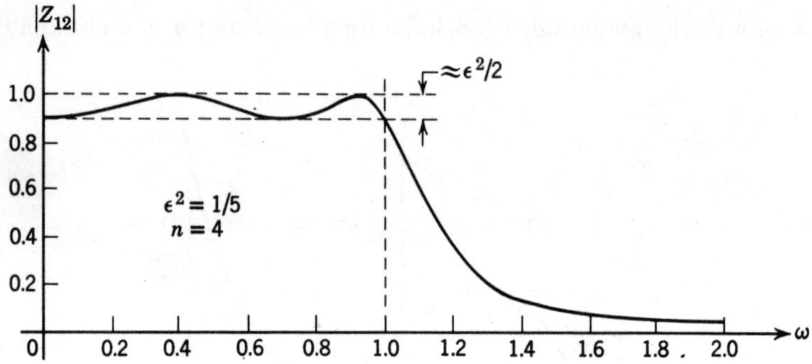

Fig. 6. Plot of the magnitude of $Z_{12}(j\omega)$ versus ω according to Eq. 28. Compare with Butterworth function of order 4 in Fig. 2.

The ripple amplitude in the pass band has the value

$$1 - \frac{1}{\sqrt{1 + \epsilon^2}} \approx \epsilon^2/2 \qquad \text{for small } \epsilon \tag{29}$$

while beyond cutoff where $\epsilon^2 V_n^2(\omega) \gg 1$ we have

$$|Z_{12}(j\omega)| \approx 1/\epsilon V_n(\omega) \tag{30}$$

As the pass-band ripple is decreased through the choice of a small ϵ value, the behavior beyond cutoff deteriorates until for $\epsilon = 0$ the function 28 exhibits no tendency to approximate the ideal rectangle. However, for moderate values of pass-band ripple that are practically tolerable, the ability of the function 28 to approximate the rectangle is better than that of a Butterworth function of like order. With two parameters ϵ and n at our disposal, we are in a very good position to meet specified requirements regarding pass-band tolerance and rate of discrimination beyond cutoff.

To generate the corresponding complex transfer impedance $Z_{12}(s)$, we must determine the left half-plane roots of the equation

$$B(-s^2) = 1 + \epsilon^2 V_n^2 = 0 \tag{31}$$

in which it is assumed that ω^2 is replaced by $-s^2$ in the polynomial $V_n^2(\omega)$. These roots are expressible in closed form in the following manner. Substituting the relation $V_n = \cos n\phi$ into Eq. 31, we must accommodate the pertinent complex ϕ values by writing $\phi = \phi_1 + j\phi_2$. Equation 31 thus yields

$$\epsilon V_n = \pm j = \epsilon \cos n(\phi_1 + j\phi_2) = \epsilon \cos n\phi_1 \cosh n\phi_2 - j\epsilon \sin n\phi_1 \sinh n\phi_2 \tag{32}$$

Equating real and imaginary parts on both sides of this equation gives

$$\cos n\phi_1 \cosh n\phi_2 = 0 \tag{33}$$

$$\sin n\phi_1 \sinh n\phi_2 = \pm 1/\epsilon \tag{34}$$

from which ϕ_1 and ϕ_2 can be determined.

Since $\cosh \phi_2$ is nonzero for all real ϕ_2 values, Eq. 33 can be satisfied only by choosing ϕ_1 values that result in $\cos n\phi_1 = 0$. The distinct values of ϕ_1 thus determined are given by

$$\phi_1 = \nu\pi/2n \qquad \text{for } \nu = 1, 3, 5, \cdots, 4n - 1 \tag{35}$$

and Eq. 34 then yields

$$\sinh n\phi_2 = \pm 1/\epsilon \qquad \text{and} \qquad \cosh n\phi_2 = \sqrt{\frac{1}{\epsilon^2} + 1} \tag{36}$$

Hence

$$e^{n\phi_2} = \sqrt{\frac{1}{\epsilon^2} + 1} \pm \frac{1}{\epsilon} \tag{37}$$

Since the $\pm$ signs merely yield reciprocal values for the exponential, we can write

$$e^{\phi_2} = \left(\sqrt{\frac{1}{\epsilon^2} + 1} + \frac{1}{\epsilon}\right)^{1/n} \tag{38}$$

and have

$$\left.\begin{matrix} \cosh \phi_2 \\ \sinh \phi_2 \end{matrix}\right\} = \frac{\left(\sqrt{\frac{1}{\epsilon^2} + 1} + \frac{1}{\epsilon}\right)^{1/n} \pm \left(\sqrt{\frac{1}{\epsilon^2} + 1} + \frac{1}{\epsilon}\right)^{-1/n}}{2} \tag{39}$$

where the $\pm$ signs respectively yield the cosh and sinh functions.

We now have for the corresponding complex s values

$$s_\nu = j \cos(\phi_1 + j\phi_2) = \sin \phi_1 \sinh \phi_2 + j \cos \phi_1 \cosh \phi_2 \tag{40}$$

in which the pertinent values of ϕ_1 are given by Eq. 35 and values for the hyperbolic sine and cosine functions are obtained from Eq. 39. We are interested only in the left half-plane roots of Eq. 31. Hence if we write

$$s_\nu = \sigma_\nu + j\omega_\nu \tag{41}$$

then

$$\left.\begin{aligned} \sigma_\nu &= -\sin \frac{\nu\pi}{2n} \sinh \phi_2 \\ \omega_\nu &= \cos \frac{\nu\pi}{2n} \cosh \phi_2 \end{aligned}\right\} \text{for } \nu = 1, 3, 5, \cdots, 2n - 1 \tag{42}$$

These roots lie upon an elliptic locus in the s-plane, as may be seen from Eqs. 42 by noting that

$$\frac{\sigma_\nu^2}{\sinh^2 \phi_2} + \frac{\omega_\nu^2}{\cosh^2 \phi_2} = 1 \tag{43}$$

Evidently, $\sinh \phi_2$ and $\cosh \phi_2$ are the semiaxes of the ellipse. In Fig. 7 this elliptic contour is drawn for $n = 6$ and a ratio of 2 for the semiaxes. This contour as well as the points on it corresponding to the roots of Eq. 31 are easily determined by first drawing circles with radii equal to the semiaxes and radial lines according to the angles 35, whereupon a familiar graphical construction indicated by the horizontal and vertical dashed lines in Fig. 7 completes the process.

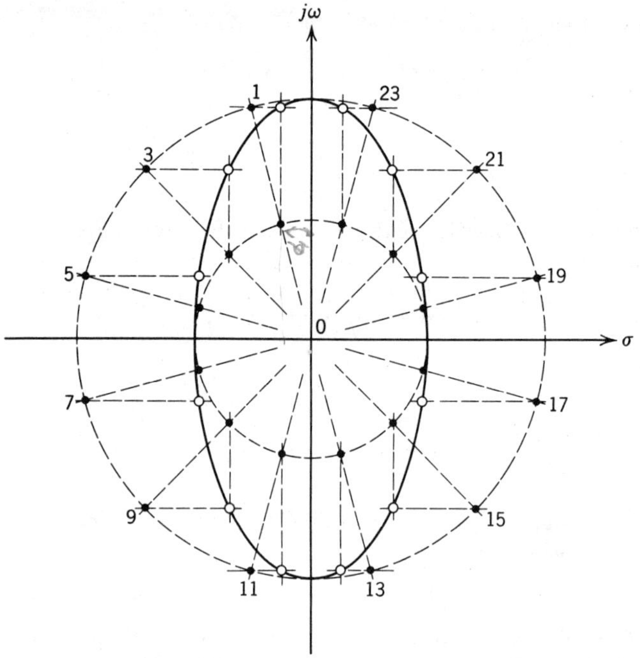

Fig. 7. Graphical construction according to Eqs. 42 and 43 yielding zeros of the polynomial 31 on an elliptic locus in the s-plane.

Because of the dependence of this locus upon the two parameters ϵ and n, it is difficult to compare it with the circular Butterworth locus in Fig. 3 with regard to either its size or shape. For values of ϵ equal to about $\frac{1}{3}$ or less we can approximate Eq. 38 by

$$e^{\phi_2} \approx (2/\epsilon)^{1/n} \tag{44}$$

The average radius of the two circles in Fig. 7 is then given by

$$r_0 \approx \tfrac{1}{2}(2/\epsilon)^{1/n} \tag{45}$$

and the ratio of the minor to the major axis of the ellipse is

$$\rho \approx \frac{(2/\epsilon)^{2/n} - 1}{(2/\epsilon)^{2/n} + 1} \tag{46}$$

For small enough values of ϵ and moderate values of n, we see that the average radius may easily be larger than unity and the ratio ρ nearly unity. The locus then is almost circular and larger than the unit circle. For moderately small values of ϵ and large values of n, on the other hand, the average radius approaches the value $\frac{1}{2}$, and the ratio ρ becomes small. The poles of our transfer function then are all inside the unit circle and lie closely along the j-axis.

By a simple modification of the function 28 we can obtain a resulting approximation of the rectangle that has equiripple character beyond cutoff rather than in the pass band. This modification reads

$$|Z_{12}(j\omega)|^2 = \frac{\epsilon^2 V_n^2(1/\omega)}{1 + \epsilon^2 V_n^2(1/\omega)} \tag{47}$$

Without ω replaced by $1/\omega$, as is done in the argument of the $V_n(\omega)$ function appearing here, this modified expression would be the complement of 28 since the sum of the two would then clearly equal unity. This complementary function, which has high-pass characteristics, exhibits an equiripple approximation to zero in its stop band $-1 < \omega < 1$ and a monotonic approximation to unity in its pass band $|\omega| > 1$. Replacing ω by $1/\omega$ evidently has the effect of interchanging the stop and pass regions, and yields the function 47 that approximates unity in a monotonic manner and zero in an equiripple manner within the ranges $-1 < \omega < 1$ and $|\omega| > 1$ respectively.

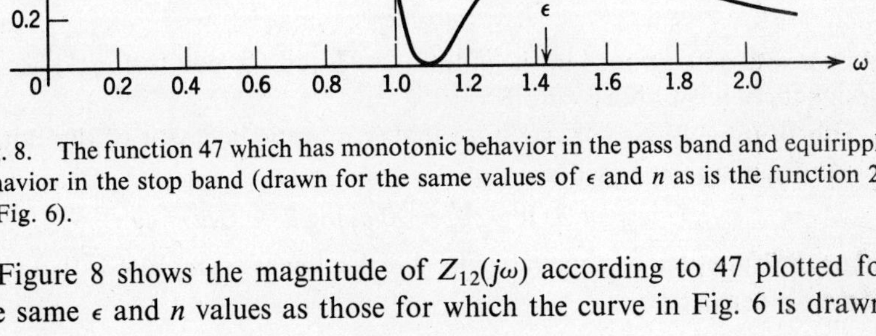

Fig. 8. The function 47 which has monotonic behavior in the pass band and equiripple behavior in the stop band (drawn for the same values of ϵ and n as is the function 28 in Fig. 6).

Figure 8 shows the magnitude of $Z_{12}(j\omega)$ according to 47 plotted for the same ϵ and n values as those for which the curve in Fig. 6 is drawn.

It is important to observe that the ripple amplitude now is ϵ instead of $\epsilon^2/2$. To get the same ripple in the stop band with the function 47 that the function 28 yields in the pass band requires a considerably smaller value of ϵ.

When the Butterworth or Tschebyscheff functions (1 or 28) are used in the design of lossless coupling networks with resistance loading at both ends (Art. 5, Ch. 11), it is worth noting that the resulting networks will always be either symmetrical or antimetrical according to the discussion in Art. 6, Ch. 11. It is pointed out there that these special forms result if the assumed squared-magnitude function has the form of Eq. 9 in which $F^2(\omega)$ is the square of an odd or even rational function respectively. We thus see that the Butterworth and Tschebyscheff functions of odd order will yield symmetrical networks and those of even order will yield antimetrical networks. The pertinent lossless networks, whether appropriate to double or single loading, are seen to be unbalanced ladders with series inductances and shunt capacitances, the total number of elements being equal to the order n of the approximating function. Butterworth and Tschebyscheff functions both lead to the same form of ladder structure; the difference between the two is in the element-value distribution only.

.3 Asymptotic Behavior of the Butterworth and Tschebyscheff Functions

It is useful to study and compare the asymptotic behaviors of the Butterworth (Eq. 1) and Tschebyscheff (Eq. 28) functions, using logarithmic scales for both the dependent and the independent variables. For $\omega \to \infty$, $|Z_{12}(j\omega)|$ according to Eq. 1 behaves like $1/\omega^n$. If we express this behavior as a loss in decibels, it is given by

$$\alpha_{db} = 20 \log \omega^n \tag{48}$$

where the notation "log" designates logarithms to the base 10 in contrast to "ln," which is the common notation for the natural logarithm function. For the independent variable it is customary to use a logarithmic scale to the base 2, and write

$$\omega = 2^u \tag{49}$$

Here $u = 0$ corresponds to the cutoff $\omega = 1$ and the points $u = 1, 2, \cdots$ designate *octaves* above cutoff.

Substituting 49 into 48, we have the asymptotic behavior of the Butterworth function expressed as

$$\alpha_{db} = 20 \log 2^{nu} = 20nu \log 2 \approx 6nu \tag{50}$$

Thus this function, as shown in Fig. 9, has a linear asymptote with a slope of $6n$ db per octave.

By the same reasoning it follows that the attentuation below cutoff has an asymptote that is constant and equal to zero. An approximate representation for the total loss function is thus given by a broken line consisting of the horizontal axis for negative values of u (octaves below cutoff) and the straight line with a 6-db-per-octave slope for positive u, the break point occurring at the cutoff $u = 0$.

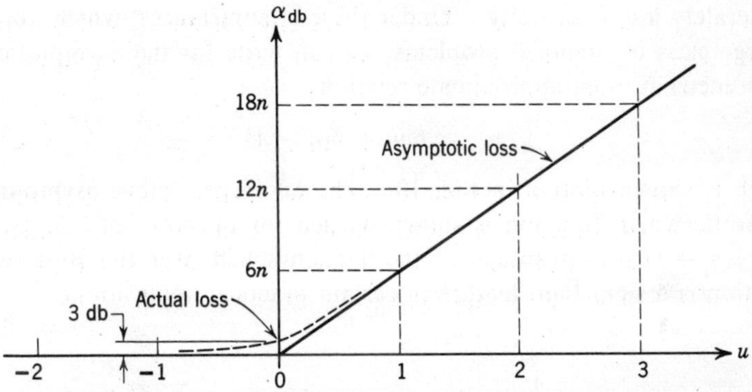

Fig. 9. Asymptotic loss of the Butterworth function in terms of a logarithmic frequency scale in which the origin corresponds to cutoff and units are octaves above or below cutoff.

This broken-line representation differs from the actual loss curve plotted to these logarithmic scales, essentially only in the vicinity of the break point where the actual loss is 3 db (independent of the order n) instead of zero. The actual loss function in this vicinity has the appearance of the dashed curve in Fig. 9 which smoothes the broken-line representation by introducing a 3-db "fillet" at the cutoff point.

A similar representation is possible for the Tschebyscheff function 28 through use of Eqs. 26 and 30 giving the behavior for large values of ω. We thus obtain for the asymptotic loss of this function in decibels

$$\alpha_{db} = 20 \log (2^{n-1}\epsilon\omega^n) \tag{51}$$

Again introducing the change of independent variable indicated in Eq. 49, we find

$$\alpha_{db} = 20 \log 2^{nu} + 20 \log 2^{n-1} + 20 \log \epsilon \tag{52}$$

or approximately

$$\alpha_{db} \approx 6nu + 6(n-1) + 20 \log \epsilon \tag{53}$$

This result differs from that given by Eq. 50 for the Butterworth function only in the addition of its second and third terms which are constants. Since the value of ϵ is unity or less, the third term in Eq. 53 is zero or negative and will tend to cancel the second term which is always positive. Since the maximum loss in the pass band, for the Tschebyscheff function, is given by 20 log $\sqrt{1 + \epsilon^2}$ db, we see that the value $\epsilon = 1$ yields a 3-db ripple. Although fairly large, this amount of pass-band ripple is tolerable in many practical applications, and it is tolerated, as we now see from Eq. 53, because of the increase in asymptotic loss which results. For a Tschebyscheff function of sufficiently high order n, the second term in Eq. 53 will swamp the effect of the third term even for values of ϵ moderately less than unity. Under these circumstances, which apply to a large class of practical problems, we can write for the asymptotic loss the somewhat more approximate relation

$$\alpha_{db} \approx 6nu + 6(n-1) \tag{54}$$

which is shown plotted in Fig. 10. The $6n$-db-per-octave asymptote of the Butterworth function is shown dashed for purposes of comparison. The $6(n-1)$ db advantage of the Tschebyscheff over the Butterworth function is a significant feature in certain practical applications.

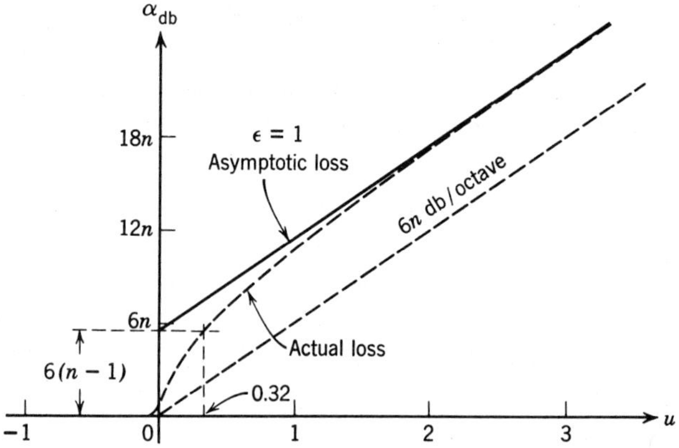

Fig. 10. Asymptotic loss of the Tschebyscheff function comparable to that shown in Fig. 9 for the Butterworth.

4 Frequency Transformations

By means of appropriate transformations of the independent variable, the rational approximants obtained for the ideal low-pass characteristic in Fig. 1 may be rendered applicable to analogous high-pass, band-pass, band-stop, and other more general classes of filter characteristics. It has, for example, already been seen that if, in a normalized low-pass approximant, we replace ω by $1/\omega$, the frequency scale is inverted about the cut-off point $\omega = 1$, and the function becomes an appropriate approximant to an ideal high-pass filter characteristic. In the same simple manner, it is possible to transform the low-pass approximants to any other desired filter class.

Since the process operates only upon the independent variable, it has no effect upon the tolerance with which the pertinent function approximates a given ideal characteristic. Thus a certain ripple amplitude in the low-pass approximant remains unchanged when that function is transformed to its band-pass counterpart by means of a suitable change in its independent variable.*

This change of variable can, moreover, be so chosen that its effect upon the given low-pass approximant may be interpreted directly in terms of the modifications it produces in the structure and element values of the low-pass network realization of that approximant (the so-called *low-pass prototype* network). In this way one can avoid altogether the need for forming the pertinent band pass or other more complicated approximant and applying to it a realization procedure which correspondingly becomes computationally more tedious to carry out.

Suppose for the purposes of this discussion we denote the complex frequency variable for the low-pass function by the symbol $\lambda = \bar{\sigma} + j\bar{\omega}$, and that for the transformed function by $s = \sigma + j\omega$. What we are looking for is a relation between λ and s that converts a given low-pass $\bar{Z}(\lambda)$ into an approximant $Z(s)$ appropriate to whatever filter class we may be interested in. The pertinent relation, as it turns out, is simply

$$\lambda = z(s) \tag{55}$$

in which $z(s)$ is a reactance function. It is not difficult to appreciate why this simple and (to us, at least) well-known function fulfills all of the requirements mentioned above.

Two points in the λ-plane, namely, $\lambda = 0$ and $\lambda = \infty$, are of particular interest. The point $\lambda = 0$ is where the pass band of the low-pass characteristic is centered, and the vicinity of $\lambda = \infty$ is where the low-pass approximating function is described by its asymptotic behavior. The function $\lambda = z(s)$ essentially is required to make these two λ values fall at certain specified points on the j-axis of the s-plane. A reactance function does just this. Thus the lowpass-highpass transformation $\lambda = 1/s$

---

* Although the attitude taken here is justified in most practical situations, there are some cases where distortion of the approximant in the abscissa direction caused by transformation of the independent variable is objectionable, and one must then solve the approximation problem for the particular filter class desired. A suitable procedure for carrying out this objective is discussed in "Equal-ripple Bandpass Amplifiers," D. L. Trautman and J. A. Aseltine, *Report* 51.9, University of California at Los Angeles Aug. 1951. Also relevant is Trautman's paper, "The Application of Conformal Mapping to the Synthesis of Band-pass Networks," *Proc. Symposium on Modern Network Synthesis*, Apr. 1952, pp. 179–192.

interchanges the origin and infinity. It makes $\lambda = 0$ fall at $s = \infty$, and $\lambda = \infty$ at $s = 0$. The pass band of the approximating function becomes centered at $s = \infty$ and its asymptotic behavior at $s = 0$; and a low-pass characteristic is thereby changed to high-pass.

If we want the low-pass characteristic changed to band-pass, then we want $z(s)$ to have a zero at a finite nonzero s value and poles (points where $\lambda = \infty$) at $s = 0$ and $s = \infty$. The reactance function appropriate to a series LC combination has precisely this character. Specifically the function

$$\lambda = \frac{s^2 + \omega_0^2}{s} \tag{56}$$

as we know from our familiarity with reactance functions, transforms the point $\lambda = 0$ into the two j-axis points $s = \pm j\omega_0$, and the point $\lambda = \infty$ into the points $s = 0$ and $s = \infty$. Hence the function 56 becomes an appropriate lowpass-bandpass transformation.

A lowpass-bandstop transformation is evidently given by the reciprocal of 56, namely

$$\lambda = \frac{s}{s^2 + \omega_0^2} \tag{57}$$

which transforms $\lambda = 0$ into $s = 0$ and $s = \infty$, and the point $\lambda = \infty$ into $s = \pm j\omega_0$. A lowpass to low-and-bandpass transformation is effected by the reactance function

$$\lambda = \frac{s(s^2 + \omega_0^2)}{(s^2 + \omega_\infty^2)} \tag{58}$$

for which the point $\lambda = 0$ is carried over into the three points $s = 0$, $s = \pm j\omega_0$; and the point $\lambda = \infty$ is likewise triplicated since it corresponds to $s = \infty$ and $s = \pm j\omega_\infty$. Transformations which are capable of carrying a low-pass approximant over into any desired filter class can thus be written down without difficulty.

All of these transformations likewise have the property that the network modifications which they imply are obvious by inspection if we are familiar with reactance function synthesis. Thus the low-pass-bandpass transformation 56 changes an inductance having the reactance $L\lambda$ into a series LC combination determined by a reactance function equal to 56, with its impedance level multiplied by L; and it changes a capacitance having the susceptance $C\lambda$ into a parallel LC combination determined by a susceptance function equal to 56, with its admittance level multiplied by C. That is to say, each inductance in the low-pass prototype network having a value of L henrys has placed in series with it a capacitance of $1/L\omega_0^2$ farads; and a capacitance in the low-pass prototype network having a value of C farads has placed in parallel with it an inductance of $1/C\omega_0^2$ henrys.

The lowpass-bandstop transformation 57 may similarly be interpreted directly in terms of modifications in the low-pass prototype network. Since this function is the reciprocal of 56, the interpretations are essentially the same as for the lowpass-bandpass situation with the roles of L and C interchanged. Specifically, a low-pass L becomes a capacitance of $1/L$ farads in parallel with an inductance of L/ω_0^2 henrys; and a low-pass C becomes an inductance of $1/C$ henrys in series with a capacitance of C/ω_0^2 farads.

More elaborate transformations like 58 and others nevertheless are readily interpreted networkwise since the pertinent functions need merely be regarded as reactances or susceptances that have their impedance or admittance levels respectively multiplied by the low-pass prototype L and C values.

It remains to see what some of the more detailed properties of these transformations are, and how a particular situation calling for specific cutoff frequencies is accommodated. If we rewrite the lowpass-bandpass transformation 56 in terms of j-axis values, it reads

$$\bar{\omega} = \frac{\omega^2 - \omega_0^2}{\omega} \tag{59}$$

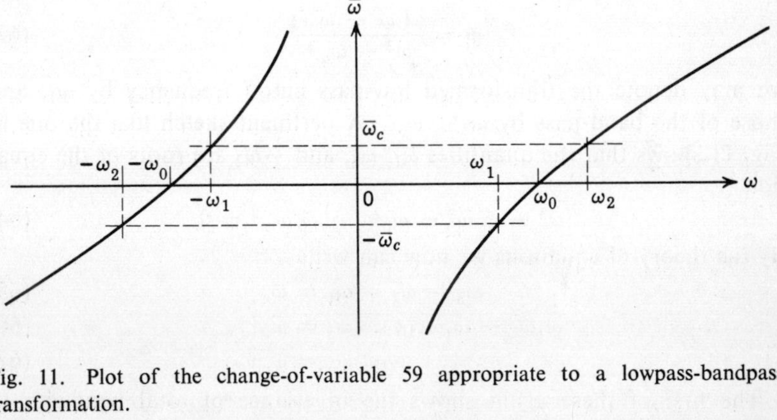

Fig. 11. Plot of the change-of-variable 59 appropriate to a lowpass-bandpass transformation.

a plot of which has the familiar form shown in Fig. 11. The cutoff frequency of the low-pass function is denoted by $\bar{\omega}_c$, and the cutoff frequencies of the corresponding bandpass by ω_1 and ω_2. These are given in terms of $\bar{\omega}_c$ and ω_0 by the roots of the following equation

$$\omega^2 - \bar{\omega}_c\omega - \omega_0^2 = 0 \tag{60}$$

which is obtained by setting $\bar{\omega} = \bar{\omega}_c$ in Eq. 59. Reference to Fig. 11

shows that the roots of Eq. 60 are $-\omega_1$ and ω_2. By the theory of equations, the sum of these roots equals the coefficient of the linear term with its sign changed, and their product equals the constant term. Hence we have

$$\omega_2 - \omega_1 = w = \bar{\omega}_c \tag{61}$$

and

$$\omega_1\omega_2 = \omega_0^2 \tag{62}$$

where we have denoted the bandwidth of the band pass by w.

Equation 61 shows that the transformation 56 or 59 has the property of keeping the bandwidth of the filter invariant; and Eq. 62 shows that the arithmetic symmetry of the low-pass filter characteristic (which is an even function of $\bar{\omega}$) is transformed into a band-pass characteristic having geometric symmetry about the center frequency ω_0, since any two frequencies whose geometric mean value is ω_0 correspond to a pair of $\bar{\omega}$ values like $\pm\bar{\omega}_c$.

In applying the transformation to a specific example, the procedure is first to frequency-scale the low-pass prototype network to have a cutoff $\bar{\omega}_c$ equal to the bandwidth of the desired band pass. Then, picking the prescribed values of ω_1 and ω_2 consistent with this bandwidth 61, compute ω_0^2 from Eq. 62. This fixes the transformation.

In the more elaborate lowpass to low-and-bandpass transformation 58, which has the j-axis form

$$\bar{\omega} = \frac{\omega(\omega^2 - \omega_0^2)}{\omega^2 - \omega_\infty^2} \tag{63}$$

we may denote the transformed low-pass cutoff frequency by ω_1, and those of the band pass by $\omega_2 < \omega_3$. A pertinent sketch like the one in Fig. 11 shows that the quantities ω_1, ω_3, and $-\omega_2$ are roots of the equation

$$\omega^3 - \bar{\omega}_c\omega^2 - \omega_0^2\omega + \bar{\omega}_c\omega_\infty^2 = 0 \tag{64}$$

By the theory of equations we now can write

$$\omega_3 - \omega_2 + \omega_1 = \bar{\omega}_c \tag{65}$$

$$\omega_1\omega_2 - \omega_1\omega_3 + \omega_2\omega_3 = \omega_0^2 \tag{66}$$

$$\omega_1\omega_2\omega_3 = \bar{\omega}_c\omega_\infty^2 \tag{67}$$

The first of these again shows the invariance of total bandwith (a property which is generally true for transformations of this type).* The

* This result, which has an interesting relation to the resistance-integral theorem (Art. 3, Ch. 8), was originally presented by R. Feldtkeller in an article entitled "Zur Frequenzabhangigkeit der Reaktanzen," *Elek. Nachr.-Tech.*, Vol. 13, No. 12, Dec. 1936, pp. 401–404. The above presentation follows a method of proof given by Hans Salinger in a "Note on the Frequency Behavior of Reactances," *Proc. I.R.E.*, Vol. 26. No. 1, Jan. 1938, pp. 107–111.

procedure for determining a specific transformation is to frequency-scale the low-pass prototype network to a value of $\bar{\omega}_c$ equal to the *combined* bandwidth of the desired transformed filter characteristic; then, picking values for the cutoff frequencies ω_1, ω_2, ω_3 consistent with this bandwidth 65, we obtain the appropriate values for the parameters $\omega_0{}^2$ and $\omega_\infty{}^2$ from Eqs. 66 and 67. The desired transformation 58 is thus established.

5 Equiripple Characteristics in Both Pass and Stop Bands

Although the Tschebyscheff polynomial affords a means for obtaining a low-pass approximant having equiripple character in either the pass or stop region, its use alone cannot provide this kind of behavior in both approximation intervals. Since equiripple, as contrasted to monotonic, behavior in one region alone results in a significant improvement in the over-all approximation ability of a rational function of given complexity, one feels intuitively that a still further improvement is achievable if such behavior can be extended over the entire frequency range. One way in which this result can be obtained is discussed in the following paragraphs.* (Another is given in Art. 10.)

A suitable form for the function $F(\omega)$ in Eq. 9 is given by either of the following two expressions

$$F(\omega) = \frac{(\omega_1{}^2 - \omega^2)(\omega_3{}^2 - \omega^2) \cdots (\omega_{2n-1}{}^2 - \omega^2)}{(1 - \omega_1{}^2\omega^2)(1 - \omega_3{}^2\omega^2) \cdots (1 - \omega_{2n-1}{}^2\omega^2)} \tag{68}$$

$$F(\omega) = \frac{\omega(\omega_2{}^2 - \omega^2)(\omega_4{}^2 - \omega^2) \cdots (\omega_{2n}{}^2 - \omega^2)}{(1 - \omega_2{}^2\omega^2)(1 - \omega_4{}^2\omega^2) \cdots (1 - \omega_{2n}{}^2\omega^2)} \tag{69}$$

of which the first is even and the second odd. Each function has zeros on the ω-axis in the range $0 < \omega < 1$ and poles in the complementary range $1 < \omega < \infty$, the zeros and poles respectively being located at reciprocal points on this axis. Both functions have the property

$$F(1/\omega) = 1/F(\omega) \tag{70}$$

according to which their values at reciprocal points on the ω-axis are themselves reciprocal. We may regard such a function as oscillating about zero in the range $0 < \omega < 1$, and about infinity in the range $1 < \omega < \infty$. If the zeros at $\omega_1, \omega_3, \cdots$ or at $\omega_2, \omega_4, \cdots$ are so located

* The mathematical basis for this procedure was first given by W. Cauer in his monograph entitled *Siebschaltungen*, V.D.I. Verlag G.M.B.H., Berlin, 1931. Details of the method given here are adapted from E. L. Norton's paper "Constant Resistance Networks with Applications to Filter Groups," *Bell System Tech. J.*, Vol. 15, Apr. 1937, pp. 178–193 and S. Darlington's "Synthesis of Reactance 4-Poles," *J. Math. and Phys.*, Vol. 18, Sept. 1939, pp. 257–353.

that the function exhibits equiripple character for $0 < \omega < 1$, then it automatically has equiripple character in the range $1 < \omega < \infty$ in the sense that its minimum values there (which can be regarded as maximum deviations from infinity) are all equal. Substituted into expression 9, such a function* clearly yields a $Z_{12}(s)$ whose magnitude exhibits the desired equiripple approximation property in both pass and stop bands.

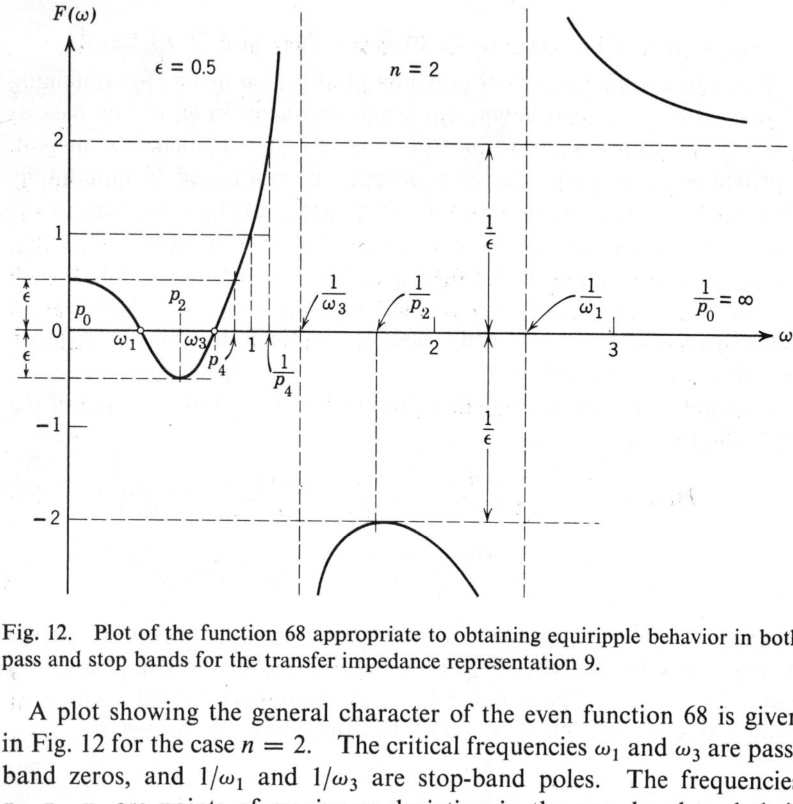

Fig. 12. Plot of the function 68 appropriate to obtaining equiripple behavior in both pass and stop bands for the transfer impedance representation 9.

A plot showing the general character of the even function 68 is given in Fig. 12 for the case $n = 2$. The critical frequencies ω_1 and ω_3 are pass-band zeros, and $1/\omega_1$ and $1/\omega_3$ are stop-band poles. The frequencies p_0, p_2, p_4 are points of maximum deviation in the pass band, and their reciprocals are minima of the function in the stop band; or we may say that the frequencies $1/p_0$, $1/p_2$, $1/p_4$ are points of maximum deviation of the function from infinity. The maximum deviation from zero in the pass band is denoted by ϵ, and is the tolerance with which the function approximates zero in this range. The minima of the function in the stop band have the common value $1/\epsilon$. The lowest frequency of maximum deviation in the pass band is $p_0 = 0$, and the highest is $p_{2n} = p_4$ for the function in Fig. 12.

* First proposed by E. L. Norton, *loc. cit.*

The odd function 69 differs from the even one only in that $\omega = 0$ is a zero instead of a point of maximum deviation. The remaining properties are identical with those pictured in Fig. 12.

Although the point $\omega = 1$ is the nominal boundary between the pass and stop bands, we observe that the range $p_{2n} < \omega < 1/p_{2n}$ is a transition interval that belongs to neither the pass nor the stop bands. Below this interval the approximating function stays within its uniform tolerance ϵ, and above it the reciprocal of the function maintains this tolerance. These are the actual ranges of stop and pass. The transition interval in general becomes narrower as we choose functions of higher order n, but it cannot become zero for a finite network.

It remains to establish a method whereby the pass-band zeros ω_ν may be found such that the function $F(\omega)$ has equiripple character for $0 < \omega < 1$. For functions of low order, one may perhaps accomplish the desired end by a trial-and-error process. An exact analytic method may be found by making a transformation of the independent variable ω that renders the function periodic, just like the substitution $\omega = \cos \phi$ transforms the polynomial $V_n(\omega)$ into the periodic function $\cos n\phi$. It is clear that some periodic function like the sinusoid with an amplitude equal to p_{2n} (the effective cutoff frequency) will accomplish the desired result.

This function, as we will presently demonstrate, turns out to be the Jacobian elliptic function *snu* where u designates the new independent variable. In form, this elliptic sine function is not unlike the trigonometric sine, the essential difference being that it is more flat-topped than the latter. The degree to which it is thus elongated and flat, as contrasted with an ordinary sinusoid, depends upon a parameter k, called the *modulus* of the elliptic function. This modulus can assume values between 0 and 1. For $k = 0$, the elliptic sine is identical with the trigonometric sine. For small values of its argument, *snu* coincides very nearly (and for $u = 0$, exactly) with the trigonometric sine function regardless of the value of the modulus k; but in the vicinity of its maxima and minima it becomes more and more elongated as k increases, and finally for $k = 1$ it becomes infinitely elongated (its period becomes infinite).

We can appreciate in a general way how this flexibility in being able to adjust the shape of the *sn* function is needed in connection with our present problem. With reference to the plot in Fig. 12, if we merely ask for that pair of values for ω_1 and ω_3 for which $F(\omega)$ has equiripple character in the range $0 < \omega < p_4$, there are an infinite number of possible solutions. If, however, we arbitrarily pick a value for ω_3 in this range, and ask for that value of $\omega_1 < \omega_3$ for which the function has equiripple character, then there is only one possible solution. Similarly, in the case of a function corresponding to a larger value of n, if we fix ω_{2n-1} (the zero nearest the point $\omega = p_{2n}$) then the positions of all the other ω_ν resulting in an equiripple behavior for $F(\omega)$ are uniquely fixed.

As we choose the largest zero ω_{2n-1} closer to the point p_{2n}, all of the oscillations in the equiripple function $F(\omega)$ for a given n become crowded toward the upper end of the pass band. In such a situation, the change-of-variable $\omega = p_{2n} snu$ needed to render the function periodic in u must evidently be more flat-topped than it would need to be if the oscillations of $F(\omega)$ were less nonuniform.

The elliptic sine function snu is expressible in terms of the trigonometric sine by

$$snu = \sin \phi \tag{71}$$

where the relation between the variables u and ϕ is given by the so-called elliptic integral of the first kind

$$u(k, \phi) = \int_0^\phi \frac{d\xi}{\sqrt{1 - k^2 \sin^2 \xi}} \tag{72}$$

The dependence of u upon ϕ afforded by this integral is more easily understood by the derivative

$$\frac{du}{d\phi} = \frac{1}{\sqrt{1 - k^2 \sin^2 \phi}} \tag{73}$$

Clearly, when $k = 0$, u and ϕ may be identified, and Eq. 71 then shows that the elliptic sine and the trigonometric sine are identical. For a non-zero k value in the range $0 < k < 1$, the derivative 73 is still equal to unity for $\phi = 0$, showing that the elliptic sine and the trigonometric sine are in any case nearly alike for small values of their arguments. As ϕ increases toward the value $\pi/2$, the derivative 73 increases, reaching the value $1/\sqrt{1 - k^2}$ for $\phi = \pi/2$, after which it decreases again to the value unity at $\phi = \pi$, and then repeats the same cycle over and over. It is easy to see, on the basis of this reasoning, that a plot of u versus ϕ has the form shown in Fig. 13 (drawn for $k = 0.9$).

The quantity K, which is the value of u for $\phi = \pi/2$, is called the *complete* elliptic integral of the first kind. It evidently equals one-fourth of a period of the *sn* function, since $4K$ corresponds to $\phi = 2\pi$. The slope of the u versus ϕ curve at the point $\phi = \pi/2$ approaches infinity as $k \rightarrow 1$. Hence for $k = 1$, K is infinite; the elliptic sine has an infinitely long period.

Values of snu for given values of u may be found in tabulations of the *sn* function for various values of the modulus k (which commonly is replaced by a parameter α related to k by $k = \sin \alpha$), or one may use tables of the elliptic integral 72 to determine ϕ for a given value of u

and modulus k, and then use the trigonometric relation 71 to compute* snu.

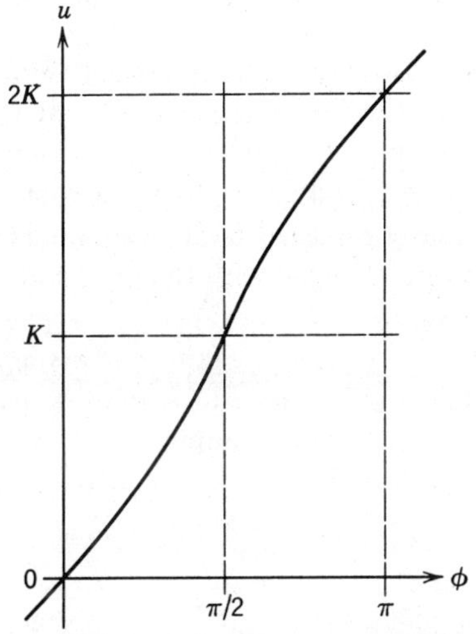

Fig. 13. Relationship (drawn for $k = 0.9$) between the independent variables u and ϕ in Eq. 71 which shows the connection between the elliptic and the trigonometric sine functions.

The appropriateness of the sn function to afford the change of variable that renders $F(\omega)$ periodic may be shown by making use of the following identity†

$$sn(u + u_\nu)sn(u - u_\nu) = \frac{sn^2u - sn^2u_\nu}{1 - k^2sn^2usn^2u_\nu} \tag{74}$$

Namely, if we let $p_{2n} = \sqrt{k}$, so that the change of variable reads

$$\omega = \sqrt{k}\,\sin\phi = \sqrt{k}\,snu \tag{75}$$

$$\omega_\nu = \sqrt{k}\,\sin\phi_\nu = \sqrt{k}\,snu_\nu \tag{76}$$

then we see that a typical pair of factors in 68 becomes

$$\frac{(\omega^2 - \omega_\nu^2)}{(1 - \omega_\nu^2\omega^2)} = \frac{k(sn^2u - sn^2u_\nu)}{(1 - k^2sn^2usn^2u_\nu)} = ksn(u + u_\nu)sn(u - u_\nu) \tag{77}$$

The total expression for $F^2(\omega)$, which is the square of 68, is thus seen to

---

* The familiar Jahnke and Emde tables can be used in this way to yield values for the Jacobian elliptic function. A table giving the latter function directly in terms of a parameter $m = k^2$ may be found in *Jacobian Elliptic Function Tables*, L. M. Milne-Thomson, Dover Publications, New York, 1950.

† See *Lectures on the Theory of Elliptic Functions*, H. Hancock, John Wiley & Sons, New York, 1910, p. 490.

be transformed into the function

$$G^2(u) = \prod_{\nu=-1,\,-3,\,\cdots,\,-(4n-1)}^{\nu=1,\,3,\,\cdots,\,4n-1} k\,sn(u + u_\nu) \tag{78}$$

in which $u_{-\nu} = -u_\nu$ so that the negative integer values of ν in this product expression account for the factors $sn(u - u_\nu)$. If the u_ν values are now chosen as indicated in Fig. 14 (for an assumed $n = 5$), we see that the totality of factors in the product 78 forms a cyclic group which is unchanged if each u value is replaced by its next succeeding one because the last merely becomes interchanged with the first, and the set of factors as a whole remains unchanged. Similarly, if u is replaced by $u + \Delta u_\nu$ with $\Delta u_\nu = u_{\nu+1} - u_\nu$, the set of factors again remains unchanged except for a rearrangement in their cyclic order, since adding this value of Δu_ν to u or to u_ν in the argument $(u + u_\nu)$ is immaterial. Thus the periodicity of $G^2(u)$ in Eq. 78 is established, and Δu_ν is revealed to be its period.

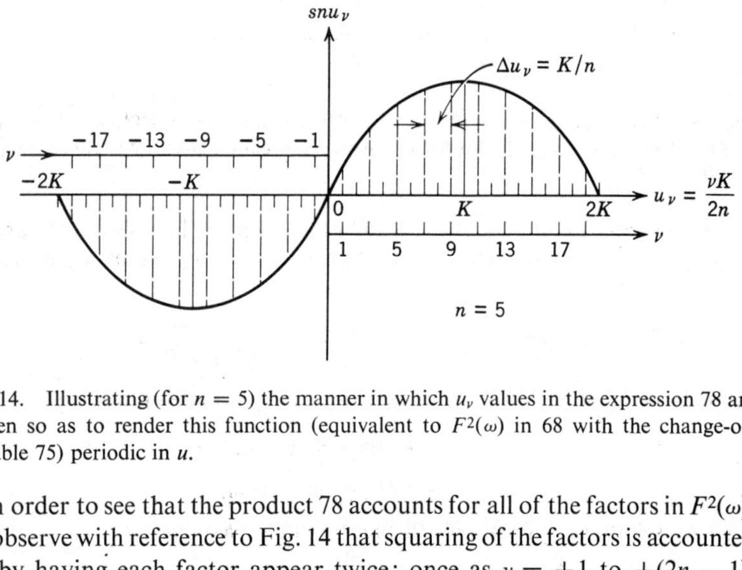

Fig. 14. Illustrating (for $n = 5$) the manner in which u_ν values in the expression 78 are chosen so as to render this function (equivalent to $F^2(\omega)$ in 68 with the change-of-variable 75) periodic in u.

In order to see that the product 78 accounts for all of the factors in $F^2(\omega)$, we observe with reference to Fig. 14 that squaring of the factors is accounted for by having each factor appear twice: once as $\nu = \pm 1$ to $\pm(2n - 1)$, and again as $\nu = \pm(2n + 1)$ to $\pm(4n - 1)$. If we observe that the square of the function in Fig. 12 (for $n = 2$) completes two periods in the interval $0 < \omega < p_4$, it is clear in general that the interval $0 < \omega < p_{2n} = \sqrt{k}$ contains n periods. Hence $G^2(u)$ has n periods in the corresponding interval $0 < u < K$, which makes each period equal to

$$\Delta u_\nu = K/n = u_{\nu+1} - u_\nu \tag{79}$$

and the u_ν values, as indicated in Fig. 14, are given by

$$u_\nu = \nu K/2n \qquad \text{for } \nu = \pm 1, \pm 3, \cdots \tag{80}$$

According to Eq. 76, the discrete ω_ν values are given by

$$\omega_\nu = \sqrt{k}\, sn(\nu K/2n) \qquad \text{for } \nu = 1, 3, 5, \cdots, 2n - 1 \tag{81}$$

The p_ν values, corresponding to points of maximum deviation, evidently occur for u values midway between the u_ν values 80. Hence we have

$$p_\nu = \sqrt{k}\, sn(\nu K/2n) \qquad \text{for } \nu = 0, 2, 4, \cdots, 2n \tag{82}$$

The first of these yields $p_0 = 0$ and the last is $p_{2n} = \sqrt{k}$ since $snK = \sin \pi/2 = 1$.

The tolerance factor ϵ, as Fig. 12 shows, is equal to $F(0)$, and hence according to Eq. 68 is given by

$$\epsilon = (\omega_1\omega_3 \cdots \omega_{2n-1})^2 \tag{83}$$

Determination of the zeros $\omega_2, \omega_4, \cdots, \omega_{2n}$ for the odd function 69 follows an entirely analogous pattern. Here the function $F^2(\omega)$ has $n + 1/2$ periods in the interval $0 < \omega < \sqrt{k}$; $G^2(u)$ therefore has $n + 1/2$ periods in the interval $0 < u < K$. Hence one period of the function $G^2(u)$ is

$$\Delta u_\nu = 2K/(2n + 1) \tag{84}$$

and the pertinent u_ν values are given by

$$u_\nu = \nu K/(2n + 1) \qquad \text{for } \nu = 0, \pm 2, \pm 4, \cdots \tag{85}$$

The discrete ω_ν values are

$$\omega_\nu = \sqrt{k}\, sn(\nu K/2n + 1) \qquad \text{for } \nu = 2, 4, \cdots, 2n \tag{86}$$

and the frequencies of maximum deviation are

$$p_\nu = \sqrt{k}\, sn(\nu K/2n + 1) \qquad \text{for } \nu = 1, 3, \cdots, 2n + 1 \tag{87}$$

the last of these again being $p_{2n+1} = \sqrt{k}$.

In order to recognize the expression for the tolerance factor ϵ in this case [since $F(0) = 0$, it is not as obvious as for the function 68], we observe first that the numerator polynomial in $F^2(\omega)$ now is one degree higher in ω^2 than the denominator polynomial, and that the coefficient of the highest power in the numerator is unity. Therefore the numerator polynomial of $(\epsilon^2 - F^2)$ has the factored form

$$(p_1{}^2 - \omega^2)^2(p_3{}^2 - \omega^2)^2 \cdots (p_{2n-1}{}^2 - \omega^2)^2(p_{2n+1}{}^2 - \omega^2) \tag{88}$$

since zeros of $(\epsilon^2 - F^2)$ are the ϵ^2 points of F^2, and all of these are double except the largest one. The zero-frequency value of $(\epsilon^2 - F^2)$, which has the same denominator as F^2, yields the value of ϵ^2. Thus we find

$$\epsilon = (p_1 \cdot p_3 \cdots p_{2n-1})^2\sqrt{k} \tag{89}$$

the factor $\sqrt{k}$, as pointed out above, being p_{2n+1}.

In a given application of these results, one must first make a choice for the value of $\sqrt{k} < 1$, which equals the effective cutoff frequency for the pass band. The closer we choose this value to unity, the larger ϵ becomes for a given n, or the larger n must be for a specified ϵ. It may be necessary to compute several trial solutions in order to make a judicious choice in this regard.

In applying the function $F^2(\omega)$ in the manner indicated in Eq. 9, a greater degree of flexibility is achieved if we introduce an additional numerical factor h as shown by

$$|Z_{12}(j\omega)|^2 = \frac{1}{1 + h^2F^2(\omega)} \tag{90}$$

This factor plays the same role that ϵ does in expression 28 involving the Tschebyscheff polynomial, and needs no further comment.

6 The Use of Fourier Methods

The Fourier series involves an approximation procedure with which most engineers are reasonably familiar. The problem of finding a rational approximant to some graphically given curve will therefore be reduced to a fairly standard procedure if we can show how to convert it to an equivalent problem in Fourier analysis. At the same time, we are thus able to provide an approximation method of more general character, since Fourier methods may be applied to functions of arbitrary shape, whereas the rather specialized procedures discussed so far have a very limited field of application.

The key relation for achieving the desired result is the Tschebyscheff polynomial 13, because it is the means for linking finite polynomials with trigonometric functions. By itself, however, it is insufficient to provide the basis for using finite trigonometric polynomial approximation methods since they require the given function to be periodic whereas the functions we wish to deal with are for the most part aperiodic. This difficulty is resolved by the choice of a suitable transformation of the independent variable, as the following discussion will show.

If in the defining equation

$$V_n(x) = \cos(n \cos^{-1} x) \tag{91}$$

we make the usual substitution

$$x = \cos\phi \tag{92}$$

and obtain for the Tschebyscheff polynomial the form

$$V_n(x) = \cos n\phi \tag{93}$$

then it becomes clear that the following trigonometric polynomial

$$f(\phi) = a_0 + a_1 \operatorname{ocs} \phi + a_2 \cos 2\phi + \cdots + a_n \cos n\phi \tag{94}$$

is equivalent to an ordinary polynomial in x that reads

$$g(x) = b_0 + b_1x + b_2x^2 + \cdots + b_nx^n \tag{95}$$

Given 94, we can convert to the form 95, or vice versa. That is to say, we can compute the coefficients $b_0 \cdots b_n$ in the polynomial 95 by substituting for $\cos \nu\phi$ in each term of 94 its polynomial in x according to 93, and collecting coefficients of like powers of x; or we can compute the coefficients $a_0 \cdots a_n$ by substituting $\cos \phi$ for x in 95, converting powers of $\cos \phi$ to cosines of multiple angles, and collecting coefficients of $\cos \nu\phi$ for $\nu = 0, 1, \cdots, n$.

The Tschebyscheff polynomial thus provides the means for uniquely converting back and forth from an ordinary polynomial to a finite trigonometric polynomial with relation 92 linking the variables in these two polynomials. The degree n of the ordinary polynomial in x becomes the highest integer multiple of the variable ϕ in the trigonometric polynomial, so that both polynomials always have the same number of terms.

Connection with the frequency variable ω is now determined by choosing

$$x = \frac{1 - \omega^2}{1 + \omega^2}, \qquad \omega^2 = \frac{1 - x}{1 + x} \tag{96}$$

yielding, with 92, the following relation between ω and ϕ

$$\omega = \tan(\phi/2), \qquad \phi = 2 \tan^{-1} \omega \tag{97}$$

As pointed out in Art. 5, Ch. 8 (Eqs. 114 and 115), this transformation is equivalent to

$$\frac{1 - j\omega}{1 + j\omega} = \frac{1 - s}{1 + s} = e^{-j\phi} = w \tag{98}$$

and maps the right half of the s-plane upon the interior of the unit circle of the w-plane (Fig. 11, Ch. 8), the periphery of this circle corresponding to the j-axis of the s-plane. Traversal once around the unit circle in a clockwise direction starting from the point $w = +1$ is equivalent to traversing the j-axis from $s = 0$ through $s = j$, through $s = \infty$, and returning to $s = 0$ via the negative branch of this axis. The variable ϕ meanwhile varies from 0 to 2π. This mapping property clearly is the means for converting from an aperiodic function in ω to a periodic function in ϕ.

Figure 12 in Ch. 8 further elucidates this transformation of the independent variable by a direct comparison of certain key points on the ω and ϕ axes. Thus we see that the origins $\omega = 0$ and $\phi = 0$ coincide; $\omega = \pm 1$ corresponds to $\phi = \pm\pi/2$, and $\omega = \pm\infty$ to $\phi = \pm\pi$.

The first step in dealing with a given approximation problem, in which the function we wish to represent by a quotient of polynomials in ω^2 is given in graphical form, is to replot this function of ω as a function of ϕ. We refer to this step as transforming the given function from the ω-domain to the ϕ-domain. The means for accomplishing it is straightforward. We simply transfer ordinates from an ω-scale to a ϕ-scale by computing corresponding values of ω and ϕ according to relation 97. Values of the ordinates are the same along both scales; only the points at which these same ordinates occur are different on the two scales. The function is distorted only in the horizontal or longitudinal direction (along the axis of the independent variable), as if by stretching or compressing it; ordinatewise nothing changes, as is always the case when making a change of the independent variable (like frequency scaling, or the low-pass–band-pass transformation, and so on).

In the ϕ-domain we now approximate our function by means of a cosine series like 94, having a finite number of terms. The function must, of course, be even about $\phi = 0$, as is also evident from the fact that the resulting approximant (according to transformations 92 and 96) is a function of ω^2 like the magnitude or real part of an impedance. If the given function of ω is odd, like the tangent of a phase angle (see Arts. 9 and 10, Ch. 8), we can divide it by ω at the outset to convert it to an even function, and restore the factor ω in the resulting rational approximant. Hence the present method can handle odd as well as even functions, notwithstanding the fact that the approximation procedure itself is restricted to even functions.

In the determination of the trigonometric polynomial 94, we can use as large a value of n as is needed to meet stated tolerance requirements, and we can avail ourselves of the many techniques that have been developed in regard to Fourier analysis and related topics. After an acceptable trigonometric approximation has thus been found, we use the Tschebyscheff polynomial to convert to the polynomial form 95 as explained above, and finally, by means of relation 96, obtain the following rational function in ω^2

$$F(\omega^2) = \frac{A_0 + A_1\omega^2 + A_2\omega^4 + \cdots + A_n\omega^{2n}}{(1 + \omega^2)^n} \tag{99}$$

which approximates the originally given function of ω with the same ordinate tolerances that are obtained in the trigonometric approximation in the ϕ-domain.

Although there is no objection, on grounds of acceptability or other theoretical considerations, to the rather special form of the denominator polynomial which this procedure yields in the rational approximant 99, there are two major reasons why we should seek to free this approximation method of the restrictions implied by this form. Firstly, if the function 99 is a real part or a magnitude, it yields a complex function having an nth order pole at $s = -1$, which is unacceptable if the pertinent function is to be representative of an *RC* network. Secondly, we have the intuitive feeling that by restricting the poles of our response function in this very special way, we are seriously limiting its approximation ability along the

j-axis for any given polynomial degree, thereby failing to realize the full potential which that function is otherwise capable of providing.

An alternate, more simple-minded, way of looking at the restrictive influence of the special denominator in Eq. 99 is to observe that if, instead, it had the general form $B_0 + B_1\omega^2 + B_2\omega^4 + \cdots + B_n\omega^{2n}$, n additional arbitrary constants would be available for augmenting the approximating ability of the rational function $F(\omega^2)$. By choosing the special denominator $(1 + \omega^2)^n$ a priori, as our present method implicitly does, we are seemingly throwing away half of the approximating ability of our function at the outset. Either we should be aware of what this sacrifice has to offer in the way of compensating advantages (so that we can capitalize upon them), or we should find a modification in our procedure that avoids it. Each of these ideas is considered in the following.

To avoid the special result in Eq. 99, suppose we denote by $G(\phi)$ the ϕ-domain plot of the given function, and give it the representation

$$G(\phi) = G_1(\phi)/G_2(\phi) \tag{100}$$

The specific determination of the functions $G_1(\phi)$ and $G_2(\phi)$, which we assume is accomplished graphically, adds to the flexibility and approximating ability of our procedure in a way that is elaborated upon presently. These two functions may now be approximated by trigonometric polynomials $f_1(\phi)$ and $f_2(\phi)$ respectively and converted, according to the above procedure, to functions $F_1(\omega^2)$ and $F_2(\omega^2)$ having the forms

$$F_1(\omega^2) = \frac{A_0 + A_1\omega^2 + \cdots + A_n\omega^{2n}}{(1 + \omega^2)^n} \tag{101}$$

and

$$F_2(\omega^2) = \frac{B_0 + B_1\omega^2 + \cdots + B_n\omega^{2n}}{(1 + \omega^2)^n} \tag{102}$$

Their ratio yields the rational approximant

$$F(\omega^2) = \frac{F_1(\omega^2)}{F_2(\omega^2)} = \frac{A_0 + A_1\omega^2 + \cdots + A_n\omega^{2n}}{B_0 + B_1\omega^2 + \cdots + B_n\omega^{2n}} \tag{103}$$

representing the desired solution.

The advantage which accrues from this modification in our procedure is due entirely to the possibilities offered by the step indicated in Eq. 100. First of all, we recognize that an unlimited number of function pairs G_1 and G_2 are possible, so that the procedure acquires a considerable degree of flexibility. Second, any characteristics of the function $G(\phi)$ that tend to make its approximation difficult (such as sudden jumps of large magnitude) can be divided between G_1 and G_2 in a way that relieves the burden on each. For example, if $G(\phi)$ is a low-pass filter characteristic that is required to drop from the value 1 at cutoff to a tolerance value of

1/100, we can have G_1 jump from 1 to $\frac{1}{10}$ at this point while G_2 jumps from 1 to 10. Thus a 100:1 ratio in G is achieved with only a 10:1 ratio in the separate functions G_1 and G_2.

Experience plus some ingenuity together with a little cut and try will be needed in order to exploit effectively the advantage inherent in the representation 100, but it should be logically clear that the approximating ability of a function of given complexity is materially increased by this device. Most important about this result, however, is the fact that it demonstrates the possibility of solving a given approximation problem in the presence of a stated constraint upon the pole positions of our network function, as is the case with *RC* impedances whose poles must lie on the negative real axis.

Fixing the pole locations at the outset fixes the function $F_2(\omega^2)$ in Eq. 102 and hence, in the ϕ-domain, it fixes $f_2(\phi)$, which in this approach to the problem is identified with $G_2(\phi)$. In terms of the desired $G(\phi)$, we can then obtain a plot of $G_1(\phi)$ from Eq. 100, which is approximated by a polynomial $f_1(\phi)$ and converted into $F_1(\omega^2)$ as described before. Since the degree of $f_1(\phi)$ cannot exceed the degree of $f_2(\phi)$ which is fixed to begin with, we do not have the freedom of choosing as many terms as may be needed to achieve the desired approximation of $G_1(\phi)$ unless we revise the number of poles chosen for the function $F_2(\omega^2)$. Although this circumstance injects some inconvenience into the use of this method, the important fact remains that presetting the poles of a response function *is* possible, and that, in spite of such a rather drastic constraint, one can still approximate arbitrarily closely to a desired characteristic.

This result is relevant to the method of precorrecting a given impedance function for the effect of incidental dissipation, as discussed in Art. 6, Ch. 12, where it is pointed out that the nearness of the poles to the *j*-axis sets a limit upon the extent to which such a precorrection scheme can be carried out. Being able to choose the distance that the poles are to the left of the *j*-axis removes this restriction completely and makes precorrection possible under all normal conditions provided we are willing to reconsider the approximation problem if necessary. Moreover, we may be required to accept a realization involving increased numbers of elements consistent with an increase in the number of poles needed because of the constraint upon their location.

An important conclusion may be drawn to the effect that the attainment of exacting requirements in the design of filters and other networks does not necessarily require inductive elements having unusually low parasitic losses (or unusually high Q's). Since a resistance may be regarded as the limiting form of an inductive element having zero Q, and since one can even in this worst case (which is the *RC* network) still approximate arbitrarily closely to any desired characteristic, it should certainly be possible to do so with inductive elements having nonzero Q's, however small. One must, of course, be willing to supply gain (by means of vacuum tubes or transistors) to offset the parasitic losses, and one must be prepared also to pay for the privilege of using arbitrarily poor quality inductances by being required to provide a larger total number of network elements.

The method possesses another degree of arbitrariness that must be resolved judiciously. It was pointed out that the frequencies $\omega = \pm 1$ correspond to $\phi = \pm\pi/2$. In the positive half of the ϕ-scale, the point $\pi/2$ is midway between the two extremities of this scale. In the positive half of the ω-scale, the point $\omega = 1$ is one unit from the left-hand end of this scale and infinitely far from the right-hand end. This situation requires that we so normalize our ω-scale to begin with that the important variations of the function under consideration occur in the vicinity of $\omega = 1$, and that such variations at values of ω much larger than unity be avoided, since this part of the ω-scale appears highly compressed on the ϕ-scale. Any moderately abrupt change in the given function at values of ω large compared with unity is converted into an extremely abrupt change in the ϕ-domain function.

In applying the method to the ideal low-pass filter characteristic, placing the cutoff at $\omega = 1$, as is done in Fig. 1, yields a uniform square wave (plus a constant) in the ϕ-domain and hence is a most advantageous choice. A function having geometric symmetry about $\omega = 1$ in the ω-domain exhibits arithmetic symmetry about $\phi = \pi/2$ when transformed to the ϕ-domain. If we plot the given function versus the logarithmic variable u in Eq. 49 so that $\omega = 1$ corresponds to $u = 0$, then an even function of u becomes an even function of ϕ about the point $\phi = \pi/2$. An ideal band-pass characteristic is advantageously normalized so as to have this type of symmetry. In an unusual situation one may need to try several normalizations of the given frequency scale before obtaining a ϕ-domain function that best lends itself to trigonometric approximation.

7 A Method Based Upon Semi-infinite Slopes*

In terms of the discussion in Art. 3 regarding the asymptotic behavior of the Butterworth function, one is able to develop an extremely simple and effective method for obtaining a rational approximant in the form of Eq. 103 from a given graphical plot of the desired impedance magnitude using logarithmic scales such as those shown in Fig. 9. If we apply the

* This method is discussed in a paper by R. F. Baum entitled "A Contribution to the Approximation Problem," *Proc. I.R.E.*, Vol. 36, No. 7, July 1948, pp. 863–869.

same procedure to the following Butterworth function of order n with cutoff at $\omega = \omega_\nu$

$$B_\nu^{(n)} = \frac{1}{\sqrt{1 + (\omega/\omega_\nu)^{2n}}} \tag{104}$$

and, by analogy to Eq. 49, write

$$\omega_\nu = 2^{u_\nu} \tag{105}$$

then the corresponding loss function, according to an extended form of Eq. 50, becomes

$$\begin{aligned} \alpha_\nu^{(n)} &\approx 6n(u - u_\nu) \qquad \text{for } u > u_\nu \\ &\approx 0 \qquad \text{for } u < u_\nu \end{aligned} \tag{106}$$

An arbitrary loss function $\alpha(u)$ may evidently be approximated by a sum of components $\alpha_\nu^{(n)}$ with suitably chosen values of u_ν and n, since any broken-line approximation to $\alpha(u)$ consisting of straight-line segments with positive or negative slopes equal to integer multiples of 6 db per octave is equivalent to a sum of functions having the form 106, a segment with negative slope corresponding to the reciprocal of the factor 104. Since a sum of functions like 106 results from the logarithm of a product of factors like 104, or their reciprocals, we see that the squared magnitude of the desired impedance is immediately obtainable from such a broken-line approximation in the form

$$|Z_{12}(j\omega)|^2 = \frac{[1 + (\omega/\omega_a)^{2p}] \times [1 + (\omega/\omega_c)^{2r}] \times \cdots}{[1 + (\omega/\omega_b)^{2q}] \times [1 + (\omega/\omega_d)^{2t}] \times \cdots} \tag{107}$$

in which $\omega_a, \omega_b, \cdots$ are obtained from the break points $u_a, u_b, \cdots$ by Eq. 105, and the integers $p, q, r, \cdots$ are the appropriate *differences* in 6-db-per-octave slopes of the adjacent segments at each break point.

An example is given by the loss curve in Fig. 15. The broken-line approximation, shown dashed, has break points at $u_1 = 1$, $u_2 = 2$, $u_3 = 3.15$, $u_4 = 4.20$, and $u_5 = 5.10$, yielding for the corresponding Butterworth factors the cutoff frequencies $\omega_1 = 2$, $\omega_2 = 4$, $\omega_3 = 8.83$, $\omega_4 = 18.3$, $\omega_5 = 34.2$. The first segment of the broken-line approximation has a slope of -12 db per octave. In expression 107 this segment, therefore, yields a denominator factor of the form $1 + (\omega/2)^4$. At the break point $u_2 = 2$, the slope of the broken line increases by 12 db per octave, calling for the numerator factor $1 + (\omega/4)^4$. At $u_3 = 3.15$ the slope suddenly increases by 18 db per octave, which yields another numerator factor equal to $1 + (\omega/8.83)^6$. At $u_4 = 4.20$ the slope decreases from 18 to 12 db per octave. The denominator factor $1 + (\omega/18.3)^2$ accomplishes this slope change of 6 db per octave. Finally, at $u_5 = 5.10$ the slope becomes zero, giving rise to a denominator factor $1 + (\omega/34.2)^4$. The complete expression for the rational approximant to the squared magnitude of this impedance becomes

$$|Z_{12}(j\omega)|^2 = \frac{\left[1 + \left(\frac{\omega}{4}\right)^4\right] \times \left[1 + \left(\frac{\omega}{8.83}\right)^6\right]}{\left[1 + \left(\frac{\omega}{2}\right)^4\right] \times \left[1 + \left(\frac{\omega}{18.3}\right)^2\right] \times \left[1 + \left(\frac{\omega}{34.2}\right)^4\right]} \tag{108}$$

The actual approximation given by this approximant is smoother than the broken line because of the 3-db fillets at the break points as discussed in Art. 3 and illustrated in Fig. 9.

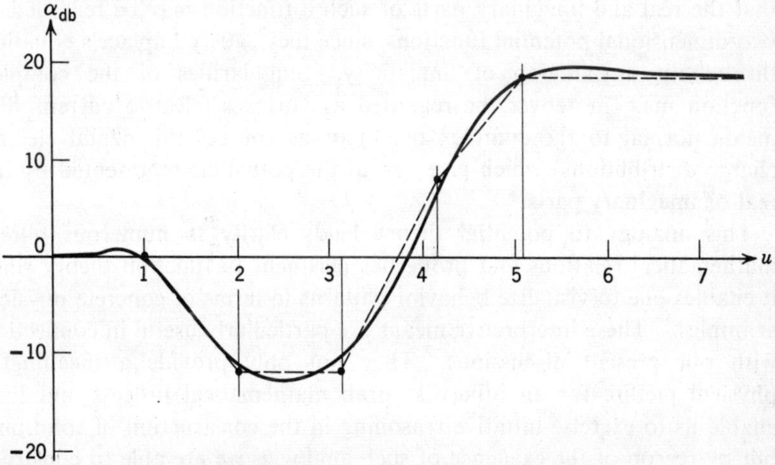

Fig. 15. Approximation of a given loss curve (as a function of the logarithmic frequency variable defined by Eq. 49) by means of straight-line segments having multiples of 6-db-per octave slopes leads directly to the rational approximant 108.

The fact that the slopes of the broken-line segments must be multiples of 6 db per octave injects a constraint into the method of carrying out the graphical approximation process which, however, becomes awkward only in certain extreme cases. For example, if the given curve has an almost constant slope over an interval of considerable extent, and if this slope is roughly midway between two integer multiples of 6 db per octave, then it becomes necessary to alternate between slopes that are equal to the nearest larger and smaller 6-db multiples. The broken line then oscillates about the required average slope. By a suitable spacing of break points, the closeness of approximation can be made arbitrarily good.

This method is superior to the brute-force procedure of determining coefficients in the rational function 107 such that it passes through the same number of selected points on the graphically given curve, since that method exercises no control over the behavior of the resulting approximant throughout the intervals between the selected points. The method illustrated in Fig. 15, on the other hand, does exercise such control. Furthermore, the complex function can readily be obtained from the squared magnitude expression 108, since each Butterworth factor readily yields its corresponding left half-plane zeros or poles.

8 Use of Potential Theory Analogies

It is a well-known fact in the theory of functions of a complex variable that the real and imaginary parts of such a function may be regarded as two-dimensional potential functions, since they satisfy Laplace's equation throughout any region of analyticity. Singularities of the complex function may, moreover, be regarded as vortexes (electric current filaments normal to the complex plane) or as sources (filamental electric charge distributions) which give rise to the potentials represented by the real or imaginary parts.*

This analogy to potential theory lends clarity to numerous purely mathematical relations and properties pertinent to function theory since it enables one to visualize behavior patterns in terms of concrete physical examples. These interpretive means are particularly useful in connection with our present discussions. They not only provide a meaningful physical picture for an otherwise drab mathematical process and thus enable us to exercise intuitive reasoning in the construction of solutions, but by reason of the existence of such analogies we are able to construct physical devices (called analog computers) that greatly aid in the actual solution of an approximation problem by the trial-and-error method.

In connection with rational functions, potential theory analogies are based upon either one of two distinctly different set of interpretations, according to whether we wish thus to obtain a representation for the real and imaginary parts of the function itself or for the logarithm of that function.†

As a representation for the real part of the function itself, one finds that simple poles may be interpreted as physical dipoles, and the residues of the complex function in these poles, as the dipole moments, their angular orientations coinciding with the angles of the respective complex residues. The real part of the function at any point in the complex s-plane equals the resulting potential at that point due to dipoles with appropriate moments located at the various poles. The corresponding imaginary part equals the resulting potential after each dipole moment is

* See *MCA*, Art. 5, Ch. VI, p. 263.

† *Ibid.*, pp. 349–360. The reader is expected to be thoroughly familiar with this material.

rotated clockwise through 90°. A pole of order n is similarly represented physically as a $2n$ pole, this being the logical extension of the dipole to a symmetrical arrangement of n pairs of oppositely charged filaments.

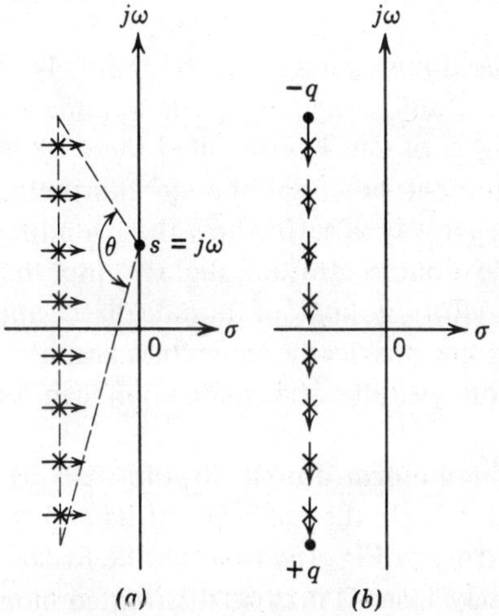

Fig. 16. Poles, regarded as electric dipoles, yield a pair of magnitude and phase functions [the j-axis potential functions of (a) and (b) respectively] that are seen by inspection to have low-pass character.

As an interesting example of how this physical analogy can help one obtain a quick grasp of some otherwise rather subtle relationships, let us consider the pole arrangement shown in Fig. 16a consisting of a row of equally spaced poles along a line parallel to the j-axis, all residues being positive real and equal, as indicated by the horizontal arrows which simultaneously can be regarded as dipole moments. To a reasonable approximation, such a dipole array may be regarded as a uniform double stratum, the potential of which at some point like $s = j\omega$ is given by the subtended angle θ. As the point s traverses the j-axis, it is clear that θ is large and nearly constant while s is adjacent to the string of poles; whereas it rapidly decreases and becomes nearly zero for values of $s = j\omega$ below or above this pole distribution. For a string of closely spaced poles very near the j-axis, the real-part function approximates very closely the ideal low-pass filter characteristic of Fig. 1. If the pertinent rational function is the input impedance to a resistance-terminated lossless network, then the discussion in Art. 1, Ch. 11, shows that the transfer impedance magnitude has this low-pass filter characteristic.

If we are interested in the corresponding imaginary part of this input impedance as a function of $s = j\omega$, we imagine all of the dipoles rotated clockwise through 90°, resulting in the dipole chain shown in Fig. 16*b*, which is equivalent simply to one negative charge $-q$ and one positive charge $+q$ at the terminations of this chain. The potential along the $j\omega$-axis produced by this pair of opposite charges is easily visualized or calculated.

If we imagine the double stratum to be infinitely close to the j-axis, it is clear that the real and imaginary parts assume the idealized limiting forms shown in Fig. 5 of Ch. 8, computed there by means of the Hilbert transform. The present potential theory interpretations not only give us a quicker and easier way of seeing how these limiting forms come about, but by moving the double stratum slightly into the left half-plane we obtain with little effort a host of nonidealized approximate low-pass characteristics that are physically realizable, and a possible pole pattern or rational function yielding this realization can be written down by inspection.

The theory of the uniform double stratum tells us, moreover, that its shape is unimportant; only the locations of its end points are significant. Hence the dipole string in Fig. 16*a* need not be linear. The poles (if they are sufficiently closely spaced) may be distributed along any curve having the same end points, the arrows indicating the residues remaining normal to this curve. Whether we wish to exploit these results practically or not, they are valuable for the insight that they give us into the directly and indirectly related network properties involved.

The loss in nepers and phase lag in radians associated with a transfer impedance are given by the real and imaginary parts of the following expression

$$\ln\,[1/Z_{12}(s)] = \ln\left[\frac{(s-s_2)(s-s_4)\cdots(s-s_{2n})}{(s-s_1)(s-s_3)\cdots(s-s_{2n-1})}\right] = \alpha + j\beta \quad (109)$$

for $s = j\omega$. Here $s_1, s_3, \cdots, s_{2n-1}$ are zeros and $s_2, s_4, \cdots, s_{2n}$ are poles of the transfer impedance. The real part yields

$$\alpha = \ln\frac{1}{|s-s_1|} + \ln\frac{1}{|s-s_3|} + \cdots + \ln\frac{1}{|s-s_{2n-1}|}$$
$$-\ln\frac{1}{|s-s_2|} - \ln\frac{1}{|s-s_4|} - \cdots - \ln\frac{1}{|s-s_{2n}|} \quad (110)$$

Except for a constant multiplier (which is of no interest at present) each term in this expression represents the potential at a variable point s due to a uniform filamental charge or current distribution normal to the s-plane at the point s_ν. If the s-plane is represented by a semiconducting sheet, one can simulate these terms by touching the sheet with pin-pointed probes at the points s_ν, feeding a fixed amount of current into each probe at the zeros $s_1 \cdots s_{2n-1}$ and withdrawing the same amount at each pole $s_2 \cdots s_{2n}$. If the number of finite zeros is less than the number of poles, then a number of zeros equal to the difference are at $s = \infty$. One must in this instance feed this difference current into the sheet at some remote point or set of points.

The potential measured at any point in the semiconducting sheet (with respect to a convenient datum—the remote point will do) by means of a roving probe is proportional to the function α. Within a constant multiplier, one may thus determine the attenuation as a function of s for any given pole-zero distribution by physical means.*

The same physical mechanism may be used for the dipole representation discussed previously. A pair of closely spaced probes, of which one feeds and the other withdraws current, simulates a dipole and represents each pole of the rational function. The intensity of the current is proportional to the magnitude of the residue, and the angular orientation of the dipole equals the angle of the residue. In this arrangement the rational function is represented by its poles and residues; in the other arrangement, by its poles and zeros.

In the dipole analog, the roving probe picks up a potential proportional to the real part of the rational function (or the imaginary part if all dipoles are rotated clockwise by 90°). In the pole-zero analog, the roving probe picks up a potential proportional to the logarithm of the magnitude of the function or its reciprocal according to whether the pole probes or the zero probes feed current into the sheet. The latter circumstance is actually of little importance, since it affects only the algebraic sign of the potential.

Phase angle, or the imaginary part of the logarithm of the rational function, is not quite as easily determined in the pole-zero analog. A basis for its determination is given by the Cauchy-Riemann equation

$$\frac{d\beta}{d\omega} = \frac{d\alpha}{d\sigma} \tag{111}$$

from which

$$\beta = \int_0^\omega \frac{d\alpha}{d\sigma}\, d\omega \tag{112}$$

---

* The literature on this subject is extensive. A representative list of papers is the following: A. R. Boothroyd, E. C. Cherry, R. Makar, "An Electrolytic Tank for the Measurement of Steady-State Response, Transient Response, and Allied Properties of Networks," *Proc. Inst. Elec. Engrs.* (London), Pt. I, Vol. 96, May 1949, pp. 163–177. S. Darlington, "The Potential Analogue of Network Synthesis," *Bell System Tech. J.*, Vol. 30, Apr. 1951, pp. 315–365. W. H. Huggins, "The Potential Analogue in Network Synthesis and Analysis," *Report 5066, Air Force Research Laboratories*, Cambridge, Mass., Mar. 1951. R. E. Scott, "Network Synthesis by the Use of Potential Analogs," *Proc. I.R.E.*, Vol. 40, Aug. 1952, pp. 970–973.

Here $d\alpha/d\sigma$ represents the potential gradient in a direction normal to the j-axis. For any j-axis point it is proportional to the current density normal to this axis at that point. The phase β for any $s = j\omega$, according to Eq. 112, is thus seen to be proportional to the total current crossing the j-axis between $s = 0$ and the pertinent point $s = j\omega$.

Since it is physically not feasible to measure this current in any analog device using a semiconducting sheet, one resorts to measuring the slope (equal to delay, or the reciprocal of group velocity) according to Eq. 111 by means of probe pairs straddling the j-axis so as to measure the potential gradient across it. By using a collateral integrating device the phase β may thus be obtained, although not in a very satisfactory manner.

In our present discussions we are not so much interested in analog devices as we are in the physical interpretations which their theoretical basis provides. For example, it is physically immediately obvious that a zero or pole pattern symmetrical about the j-axis produces no net current across that axis because the correspondingly symmetrically located probes feed equal currents into the sheet. Hence it is at once clear that such a pattern produces no phase shift and affects only the loss function. If in such a pattern all the probes in one half-plane are removed, the loss function is unaffected except for a factor $\frac{1}{2}$, but a phase shift now takes place.

Analogously, if we have poles on one side of the j-axis and zeros in symmetrically located positions, the j-axis is an equipotential locus and hence the loss function is obviously a constant. We do, however, have current crossing the j-axis, and so this pattern produces phase shift only. If now we remove all the probes on one side of the j-axis, the current crossing it is merely halved; its distribution along the axis is not changed. Hence such an "all-pass" pole-zero pattern clearly yields the same phase function, except for a factor $\frac{1}{2}$ when all of the critical frequencies in one half-plane are removed. However, the j-axis does not remain an equipotential locus, and hence the associated loss function does not remain constant.

We can, of course, obtain these same results without the aid of the physical picture lent by these potential theory interpretations, but they are so much more meaningful, and evident by inspection, when seen in this light. We can furthermore make use of certain well-known experimental evidence in potential theory to establish useful principles and points of view. Faraday's famous ice pail experiment, for example, gives us a basic principle that is useful in solving the approximation problem for filter networks. According to this experiment the potential inside a conducting vessel is constant; or we can say that the potential inside a closed (or substantially closed) equipotential surface is constant, since the conducting vessel merely serves as a convenient mechanism for providing the almost closed equipotential surface.

With reference to Fig. 17*a*, the cylindrical surface axially normal to the s-plane is assumed to provide such an equipotential locus. The cylinder is assumed to carry some total charge Q per unit of axial length, which distributes itself uniformly over the surface and produces a potential outside that may be computed as though Q were a filamental distribution normal to the s-plane at $s = 0$. If we assume Q to be negative and regard the potential along $s = j\omega$ as a loss function α, we have

$$\alpha = -K \ln (1/\omega) = K \ln \omega \qquad \text{for } \omega > 1 \tag{113}$$

where K is an appropriate constant multiplier. For $|\omega| < 1$, Faraday's ice pail experiment tells us that the potential remains constant at the value it has for $\omega = 1$ where the variable point enters the cylinder. A plot of this loss function is shown in Fig. 18, in which the scale factor for α is proportional to the charge Q.

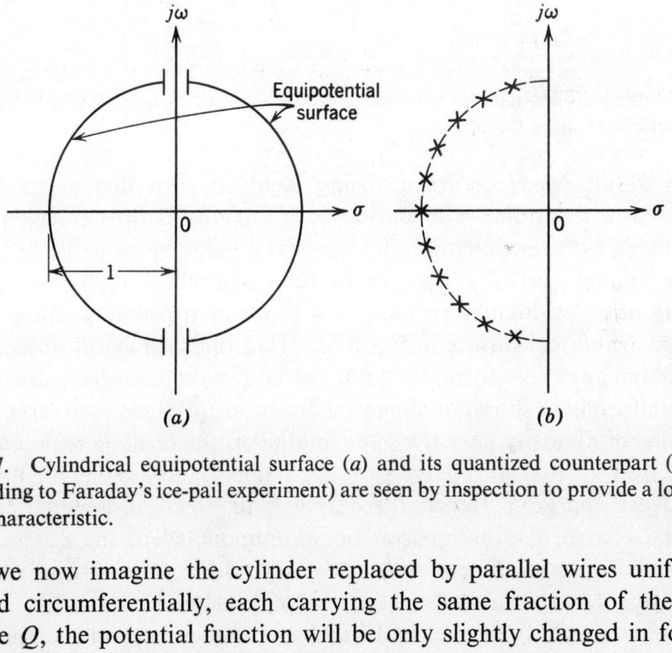

Fig. 17. Cylindrical equipotential surface (*a*) and its quantized counterpart (*b*) that (according to Faraday's ice-pail experiment) are seen by inspection to provide a low-pass filter characteristic.

If we now imagine the cylinder replaced by parallel wires uniformly spaced circumferentially, each carrying the same fraction of the total charge Q, the potential function will be only slightly changed in form if the spacing between wires is small. The total charge Q, instead of being uniformly distributed circumferentially, is divided into a finite number of equal lumps concentrated at equally spaced points. This process is spoken of as "quantization," the "quanta" being the equal portions of charge carried by the single filaments or wires.

Networkwise this process of quantization is equivalent to the transition from a distributed parameter network characterized by transcendental impedance functions to a lumped network characterized by rational impedance functions. The poles of the latter may now be identified, according to the potential analog previously discussed, with the filaments carrying lumped charges. If we do this, and subsequently discard the right half-plane poles (which, as pointed out above, merely changes the potential or the loss function α by a factor $\frac{1}{2}$), we get the picture in Fig. 17*b*, which still has essentially the loss function shown in Fig. 18.

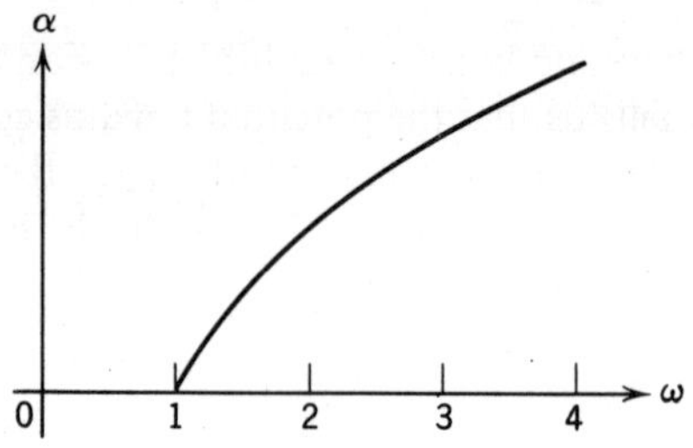

Fig. 18. The *j*-axis potential of the arrangement in Fig. 17*a*. It may be regarded as $1/n$ times the loss function of a low-pass filter having the limiting form of a Butterworth characteristic of infinite order.

This result we recognize as being identical with that given by the Butterworth function. The Butterworth function is thus revealed to be identifiable with the potential of a quantized Faraday ice pail!

One should carefully observe in this connection that the uniform spacing of poles circumferentially is a result of the circular form of the original cylindrical surface in Fig. 17*a*. Had this cylindrical surface been assumed in any other form, the total charge Q would not have distributed itself uniformly in the circumferential dimension. Since each term in Eq. 110 must of necessity have the *same* multiplier, the analogy between poles (respectively, zeros) and charge-carrying filaments implies that all filaments carry *equal* charges. Hence, the only way in which an unequal circumferential charge distribution can be accommodated in the quantization process is to make the circumferential pole density (reciprocal spacing) proportional to the circumferential charge density.

We thus see that our "ice pail" in Fig. 17*a* may have any shape other than that of a circular cylinder provided we determine, for that shape, the corresponding charge density distribution, and quantize accordingly. Thus the Butterworth function is not the only one capable of yielding a Butterworth type of loss function, although it is analytically the simplest. Finding the charge distribution on an arbitrarily shaped cylindrical surface presents a boundary value problem that is much more difficult to solve. In fact, any boundary value problem (even the two-dimensional one considered here) that cannot be solved by inspection presents considerable difficulty.

Computationally, we thus recognize that the use of potential theory does not make our problem easier to solve unless it yields a situation for which the solution is recognizable by inspection. Otherwise we merely find ourselves trading one computationally awkward problem for another equally awkward one. If the assumed equipotential cylindrical surfaces have s-plane cross sections that are polygonal in shape (or can reasonably be approximated by polygons), then we can always construct a solution to the boundary value problem by making use of the Schwarz-Christoffel formula (see *MCA*, Art. 25, pp. 378–395). It is rather doubtful, however, whether potential theory can still be regarded as an "aid" in the solution to an approximation problem under these circumstances.

Primarily, potential theory provides a useful mechanism for intuitive reasoning and mental visualization. It also provides a number of specific solutions in terms of certain classical boundary value problems for which solutions are available in the rather extensive literature on potential theory and conformal mapping. An example of this sort, appropriate to our present discussion, is given by the well-known conformal transformation

$$w = s \pm \sqrt{s^2 + 1} \qquad \text{or} \qquad s = \frac{1}{2}\left(w - \frac{1}{w}\right) \tag{114}$$

which maps the s-plane doubly upon the w-plane, once upon the interior of the unit circle of that plane and again upon the exterior of this unit circle. The unit circle itself becomes the doubly traversed interval of the j-axis corresponding to $-1 < \omega < 1$.

Concentric circles and radial lines in the w-plane become ellipses and hyperbolas respectively in the s-plane. Letting

$$w = re^{j\beta} \tag{115}$$

the family of confocal and orthogonal ellipses and hyperbolas in the s-plane result for constant values of r and β respectively, as shown in Fig. 19. The s-plane is a two-sheeted Riemann surface; the points $s = \pm j$ are branch points, and the j-axis interval connecting them may be interpreted as a branch cut.

If we regard the circular cylinder designated by $r = 1$ in the w-plane as an equipotential surface, then the corresponding surface in the s-plane becomes a sheet whose normal intersection with this plane coincides with the branch cut. A net charge Q on the circular cylinder of the w-plane distributes uniformly in a circumferential direction, as is also indicated in this field map by drawing the flux lines $\beta =$ constant with uniform spacing. On the corresponding equipotential surface in the s-plane, the total charge Q distributes nonuniformly, the charge density for any ω value being proportional to the density of the flux lines $\beta =$ constant at that point.

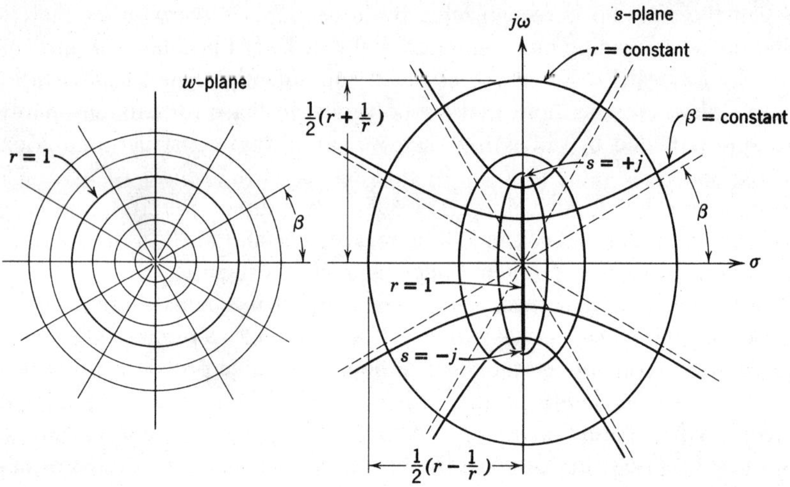

Fig. 19. Graphical interpretation of the conformal transformation 114 leading to a potential theory analog of the Tschebyscheff low-pass characteristic according to Eqs. 116 through 124.

Variation of the potential along the j-axis of the s-plane may be obtained by noting first that, for a negative charge on the cylinder of the w-plane, we have

$$\alpha = -K \ln (1/r) = K \ln r \tag{116}$$

or

$$r = e^{\alpha/K} \tag{117}$$

Equations 114 and 115 thus yield

$$s = \sinh (\alpha/K + j\beta) \tag{118}$$

giving

$$\sigma + j\omega = \sin (\alpha/K) \cos \beta + j \cosh (\alpha/K) \sin \beta \tag{119}$$

whence

$$\sigma = \sinh (\alpha/K) \cos \beta \tag{120}$$

$$\omega = \cosh (\alpha/K) \sin \beta \tag{121}$$

For $\sigma = 0$ ($s = j\omega$) we have

either

$$\alpha = 0, \quad \beta = \sin^{-1} \omega \qquad \text{for } -1 < \omega < 1 \tag{122}$$

or

$$\beta = \pm\pi/2, \quad \alpha = K \cosh^{-1} \omega \qquad \text{for } |\omega| > 1 \tag{123}$$

These two relations are pertinent respectively to the pass and stop bands. The characteristics which they yield are shown in Fig. 20. They are unattainable in a lumped network since they would require a continuous pole distribution having the density variation β in the interval $-1 < \omega < 1$ of the j-axis. A lumped network impedance cannot have a singularity of this sort; it can have only poles, and these must not lie upon the j-axis if the impedance is to remain bounded for $s = j\omega$.

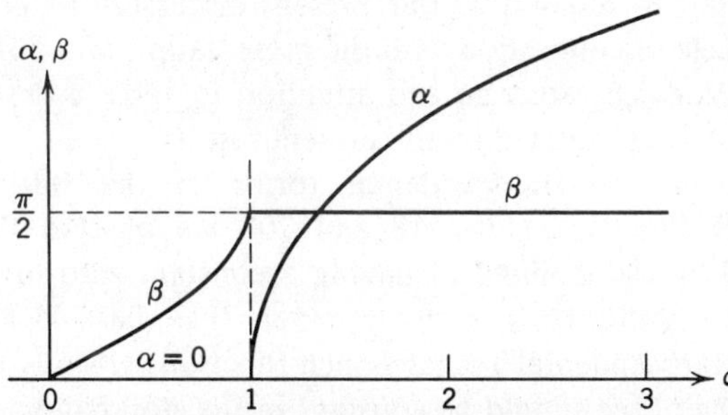

Fig. 20. Loss and phase characteristics derived from the potential analog based upon the conformal representation in Fig. 19. They may be interpreted as the limiting form assumed by a Tschebyscheff function for $\epsilon \to 0$ and $n \to \infty$.

We can, of course, approximate these characteristics in a lumped network by placing poles along some other equipotential contour, that is to say, upon an elliptic contour in the left half-plane. Such a contour in the s-plane corresponds to a circle in the w-plane for r larger or smaller than unity. Since circumferential charge distribution on a circle in the w-plane is uniform, quantization there is simple. We may pick a circle for r slightly larger than 1, for example, and place poles upon it with uniform spacing.

On the circle $r = 1$, corresponding to the pass band $-1 < \omega < 1$, we no longer have a constant potential. It is not difficult to see that the potential, and hence the loss function, has equiripple character in the pass band. This is the result we obtain in Art. 2 through use of the Tschebyscheff polynomial. What we are discovering is that a rational approximant to the transcendental loss and phase functions shown in Fig. 20 is none other than the Tschebyscheff function obtained earlier through a completely different approach.

Placing $2n$ uniformly spaced poles on a circle in the w-plane defined by $r = e^{\delta}$ yields poles in the s-plane at points that can be calculated from Eqs. 117, 120, and 121 by letting $\beta = (\pi/2) + \nu\pi/2n$ for ν equal to odd integer values. Left half-plane poles at the points $s_\nu = \sigma_\nu + j\omega_\nu$ are thus determined by the relations

$$\left.\begin{aligned}\sigma_\nu &= \sinh \delta \cos \frac{(n+\nu)\pi}{2n} \\ \omega_\nu &= \cosh \delta \sin \frac{(n+\nu)\pi}{2n}\end{aligned}\right\} \text{for } \nu = 1, 3, \cdots, 2n-1 \tag{124}$$

These are entirely equivalent to Eqs. 42. We lack in the present formulation only a correlation between the quantity δ and a corresponding ripple amplitude, which we shall not evaluate at the moment since it is an irrelevant detail in relation to our present discussion (it is given in the following article in connection with the presentation of a different type of problem). We wish rather to call attention to some more fundamental principles that are suggested by the present results.

In comparing the transcendental forms of the Butterworth and Tschebyscheff functions (Figs. 18 and 20), we observe that both are characterized by the demand of having a constant zero loss in the pass band. In the Butterworth case, however, the phase is left arbitrary, whereas the transcendental Tschebyscheff function evidently adds the constraint that the phase should be constant in the stop band. As has been pointed out, the single demand of $\alpha = 0$ in the pass band can be met in an endless variety of ways. When we add the phase constraint, $\beta =$ constant, in the stop band, there exists only one solution, namely, that provided by the conformal representation in Fig. 19, yielding the characteristics in Fig. 20.

Finally, it should be mentioned that the potential theory approach to the equiripple pass-band behavior yields a larger variety of possible rational approximants than does the approach by use of the Tschebyscheff polynomial. Thus we can vary the method of quantization by placing poles along more than one equipotential contour.* We can, for example, use a double string of poles or multiple-order poles. Such schemes allow us to obtain nearly the same results with poles farther to the left of the j-axis, which (as we have seen in Art. 6, Ch. 12) allows the absorption of a larger amount of parasitic loss.

9 Linear Phase Approximation

In the design of delay networks one is required to determine a transfer impedance whose phase angle is an approximately linear function of the

* For a further elaboration of these ideas and a more detailed potential theory interpretation of the use of the Jacobian elliptic function as well as the Tschebyscheff polynomial in filter design, see "A Note on the Solution of Certain Approximation Problems in Network Synthesis" by R. M. Fano, Tech. Rept. 62, Mass. Inst. Technol., Research Lab. Electronics, Apr. 16, 1948.

frequency ω from $\omega = 0$ to some specified upper limit depending upon the type of signal that the network is required to transmit with a minimum of distortion. Networks of this kind fall into two categories according to whether they are of the all-pass variety or not. In the transfer impedance

$$Z_{12}(s) = \frac{P(s)}{Q(s)} = \frac{m_1 + n_1}{m_2 + n_2} \tag{125}$$

of an all-pass network, $P(s) = Q(-s)$, so that $|Z_{12}(j\omega)| \equiv 1$.

Signal distortion in such a network depends solely upon the character of the phase function. We shall show in the next article, however, that one can modify an all-pass function to have a wide variety of amplitude characteristics without affecting the associated phase. By this means it is possible in a large measure to control the transient response of the pertinent network so as to meet the requirements of a particular application.

The lossless constant-resistance lattice is automatically an all-pass network, since for $z_a = jx_a$ we have (for $s = j\omega$)

$$Z_{12} = \frac{1 - z_a}{1 + z_a} = \frac{1 - jx_a}{1 + jx_a} = e^{-j\beta} \tag{126}$$

The relationship between the phase β and the reactance function x_a reads

$$x_a = \tan(\beta/2) \qquad \text{or} \qquad \beta = 2\tan^{-1} x_a \tag{127}$$

Since a reactance function must be a continuously increasing function of ω (Eq. 9, Ch. 2), we see that the phase of an all-pass network has this property also. This much is, of course, obvious from the pole-zero pattern of the all-pass function. In fact, it is thus clear that a monotonically increasing phase function (defined as phase *lag*, of course) must result for any transfer function having no zeros in the finite left half-plane.

We may obtain a somewhat more precise condition on the phase of an all-pass network from the following derivative of 127

$$\frac{dx_a}{d\omega} = \frac{1}{2\cos^2(\beta/2)}\frac{d\beta}{d\omega} \tag{128}$$

by making use of the property (Eq. 22, Ch. 2) of reactance functions

$$\frac{dx_a}{d\omega} > \frac{x_a}{\omega} \tag{129}$$

Thus we find

$$\frac{d\beta}{d\omega} > \left(\frac{\sin\beta}{\beta}\right)\frac{\beta}{\omega} \tag{130}$$

which should be regarded as supplementing the condition $d\beta/d\omega > 0$; the stronger one in any situation is the ruling condition. For small values of β for which $\sin \beta/\beta$ is near unity, the phase function must evidently fulfill a condition as stringent as 129 pertinent to reactance functions, whereas for large β, where $\sin \beta/\beta$ is practically zero, the slope of the phase may have any positive value, however small.

If the desired phase variation is to be free from these restrictions, the pertinent transfer impedance cannot be all pass. We may regard any arbitrary phase function as the sum of two components of which one is that associated with the amplitude of a minimum-phase function and the other is an all-pass phase function. In any given situation, the above conditions on the all-pass function enable one to separate the component (if any) that is associated with the amplitude function of a minimum-phase network.

In the design of a network for prescribed phase shift, it is sometimes advantageous to use a procedure in which the phase slope (delay) rather than the phase itself is controlled. To formulate this approach we begin with the familiar separation of the impedance 125 into even and odd parts

$$Z_{12}(s) = \frac{P(s)Q(-s)}{Q(s)Q(-s)} = \frac{M + N}{m_2^2 - n_2^2} \tag{131}$$

with

$$M = \tfrac{1}{2}[P(s)Q(-s) + P(-s)Q(s)] \tag{132}$$

$$N = \tfrac{1}{2}[P(s)Q(-s) - P(-s)Q(s)] \tag{133}$$

from which the lag angle of $Z_{12}(j\omega)$ is given by

$$-j \tan \beta = (N/M)_{s=j\omega} \qquad \text{or} \qquad -\beta = \tan^{-1} (N/jM)_{s=j\omega} \tag{134}$$

The derivative of this equation yields

$$-\frac{d\beta}{d\omega} = \frac{M^2}{M^2 - N^2} \times \frac{d}{ds}(N/M) = \frac{MN' - NM'}{M^2 - N^2} \tag{135}$$

where the prime denotes differentiation with respect to s and evaluation of the resulting expression for $s = j\omega$ is tacitly understood. The expression obtained here is recognized as being the even part of a rational function $(M' + N')/(M + N)$. Hence we see that the phase slope of a network may be written in the form*

$$\frac{d\beta}{d\omega} = -\text{Ev}\left(\frac{M' + N'}{M + N}\right) = +\text{Ev}\left[\frac{\phi'(s)}{\phi(s)}\right] \tag{136}$$

in which

$$\phi(s) = P(-s)Q(s) \tag{137}$$

and evaluation for $s = j\omega$ is again understood. A possible approach to delay network synthesis is to construct a polynomial $\phi(s)$ that meets specifications on $d\beta/d\omega$ according to relation 136. The polynomials $P(s)$

* Ev again means "even part of."

and $Q(s)$ are then found from Eq. 137 by the method discussed in Art. 10, Ch. 8.

If the network is to be all pass, $P(-s) = Q(s)$, and Eq. 137 gives

$$\phi(s) = Q^2(s), \qquad \phi'(s) = 2Q(s)Q'(s) \tag{138}$$

Equation 136 then yields

$$\frac{d\beta}{d\omega} = 2\text{Ev}\left[\frac{Q'(s)}{Q(s)}\right] \tag{139}$$

On the other hand, if the transfer impedance has all zeros at infinity, $P(s) = 1$ and $\phi(s) = Q(s)$, so that

$$\frac{d\beta}{d\omega} = \text{Ev}\left[\frac{Q'(s)}{Q(s)}\right] \tag{140}$$

which is the same as 139 pertinent to the all-pass network except for a factor of 2. This result we could, of course, have predicted on the basis of the interpretations given in the previous article.

A linear phase-shift network is one for which the transfer impedance has a lag angle β such that $d\beta/d\omega$ approximates the same ideal rectangle as does the magnitude of this impedance (Fig. 1) in the design of a low-pass filter. By analogy to the solution of that problem in terms of the Butterworth function, we may seek a design in which $d\beta/d\omega$ approximates the ideal rectangle in a maximally flat manner. Because of the special form of the rational function 136, however, it is not simple to see how such a solution may be obtained. In order to develop a better understanding of the problem, suppose we consider first a few particular examples.

Assume for $\phi(s)$ the second-degree polynomial

$$\phi(s) = 1 + a_1 s + a_2 s^2 \tag{141}$$

We have

$$\phi'(s) = a_1 + 2a_2 s \tag{142}$$

and for $s = j\omega$

$$\text{Ev}\left[\frac{\phi'(j\omega)}{\phi(j\omega)}\right] = \frac{a_1 + a_1 a_2 \omega^2}{1 + (a_1^2 - 2a_2)\omega^2 + a_2^2\omega^4} \tag{143}$$

A flat-flat result is obtained if the terms in the numerator polynomial are proportional to the respective ones in the denominator. This condition is expressed by

$$a_2 = a_1^2 - 2a_2 \qquad \text{or} \qquad a_2 = a_1^2/3 \tag{144}$$

for which we get the phase-slope function

$$\frac{d\beta}{d\omega} = a_1 \times \frac{1 + \dfrac{a_1^2\omega^2}{3}}{1 + \dfrac{a_1^2\omega^2}{3} + \dfrac{a_1^4\omega^4}{9}} \tag{145}$$

The range of ω over which this function approximates the constant value a_1 with a given tolerance evidently decreases as a_1 increases. We can have a large delay over a narrow band or a small delay over a wider band. We can have both, if we cascade identical constant-resistance lattices realizing the transfer impedance implied by 145, since the value of a_1 can be chosen to yield the required bandwith, and the total delay provided by n cascaded sections equals $n \times a_1$.

Alternately, one can assume a polynomial $\phi(s)$ of higher degree. For example

$$\phi(s) = 1 + a_1 s + a_2 s^2 + a_3 s^3 \tag{146}$$

for which

$$\phi'(s) = a_1 + 2a_2 s + 3a_3 s^2 \tag{147}$$

and for $s = j\omega$

$$\mathrm{Ev}\left[\frac{\phi'(j\omega)}{\phi(j\omega)}\right] = \frac{a_1 + (a_1 a_2 - 3a_3)\omega^2 + a_2 a_3 \omega^4}{1 + (a_1^2 - 2a_2)\omega^2 + (a_2^2 - 2a_1 a_3)\omega^4 + a_3^2 \omega^6} \tag{148}$$

The maximally flat requirement yields the equations

$$a_1 a_2 - 3a_3 = a_1(a_1^2 - 2a_2), \qquad a_2 a_3 = a_1(a_2^2 - 2a_1 a_3) \tag{149}$$

giving

$$a_2 = 2a_1^2/5 \qquad \text{and} \qquad a_3 = a_1^3/15 \tag{150}$$

The desired phase-slope function in this case reads

$$\frac{d\beta}{d\omega} = a_1 \times \frac{1 + \dfrac{a_1^2\omega^2}{5} + \dfrac{2a_1^4\omega^4}{75}}{1 + \dfrac{a_1^2\omega^2}{5} + \dfrac{2a_1^4\omega^4}{75} + \dfrac{a_1^6\omega^6}{225}} \tag{151}$$

Again, a desired tolerance and bandwith fixes a_1, equal to the resulting delay. This method is very straightforward in its application, although it becomes computationally tedious for polynomials $\phi(s)$ of higher degree.

An approach to this problem that yields approximately the same results and is computationally very much simpler is based upon the observation that if $\phi(s)$ approximates e^s, then $\phi'(s) \approx \phi(s)$, and Eq. 136 yields $d\beta/d\omega \approx 1$ over the approximation interval. More specifically, if we consider the following portion of a Maclaurin expansion of e^s as representing the polynomial $\phi(s)$

$$\phi(s) = 1 + s + \frac{s^2}{2} + \frac{s^3}{6} + \cdots + \frac{s^n}{n!} \tag{152}$$

then

$$\phi'(s) = 1 + s + \frac{s^2}{2} + \frac{s^3}{6} + \cdots + \frac{s^{n-1}}{(n-1)!} \tag{153}$$

The rational function $\phi'(s)/\phi(s)$ has a saddle point of order $n - 1$ at $s = 0$; the function $1 - (\phi'/\phi)$ has a zero of order n at $s = 0$, or the first $n - 1$ derivatives of ϕ'/ϕ are zero at $s = 0$. The even part of ϕ'/ϕ, which has the form

$$\mathrm{Ev}\left[\frac{\phi'(s)}{\phi(s)}\right] = \frac{1 + \alpha_1 s^2 + \alpha_2 s^4 + \cdots + \alpha_{n-1} s^{2(n-1)}}{1 + \beta_1 s^2 + \beta_2 s^4 + \cdots + \beta_n s^{2n}} \tag{154}$$

has a saddle point of order n (resp. $n - 1$) at $s = 0$ for n odd (resp. even). Hence we have

$$\alpha_k = \beta_k \qquad \text{for } k = 1, 2, \cdots \begin{cases} (n-1)/2 \text{ for } n \text{ odd} \\ (n-2)/2 \text{ for } n \text{ even} \end{cases} \tag{155}$$

which falls considerably short of fulfilling the maximally flat condition, although it yields one that is reasonably flat.

It is of collateral interest to point out that $\phi(s)$ in Eq. 152 fails to be a Hurwitz polynomial for $n > 4$. In this event its right half-plane zeros must, according to Eq. 137, be allotted to $P(-s)$, and we have no choice but to accept a transfer impedance 125 with (left half-plane) zeros as well as poles. So long as $\phi(s)$ is Hurwitz, we are free to construct a transfer impedance with or without finite (right half-plane) zeros, but any right half-plane zeros of $\phi(s)$ must become left half-plane zeros of $Z_{12}(s)$.

Although these considerations do not yield the desired solution, they point the way toward the following results. Since $\phi(s)$ in Eq. 136 is $(M - N)$, we see that requiring $\phi(s)$ to approximate e^s is, according to Eq. 134, equivalent to having the rational function $(-N/M) = \tanh j\beta$ approximate $\tanh s$. Since the latter function has simple poles with positive real residues restricted to the j-axis of the s-plane, it follows that the desired approximant $(-N/M)$ is a reactance function (LC realizable impedance).

The flat-flat requirement on the phase β may be interpreted as stipulating that we want the reactance function $(-N/M)$ to approximate $\tanh s$ in a manner that places all the emphasis upon the point $s = 0$. Our familiarity with the synthesis of reactance functions tells us how to proceed in order to accomplish this objective. Namely, we should manipulate $\tanh s$ as we would a reactance function when developing it into a ladder network by removing poles at $s = 0$ from this function (or its reciprocal) and from its successively formed reciprocal remainder functions.

Expressed in another way, we should manipulate $\tanh s$ as we would a reactance function when making the canonic Cauer development leading to a ladder network with series C's and shunt L's. That is to say, we should develop $\tanh s$ into a continued fraction in terms of the variable $1/s$. We can readily find this development by starting from the Maclaurin expansions in

$$\tanh s = \frac{\sinh s}{\cosh s} = \frac{s + \dfrac{s^3}{3!} + \dfrac{s^5}{5!} + \dfrac{s^7}{7!} + \cdots}{1 + \dfrac{s^2}{2!} + \dfrac{s^4}{4!} + \dfrac{s^6}{6!} + \cdots} \tag{156}$$

The familiar process of continued division and inversion yields

$$\tanh s = \cfrac{1}{(1/s) + \cfrac{1}{(3/s) + \cfrac{1}{(5/s) + \cfrac{1}{(7/s) + \cdots}}}} \tag{157}$$

In the first step of this process we remove the pole $(1/s)$ from $\coth s$, leaving a remainder $(\coth s - 1/s)$ that is zero at $s = 0$. Its reciprocal has the pole $(3/s)$ at $s = 0$, which is removed in the second step, and so forth. It is not hard to see that this development accents the point $s = 0$ exclusively.

If we terminate this development after n terms, we get a rational approximant that we can identify with $(-N/M)$, yielding a polynomial $\phi(s) = (M - N)$ of degree n. Observe, however, that the polynomials $-N$ and M are not the corresponding finite portions of the Maclaurin expansions in Eq. 156, as is already evident from the consideration that $\phi(s)$ formed from expansion 157 is always a Hurwitz polynomial (since the coefficients in 157 are all positive). Moreover, the Maclaurin expansion of $\tanh s$ which reads

$$\tanh s = s - \frac{s^3}{3} + \frac{2s^5}{15} - \frac{17s^7}{315} + - \cdots \tag{158}$$

is correctly obtained for n terms by an approximant $\phi(s)$ formed from n terms of expansion 157, whereas an approximant formed from 156 by considering terms through s^n yields a Maclaurin expansion that agrees with 158 for roughly only half as many terms, as may be expected also from the results expressed in Eqs. 152 through 155.

By use of the algebraic properties of continued fractions* one can obtain the following recursion formula for the construction of polynomials $\phi_n(s)$ corresponding to n terms of 157

$$\phi_n = (2n - 1)\phi_{n-1} + s^2\phi_{n-2} \tag{159}$$

Together with the particular ones

$$\phi_0 = 1 \qquad \text{and} \qquad \phi_1 = 1 + s \tag{160}$$

one can readily find those of higher degree.

---

* See Chrystal's *Text Book of Algebra*, Part II, Ch. 34, Adam & Charles Black, London, 1900, and Muir's *Theory of Determinants*, Ch. 3, Macmillan, New York, 1882.

It is further possible to express this polynomial explicitly in the form*

$$\phi_n(s) = \sum_{r=0}^{n} a_r s^r \tag{161}$$

with

$$a_r = \begin{bmatrix} 2n - r \\ r \end{bmatrix} \times \begin{bmatrix} 1 \cdot 3 \cdot 5 \cdots (2n - 2r - 1) \end{bmatrix} \tag{162}$$

The first factor is the familiar binomial coefficient

$$\binom{k}{t} = \frac{k(k-1)\cdots(k-t+1)}{t!} \tag{163}$$

which has the value unity for either $t = 0$ or $k = t$.

A more practical approach to the construction of a transfer impedance having an approximately linear phase is to consider an s-plane pole pattern like that shown in Fig. 21 where the total number of poles is, for

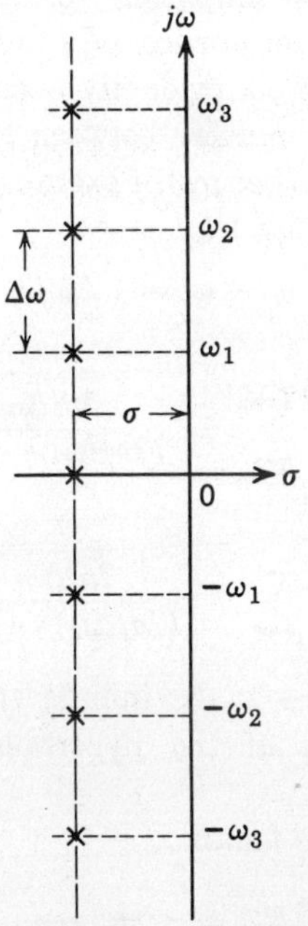

Fig. 21. Pole pattern yielding an approximately linear phase.

---

* See "An Application of Modern Network Synthesis to the Design of Constant-Time-Delay Networks with Low-Q Elements," Leo Storch, 1954, *IRE Convention Record, Part* 2, *Circuit Theory*, pp. 105–117. It is incidentally shown there that $\phi(s)$ is the so-called *Bessel* polynomial, and a rigorous proof of the maximally flat character of the phase β is given in terms of properties of these polynomials.

the moment, left unspecified. If a symmetrically placed linear array of zeros is added in the right half-plane, the total phase shift merely doubles, so that we need not separately consider this all-pass arrangement in the present analysis.

The net phase angle for this uniform pole pattern is evidently given by

$$\beta = \sum_{\nu} \tan^{-1} \frac{\omega - \omega_\nu}{\sigma} \tag{164}$$

where the summation extends over all poles. The phase slope is

$$\frac{d\beta}{d\omega} = \sum_{\nu} \frac{\sigma}{\sigma^2 + (\omega - \omega_\nu)^2} \tag{165}$$

If we consider the pole pattern to be infinite in extent, then the phase slope 165 has equiripple character. For a finite pattern it likewise will have an approximately equiripple variation except near the ends of the array. The resulting ripple amplitude for an infinite array will apply very nearly also to the central portion of a finite array.

A maximum value of $d\beta/d\omega$ evidently occurs for any $\omega = \omega_\nu$, and a minimum value for any ω midway between two successive ω_ν's. These two extreme values of $d\beta/d\omega$ occur, for example, for $\omega = 0$ and $\omega = \Delta\omega/2$ respectively. If in Eq. 165 we let

$$\omega - \omega_\nu = \nu \, \Delta\omega/2 \tag{166}$$

then we find in this way

$$\left(\frac{d\beta}{d\omega}\right)_{\max} = \frac{2}{\Delta\omega} \sum_{\nu=0,-2,-4,\cdots}^{2,4,\cdots} \frac{2\sigma/\Delta\omega}{(2\sigma/\Delta\omega)^2 + \nu^2} = \frac{\pi}{\Delta\omega} \coth \frac{\pi\sigma}{\Delta\omega} \tag{167}$$

and

$$\left(\frac{d\beta}{d\omega}\right)_{\min} = \frac{2}{\Delta\omega} \sum_{\nu=-1,-3,\cdots}^{1,3,\cdots} \frac{2\sigma/\Delta\omega}{(2\sigma/\Delta\omega)^2 + \nu^2} = \frac{\pi}{\Delta\omega} \tanh \frac{\pi\sigma}{\Delta\omega} \tag{168}$$

where except for a factor $\pi/2$, the infinite sums are recognized as the partial-fraction expansions of the hyperbolic cotangent and tangent functions respectively.

We thus obtain the ripple factor

$$\frac{(d\beta/d\omega)_{\max} - (d\beta/d\omega)_{\min}}{(d\beta/d\omega)_{\max} + (d\beta/d\omega)_{\min}} = \rho_p = \frac{1}{\cosh 2\pi\sigma/\Delta\omega} \approx 2e^{-2\pi\sigma/\Delta\omega} \tag{169}$$

while the arithmetic mean of the values 167 and 168 yields the average slope

$$t_d = \left(\frac{d\beta}{d\omega}\right)_{\text{av}} = \frac{\pi}{\Delta\omega} \coth 2\pi\sigma/\Delta\omega \approx \frac{\pi}{\Delta\omega} \tag{170}$$

which equals the delay produced by the pertinent network. The fact that the average slope is essentially independent of σ results from the assumption of an infinite array of poles. For a finite number of poles this event is still very nearly true so long as σ is small compared with the total bandwidth, a condition that is practically always fulfilled.

For a chosen average slope, fixing $\Delta\omega$, the ripple factor is controlled by the value of σ. In this regard it would seem offhand that we can reduce the ripple to any desired value without cost, since σ may be increased, within reasonable limits, as much as we wish. The fallacy in this viewpoint lies in the fact that it disregards the effect of σ upon the behavior of the phase near the band limits of the finite array of poles. As the following analysis shows, the average slope remains very nearly constant throughout most of the frequency range covered by the pole pattern, but decreases as ω nears the limits of this range, and equals half its normal value at these limits. The "end zone" throughout which this significant variation in the phase slope occurs becomes a larger fraction of the total bandwidth as the ratio $\sigma/\Delta\omega$ is increased.

Computation of the function $d\beta/d\omega$ so as to portray this end-zone behavior is facilitated by assuming a continuous pole distribution parallel to the ω-axis as shown in Fig. 22. The sum 165 is then replaced by the integral

$$\frac{d\beta}{d\omega} = \frac{\sigma}{\Delta\omega}\int_{-(\omega_c+\omega)}^{\omega_c-\omega} \frac{dx}{\sigma^2 + x^2} = \frac{1}{\Delta\omega}\int_{-(\omega_c+\omega)/\sigma}^{(\omega_c-\omega)/\sigma} \frac{d(x/\sigma)}{1 + (x/\sigma)^2} \tag{171}$$

Here ω is a parameter and x or x/σ is the variable of integration. We thus obtain

$$\frac{d\beta}{d\omega} = \frac{1}{\Delta\omega}\left(\tan^{-1}\frac{\omega_c + \omega}{\sigma} + \tan^{-1}\frac{\omega_c - \omega}{\sigma}\right) \tag{172}$$

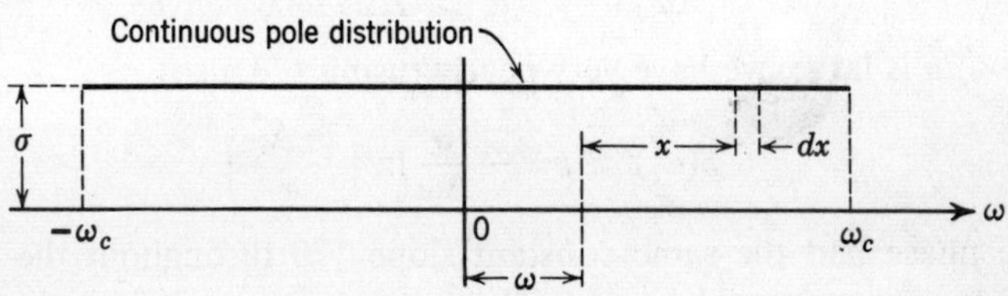

Fig. 22. Continuous pole distribution assumed for the purpose of calculating the variation in phase slope in the "end zones" near $\omega = \pm\omega_c$.

If the total number of discrete poles is denoted by $2n$, we may write

$$\omega_c = n\,\Delta\omega, \qquad \omega = \nu\,\Delta\omega \tag{173}$$

and have

$$\frac{d\beta}{d\omega} = \frac{1}{\Delta\omega}\left[\tan^{-1}(n + \nu)\frac{\Delta\omega}{\sigma} - \tan^{-1}(n - \nu)\frac{\Delta\omega}{\sigma}\right] \tag{174}$$

At the band center ($\omega = 0$ or $\nu = 0$) this slope is essentially equal to $\pi/\Delta\omega$, which agrees with 170. At the band limits ($\omega = \omega_c$ or $\nu = n$), it equals half this value.

Since the value of $\Delta\omega/\sigma$, according to the ripple factor 169, will in most cases lie between 1 and 2, Eq. 174 shows that the region of major variation in $d\beta/d\omega$ involves only the last two or three poles in the array. A reasonably satisfactory result is therefore obtained by simply extending the pole pattern this much beyond the desired band limits. A more exacting requirement can be met by inserting one or two extra poles near the end of the array so as to supplement the sum 165 by appropriately chosen additional terms. This process is best carried out by plotting the function 174 in the end zone and then adding terms like those in 165 with σ and ω_ν values selected on a cut-and-try basis. Since one or two extra terms will usually suffice, the computational work involved is very reasonable, and no elaborate analytical procedure is called for.

If the behavior of β versus ω is desired, one can readily obtain it through integration of expression 172. This yields

$$\beta(\omega) = \frac{1}{\Delta\omega}\int_0^\omega \tan^{-1}\frac{\omega_c + \omega}{\sigma}\,d\omega + \frac{1}{\Delta\omega}\int_0^\omega \tan^{-1}\frac{\omega_c - \omega}{\sigma}\,d\omega \tag{175}$$

or

$$\beta(\omega) = \frac{\omega_c + \omega}{\Delta\omega}\tan^{-1}\frac{\omega_c + \omega}{\sigma} - \frac{\omega_c - \omega}{\Delta\omega}\tan^{-1}\frac{\omega_c - \omega}{\sigma}$$
$$- \frac{\sigma}{2\Delta\omega}\ln\left[1 + \left(\frac{\omega_c + \omega}{\sigma}\right)^2\right] + \frac{\sigma}{2\Delta\omega}\ln\left[1 + \left(\frac{\omega_c - \omega}{\sigma}\right)^2\right] \tag{176}$$

At the band limit $\omega = \omega_c$, the value of β is given by

$$\beta(\omega_c) = \frac{2\omega_c}{\Delta\omega}\tan^{-1}\frac{2\omega_c}{\sigma} - \frac{\sigma}{2\Delta\omega}\ln\left(1 + \frac{4\omega_c^2}{\sigma^2}\right) \tag{177}$$

Since ω_c/σ is large, we have very nearly (using 173 also)

$$\beta(\omega_c) \approx n\pi - \frac{\sigma}{\Delta\omega}\ln\left(\frac{2n\,\Delta\omega}{\sigma}\right) \tag{178}$$

If the phase had the same constant slope 170 throughout the band, its value at $\omega = \omega_c$ would be $n\pi$. Hence, the amount by which the phase drops below a linear variation at the band limit is

$$n\pi - \beta(\omega_c) \approx \frac{\sigma}{\Delta\omega}\ln\left(\frac{2n\,\Delta\omega}{\sigma}\right) \tag{179}$$

which is a small fraction of $n\pi$. In fact, as a fraction of $n\pi$, the value 179 tends toward zero with increasing n.

If the pole pattern in Fig. 21 is supplemented by a symmetrically placed linear array of right half-plane zeros, then, as mentioned above, the phase merely doubles, and in that case the associated amplitude is constant, for the transfer function has all-pass character. If the pole pattern in Fig. 21 is not so supplemented, then the transfer function has an amplitude variation which, for an infinite array, is evidently constant except for an equiripple variation. The pertinent ripple factor may be evaluated in the following manner.

The squared magnitude function for the pole pattern in Fig. 21 is given by

$$f(\omega) = \frac{K}{\prod_{\nu} [\sigma^2 + (\omega - \omega_\nu)^2]} \tag{180}$$

in which the product extends over all poles in the array. We will assume tentatively that the array consists of a total of $2n + 1$ poles, one corresponding to $\omega = 0$, and n in each of the positive and negative branches. The ripple factor for an infinite array will then be obtained for the limit $n \to \infty$.

Anticipating this limit process, a maximum value of $f(\omega)$ occurs for $\omega = 0$, and a minimum value for $\omega = \omega_1/2$. Again making use of 166, these values, according to Eq. 180, are given by

$$f_{\max} = \frac{K}{\sigma^2 \prod_{\nu=2,4,\ldots}^{2n} \left(\frac{\nu\,\Delta\omega}{2}\right)^4 \prod_{\nu=2,4,\ldots}^{2n} \left[1 + \frac{(2\sigma/\Delta\omega)^2}{\nu^2}\right]^2} \tag{181}$$

and

$$f_{\min} = \frac{K}{[\sigma^2 + (2n+1)^2\,\Delta\omega^2/4] \prod_{\nu=1,3,\ldots}^{2n-1} \left(\frac{\nu\,\Delta\omega}{2}\right)^4 \prod_{\nu=1,3,\ldots}^{2n-1} \left[1 + \frac{(2\sigma/\Delta\omega)^2}{\nu^2}\right]^2} \tag{182}$$

Making use of the infinite product expansion of the hyperbolic cotangent function, we find for $n \to \infty$

$$\sqrt{f_{\max}/f_{\min}} = m\frac{\pi}{2} \coth \frac{\pi\sigma}{\Delta\omega} \tag{183}$$

in which the factor m is the limit for $n \to \infty$ of the following expression

$$m = \frac{(2n+1) \prod_{\nu=1,3,\ldots}^{2n-1} \nu^2}{\prod_{\nu=2,4,\ldots}^{2n} \nu^2} \tag{184}$$

It can be shown that this factor approaches the limit $2/\pi$, as may crudely be appreciated from specific examples by noting that for any finite n, however large, we have

$$m = \frac{1\cdot 3}{2\cdot 2} \times \frac{3\cdot 5}{4\cdot 4} \times \frac{5\cdot 7}{6\cdot 6} \times \frac{7\cdot 9}{8\cdot 8} \times \frac{9\cdot 11}{10\cdot 10} \cdots \tag{185}$$

The magnitude ratio for an infinite pole array is, therefore, seen to be

$$\sqrt{f_{\max}/f_{\min}} = \coth \frac{\pi\sigma}{\Delta\omega} \tag{186}$$

and the ripple factor becomes

$$\frac{\sqrt{f_{\max}/f_{\min}} - 1}{\sqrt{f_{\max}/f_{\min}} + 1} = \rho_m = e^{-2\pi\sigma/\Delta\omega} \tag{187}$$

which is just half the ripple factor for the phase slope as given by Eq. 169.

By itself this result is not particularly useful since the average value of the magnitude throughout the frequency range covered by a finite pole array is far from being constant. In terms of the continuous pole distribution in Fig. 22 one can evaluate the logarithm of 180 by an integral and find the loss at the band limits relative to that at $\omega = 0$ to be

$$\alpha(\omega_c) \approx (4\omega_c/\Delta\omega) \ln 2 - (\pi\sigma/\Delta\omega) \text{ nepers} \tag{188}$$

which is so considerable that one can hardly refer to this range as a "pass band." However, if the pole array in Fig. 21 is supplemented by a similar distribution of right half-plane zeros with the same spacing $\Delta\omega$ but for a different value of σ, then the magnitude of the transfer function, although not identically equal to unity, is substantially constant and nearly unity throughout the pass band. Such a situation may arise if we design for an all-pass function but fail to precorrect for incidental losses (see Art. 6, Ch. 12), so that the resulting poles are farther to the left of the j-axis than the zeros are to the right of it.

The above analysis enables us at once to state that if the pole distance is σ_1 and the zero distance σ_2, then we will have a max/min amplitude ratio given by

$$\sqrt{f_{\max}/f_{\min}} = \frac{\coth(\pi\sigma_1/\Delta\omega)}{\coth(\pi\sigma_2/\Delta\omega)} \tag{189}$$

For this same situation, Eqs. 167 and 168 enable us to write

$$\left(\frac{d\beta}{d\omega}\right)_{\max} = \frac{\pi}{\Delta\omega}\left(\coth\frac{\pi\sigma_1}{\Delta\omega} + \coth\frac{\pi\sigma_2}{\Delta\omega}\right) \tag{190}$$

and

$$\left(\frac{d\beta}{d\omega}\right)_{\min} = \frac{\pi}{\Delta\omega}\left(\tanh\frac{\pi\sigma_1}{\Delta\omega} + \tanh\frac{\pi\sigma_2}{\Delta\omega}\right) \tag{191}$$

from which we get

$$\left(\frac{d\beta}{d\omega}\right)_{\max} \Big/ \left(\frac{d\beta}{d\omega}\right)_{\min} = \coth\frac{\pi\sigma_1}{\Delta\omega}\coth\frac{\pi\sigma_2}{\Delta\omega} \tag{192}$$

The ripple factor may readily be computed from this result.

10 An Alternate Derivation of the Equiripple Approximant and Some Useful Applications

If we are interested in a rational function having equiripple character, we do not need to look far to find one. The familiar circuit in Fig. 23 consisting of a resistance of 1 ohm in parallel with a reactance jx yields an impedance $Z(j\omega)$ whose magnitude or real part oscillates between 0 and 1 as ω varies from 0 to ∞. The zeros occur where the reactance has zeros and the "ones" where the reactance has poles. Our knowledge of reactance functions and their synthesis enables us at once to construct a function $Z(s)$ whose j-axis magnitude or real part has zeros and "ones" at prescribed points.

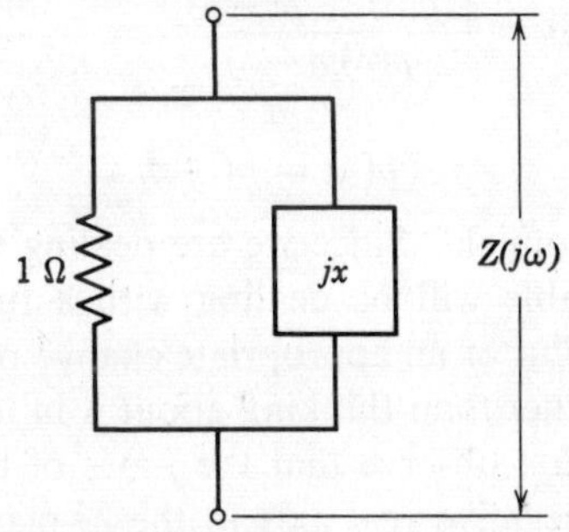

Fig. 23. This impedance magnitude or real part has equiripple character over the entire j-axis (jx is any reactance function).

If $p(s) = m + n$ is a Hurwitz polynomial, then the pertinent impedance may be written

$$Z(s) = \frac{m}{m+n} \tag{193}$$

if we want $\omega = 0$ to be a "one." If the numerator is chosen to be n instead of m, then $\omega = 0$ becomes a zero. From 193 it follows that

$$\text{Re}\,[Z(j\omega)] = |Z(j\omega)|^2 = \left(\frac{m^2}{m^2 - n^2}\right)_{s=j\omega} = \left(\frac{1}{1 - n^2/m^2}\right)_{s=j\omega} \tag{194}$$

Since the ratio of the even and odd parts of a Hurwitz polynomial is a reactance function, we may write

$$jx = (m/n)_{s=j\omega} \tag{195}$$

and have

$$Z(j\omega) = \frac{jx}{1 + jx} \tag{196}$$

whence

$$\text{Re}\,[Z(j\omega)] = |Z(j\omega)|^2 = \frac{x^2}{1 + x^2} \tag{197}$$

This function, of course, has equiripple character over the entire j-axis, whereas the usual approximation problem calls for this kind of behavior over only a finite interval on the j-axis. Accommodation of this requirement is readily achieved through an appropriate transformation of the independent variable that maps the entire j-axis in the complex plane of that variable upon the desired portions of the j-axis in the plane of the new variable. As we shall show, such a transformation is not difficult to construct.

In order to avoid confusion in notation, we shall denote by $\lambda = \bar{\sigma} + j\bar{\omega}$ the original independent variable, and by $s = \sigma + j\omega$ the new one. The rational function having equiripple character throughout its entire j-axis may be written

$$f(\lambda^2) = \frac{\frac{1}{4}[p(\lambda) + p(-\lambda)]^2}{p(\lambda)p(-\lambda)} = \frac{m^2}{m^2 - n^2} \tag{198}$$

where

$$p(\lambda) = m + n \tag{199}$$

is any Hurwitz polynomial. Since we are dealing with a function of λ^2, and in the new variable will be dealing with a function of s^2, we can simplify the construction of an appropriate change of variable by expressing it as $\lambda^2 = t(s^2)$ rather than thinking about it in terms of λ and s.

In this regard we first observe that the j-axis of the λ-plane or s-plane is equivalent to the negative real axis in the λ^2-plane or in the s^2-plane. Hence, if we wish to map the entire j-axis of the λ-plane upon a portion of the j-axis in the s-plane, we need merely superimpose the λ^2-plane upon the s^2-plane so that their real axes coincide, but leave part of the negative real axis of the s^2-plane untouched by the negative real axis of the λ^2-plane. Only that portion of the negative real axis of the s^2-plane coinciding with the negative real axis of the λ^2-plane will, in the corresponding s-plane, become the j-axis region in which the transformed rational function exhibits equiripple behavior.

In our initial thinking about the construction of this transformation $\lambda^2 = t(s^2)$, we need not be particularly concerned as to where on the i-axis of the s-plane the equiripple behavior takes place, since we can always adjust this item subsequently through use of a further transformation of the sort described in Art. 4. Figure 24, for example, shows a superposition of the λ^2- and s^2-planes that results in the entire j-axis of the λ-plane being mapped upon that portion of the j-axis of the s-plane corresponding to $s^2 < -1$, which is $|\omega| > 1$. The transformation appropriate to this mapping property evidently reads

$$\lambda^2 = s^2 + 1 \tag{200}$$

If, instead, we wish to have the j-axis of the λ-plane mapped upon the interval $-1 < \omega < 1$ on the j-axis of the s-plane, we have but to replace s by $1/s$ in Eq. 200. The appropriate change of variable for the rational function 198 then reads

$$\lambda^2 = \frac{s^2 + 1}{s^2} \tag{201}$$

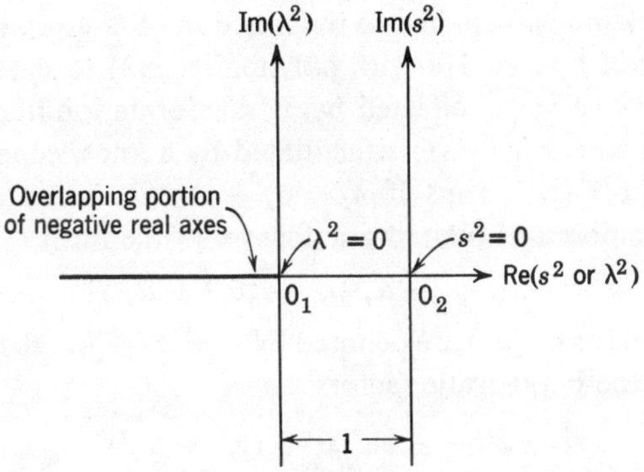

Fig. 24. Superposition of λ^2- and s^2-planes such that the entire j-axis in the λ-plane maps upon the portion $|\omega| > 1$ of the j-axis in the s-plane. The points 0_1 and 0_2 are respectively the origin of the λ^2-plane and that of the s^2-plane.

Significant for our purposes is the fact that the desired transformation is rational, so that the rational function in λ^2 (Eq. 198) remains rational when expressed as a function of s^2. In terms of λ or s, the transformation 200 or 201 is, of course, irrational and hence multivalued in both λ and s, but fortunately the functions that we are concerned with in our network approximation problem are almost exclusively functions of the frequency variable squared.

The function 198 has double zeros on the negative real axis of the λ^2-plane, and poles symmetrically disposed about the real axis of that plane. Pertinent to the transformation 201, these critical frequencies have precisely the same disposition in the $(1/s^2)$-plane, and hence the transformed function $F(s^2)$ for either transformation 200 or 201 has the form

$$F(s^2) = \frac{P(s^2)}{Q(s)Q(-s)} \tag{202}$$

Specifically, the change-of-variable 200 transforms $f(\lambda^2)$ in Eq. 198 into $F(s^2)$ such that

$$\tfrac{1}{4}[p(\lambda) + p(-\lambda)]^2 \rightarrow P(s^2) \tag{203}$$

and

$$p(\lambda)p(-\lambda) \rightarrow Q(s)Q(-s) \tag{204}$$

but one must not conclude that $p(\lambda)$ is carried directly over into $Q(s)$, for the transformation 200 can uniquely carry only a function of λ^2 over into a function of s^2, and $p(\lambda)$ and $Q(s)$ are not polynomials in λ^2 or s^2. Nevertheless, the polynomial $p(\lambda)$, which completely fixes $f(\lambda^2)$ in Eq. 198, uniquely determines $Q(s)$ if we specify that it shall be a Hurwitz polynomial. In this event $Q(s)$ is constructed from the left half-plane zeros of the function of s^2 represented by $Q(s)Q(-s)$, the latter resulting from the substitution 200 applied to the function of λ^2 given by $p(\lambda)p(-\lambda)$.

This conversion from a Hurwitz polynomial $p(\lambda)$ to a Hurwitz polynomial $Q(s)$, or vice versa, effected by a transformation like 200 applied to $p(\lambda)p(-\lambda)$ or to $Q(s)Q(-s)$, is facilitated by a knowledge of the zeros of either $p(\lambda)$ or $Q(s)$. Thus if $\lambda_\nu = \bar{\sigma}_\nu \pm j\bar{\omega}_\nu$ are zeros of $p(\lambda)$, then $p(\lambda)p(-\lambda)$ is composed of biquadratic factors of the form

$$\lambda^4 - 2(\bar{\sigma}_\nu^2 - \bar{\omega}_\nu^2)\lambda^2 + (\bar{\sigma}_\nu^2 + \bar{\omega}_\nu^2)^2 \tag{205}$$

Similarly, if the zeros of $Q(s)$ are denoted by $s_\nu = \sigma_\nu \pm j\omega_\nu$, then $Q(s)Q(-s)$ is composed of the biquadratic factors

$$s^4 - 2(\sigma_\nu^2 - \omega_\nu^2)s^2 + (\sigma_\nu^2 + \omega_\nu^2)^2 \tag{206}$$

The substitution 200 converts the biquadratic factor 205 into a biquadratic factor in s^2 which we identify with 206, and thus solve for σ_ν and ω_ν in terms of $\bar{\sigma}_\nu$ and $\bar{\omega}_\nu$ or vice versa. From these s_ν values we can then construct $Q(s)$; and if we need the substitution 201 instead of 200, we simply replace s by $1/s$ afterward. Any other transformation described in Art. 4 may be used in this final step to obtain a polynomial $Q(s)$ appropriate to the application in hand.

Identification of 205 and 206 through the substitution 200 thus gives

$$\begin{aligned} \sigma_\nu^2 - \omega_\nu^2 &= \bar{\sigma}_\nu^2 - \bar{\omega}_\nu^2 - 1 \\ \sigma_\nu^2 + \omega_\nu^2 &= \sqrt{(\bar{\sigma}_\nu^2 - \bar{\omega}_\nu^2 - 1)^2 + 4\bar{\sigma}_\nu^2\bar{\omega}_\nu^2} \end{aligned} \tag{207}$$

Solution for σ_ν and ω_ν values from given $\bar{\sigma}_\nu$ and $\bar{\omega}_\nu$ values is straightforward. There can be no ambiguity about algebraic signs since we know that the desired polynomial $Q(s)$ is Hurwitz.

Construction of $Q(s)$ by this process fails if the function $Q(s)Q(-s)$ resulting from $p(\lambda)p(-\lambda)$ has j-axis zeros of odd multiplicity, for then it is impossible for $Q(s)$ and $Q(-s)$ to share these equally. Occurrence of this situation requires that $p(\lambda)p(-\lambda)$ have zeros of odd multiplicity on the real axis of the λ^2-plane in the range $\lambda^2 < 1$. Although the postulated separate existence of $p(\lambda)$ precludes the possibility of $p(\lambda)p(-\lambda)$ having zeros of odd multiplicity on the j-axis of the λ-plane (corresponding to the negative real axis of the λ^2-plane), this polynomial can have such zeros in the range $0 < \lambda^2 < 1$. In this event, $p(\lambda)$ has odd-multiplicity zeros on the real axis of the λ-plane in the range $-1 < \lambda < 1$.

We can, of course, still form the polynomial $Q(s)Q(-s)$ in situations of this sort, and hence construct the desired approximant, but the polynomial $Q(s)$ has no separate existence. This fact may be an important consideration in certain applications, and the pertinent restriction on the zeros of $p(\lambda)$ is then significant.

Conversion in the opposite direction from a given polynomial $Q(s)$, where we are interested in the construction of $p(\lambda)$, is not subject to a similar restriction since no part of the positive real axis of the s^2-plane overlaps with the negative real axis of the λ^2-plane (see Fig. 24). In this situation the substitution 200 together with 205 and 206 yields

$$\begin{aligned} \bar{\sigma}_\nu^2 - \bar{\omega}_\nu^2 &= \sigma_\nu^2 - \omega_\nu^2 + 1 \\ \bar{\sigma}_\nu^2 + \bar{\omega}_\nu^2 &= \sqrt{(\sigma_\nu^2 - \omega_\nu^2 + 1)^2 + 4\sigma_\nu^2\omega_\nu^2} \end{aligned} \tag{208}$$

If here the substitution 201 rather than 200 is desired, then the pertinent s_ν values must be understood to be zeros of $Q(1/s)$, which are reciprocals of the zeros of $Q(s)$.

In the s-plane the function $F(s^2)$ has a pole-zero pattern with symmetry about both the real and imaginary axes (also called "quadrantal" symmetry) as is likewise true of the pole-zero pattern of $f(\lambda^2)$ in the λ-plane.

For the substitution 201, $F(-\omega^2)$ has equiripple character in the range $-1 < \omega < 1$. At the points $\omega = \pm 1$, corresponding to $\lambda = 0$, $F(-\omega^2)$ always equals $+1$, as may best be seen from Eq. 198 by noting that $n = 0$ for $\lambda = 0$. In the range $1 < |\omega| < \infty$, corresponding to $0 < \lambda^2 < 1$, the behavior of $F(-\omega^2)$ may be recognized by observing from Fig. 25 that the squared reactance function n^2/m^2 in most cases

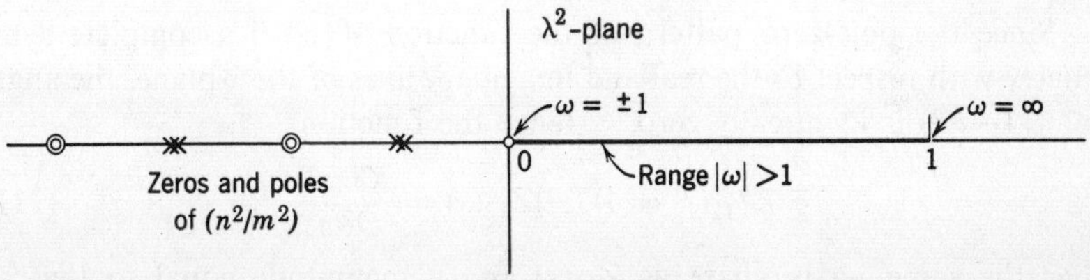

Fig. 25. Pertinent to evaluation of the behavior of $F(-\omega^2)$ in the range $|\omega| > 1$ when the transformation 201 applies. (Note that the function n^2/m^2 has all double-order zeros and poles except at $\lambda^2 = 0$, which point is a simple zero.)

increases monotonically (starting from zero at $\omega = \pm 1$) throughout this range, and hence that $f(\lambda^2)$, which assumes the same values as $F(-\omega^2)$, increases monotomically also and will pass through the value ∞ if n^2/m^2 surpasses the value unity. The precise behavior of $F(-\omega^2)$ in this range, therefore, depends upon the polynomial $p(\lambda)p(-\lambda)$ or its correspondent $Q(s)Q(-s)$ in the s-plane.

If n^2/m^2 passes through the value unity at some positive $\lambda^2 < 1$, then

$p(\lambda)p(-\lambda)$ has a simple zero in this range, and we have the situation described above in which $Q(s)$ has no separate existence. In an application in which we begin with the polynomial $Q(s)$, we see, therefore, that the quantity n^2/m^2 cannot surpass unity in the range $0 < \lambda^2 < 1$, although it may equal this value at discrete points.

An interesting application of the^e results arises if we are required to construct the transfer impedance for a low-pass filter that is to have an equiripple amplitude response combined with a linear or arbitrary phase behavior in the pass band.† We begin by constructing the all-pass function

$$Z_{12}(s) = \frac{Q(-s)}{Q(s)} \tag{209}$$

so as to meet the desired phase characteristic. For the polynomial $Q(s)$ thus determined, we construct $p(\lambda)$ according to the above process, and from its even part obtain $P(s^2)$ in Eq. 203 and hence $F(s^2)$ given by Eq. 202. We then form the function

$$W(s^2) = K[1 - \epsilon F(s^2)] = \frac{KT(s^2)}{Q(s)Q(-s)} \tag{210}$$

choosing values for ϵ and K that yield a function $W(-\omega^2)$ with an acceptable behavior so far as the desired amplitude response of the filter is concerned. In this connection it should be observed that $F(-\omega^2)$ will remain bounded in the range $|\omega| > 1$ since the polynomial $Q(s)$, in this application, has no j-axis zeros. Hence a value for ϵ can be chosen such that $\epsilon F(-\omega^2) \leq 1$. The value of K is not critical, it being simply an arbitrary multiplier.

Since the pole-zero pattern of the function $W(s^2)$ has complete symmetry with respect to the real and imaginary axes of the s-plane, the angle of $W(-\omega^2)$ is identically zero. Hence the function

$$Z^*{}_{12}(s) = W(s^2)Z_{12}(s) = \frac{KT(s^2)}{Q^2(s)} \tag{211}$$

has the same j-axis phase as $Z_{12}(s)$ and a magnitude equal to that of $W(-\omega^2)$. It is, therefore, the desired approximant.

Another interesting application of this procedure is recognized if we split the symmetrical zero pattern of $T(s^2)$ by writing

$$T(s^2) = t(s)t(-s) \tag{212}$$

in which $t(s)$ and $t(-s)$ represent respectively the left and right half-plane

---

† See "Synthesis of Electric Filters with Arbitrary Phase Characteristics," B. J. Bennett, 1953 *IRE Convention Record, Part 5, Circuit Theory*, pp. 19–26. Also *IRE PGCT Trans.*, Vol. CT-2, No. 2, June 1955, pp. 217–218. Although based upon the method given in this paper, the discussion in this text presents the method in a somewhat different and (we hope) clearer manner.

zeros. We can then split off the left half-plane part of $W(s^2)$, Eq. 210, as the minimum-phase transfer function

$$H(s) = \frac{Kt(s)}{Q(s)} \tag{213}$$

having a squared j-axis magnitude equal to the magnitude of $W(-\omega^2)$.

Since the polynomial $Q(s)$ may be chosen arbitrarily, we have here another method (alternate to that discussed in Art. 6) for constructing the approximant appropriate to a desired filter design with prechosen pole locations. An important application of this method* is concerned with the approximation problem arising in the design of RC networks for which the poles are restricted to the negative real axis.

If we use the method of Art. 6 to construct $H(s)$ to have a desired j-axis magnitude for prescribed poles determined by $Q(s)$, we see that it is possible to associate a wide variety of amplitude functions with a given phase. These results are thus seen to supplement the discussion in Art. 11, Ch. 8, regarding the extent to which amplitude and phase characteristics can be controlled independently.

The process discussed in this article may also be used as an alternate means for the construction of an impedance magnitude having equiripple character in both the stop and pass bands. The solution to this problem given in Art. 5 above makes use of the properties of Jacobian elliptic functions. The present method† yields the same results in a less roundabout manner, and computationally without recourse to elliptic function tables.

Consider the function $F(\omega)$ given by Eq. 68 and illustrated for $n = 2$ in Fig. 12. The zeros and poles of this function are located exclusively on the j-axis of the complex frequency plane. The zeros are confined to the pass band ($-1 < \omega < 1$) and the poles to the stop band ($|\omega| > 1$). Locations of the zeros, moreover, are reciprocal to the locations of the poles. With respect to the squared impedance magnitude given by Eq. 9, a function having this special form will (as pointed out in Art. 5) *automatically* yield equiripple character in the stop band if the zeros in the pass band are properly located to yield an equiripple response there for the function $F(\omega)$. In other words, the problem of constructing a low-pass rational approximant having equiripple properties in both pass and stop bands, according to the expression for this function given in Eq. 9,

* First suggested by G. L. Matthaei in "A General Method for Synthesis of Filter Transfer Functions," Stanford University Dissertation, 1952, and Office of Naval Research Rep. No. 39.

† Based upon the paper: "A Note on Filter Synthesis," B. J. Bennett, *IRE PGCT Trans.*, Vol. CT-1, No. 3, Sept. 1954, pp. 61–64.

reduces to that of determining the critical frequencies in $F(\omega)$, described in Eq. 68 or 69, so that it exhibits equiripple behavior in the pass band alone.

The method discussed in this article enables one to construct a function like $F(\omega)$ having equiripple pass-band behavior for any specified set of poles. By specifying that the poles shall lie upon the j-axis at points that are reciprocal to the resulting zero locations, we constrain $F(\omega)$ to have the desired form. A computational hazard is encountered in carrying out this scheme because of the fact that we cannot compute the zero locations without specifying a set of poles, and we cannot specify the poles to be at points reciprocal to the zero locations until we have determined the zeros. We can, of course, write analytic expressions demanding this condition, and hope to solve for the unknown parameters, but this process leads to a set of highly nonlinear simultaneous equations whose straightforward solution is practically hopeless. It is more effective to assume some pole locations to start with, calculate zeros by the method of the present procedure, and from these determine a set of revised pole locations. Repetition of this computational process is straightforward, and converges rapidly upon the desired solution.

In the following detailed discussion of this procedure we shall consider the high-pass instead of the low-pass problem because the substitution 200 is a little easier to work with than 201, and one can always convert to the pertinent low-pass function afterward if necessary. It is further important to observe that the process described in this article yields a function for which the boundaries of the equiripple region are the points $\omega = \pm 1$ whereas the functions 68 and 69 are frequency-scaled in such a way that the points $\omega = \pm\sqrt{k}$ are the boundaries of the equiripple region (these become the points $\omega = \pm 1/\sqrt{k}$ for the corresponding high-pass functions obtained through replacing ω by $1/\omega$). Before comparing the present results with those obtained by the method of Art. 5, it is necessary to revise the frequency scale accordingly.

For a high-pass filter the range $-1 < \omega < 1$ is the stop band. The assumed pole positions $\omega_1, \omega_3, \omega_5, \cdots$ for a function like 68 with ω replaced by $1/\omega$, are assumed to be located in this interval. These are the poles $s_\nu = j\omega_\nu$ for which we must compute the corresponding critical frequencies λ_ν in the λ-plane through use of Eqs. 208. Since $\sigma_\nu = 0$, these equations are seen to yield $\bar{\omega}_\nu = 0$ and

$$\bar{\sigma}_\nu = \pm\sqrt{1 - \omega_\nu^2} \tag{214}$$

The pair of left-half λ-plane points corresponding to $s_\nu = \pm j\omega_\nu$ are $\lambda_\nu = -\bar{\sigma}_\nu \pm j\bar{\omega}_\nu$, but since $\bar{\omega}_\nu = 0$, these are coincident and hence produce a quadratic factor of the form $(\lambda + \bar{\sigma}_\nu)^2$. The Hurwitz polynomial $p(\lambda)$

is, therefore, given by

$$p(\lambda) = (\lambda + \bar{\sigma}_1)^2(\lambda + \bar{\sigma}_3)^2 + \cdots + (\lambda + \bar{\sigma}_{2n-1})^2 \qquad (215)$$

in which the $\bar{\sigma}_\nu$ values are computed from given or assumed values ω_1, $\omega_3, \cdots, \omega_{2n-1}$ by Eq. 214.

The λ^2-plane zeros of our equiripple function, according to Eqs. 202 and 203, are those of the even part of $p(\lambda)$, which we can readily evaluate from expression 215, and the corresponding s^2-plane or s-plane zeros are then easily computed through use of the substitution 200. The basis for the pertinent iterative process is thus established.

Observe that this procedure yields an even function like 68 because the point $s = \infty$ (corresponding to $\lambda = \infty$) yields $f(\lambda^2) = +1$ in Eq. 198 (since the even part of 215 is always one degree higher than the odd part) and hence we have $F(-\omega^2) = +1$ for $\omega = \infty$. In other words, the "mid-band value" of our equiripple function is a point of maximum departure from zero. For our present high-pass function "mid-band" is the point $\omega = \infty$; for the low-pass function 68 it is $\omega = 0$.

In order to obtain an odd function like 69, we must start out with one of the assumed poles on the j-axis of the s-plane at $\omega = 0$, and the rest at $\omega_2, \omega_4, \cdots, \omega_{2n}$. For the pole at $\omega = 0$, Eq. 214 gives $\bar{\sigma} = \pm 1$, and the Hurwitz polynomial $p(\lambda)$ correspondingly contains the simple factor $(\lambda + 1)$. Specifically, we have in place of 215

$$p(\lambda) = (\lambda + 1)(\lambda + \bar{\sigma}_2)^2(\lambda + \bar{\sigma}_4)^2 \cdots (\lambda + \bar{\sigma}_{2n})^2 \qquad (216)$$

where the $\bar{\sigma}_2, \bar{\sigma}_4, \cdots, \bar{\sigma}_{2n}$ are again computed from the given or assumed $\omega_2, \omega_4, \cdots, \omega_{2n}$ by Eq. 214.

We again form the even part of $p(\lambda)$, find its λ^2 zeros, convert to s^2 values by Eq. 200, and thus compute the s-plane zeros. The reciprocals of these values are the revised pole positions. These are not reciprocals with respect to unity, however, since unity marks the boundary of the equiripple range (as pointed out above) and does not correspond to unity in the frequency scale of the functions 68 and 69, or in Fig. 12. We must reciprocate with respect to a constant k, which plays the same role as in the discussion given in Art. 5. Some examples will now serve best to fill in the missing details.

Suppose we begin with the construction of an odd function (the high-pass equivalent to 69) for $n = 1$. In the s-plane we have one pole at $\omega = 0$ and a pair at $s = \pm j\omega_2$ with $0 < \omega_2 < 1$. Equation 214 gives

$$\bar{\sigma}_2 = \sqrt{1 - \omega_2{}^2} \qquad (217)$$

and according to Eq. 216

$$p(\lambda) = (\lambda + 1)(\lambda + \bar{\sigma}_2)^2 = \lambda^3 + (1 + 2\bar{\sigma}_2)\lambda^2 + (2\bar{\sigma}_2 + \bar{\sigma}_2{}^2)\lambda + \bar{\sigma}_2{}^2 \qquad (218)$$

The even part

$$m(\lambda) = (1 + 2\bar{\sigma}_2)\lambda^2 + \bar{\sigma}_2^2 \tag{219}$$

has the λ^2 zero

$$\lambda_2^2 = \frac{-\bar{\sigma}_2^2}{1 + 2\bar{\sigma}_2} \tag{220}$$

Let us assume $\omega_2 = 0.8$. From 217 we compute $\bar{\sigma}_2 = 0.6$, and from Eq. 220 we get $\lambda_2^2 = -0.36/2.2 = -0.1635$. The corresponding s-plane zeros are obtained from the substitution 200, giving $s_2^2 = \lambda_2^2 - 1 = -1.1635 = -u_2^2$, where we are denoting the zeros as u_ν so as not to confuse them with the poles ω_ν. We now introduce the reciprocal relationship between the s-plane zeros and poles*

$$u_\nu = k/\omega_\nu \qquad \text{or} \qquad u_\nu^2 = k^2/\omega_\nu^2 \tag{221}$$

In our present example the relation reads

$$1.1635 = k^2/0.64 \tag{222}$$

from which we get

$$k^2 = 0.745 \tag{223}$$

This example is too simple to require iteration. For each assumed ω_2 value, we obtain a corresponding value of k which is the radian frequency below which the value of our equiripple function is consistently greater than its guaranteed equiminimum value. The function resulting from this example (provided with a suitable multiplier to normalize the amplitude at unity for $\omega = \sqrt{k}$) reads

$$F(\omega) = \frac{(u_2^2 - \omega^2)}{\sqrt{k}\omega(1 - \omega^2/\omega_2^2)} = \frac{1.076(1.1635 - \omega^2)}{\omega(1 - 1.563\omega^2)} \tag{224}$$

If we shrink the frequency scale through replacing ω by $\omega\sqrt{k}$, it becomes comparable with the scale used in the functions 68 and 69. We then have

$$F(\omega) = \frac{(1.345 - \omega^2)}{\omega(1 - 1.345\omega^2)} \tag{225}$$

which becomes converted to a low-pass function upon replacing ω by $1/\omega$, equivalent to turning the function upside down.

Suppose we consider next an even function for $n = 2$. We choose $k = 0.75$, and as a first guess try

$$\omega_1 = 0.5, \qquad \omega_3 = 0.707 \tag{226}$$

The $\bar{\sigma}$ values computed from Eq. 214 are

$$\bar{\sigma}_1 = 0.866, \qquad \bar{\sigma}_3 = 0.707 \tag{227}$$

* In the frequency scale of the functions 68 and 69 we would have $u_\nu'\omega_\nu' = 1$. Our present scale is related to this one by the expression $u = u'\sqrt{k}$ or $\omega = \omega'\sqrt{k}$. Hence $u_\nu\omega_\nu = u_\nu'\omega_\nu'k = k$.

and the corresponding quadratic factors in $p(\lambda)$ become

$$\begin{aligned}(\lambda + 0.866)^2 &= \lambda^2 + 1.732\lambda + 0.75\\(\lambda + 0.707)^2 &= \lambda^2 + 1.414\lambda + 0.50\end{aligned} \tag{228}$$

The even part of their product reads

$$m(\lambda) = \lambda^4 + 3.70\lambda^2 + 0.375 \tag{229}$$

and its zeros are

$$\lambda_1^2 = -3.60, \qquad \lambda_3^2 = -0.10 \tag{230}$$

giving

$$s_1^2 = -4.60 = -u_1^2, \qquad s_3^2 = -1.10 = -u_3^2 \tag{231}$$

from which

$$u_1 = 2.15, \qquad u_3 = 1.05 \tag{232}$$

Equation 221 now yields the revised pole positions

$$\omega_1 = \frac{0.75}{2.15} = 0.349, \qquad \omega_3 = \frac{0.75}{1.05} = 0.715 \tag{233}$$

For the first iteration we compute in succession

$$\bar{\sigma}_1 = 0.937, \qquad \bar{\sigma}_3 = 0.700 \tag{234}$$

$$\begin{aligned}(\lambda + 0.937)^2 &= \lambda^2 + 1.874\lambda + 0.877\\(\lambda + 0.700)^2 &= \lambda^2 + 1.40\lambda + 0.49\end{aligned} \tag{235}$$

$$m(\lambda) = \lambda^4 + 3.987\lambda^2 + 0.43 = 0 \tag{236}$$

$$\lambda_1^2 = -3.878, \qquad \lambda_3^2 = -0.108 \tag{237}$$

$$s_1^2 = -4.878 = -u_1^2, \qquad s_3^2 = -1.108 = -u_3^2 \tag{238}$$

$$u_1 = 2.21, \qquad u_3 = 1.051 \tag{239}$$

and finally

$$\omega_1 = \frac{0.75}{2.21} = 0.340, \qquad \omega_3 = \frac{0.75}{1.051} = 0.713 \tag{240}$$

The process is evidently converging very well.

Amplitude and frequency scaling, to bring the result into the normal form of Eq. 68, follows the same pattern as before. Construction of either function 68 or 69 is thus seen to be computationally possible in a straightforward manner without the aid of elliptic function tables.

PROBLEMS

1. Design a prototype low-pass filter to have Tschebyscheff behavior in the pass band and monotonic behavior in the stop band according to the squared transfer function

$$|Z_{12}|^2 = \frac{1}{1 + \epsilon^2 V_n^2(\omega)}$$

The network is to be lossless and terminated in a 1-ohm resistance. It is to be driven by a pentode so that it is not appropriate to have any resistance loading on the input side. In the pass band the voltage ripples are not to exceed 5 per cent, and at least 40 db of attenuation are to be reached at the point $\omega = 1.5$.

2. Adapt the design in the previous problem to a situation where the lossless network is to be associated with 1-ohm resistances at both ends.

3. Transform the low-pass prototype designs found in the two foregoing problems into band-pass designs for a mid-band frequency of 100 kilocycles/sec, a bandwidth of 10 kilocycles/sec, and an impedance level of 500 ohms.

4. Design a low-pass prototype filter of the type described in Prob. 1 for $\epsilon = 1$ and $n = 4$, the network to be associated with resistances at both ends. Assume the load resistor R_2 for the output end to be 1 ohm and find the corresponding value of the input resistor R_1. Note the latter cannot also have a 1-ohm value since $|Z_{12}| \neq 1$ for $\omega = 0$.

5. A low-pass filter with cutoff normalized at 1 radian/sec is to be realized with inductances having a loss factor (R/L ratio) of 0.1, ideal capacitances, and ideal resistances. It is to have the Tschebyscheff behavior described by

$$|Z_{12}(j\omega)|^2 = \frac{1}{1 + \frac{1}{4}V_3^2(\omega)}$$

The impedance level is to be 1 ohm. Determine a network which approximately meets these specifications.

6. A lossless two terminal-pair network terminated in a 1-ohm resistance is to have a transfer impedance whose magnitude squared for $s = j\omega$ is approximated by the straight lines shown in the accompanying logarithmic plot.

(*a*) Write the appropriate transfer impedance $Z_{12}(s)$ as a rational function, assuming $Z_{12}(0) = 1$.

(*b*) Synthesize a symmetrical lattice lossless network which, when terminated in 1 ohm, will realize the specified transfer impedance. Choose the constant multiplier of Z_{12} so as to make the network simple.

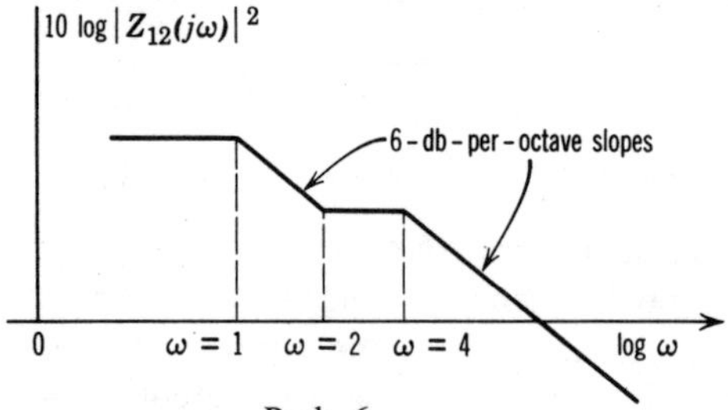

Prob. 6.

7. Design a low-pass filter of ladder form having the transfer impedance

$$|Z_{12}|^2 = \frac{1}{1 + \epsilon^2 V_n^2(\omega)}$$

for $\epsilon = 1$ and $n = 4$. Find a complementary network which connected as shown with the filter designed provides a constant driving-point impedance. Sketch $|E_c/I|$ for $s = j\omega$.

8. The transfer impedance of a lossless network terminated in a 1-ohm load is in magnitude given by

$$\left|\frac{E_2}{I_1}\right| = \frac{1}{\sqrt{1 + \omega^8}}$$

Determine the network.

9. Determine the network complementary to that of Prob. 8, and by combining the two obtain a low-pass, high-pass constant-resistance filter group. Reciprocate the result so as to obtain the dual filter group.

10. Consider the function of Prob. 8 as representing the voltage ratio $|2E_2/E_1|$ for a lossless network excited by the source E_1 with an internal resistance of 1 ohm and terminated in a 1-ohm load. Synthesize the network.

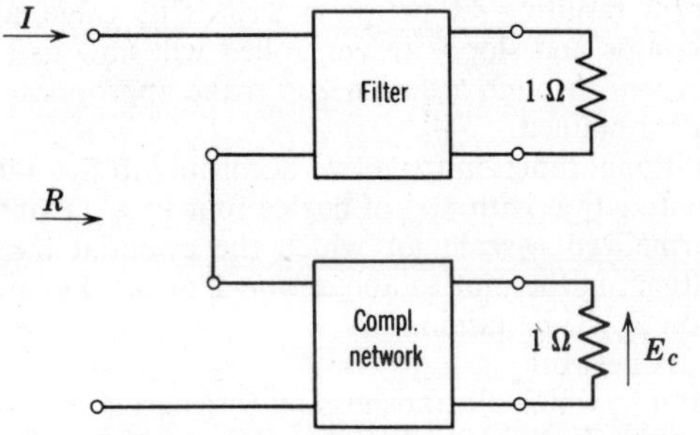

Prob. 7.

11. Through use of Tschebyscheff polynomials in a function of the form

$$|Z_{12}|^2_{s=j\omega} = \frac{\epsilon^2 V_n^2(1/\omega)}{1 + \epsilon^2 V_n(1/\omega)}$$

design a low-pass filter (develop into ladder form) with monotonic behavior in the pass band and with $|Z_{12}|$ having equal tolerance behavior for $\omega > 1$ with $|Z_{12}|_{\max}$ in this range being 0.1. Choose the order of the polynomial such that infinite attenuation results at $\omega = \infty$ and at one other point beyond $\omega = 1$. Utilizing the low-to-bandpass transformation, obtain a band-pass filter of 1 radian/sec bandwidth having the same quality as the low-pass filter but with its pass band centered at $\omega = 10$.

12. Making use of Eq. 127 and the infinite product expansion of the tangent function

$$\tan x = x\,\frac{[1 - (x^2/x_2^2)][1 - (x^2/x_4^2)]\cdots}{[1 - (x^2/x_1^2)][1 - (x^2/x_3^2)]\cdots}$$

with $x_k = k\pi/2$, show that a linear phase-shift constant-resistance lattice can be designed by letting the reactance x_a be represented by the truncation of this expansion through n factors. Choose $\beta = 2\omega$ and $n = 4$. Plot the ratio $x_a/\tan\omega$ for the range $0 < \omega < 5\pi/2$ and interpret the result in terms of the quality achievable by this method.

13. Modify the design in the previous problem by inserting into the expression for x_a an additional factor $(1 - x^2/x_5^2)$ with x_5 between the values $5\pi/2$ and $6\pi/2$. Make plots of the ratio $x_a/\tan\omega$ for several values of x_5 and show that the results of Prob. 12 can be materially improved by this method.

14. A monotonic phase function β is desired to be a smooth curve passing through the points:

$$\beta = 0,\ \pi,\ 2\pi,\ \ 3\pi,\ 4\pi$$

for

$$\omega = 0,\ 1,\ 2.5,\ \ 5,\ \ 10$$

Making use of Eq. 127 determine the simplest reactance function x_a which will yield a β-function passing through the desired points, and plot the actual β-function obtained by this method.

By extending the desired β-curve smoothly beyond the specified range and inserting one additional critical frequency into x_a, note the improvement obtainable.

15. Reconsider Prob. 14 by considering additionally Eq. 128 enabling control of the slope of the resulting β-function as well as its value, although the points at which both value and slope are controlled will now be spaced twice as far apart. Again carry through a design and make appropriate plots showing the quality of results obtained.

16. Find a rational function for $d\beta/d\omega$ according to Eq. 136 corresponding to a flat-flat phase function with $\phi(s)$ of degree four in s. Plot the resulting phase slope to a normalized ω-scale for which the essential range corresponds to $0 < \omega < 1$, extending the plot to about $\omega = 2$ or 3. Form the corresponding transfer function $Z_{12}(s)$ pertaining to:

(*a*) An all-pass network.

(*b*) A situation in which all transmission zeros occur at $s = \infty$.

For the latter, plot $|Z_{12}(j\omega)|$ for the interval $0 < \omega < 2$.

17. Determine a rational approximant $|Z_{12}(j\omega)|^2$ appropriate to a low-pass filter having equiripple characteristics in both pass and stop regions. The pass-band ripple shall not exceed 0.5 db for $0 < \omega < \sqrt{k}$ and the stop-band loss shall be at least 50 db for the range $(1/\sqrt{k}) < \omega < \infty$ with $k = 0.9$.

18. Assume the negative real poles $s = -1, -2, -3, -4$, and construct an equiripple low-pass *RC* transfer impedance according to the method given in Art. 10 for a normalized pass band defined by $-1 < \omega < 1$. Choose a pass-band ripple amplitude equal to 0.1, and plot the magnitude of the resulting transfer function for the range $0 < \omega < 4$.

19. Show by means of the results given in Art. 10 that one can devise a simple lowpass-bandpass transformation applicable to *RC* networks permitting a band-pass design to be obtained directly from the solution of the pertinent low-pass approximation problem. Does this process place any constraints upon the pole locations?

20. From Eqs. 127 and 128 derive the following relations:

$$(dx_a/d\omega)_{x_a=0} = \tfrac{1}{2}(d\beta/d\omega)_{\beta=2\pi\nu}$$

$$(db_a/d\omega)_{b_a=0} = \tfrac{1}{2}(d\beta/d\omega)_{\beta=2\pi(\nu+\frac{1}{2})}$$

for $\nu = 1, 2, \cdots$, in which b_a is a susceptance function corresponding to the reactance x_a.

Through use of these results plus Eqs. 127 and 128, show that one can design a lossless constant-resistance lattice to approximate any graphically given phase characteristic either by assuring that it have the proper phase shift at points where β equals integer multiples of π, or by assuring that both the value and the slope of the resulting phase be correct where β equals only odd or only even multiples of π.

CHAPTER FIFTEEN

Time-Domain Synthesis

1 Introduction

In time-domain synthesis (sometimes also referred to as *transient* synthesis) we are interested in determining a network whose transient response is specified in terms of given input and output time functions. Thus the important feature of this problem resides in the fact that the specified input and output are time functions rather than functions of frequency as has been almost exclusively the case in all of the foregoing discussions.

In view of the fact that Fourier or Laplace transform methods enable us to convert the given input and output time functions to equivalent functions of the complex frequency variable s, our attitude initially may be that there is essentially nothing new or different or unique about this form of the general synthesis problem. Having transformed the given functions (graphically or analytically) to the frequency domain, the methods already discussed should afford the means for constructing a solution.

They do, except that in the solution to the pertinent approximation problem (in which we construct an acceptable rational approximant to the required frequency response function) we cannot limit our considerations to approximation tolerances in the frequency domain since our ultimate interest lies in the quality of the approximation in the time domain where the given data are initially specified. In dealing with the pertinent approximation problem we become involved in a simultaneous consideration of both the time and frequency domains and in the interrelation of errors in these two domains.

Thus time-domain synthesis presents new considerations only so far as a discussion of the associated approximation problem is concerned. Once an acceptable rational approximant is found, there is no difference between time- or frequency-domain synthesis. The two differ only in the aspects presented by their respective approximation problems, and hence a consideration of time-domain synthesis amounts essentially to an enlargement of the discussions of the previous chapter so as to include related time-domain interpretations.

It is, of course, conceivable that we may devise methods of constructing networks to have desired forms of transient response without recourse to Laplace transforms or frequency-domain considerations at all. In fact, our familiarity with transient responses of circuits consisting of a variety of simple element combinations enables us to solve a number of problems of this sort by inspection, especially if the required tolerances are lenient and we are gifted with intuitive reasoning and experimental manipulation, so that cut-and-try procedures are capable of yielding satisfactory results. Where we seek straightforwardly obtainable solutions with carefully controllable tolerances, however, a direct time-domain approach as yet is not available, although we come pretty close to achieving this kind of result when we devise methods of solving the approximation problem (finding rational approximants) in which the process of error control involves time-domain considerations only.*

As has been mentioned in the preceding chapter, the approximation problem is concerned essentially with two things: fulfillment of the desired response characteristics on the one hand, and satisfaction of the necessary conditions for physical realizability of the resulting rational approximant on the other. The latter item injects into the time-domain synthesis problem an additional difficulty since all previously discussed realizability conditions are pertinent to functions of frequency, whereas we now need realizability conditions that can be applied directly to the specified time functions. In other words, we must be able to determine at the outset whether an appropriate rational approximant exists.

This existence question is, of course, not clear cut since the answer depends, in some measure at least, upon the tolerances that we are willing to accept in the solution of our approximation problem. If we are

* A method which accomplishes this end was first used for the synthesis of networks to produce rectangular pulses for radar applications; see "An Historical Account of the Development of a Design Procedure for Pulse-Forming Networks," *Report 43*, Mass. Inst. Technol., Radiation Lab., Oct. 16, 1944. An extension of this method to the production of arbitrary pulse shapes was given later in the *Quart. Prog. Rept.*, Mass. Inst. Technol., Research Lab. Electronics, Jan. 15, 1952, pp. 66–71. This method is discussed in Arts. 6 and 7 of this chapter.

Another one that also avoids correlation of frequency- and time-domain errors is given by A. Papoulis, "Network Response in Terms of Behavior at Imaginary Frequencies," presented at the Symposium on Modern Network Synthesis, Apr. 13–15, 1955, and published in the *Transactions*, Polytechnic Institute of Brooklyn, 1956. This procedure involves a transformation of the independent variable and requires the preselection of poles, so that the character of the approximation as well as the relation between tolerances and complexity of the resulting rational approximant are difficult to assess.

willing to depart arbitrarily far from the initially stated input and output time functions, then an approximant fulfilling these demands as well as realizability conditions always exists. Although it is, therefore, not practically sensible to ask for too precise a formulation of existence criteria, we do need to clarify the relationship between nature and order of magnitude of tolerances on the one hand, and realizability conditions on the other. Once we have an appreciation of the time-domain properties upon which realizability depends, then we will be able to cope with this additional consideration which time-domain synthesis injects into the approximation problem.

If the specified input function is an impulse or a step occurring at $t = 0$, and the output is a bounded time function of finite duration, zero before $t = 0$, then a realizable frequency-domain approximant always exists (as a transfer function). If the input time function is also bounded, zero before $t = 0$, of finite duration, but arbitrary in form, then a realizable frequency-domain approximant may not exist because the ratio (output over input) of the Laplace transforms of these time functions (or reasonable approximations thereof) may contain right half-plane poles. In fact, such poles will be present whenever the transform of the input time function has right half-plane zeros not also contained in the transform of the output time function.

We shall develop simple numerical methods, applicable directly to the given time functions, that enable us to see at the outset what the situation is in this regard. In general it can be said that any output wave shape can be had from any input wave shape (within reason) if we are willing to accept a sufficient delay in the output time function.

In contemplating the general time-domain synthesis problem, we are naturally led to consider the possibility that the Laplace transforms of the stated input and output time functions are rational. In that event no approximation problem would need to be solved provided the right-half-plane-pole difficulty mentioned above did not occur. The only time functions whose transforms are rational functions, however, are sums of exponentials of the form $A_\nu e^{s_\nu t}$, and it is hardly likely that the specified time functions will be expressed in this form. If a method were available whereby arbitrary time functions could systematically be approximated with any desired tolerance and in a specified manner by sums of exponentials, then our problem would be solved. The Fourier expansion unfortunately represents such a solution only for periodic time functions.*

* It is, of course, theoretically possible to construct a set of orthogonal normal functions as linear combinations of exponentials and then apply well-known methods to carry out the desired approximation. This scheme fails practically because of the huge computational labor involved and because the s_ν values must be known beforehand.

In most situations, for analytically as well as graphically specified time functions, the transforms and hence their ratio (the required frequency-domain response function) turn out to be transcendental functions of the complex frequency variable. For example, if the output time function is identical in form with the input time function except for a relative delay of 1 second, then the response function is recognized to be simply e^{-s}. For synthesis purposes this result is useless unless we can construct a rational approximant to this transcendental function in such a way that the quality of the approximation can be judged in the time domain, for experience has shown that what seems to be a good approximation in the frequency domain can lead to a totally unacceptable result in the time domain.*

If, for one reason or another, the approximation is tentatively carried out in the frequency domain, two major questions arise: (1) Over what region of the s-plane and in what manner should the approximation of a complex transcendental function be done? (2) How can the error committed in the frequency domain by this approximation process be transformed to the time domain with sufficient ease to facilitate control of the resultant error in that domain?

A reasonable answer to both of these questions is suggested by the sufficiency of the j-axis real part of a driving-point or transfer function in completely and uniquely specifying not only that complex function but also its inverse transform in the time domain (as demonstrated in Ch. 8). Thus we can approximate the j-axis real part of the desired transcendental function by means of a quotient of finite polynomials in ω^2, with the assurance that a corresponding complex response function (having only left half-plane poles) is always uniquely constructible therefrom. Moreover, since the inverse transform (time function) is uniquely and *linearly* related to the j-axis real part of the complex frequency response function, error correlation in the two domains is straightforwardly possible because additive terms in the j-axis real part are separately convertible into additive terms in the corresponding time function, and no consideration need be given to the possible minimum-phase or nonminimum-phase character of the pertinent response function.†

Since the inverse transform of a given complex transfer function is the unit impulse response of the pertinent network, a time-domain synthesis problem is most easily dealt with when the given data call for a specified output time function resulting from an impulse excitation. If the input

* See, for example, "What is Network Synthesis?", E. A. Guillemin, *IRE Trans.*, Vol. PGCT–1, Dec. 1952, pp. 4–19.

† These advantages of working with the j-axis real-part function in time-domain synthesis were initially explored at the writer's suggestion by R. Marston, *Master's thesis*, Massachusetts Institute of Technology, 1946.

time function likewise has some arbitrary form, our procedure is first to determine the impulse response from the two given time functions and then treat the problem as we would the simpler one for which the impulse response is specified directly.

To implement this approach we must develop methods for determining the impulse response for arbitrarily given input and output time functions (including existence criteria); and to facilitate the synthesis in terms of a specified impulse response, we need to develop computational procedures which will expedite transformation between time and frequency domains in order to deal effectively with error control. These matters are discussed in the immediately following articles.

2 Computational Techniques Which Simplify Correlation Between Time and Frequency Domains*

Our point of departure is given by the familiar Fourier integrals

$$f(t) = \frac{1}{2\pi}\int_{-\infty}^{\infty} F(j\omega)e^{j\omega t}\, d\omega \tag{1}$$

$$F(j\omega) = \int_{-\infty}^{\infty} f(t)e^{-j\omega t}dt \tag{2}$$

in which we may regard $F(j\omega)$ as the conventional response function (driving-point or transfer) for the substitution $s = j\omega$, and $f(t)$ as the corresponding unit impulse response of the pertinent network.

If we separate $f(t)$ into its even and odd parts by writing

$$f(t) = f_e(t) + f_o(t) \tag{3}$$

and indicate the complex character of $F(j\omega)$ in the form

$$F(j\omega) = F_1(\omega) + jF_2(\omega) \tag{4}$$

then it follows rather straightforwardly that

$$F_1(\omega) = \int_{-\infty}^{\infty} f_e(t) \cos \omega t\, dt, \quad F_2(\omega) = -\int_{-\infty}^{\infty} f_o(t) \sin \omega t\, dt \tag{5}$$

from which we note that the real and imaginary parts of $F(j\omega)$ are respectively even and odd functions of ω.

Substitution of 4 into 1 and separation into two integrals yields the respective inverse transforms to those stated in 5, namely

$$f_e(t) = \frac{1}{2\pi}\int_{-\infty}^{\infty} F_1(\omega) \cos \omega t\, d\omega, \qquad f_o(t) = \frac{-1}{2\pi}\int_{-\infty}^{\infty} F_2(\omega) \sin \omega t\, d\omega \tag{6}$$

Here, as in 5, the limits of integration may be extended only over positive

* Based upon *Tech. Rept. 268*, Mass. Inst. Technol., Research Lab. Electronics, Sept. 2, 1953, presented at the National Electronics Conference (Chicago) of that date, and published in their *Transactions*, Vol. 9, Feb. 1954.

values of the independent variable and the result multiplied by 2, since the integrands are even functions.

If the time function is tacitly assumed zero for $t < 0$, then it is clear that $f_e(t)$ and $f_o(t)$ cancel for $t < 0$, and hence must be equal for $t > 0$. Therefore, such an $f(t)$ for $t > 0$ is given by either twice $f_e(t)$ or twice $f_o(t)$ for $t > 0$, and we have from 6

$$f(t) = \frac{2}{\pi}\int_0^\infty F_1(\omega)\cos\omega t\,d\omega = \frac{-2}{\pi}\int_0^\infty F_2(\omega)\sin\omega t\,d\omega \tag{7}$$

Since the impulse response of a passive network must be dead before the advent of the excitation (assumed to occur at $t = 0$), we see that the real or imaginary part of the response function $F(j\omega)$ is sufficient to characterize this impulse response (other aspects of this theme have already been elaborated upon in Ch. 8).

The companion integrals to 7 are obtained from 5 on the tacit assumption that twice $f_e(t)$ or twice $f_o(t)$ equals $f(t)$ for $t > 0$. Thus we get

$$F_1(\omega) = \int_0^\infty f(t)\cos\omega t\,dt, \qquad F_2(\omega) = -\int_0^\infty f(t)\sin\omega t\,dt \tag{8}$$

If we use the integrals 7 and 8 instead of 1 and 2, then the function which uniquely characterizes $f(t)$ in the frequency domain may be regarded as given by either $F_1(\omega)$ or $F_2(\omega)$. Like $f(t)$, it is also a real function of a real variable, and we are rid of the nuisance of having to deal with a complex frequency function while we are concerned with the correlation of time- and frequency-domain tolerances. The rational function in the frequency domain with which we seek to fulfill the requirements on a stated $f(t)$ in the time domain is $F_1(\omega)$ instead* of the complex $F(j\omega)$, which we can readily generate (by the methods discussed in Ch. 8) once we have an acceptable rational function $F_1(\omega)$.

Of particular importance in the present discussions is the inclusion of impulses in the real-part function $F_1(\omega)$. As pointed out in Art. 2, Ch. 8, these are associated with j-axis poles of the complex function $F(j\omega)$, the value of a particular impulse being equal to π times the residue of this function in the respective pole (see Eqs. 57 through 61 in Ch. 8).

The method of carrying out a transformation from a given real-part function $F_1(\omega)$ to its associated $f(t)$, or vice versa, is based upon an approximation of the given function in such a way that either the approximating function or its first or higher derivative consists of an impulse train† (succession of impulses at regular or irregular intervals). For

* Alternately we may use the imaginary part $F_2(\omega)$. A choice is sometimes indicated by the nature of $f(t)$.

† This impulse-train method of approximation was first used by W. K. Linvill in "Analysis and Design of Sampled Data Control Systems," *Sc.D. thesis*, Massachusetts Institute of Technology, 1949.

example, if the given function is approximated by rectangular elements or pulses of finite width, and each of these is approximated by an impulse whose value equals the respective pulse area, then we regard the impulses as "samples" of the function and the impulse train is the given function in so-called sampled form. This form of representation is approximate because not only does the succession of rectangular pulses yield a step approximation to the given function but an additional source of error is introduced by replacing each pulse by an impulse. This latter approximation introduces a time-dependent error into the evaluation of $f(t)$ in the first integral 7 [or a frequency-dependent error into the evaluation of $F_1(\omega)$ in the first of the integrals 8] inasmuch as $\cos \omega t$ actually varies throughout an element of finite width whereas substitution of an impulse for that element suppresses this variation.

The latter error is completely removed, if the step approximation to the given function is differentiated, since this process converts it into an impulse train without injecting any additional errors. The effect of this differentiation upon the associated function $f(t)$ or $F_1(\omega)$, as the case may be, amounts essentially to multiplication respectively by t or ω (an algebraic sign is unimportant at the moment) and can readily be undone. The only remaining error is that due to the stepped character of the approximating function. Reduction of this error involves an increase in the total number of terms in the approximation.

In this regard we can obtain a more advantageous result by use of a piecewise-linear or broken-line approximation to the given function. Two successive differentiations then are needed to reduce the approximant to a sum of impulses, their number being equal to the number of break points in the piecewise-linear approximation, which, for a given tolerance, is appreciably smaller than the number of terms resulting from a step approximation.

Logically a further reduction in the number of terms for a given error, or in the error for a given number of terms, should be possible if we use a piecewise parabolic arc approximation to the given function, corresponding to a piecewise-linear approximation of its first derivative. The latter, however, may require more break points than a comparable piecewise-linear approximation of the given function, so that this procedure does not necessarily lead to a further reduction in the number of terms although it will surely decrease the resulting error. We will return to a detailed consideration of these matters after the Fourier integral evaluation in terms of an impulse-train representation of this sort has been formulated.

First we remark that in these approximate methods of integration we always consider the function $F_1(\omega)$ to extend over a finite frequency interval, which we can define as the normalized range $0 < \omega < 1$. The effect of chopping off an asymptote in order to achieve this representation, and the way in which one may nevertheless properly take account of the relationship between the behavior of $F(j\omega)$ for $\omega \to \infty$ and that of $f(t)$ for $t \to 0$, will receive separate attention.

In an analogous transformation from $f(t)$ to $F_1(\omega)$, the time function is likewise considered to have nonzero values over only a finite interval through an appropriate compensation for the neglect of its asymptotic behavior.

If we observe that the function $F_1(\omega)$ is even but that its derivatives of odd order are odd while those of even order are even, then a piecewise-linear approximation of the $(\nu-2)$th derivative with n break points is seen to yield for the derivative of order ν the impulse representation

$$F_1^{(\nu)}(\omega) = \sum_{k=1}^{n} a_k[u_0(\omega - \omega_k) + (-1)^\nu u_0(\omega + \omega_k)] \tag{9}$$

By use of the reasoning that relates Eqs. 59 and 60 in Ch. 8, noting also that $d/ds = d/j\,d\omega$, we see that the complex response function $F(s)$ having the real part 9 is given by

$$F^{(\nu)}(s) = \frac{1}{\pi(j)^\nu}\sum_{k=1}^{n} a_k\left[\frac{1}{s - j\omega_k} + (-1)^\nu \frac{1}{s + j\omega_k}\right] \tag{10}$$

where the superscript (ν) here indicates the νth derivative with respect to s. The coefficients a_k in Eqs. 9 and 10 equal the changes in slope of the broken-line approximation at its break points which occur at the frequencies $\omega = \omega_k$.

Since the ν-fold differentiation of $F(s)$ corresponds to multiplication of $f(t)$ by $(-t)^\nu$, and the time function resulting from the function within the bracket of Eq. 10 is obvious by inspection, we see that the time function corresponding to the pertinent real part $F_1(\omega)$ is given by

$$f(t) = \frac{2(-1)^{\nu/2}}{\pi t^\nu}\sum_{k=1}^{n} a_k \cos \omega_k t \qquad \text{for } \nu \text{ even} \tag{11}$$

and

$$f(t) = \frac{2(-1)^{(\nu+1)/2}}{\pi t^\nu}\sum_{k=1}^{n} a_k \sin \omega_k t \qquad \text{for } \nu \text{ odd} \tag{12}$$

As an illustration of how these relations are used, consider the trapezoidal real-part function shown in the top part of Fig. 1. Directly below it is drawn the first derivative and below this we have the second derivative with the impulse representation

$$F_1^{(2)}(\omega) = -5u_0(\omega - 0.8) + 5u_0(\omega - 1) \tag{13}$$

Substitution into 11 gives the associated time function

$$f(t) = \frac{10}{\pi t^2}(\cos 0.8t - \cos t) \tag{14}$$

having the initial value $f(0) = 9/5\pi$ as it should since this value, by the first of the integrals 7, equals $2/\pi$ times the net area under the real-part function.

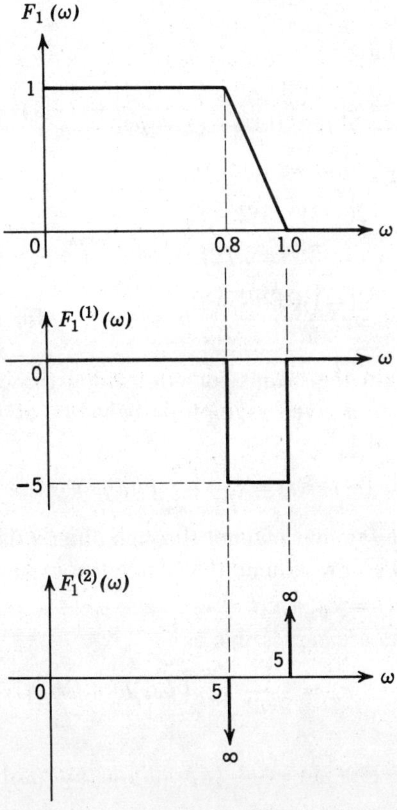

Fig. 1. Assumed real part of a frequency response function and its first and second derivatives, yielding (according to Eq. 11) the associated time function 14.

Incidentally, the time function does not "blow up" for $t = 0$ as a first reaction to the form of the result in Eq. 14 might lead us to suspect. The reason for the proper behavior of $f(t)$ for $t = 0$ owes to a condition which the coefficients a_k automatically fulfill as a result of the implied asymptotic behavior of $F(s)$. Since $F_1(\omega)$ in Fig. 1 is nonnegative, the angle of $F(s)$ in absolute value does not exceed $\pi/2$, which implies that $F(s)$ behaves like

$1/s$ for large s and $f(t)$ behaves like a step function for $t = 0$. These matters will presently be discussed in greater detail.

We can in an analogous fashion obtain $f(t)$ from the imaginary part $F_2(\omega)$. Since it is an odd function, whereas its odd-order derivatives are even, the impulse representation of its νth derivative, similar to that in Eq. 9 for the real part, reads

$$F_2^{(\nu)}(\omega) = \sum_{k=1}^{n} b_k[u_0(\omega - \omega_k) - (-1)^\nu u_0(\omega + \omega_k)] \tag{15}$$

In place of 10 we have

$$F^{(\nu)}(s) = \pi \frac{1}{(j)^{(\nu-1)}} \sum_{k=1}^{n} b_k \left[\frac{1}{s - j\omega_k} - (-1)^\nu \frac{1}{s + j\omega_k} \right] \tag{16}$$

and for the time function we find

$$f(t) = \frac{2(-1)^{(\nu+1)/2}}{\pi t^\nu} \sum_{k=1}^{n} b_k \cos \omega_k t \qquad \text{for } \nu \text{ odd} \tag{17}$$

$$f(t) = \frac{2(-1)^{(\nu+2)/2}}{\pi t^\nu} \sum_{k=1}^{n} b_k \sin \omega_k t \qquad \text{for } \nu \text{ even} \tag{18}$$

In order to obtain the means for controlling the behavior of $f(t)$ for $t = 0$ appropriate to a given asymptotic behavior of $F(s)$, we begin with the modified form of 1

$$(-t)^\nu f(t) = \frac{1}{2\pi} \int_{-\infty}^{\infty} F^{(\nu)}(j\omega) e^{j\omega t} \, d\omega \tag{19}$$

which is gotten in a familiar manner through differentiation of 2 under the integral sign. If we now assume the Maclaurin expansion

$$(-t)^\nu f(t) = c_0 + c_1 t + c_2 t^2 + \cdots + c_\mu t^\mu + \cdots \tag{20}$$

then the coefficients are expressible as

$$c_\mu = \frac{1}{2\pi\mu!} \int_{-\infty}^{\infty} (j\omega)^\mu F^{(\nu)}(j\omega) \, d\omega \tag{21}$$

Noting that

$$F^{(\nu)}(j\omega) = \frac{1}{(j)^\nu} [F_1^{(\nu)}(\omega) + jF_2^{(\nu)}(\omega)] \tag{22}$$

because the superscript (ν) on $F(j\omega)$ denotes differentiation with respect to $j\omega$ while on $F_1(\omega)$ or $F_2(\omega)$ it denotes differentiation with respect to ω, and substituting from Eqs. 9 and 15, evaluation of 21 leads to

$$c_\mu = \frac{(-1)^\nu (j)^{\nu+\mu}}{\pi\mu!} \sum_{k=1}^{n} a_k \omega_k^\mu \qquad \text{for } (\nu + \mu) \text{ even} \tag{23}$$

and

$$c_\mu = \frac{(-1)^\nu (j)^{\nu+\mu+1}}{\pi\mu!} \sum_{k=1}^{n} b_k \omega_k^\mu \qquad \text{for } (\nu + \mu) \text{ odd} \tag{24}$$

Now if the asymptotic behavior of $F(s)$ is described by

$$F(s) \to 1/s^q \qquad \text{for } s \to \infty \tag{25}$$

then the associated time function should exhibit a behavior like

$$f(t) \to t^{q-1} \qquad \text{for } t \to 0 \tag{26}$$

In the Maclaurin expansion 20 we should, therefore, have

$$c_\mu = 0 \qquad \text{for } \mu = 0, 1, 2, \cdots, q + \nu - 2 \tag{27}$$

Expressions 23 and 24 thus yield the conditions

$$\sum_{k=1}^{n} a_k \omega_k^{\mu} = 0 \begin{cases} \text{for } \mu = 0, 2, 4, \cdots, \leq q + \nu - 2 \text{ when } \nu \text{ is even} \\ \text{for } \mu = 1, 3, 5, \cdots, \leq q + \nu - 2 \text{ when } \nu \text{ is odd} \end{cases} \tag{28}$$

and

$$\sum_{k=1}^{n} b_k \omega_k^{\mu} = 0 \begin{cases} \text{for } \mu = 0, 2, 4, \cdots, \leq q + \nu - 2 \text{ when } \nu \text{ is odd} \\ \text{for } \mu = 1, 3, 5, \cdots, \leq q + \nu - 2 \text{ when } \nu \text{ is even} \end{cases} \tag{29}$$

In the above example, $\nu = 2$ and $q = 1$. Conditions 28 apply only for $\mu = 0$ and state that the pertinent behavior of $f(t)$ for $t = 0$ is assured if the sum of the a_k's equals zero. By inspection of Fig. 1 we see that this requirement is met.

In order to see from another point of view what these conditions mean, let us consider the rth repeated integral of Eq. 9, which reads

$$F_1^{(\nu-r)}(\omega) = \sum_{k=1}^{n} a_k[u_{-r}(\omega - \omega_k) + (-1)^\nu u_{-r}(\omega + \omega_k)] \tag{30}$$

in which $u_{-r}(x)$ denotes the singularity function of order $-r$ given by

$$u_{-r}(x) = \frac{x^{r-1}}{(r-1)!} \qquad \text{for } x > 0 \tag{31}$$

and equal to zero for $x < 0$. From 30 and 31 we get for $\omega = 0$

$$F_1^{(\nu-r)}(0) = \sum_{k=1}^{n} \frac{a_k \omega_k^{r-1}}{(r-1)!}(-1)^\nu \tag{32}$$

or, if we let $r - 1 = \mu$, we may write

$$\sum_{k=1}^{n} a_k \omega_k^{\mu} = (-1)^\nu \mu! F_1^{(\nu-\mu-1)}(0) \tag{33}$$

Since the odd-order derivatives of $F_1(\omega)$ are zero for $\omega = 0$, we again have conditions 28, without, however, a knowledge of the upper limit in the range of μ values to which they apply, or the relationship of this upper limit to the implied asymptotic behavior of $F(s)$. The result 33, on the other hand, shows that fulfillment of conditions 28 for μ values up to and including the integer $\nu - 2$ has nothing directly to do with asymptotic conditions* but only with the requirement that odd-order derivatives of $F_1(\omega)$ must pass through the value zero at $\omega = 0$. The additional conditions, for μ values beyond $\nu - 2$, are the ones whereby a particular asymptotic behavior on the part of $F(s)$ is accounted for.

In this connection it should be observed that $f(t)$, determined from the real part $F_1(\omega)$, is even and hence can behave only like some even power of t for $t \to 0$. The asymptotic behavior of $F(s)$ according to 25 can in this situation be accommodated only for odd integer values of q. When an even integer value of q is called for, and $f(t)$ is required to behave like an odd power of t for $t \to 0$, then the imaginary part $F_2(\omega)$ should be used to generate the time function.

Actually these matters are not too significant unless we insist that the behavior of $f(t)$ for $t \to 0$ must be just right. Since in any method of approximate integration a certain error is tolerated anyway, the precise behavior at the point $t = 0$ is in most situations not so rigid that one cannot freely choose $F_1(\omega)$ or $F_2(\omega)$ for the determination of $f(t)$ since either one of these completely fixes the time function everywhere except at the point $t = 0$.

The result 33 for $\nu - \mu - 1 = 0$ yields the condition

$$\sum_{k=1}^{n} a_k \omega_k^{\nu-1} = (-1)^\nu (\nu - 1)! F_1(0) \tag{34}$$

which must be satisfied by the a_k values in order to insure zero error in the approximation of $F_1(\omega)$ at $\omega = 0$. For the trapezoidal function in Fig. 1 we see that this relation also is automatically fulfilled.

Relations 28 and 29 are commonly referred to as *moment* conditions. They are equivalent to corresponding integral relations referred to in Fourier or Laplace transform theory by the same name.

As an example of the process of meeting certain desired asymptotic behavior, suppose we assume that $F(s) \to 1/s^3$ for $s \to \infty$, and ask to construct a pertinent real part $F_1(\omega)$ together with its associated time function. We will assume that the third derivative of $F_1(\omega)$ is represented by a sum of impulses, or that $F_1(\omega)$ is approximated by confluent parabolic arcs of second degree. Regarding conditions 28, we then have $\nu = 3$ and $q = 3$, so that these conditions apply for $\mu = 1, 3$.

If we normalize $F_1(\omega)$ by letting $F_1(0) = 1$, then conditions 28 together

---

* Indirectly it does, of course, since the proper behavior of $f(t)$ for $t \to 0$ can certainly not result if the implied F_1 function contains drift error.

with relation 34 yield

$$\sum_{k=1}^{n} a_k \omega_k = 0 \tag{35}$$

$$\sum_{k=1}^{n} a_k \omega_k^2 = -2 \tag{36}$$

$$\sum_{k=1}^{n} a_k \omega_k^3 = 0 \tag{37}$$

The simplest approximation will therefore involve three arcs, and in that event the a_k values must fulfill the equations

$$\begin{aligned} a_1\omega_1 + a_2\omega_2 + a_3\omega_3 &= 0 \\ a_1\omega_1^2 + a_2\omega_2^2 + a_3\omega_3^2 &= -2 \\ a_1\omega_1^3 + a_2\omega_2^3 + a_3\omega_3^3 &= 0 \end{aligned} \tag{38}$$

Normalization of the frequency scale amounts to choosing the break point $\omega_3 = 1$. An $F_1(\omega)$ function appropriate to this situation is now constructed as soon as we choose values for the break points ω_1 and ω_2. Suppose we let $\omega_1 = \frac{1}{3}$ and $\omega_2 = \frac{2}{3}$. Equations 38 then yield

$$a_1 = 45, \qquad a_2 = -36, \qquad a_3 = 9 \tag{39}$$

The resulting real part $F_1(\omega)$ is shown at the top of Fig. 2 and its first, second, and third derivatives in the successive sketches below it. The corresponding time function according to Eq. 12 reads

$$\begin{aligned} f(t) &= \frac{2}{\pi t^3}[45 \sin (t/3) - 36 \sin (2t/3) + 9 \sin t] \\ &= \frac{2}{\pi}[(t^2/27) - (t^4/729) + \cdots] \end{aligned} \tag{40}$$

from which we see that the proper behavior at $t = 0$ is obtained.

In situations where the function $F_1(\omega)$ is graphically given, and one wishes to apply the piecewise-linear approximation to its first or higher derivative [equivalent to approximating $F_1(\omega)$ itself by second-degree or higher degree parabolic arcs], the problem of constructing these derivatives with the required accuracy arises. They obviously cannot be constructed with the usual attitude of regarding the value of the derivative at any point as being equal to the slope of the given function at that point, nor can one use a numerical method based upon tables of finite differences. If we were to construct a second or third derivative by such means and then undo the process through repeated graphical integration, the errors in the derivative functions would be found to cause excessive "drift" error in the integration process, with the result that the original function would not be regained.

In this connection it is important to recognize that fulfillment of the moment conditions 28 for μ values up to and including $\nu - 2$, as pointed out above, assures that odd-order derivatives of $F_1(\omega)$ will pass through

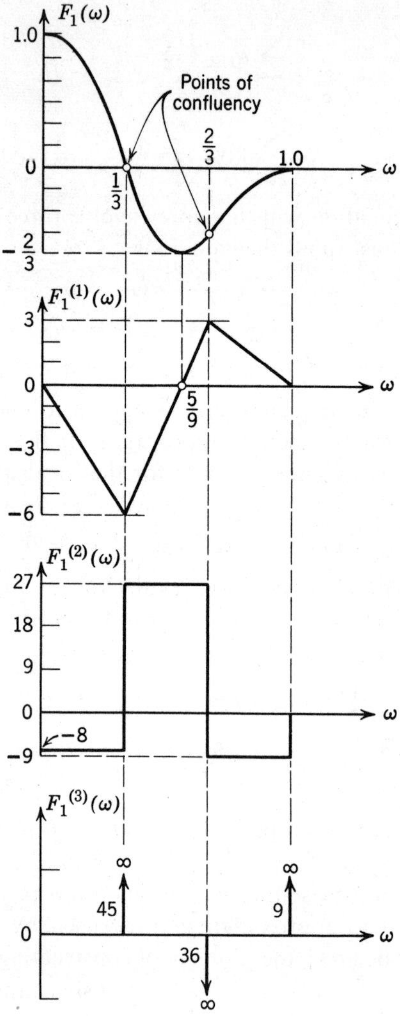

Fig. 2. Real-part function yielding the a_k values 39 in the representation 9 appropriate to the asymptotic behavior $F(s) \to 1/s^3$ for $s \to \infty$.

the value zero at $\omega = 0$, and hence (together with condition 33) they guarantee that there will be no drift error resulting from successive integrations, which must in these considerations be thought of as extending over the entire normalized frequency interval $-1 < \omega < 1$ so that conditions assuring zero drift at $\omega = 0$ automatically assure zero drift throughout this interval. We thus see that a set of a_k values based upon a broken-line approximation to an inaccurate higher derivative (one whose subsequent integration causes an intolerable amount of drift) would fail to fulfill conditions 28 and 33.

Such failure can be avoided if we use a graphical construction process that defines successive ordinates in the derivative function by requiring that the accumulated area under this function from $\omega = -1$ to any particular ω value shall equal the ordinate of the given function at the same point, or that the incremental area under the derivative function for a given $\Delta\omega$ shall equal the ordinate increment in the given function for the same $\Delta\omega$. With this attitude, one can construct derivative functions whose subsequent integration regains the given function with the desired accuracy.

If the piecewise-linear approximation is to be applied to the first derivative, one can alternately devise a straightforward method of approximating the given function by parabolic arcs directly and thus save some manipulative work. The geometric principles found useful in this connection may be seen by reference to Fig. 2. If in the graph of $F_1(\omega)$ we join maxima and minima by straight lines, these intersect the curve for $F_1(\omega)$ at points where the parabolic arcs are confluent; and the ordinate difference $F_1(0) - F_1(\frac{5}{9})$, for example, is given by the area of the negative triangle in $F_1^{(1)}(\omega)$. The altitude of this triangle is fixed by the desired ordinate difference and the location of its tip by the desired point of confluency.

We are now in a position to discuss more adequately the question of error estimation and its relation to the total number of impulses needed in the representation of $F_1^{(\nu)}(\omega)$. Fundamental to this discussion is the normalization of the pertinent frequency interval as shown in Figs. 1 and 2. Since the method of taking into account the proper asymptotic behavior of $F(s)$, in spite of the chopping off of an asymptote in order to accommodate this normalization of the frequency scale, is now clear, we are in a position to interrelate the following three things: the closeness with which $F_1(\omega)$ is approximated, the order ν of the derivative for which the approximation is given by a sum of impulses, and the number of these impulses (equal to the number of terms needed to represent the associated time function).

Imagine the $(\nu - 2)$th derivative of $F_1(\omega)$ to be approximated by a broken line in such a way that the error (difference between the broken line and the actual curve) changes sign three times within each linear interval, which it normally will if a line is passed through each inflection point. Within the interval $0 < \omega < 1$ there then are $3(n - 1)$ changes of sign if n equals the total number of break points in the piecewise-linear approximation.

Hence there are $3(n - 1)/2$ cycles of error oscillation in the interval $0 < \omega < 1$.

Let $3(n - 1)/2 = \tau/2\pi$, and assume that the error variation can roughly be regarded as sinusoidal with an amplitude ϵ. Then we have an error

$$\begin{aligned} &\text{in } F_1^{(\nu-2)}(\omega) \approx \epsilon \sin \omega\tau \\ &\text{in } F_1^{(\nu-3)}(\omega) \approx -(\epsilon/\tau) \cos \omega\tau \\ &\text{in } F_1^{(\nu-4)}(\omega) \approx -(\epsilon/\tau^2) \sin \omega\tau \end{aligned} \tag{41}$$

and so forth. The approximate maximum error in the approximation of of $F_1(\omega)$ itself becomes

$$\frac{\epsilon}{\tau^{\nu-2}} = \frac{\epsilon}{[3\pi(n - 1)]^{\nu-2}} \tag{42}$$

Although somewhat crude in its derivation, this result is nevertheless useful in estimating the error of approximation for particular values of n and ν. For example, if the first derivative of $F_1(\omega)$ is approximated in a piecewise-linear manner involving four break points and a maximum error of 10 per cent, then Eq. 42 for $\nu = 3$, $n = 4$, and $\epsilon = 0.1$ shows that $F_1(\omega)$ is approximated by parabolic arcs with a maximum error of less than 0.5 per cent, while the associated time function is given by four terms.

Essentially the same methods may be used to find the real and imaginary parts of $F(j\omega)$ from a given time function. Thus if the νth derivative of $f(t)$ for $t > 0$ is a sum of impulses, then $f(t)$ itself has the representation

$$f(t) = \sum_{k=1}^{n} a_k u_{-\nu}(t - t_k) \tag{43}$$

Since $u_{-\nu}(t)$ has the transform $s^{-\nu}$, and delay of a time function by t_k seconds amounts to multiplication by $e^{-j\omega t_k}$ in the frequency domain, we see that the transform of the function in Eq. 43 is

$$F(j\omega) = \sum_{k=1}^{n} a_k \frac{e^{-j\omega t_k}}{(j\omega)^\nu} \tag{44}$$

having the real and imaginary parts

$$F_1(\omega) = \frac{(-1)^{\nu/2}}{\omega^\nu} \sum_{k=1}^{n} a_k \cos \omega t_k \qquad \text{for } \nu \text{ even} \tag{45}$$

$$F_1(\omega) = \frac{(-1)^{(\nu+1)/2}}{\omega^\nu} \sum_{k=1}^{n} a_k \sin \omega t_k \qquad \text{for } \nu \text{ odd} \tag{46}$$

and

$$F_2(\omega) = \frac{(-1)^{(\nu+2)/2}}{\omega^\nu} \sum_{k=1}^{n} a_k \sin \omega t_k \qquad \text{for } \nu \text{ even} \tag{47}$$

$$F_2(\omega) = \frac{(-1)^{(\nu+1)/2}}{\omega^\nu} \sum_{k=1}^{n} a_k \cos \omega t_k \qquad \text{for } \nu \text{ odd} \tag{48}$$

3 Impulse Response Determination

When we are asked to find a network that will convert a given input time function into a specified output time function, it is effective to determine first the required impulse response of the network and then construct a rational function of s whose inverse transform is an acceptable approximation to this impulse response. Stability evidently requires that the pertinent impulse response be a bounded function of finite duration, or, if the duration is not finite, the function should settle down to a zero asymptote. Our first concern is with the determination of methods by which the stated input and output time functions may be tested to see whether these stability conditions are met, for no solution to the given problem exists unless they are.

Suppose we denote the input and output time functions respectively by $a(t)$ and $b(t)$, and the impulse response by $c(t)$. These are related by the familiar convolution integral

$$b(t) = \int_{-\infty}^{\infty} a(\xi)c(t - \xi)\, d\xi = \int_{-\infty}^{\infty} a(t - \xi)c(\xi)\, d\xi \tag{49}$$

If one of the functions $a(t)$ or $c(t)$ is replaced by its derivative, then $b(t)$ likewise becomes replaced by its derivative; if one of the functions $a(t)$ or $c(t)$ is differentiated and the other integrated, then $b(t)$ is unaffected. The fact that such operations may (if necessary) be applied repeatedly allows us to take advantage of the possibility of constructing $b(t)$ by inspection if either $a(t)$ or $c(t)$, or both, are in the form of impulse trains, for it is well known that the convolution of anything with a unit impulse equals that thing, whatever it is. Since the discussions of the preceding article have shown that some higher derivative of a function can, with remarkably good accuracy, be regarded as an impulse train, we see that an evaluation of the convolution integral by numerical methods is always possible with a precision that is subject only to limitations imposed by the pertinent graphical procedure.

Although in our present problem we are concerned with the evaluation of $c(t)$ from given functions $a(t)$ and $b(t)$ rather than with an evaluation of $b(t)$ from given functions $a(t)$ and $c(t)$, an example of the latter, illustrating the significance of these ideas, may be helpful in clarifying their implementation.

Accordingly, suppose we wish to determine the shape of pulse that results from convolving the rectangle shown in the top part of Fig. 3 with the triangle in the sketch below it. Following the process indicated in the first of the two integrals 49 leads to the graphical interpretation illustrated in the bottom part of Fig. 3 where the function $c(t - \xi)$ is drawn for a specific value of the parameter t, and the corresponding value of the output function $b(t)$ is indicated as the area (shown crosshatched) under the pertinent product $a \cdot c$ as a function of ξ.

Although the method of evaluating $b(t)$ for various values of t is thus very clear, the mechanics of carrying out this evaluation and obtaining a plot of $b(t)$ versus t appears to be quite a task if approached in this straightforward manner. We can materially simplify this problem by taking advantage of the principles pointed out above.

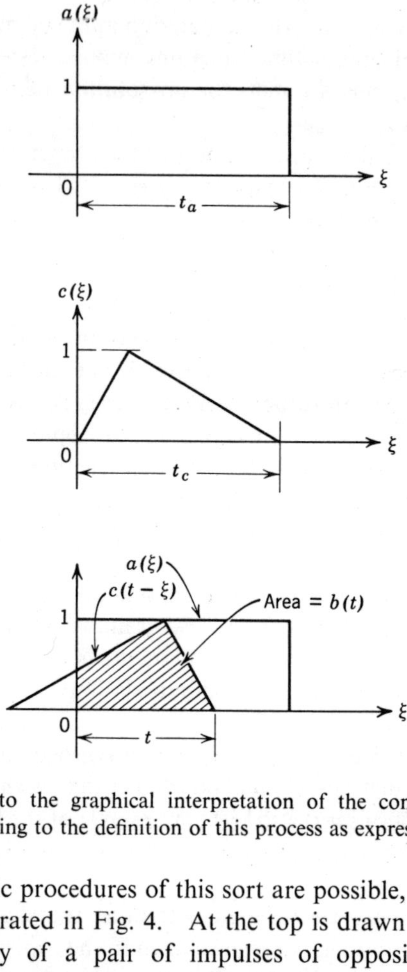

Fig. 3. Pertinent to the graphical interpretation of the convolution of functions $a(\xi)$ and $c(\xi)$ according to the definition of this process as expressed by the integral 49.

Various specific procedures of this sort are possible, a particular one of these being illustrated in Fig. 4. At the top is drawn the function da/dt, consisting simply of a pair of impulses of opposite sign and equal value unity. Knowing that the convolution or "scanning" of any function by a unit impulse reproduces that function exactly (see *MCA*, p. 488), we observe by inspection that scanning the triangle $c(\xi)$ with the pair of impulses given by da/dt yields the pair of alternately positive and negative triangles shown in the center of Fig. 4. Since this result is the derivative of the desired function $b(t)$, we have but to integrate it to obtain the solution to our problem, which is shown in the bottom part of Fig. 4. It consists of a set of confluent parabolic arcs connected by a flat portion of rather short duration at the center. These arcs are rather easily drawn knowing the location of each apex, the points of confluency, and the slopes at these points.

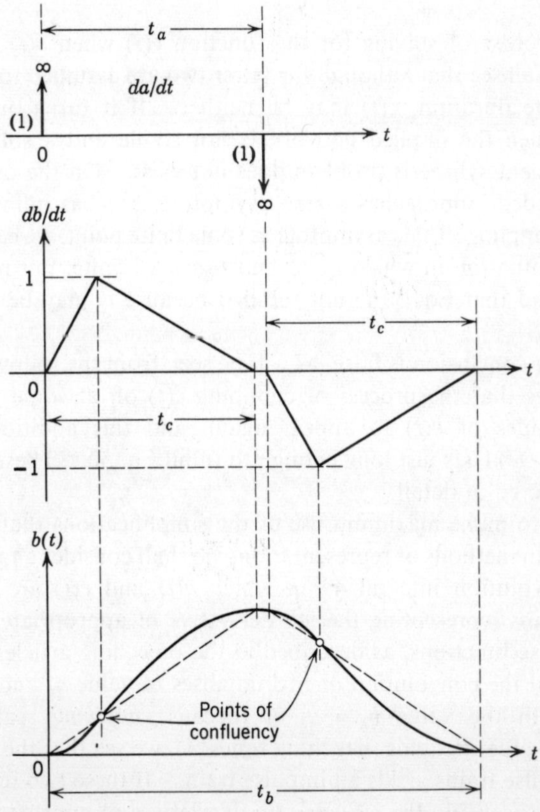

Fig. 4. Showing how the process of convolution in Fig. 3 may be more easily evaluated through initial differentiation of the function $a(t)$.

It is clear from this example that duration t_b of the resultant output equals the sum of durations t_a and t_c of the input and the impulse response functions respectively, that is

$$t_b = t_a + t_c \tag{50}$$

This situation, of course, assumes that all three functions $a(t)$, $b(t)$, and

$c(t)$ are of finite duration. In connection with physical network problems we know that although an input may be chosen to have finite duration, the impulse response and the output functions are theoretically infinite in extent. Nevertheless, in any stable situation, these response functions become substantially zero in a reasonably short finite interval, so that the method of chopping off asymptotes discussed in the previous article can be used to obtain finite intervals t_a, t_b, and t_c for all three functions, and under these circumstances the relation 50 between them must always be fulfilled.

In the process of solving for the function $c(t)$ when $a(t)$ and $b(t)$ are given, we shall see that although the latter two are assumed to be bounded and of finite duration, $c(t)$ may be neither. If it turns out not to be bounded, then the implied network is not stable and a solution to the stated transient synthesis problem does not exist. On the other hand, if $c(t)$ is bounded, approaches a zero asymptote, but has infinite duration, then, by chopping off this asymptote at some finite point, we can essentially regain the situation in which t_a, t_b, and t_c are all finite, but now it would seem offhand that Eq. 50 is not fulfilled because t_c may be much larger than t_a or t_b.

The latter conclusion is false, as will be seen from the following analysis which shows that the process of chopping $c(t)$ off at some point causes nonzero values of $b(t)$ to appear again, and this additional nonzero interval for $b(t)$ lasts just long enough to fulfill Eq. 50. These matters are now considered in detail.

In order to make maximum use of the simplifications that derive from impulse-train methods of representation, we shall consider a modified form of the convolution integral 49 in which $a(t)$ and $c(t)$ are replaced by impulse trains representing the νth derivative of appropriate approximations to these functions, as described in the preceding article. Since it is obvious that the convolution of two impulses of value a_k and c_k yields an impulse with the value $b_k = a_k \cdot c_k$ (because *anything* scanned by an impulse of value A equals that thing times A), we see that the convolution of two impulse trains yields an impulse train. If these two impulse trains represent respectively the νth and μth derivatives of approximants to $a(t)$ and $c(t)$, then the resulting impulse train represents the $(\nu + \mu)$th derivative of an approximant to $b(t)$.

To keep the following detailed discussion as simple as possible, we make the following two choices: (1) We use step approximations to the functions $a(t)$ and $c(t)$ so that their *first* derivatives become impulse trains, and (2) we choose a uniform time increment Δt between steps so that all relevant impulse trains become uniform and involve this same spacing. Neither the purpose of the following analysis nor the validity of the conclusions drawn from it is marred by this specialization, and at the same time greater clarity and simplicity are thereby achieved.

Convolution of the impulse trains representing da/dt and dc/dt thus yields a resulting impulse train which we must interpret as representing the second derivative of the desired function $b(t)$. The latter therefore should be thought of as approximated by a broken line with uniform spacing Δt between break points. The character of these approximations to the functions $a(t)$, $c(t)$ and $b(t)$ is shown in Fig. 5.

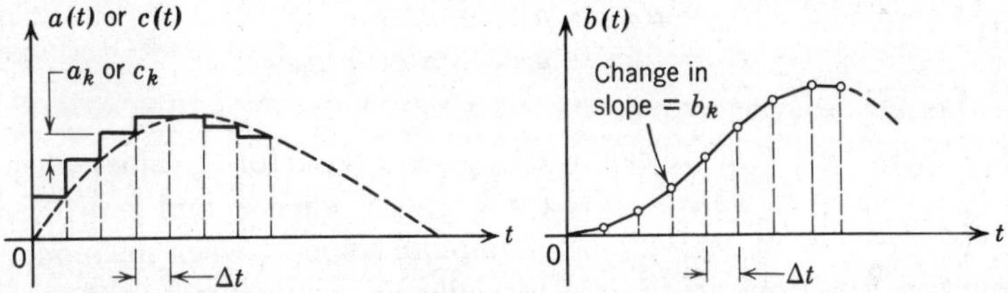

Fig. 5. Choosing a step approximation to both $a(t)$ and $c(t)$ yields a broken-line or piecewise linear approximation to $b(t)$ in the integral 49.

If we indicate differentiation by means of a prime, the functions pertinent to the convolution integral 49 then have the following representations

$$a'(t) = \sum_{k=0}^{n_a} a_k u_0(t - k\,\Delta t) \tag{51}$$

$$c'(t) = \sum_{k=0}^{n_c} c_k u_0(t - k\,\Delta t) \tag{52}$$

$$b''(t) = \sum_{k=0}^{n_b} b_k u_0(t - k\,\Delta t) \tag{53}$$

in which the integers n_a, n_b, n_c fulfill the relations

$$t_a = n_a\,\Delta t, \qquad t_b = n_b\,\Delta t, \qquad t_c = n_c\,\Delta t \tag{54}$$

and Eq. 50 demands that

$$n_b = n_a + n_c \tag{55}$$

The chief advantage of these impulse-train representations may now be seen to accrue from the fact that the value of the convolution integral

$$b''(t) = \int_{-\infty}^{\infty} a'(\xi)c'(t - \xi)\,d\xi \tag{56}$$

yields nonzero values only for t equal to an integer multiple of Δt (including zero), because the integrand is identically zero except when the impulses of $a'(\xi)$ and of $c'(t - \xi)$ fall upon each other. Being mindful of the reversed orientation of the function $c'(t - \xi)$ relative to $c'(\xi)$, as shown in Fig. 3, we see that for $t = 0$ only the first impulses in the two trains $a'(\xi)$ and $c'(\xi)$ coincide, giving $b_0 = a_0c_0$. For $t = \Delta t$, the first impulse in $c'(\xi)$ coincides with the second in $a'(\xi)$ and the second in $c'(\xi)$, with the first in a$'(\xi)$, giving $b_1 = a_1c_0 + a_0c_1$, and so forth. We thus obtain the sequence of relations

$$\begin{aligned} b_0 &= a_0c_0 \\ b_1 &= a_1c_0 + a_0c_1 \\ b_2 &= a_2c_0 + a_1c_1 + a_0c_2 \\ b_3 &= a_3c_0 + a_2c_1 + a_1c_2 + a_0c_3 \\ &\cdots\cdots\cdots\cdots\cdots \end{aligned} \tag{57}$$

It is quite clear that we can as easily solve these for c_k values when a_k and b_k are given as we can solve for b_k values when a_k and c_k are given. Thus the problem of determining an impulse response when the input and output time functions are given is formally, at least, reduced to the simple algebraic process of solving the system of triangular Eqs. 57.

In this connection it is significant to observe that in any specific situation in which the number of impulses in $a'(t)$ and in $c'(t)$ is finite, the number in $b''(t)$ is likewise finite, and so is the equation set 57. To illustrate the form of a complete set of these equations, suppose we consider the simple case defined by $n_a = 2$, $n_c = 4$, and $n_b = 6$, which yields

$$\begin{aligned} b_0 &= a_0c_0 \\ b_1 &= a_1c_0 + a_0c_1 \\ b_2 &= a_2c_0 + a_1c_1 + a_0c_2 \\ b_3 &= \phantom{a_2c_0 + {}} a_2c_1 + a_1c_2 + a_0c_3 \\ b_4 &= \phantom{a_2c_0 + a_2c_1 + {}} a_2c_2 + a_1c_3 + a_0c_4 \\ b_5 &= \phantom{a_2c_0 + a_2c_1 + a_2c_2 + {}} a_2c_3 + a_1c_4 \\ b_6 &= \phantom{a_2c_0 + a_2c_1 + a_2c_2 + a_2c_3 + {}} a_2c_4 \end{aligned} \tag{58}$$

Several things are of interest about this result. First, we see that the equations are solvable (for c_k's) with equal ease, starting from the top or from the bottom. Second, we see that if *all* the a_k's and b_k's are given, the problem is overspecified. Starting from the top, we have determined all the c_k's after considering the equation with b_4, and unless the last two equations are then found to be satisfied, the set of Eqs. 58 are inconsistent. Similarly, if we start from the bottom, all c_k's are computed after consideration of the equation with b_2, and the first two equations are required to be consistent with these results.

This situation could have been seen at the outset from the condition expressed by Eq. 50 or its equivalent form 55 which shows that if $a(t)$ is specified over its entire interval t_a, then $b(t)$ can be specified independently only over an interval $t_b - t_a = t_c$. Throughout the remainder of its interval, $b(t)$ must then assume values that are consistent with the specifications already made, which include the a priori stipulation regarding a finite duration of the function $c(t)$.

Before considering this question in greater detail, we must pay some attention also to stability requirements. In this regard we shall ask that all three functions $a(t)$, $b(t)$, and $c(t)$ begin and end with zero values. So far as the impulse-train representations 51 and 52 are concerned, this requirement leads to the conditions

$$\sum_{k=0}^{n_a} a_k = 0 \tag{59}$$

and

$$\sum_{k=0}^{n_c} c_k = 0 \tag{60}$$

With regard to the impulse train 53 we note that if $b(t)$ encloses a nonzero area (the usual situation), then the transform of $b''(t)$ has a double-order zero at $s = 0$ and hence the area enclosed by $tb''(t)$ as well as that enclosed by $b''(t)$ itself must be zero. This result leads to the pair of conditions*

$$\sum_{k=0}^{n_b} b_k = 0 \quad \text{and} \quad \sum_{k=0}^{n_b} kb_k = 0 \tag{61}$$

Actually, one does not need additionally to be concerned with the latter conditions on the b_k's if those on the a_k's and c_k's expressed by Eqs. 59 and 60 are fulfilled, since the convolution process in that event automatically yields b_k's satisfying conditions 61. Condition 59 presumably will also be fulfilled in every case because the given input function $a(t)$ is specified to have zero values at both ends. However, condition 60 involving the impulse response which we are computing is neither automatically fulfilled nor fixed by specification. Its fulfillment adds one more equation to those in the set 58 which we are using for the determination of the c_k's; and thus it subtracts one more b_k value from the independently specifiable number.

Since the total number of c_k's is $n_c + 1$, the number determined by Eqs. 58 alone is n_c, which is, therefore, also equal to the number of inde-

---

* If $b(t)$ encloses *zero* area, then an additional moment condition involving the summation of $k^2 b_k$ is required. Conditions 61 therefore represent a minimum requirement.

pendently specifiable b_k's if all $n_a + 1$ of the a_k's are assumed to be given. In that event, we see that $n_b - n_c + 1$ of the b_k's must remain unspecified, or that a portion of the output function having the duration of $a(t)$ is unspecifiable. Let us consider a few specific examples illustrating the significance of these results.

Suppose we begin by stipulating that we want the pair of input and output functions shown in Fig. 6. Since we know that we cannot completely specify the output, we shall assume that an unspecified portion of duration $2\Delta t$ exists beyond the point $t = 6\Delta t$, and that we are going to let the computations determine what this portion looks like.

The given data enable us to write down the impulse values

$$a_k = 1, 1, -2 \qquad \text{for } k = 0, 1, 2 \tag{62}$$

$$b_k = 1, -\tfrac{1}{2}, -\tfrac{1}{2}, -\tfrac{1}{2}, -\tfrac{1}{2}, 1 \qquad \text{for } k = 0, 1, 2, 3, 4, 5 \tag{63}$$

Additional b_k values for $k = 6, 7, 8$ remain unspecified. These, together with the values of $c_0, c_1, \cdots, c_6$ are to be determined from condition 60 and the pertinent equations like 58, which read

$$\begin{aligned}
1 &= c_0 \\
-\tfrac{1}{2} &= c_0 + c_1 \\
-\tfrac{1}{2} &= -2c_0 + c_1 + c_2 \\
-\tfrac{1}{2} &= -2c_1 + c_2 + c_3 \\
-\tfrac{1}{2} &= -2c_2 + c_3 + c_4 \\
1 &= -2c_3 + c_4 + c_5 \\
b_6 &= -2c_4 + c_5 + c_6 \\
b_7 &= -2c_5 + c_6 \\
b_8 &= -2c_6
\end{aligned} \tag{64}$$

Together with condition 60, these yield the values

$$c_k = 1, -\tfrac{3}{2}, 3, -\tfrac{13}{2}, 12, -24, 16 \qquad \text{for } k = 0, \cdots, 6 \tag{65}$$

and

$$b_k = -32, 64, -32 \qquad \text{for } k = 6, 7, 8 \tag{66}$$

The last three values indicate that the unspecified portion of the output function consists of a negative triangular spike, 32 units tall. Compared with the desired portion of $b(t)$ shown in Fig. 6, this spike almost dwarfs the rest of the output. We may interpret it as being called for in order to permit the impulse response (shown in Fig. 7) to remain bounded and have a finite duration.

Now let us consider the unspecified portion of the output function $b(t)$ to be located at the front end of this pulse. The b_k values 63 then are considered to pertain respectively to $k = 3, 4, 5, 6, 7, 8$, and b_0, b_1, b_2 as well as the c_k's are regarded as unknown. Equations 64 are altered only in that the numerical values on the left are moved bodily down by three lines, thus displacing the quantities b_6, b_7, b_8 at the bottom and leaving room for b_0, b_1, b_2 at the top. Together with condition 60, these modified equations yield

$$c_k = \tfrac{1}{16}, -\tfrac{3}{16}, \tfrac{3}{8}, \tfrac{1}{4}, 0, 0, -\tfrac{1}{2} \qquad \text{for } k = 0, \cdots, 6 \tag{67}$$

and

$$b_k = \tfrac{1}{16}, -\tfrac{1}{8}, \tfrac{1}{16} \qquad \text{for } k = 0, 1, 2 \tag{68}$$

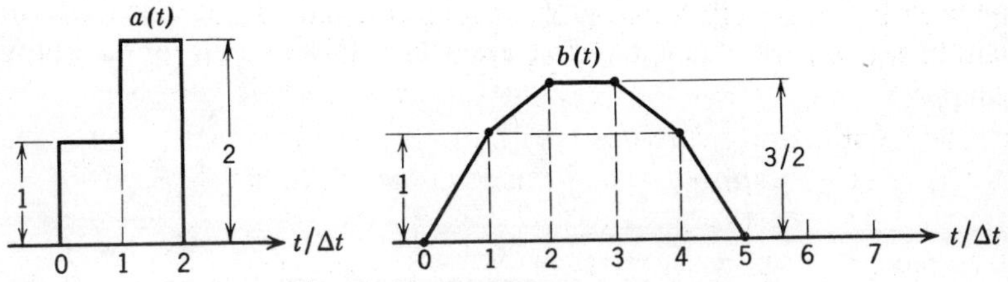

Fig. 6. Pair of input and output time functions for which the appropriate impulse response is to be calculated. An unspecified portion of $b(t)$ remains to be determined from the condition that the impulse response be bounded and of finite duration.

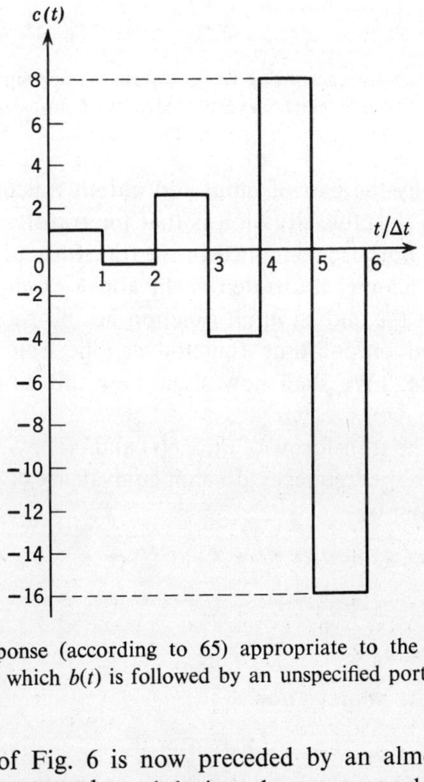

Fig. 7. Impulse response (according to 65) appropriate to the input and output functions in Fig. 6 in which $b(t)$ is followed by an unspecified portion defined by the values 66.

The pulse $b(t)$ of Fig. 6 is now preceded by an almost insignificant triangular bump of height $\frac{1}{16}$, and the impulse response relating this output to the input $a(t)$ is shown in Fig. 8. It is quite apparent that this situation is far less critical than the previous one in which a threatening instability is held in check only by acceptance of an excessively large quasi impulse in the final stage of the output function. In contrast, the impulse response now is very mild in character, and no violent departure from the desired form of output pulse is called for. We begin to see the truth of a statement made earlier to the effect that any pair of input and output time functions (within reason) are obtainable if we are willing to wait a bit for the output to appear.

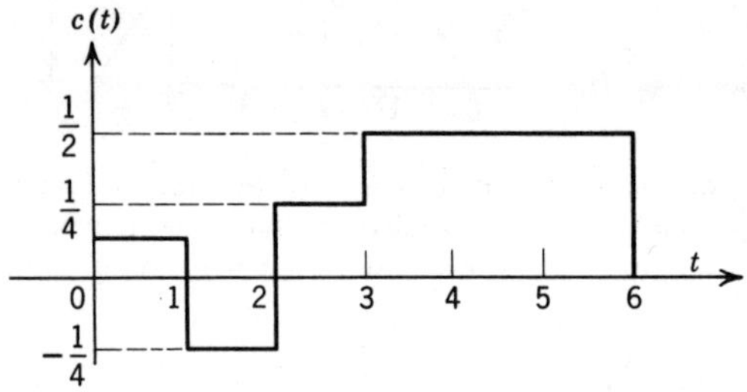

Fig. 8. Impulse response (according to the values 67) appropriate to the input and output functions in Fig. 6 in which $b(t)$ is preceded by an unspecified portion defined by the values 68.

The reason why the pair of input and output functions in Fig. 6 leads essentially to an unstable situation is that the transform of $a(t)$ has right half-plane zeros not also contained in the transform of $b(t)$ unless its form is modified in a manner illustrated in the above examples. Evidently, if the transform of the chosen input function has no right half-plane zeros, then any desired output time function can be obtained with a stable impulse response. We shall now show how this situation can be controlled.

If we denote the transforms of $a(t)$, $b(t)$ and $c(t)$ by $A(s)$, $B(s)$, and $C(s)$ respectively, then the frequency-domain equivalents of Eqs. 51, 52, and 53 are evidently given by

$$sA(s) = a_0 + a_1 e^{-\Delta ts} + a_2 e^{-2\Delta ts} + \cdots + a_{n_a} e^{-n_a \Delta ts} \tag{69}$$

$$sC(s) = c_0 + c_1 e^{-\Delta ts} + c_2 e^{-2\Delta ts} + \cdots + c_{n_c} e^{-n_c \Delta ts} \tag{70}$$

$$s^2 B(s) = b_0 + b_1 e^{-\Delta ts} + b_2 e^{-2\Delta ts} + \cdots + b_{n_b} e^{-n_b \Delta ts} \tag{71}$$

Making use of the abbreviation

$$p = e^{-\Delta ts} \tag{72}$$

these frequency-domain functions become polynomials in the variable p.

We will write them

$$A^*(p) = a_0 + a_1 p + a_2 p^2 + \cdots + a_{n_a} p^{n_a} \tag{73}$$

$$C^*(p) = c_0 + c_1 p + c_2 p^2 + \cdots + c_{n_c} p^{n_c} \tag{74}$$

$$B^*(p) = b_0 + b_1 p + b_2 p^2 + \cdots + b_{n_b} p^{n_b} \tag{75}$$

Since the change-of-variable 72 maps the right half of the s-plane upon the interior of the unit circle in the p-plane, we can say that the transform $A(s)$ of the input time function $a(t)$ has no right half-plane zeros if the polynomial $A^*(p)$ has no zero with an absolute value less than unity. If we make an additional change-of-variable by means of the well-known transformation

$$p = \frac{1 - \lambda}{1 + \lambda} \quad \text{or} \quad \lambda = \frac{1 - p}{1 + p} \tag{76}$$

then the left half of the s-plane is mapped upon the left half of the λ-plane. The polynomial $A^*(p)$ becomes

$$A^*(p) = P_a(\lambda)/(1 + \lambda)^{n_a} \tag{77}$$

where

$$P_a(\lambda) = a_0(1 + \lambda)^{n_a} + a_1(1 + \lambda)^{n_a - 1} \times (1 - \lambda) + \cdots + a_{n_a}(1 - \lambda)^{n_a} \tag{78}$$

and we may conclude that the transform $A(s)$ of $a(t)$ has no right half-plane zeros if $P_a(\lambda)$ is a Hurwitz polynomial. Since the latter can readily be formed from the a_k values characterizing $a(t)$, we have a direct way of determining whether this time function leads to a transform with right half-plane zeros.

For the function $a(t)$ in Fig. 6, with the a_k values 62, we find

$$P_a(\lambda) = 2\lambda(3 - \lambda) \tag{79}$$

which clearly is not Hurwitz. The origin of the λ-plane, incidentally, corresponds to the origin of the s-plane. That is why the polynomial $P_a(\lambda)$, corresponding to $sA(s)$ in Eq. 69, contains the factor λ. Its presence, therefore, has nothing to do with the stability considerations under discussion. If $a(t)$, like $b(t)$ in Fig. 6, is given by a broken-line approximation, then 69 becomes a representation for $s^2A(s)$ and $P_a(\lambda)$ contains the factor λ^2. In general, if the νth derivative of $a(t)$ is represented by a sum of impulses, then $P_a(\lambda)$ contains λ^ν as a factor. In any case, it is the rest of $P_a(\lambda)$ to which the pertinent Hurwitz test is applied.

Significant in connection with the present topic is the effect upon the polynomial $P_a(\lambda)$ of turning the time function $a(t)$ end for end, that is to say, of replacing $a(t)$ by $a(n_a\,\Delta t - t)$. Since this process replaces the sequence of values $a_0, a_1, \cdots, a_{n_a}$ respectively by the sequence $a_{n_a}, a_{n_a - 1}, \cdots, a_0$, Eq. 73 shows that $A^*(p)$ is replaced by $p^{n_a}A^*(1/p)$, and hence, by Eqs. 76 and 77, $P_a(\lambda)$ becomes replaced by $P_a(-\lambda)$. Thus we see that if a given pulse $a(t)$ has a transform $A(s)$ with right half-plane zeros exclusively, then the transform of that pulse turned end for end has no right half-plane zeros. A corollary to this result is the statement that the transform of a symmetrical pulse has a zero distribution that is symmetrical with respect to both the real and imaginary axes of the s-plane [the polynomial $P_a(\lambda)$ involves only even or only odd powers of λ].

As an illustration of these results let us consider the output function $b(t)$ as given in Fig. 6, but turn the input function around so that the values 62 become replaced by

$$a_k = 2, -1, -1 \qquad \text{for } k = 0, 1, 2 \tag{80}$$

The b_k values remain as in 63. Equations 73 and 75 then yield

$$A^*(p) = 2 - p - p^2 \tag{81}$$

$$B^*(p) = 1 - \tfrac{1}{2}p - \tfrac{1}{2}p^2 - \tfrac{1}{2}p^3 - \tfrac{1}{2}p^4 + p^5 \tag{82}$$

Since in the frequency domain we have

$$C^*(p) = B^*(p)/A^*(p) \tag{83}$$

we can find the c_k values determining the associated impulse response by the simple process of long division, as follows

$$\begin{array}{r|l}
 & \tfrac{1}{2} + 0p + 0p^2 - \tfrac{1}{4}p^3 - \tfrac{3}{8}p^4 + \tfrac{3}{16}p^5 - \tfrac{3}{32}p^6 \\
\hline
2 - p - p^2 & 1 - \tfrac{1}{2}p - \tfrac{1}{2}p^2 - \tfrac{1}{2}p^3 - \tfrac{1}{2}p^4 + p^5 \\
 & 1 - \tfrac{1}{2}p - \tfrac{1}{2}p^2 \\
\hline
 & \qquad - \tfrac{1}{2}p^3 - \tfrac{1}{2}p^4 + p^5 \\
 & \qquad - \tfrac{1}{2}p^3 + \tfrac{1}{4}p^4 + \tfrac{1}{4}p^5 \\
\hline
 & \qquad\qquad - \tfrac{3}{4}p^4 + \tfrac{3}{4}p^5 \\
 & \qquad\qquad - \tfrac{3}{4}p^4 + \tfrac{3}{8}p^5 + \tfrac{3}{8}p^6 \\
\hline
 & \qquad\qquad\qquad \tfrac{3}{8}p^5 - \tfrac{3}{8}p^6 \\
 & \qquad\qquad\qquad \tfrac{3}{8}p^5 - \tfrac{3}{16}p^6 - \tfrac{3}{16}p^7 \\
\hline
 & \qquad\qquad\qquad\quad - \tfrac{3}{16}p^6 + \tfrac{3}{16}p^7
\end{array} \tag{84}$$

The resulting impulse response is theoretically infinite in duration, having the sequence of c_k values

$$c_k = \tfrac{1}{2}, 0, 0, -\tfrac{1}{4}, -\tfrac{3}{8}, \tfrac{3}{16}, -\tfrac{3}{32}, \tfrac{3}{64}, -\tfrac{3}{128}, \cdots \tag{85}$$

The corresponding time function is shown in Fig. 9.

As mentioned above, one might in a situation of this sort be tempted to conclude that the basic relation 50 is violated, since t_a and t_b are presumably both finite while t_c is infinite. However, the conclusion that t_b consists only of the five Δt intervals shown in Fig. 6 is incorrect. To be sure, $b(t)$ remains zero after $t = 5\Delta t$, but only because $c(t)$ keeps on having the proper nonzero values to keep $b(t)$ zero. These zero values of $b(t)$ should, strictly speaking, be regarded as a legitimate part of this function, for, should $c(t)$ be cut off at some finite point, $b(t)$ would become nonzero again, and for precisely the number of Δt intervals beyond the cutoff point of $c(t)$ that equal the duration of $a(t)$. Whereupon Eq. 50 is again fulfilled.

Considering the manner in which $c(t)$ in Fig. 9 is approaching zero with increasing t, we may expect that the act of cutting this function off after $t = 9$ or 10 times Δt will cause rather negligible values of $b(t)$ to appear beyond this point, so that for all practical purposes such a reappearance of output values may be ignored.

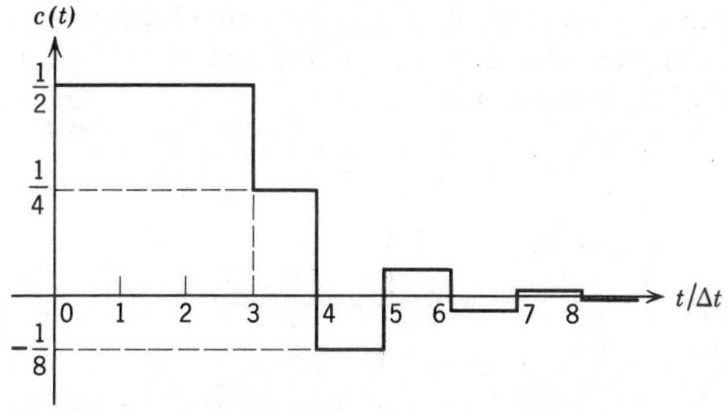

Fig. 9. Impulse response (according to the values 85) appropriate to the functions in Fig. 6 in which $a(t)$ is turned end for end.

We see then that the impulse response needed to relate a pair of given input and output time functions may always be computed in a straightforward manner, and that the stability of that impulse response can readily be predicted directly from the given time-domain data. It is sufficient, therefore, to consider only the construction of an appropriate rational response function for a specified time function which is to be the unit impulse response of a desired network.

4 Impulse Response Synthesis; General Considerations

The problem is to construct a rational function (satisfying appropriate realizability conditions) whose inverse transform is an acceptable approximation to the desired impulse response $f(t)$. More specifically, if $F(s)$ denotes the transform of $f(t)$, and if this transform is not already a realizable rational function (the usual situation), we are required to construct a rational approximant $F^*(s)$ such that its inverse transform $f^*(t)$ approximates $f(t)$ in a satisfactory manner.

In view of considerations presented in the preceding articles, it seems best to approach this problem through the formulation of a rational approximant to either the real or imaginary part of $F(j\omega)$, since correlation of errors committed in the frequency domain with resulting errors in the time domain can more readily be accomplished by straightforward numerical computation. These errors are for convenience placed into two categories: (1) errors due to the chopping off of the desired $F_1(\omega)$ or $F_2(\omega)$ beyond some finite frequency ω_c, and (2) errors arising from the imperfect manner in which the real or imaginary part is approximated within the finite interval $0 < \omega < \omega_c$.

The first of these, which is commonly referred to as a "truncation" error, is easily formulated analytically by noting (according to well-known transform theory) that multiplication of $F(j\omega)$ by some cutoff function $H(j\omega)$ amounts to convolution of $f(t)$ with $h(t)$, the inverse transform of $H(j\omega)$. If, for example, the latter is real and has the trapezoidal form shown in Fig. 10, then we find

$$h(t) = \frac{\omega_m}{\pi} \times \frac{\sin \Delta t}{\Delta t} \times \frac{\sin \omega_m t}{\omega_m t} \tag{86}$$

which, for $\Delta = 0$, becomes

$$h(t) = \frac{\omega_c}{\pi} \times \frac{\sin \omega_c t}{\omega_c t} \tag{87}$$

for $\Delta = \omega_m = \omega_c/2$ is given by

$$h(t) = \frac{\omega_c}{2\pi} \times \left(\frac{\sin \omega_c t/2}{\omega_c t/2}\right)^2 \tag{88}$$

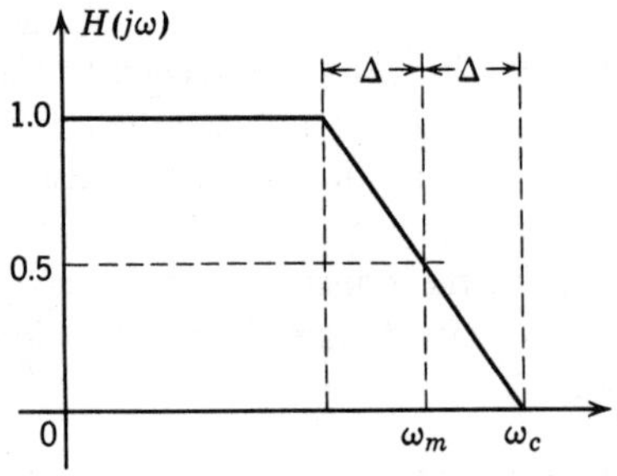

Fig. 10. Trapezoidal form of truncation function whose inverse transform is given by Eq. 86.

and, for intermediate Δ values, has a behavior that is somewhere between these two extremes. Thus, if $f(t)$ has discontinuities, truncation with a rectangular cutoff function (amounting to convolution with 87) produces the familiar Gibbs phenomenon, and truncation with a triangular cutoff function (corresponding to convolution with 88) yields the ultraconservative behavior of a Cesàro sum (see *MCA*, pp. 285, 496–501). A value of Δ yielding a proper amount of ripple can in such a situation readily be determined, whereas if $f(t)$ has no discontinuities, truncation with a rectangular cutoff function (which is simplest to consider) yields no serious distortion provided the cutoff frequency ω_c is large enough.

In any event, the nature and amount of distortion in the time domain caused by truncation of the frequency-domain function can easily be estimated, and an appropriate cutoff function determined (as is illustrated in detail in the following article).

The second group of time-domain errors arising from imperfect approximation within the interval $0 < \omega < \omega_c$ may be held within negligible limits or can, in any doubtful situation, be computed by the methods of Art. 2.

The practical consideration that aggravates control of errors is a desire to achieve an adequate result with a rational function having as few poles as possible. If the resulting number of poles were of no moment, error control would be easy; but as a practical matter we never find ourselves in such a simple situation. On the contrary, we try in every conceivable manner to squeeze the last ounce of approximation ability out of a rational approximant of given complexity.

The major problem, however, is that of finding a good way of approximating a given $F_1(\omega)$ or $F_2(\omega)$ over a finite frequency interval by means of a rational function. Strange as it may seem, mathematicians have not, so far at least, produced a satisfactory solution to this problem, and we find ourselves devising all sorts of makeshift methods which are reasonably satisfactory in special situations but none are uniformly applicable, so that a certain amount of experience and judgment is needed in their selection and use. Thus if $F_1(\omega)$ has its zeros (or most of them) on the real frequency axis, we can construct a numerator polynomial from these known zeros, and then determine a denominator polynomial by demanding that the rational approximant shall pass through selected points on the desired $F_1(\omega)$ curve. As is well known, success of this method depends upon the proper choice of these points, for which there is no guide except that provided by trial and error.

Instead of trying to select such a set of points we may expand $1/F_1(\omega)$ into partial fractions since its poles are the zeros of $F_1(\omega)$ and its residues there are given by the slopes of $F_1(\omega)$ at these same points. Not only is this scheme likewise inconsistent in the results obtained but it leads to considerable computational difficulty regarding control of the asymptotic behavior of the resulting approximant and hence of the behavior of $f^*(t)$ at or near $t = 0$.

A procedure suggested by the methods discussed in Art. 2 is to plot the function

$$\phi(\omega) = \int_{-\infty}^{\omega} F_1(u)\, du \tag{89}$$

and approximate it within the interval $0 < \omega < \omega_c$ by a step curve, whereupon the derivative yields an impulse-train representation for $F^*{}_1(\omega)$.

Interpretation of each impulse of this real part as a pole of $F^*(s)$ enables one to construct this rational approximant. However, assuming a uniform increment $\Delta\omega$ between impulses or between poles, Fourier theory tells us that the associated $f^*(t)$ is periodic with a period $\tau = 2\pi/\Delta\omega$. Hence we can obtain a desired aperiodic $f^*(t)$ only by letting the poles become infinitely dense on the j-axis, which is practically unacceptable.

We can attempt to salvage this method by predistortion of the desired $f(t)$ through multiplication by $e^{\alpha t}$ and subsequent multiplication of the periodic $f^*(t)$ by $e^{-\alpha t}$ so as to damp out essentially all but the first period of that function. This process leads to rather uneconomical results.

A vastly more satisfactory solution is obtained by noting that the graphical method in Art. 2 for the transformation from $F_1(\omega)$ or $F_2(\omega)$ to $f(t)$, leading to Eqs. 11 and 12 or 17 and 18, yields a time function in the form of a sum of damped sinusoids, albeit the damping factor is $1/t^\nu$ instead of $e^{-\alpha t}$. If, for the moment, we consider an $f(t)$ that is nonzero only over an interval throughout which these two damping factors can, except for a constant multiplier, be regarded as essentially equal, then the transform of $f(t)$ with $1/t^\nu$ replaced by $e^{-\alpha t}$ is the rational approximant desired.

We can remove the approximations involved in this reasoning by predistorting the desired $f(t)$ through multiplication by $t^{-\nu}e^{\alpha t}$ before computing its transform $F(s)$ having the real part $F_1(\omega)$. The time function obtained from this real part by the method of Art. 2, when multiplied by $t^\nu e^{-\alpha t}$, is the desired time-domain approximant having a rational transform.

Unless $f(t)$ has at least a νth-order zero at $t = 0$, multiplication by the predistortion factor $t^{-\nu}e^{\alpha t}$ yields a time function that is no longer bounded at $t = 0$. When this difficulty arises, it can be met by subtracting from $f(t)$ a component of the form

$$A_1e^{-\alpha_1 t} + A_2e^{-\alpha_2 t} + \cdots + A_ke^{-\alpha_k t} \tag{90}$$

so chosen that the first ν terms of its Maclaurin expansion coincide with the corresponding ones in a similar expansion of $f(t)$. The transform of this component is rational, and the remainder function fulfills the stated necessary condition (namely, that it have at least a νth-order zero at $t = 0$).†

5 Impulse Response Synthesis; a "Standard Parts" Method

Although the preceding considerations tend to establish the thought that the solution to a time-domain synthesis problem is far from straight-

† For a detailed discussion of this method, the reader is referred to "Network Synthesis for a Prescribed Impulse Response," R. A. Pucel, *Sc.D. thesis*, Massachusetts Institute of Technology, June 1955.

forward and involves the exercise of judgment based upon experience, only a small amount of additional sifting and organization of these methods is required to yield an approach that not only is extremely simple and infallible in principle but has the further advantage that results of the time-consuming computational part of the work can be tabulated and plotted for reuse. A sufficient store of such tabulated material may then serve as a set of "standard parts" out of which the solution to any given problem can be constructed with a minimum of effort.

The basis for such thinking is supplied by the numerical analysis method described in Art. 2 where a given function is represented as the νth repeated integral of an impulse train. Thus we see that the delayed impulse (supplemented by integration) may be regarded as a building block for the construction of arbitrary functions. Once we have rational approximants for the transforms of an appropriate set of delayed impulses or reasonable facsimiles thereof), we are in a position to construct rational approximants for arbitrary functions through their linear combination.

The reasonable facsimile of a delayed impulse is evidently a delayed pulse. On a normalized basis we need consider only a pulse-enclosing unit area, centered at the point $t = 1$. The degree of simulation is inversely proportional to the pulse width which must be defined in a manner consistent with the pulse shape chosen. A tabulation of rational approximants for various pulse widths provides the stock of standard parts needed for any given problem.

The location and number of poles required in its solution are determined by the specified approximation tolerance and by the amount of delay called for by the remotest member in the impulse-train representation of the νth derivative of the desired impulse response. As is shown in detail below, the required approximation tolerance determines the pulse width, which we may assume to be the same for any one of the simulated impulses in the train. This width divided by the delay of the remotest impulse then equals the width per unit delay characterizing the pertinent rational function in the standard parts tabulation. Appropriately scaled in magnitude and frequency, that function provides the final quasi impulse in the train; and the rational approximants for all other members of the train are formed from this one. Thus, if this most remote pulse (per unit area) is denoted by $p(t)$ and its transform by $P(s)$, we have

$$p(t) = \sum k_i e^{s_i t} \tag{91}$$

$$P(s) = \sum \frac{k_i}{s - s_i} \tag{92}$$

from which we see that an earlier pulse $p(t + \tau)$ has a transform 92 in which the residues k_i are multiplied respectively by the factors $e^{s_i \tau}$.

Important here is the result that the transforms for all pulses have the same poles, so that the number of poles in the resultant rational approximant is (as stated above) fixed by the amount of delay required by the remotest impulse and the tolerance with which it is to be reproduced. We may say that the remotest simulated impulse costs a certain price in terms of the number of poles required for its realization; all other pulses, regardless of their number, add nothing to the over-all cost.* Their presence merely affects the values of the residues in the resultant rational approximant.

In this connection it is significant to observe that the width of all pulses produced by this method is the same, which would not be so if the pulse locations were controlled by scaling. In that case the ratio of width to delay remains constant (earlier pulses being correspondingly narrower) and, of course, each pulse requires the same number of poles and not the same ones as are required by the other pulses. Hence scaling, as a means for obtaining the rational approximants of all pulses from that of a single one, does not lead to a uniform tolerance or quality factor in the over-all approximation, and furthermore yields a comparatively very poor economy in the resultant number of poles required.

It remains to determine the rational transform of a simulated impulse having unit delay, and to show how the design procedure may be conducted so as to meet the stated time-domain tolerance in the approximation of a desired impulse response.

The transform of a unit impulse occurring at $t = 1$ is e^{-s} and its j-axis real part is $\cos \omega$. If we truncate this real part by cutting it off abruptly at ω_c, leaving it otherwise intact, then (since the inverse Fourier transform of $\cos \omega$ is a pair of $\frac{1}{2}$-value impulses at $t = \pm 1$) the resultant impulse response (equal to twice the resultant inverse Fourier transform for $t > 0$) becomes $h(t - 1) + h(t + 1)$ with $h(t)$ given by Eq. 87. For any reasonable truncation frequency ω_c, this result equals essentially the function 87 delayed by 1 second.

Although this well-known function has a considerable ripple in its precursor and coda regions, it represents an economical choice, for it does yield a relatively narrow pulse width and the fewest poles for a given bandwidth or cutoff frequency ω_c. We shall show that the ripple amplitude becomes negligibly small after two or more successive integrations, and hence presents no objectionable feature so long as the desired impulse response is approximated in such a way that its second or higher derivative is the impulse train to be simulated. In other words, the quasi-impulse

* This important feature and the method of obtaining it were suggested by T. G. Stockham, Jr., in his *S.M. thesis*, Massachusetts Institute of Technology, 1956.

function 87 is adequate if we use it to approximate the second or higher derivative of our desired time function, since subsequent integrations suppress ripples sufficiently to render them unimportant.

In this connection it must not be overlooked that integration as a means for ripple suppression is not a cost-free procedure. As we go from a piecewise-linear approximation to one involving second order and higher order parabolic arcs, the spacing between impulses (in general) decreases, thus requiring narrower quasi impulses, an increased bandwidth, and hence more poles. Nevertheless, we will in most cases arrive at as good or better economy by proceeding from the function 87 as a basic quasi impulse and relying upon integration as a means for rendering its ripples insignificant, as we would through the use of a somewhat more costly (and less ripply) approximation to the impulse function in conjunction with a procedure involving fewer subsequent integrations.

We conclude that abrupt truncation of the real part cos ω, leading to the familiar function 87 as a basic simulated impulse, permits the same ultimate tolerance control, subject to pole requirements on the part of the resultant rational approximant, as would be provided by some less ripply simulated impulse obtainable from an appropriately modified form of truncation function. Choice of the function 87, moreover, permits maximum flexibility in tolerance control since it initially sets no arbitrary upper bound on the suppression of ripples, while allowing, through its use, any ultimate degree of ripple suppression in the final function synthesized.

These matters are now brought into sharper focus through the following analytical comparison of the true impulse and its simulation by the function 87. The difference between these two, which may be denoted as an error function

$$\epsilon_0(t) = \frac{\omega_c}{\pi} \times \frac{\sin \omega_c t}{\omega_c t} - u_0(t) \tag{93}$$

after one integration becomes

$$\epsilon_{-1}(t) = \int_{-\infty}^{t} \epsilon_0(\xi)\, d\xi = \frac{1}{2} + \frac{1}{\pi} Si(\omega_c t) - u_{-1}(t) \tag{94}$$

in which $u_{-1}(t)$ is the unit step and $Si(x)$ is the so-called sine-integral function.* Although its integral is not tabulated, we can compute and plot the second successive integral of $\epsilon_0(t)$ indicated by

$$\epsilon_{-2}(t) = \int_{-\infty}^{t} \int_{-\infty}^{\tau} \epsilon_0(\xi)\, d\xi\, d\tau \tag{95}$$

The resulting functions 94 and 95 are shown plotted in Figs. 11 and 12.

* See Jahnke and Emde Tables of Functions, Dover Publications, New York, 1943, pp. 1–4.

Figure 11 shows the error in the vicinity of a unit step (the familiar Gibbs phenomenon); Fig. 12, the error with which the unit ramp function is approximated. Except for some very minor ripples, the essential error here consists of the appearance of a fillet at the break point having a value $1/\pi\omega_c$. One can easily obtain this value by observing that the transform of $\epsilon_0(t)$ in Eq. 93 is $H(j\omega) - 1$ with $H(j\omega)$ given by Fig. 10 for $\Delta = 0$. The transform of $\epsilon_{-2}(t)$ is obtained through division by $(j\omega)^2$, and hence we have

$$\epsilon_{-2}(0) = \frac{1}{2\pi}\int_{-\infty}^{\infty} \frac{1 - H(j\omega)}{\omega^2}\, d\omega = \frac{1}{\pi}\int_{\omega_c}^{\infty} \frac{d\omega}{\omega^2} = \frac{1}{\pi\omega_c} \tag{96}$$

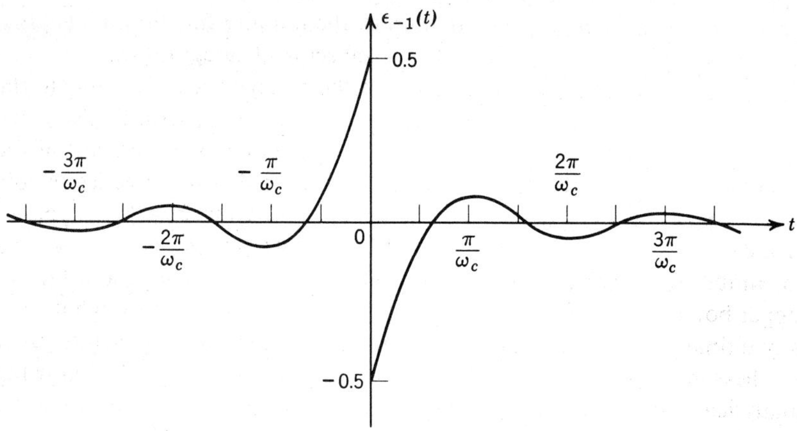

Fig. 11. The once-integrated error 93 as given in Eq. 94.

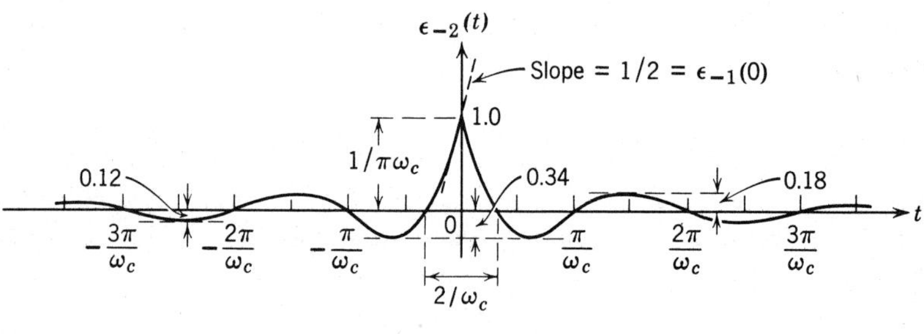

Fig. 12. The twice-integrated error 93 as represented by Eq. 95.

We will define the "width" of the pulse 87 by

$$\delta = \pi/\omega_c \tag{97}$$

which is the width of a rectangular pulse of equal height and equal (unit) area. In terms of this pulse width, the fillet 96 has the value

$$\epsilon_{-2}(0) = \delta/\pi^2 \tag{98}$$

which is roughly one-tenth of δ. At any break point involving unit change in slope, the resulting error amounts essentially to the insertion of a fillet having the magnitude 96 or 98. For a slope change other than unity, the size of the fillet is proportionally modified; and for a specified maximum allowable fillet size, the necessary pulse width δ or truncation frequency ω_c is immediately determined.

It is pretty evident that an additional integration reduces the error function to much smaller values which in most practical cases can be neglected entirely. Hence if the given impulse response is approximated by parabolic arcs, the pulse width δ is (most likely) no longer determined by a ripple requirement but rather by the condition that it must be at least as small as the closest spacing between impulses in the νth derivative.

The rational transform of the desired simulated impulse is one whose real part approximates $\cos \omega$ or whose imaginary part approximates $-\sin \omega$ over an interval $-\omega_c < \omega < \omega_c$, where the choice of ω_c is now known except for some minor revision having to do with the necessity of imitating an asymptotic behavior on the part of this transform appropriate to having the associated time function begin with at least a zero value at $t = 0$.

As a basis for constructing the rational transform of our simulated impulse we shall use the method suggested in the previous article in which we begin by constructing an $f(t)$ from the pertinent $F_1(\omega)$ or $F_2(\omega)$ function by the procedure of Art. 2, and then replace $1/t^\nu$ by $e^{-\alpha t}$. The latter step amounts to multiplying $f(t)$ by $t^\nu e^{-\alpha t}$, which is itself a pulselike function having a maximum at $t = \nu/\alpha$. If we let $\alpha = \nu$, and multiply by an amplitude factor e^ν, this function has a maximum value of unity at $t = 1$ for all integer values of ν. It reads

$$g(t) = [te^{-(t-1)}]^\nu \tag{99}$$

and is shown plotted for $\nu = 2$ and $\nu = 3$ in Fig. 13. Multiplication of our simulated impulse by this function evidently has a beneficial effect since it diminishes ripples in the precursor and coda regions without essentially affecting the pulse shape otherwise.

Regarding the question as to whether we should consider a broken-line or a parabolic-arc approximation to $\cos \omega$ or $-\sin \omega$, we shall see that the former is in most cases adequate and leads to slightly better economy. Since we prefer to have the associated time function begin with a zero value for $t = 0$, we recognize that the net area enclosed by the broken-line approximation should be zero, which requires that the cutoff frequency ω_c should equal an integer multiple of π. If we approximate $\cos \omega$ in this manner, the broken line does not terminate at a zero value for $\omega = \omega_c$, and hence the second derivative contains a doublet at this point. Since a doublet leads to a double-order pole, the resulting economy is not as good as that obtainable from a broken-line approximation of $-\sin \omega$ which ends in a zero value at $\omega = \omega_c$.

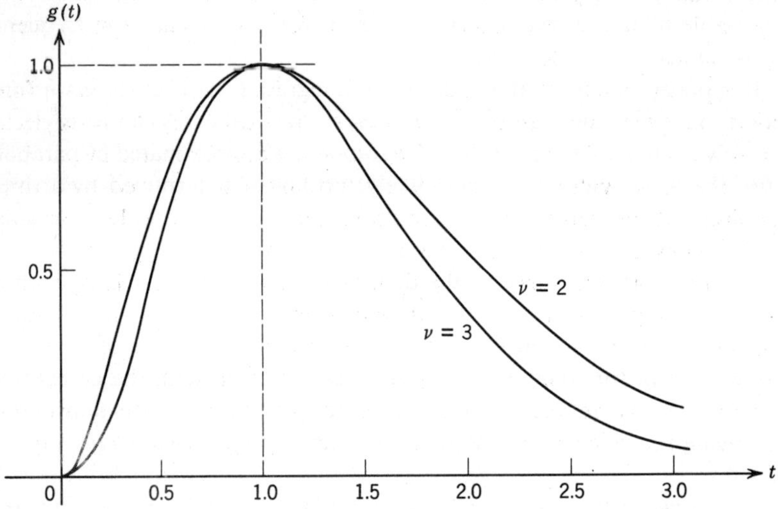

Fig. 13. The function 99 by which the normalized quasi impulse becomes multiplied.

Such an approximation is shown in the top sketch of Fig. 14 for $\omega_c = 2\pi$; below it are respectively the first and second derivatives. The amplitude, of this triangular approximation is chosen so that the areas enclosed by half cycles are the same as for the sinusoid which is being approximated, although this condition is actually not significant, and we shall have to introduce appropriate constant multipliers later on anyway so as to adjust the net area enclosed by the resultant time function to a normalized value.

In terms of the impulse values in the bottom sketch of Fig. 14, Eq. 18 yields for the associated time function

$$f(t) = \frac{32}{\pi^3 t^2}\left(\sin\frac{\pi t}{2} - \sin\frac{3\pi t}{2} + \frac{1}{2}\sin\frac{4\pi t}{2}\right) \tag{100}$$

The solid curve in Fig. 15 shows a plot of this function (with normalized amplitude) while the dashed curve indicates its modification resulting from multiplication by $g(t)$ of Eq. 99 for $\nu = 2$ (Fig. 13). The resulting ripple amplitudes are smaller than those for the function 87 plotted in Fig. 16, while the shape of the main hump is very nearly the same. For

$\omega_c = 2\pi$, Eq. 97 yields $\delta = \frac{1}{2}$ which is remarkably close to the actual width at the $(2/\pi)$ points as defined for the pulse function 87. We may conclude that, even for this rather small value of ω_c, the tolerance criteria based upon the function 87 are conservatively applicable.

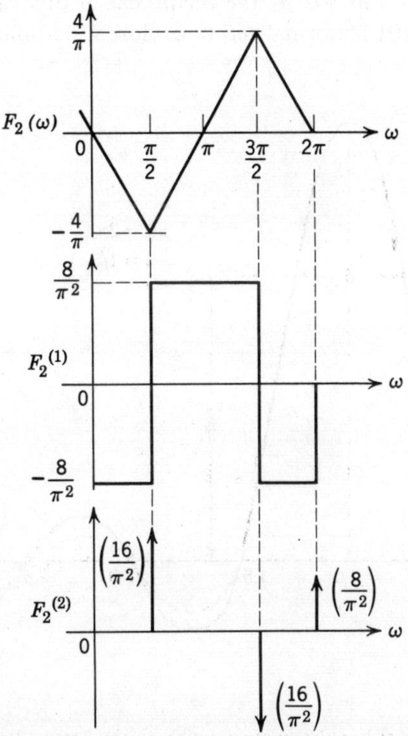

Fig. 14. Broken-line approximation of $-\sin\omega$ for $\omega_c = 2\pi$, leading to the normalized quasi impulse shown in Fig. 15.

The modified time function to which the dashed curve applies (except for the main hump which is inappreciably affected through multiplication by 99) is given by

$$f(t) = \frac{32e^{-2(t-1)}}{\pi^3}\left(\sin\frac{\pi t}{2} - \sin\frac{3\pi t}{2} + \frac{1}{2}\sin\frac{4\pi t}{2}\right) \tag{101}$$

The corresponding rational transform is readily found to be

$$F(s) = 11.98\left(\frac{1}{s^2 + 4s + 6.47} - \frac{3}{s^2 + 4s + 26.21} + \frac{2}{s^2 + 4s + 43.48}\right) \tag{102}$$

Its zero-frequency value equals the net area enclosed by the pulse function 101 thus

$$F(0) = 1.03 = \int_0^\infty f(t)\,dt \tag{103}$$

Multiplication of $F(s)$ in 102 by the reciprocal of this value is the means whereby the pulse 101 is normalized to enclose a net unit area.

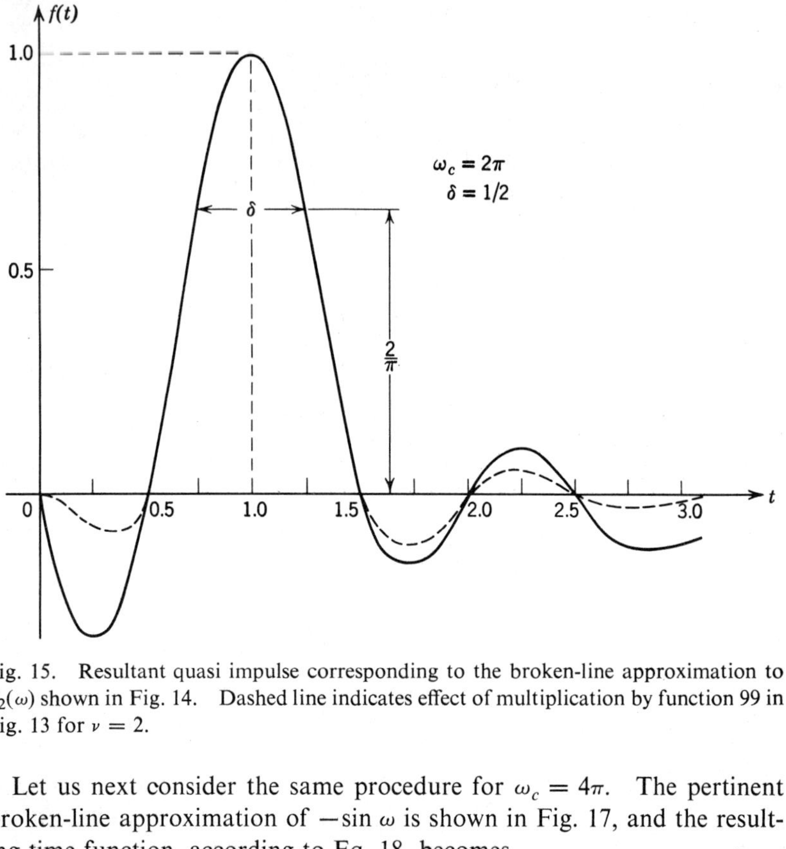

Fig. 15. Resultant quasi impulse corresponding to the broken-line approximation to $F_2(\omega)$ shown in Fig. 14. Dashed line indicates effect of multiplication by function 99 in Fig. 13 for $\nu = 2$.

Let us next consider the same procedure for $\omega_c = 4\pi$. The pertinent broken-line approximation of $-\sin\omega$ is shown in Fig. 17, and the resulting time function, according to Eq. 18, becomes

$$f(t) = \frac{32}{\pi^3 t^2}\left(\sin\frac{\pi t}{2} - \sin\frac{3\pi t}{2} + \sin\frac{5\pi t}{2} - \sin\frac{7\pi t}{2} + \frac{1}{2}\sin\frac{8\pi t}{2}\right) \tag{104}$$

It is shown by the solid curve in Fig. 18, and the effect of multiplying by $g(t)$ in Eq. 99 for $\nu = 2$ is indicated by the dashed curve. We see that the resulting ripple amplitudes still are smaller than for the prototype pulse 87 in Fig. 16, so that error criteria based upon it are again applicable.

After multiplying the function 104 by $g(t)$ in Eq. 99 for $\nu = 2$, its transform is given by

$$F(s) = 11.98\left(\frac{1}{s^2 + 4s + 6.47} - \frac{3}{s^2 + 4s + 26.2} + \frac{5}{s^2 + 4s + 65.7} - \frac{7}{s^2 + 4s + 124.9} + \frac{4}{s^2 + 4s + 161.9}\right) \quad (105)$$

The net area enclosed by the resultant pulse in this case is found to be

$$F(0) = 1.016 \quad (106)$$

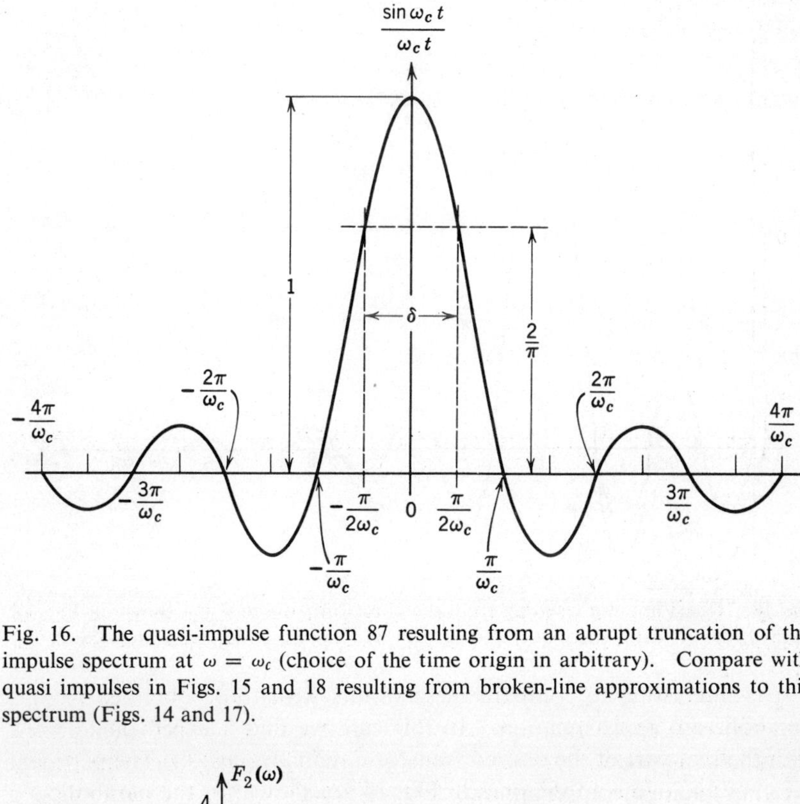

Fig. 16. The quasi-impulse function 87 resulting from an abrupt truncation of the impulse spectrum at $\omega = \omega_c$ (choice of the time origin in arbitrary). Compare with quasi impulses in Figs. 15 and 18 resulting from broken-line approximations to this spectrum (Figs. 14 and 17).

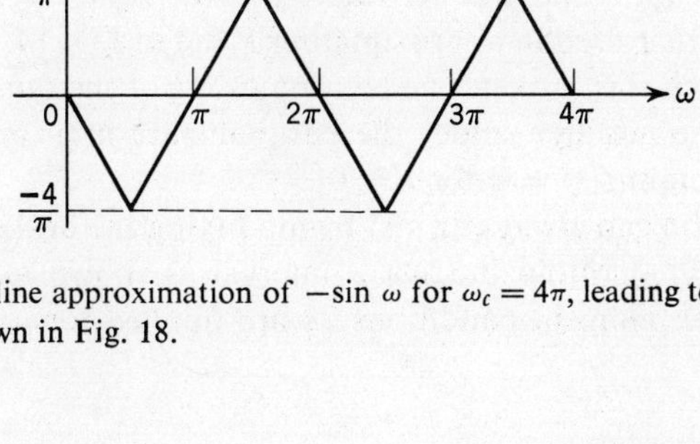

Fig. 17. Broken-line approximation of $-\sin\,\omega$ for $\omega_c = 4\pi$, leading to the normalized quasi impulse shown in Fig. 18.

The formation of rational transforms corresponding to larger values of the cutoff frequency ω_c and smaller values of the width δ is straightforward. We see that the number of pole-pairs is given by the relation

$$n = (1/\delta) + 1 = (\omega_c/\pi) + 1 \tag{107}$$

Hence, for any desired value of δ, the number of poles required by the resulting response function is easily obtained.

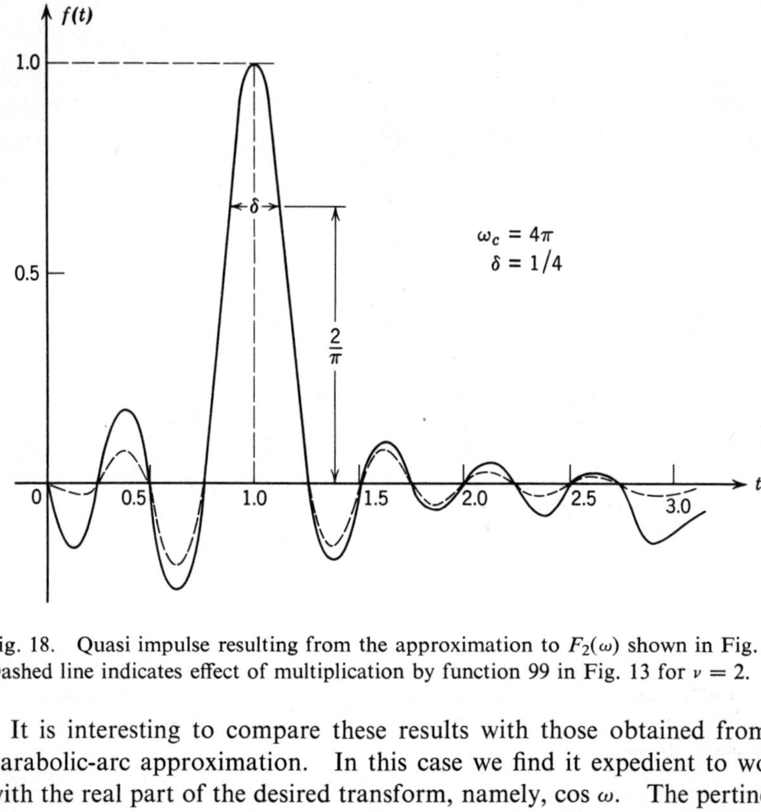

Fig. 18. Quasi impulse resulting from the approximation to $F_2(\omega)$ shown in Fig. 17. Dashed line indicates effect of multiplication by function 99 in Fig. 13 for $\nu = 2$.

It is interesting to compare these results with those obtained from a parabolic-arc approximation. In this case we find it expedient to work with the real part of the desired transform, namely, cos ω. The pertinent sketches analogous to the ones in Fig. 14 are shown for the parabolic-arc method in Fig. 19. The first derivative $F_1^{(1)}(\omega)$ is now given by a broken line similar to that used in approximating $F_2(\omega)$ in Fig. 14. However, the manner in which this broken line terminates is not the same, if we wish to require that the net area under the parabolic-arc approxmation be zero [in order to obtain $f(t) = 0$ for $t = 0$].

This condition can always be met by modifying the broken-line approximation of $F_1^{(1)}(\omega)$ within the last π interval as shown in Fig. 19, which assures that the moment conditions 28 are fulfilled for $\mu = 1, 3$ [as they should be since $\nu = 3$ and we want to simulate the asymptotic behavior expressed by $F(s) \rightarrow 1/s^2$ for $s \rightarrow \infty$]. The effect of thus terminating the approximation is to deviate slightly from an abrupt cutoff, and the corresponding value of ω_c is not so clearly defined. We shall see that

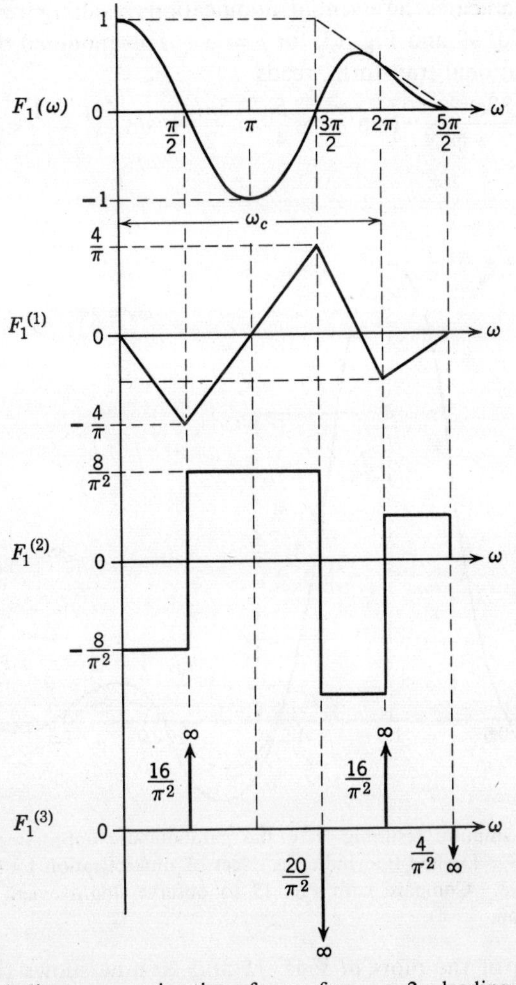

Fig. 19. Parabolic-arc approximation of cos ω for $\omega_c = 2\pi$, leading to the normalized quasi impulse shown in Fig. 20.

choosing for ω_c the mean value of the last π interval, as is indicated in Fig. 19, yields correctly the base width $2\pi/\omega_c$ of the resulting pulse (according to Fig. 16), although the width δ at the $(2/\pi)$ points is somewhat on the conservative side.

In terms of the impulse values in the bottom sketch of Fig. 19, Eq. 12 yields the associated time function

$$f(t) = \frac{32}{\pi^3 t^3}\left(\sin\frac{\pi t}{2} - \frac{5}{4}\sin\frac{3\pi t}{2} + \sin\frac{4\pi t}{2} - \frac{1}{4}\sin\frac{5\pi t}{2}\right) \tag{108}$$

Its normalized plot is shown by the solid curve of Fig. 20, in which the dashed curve indicates the essential modifications resulting from multiplication by $g(t)$, Eq. 99 and Fig. 13, for $\nu = 3$. This modified time function, which has a rational transform, reads

$$f(t) = \frac{32e^{-3(t-1)}}{\pi^3}\left(\sin\frac{\pi t}{2} - \frac{5}{4}\sin\frac{3\pi t}{2} + \sin\frac{4\pi t}{2} - \frac{1}{4}\sin\frac{5\pi t}{2}\right) \tag{109}$$

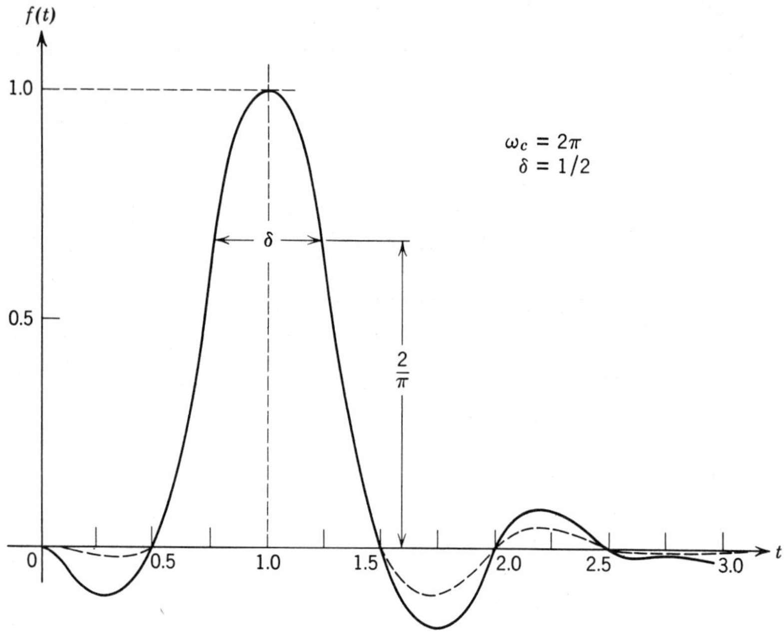

Fig. 20. Quasi impulse resulting from the parabolic-arc approximation to $F_1(\omega)$ shown in Fig. 19. Dashed line indicates effect of multiplication by function 99 in Fig. 13 for $\nu = 3$. Compare with Fig. 15 to observe improvement over broken-line approximation.

Comparison of the plots in Figs. 15 and 20 now shows the difference between the broken-line and parabolic-arc approximations. The latter yields a considerably smaller precursor ripple and a slightly smaller δ value for the same ω_c. Correspondingly, we note that the number of pole-pairs required by the parabolic-arc approximation is given by

$$n = (1/\delta) + 2 = (\omega_c/\pi) + 2 \tag{110}$$

which is one pole-pair more than required by the broken-line approximation according to Eq. 107. If the necessary δ value is small, this difference is insignificant, and the more conservative aspect of the parabolic-arc method easily justifies this slight increase in the expenditure of poles.

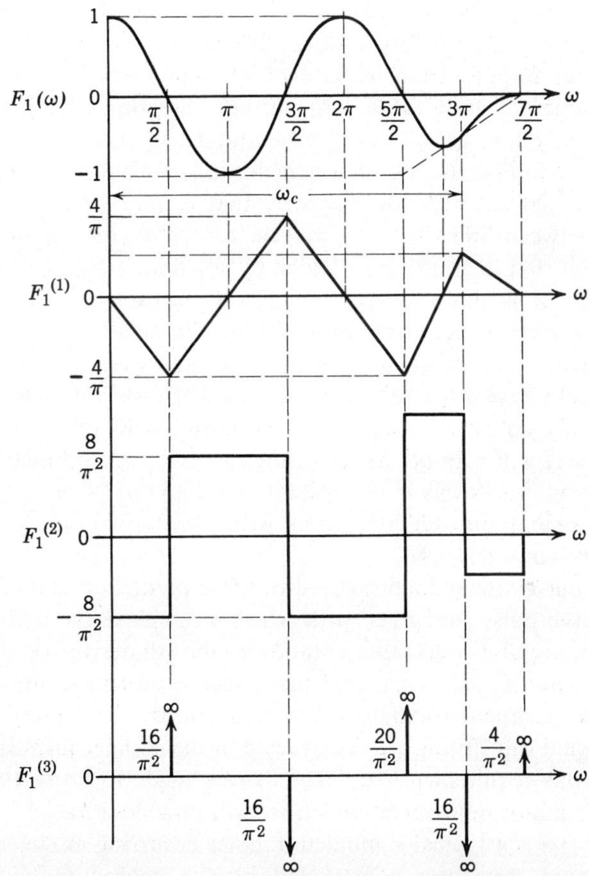

Fig. 21. Parabolic-arc approximation of cos ω for $\omega_c = 3\pi$.

The sketches analogous to those in Fig. 19 pertaining to a parabolic-arc approximation for $\omega_c = 3\pi$ are shown in Fig. 21, and the corresponding time function according to Eq. 12 reads

$$f(t) = \frac{32}{\pi^3 t^3}\left(\sin\frac{\pi t}{2} - \sin\frac{3\pi t}{2} + \frac{5}{4}\sin\frac{5\pi t}{2} - \sin\frac{6\pi t}{2} + \frac{1}{4}\sin\frac{7\pi t}{2}\right) \quad (111)$$

Multiplied by $g(t)$ of Eq. 99 for $\nu = 3$, we obtain the modified time function having a rational transform and reduced precursor and coda ripples.

With the pattern thus established, it is a simple matter to construct additional time functions and rational transforms for any chosen value of δ or ω_c. Each resulting pulse simulates the prototype in Fig. 16 conservatively in the sense that the ripple amplitudes and actual tolerances turn out to be at least as small as the predicted ones.

We come now to the consideration of a rather detailed item in the construction of rational transforms for prototype quasi impulses (the building blocks in our impulse-train method of synthesis) which, however, is of quite fundamental importance. In our consideration of the error functions $\epsilon_0(t)$, $\epsilon_{-1}(t)$, $\epsilon_{-2}(t)$ of Eqs. 93, 94, and 95, the last two of which are shown plotted in Figs. 11 and 12, we are dealing with a somewhat fortuitous situation in that $\epsilon_0(t)$ has the property that it, and *all* of its successive integrals (between limits $-\infty$ to ∞), enclose zero area [as may be seen from the fact that its transform divided by any finite power of $(j\omega)$ is zero for $\omega = 0$]. It is this property that results in each successive integral oscillating about the zero axis with smaller and smaller amplitudes. If, for one of these functions, the net enclosed area were not zero, then its integral would have a nonzero asymptote. The following integral would asymptotically equal a linear ramp; the next would follow a parabolic ramp, and so forth. In other words, the successive error functions would drift more and more away from the zero axis; and this "drift error" would completely swamp the oscillatory error with which we are solely concerned in the above considerations.

Clearly, our contemplated method of error prediction and control will fail unless the pulse functions, with which we replace the individual impulses in an impulse-train representation of the νth derivative of a desired transient response, are repeatedly integrable ν successive times without drift in the pertinent oscillatory error functions. The pulse functions whose rational transforms are constructed in the manner just discussed do not fulfill this requirement, but they can be made to do so through an appropriate minor adjustment which we will now describe.

Suppose $f(t)$ is a typical simulated impulse occurring at $t = \tau$, and $F(s)$ is its rational transform constructed by the procedure given above. Assume $F(s)$ in the form

$$F(s) = \frac{p(s)}{q(s)} = \frac{a_0 + a_1 s + a_2 s^2 + \cdots + a_m s^m}{b_0 + b_1 s + b_2 s^2 + \cdots + b_n s^n} \qquad (n > m) \tag{112}$$

If we want to integrate $f(t)$ ν times without drift error, then the error function $f(t) - u_0(t - \tau)$, after ν successive integrations, should enclose a net zero area. This condition is met if $F(s) - e^{-s\tau}$ (the transform of this error) has at least a $(\nu + 1)$-order zero at $s = 0$, or if at least the first $\nu + 1$ coefficients in the Maclaurin expansion of this transform are zero. Since

$$F(s) - e^{-s\tau} = \frac{p(s) - q(s)e^{-s\tau}}{q(s)} \tag{113}$$

and

$$q(s)e^{-s\tau} = b_0 + (b_1 - \tau b_0)s + \left(b_2 - \tau b_1 + \frac{\tau^2}{2}b_0\right)s^2 + \left(b_3 - \tau b_2 + \frac{\tau^2}{2}b_1 - \frac{\tau^3}{6}b_0\right)s^3 + \cdots \quad (114)$$

the necessary and sufficient conditions insuring zero drift error read

$$\begin{aligned} a_0 &= b_0 \\ a_1 &= b_1 - \tau b_0 \\ a_2 &= b_2 - \tau b_1 + \frac{\tau^2}{2}b_0 \\ a_3 &= b_3 - \tau b_2 + \frac{\tau^2}{2}b_1 - \frac{\tau^3}{6}b_0 \\ &\cdots\cdots\cdots\cdots \\ a_\nu &= b_\nu - \tau b_{\nu-1} + \frac{\tau^2}{2}b_{\nu-2} - \cdots + (-1)^\nu \frac{\tau^\nu}{\nu!}b_0 \end{aligned} \quad (115)$$

The first of these is equivalent to demanding $F(0) = 1$, and hence amounts to normalization of the area enclosed by the pulse $f(t)$ to the value unity. If the desired transient response function is approximated by a broken line so that $\nu = 2$, then two additional relations in the group 115 must be fulfilled; if a parabolic-arc approximation is used, three additional relations must be satisfied.

In the construction of $F(s)$ described above we make either the real or the imaginary parts of $F(j\omega)$ and $e^{-j\omega\tau}$ approximately equal over a frequency interval from $\omega = 0$ to some multiple of π. Hence we can expect $F(s)$ to behave reasonably nearly like $e^{-s\tau}$ near $s = 0$, so that conditions 115, although not exactly fulfilled, will in any reasonable situation already be nearly fulfilled. The function 105, for example, which is found from the broken-line approximation to $-\sin\omega$ shown in Fig. 17, after division by the value 106, yields $F(0) = 1$, $F'(0) = -0.985$, $F''(0) = 0.810$ in place of the exact values 1, -1, 1 respectively. For a parabolic-arc approximation these results are much closer.

We conclude that only minor adjustments in the values of the coefficients $a_0 \cdots a_\nu$ are called for in order to fulfill exactly the pertinent conditions 115. This "drift adjustment" of the transforms of quasi-impulse functions, since it affects slightly essentially only the low-frequency behavior, leaves the form of the pulse unchanged except for some modification of the ripples in the precursor and coda regions.

In any doubtful situation one can readily estimate the effect of a drift adjustment. Suppose $\nu = 2$, and the necessary changes in the coefficients a_1 and a_2 are denoted by α_1 and α_2. Then

$$D(s) = \frac{\alpha_1 s + \alpha_2 s^2}{q(s)} \tag{116}$$

is the transform of the change in $f(t)$ due to drift adjustment. Since the poles of this transform are known, it is an easy matter to compute roughly the magnitudes of its residues and compare them with the known residues of $F(s)$.

The total procedure in the synthesis of a desired impulse response is now straightforwardly describable. Suppose we start from a broken-line approximation to this given time function so that its second derivative has an impulse-train representation. It is convenient to normalize the time scale so that the last impulse falls at $t = 1$. Next we decide upon an approximation tolerance in terms of a largest allowable fillet at the break points. This fillet is converted into an allowable size per unit impulse, which is done through division by the value of the largest impulse in the train. It then represents the value of $\epsilon_{-2}(0)$ in Eq. 96 or 98 for the computation of the pertinent truncation frequency ω_c or pulse width δ.

We now construct a rational transform $F(s)$ for a quasi impulse having this width δ by the above methods (preferably using a parabolic-arc approximation), and from it obtain transforms (having the same poles) for pulses occurring at each impulse in the train, using the method discussed in connection with Eqs. 91 and 92. After each of these transforms is adjusted for zero drift error and multiplied by an appropriate (positive or negative) factor according to the values of the respective components in the pertinent impulse train, their sum, divided by s^2, is the desired response function.

Since the impulse train fulfills moment conditions which assure that its twofold integration regains the originally specified time function without drift error, and each quasi impulse likewise is adjusted for zero drift error, it follows that the sum of transforms constructed in the above manner must contain the factor s^2, which cancels out through division by s^2 as just described. The resultant response function then has a finite nonzero value for $s = 0$, equal to the net area enclosed by the specified impulse response.

Except in some special situations, a broken-line approximation to the given time function (impulse response) can be made in a variety of ways, according to the number and distribution of break points chosen. Since this kind of an approximation becomes better as the number of break points is increased, and since the cost in terms of the total number of required poles is, for a given pulse width δ, independent of the number of break points, it would seem that there is some sort of optimum spacing of break points. Two considerations primarily influence this situation: (1) As break points are spaced more closely, the resulting values of the impulses in the impulse-train representation of the second derivative decrease, and so do the amplitude factors for the error functions $\epsilon_{-2}(t)$, Fig. 12, centered at these break points. (2) The decreased spacing between such error functions causes these to overlap, with the resultant possibility of their ripple amplitudes becoming additive. A compromise between these two oppositely slanted tendencies, although not clearly defined, can satisfactorily be reached for a given situation without great difficulty.

The question as to whether a broken-line or parabolic-arc approximation should be used involves similar reasoning. In this connection we observe first of all that an impulse-train representation for the third and higher derivatives leads to error functions $\epsilon_{-3}(t)$, $\epsilon_{-4}(t)$, and so forth, which are obtained through continued integration of the function in Fig. 12. We find that the maximum ripple amplitudes of these functions, at or adjacent to the respective impulses that generate them, have the values $1/2\pi\omega_c^2 = \delta^2/2\pi^3$, $1/3\pi\omega_c^3 = \delta^3/3\pi^4$, and so forth. Although these are impressively small, it is not necessarily advantageous to adopt a procedure that leads to their consideration, as the example in Fig. 22 illustrates.

In Fig. 22*a* we approximate a desired transient pulse by parabolic arcs and in Fig. 22*b* by a broken line. The resulting error ripples for the former have the order of magnitude $(1/\Delta^2)(\delta^2/2\pi^3)$, while for the latter these are given by $(2/\Delta)(\delta/\pi^2)$ and take the form essentially of fillets at the break points. As Δ is chosen about as small as δ, there ceases to be much, if any, advantage in the parabolic-arc approximation. As the dashed curve in Fig. 22*b* indicates, the end result in using the broken-line approximation is essentially the same.

There will, of course, be some situations in which the parabolic-arc approximation has advantages. For example, it does in the construction of transforms for the quasi impulses (the building-block functions) since it adds only slightly to the cost and reduces materially the ripples, primarily in the precursor region, as has been mentioned previously.

6 Impulse Response Synthesis; a Fourier Method

This method of synthesis is based upon the following thought. If the desired impulse response were a periodically repeated transient, representable with any required degree of approximation by a finite trigonometric polynomial, then a corresponding rational transform could be obtained at once, and no correlation of errors in the time and frequency domains would be needed since the pertinent approximation problem is concerned solely with the time domain. Moreover, the techniques for dealing with this approximation problem are all available and well understood. All that is needed is some artifice whereby a desired transient time function is expressible in terms of its periodic repetition. A scheme for accomplishing this end may be described as follows.*

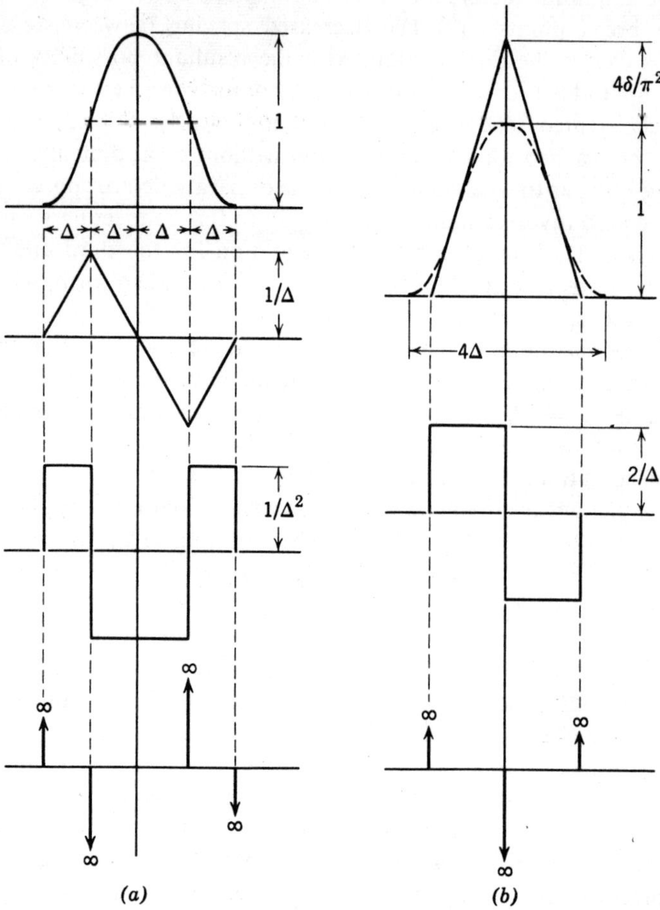

Fig. 22. A situation in which a parabolic-arc type of approximation yields essentially the same result as a broken-line approximation. The error function associated with the latter introduces the fillets shown dashed in (*b*).

Suppose a desired $f(t)$ is the transient pulse shown in Fig. 23. We consider two periodic repetitions of this function as shown in Fig. 24 and observe that their corresponding transforms $h_1(s)$ and $h_2(s)$ are readily

* This method was first presented in the *Quart. Prog. Rept.*, Mass. Inst. Technol., Research Lab. Electronics, Jan. 15, 1952, pp. 66–71.

constructible from the appropriate Fourier series for $f_1(t)$ and $f_2(t)$. If we could synthesize a pair of two-terminal networks N_1 and N_2 having the driving-point admittances* $h_1(s)$ and $h_2(s)$, it is clear that their responses to an applied unit voltage impulse would be currents given by the time functions $f_1(t)$ and $f_2(t)$ respectively. Or we can say that if we were to apply a current excitation $f_1(t)$ to N_1 or a current excitation $f_2(t)$ to N_2, the response in each case would be a unit voltage impulse.

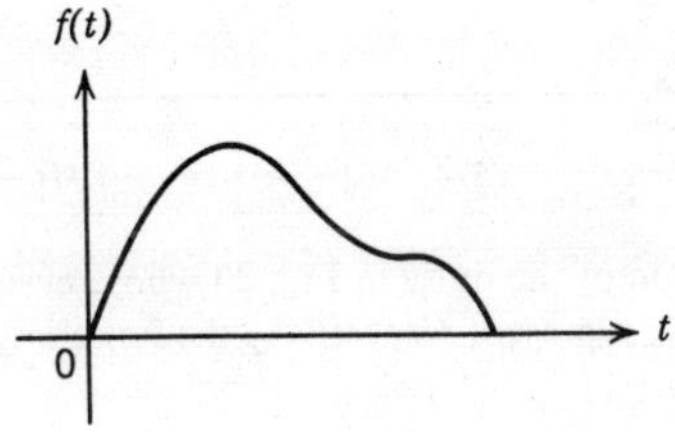

Fig. 23. Transient pulse which is to be the impulse response of a desired network.

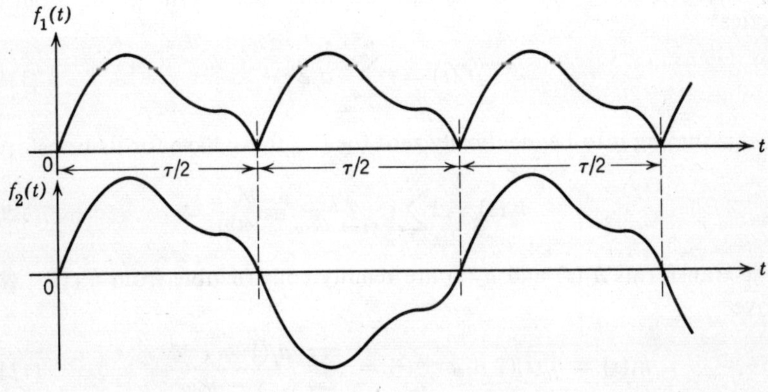

Fig. 24. Respectively even and odd repetitions of the pulse of Fig. 23.

If we now observe that

$$f(t) = f_1(t) - f_1(t - \tau/2) \tag{117}$$

and

$$f(t) = f_2(t) + f_2(t - \tau/2) \tag{118}$$

we can conclude that if we apply a current excitation $f(t)$ to N_1 the response is a unit voltage impulse at $t = 0$ followed by a negative unit impulse at $t = \tau/2$; if we apply a current excitation $f(t)$ to N_2 the response is a unit

* In general it will turn out that these are not p.r. functions and hence not realizable in terms of passive elements, but this fact is of no consequence in the present argument, which is offered merely as a simple means for visualizing the mechanism of arriving at the desired result.

voltage impulse at $t = 0$ followed by another at $t = \tau/2$. Therefore, a current $f(t)$ applied to N_1 and N_2 in series produces simply a voltage impulse at $t = 0$, or the latter produces $f(t)$, which is the desired result. It remains to fill in the details with appropriate mathematical symbols.

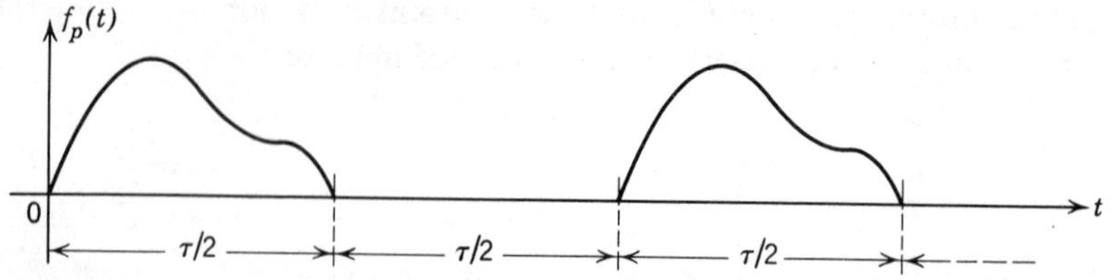

Fig. 25. Periodic repetition of the pulse in Fig. 23 whose even- and odd-order Fourier terms (except for a factor 2) represent respectively the functions $f_1(t)$ and $f_2(t)$ in Fig. 24. (See Eqs. 119 through 124.)

We begin by considering the periodic function $f_p(t)$ in Fig. 25, consisting of a repetition of $f(t)$ at half-period intervals. Writing the Fourier zeries

$$f_p(t) = \sum_{k=-\infty}^{\infty} a_k e^{jk\omega t} \tag{119}$$

and assuming it to be identically zero for $t < 0$, we have for its transform

$$h_p(s) = \sum_{k=-\infty}^{\infty} \frac{a_k}{s - jk\omega} = \frac{p(s)}{q(s)} \tag{120}$$

The transforms $h_1(s)$ and $h_2(s)$ are readily constructible from $h_p(s)$. We have

$$h_1(s) = h_p(s)(1 + e^{-s\tau/2}) = \sum_{k=-\infty}^{\infty} \frac{a_k(1 + e^{-jk\pi})}{s - jk\omega} \tag{121}$$

and

$$h_2(s) = h_p(s)(1 - e^{-s\tau/2}) = \sum_{k=-\infty}^{\infty} \frac{a_k(1 - e^{-jk\pi})}{s - jk\omega} \tag{122}$$

Noting that the factor $(1 + e^{-jk\pi})$ equals 2 for k even and zero for k odd, while the factor $(1 - e^{-jk\pi})$ equals 2 for k odd and zero for k even, we have more simply

$$h_1(s) = 2 \sum_{k=0\pm(2,4,\,\cdots)}^{\infty} \frac{a_k}{s - jk\omega} = \frac{p_1(s)}{q_1(s)} \tag{123}$$

and

$$h_2(s) = 2 \sum_{k=\pm(1,3,\,\cdots)}^{\infty} \frac{a_k}{s - jk\omega} = \frac{p_2(s)}{q_2(s)} \tag{124}$$

That is to say, $h_1(s)$ equals twice the sum of the even-order terms in Eq. 120, while $h_2(s)$ equals twice the sum of the odd-order terms.

As long as the sums in the expressions for $h_p(s)$, $h_1(s)$, and $h_2(s)$ are regarded as having an infinite number of terms, these transforms are, of course, not rational and the polynomials $p(s)$, $q(s)$, $p_1(s)$, $q_1(s)$, $p_2(s)$, and $q_2(s)$ are not finite. Since we anticipate replacing these expressions by their partial sums, it is appropriate even at this stage to think of the polynomials as though they were finite. They obviously have real coefficients even though the Fourier coefficients a_k (which are residues of the h-functions in their j-axis poles) are complex.

If we think of networks N_1 and N_2 as having the driving-point admittance functions $h_1(s)$ and $h_2(s)$, then a unit voltage impulse $u_0(t)$ applied to N_1 produces a current $f_1(t)$, and if applied to N_2, it produces a current $f_2(t)$. If we now interchange the roles of excitation and response, and say that a current $f_1(t)$ applied to N_1 or $f_2(t)$ applied to N_2 produces a voltage $u_0(t)$, then the driving-point impedance functions for these networks are the reciprocals $1/h_1(s)$ and $1/h_2(s)$. Noting Eqs. 117 and 118, we may then state that a current $f(t)$ applied to N_1 having the impedance $1/h_1(s)$ produces the voltage

$$u_0(t) - u_0(t - \tau/2) \tag{125}$$

and a current $f(t)$ applied to N_2 having the impedance $1/h_2(s)$ produces the voltage

$$u_0(t) + u_0(t - \tau/2) \tag{126}$$

Therefore, the current $f(t)$ applied to N_1 and N_2 in series with the resultant impedance $(1/h_1) + (1/h_2)$ produces a voltage $2u_0(t)$; or if we again interchange the roles of excitation and response and consider the series combination of N_1 and N_2 as having the resultant admittance

$$h(s) = \frac{1}{(1/h_1) + (1/h_2)} = \frac{h_1 h_2}{h_1 + h_2} \tag{127}$$

then its current response to an applied unit voltage impulse becomes $\frac{1}{2}f(t)$, as may readily be verified by substituting Eqs. 121 and 122 into 127, which yields

$$h(s) = \tfrac{1}{2}h_p(s)(1 - e^{-s\tau}) = \frac{p_1 p_2}{p_1 q_2 + p_2 q_1} = \frac{p_1(s)p_2(s)}{2p(s)} \tag{128}$$

Since the time-domain equivalent of this equation reads

$$f(t) = \tfrac{1}{2}[f_p(t) - f_p(t - \tau)] \tag{129}$$

the above statement is seen to be true.

Neglecting second-order effects, it is easy to show that when one replaces the above infinite series by their partial sums, the resulting transient response is as good an approximation to $f(t)$ as the partial sum of the Fourier series (or its modification to obtain any desired approximation properties) given in Eq. 119 is to $f_p(t)$ over any one of its periods. Thus, if we indicate any approximate function by attaching an asterisk to the corresponding exact function, we may write

$$f^*{}_p(t) = \sum_{k=-n}^{n} a_k e^{jk\omega t} \tag{130}$$

$$h^*{}_p(s) = \sum_{k=-n}^{n} \frac{a_k}{s - jk\omega} \tag{131}$$

and

$$h^*{}_p(s)e^{-s\tau/2*} = \sum_{k=-n}^{n} \frac{a_k e^{-jk\pi}}{s - jk\omega} \tag{132}$$

where n is a finite integer.

Figure 26 shows how the approximation to $f_p(t)$ given by the partial sum 130 might look, and correspondingly how the inverse transform of the frequency function 132 will appear. It is important to note that this time function may be regarded as an imperfect version of $f^*{}_p(t)$ delayed by a half period. Thus, the perfect version of $f^*{}_p(t)$ delayed by a half period

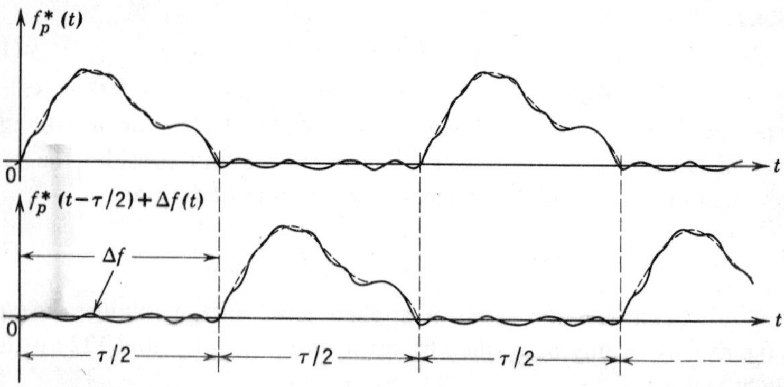

Fig. 26. Approximation to the semiperiodic function in Fig. 25 afforded by the partial sum 130, and the imperfectly delayed version of this approximation as given by the inverse transform of 132.

differs from the function shown in the bottom half of Fig. 26 in that it is *exactly* zero throughout the first half period. The imperfect version, by contrast, has the same nonzero ripply variations here that it has in any of the corresponding subsequent half-period intervals. Observe, however, that the imperfect and perfect versions are exactly alike everywhere except during the first half period where they differ by the residual ripple which in Fig. 26 is denoted by $\Delta f(t)$. That is to say, the net difference $\Delta f(t)$ between the imperfectly and the perfectly delayed functions $f^*_p(t)$ is nonzero only for the interval $0 < t < \tau/2$.

In the frequency domain we can express this result by writing

$$h^*_p(s)e^{-s\tau/2*} = h^*_p(s)e^{-s\tau 2} + \Delta h(s) \tag{133}$$

and interpreting $\Delta h(s)$ as a frequency function whose inverse transform is $\Delta f(t)$. From Eq. 133 we may now form

$$e^{-s\tau/2*} = e^{-s\tau/2} + \frac{\Delta h}{h^*_p} \tag{134}$$

By squaring and neglecting the term involving $(\Delta h)^2$, we have

$$e^{-s\tau *} = e^{-s\tau} + \frac{2\Delta h}{h^*_p} e^{-s\tau/2} \tag{135}$$

whereupon the approximate relation appropriate to Eq. 128 becomes

$$h^*(s) = \tfrac{1}{2}h^*_p(s)(1 - e^{-s\tau *}) = \tfrac{1}{2}h^*_p(s)(1 - e^{-s\tau}) - \Delta h e^{-s\tau/2} \tag{136}$$

and the corresponding time-domain relation reads

$$f^*(t) = \tfrac{1}{2}[f^*_p(t) - f^*_p(t - \tau)] - \Delta f(t - \tau/2) \tag{137}$$

The first term in this last expression equals one-half the function $f^*_p(t)$ for the interval $0 < t < \tau$ and is zero otherwise. The second term is $f^*_p(t)$ for $\tau/2 < t < \tau$ with its sign changed. The net result (except for a factor $\frac{1}{2}$) equals $f^*_p(t)$ for $0 < t < \tau/2$, $-f^*_p(t)$ for $\tau/2 < t < \tau$, and is zero otherwise.

The second-order effects, which have been neglected, alter slightly the form of $f^*(t)$ during the interval $0 < t < \tau$, and prolong the error in an asymptotically decreasing fashion thereafter.

It is important to note that the trigonometric polynomial $f_p(t)$ used as a starting point in these calculations need not be the partial sum of a Fourier series, but can be any modification thereof that more adequately fulfills the approximation properties desired for a specific problem. For example, if a pulse of rectangular shape is wanted, use of the partial sum of a Fourier series would lead to the undesirable effects caused by the Gibbs phenomenon, and this can easily be avoided by an appropriate modification in the coefficients of the pertinent trigonometric polynomial.

The rectangular pulse is of special interest since the network realization becomes particularly simple if we assume the excitation to be a unit step voltage. Thus, if $f(t)$ in Fig. 23 is a rectangular pulse of unit height, then in Fig. 24, $f_1(t)$ is a unit step function and $f_2(t)$ is a square wave of unit height (dead before $t = 0$, of course). The transform $h_1(s) = 1/s$, and since the a_k's in the series 124 are negative real (the square wave is given by a sum of sine functions), we find

$$h_2(s) = \sum_{k=1,3,\cdots}^{n} \frac{4k\omega|a_k|}{s^2 + k^2\omega^2} \tag{138}$$

Corresponding admittances of the networks N_1 and N_2 now are $s\,h_1(s)$ and $s\,h_2(s)$ respectively (because they yield responses when multiplied by $1/s$). Hence we see that N_1 is a 1-ohm resistance and N_2 is given by series LC branches in parallel, or any equivalent to this Foster form which is immediately obtainable from 138 by the methods of Ch. 3.

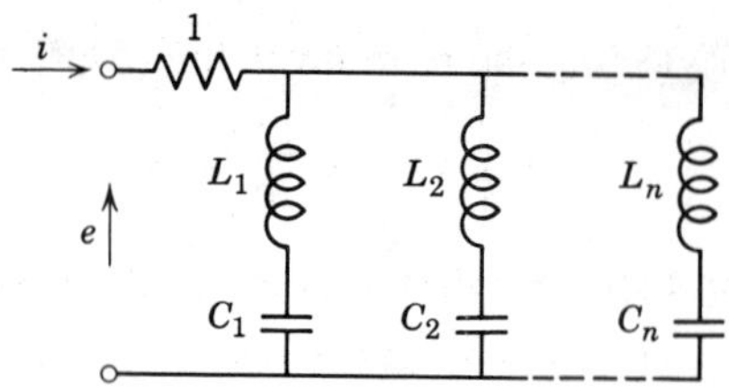

Fig. 27. Network for the production of a single rectangular current pulse $i(t)$ when $e(t)$ is a unit step.

The result is shown in Fig. 27. The current i (or voltage across the 1-ohm resistance) is an approximation to the assumed square pulse when e is a unit step voltage. This circuit can, of course, be transformed into numerous other forms, some of which are preferable for various practical reasons.*

When the functions $h_1(s)$ and $h_2(s)$, or their s-multiplied versions, are not separately realizable driving-point admittances, the pertinent network is not so readily constructible. However, the resulting response function $h(s)$ given by Eq. 128 is always realizable as a transfer impedance or admittance so long as its poles are restricted to the left half-plane. Since this function represents a stable impulse response, we expect that its right half-plane analyticity will automatically be assured. That is to say, we expect the polynomial $p(s)$ in the numerator of $h_p(s)$, Eq. 120, or in the denominator of Eq. 128 to be Hurwitz, no matter what pulse shape is assumed.

Surprisingly enough, it turns out that this expectation is not fulfilled. In fact, one finds that $p(s)$ in general has some right half-plane zeros in all but a very few special cases, and there appears to be no simple correlation

* For a more detailed discussion see "A Historical Account of the Development of a Design Procedure for Pulse-Forming Networks," *Report 43*, Mass. Inst. Technol., Radiation Lab., Oct. 16, 1944.

between this phenomenon and the pulse shape. There is actually nothing inconsistent about this result since the preceding analysis guarantees a response of finite duration only in the limit in which the pertinent Fourier series are infinite. In any finite case we have the right to expect, however, that terms for right half-plane poles, if any, will involve relatively small residues so that their contribution to the resulting function throughout the interval $0 < t < \tau$ is negligible. Numerous examples have shown that this interpretation is correct, and that in any situation in which $h(s)$, Eq. 128, has right half-plane poles, we simply discard the corresponding terms in a partial-fraction expansion.* The remaining ones constitute the desired solution. The approximation tolerance is of the order exhibited by the initial trigonometric polynomial representation for the given pulse shape.

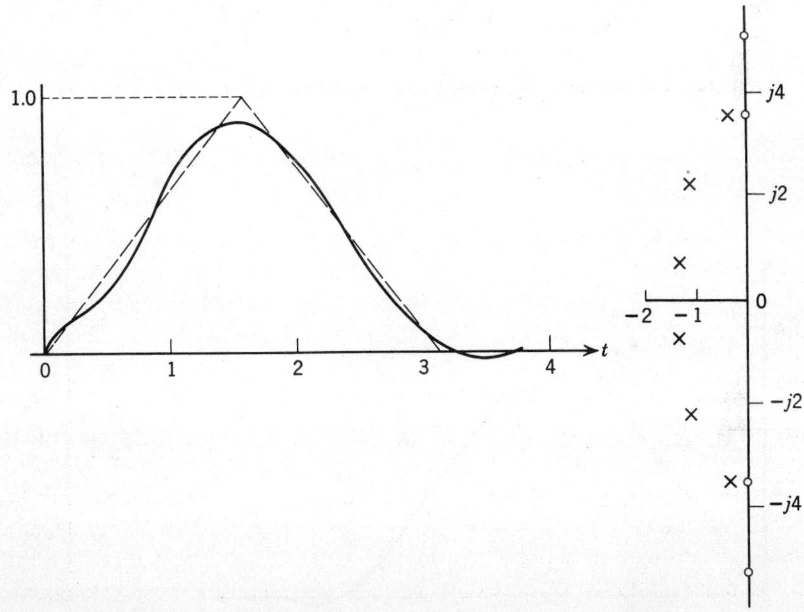

Fig. 28. Example of transient synthesis by Fourier methods. The desired pulse shape (shown dashed) is triangular. The pertinent pole-zero configuration is on the right.

An example illustrating the results of this method of synthesis is shown in Fig. 28 where the dashed curve indicates the desired triangular pulse

* These matters are discussed in detail in "Time Domain Synthesis by Means of Trigonometric Polynomial Approximations," Michael Strieby, *Sc.D. thesis*, Massachusetts Institute of Technology, July 1955. In this connection the discussions in the succeeding article are also relevant.

shape and the solid one is the result obtained from a six-pole rational approximant to its transform. On the right-hand side of the same figure is sketched the pole-zero pattern of this approximant. No right half-plane poles were encountered in this derivation and so no poles needed to be discarded. Figure 29 shows a comparison of the actual error with that predicted from the original trigonometric polynomial approximation.

A closer approximation for this same pulse shape is illustrated in Fig. 30. This example* began as a 12-pole approximation but became abbreviated

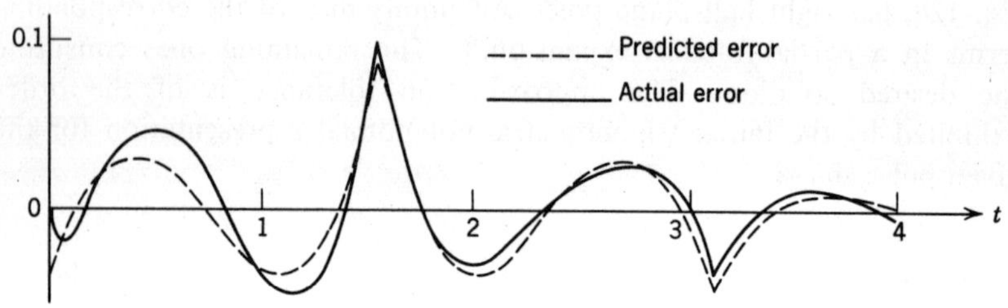

Fig. 29. Comparison of predicted and actual error for the pulse in Fig. 28.

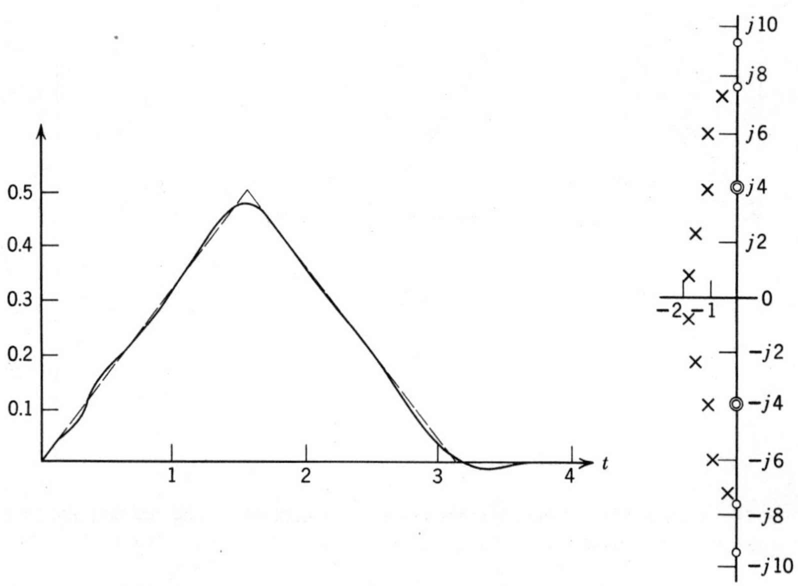

Fig. 30. A closer approximation to the triangular pulse obtained by considering a larger number of terms in the Fourier series for its periodic repetition. Solid curve is actual response computed from rational function with the pole-zero pattern shown on the right.

* Examples in the last three figures taken from the *Sc.D. thesis* by M. Strieby.

to ten poles after a pair of terms in the partial-fraction expansion of $h(s)$ representing right half-plane poles were discarded. The resulting time function is given by

$$\begin{aligned} f(t) = & \ 0.600e^{-1.112t} \cos (3.97t + 2.496) \\ & + 0.303e^{-0.981t} \cos (5.91t + 3.01) \\ & + 21.78e^{-1.799t} \cos (0.745t + 4.55) \\ & + 7.35e^{-1.602t} \cos (2.259t + 0.959) \\ & + 0.0204e^{-0.381t} \cos (7.42t + 5.13) \end{aligned} \tag{139}$$

7 Alternate Interpretation of the Fourier Method

Suppose we consider a current $i(t)$ through the series combination of boxes N_1 and N_2, as shown in Fig. 31, to have the form of the periodic

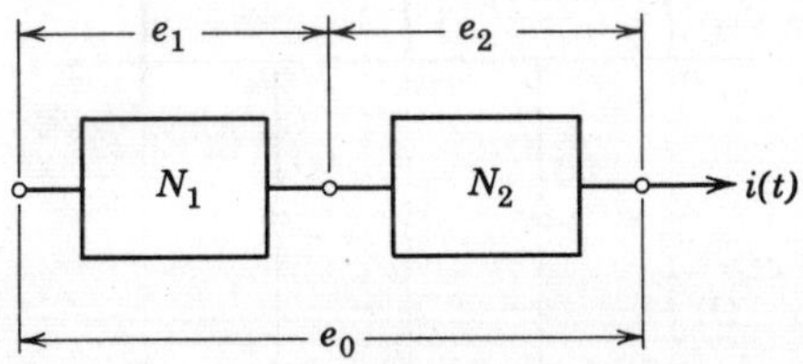

Fig. 31. Schematic circuit for the physical interpretation of the Fourier method.

function $f_2(t)$ in Fig. 24. The networks N_1 and N_2 are so constituted that the voltages $e_1(t)$ and $e_2(t)$ produced by this common current have the forms shown in Fig. 32. Observe that $e_1(t)$ is the derivative of a unit square wave for $t > 0$, and $e_2(t)$ is the derivative of a unit step. We may say that a periodic current $i(t)$, like the function $f_2(t)$ in Fig. 24, is the response of box N_1 to a differentiated square wave of voltage, and it is the response of box N_2 to a differentiated step, or unit voltage impulse. The series combination of N_1 and N_2 has this same current response for an applied voltage $e_0 = e_1 + e_2$, also shown in Fig. 32.

Because of the assumed odd character of the periodic current (its Fourier series is given by odd-order terms only), we recognize that a voltage excitation $e_0(t) + e_0(t - \tau/2)$ yields a current having the form of a single pulse like $f(t)$ in Fig. 23. Since this voltage excitation is simply an impulse function (of value 2), the desired result is again achieved.

A voltage $e_1(t) + e_1(t - \tau/2)$ applied to the box N_1 alone evidently also produces a current having the form of a single pulse like $f(t)$ in Fig. 23. This voltage is given by a unit impulse at $t = 0$ plus a negative unit impulse at $t = \tau/2$. An infinite sum of such voltages at regular intervals of $\tau/2$ seconds adds up to a single unit impulse at $t = 0$, and evidently produces an infinite sum of pulses $f(t)$ at the same regular intervals, amounting to the periodic function $f_1(t)$ in Fig. 24. That is to say, the unit impulse response of box N_1 is a periodic current like the function $f_1(t)$ in Fig. 24.

Boxes N_1 and N_2 are thus seen to have response characteristics identical with those described in the discussion given in the previous article, and so we see that our present results are in complete agreement with those already given. Although it might seem to the reader at this point that nothing has been gained, such is not the case, for the alternate physical interpretations contained in the present approach lead us to recognize a most useful property of our solution that so far has not been evident.

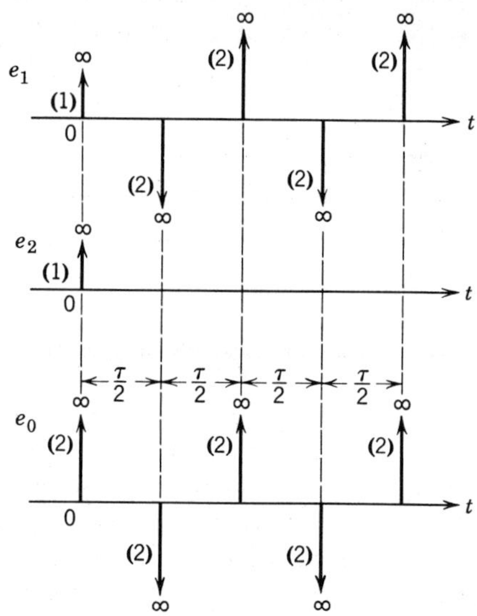

Fig. 32. The voltages in Fig. 31 as functions of time when the current $i(t)$ is identfied with the periodic function $f_2(t)$ in Fig. 24.

Suppose we let $w(t)$ denote a unit square wave with the Fourier representation

$$w(t) = \sum_{k=\pm(1,3,\cdots)}^{\infty} \alpha_k e^{jk\omega t} = \sum_{k=1,3,\cdots}^{\infty} \beta_k \sin k\omega t \tag{140}$$

in which $\beta_k = 2j\alpha_k$ since $\alpha_{-k} = -\alpha_k$. The transform of $w(t)$ for $t > 0$ is given by

$$W(s) = \sum_{k=\pm(1,3,\cdots)}^{\infty} \frac{\alpha_k}{s - jk\omega} = \sum_{k=1,3,\cdots}^{\infty} \frac{k\omega\beta_k}{s^2 + k^2\omega^2} \tag{141}$$

and hence the transform of the voltage $e_1(t)$ in Fig. 32 becomes

$$E_1(s) = sW(s) = \sum_{k=1,3,\ldots}^{\infty} \frac{sk\omega\beta_k}{s^2 + k^2\omega^2} = \frac{g_1(s)}{q_2(s)} \tag{142}$$

Here, as in expressions 123 and 124, we are anticipating a replacement of the infinite sum by a finite number of terms, so that $g_1(s)$ and $q_2(s)$ may be thought of as being finite polynomials.

We observe that the poles of $E_1(s)$ are the same as those of $h_2(s)$ in Eq. 124, and that (since the β_k's are positive real) $E_1(s)$ has the form of a realizable reactance function (a ψ-function, according to the discussion in Ch. 1, Art. 4, and Ch. 3, Art. 4). The polynomials $g_1(s)$ and $q_2(s)$ are respectively odd and even; and they have simple zeros alternating on the j-axis.

The transform for the current $i(t)$, having the form of $f_2(t)$ in Fig. 24, is evidently identical with the function $h_2(s)$ in Eq. 124, and hence if we denote it for the present interpretations as $I(s)$ we have

$$I(s) = h_2(s) = \frac{p_2(s)}{q_2(s)} \tag{143}$$

The admittance of box N_1 is thus given by

$$y_1(s) = \frac{I(s)}{E_1(s)} = \frac{p_2(s)}{g_1(s)} \tag{144}$$

that of box N_2 by

$$y_2(s) = h_2(s) = \frac{p_2(s)}{q_2(s)} \tag{145}$$

and hence we have for the admittance of their series combination

$$Y(s) = \frac{1}{(1/y_1) + (1/y_2)} = \frac{p_2(s)}{q_2(s) + g_1(s)} \tag{146}$$

This result differs in appearance from the one stated by Eq. 128. For any situation in which the pertinent infinite series are replaced by a finite number of terms, the solutions given by Eqs. 128 and 146 do yield distinctly different although not substantially different results. As we shall show, their differences become less and less significant as more and more terms of the infinite series are considered; and they coincide identically as the number of terms considered becomes unlimited. We may say that the present result 146 for the desired system function and that stated by Eq. 128 are asymptotically equivalent.

In order to demonstrate the truth of this statement we need merely show that the admittance $y_1(s)$ of box N_1 in our present approach becomes identical with the function $h_1(s)$ in the former one, since the identity of $y_2(s)$ with $h_2(s)$ is already recognized by Eq. 145. According to Eqs. 143

and 144, we must show that

$$y_1(s) = \frac{h_2(s)}{E_1(s)} \rightarrow h_1(s) \qquad \text{or} \qquad E_1(s) \rightarrow \frac{h_2(s)}{h_1(s)} \tag{147}$$

as the number of terms in the pertinent trigonometric expansions become infinite.

With regard to the time function $e_1(t)$ in Fig. 32 we see that its exact transform can be written

$$E_1(s) = 1 - 2e^{-s\tau/2} + 2e^{-s\tau} - 2e^{-3s\tau/2} + - \cdots$$
$$= \frac{2}{1 + e^{-s\tau/2}} - 1 = \frac{1 - e^{-s\tau/2}}{1 + e^{-s\tau/2}} \tag{148}$$

If we regard $h(s)$ as the exact transform of the single pulse $f(t)$ in Fig. 23, then the corresponding transforms of $f_1(t)$ and $f_2(t)$ in Fig. 24 are given by

$$h_1(s) = h(s)(1 + e^{-s\tau/2} + e^{-s\tau} + e^{-3s\tau/2} + \cdots) = \frac{h(s)}{1 - e^{-s\tau/2}} \tag{149}$$

and

$$h_2(s) = h(s)(1 - e^{-s\tau/2} + e^{-s\tau} - e^{-3s\tau/2} + - \cdots) = \frac{h(s)}{1 + e^{-s\tau/2}} \tag{150}$$

From these results, relation 147 follows.

We may conclude that asymptotically the transform $h_1(s)$ of the evenly repeated function $f_1(t)$ has the same zeros as the transform $h_2(s)$ of the oddly repeated function $f_2(t)$, these being the zeros of the transform $h(s)$ of the resultant pulse $f(t)$. Further, the relation

$$y_1(s) = \frac{p_2(s)}{g_1(s)} \rightarrow \frac{p_1(s)}{q_1(s)} = h_1(s) \qquad \text{or} \qquad g_1(s) \rightarrow q_1(s) \times \frac{p_2(s)}{p_1(s)} \tag{151}$$

then shows that the zeros of the transform $E_1(s)$ of the square wave coincide with the poles of the transform $h_1(s)$ of the evenly repeated function $f_1(t)$. Thus the zeros of $E_1(s)$, according to Eq. 148, are determined by the equation

$$1 - e^{-s\tau/2} = 0 \tag{152}$$

which yields (because of $\tau = 2\pi/\omega$) the values

$$s_k = jk\omega \qquad \text{for } k = 0, \pm 2, \pm 4, \cdots \tag{153}$$

Equation 123 shows that these are precisely the poles of $h_1(s)$.

Having thus seen that $p_1(s)$ and $p_2(s)$ coincide asymptotically, as do the odd polynomials $q_1(s)$ and $g_1(s)$, it follows that the results 128 and 146 are equivalent in the same sense.

Since 142 has the form of an LC realizable impedance, the denominator in 146 is a Hurwitz polynomial and the realizability of $Y(s)$ as a transfer function is assured.

Further, it is interesting to observe that this Hurwitz polynomial is completely determined by the transform $E_1(s)$ of a differentiated unit square wave. It is independent of the particular pulse shape for which the synthesis is done. Hence a family of such polynomials can be constructed for various integer values of n (the finite number of terms considered in the pertinent trigonometric expansions) and reused in every new synthesis problem. Only the numerator polynomial $p_2(s)$ depends upon the particular pulse shape desired.

It might also be mentioned that the transform $E_1(s)$ may be derived from any trigonometric approximation to a square wave other than that given by the partial sum of a Fourier series if the approximation properties of the latter are undesirable. We thus see that the poles of the resulting function 146, while determined by this trigonometric approximation alone, are not uniquely fixed.

For any finite number of terms in the trigonometric expansions, the results expressed by Eqs. 128 and 146 are, of course, not identical. The former contains some redundancies, as is evident from the fact that it may produce some right half-plane poles with relatively small residues. Some left half-plane poles may likewise be redundant as, for example, the pair yielding the last term in expressions 139 for the triangular pulse shown in Fig. 30. Such poles are very nearly canceled by neighboring zeros. In this regard the formulation expressed by Eq. 146 is more satisfactory since it yields only left half-plane poles, and none that are unessential.

What is needed at this point is an evaluation of the approximate time-domain error committed in the present formulation. We shall see that, with respect to its order of magnitude, the error is the same as for the previous formulation, although specifically it may be somewhat larger. Roughly, we may still adhere to the simple rule that the resultant tolerance is about the same (within a factor of 2) as for the pertinent trigonometric approximation, which in the present method need be carried out only for the oddly repeated function $f_2(t)$ in Fig. 24.

With reference to Fig. 31, suppose we assume that the impulse response of box N_2 is some ripply approximation to the desired function $f_2(t)$ in Fig. 24, similar to the approximate plots shown in Fig. 26. The error thus placed in evidence results from the truncation of an infinite series representation for the periodic function $f_2(t)$ by considering only its terms through $k = n$. If box N_1 were now so constituted that this current passing through it produced a perfect impulse train like $e_1(t)$ in Fig. 32 (the derivative of a perfect square wave), then the net error would be just the ripply one already in evidence. Because the series 142 for $E_1(s)$ is also terminated after n terms, however, $e_1(t)$ is a train of imperfect impulses, and this fact introduces an additional error increment which can be assessed in the following manner.

In order to compute this error increment, we evaluate first the excitation that causes it, which is the difference between what the voltage $e_0(t) + e_0(t - \tau/2)$ actually is and what it should be. Thus, if $e_1(t)$ is a perfect impulse train as shown in Fig. 32, then this voltage (as described above) is an impulse of value 2, occurring at $t = 0$. We must see what it becomes when $e_1(t)$ is a train of imperfect impulses.

To this end, we first sketch the appearance of this imperfect version of $e_1(t)$ as is done in Fig. 33. Its exact appearance will, of course, depend upon the kind of approximating polynomial used, but this is a minor detail that is unimportant at the moment. More significant is a recognition of the fact that the first quasi impulse in the train is like all the others except that it is halved through its vertical center line.

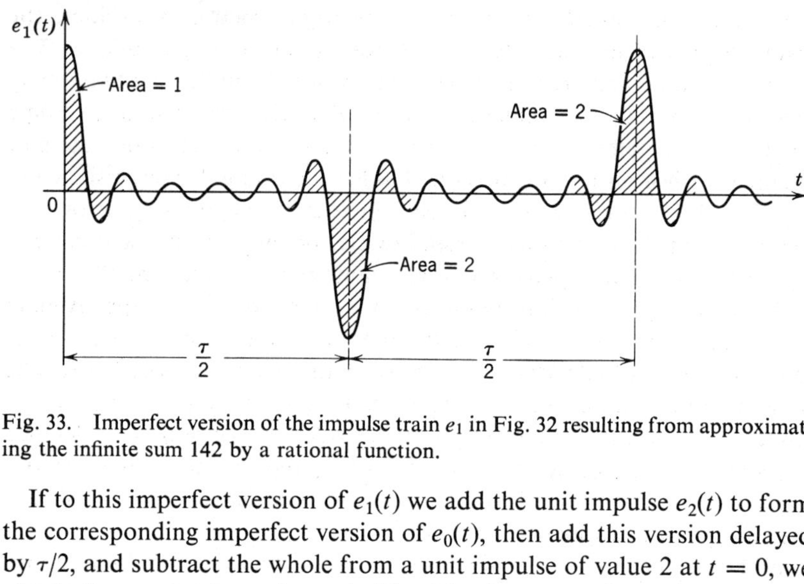

Fig. 33. Imperfect version of the impulse train e_1 in Fig. 32 resulting from approximating the infinite sum 142 by a rational function.

If to this imperfect version of $e_1(t)$ we add the unit impulse $e_2(t)$ to form the corresponding imperfect version of $e_0(t)$, then add this version delayed by $\tau/2$, and subtract the whole from a unit impulse of value 2 at $t = 0$, we obtain the error voltage shown in Fig. 34. Important is the fact that this error voltage is identically zero for $t > \tau/2$.

The resultant error in the desired current pulse is larger than the tolerance indicated in the trigonometric approximation of $f_2(t)$ in Fig. 24 by the incremental response resulting from this error voltage. Except for second-order effects, this response may be found through convolution of the error function in Fig. 34 with the desired pulse shape. But this is precisely the sort of error present in the trigonometric approximation of $f_2(t)$ which is found through convolution of the function $f_2(t)$ with an error function given by the difference between a full quasi impulse of area 2 and a centrally placed perfect impulse of value 2. Hence we may at most expect that *that* error is doubled because of the perturbations presently considered. In other words, the resultant error, within a factor of at most 2, is given by the one observed in the trigonometric approximation of the oddly repeated pulse function $f_2(t)$.

Returning now to the admittance $Y(s)$ given by Eq. 146, we may point out that if we wish to produce the desired current pulse $i(t)$ in Fig. 31 by an applied voltage $e_0(t)$ which is a step instead of an impulse, we have but to construct an admittance $sY(s)$. The voltage $e_1(t)$ appearing across box N_1 is then a rectangular pulse of unit height and duration $\tau/2$ (neglecting errors) regardless of the shape of the current pulse $i(t)$. Hence, if we wish to design a network that converts a voltage step into a rectangular voltage pulse, we can choose any current pulse shape that makes realization of the boxes N_1 and N_2 easiest.

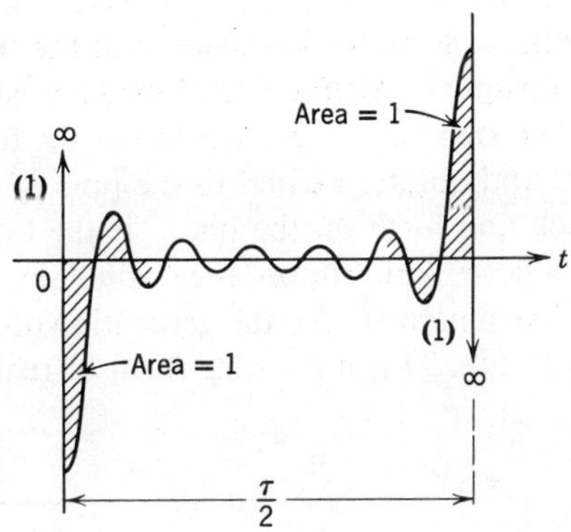

Fig. 34. Difference between what the voltage $e_0(t) + e_0(t - \tau/2)$ actually is and what it should be [namely $2u_0(t)$ according to Fig. 32]. This is the error voltage that causes additional error in the response.

Regarding realization of the function 146, which may at this stage be regarded as either an impedance or admittance, it is useful to observe that if the specified pulse shape has symmetry about a vertical center line, then the trigonometric approximation of $f_2(t)$ in Fig. 24 contains sine terms only and $p_2(s)$ becomes an even polynomial. On the other hand, if the pulse is antisymmetrical about a vertical center line, then it is approximated by cosine terms and $p_2(s)$ is an odd polynomial. In either of these special cases, realization of the resultant transfer function 146 is possible (according to the discussion in Art. 1, Ch. 11) as a lossless two terminal-pair network terminated in a resistance. The pertinent lossless network takes the form of an unbalanced ladder only if the zeros of $p_2(s)$ are simple and restricted to the j-axis, but an unbalanced realization (perhaps involving mutual inductive coupling) is possible in any case according to the procedure discussed in Art. 4, Ch. 7.

In a general situation in which the pulse has symmetrical and antisymmetrical components about a vertical center line, $p_2(s)$ correspondingly has even and odd components. Of the various ways in which the pertinent transfer function can be realized (according to methods discussed in Chs. 11 and 12), the one given in Art. 5, Ch. 12, is particularly interesting because of a rather lucid physical interpretation of the mechanism responsible for termination of the transient after a single pulse in the periodic function $f_2(t)$ has been generated.

Consider first the situation shown in Fig. 35 in which the resistance R has a value equal to the characteristic impedance of the transmission line (assumed lossless). Upon closure of the switch S, a rectangular voltage wave of value $E/2$ travels down the line and is reflected without change of sign at its open end. When the reflected wave reaches the sending end of the line, no further reflection takes place, and the transient response of the system ceases abruptly. Voltage developed across the resistance R as a result of the applied step voltage excitation has the form of a rectangular pulse of height $E/2$ and duration equal to the time of travel of an electromagnetic wave back and forth on the line. If the latter is replaced by a lumped network whose input impedance simulates that of the open-circuited line, we have a network for the generation of rectangular pulses.* The circuit shown in Fig. 27 is a possible form of realization.

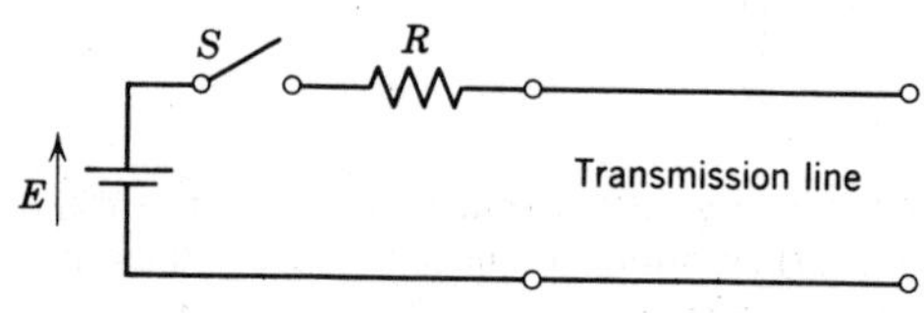

Fig. 35. A situation that generates a rectangular current pulse if R equals the characteristic impedance of the lossless transmission line.

If the line length is l meters and the velocity of electromagnetic wave propagation on the lossless line is denoted by v, then the pulse length $\tau/2 = 2l/v$. The input impedance of the line (with its characteristic impedance normalized at 1 ohm) is therefore equal to coth $s\tau/4$, which is the reciprocal of expression 148 for $E_1(s)$, and hence (by Eq. 142) equal to $q_2(s)/g_1(s)$—the ratio of even-to-odd parts in the denominator of $Y(s)$, Eq. 146.

When $p_2(s)$ is even, and we realize the transfer function 146 (regarded as

* This is discussed in the MIT *Report 43* referred to on p. 417.

an impedance) in the form of a lossless network terminated in a 1-ohm load (Art. 1, Ch. 11), then this network, viewed from its terminal pair 2, simulates an open-circuited transmission line. Driven from this end, the loaded network approximates the situation pictured in Fig. 35, and hence the transient terminates for the reason described above.*

The voltage across the load resistance is a rectangular pulse, but that appearing at the opposite end of the lossless network (terminal pair 1) has the form of the desired pulse, because this network does *not* simulate an open-circuited transmission line so far as its transfer impedance is concerned. By the reciprocity theorem we can alternately drive the network from its end 1 and observe the desired pulse as a voltage across the load resistance.

When $p_2(s)$ is odd (the desired pulse is antisymmetrical about a vertical center line), then z_{22} for the lossless network becomes $g_1(s)/q_2(s)$—the reciprocal of its form when $p_2(s)$ is even. On a 1-ohm impedance level, this function simulates the input impedance of a short-circuited transmission line. A corresponding situation to that pictured in Fig. 35 again yields a transient that ceases abruptly after a wave travels down and back on the line, although a steady current equal to E/R persists. The voltage across the input terminals of the line (instead of across R) now has the form of a rectangular pulse. If the lumped lossless network is synthesized to have the proper transfer impedance with the zeros of $p_2(s)$, then the voltage developed at its end 1 again has the form of the desired pulse.

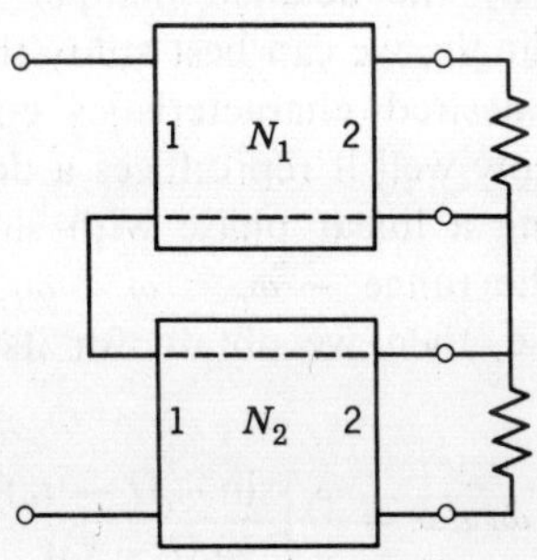

Fig. 36. Realization of the transfer function 146 (regarded as an impedance) according to the method of Art. 5, Ch. 12.

Finally, when $p_2(s)$ has even and odd components, realization of the transfer function according to the method of Art. 5, Ch. 12, yields a result in the form shown in Fig. 36. The network N_1 has an impedance

* This interpretation was pointed out to the writer by P. M. Lewis.

$z_{22}^{(1)} = q_2(s)/g_1(s)$, and for N_2 we have $z_{22}^{(2)} = g_1(s)/q_2(s)$. The first of these networks has the proper transfer impedance to generate the symmetrical component of the desired pulse and the second generates the antisymmetrical component. Finite duration of the total transient response may still be ascribed to the same phenomenon characterizing the behavior of the simple system in Fig. 35.

8 Time-Domain Synthesis of Delay Networks

As mentioned in Art. 9 of the preceding chapter, the design of a delay network (often called a delay line since it simulates a transmission line of finite length) is usually approached by attempting to construct a rational transfer function having an approximately linear phase characteristic throughout a frequency interval $0 < \omega < \omega_c$ with ω_c chosen large enough to accommodate the frequency spectrum of the signal which the contemplated network is intended to transmit. The magnitude of the transfer function is given little consideration except to see that it is essentially constant over most of the pertinent frequency interval. If an all-pass type of function is chosen, this latter item takes care of itself, and so it is felt that the essential features of an adequate delay system are provided for.

In certain situations, particularly those in which the control of specific signal wave-form characteristics is important, this attitude may lead to unacceptable results. The previous discussions of this chapter have shown that a convenient and adequate test for the transient response of a network is provided by the detailed manner in which it transmits a unit impulse. Accordingly, we can best study the adequacy of a delay network in preserving desired characteristics concerning signal wave shape by determining how well it reproduces a delayed impulse. If we imagine a network having a linear phase with slope t_0 and a constant amplitude throughout the range $-\omega_c < \omega < \omega_c$, while its response is essentially zero otherwise, then we obtain for its impulse response the function

$$f_1(t) = \frac{2}{\pi}\int_0^{\omega_c} \cos \omega t_0 \cos \omega t \, d\omega = \frac{\omega_c}{\pi}\left[\frac{\sin \omega_c(t - t_0)}{\omega_c(t - t_0)} + \frac{\sin \omega_c(t + t_0)}{\omega_c(t + t_0)}\right] \tag{154}$$

for $t > 0$. For reasonably large values of ω_c, this function is essentially given by the first term of the bracket expression, and hence (except for the delay t_0) is seen to have the familiar form of the quasi impulse shown in Fig. 16.

If the transfer function has the same phase (having essentially zero slope for $|\omega| > \omega_c$), but is all pass in its amplitude response, then we must add to the impulse response 154 the function

$$f_2(t) = \frac{2}{\pi}\int_{\omega_c}^{\infty} \cos \omega t \, d\omega = \frac{2}{\pi}\left(\int_0^{\infty} \cos \omega t \, d\omega - \int_0^{\omega_c} \cos \omega t \, d\omega\right)$$

$$= u_0(t) - \frac{2}{\pi}\int_0^{\omega_c} \cos \omega t \, d\omega = u_0(t) - \frac{2\omega_c}{\pi}\frac{\sin \omega_c t}{\omega_c t} \qquad \text{for } t > 0 \quad (155)$$

This component in the impulse response of an all-pass delay network is rather undesirable. First of all, it contains the undelayed impulse with which the network is excited. In addition, it contains a term which, for large values of ω_c, consists of high-frequency oscillations of tremendous amplitude occurring near $t = 0$.

How severe the signal distortion will be that is caused by an impulse response equal to the sum of 154 and 155 depends, of course, upon the detailed character of that signal wave form. For example, if it is very smooth so that it can be approximated by parabolic arcs of rather high degree and hence can be expressed as the third, fourth, or higher successive integral of an impulse train, then the ripply character of the quasi impulse 154 will cause no noticeable resultant ripples, and even the second term in the function 155 will be rendered unimportant. The first term in 155 yields the undelayed input at the output, and hence is objectionable no matter what the signal wave shape may be.

On the other hand, if the signal contains discontinuities, then even the function 154 introduces objectionable ripples, and 155 causes so much distortion as to practically obliterate the desired signal altogether. It is clear that an all-pass amplitude characteristic is practically never tolerable in a delay network, and that a band-pass amplitude characteristic coupled with linear phase response throughout the band may likewise be unacceptable unless the amplitude departs from constancy in just the correct manner within the pass band.

It is important in this connection to recognize that the response function referred to in these considerations is the *total* function for the pertinent system. Sometimes the delay network per se is associated in cascade with other networks which provide a band-pass character for the total system, so that an all-pass function for the delay component alone becomes tolerable.

In many practical applications a delay network is sought in the form of identical components in cascade so that a variable amount of delay can be had simply by tapping the cascade at an appropriate point. In such a situation the transient response, or signal wave-form distortion, cannot be independent of the amount of delay unless each component in the cascade provides a linear phase and an ideal band-pass amplitude characteristic over the appropriate frequency range. Since, for economical reasons, each component must be relatively simple, its phase is only moderately linear and its amplitude only moderately ideal band pass, with the result that such a delay system has a widely varying transient response as a function of the amount of delay unless the maximum required delay is small.

Exact control of the transient response of a delay network (with a reasonable number of total circuit elements) is possible only when the desired delay is a fixed amount. The design procedure is precisely that appropriate to the construction of a rational approximant to the transform of a prescribed impulse response, as discussed in the foregoing articles of this chapter. Particularly interesting in this connection is the Fourier method given in the preceding article. That method essentially converts a periodic function (Fig. 25) into an aperiodic one (Fig. 23) through delaying the former by a period and subtracting it from its undelayed version. We may say that transient synthesis of any desired pulse shape is obtainable as soon as we are able to produce a sufficiently good delay; or the latter may be regarded as accomplished if we have found an acceptable solution to the transient synthesis problem. We shall leave the reader to ponder the practical implications of this observation for himself.

PROBLEMS

1. Given the real parts of Fourier transforms shown, compute and plot the corresponding time functions.

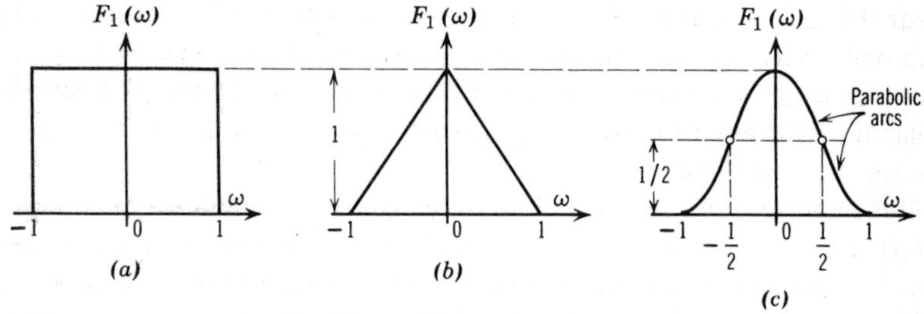

Prob. 1.

2. Consider the plots in the preceding problem as representing time functions $f(t)$ versus t. Compute and plot the corresponding Fourier transforms.

3. Consider the time functions of Prob. 2 delayed by 10 seconds. Find the corresponding real parts of their Fourier transforms.

4. Using the standard parts method and some of the quasi impulses for which rational approximants are found in the text, determine the transfer impedance of a network whose unit step response is an approximation to the time function $f(t)$ shown. Plot the results with sufficient care to illustrate the tolerances obtained.

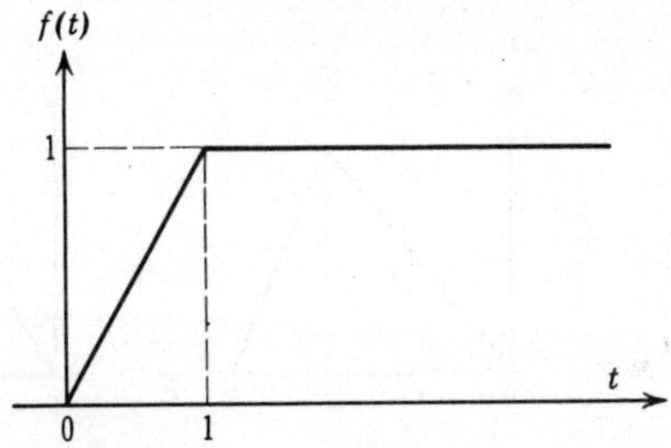

Prob. 4.

5. Reconsider Prob. 4 for the time function shown here.

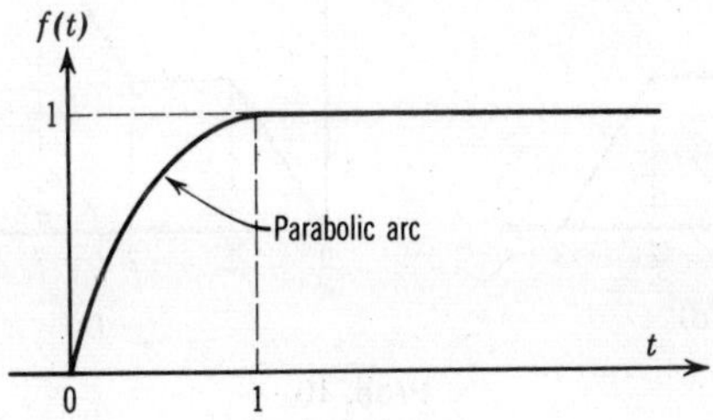

Prob. 5.

6. Solve Prob. 4 by the Fourier method.
7. Solve Prob. 5 by the Fourier method.
8. It is desired to find a rational function whose inverse transform approximates the time function $f(t) = 1 - \cos 4\pi t$ for $0 < t < 1$ and is zero otherwise. Outline the standard parts method of solution for each of three procedures according to which the desired function $f(t)$ is approximated:

(*a*) By rectangles.
(*b*) By triangles.
(*c*) By parabolic arcs.

9. Outline a Fourier method of solution to Prob. 8 comparable tolerance-wise to one of the cases considered there, and compare the number of poles in the resulting rational function.

10. Determine whether any of the time functions shown in the figure on page 730 have transforms with right half-plane zeros.

11. The input and output time functions $a(t)$ and $b(t)$ respectively have the forms shown on the next page. Determine and sketch the corresponding impulse response, and tell whether the implied network is stable or unstable. If unstable compute:

(*a*) A precursor to $b(t)$ of approximate duration such that the impulse response is stabilized.
(*b*) A trailing portion of appropriate duration to stabilize the impulse response.

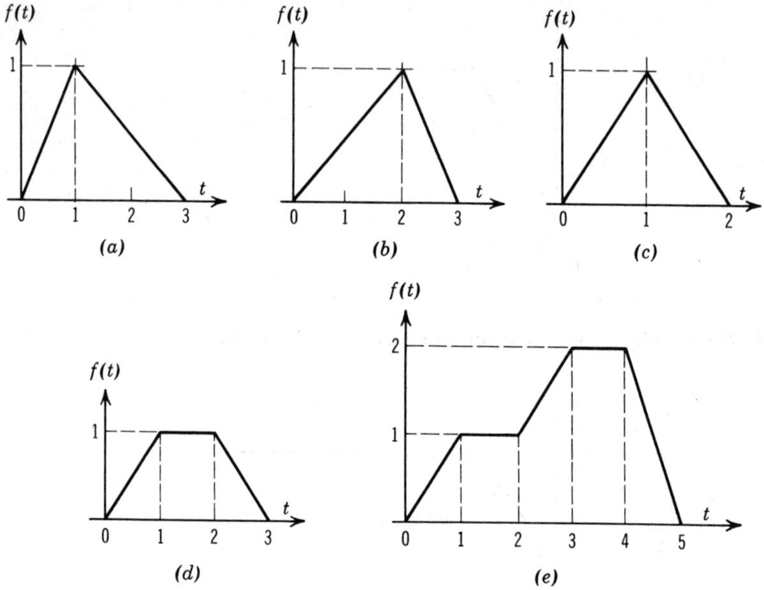

Prob. 10.

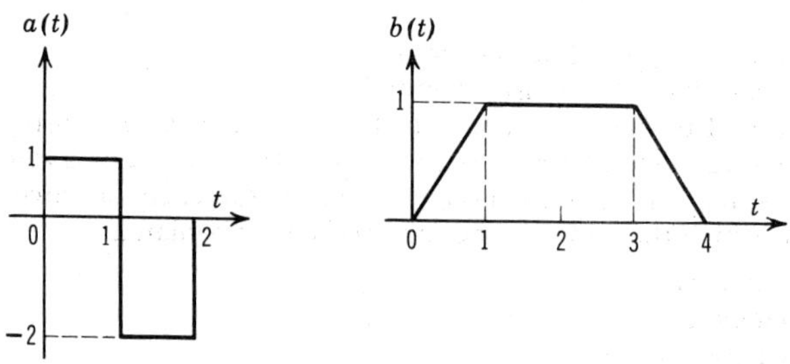

Prob. 11.

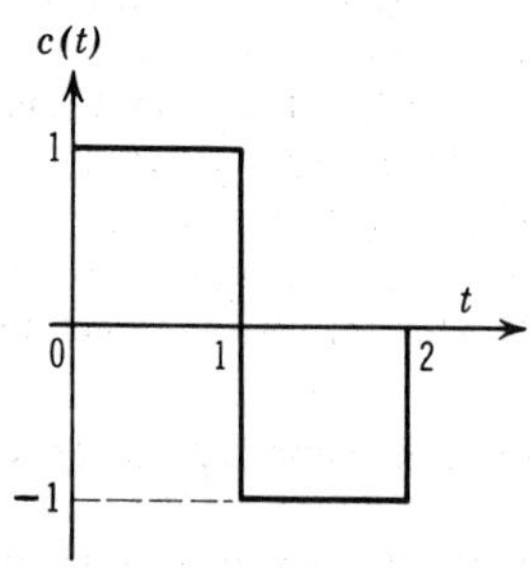

Prob. 12.

12. A given network has the impulse response $c(t)$ shown here. Compute the input $a(t)$ that yields the output $b(t)$ shown in Prob. 11.

13. By the Fourier method of Art. 7, find a rational approximant that will yield the pulse shape defined by

$$f(t) = 1 + \cos \pi t \qquad \text{for } -1 < t < 1$$

with a tolerance of about 3 per cent.

Regarding this approximant as an open-circuit impedance, find its realization in the form of a lossless network loaded by a 1-ohm resistance at the input end, and synthesize this lossless network by the method of Art. 1, Ch. 7 (as illustrated there in Figs. 5 and 8).

INDEX